Operator Theory
Advances and Applications
Vol. 100

# Metric Constrained Interpolation, Commutant Lifting and Systems

C. Foias
A.E. Frazho
I. Gohberg
M.A. Kaashoek

Springer Basel AG

Authors:

C. Foias
Department of Mathematics
Indiana University
Rawles Hall
Bloomington, IN 47405-5701
USA

A.E. Frazho
Department of Aeronautics
Purdue University
Main Campus
West Lafayette, IN 47907
USA

I. Gohberg
School of Mathematical Sciences
Raymond and Beverly Sackler
Faculty of Exact Sciences
Tel Aviv University
Ramat Aviv 69978
Israel

M.A. Kaashoek
Dept. of Mathematics and Computer Science
Vrije Universiteit Amsterdam
De Boelelaan 82
1081 HV Amsterdam
The Netherlands

1991 Mathematics Subject Classification 47A57, 47A20, 47B35, 93A25, 93B36, 30E05

A CIP catalogue record for this book is available from the
Library of Congress, Washington D.C., USA

Deutsche Bibliothek Cataloging-in-Publication Data
**Metric constrained interpolation, commutant lifting systems** / C. Foias ...
– Basel ; Boston ; Berlin : Birkhäuser, 1998
(Operator theory ; Vol. 100)
ISBN 978-3-0348-9775-4 ISBN 978-3-0348-8791-5 (eBook)
DOI 10.1007/978-3-0348-8791-5


Originally published by Birkhäuser Verlag in 1998
Softcover reprint of the hardcover 1st edition 1998
Printed on acid-free paper produced from chlorine-free pulp. TCF ∞
Cover design: Heinz Hiltbrunner, Basel

ISBN 978-3-0348-9775-4

9 8 7 6 5 4 3 2 1

# PREFACE

This book presents a unified approach for solving both stationary and nonstationary interpolation problems, in finite or infinite dimensions, based on the commutant lifting theorem from operator theory and the state space method from mathematical system theory.

Initially the authors planned a number of papers treating nonstationary interpolation problems of Nevanlinna-Pick and Nehari type by reducing these nonstationary problems to stationary ones for operator-valued functions with operator arguments and using classical commutant lifting techniques. This reduction method required us to review and further develop the classical results for the stationary problems in this more general framework. Here the system theory turned out to be very useful for setting up the problems and for providing natural state space formulas for describing the solutions. In this way our work involved us in a much wider program than original planned. The final results of our efforts are presented here.

The financial support in 1994 from the "NWO-stimulansprogramma" for the Thomas Stieltjes Institute for Mathematics in the Netherlands enabled us to start the research which lead to the present book. We also gratefully acknowledge the support from our home institutions: Indiana University at Bloomington, Purdue University at West Lafayette, Tel-Aviv University, and the Vrije Universiteit at Amsterdam. We warmly thank Dr. A.L. Sakhnovich for his carefully reading of a large part of the manuscript. Finally, Sharon Wise prepared very efficiently and with great care the troff file of this manuscript; we are grateful for her excellent typing.

May, 1997 The authors

# TABLE OF CONTENTS

# INTRODUCTION

The theory of interpolation with metric constraints started in the beginning of this century with papers of Carathéodory [1], [2] and Schur [1], [2], and was continued by Nevanlinna [1] and Pick [1], [2], and then later in the fifties by Nehari [1]. At the end of the sixties and in the beginning of the seventies operator theoretical methods for solving these classical function theory interpolation problems were discovered. The most important developments started with the papers Adamjan-Arov-Krein [1], Sarason [1], and Sz.-Nagy-Foias [1], [2]. In particular, in 1967 Sarason [1] encompassed these classical interpolation problems in a representation theorem of operators commuting with special contractions. Shortly after that, in 1968, Sz.-Nagy-Foias [1], [2] derived a purely geometrical extension of Sarason's results. Actually, their result states that operators intertwining restrictions of co-isometries can be extended, by preserving their norm, to operators intertwining these co-isometries, and this work formed the basis of a new method to deal with metric constrained interpolation problems, which usually is referred to as the commutant lifting approach.

In the seventies and the eighties the commutant lifting approach was extended and enriched with important applications to different branches of sciences; in particular, to control theory, prediction theory and geophysics. Especially we would like to mention a new branch of control theory, called $H^\infty$ control, which was pioneered by Zames [1]-[3], Helton [1], [2], and Glover [1]. The book Foias-Frazho [4] presents these developments systematically with an emphasis on choice sequences, Schur type algorithm and inverse scattering techniques.

In recent years the commutant lifting approach was further developed, for example, to solve more complex metric constrained interpolation problems (like the ones involving operator-valued functions with operator arguments and the two-sided Nudelman problem), to understand better the maximum entropy properties and the role of the central solution, and to obtain a more detailed description of the linear fractional representation of the solutions. Moreover, a new area of applications appeared in connection with $H^\infty$-control problems for the time-variant systems, which led in a natural way to new interpolation problems with metric constraints in which the role of functions is taken over by input-output operators of time-dependent systems. In the discrete time case this means that bounded analytic functions on the open unit disc, which may be viewed as lower triangular Toeplitz or Laurent operators, are replaced by arbitrary lower triangular bounded linear operators. A simple example of an interpolation problem that emerges in this way is the generalization of the Nehari problem to a time-variant setting, which for the discrete case has the following form. Given an array of complex numbers $f_{j,k}$, $-\infty < k \leq j < \infty$, determine a bounded linear operator $G = (g_{j,k})_{j,k=-\infty}^{\infty}$

acting on $\ell^2$, the Hilbert space of doubly infinite square summable sequences, such that the lower triangular part of G coincides with the given array of complex numbers, that is,

$$g_{j,k} = f_{j,k} \qquad (-\infty < k \leq j < \infty)$$

and the operator norm of G is as small as possible (see Arveson [1], Feintuch-Francis [1], and Gohberg-Kaashoek-Woerdeman [1]-[4]).

Recently the concept of point evaluation has been extended to a time-variant setting too. For the discrete time case the definition goes back to Dewilde, Alpay, and Dym (Dewilde [1], Alpay-Dewilde [1] and Dewilde-Dym [1]), and its continuous time version is due to Ball-Gohberg-Kaashoek [2], [5]. The new concept of time-variant point evaluation led to the formulation of various nonstationary interpolation problems of Nevanlinna-Pick type, which appear in the following way. One first replaces a point z in the open unit disc by a weighted shift

$$\begin{bmatrix} \ddots & \ddots & & & & \\ & 0 & z_{-1} & & & \\ & & \underline{0} & z_0 & & \\ & & & 0 & z_1 & \\ & & & & \ddots & \ddots \end{bmatrix} \tag{0.1}$$

acting on $\ell^2$ with spectral radius strictly less than one. (The underline indicates the 0-0 position. The blank spots in the definition of the weighted shift in (0.1) are zero.) Next, an $H^\infty$ function f is replaced by a lower triangular operator

$$F = \begin{bmatrix} \ddots & & & & \\ \ddots & f_{-1,-1} & & & \\ \ddots & f_{0,-1} & \underline{f_{0,0}} & & \\ \ddots & f_{1,-1} & f_{1,0} & f_{1,1} & \\ & \ddots & \ddots & \ddots & \ddots \end{bmatrix}. \tag{0.2}$$

The value of the operator F in (0.2) at the weighted shift (0.1) is defined to be the diagonal operator diag $(y_k)_{k=-\infty}^{\infty}$, where

$$y_k = f_{k,k} + \sum_{n=1}^{\infty} z_k z_{k+1} \cdots z_{k+n-1} f_{k+n,k} \qquad (-\infty < k < \infty).$$

In the time-invariant case, when in (0.1) the entries $z_j$ do not depend on j (and are equal to z say) and the operator F is a Laurent operator (i.e., $f_{j,k} = f_{j-k}$ for all j and k), the diagonal entries $y_k$ do not depend on k and are all equal to f(z), where $f(\lambda) = \sum_0^\infty \lambda^n f_n$. The definition of nonstationary point evaluation given above leads in a natural way to nonstationary versions of the classical Nevanlinna-Pick type interpolation problem.

In the last six/seven years almost all nonstationary versions of the classical interpolation problems have been solved, including the basic applications to $H^\infty$ control (see, e.g., the papers in Gohberg [1], Dewilde-Kaashoek-Verhaegen [1], and Helmke-Mennicken-Saurer [1]). Also, recursive methods for obtaining solutions have been derived (see Sayed-Constantinescu-Kailath [1]).

Several methods to solve norm constrained interpolation problems, stationary as well as nonstationary, have been developed. The literature on the subject is very rich, and it is impossible to mention all the related publications. For recent accounts and extended lists of references we refer the reader to the monographs Rosenblum-Rovnyak [3], Francis [1], Helton [3], Dym [1], Ball-Gohberg-Rodman [1], Foias-Frazho [4], Bakonyi-Constantinescu [1], Dubovoj-Fritzsche-Kirstein [1], Gohberg-Goldberg-Kaashoek [2], Halany-Ionescu [1], Constantinescu [3] as well as to the review paper Kailath-Sayed [1], to the collection of papers in Gohberg [1] and Dewilde-Kaashoek-Verhaegen [1] and the references mentioned in the notes to the chapters in this book.

The present book treats both stationary and nonstationary interpolation problems. Originally it aimed at the nonstationary theory only, but when working on the first draft the authors understood that the connection between the stationary and nonstationary problems are very strong, and that the nonstationary theory can be based entirely on an advanced operator oriented version of the stationary theory. The latter is also developed in this book in full detail. So the fact that nonstationary problems can be reduced in a canonical way to classical interpolation problems involving operator-valued functions became one of the leading ideas of the present book. In this context, the point evaluation, whenever it appears in the formulation of the interpolation problem, can be understood as an evaluation for the case when the argument is an operator. It turns out that in this operator-valued function setting only one operator point evaluation has to be considered (or at most two for bitangential problems).

The reduction from time-variant to time-invariant and the corresponding inverse operation led us to the following norm bound completion theorem. Let $\mathcal{U}$ and $\mathcal{Y}$ be Hilbert Spaces, and let $\mathcal{H}_k \subset \mathcal{U}$ and $\mathcal{M}_k \subset \mathcal{K}_k \subset \mathcal{Y}$, where $k \in \mathbb{Z}$, be subspaces satisfying

$$\mathcal{H}_{k-1} \subset \mathcal{H}_k\,,\ \mathcal{M}_{k-1} \subset \mathcal{M}_k\,,\ \mathcal{K}_{k-1} \subset \mathcal{K}_k \qquad (k \in \mathbb{Z})\,.$$

Furthermore, for each $k \in \mathbb{Z}$ let $A_k$ from $\mathcal{H}_k$ to $\mathcal{K}_k \ominus \mathcal{M}_k$ be a bounded linear operator, and let $P_{\mathcal{H}_k}$ denote the orthogonal projection onto $\mathcal{M}_k$. In order that there exists a bounded linear operator B from $\mathcal{U}$ to $\mathcal{Y}$ such that

$$B\mathcal{H}_k \subset \mathcal{K}_k \text{ and } (I - P_{\mathcal{M}_k})B \mid \mathcal{H}_k = A_k \qquad (k \in \mathbb{Z}) \tag{0.3}$$

it is necessary and sufficient that

(i) $(I - P_{\mathcal{M}_k})A_{k-1} = A_k \mid \mathcal{H}_{k-1}$ for all $k \in \mathbb{Z}$,

(ii) $\mu := \sup\{\|A_k\| : k\in\mathbb{Z}\} < \infty$.

Moreover, in this case there exists a B from $\mathcal{U}$ to $\mathcal{Y}$ satisfying (0.3) with the additional property that $\|B\| = \mu$. The above three chains completion theorem can be viewed as a time-variant version of the commutant lifting theorem. In fact, we show that there are direct ways to derive one theorem from the other and vice versa.

The present monograph combines the commutant lifting theorem from operator theory and the state space method from system theory to provide a unified approach for solving both stationary and nonstationary interpolation problems with norm constraints. This approach allows us to clarify the Euclidean infinite dimensional geometric properties of these problems. We concentrate on the existence of a solution, the explicit construction of the central solution (which corresponds to the maximum entropy solution in each of the interpolation problems), and the parameterization of all solutions. On the way we also display direct connections between the various interpolation problems.

The book consists of two parts. In Part A a general theory is developed for modern $H^\infty$ interpolation problems for operator-valued functions with operator arguments. Included are the operator-valued versions of the left tangential Nevanlinna-Pick problem, the tangential Hermite-Fejér problem, the Nehari problem, the Sarason problem, the two-sided Nudelman problem and the two-sided Sarason problem. The results include necessary and sufficient conditions for the existence of solutions, central state space solutions, and state space formulas for the description of all solutions. Special attention is payed to the central solution, its maximum entropy property, and its mixed $H^\infty$ and $H^2$ bounds. Also applications to control problems are given.

Part B treats nonstationary analogues of the $H^\infty$-interpolation problems studied in Part A. Two methods are employed. One is based on reduction to time-variant problems, and the other uses the three chains completion theorem. Part B also contains descriptions of the time-variant versions of the central solutions, the description of all solutions of the three chains completion problem, and applications to time-dependent $H^\infty$ control problems.

Although the present book may be viewed as a natural continuation to the monograph Foias-Frazho [4], it can be read independently of the latter. In fact, the prerequisites for this book are rather modest; they include the regular courses in linear algebra and complex analysis, and a first course in Hilbert space operator theory. In order to avoid too many cross references we often recall definitions and results and even repeat appropriate parts of proofs when necessary. We also tried to state the theorems with potential applications as self contained as possible. In particular, the state space solutions to the various interpolation problems presented in the theorems in Chapter V and Sections 7-9 of Chapter VI can be implemented without reading the proofs.

PART A

# Interpolation and Time-Invariant Systems

In this part a general theory is developed for modern $H^\infty$ interpolation problems for operator-valued functions with operator argument. System theory in its state space form and the commutant lifting theorem from operator theory are combined to solve metric constrained interpolation problems. The results include explicit state space formulas for the solutions and applications to $H^\infty$ and mixed $H^\infty$ and $L^2$ control problems.

## CHAPTER I

# INTERPOLATION PROBLEMS FOR OPERATOR-VALUED FUNCTIONS

This chapter concentrates on the necessary and sufficient conditions for the existence of solutions for classical and modern $H^\infty$ interpolation problems. These interpolation problems have their roots deep in classical analysis, and they are introduced here in a sophisticated way for operator valued functions with operator arguments. Included are the Nevanlinna-Pick, Hermite-Fejér, Nehari, Sarason and Nudelman interpolation problems, in their classical and tangential form. Both one sided and two-sided versions are considered. Proofs of the main existence results will be given in the next chapter, based on the commutant lifting theorem.

## I.1. PRELIMINARIES ABOUT NOTATION AND TERMINOLOGY

Throughout this work $\mathcal{X}$, $\mathcal{U}$ and $\mathcal{Y}$ are Hilbert spaces. One may assume that these spaces are always separable, although whenever this assumption is used it will be mentioned explicitly. The Hilbert space direct sum of n copies of $\mathcal{X}$ will be denoted by $\ell^2_n(\mathcal{X})$, that is, $\ell^2_n(\mathcal{X})$ is the Hilbert space consisting of all column vectors of the form $x = [x_1, x_2, ..., x_n]^{tr}$ where tr denotes the transpose, $x_j$ is in $\mathcal{X}$ for all j and $\|x\|^2 = \|x_1\|^2 + \cdots + \|x_n\|^2$. We write $\ell^2_+(\mathcal{X})$ for the Hilbert space of all square norm summable sequences $(x_0, x_1, x_2, ...)$ with elements in $\mathcal{X}$. It will be convenient to deal with the elements in $\ell^2_+(\mathcal{X})$ as infinite columns rather than sequences. To simplify the notation we write $[x_0, x_1, x_2, ...]^{tr}$ for the infinite column vector whose j-th entry is $x_j$. Similarly, $\ell^2(\mathcal{X})$ denotes the Hilbert space of all square norm summable doubly infinite sequences with entries in $\mathcal{X}$. We will often identify the column vector

$$\begin{bmatrix} \vdots \\ x_{-1} \\ x_0 \\ x_1 \\ \vdots \end{bmatrix} = [..., x_{-1}, \underline{x_0}, x_1, ...]^{tr}$$

in $\ell^2(\mathcal{X})$ with the corresponding sequence $(x_j)_{-\infty}^{\infty}$. The underline $\underline{x_0}$ means that $x_0$ appears in the zero position. Often we view $\ell^2_+(\mathcal{X})$ as the subspace of $\ell^2(\mathcal{X})$ consisting of all column vectors of the form $[..., x_{-1}, \underline{x_0}, x_1, ...]^{tr}$ where $x_j = 0$ for $j < 0$. Throughout $\ell^2_-(\mathcal{X})$ is the orthogonal complement of $\ell_+^2(\mathcal{X})$ in $\ell^2(\mathcal{X})$, that is, the space $\ell_-^2(\mathcal{X})$ consists of all vectors in $\ell^2(\mathcal{X})$ of the form $[\ldots, x_{-1}, \underline{x_0}, x_1, \ldots]^{tr}$, where $x_j = 0$ for all $j \geq 0$. By $H^2(\mathcal{X})$ we denote the Hardy space consisting of all $\mathcal{X}$-valued analytic functions f on the open unit disc $\mathbb{D}$ with the property that

$$\|f\|_2 := \sup_{0 \leq r < 1} \left[ \frac{1}{2\pi} \int_0^{2\pi} \|f(re^{it})\|^2 dt \right]^{1/2} < \infty .$$

As usual, we identify $\ell_+^2(\mathcal{X})$ with $H^2(\mathcal{X})$ via the Fourier transform

$$\mathcal{F}_{\mathcal{X}+}(x_0, x_1, x_2, \ldots) = f(\lambda), \quad \text{where} \quad f(\lambda) = \sum_{j=0}^{\infty} \lambda^j x_j \quad (\lambda \in \mathbb{D}) . \tag{1.1}$$

The set of all bounded linear operators from $\mathcal{X}$ into $\mathcal{Y}$ is denoted by $\mathcal{L}(\mathcal{X}, \mathcal{Y})$. Unless stated explicitly otherwise, an operator will always be a bounded linear operator. By $H^\infty(\mathcal{X}, \mathcal{Y})$ we denote the Hardy class of all uniformly bounded analytic functions on $\mathbb{D}$ with values in $\mathcal{L}(\mathcal{X}, \mathcal{Y})$. By definition, the $H^\infty$ norm of such a function G is given by

$$\|G\|_\infty := \sup \left\{ \|G(\lambda)\| : \lambda \in \mathbb{D} \right\}, \tag{1.2}$$

where $\|\cdot\|$ is the operator norm on $\mathcal{L}(\mathcal{X}, \mathcal{Y})$. If G is a function in $H^\infty(\mathcal{X}, \mathcal{Y})$, then $M_G^+$ denotes the multiplication operator from $H^2(\mathcal{X})$ into $H^2(\mathcal{Y})$ defined by $M_G^+ h = Gh$ for h in $H^2(\mathcal{X})$. It is well known that the operator norm $\|M_G^+\|$ of $M_G^+$ coincides with the $H^\infty$ norm $\|G\|_\infty$, Chapter XXIII in Gohberg-Goldberg-Kaashoek [2], Chapter V in Sz.-Nagy-Foias [3], or Chapter IX in Foias-Frazho [4].

An operator R from $\ell^2(\mathcal{X})$ to $\ell^2(\mathcal{Y})$ is represented in a natural way by a doubly infinite operator matrix with entries in $\mathcal{L}(\mathcal{X}, \mathcal{Y})$, that is, R admits a matrix representation of the form

$$R = \begin{bmatrix} & \vdots & \vdots & \vdots & \\ \cdots & R_{-1,-1} & R_{-1,0} & R_{-1,1} & \cdots \\ \cdots & R_{0,-1} & \underline{R_{0,0}} & R_{0,1} & \cdots \\ \cdots & R_{1,-1} & R_{1,0} & R_{1,1} & \cdots \\ & \vdots & \vdots & \vdots & \end{bmatrix}, \tag{1.3}$$

where $R_{i,j} \in \mathcal{L}(\mathcal{X}, \mathcal{Y})$ for all i and j. The underline indicates the entry in the zero row and zero column of R. We call R *(block) lower triangular* if $R_{i,j} = 0$ for all $i < j$.

An operator L from $\ell^2(\mathcal{X})$ to $\ell^2(\mathcal{Y})$ is a *(block) Laurent operator* if the entries $L_{i,j}$ in its matrix representation depend only on $i - j$, that is, $L_{i,j} = G_{i-j}$. So a Laurent operator admits a matrix representation of the form

$$L = \begin{bmatrix} & \vdots & \vdots & \vdots & \\ \cdots & G_0 & G_{-1} & G_{-2} & \cdots \\ \cdots & G_1 & \underline{G_0} & G_{-1} & \cdots \\ \cdots & G_2 & G_1 & G_0 & \cdots \\ & \vdots & \vdots & \vdots & \end{bmatrix} \tag{1.4}$$

In this case we refer to L as the block Laurent operator *generated by* the sequence of operators $\{G_n\}_{-\infty}^{\infty}$. Recall that V is a *bilateral forward shift* on $\ell^2(\mathcal{E})$ if V is the unitary operator on $\ell^2(\mathcal{E})$ defined by

$$V[\ldots, f_{-1}, \underline{f_0}, f_1, \ldots]^{tr} = [\ldots, f_{-2}, \underline{f_{-1}}, f_0, \ldots]^{tr} . \tag{1.5}$$

Now let $V_1$ on $\ell^2(\mathcal{X})$ and $V_2$ on $\ell^2(\mathcal{Y})$ be forward shifts, and L an operator from $\ell^2(\mathcal{X})$ to $\ell^2(\mathcal{Y})$. Then it is easy to verify that L is a Laurent operator if and only if $V_2L = LV_1$.

A Laurent operator L from $\ell^2(\mathcal{X})$ into $\ell^2(\mathcal{Y})$ is lower triangular if and only if L maps $\ell_+^2(\mathcal{X})$ into $\ell_+^2(\mathcal{Y})$. In this case there exists a unique $G \in H^\infty(\mathcal{X}, \mathcal{Y})$, called the *symbol* of L, such that $L_{j,k} = G_{j-k}$ for $j \geq k$ (and $L_{j,k} = 0$ for $j < k$). Here $G_0, G_1, G_2, \ldots$ are the Taylor coefficients of G, that is,

$$G(\lambda) = \sum_{j=0}^{\infty} \lambda^j G_j \qquad (\lambda \in \mathbb{D}) . \tag{1.6}$$

It is well-known that the operator norm of a lower triangular block Laurent operator coincides with the $H^\infty$ norm of its symbol. Conversely, any G in $H^\infty(\mathcal{X}, \mathcal{Y})$ defines in this way a lower triangular block Laurent operator.

An operator T from $\ell_+^2(\mathcal{X})$ into $\ell_+^2(\mathcal{Y})$ is called a *(block) Toeplitz operator* if T is the natural compression of a Laurent operator L from $\ell_+^2(\mathcal{X})$ into $\ell_+^2(\mathcal{Y})$, or equivalently, if T admits a block matrix representation of the form:

$$T = \begin{bmatrix} G_0 & G_{-1} & G_{-2} & \cdots \\ G_1 & G_0 & G_{-1} & \cdots \\ G_2 & G_1 & G_0 & \cdots \\ \vdots & \vdots & \vdots & \end{bmatrix} . \tag{1.7}$$

If T is given by (1.7) and L by (1.4), then T and L have the same operator norms. Recall that the *unilateral forward shift* S is the isometry on $\ell_+^2(\mathcal{E})$ given by

$$S[f_0, f_1, f_2, \ldots]^{tr} = [0, f_0, f_1, f_2, \ldots]^{tr} . \tag{1.8}$$

Now let $S_2$ be the forward shift on $\ell_+^2(\mathcal{Y})$ and $S_1$ be the forward shift on $\ell_+^2(\mathcal{X})$. Then it is easy to show that an operator T from $\ell^2_+(\mathcal{X})$ to $\ell^2_+(\mathcal{Y})$ is Toeplitz if and only if $S_2^* T S_1 = T$.

If $G \in H^\infty(\mathcal{X}, \mathcal{Y})$, then the block Toeplitz operator

$$T_G = \begin{bmatrix} G_0 & 0 & 0 & \dots \\ G_1 & G_0 & 0 & \dots \\ G_2 & G_1 & G_0 & \dots \\ \vdots & \vdots & \vdots & \dots \end{bmatrix}. \tag{1.9}$$

is unitarily equivalent to the multiplication operator $M_G^+$ via the Fourier transform. To be precise, $\mathcal{F}_{\mathcal{Y}+} T_G = M_G^+ \mathcal{F}_{\mathcal{X}+}$ where $\mathcal{F}_{\mathcal{Y}+}$ is the Fourier transform from $\ell^2_+(\mathcal{Y})$ onto $H^2(\mathcal{Y})$ and $\mathcal{F}_{\mathcal{X}+}$ is given by (1.1). It follows that $\|T_G\| = \|M_G^+\| = \|G\|_\infty$. Let T be an operator from $\ell^2_+(\mathcal{X})$ into $\ell^2_+(\mathcal{Y})$, and let $S_1$ and $S_2$ be the (block) forward shifts on $\ell^2_+(\mathcal{X})$ and $\ell^2_+(\mathcal{Y})$, respectively. Then it is easy to show that T is a block lower triangular Toeplitz operator (i.e., T admits a matrix representation of the form (1.9)) if and only if $S_2 T = T S_1$.

We let $L^2(\mathcal{X})$ denote the set of all Lebesgue measurable functions f on the unit circle with values in $\mathcal{X}$ such that

$$\|f\|_2 = \left[\frac{1}{2\pi}\int_0^{2\pi} \|f(e^{it})\|^2 dt\right]^{1/2} < \infty .$$

Here, and in the sequel, whenever we write $L^2(\mathcal{X})$ we assume that the underlying Hilbert space $\mathcal{X}$ is separable. In this case $H^2(\mathcal{X})$ may be identified with the subspace of $L^2(\mathcal{X})$ consisting of all functions f of which the negative Fourier coefficients are zero, that is,

$$f_n = \frac{1}{2\pi}\int_0^{2\pi} e^{-int} f(e^{it}) dt = 0 \quad (\text{for } n < 0) .$$

The spaces $L^2(\mathcal{X})$ and $\ell^2(\mathcal{X})$ may be identified by the Fourier transform $\mathcal{F}_\mathcal{X}$, which is the unitary map from $\ell^2(\mathcal{X})$ onto $L^2(\mathcal{X})$ defined by

$$\mathcal{F}_\mathcal{X}[\dots, x_{-1}, \underline{x_0}, x_1, \dots]^{tr} = f \quad \text{where} \quad f(e^{it}) = \sum_{n=-\infty}^{\infty} x_n e^{int} . \tag{1.10}$$

For separable Hilbert spaces $\mathcal{X}$ and $\mathcal{Y}$ we let $L^\infty(\mathcal{X}, \mathcal{Y})$ denote the set of all strongly measurable functions F on the unit circle **T** with values in $\mathcal{L}(\mathcal{X}, \mathcal{Y})$ such that

$$\|F\|_\infty := \text{ess sup}\left\{\|F(e^{it})\| : 0 \le t \le 2\pi\right\} < \infty . \tag{1.11}$$

In this case $H^\infty(\mathcal{X}, \mathcal{Y})$ may be identified with all F in $L^\infty(\mathcal{X}, \mathcal{Y})$ such that

$$\frac{1}{2\pi}\int_0^{2\pi} e^{-int}F(e^{it})dt = 0 \quad (\text{for } n < 0)\,.$$

Given $F \in L^\infty(\mathcal{X}, \mathcal{Y})$ we let $M_F$ denote the operator of multiplication by F from $L^2(\mathcal{X})$ into $L^2(\mathcal{Y})$. Thus

$$(M_F h)(e^{it}) = F(e^{it})h(e^{it}) \quad (h \in L^2(\mathcal{X}))\,. \tag{1.12}$$

It is well-known that the operator norm $\|M_F\|$ and the function norm $\|F\|_\infty$ coincide. We write $T_F$ and $L_F$ for the Toeplitz operator and Laurent operator, respectively, defined by F. Thus $T_F$ and $L_F$ are defined by convolution; that is,

$$\begin{aligned} &T_F : \ell_+^2(\mathcal{X}) \to \ell_+^2(\mathcal{Y})\,, \ (T_F x)_j = \sum_{k=0}^{\infty} F_{j-k}x_k \quad (j \geq 0)\,, \\ &L_F : \ell^2(\mathcal{X}) \to \ell^2(\mathcal{Y})\,, \ (L_F x)_j = \sum_{k=-\infty}^{\infty} F_{j-k}x_k \quad (j \in \mathbb{Z})\,. \end{aligned} \tag{1.13}$$

Here $F_n$ is the n-th Fourier coefficients of F, and $\mathbb{Z}$ denotes the integers ..., – 2, – 1, 0, 1, 2, .... The operator norms of $T_F$ and $L_F$ coincide and are equal to $\|F\|_\infty$. Moreover, F is usually referred to as the *symbol* of $T_F$ and $L_F$. The operators $L_F$ and $M_F$ are unitarily equivalent via the Fourier transform, in fact $\mathcal{F}_\mathcal{Y} L_F = M_F \mathcal{F}_\mathcal{X}$. In the sequel we shall often identify the operators $M_F$ and $L_F$ with the function F, and simply write F instead of $M_F$ and $L_F$. If $F \in H^\infty(\mathcal{X}, \mathcal{Y})$, then $T_F$ and $L_F$ are well-defined operators also for the case when $\mathcal{X}$ and $\mathcal{Y}$ are non-separable. For a more in-depth discussion on Toeplitz and Laurent operators, see Chapter VIII in Bercovici [1], Chapter XXIII in Gohberg-Goldberg-Kaashoek [2], Chapter V in Sz.-Nagy-Foias [3] or Chapter IX in Foias-Frazho [4]. Finally, if $\mathcal{X} = \mathcal{Y} = \mathbb{C}^1$, then the spaces $\ell^2(\mathcal{Y})$, $\ell_+^2(\mathcal{Y})$, $L^2(\mathcal{Y})$, $H^\infty(\mathcal{X}, \mathcal{Y})$ are denoted by $\ell^2$, $\ell_+^2$, $L^2$, $H^\infty$, respectively.

The word *subspace* stands for closed linear manifold. Given a subspace $\mathcal{H}$ of a Hilbert space $\mathcal{X}$, we write $P_\mathcal{H}$ for the orthogonal projection of $\mathcal{X}$ onto $\mathcal{H}$. Moreover, $\Pi_\mathcal{H}$ denotes the orthogonal projection of $\mathcal{X}$ onto $\mathcal{H}$ viewed as a map from $\mathcal{X}$ onto $\mathcal{H}$. Clearly $P_\mathcal{H} = \Pi_\mathcal{H}^* \Pi_\mathcal{H}$. The identity operator on $\mathcal{X}$ will be denoted by $I_\mathcal{X}$ or just by I if the underlying space is clear from the context. The range of an operator T is denoted by ran T and ker T is the kernel of T. An operator A on $\mathcal{X}$ is *positive* if $(Ax, x) \geq 0$ for all x in $\mathcal{X}$. In this case we write $A \geq 0$. An operator A on $\mathcal{X}$ is *strictly positive* if there exists a $\delta > 0$ such that $(Ax, x) \geq \delta\|x\|^2$ for all x in $\mathcal{X}$. Notice that A is strictly positive if and only if A is positive and invertible. The symbol $\bigvee\{\mathcal{M}_\alpha : \alpha \in \mathcal{A}\}$ denotes the closed linear span of the subspaces $\mathcal{M}_\alpha$ for $\alpha$ in $\mathcal{A}$. Given the Hilbert spaces $\ldots \mathcal{X}_{-1}, \mathcal{X}_0, \mathcal{X}_1, \ldots$ we write $\vec{\mathcal{X}}$ for the Hilbert space direct sum generated by $\{\mathcal{X}_j\}$, that is,

$$\vec{\mathcal{X}} = \bigoplus_{j-\infty}^{\infty} \mathcal{X}_j = \oplus \{\mathcal{X}_j : -\infty < j < \infty\} . \tag{1.14}$$

We will also use the notation

$$\vec{\mathcal{X}}_{(m,\, n)} = \bigoplus_{j=m}^{n} \mathcal{X}_j = \oplus \{\mathcal{X}_j : m \leq j \leq n\} . \tag{1.15}$$

Obviously $\vec{\mathcal{X}} = \vec{\mathcal{X}}_{(-\infty,\, \infty)}$.

We conclude this section with a discussion of general isometries. An operator T from $\mathcal{H}$ into $\mathcal{K}$ is an *isometry* if $\|Tx\| = \|x\|$ for all x in $\mathcal{H}$, or equivalently, $T^*T = I$. If both T and $T^*$ are isometries, then T is called *unitary*. We say that T on $\mathcal{H}$ is a *unilateral shift* (respectively a *bilateral shift*) if T is unitarily equivalent to the unilateral forward shift S on some $\ell_+^2(\mathcal{E})$ space (respectively the bilateral forward shift V on some $\ell^2(\mathcal{E})$ space). Recall that two operators $T_1$ and $T_2$ are *unitarily equivalent* if there exists a unitary operator W satisfying $T_2W = WT_1$. Now let T be an isometry on $\mathcal{H}$, a subspace $\mathcal{L}$ of $\mathcal{H}$ is *wandering* for T if $T^n\mathcal{L}$ is orthogonal to $T^m\mathcal{L}$ for all $n \neq m$, where m and n are nonnegative integers. Clearly $\mathcal{L} = \ker T^*$ is a wandering subspace for T. The classical Wold decomposition shows that any isometry T on $\mathcal{H}$ admits a unique reducing decomposition of the form $T = T_o \oplus T_u$ on $\mathcal{H} = \mathcal{H}_o \oplus \mathcal{H}_u$, where $T_o$ on $\mathcal{H}_o$ is a unilateral shift and $T_u$ on $\mathcal{H}_u$ is unitary. In this decomposition the spaces $\mathcal{H}_o$ or $\mathcal{H}_u$ can be zero. Moreover, $\mathcal{H}_o$ and $\mathcal{H}_u$ are given by

$$\mathcal{H}_o = \bigoplus_{0}^{\infty} T^n\mathcal{L} \text{ and } \mathcal{H}_u = \bigcap_{0}^{\infty} T^n\mathcal{H} \tag{1.16}$$

where $\mathcal{L} = \ker T^*$; see Chapter I of Sz.-Nagy-Foias [3] or Chapter VI in Foias-Frazho [4].

In this book the reader will often encounter a phrase like: ... the optimization problem

$$d = \inf \{\|f(x)\| : x \in \Omega\} , \tag{1.17}$$

where f is some operator-valued function and the parameter x runs over a certain set $\Omega$. Here we would like to mention that this phrase actually refers to a set of problems. In fact, the optimization problem (1.17) includes the problem to determine and compute the value of the number d, the problem of finding elements x in $\Omega$ for which the infimum in (1.17) is attained (such an element x will be called an *optimal solution* to the problem (1.17)), and also the problem to describe the set of all x in $\Omega$ such that $\|f(x)\| \leq \gamma$, where $\gamma \geq d$ is a certain prespecified bound (which is usually referred to as a *tolerance*).

## I.2. NEVANLINNA-PICK INTERPOLATION

Let G be an analytic function on $\mathbb{D}$ with values in $\mathcal{L}(\mathcal{Y}, \mathcal{X})$, and let Z on $\mathcal{X}$ be a bounded linear operator with spectral radius $r_{spec}(Z) < 1$. We define the *left value of* G *at* Z (notation: $G(Z)_{left}$) to be the operator $\tilde{B} \in \mathcal{L}(\mathcal{Y}, \mathcal{X})$ given by

$$\tilde{B} = \sum_{\nu=0}^{\infty} Z^{\nu} G_{\nu} , \tag{2.1}$$

where $G_0, G_1, G_2, \ldots$ are the Taylor coefficients of G as in (1.6). Since G is analytic on $\mathbb{D}$, we have

$$\limsup_{n\to\infty} \|G_n\|^{1/n} \leq 1 .$$

This inequality, together with the fact that $r_{spec}(Z) < 1$, implies that the series in the right hand side of (2.1) converges in the operator norm. By interchanging in (2.1) the positions of $Z^{\nu}$ and $G_{\nu}$ one obtains the definition of the right value $G(Z)_{right}$ of G at Z (for $Z \in \mathcal{L}(\mathcal{Y}, \mathcal{Y})$).

For $j = 1, \ldots, N$, let $Z_j \in \mathcal{L}(\mathcal{X}, \mathcal{X})$ with $r_{spec}(Z_j) < 1$ and $\tilde{B}_j \in \mathcal{L}(\mathcal{Y}, \mathcal{X})$ be given operators. The *left Nevanlinna-Pick interpolation problem* for the data $Z_j, \tilde{B}_j$, where $j = 1, \ldots, N$, with predetermined specified bound (tolerance) $\gamma$, is to find (if possible) a G in $H^{\infty}(\mathcal{Y}, \mathcal{X})$ such that

$$G(Z_j)_{left} = \tilde{B}_j \quad (\text{for } j = 1, \ldots, N) \quad \text{and} \quad \|G\|_{\infty} \leq \gamma . \tag{2.2}$$

Notice that this is precisely the classical Nevanlinna-Pick interpolation problem when $\mathcal{Y} = \mathcal{X} = \mathbb{C}$ and $\gamma = 1$. The Nevanlinna-Pick interpolation problem may not have a solution. The following result which we will prove (in Chapter II), by using the commutant lifting theorem, gives necessary and sufficient conditions for the existence of a solution to this problem.

**THEOREM 2.1.** *The left Nevanlinna-Pick interpolation problem for the data* $Z_j, \tilde{B}_j$, *where* $j = 1, \ldots, N$ *and with tolerance* $\gamma$ *has a solution if and only if the* $N \times N$ *operator matrix with entries*

$$\left[ \sum_{\nu=0}^{\infty} Z_j^{\nu} [\gamma^2 I - \tilde{B}_j \tilde{B}_k^*] (Z_k^{\nu})^* \right]_{j,k=1}^{N} \tag{2.3}$$

*is positive* ($\geq 0$).

In an analogous way one may formulate the right Nevanlinna-Pick interpolation problem for operator-valued functions and the conditions under which it is solvable.

**REMARK 2.2.** Recall that the classical block Nevanlinna-Pick interpolation problem is, given a set of points $z_1, z_2, \ldots, z_N$ in the open unit disc and a set of operators $\tilde{B}_j$ from $\mathcal{Y}$ to $\mathcal{X}$ for $1 \leq j \leq N$, to find a function G in $H^\infty(\mathcal{Y}, \mathcal{X})$ (if one exists) such that

$$G(z_j) = \tilde{B}_j \quad (\text{for } j = 1, \ldots N) \quad \text{and} \quad \|G\|_\infty \leq \gamma. \tag{2.4}$$

Notice that in this case the j,k entry of the matrix (2.3) is given by

$$\sum_{\nu=0}^{\infty} z_j^\nu (\gamma^2 I - \tilde{B}_j \tilde{B}_k^*) \bar{z}_k^\nu = \frac{\gamma^2 I - \tilde{B}_j \tilde{B}_k^*}{1 - z_j \bar{z}_k}. \tag{2.5}$$

Therefore there exists a solution to the classical block Nevanlinna-Pick interpolation problem if and only if the block matrix

$$\left[ \frac{\gamma^2 I - \tilde{B}_j \tilde{B}_k^*}{1 - z_j \bar{z}_k} \right]_{j=1,k=1}^{N} \tag{2.6}$$

is positive ($\geq 0$). For the scalar case (i.e., when $\mathcal{Y} = \mathcal{X} = \mathbb{C}$) the matrix (2.6) is the classical Pick matrix.

For later purposes we mention the following product rule

$$(GF)(Z)_{\text{left}} = (G(Z)_{\text{left}} F)(Z)_{\text{left}}. \tag{2.7}$$

Here $F : \mathbb{D} \to \mathcal{L}(\mathcal{E}, \mathcal{Y})$ and $G : \mathbb{D} \to \mathcal{L}(\mathcal{Y}, \mathcal{X})$ are analytic operator-valued functions (with $\mathcal{E}$, $\mathcal{Y}$ and $\mathcal{X}$ Hilbert spaces), and $Z \in \mathcal{L}(\mathcal{X}, \mathcal{X})$ has spectral radius strictly less than one. To prove (2.7) let $F_0, F_1, F_2, \ldots$ and $G_0, G_1, G_2, \ldots$ be the Taylor coefficients of F and G, respectively. Then $\sum_{j=0}^{n} G_{n-j} F_j$ is the n-th Taylor coefficient of GF, and hence

$$\begin{aligned}(GF)(Z)_{\text{left}} &= \sum_{n=0}^{\infty} Z^n \left[ \sum_{j=0}^{n} G_{n-j} F_j \right] = \sum_{j=0}^{\infty} Z^j \left[ \sum_{n=j}^{\infty} Z^{n-j} G_{n-j} \right] F_j \\ &= \sum_{j=0}^{\infty} Z^j (G(Z)_{\text{left}}) F_j = (G(Z)_{\text{left}} F)(Z)_{\text{left}},\end{aligned}$$

which proves (2.7). The change of the order of summation in the above calculation is justified by the fact that $r_{\text{spec}}(Z) < 1$.

## I.3. TANGENTIAL NEVANLINNA-PICK INTERPOLATION

To introduce the *data* for the left tangential Nevanlinna-Pick problem, for $j = 1, \ldots, N$ let

$$Z_j : \mathcal{X}_j \to \mathcal{X}_j, \quad B_j : \mathcal{U} \to \mathcal{X}_j \text{ and } \tilde{B}_j : \mathcal{Y} \to \mathcal{X}_j \tag{3.1}$$

be bounded linear operators acting between Hilbert spaces, and assume that

$$r_{spec}(Z_j) < 1 \qquad (\text{for } j = 1, \ldots, N)\,. \tag{3.2}$$

The *left tangential Nevanlinna-Pick interpolation problem* for the data $Z_j$, $B_j$ and $\tilde{B}_j$, where $j = 1, \ldots, N$, and tolerance $\gamma$ is to find (if possible) a $G \in H^\infty(\mathcal{Y}, \mathcal{U})$ such that

$$(B_j G)(Z_j)_{\text{left}} = \tilde{B}_j \qquad (\text{for } j = 1, \ldots, N) \text{ and } \|G\|_\infty \le \gamma\,. \tag{3.3}$$

Here $B_jG$ denotes the function in $H^\infty(\mathcal{Y}, \mathcal{X}_j)$ whose n-th Taylor coefficient is equal to $B_jG_n$, where $G_n$ is the n-th Taylor coefficient of G. Thus

$$(B_j G)(Z_j)_{\text{left}} = \sum_{n=0}^{\infty} Z_j^n B_j G_n\,. \tag{3.4}$$

The following result provides a necessary and sufficient condition for the existence of a solution to the tangential Nevanlinna-Pick interpolation problem.

**THEOREM 3.1**. *The left tangential Nevanlinna-Pick interpolation problem for the data* $Z_j$, $B_j$, $\tilde{B}_j$ *for* $j = 1, \ldots, N$ *and tolerance* $\gamma$ *has a solution if and only if the* $N \times N$ *operator matrix*

$$\left[\sum_{\nu=0}^{\infty} Z_j^\nu [\gamma^2 B_j B_k^* - \tilde{B}_j \tilde{B}_k^*](Z_k^\nu)^*\right]_{j,k=1}^{N} \tag{3.5}$$

*is positive.*

Theorem 3.1 has an obvious right tangential version which we shall not formulate explicitly. If in Theorem 3.1 the operators $B_1, \ldots, B_N$ are taken to be the identity operator on $\mathcal{U}$, then Theorem 3.1 reduces to Theorem 2.1, and hence the latter is a corollary of the former.

**REMARK 3.2**. Recall that the classical block left tangential Nevanlinna-Pick interpolation problem is: Given a set of complex numbers $z_1, \ldots, z_N$ in the open unit disc along with the operators $B_j$ from $\mathcal{U}$ to $\mathcal{X}_j$ and $\tilde{B}_j$ from $\mathcal{Y}$ to $\mathcal{X}_j$, find (if possible) a function G in $H^\infty(\mathcal{Y}, \mathcal{U})$ satisfying

$$B_jG(z_j) = \tilde{B}_j \quad (\text{for } j = 1, \ldots, N) \quad \text{and} \quad \|G\|_\infty \le \gamma \tag{3.6}$$

where $\gamma$ is a prespecified bound. According to Theorem 3.1, there exists a solution to this Nevanlinna-Pick interpolation problem if and only if the $N \times N$ block matrix

$$\left[\frac{\gamma^2 B_j B_k^* - \tilde{B}_j \tilde{B}_k^*}{1 - z_j \bar{z}_k}\right]_{j,k=1}^{N} \tag{3.7}$$

is positive.

**REMARK 3.3**. Let us associate with the interpolation data $Z_j$, $B_j$, $\tilde{B}_j$ for $j = 1, \ldots, N$, the following operator matrices

$$Z = \begin{bmatrix} Z_1 & & & \\ & Z_2 & & \\ & & \ddots & \\ & & & Z_N \end{bmatrix}, \quad B = \begin{bmatrix} B_1 \\ B_2 \\ \vdots \\ B_N \end{bmatrix}, \quad \tilde{B} = \begin{bmatrix} \tilde{B}_1 \\ \tilde{B}_2 \\ \vdots \\ \tilde{B}_N \end{bmatrix}. \tag{3.8}$$

The blank spots in the definition of $Z$ denote zero operator entries and thus $Z$ is a block diagonal matrix. Obviously,

$$Z : \mathcal{X} \to \mathcal{X}, \; B : \mathcal{U} \to \mathcal{X} \text{ and } \tilde{B} : \mathcal{Y} \to \mathcal{X}, \tag{3.9}$$

where $\mathcal{X}$ stands for the Hilbert space direct sum $\mathcal{X}_1 \oplus \mathcal{X}_2 \oplus \cdots \oplus \mathcal{X}_N$. Because of (3.2) we have $r_{spec}(Z) < 1$. Note that the first N conditions in (3.3) are equivalent to the requirement that $(BG)(Z)_{left} = \tilde{B}$. Furthermore, the $N \times N$ operator matrix in (3.5) is now given by

$$\sum_{\nu=0}^{\infty} Z^\nu [\gamma^2 BB^* - \tilde{B}\tilde{B}^*](Z^\nu)^*, \tag{3.10}$$

and hence the operator matrix in (3.5) is positive if and only if the operator in (3.10) is positive. Thus we may conclude that it suffices to prove Theorem 3.1 for the case when $N = 1$. In particular, G is a solution to the classical left tangential Nevanlinna-Pick interpolation problem with data $Z_j$, $B_j$, $\tilde{B}_j$ and tolerance $\gamma$ if and only if G is a solution to the left tangential Nevanlinna-Pick interpolation problem with one data point $Z$, $B$, $\tilde{B}$ defined by (3.8) and tolerance $\gamma$. Motivated by this reduction we call the left tangential Nevanlinna-Pick problem for the case when $N = 1$ the *standard left Nevanlinna-Pick problem.* It turns out that many important $H^\infty$ interpolation problems, such as the Hermite-Fejér interpolation problem, can be converted to the standard left Nevanlinna-Pick problem.

As a first illustration of Theorem 3.1, let us solve the following operator-valued version of the Schur interpolation problem. *Given operators* $A_0, \ldots, A_{r-1}$ *from* $\mathcal{Y}$ *into* $\mathcal{U}$ *determine* $G \in H^\infty(\mathcal{Y}, \mathcal{U})$ *such that* $A_0, \ldots, A_{r-1}$ *are the first r Taylor coefficients of* G *and* $\|G\|_\infty \le 1$. To obtain necessary and sufficient conditions for the existence of such a G let

$$Z = \begin{bmatrix} 0 & \ldots & 0 & 0 \\ I & & & 0 \\ \vdots & \ddots & & \vdots \\ 0 & \ldots & I & 0 \end{bmatrix}, \quad B = \begin{bmatrix} I \\ 0 \\ \vdots \\ 0 \end{bmatrix}, \quad \tilde{B} = \begin{bmatrix} A_0 \\ A_1 \\ \vdots \\ A_{r-1} \end{bmatrix}, \tag{3.11}$$

where I denotes the identity operator on $\mathcal{U}$. Here Z is a block lower shift on $\mathcal{X} = \ell_r^2(\mathcal{U})$ where the identity operators appear immediately below the main diagonal and zeros elsewhere. The operator B maps $\mathcal{U}$ into $\mathcal{X}$ while $\tilde{B}$ maps $\mathcal{Y}$ into $\mathcal{X}$. It is straightforward to check that $G \in H^\infty(\mathcal{Y}, \mathcal{U})$ has $A_0, \ldots, A_{r-1}$ as its first r Taylor coefficients if and only if $(BG)(Z)_{\text{left}} = \tilde{B}$. It follows that the Schur interpolation problem for the data $A_0, \ldots, A_{r-1}$ is solvable if and only if the left tangential Nevanlinna-Pick interpolation problem for the data Z, B, $\tilde{B}$ and tolerance $\gamma = 1$ is solvable. Furthermore, in this case both problems have the same set of solutions. Now let W from $\ell_r^2(\mathcal{U})$ into $\mathcal{X}$ and $\tilde{W}$ from $\ell_r^2(\mathcal{Y})$ into $\mathcal{X}$ be the operators defined by

$$W = [B, ZB, \ldots, Z^{r-1}B] \text{ and } \tilde{W} = [\tilde{B}, Z\tilde{B}, \ldots, Z^{r-1}\tilde{B}]. \tag{3.12}$$

A simple calculation readily shows that $W = I$ and $\tilde{W} = T$, where I is the identity operator on $\ell_r^2(\mathcal{U})$ and T is the operator given by the following block lower triangular operator Toeplitz matrix

$$T = \begin{bmatrix} A_0 & & & \\ A_1 & A_0 & & \\ \vdots & \vdots & \ddots & \\ A_{r-1} & A_{r-2} & \ldots & A_0 \end{bmatrix} : \ell_r^2(\mathcal{Y}) \to \ell_r^2(\mathcal{U}). \tag{3.13}$$

Using the fact that $Z^\nu = 0$ for all $\nu \ge r$ and $\gamma = 1$, it readily follows that the operator in (3.10) is given by

$$\sum_{\nu=0}^{\infty} Z^\nu [BB^* - \tilde{B}\tilde{B}^*] Z^{*\nu} = WW^* - \tilde{W}\tilde{W}^* = I - TT^*. \tag{3.14}$$

Therefore the operator-valued Schur interpolation problem for the data $A_0, \ldots, A_{r-1}$ is solvable if and only if the operator T in (3.13) is a contraction, that is, $\|T\| \le 1$. Finally, it is noted that the above Schur interpolation problem may also be seen as a special case of the left tangential Hermite-Fejér interpolation problem that we shall consider in Section I.5.

## I.4. CONTROLLABILITY OPERATORS AND INTERPOLATION

In this section we will present a systems interpretation for the standard left Nevanlinna-Pick interpolation problem. To this end, let Z be an operator on $\mathcal{X}$ whose spectral radius is strictly less than one and B an operator from $\mathcal{U}$ to $\mathcal{X}$. The *controllability operator* W corresponding to the pair {Z, B} is the linear operator from $\ell_+^2(\mathcal{U})$ into $\mathcal{X}$ defined by

$$W = [B, ZB, Z^2B, Z^3B, \dots] . \tag{4.1}$$

The right hand side of (4.1) denotes the one row operator matrix which assigns to the vector $[y_o, y_1, y_2, \dots]^{tr}$ from $\ell_+^2(\mathcal{U})$ the vector $\Sigma_{n=0}^{\infty} Z^n B y_n$ in $\mathcal{X}$. Because $r_{spec}(Z) < 1$, it follows that W is a well-defined bounded operator. The positive operator $P = WW^*$ on $\mathcal{X}$ is called the *controllability grammian* for {Z, B}. Notice that

$$P = WW^* = \sum_{n=0}^{\infty} Z^n BB^* Z^{*n} . \tag{4.2}$$

The series on the right converges in the operator topology because $r_{spec}(Z) < 1$. The pair {Z, B} is *controllable* if its controllability grammian P is strictly positive. In Chapter III we will present a system theory motivation behind the controllability operator.

The controllability grammian P can be obtained by solving the controllability Lyapunov equation

$$P = ZPZ^* + BB^* . \tag{4.3}$$

Moreover, there is only one solution to this Lyapunov equation and this solution is given by (4.2). Indeed, direct substitution shows that $P = \Sigma_0^\infty Z^n BB^* Z^{*n}$ is a solution to (4.3). On the other hand, if P is any solution to this Lyapunov equation, then by recursive substitution

$$P = ZPZ^* + BB^* = Z^2PZ^{*2} + ZBB^*Z^* + BB^* = Z^m PZ^{*m} + \sum_{n=0}^{m-1} Z^n BB^* Z^{*n} .$$

Since $r_{spec}(Z) < 1$, this converges to $P = \Sigma_0^\infty Z^n BB^* Z^{*n}$. Hence this P is the only solution to the Lyapunov equation in (4.3).

Now let $\tilde{B}$ be an operator from $\mathcal{Y}$ into $\mathcal{X}$ and $\tilde{W}$ the controllability operator from $\ell_+^2(\mathcal{Y})$ to $\mathcal{X}$ corresponding to the pair $\{Z, \tilde{B}\}$, that is, $\tilde{W}$ is given by (4.1) where $\tilde{B}$ replaces B. Let G be a function in $H^\infty(\mathcal{Y}, \mathcal{U})$ and $T_G$ the corresponding (lower triangular) Toeplitz operator from $\ell_+^2(\mathcal{Y})$ into $\ell_+^2(\mathcal{U})$ given by (1.9). Then it is easy to verify that

$$(BG)(Z)_{left} = \tilde{B} \text{ if and only if } WT_G = \tilde{W} . \tag{4.4}$$

Therefore the standard left tangential Nevanlinna-Pick interpolation problem is equivalent to finding a function G in $H^\infty(\mathcal{Y}, \mathcal{U})$ satisfying

$$WT_G = \tilde{W} \quad \text{and} \quad \|T_G\| = \|G\|_\infty \le \gamma . \tag{4.5}$$

where $\gamma$ is a predetermined specified bound. For the moment assume that G is a solution to the left tangential Nevanlinna-Pick interpolation problem. Let $\tilde{P} = \tilde{W}\tilde{W}^*$ be the controllability grammian for the pair $\{Z, \tilde{B}\}$. Then using $WT_G = \tilde{W}$ we have

$$\tilde{P} = \tilde{W}\tilde{W}^* = WT_G T_G^* W^* \le \gamma^2 WW^* = \gamma^2 P .$$

This proves the necessary part of the following result which is equivalent to Theorem 3.1.

**THEOREM 4.1**. *Let* Z *be an operator on* $\mathcal{X}$ *whose spectral radius is strictly less than one,* B *an operator from* $\mathcal{U}$ *to* $\mathcal{X}$ *and* $\tilde{B}$ *an operator from* $\mathcal{Y}$ *to* $\mathcal{X}$. *Let* P *and* $\tilde{P}$ *be the controllability grammians for the pairs* $\{Z, B\}$ *and* $\{Z, \tilde{B}\}$, *respectively. Then there exists a solution to the standard left tangential Nevanlinna-Pick interpolation problem with data* Z, B, $\tilde{B}$ *and tolerance* $\gamma$ *if and only if* $\tilde{P} \le \gamma^2 P$.

Let Z be the block diagonal operator on $\mathcal{X} = \mathcal{X}_1 \oplus \ldots \oplus \mathcal{X}_n$ and B the operator from $\mathcal{U}$ to $\mathcal{X}$ defined by

$$Z = \begin{bmatrix} z_1 I & & & \\ & z_2 I & & \\ & & \ddots & \\ & & & z_N I \end{bmatrix} \quad \text{and} \quad B = \begin{bmatrix} B_1 \\ B_2 \\ \vdots \\ B_N \end{bmatrix} \tag{4.6}$$

where $z_i$ are scalars in the open unit disc for all i and the blank spots in the matrix for Z are zero. Then using (4.2) it is easy to show that the controllability grammian P for the pair {Z, B} is given by the $N \times N$ block matrix

$$P = \left\{ \frac{B_j B_k^*}{1 - z_j \bar{z}_k} \right\}_{j,k=1}^{N} . \tag{4.7}$$

So if $\tilde{B}$ is the matrix from $\mathcal{Y}$ to $\mathcal{X}$ given in (3.8), then the controllability grammian $\tilde{P}$ for the pair $\{Z, \tilde{B}\}$ is given by the $N \times N$ block matrix in (4.7) where $\tilde{P}$ replaces P and $\tilde{B}$ replaces B. Therefore $\gamma^2 P - \tilde{P}$ is precisely the $N \times N$ block matrix in (3.7). Hence there exists a solution to the classical block Nevanlinna-Pick problem if and only if $\gamma^2 P - \tilde{P} \ge 0$, or equivalently, the matrix in (3.7) is positive. Finally, G is a solution to the classical Nevanlinna-Pick interpolation problem with data $z_j$, $B_j$, $\tilde{B}_j$ and tolerance $\gamma$ if and only if G is a solution to the standard Nevanlinna-Pick problem with data Z, B, $\tilde{B}$ defined in (4.6), (3.8) and tolerance $\gamma$.

As before, assume that Z B and $\tilde{B}$ are the data for the standard left Nevanlinna-Pick problem. Now consider the following $H^\infty$ optimization problem:

$$d_\infty = \inf \{ \|G\|_\infty : G \in H^\infty(\mathcal{Y}, \mathcal{U}) \text{ and } WT_G = \tilde{W} \} . \tag{4.8}$$

Let $\mathcal{H}'$ be the closure of the range of $W^*$. Then we claim that $d_\infty$ is finite if and only if there exists an operator A from $\ell_+^2(\mathcal{Y})$ into $\mathcal{H}'$ satisfying $WA = \tilde{W}$. Moreover, in this case $d_\infty = \|A\|$. Notice that if $d_\infty$ is finite, then for any $\gamma > d_\infty$ there exists a G in $H^\infty(\mathcal{Y}, \mathcal{U})$ such that $WT_G = \tilde{W}$ where $\|T_G\| = \gamma$. Hence $\tilde{W}\tilde{W}^* = WT_G T_G^* W^* \le \gamma^2 WW^*$. So there exists an operator C from $\mathcal{H}'$ to $\ell_+^2(\mathcal{Y})$ satisfying $CW^* = \tilde{W}^*$ and $\|C\| \le \gamma$. Choosing $A = C^*$ gives $WA = \tilde{W}$. Since $\mathcal{H}'$ is orthogonal to ker W, the operator A is uniquely determined by $WA = \tilde{W}$, and A does not depend on the choice of $\gamma$. Because we can choose $\gamma$ arbitrarily close to $d_\infty$ we see that $\|A\| \le d_\infty$. On the other hand, if $WA = \tilde{W}$, then clearly $\tilde{P} = \tilde{W}\tilde{W}^* \le \|A\|^2 \tilde{W}\tilde{W}^* = \|A\|^2 P$. By Theorem 4.1 there exists a G in $H^\infty(\mathcal{Y}, \mathcal{U})$ satisfying $WT_G = \tilde{W}$ and $d_\infty \le \|G\|_\infty \le \|A\|$. Therefore $d_\infty = \|A\|$. This proves our claim. Furthermore, we see that the infimum in (4.8) is attained.

## I.5. TANGENTIAL HERMITE-FEJER INTERPOLATION

To introduce the data for the left tangential Hermite-Fejér problem, for $i = 0, \ldots, r_j - 1$ and $j = 1, \ldots, N$, let

$$Z_j : \mathcal{X}_j \to \mathcal{X}_j \, , \; B_{i,j} : \mathcal{U} \to \mathcal{X}_j \quad \text{and} \quad \tilde{B}_{i,j} : \mathcal{Y} \to \mathcal{X}_j \tag{5.1}$$

be bounded linear operators acting between Hilbert spaces, and assume that

$$r_{spec}(Z_j) < 1 \qquad (\text{for } j = 1, \ldots, N) . \tag{5.2}$$

The *left tangential Hermite-Fejér interpolation problem* for the data $Z_j$, $B_{i,j}$, $\tilde{B}_{i,j}$, where $i = 0, \ldots, r_j - 1$ and $j = 1, \ldots, N$, with tolerance $\gamma$ is to find $G \in H^\infty(\mathcal{Y}, \mathcal{U})$ such that

$$\sum_{k=0}^{i} \frac{1}{k!} (B_{i-k,j} G)^{(k)} (Z_j)_{\text{left}} = \tilde{B}_{i,j} \qquad (i = 0, \ldots r_j - 1 \text{ and } j = 1, \ldots, N) \text{ and } \|G\|_\infty \le \gamma . \tag{5.3}$$

Here $(B_{q,j}G)^{(k)}$ denotes the k-th derivative of the function $B_{q,j}G$. If $F_0, F_1, F_2, \ldots$ are the Taylor coefficients for a function F in $H^\infty(\mathcal{Y}, \mathcal{U})$, then the k-th derivative of F is the function

$$F^{(k)}(\lambda) = k! \sum_{n=k}^{\infty} \begin{bmatrix} n \\ k \end{bmatrix} \lambda^{n-k} F_n \qquad (\lambda \in \mathbb{D}) . \tag{5.4}$$

(The n over k in the brackets is the binomial coefficient $n!/k!(n-k)!$) It follows that the operators appearing in the left hand side of (5.3) are defined by

$$\frac{1}{k!}(B_{i-k,j}G)^{(k)}(Z_j)_{\text{left}} = \sum_{n=k}^{\infty} \begin{bmatrix} n \\ k \end{bmatrix} Z_j^{n-k} B_{i-k,j} G_n , \tag{5.5}$$

where $G_0, G_1, G_2, \ldots$ are the Taylor coefficients of G. Since $(B_{i-k,j}G)^{(k)}(\lambda)$ is analytic on $\mathbb{D}$, the spectral radius condition (5.2) implies that the series in the right hand side of (5.5) converges in the operator norm.

If the operators $B_{i,j}$ are taken to be zero operators for $i = 1, \ldots, r_j - 1$ and $j = 1, \ldots, N$, then the previous Hermite-Fejér problem will be called the *left tangential Schur-Fejér interpolation problem*. In particular, by choosing $Z_1 = 0$ and $N = 1$ this contains the following tangential version of the classical block Schur interpolation problem considered at the end of Section 3. Given an operator B from $\mathcal{U}$ to $\mathcal{X}_1$ and a set of operators $\tilde{B}_i$ from $\mathcal{Y}$ to $\mathcal{X}_1$ for $i = 0, 1, \ldots, r_1 - 1$, find (if possible) a function G in $H^\infty(\mathcal{Y}, \mathcal{U})$ satisfying $\|G\|_\infty \leq \gamma$ for some specified bound $\gamma$ and $BG_i = \tilde{B}_i$ for $i = 0, 1, 2, \ldots, r_1 - 1$, where $G_0, G_1, \ldots, G_{r_1-1}$ are the first $r_1$ Taylor coefficients of G.

The following result gives necessary and sufficient conditions for the existence of a solution to the left tangential Hermite-Fejér interpolation problem.

**THEOREM 5.1.** *The left tangential Hermite-Fejér interpolation problem for the data* $Z_j, B_{i,j}, \tilde{B}_{i,j}$, *where* $i = 0, \ldots, r_j - 1$ *and* $j = 1, \ldots, N$, *with tolerance* $\gamma$ *has a solution if and only if the* $N \times N$ *operator matrix*

$$\left[ \sum_{v=0}^{\infty} M_j^v [\gamma^2 R_j R_k^* - \tilde{R}_j \tilde{R}_k^*](M_k^v)^* \right]_{j,k=1}^{N} \tag{5.6}$$

*is positive. Here, for* $j = 1, \ldots, N$,

$$M_j = \begin{bmatrix} Z_j & & & \\ I & Z_j & & \\ & \ddots & \ddots & \\ & & I & Z_j \end{bmatrix} : \ell^2_{r_j}(\mathcal{X}_j) \to \ell^2_{r_j}(\mathcal{X}_j) , \tag{5.7}$$

$$R_j = \begin{bmatrix} B_{0,j} \\ B_{1,j} \\ \vdots \\ B_{r_j-1,j} \end{bmatrix} : \mathcal{U} \to \ell^2_{r_j}(\mathcal{X}_j) \quad \textit{and} \quad \tilde{R}_j = \begin{bmatrix} \tilde{B}_{0,j} \\ \tilde{B}_{1,j} \\ \vdots \\ \tilde{B}_{r_j-1,j} \end{bmatrix} : \mathcal{Y} \to \ell^2_{r_j}(\mathcal{X}_j), \tag{5.8}$$

*and the blank spots in the operator matrix representing* $M_j$ *denote zero entries.*

Theorem 5.1 has an obvious right tangential version which we shall not formulate here. If in Theorem 5.1 one takes $r_j = 1$ and $j = 1, \ldots, N$, then Theorem 5.1 reduces to Theorem 3.1. For the next theorem we need the following notation. Given a set of operators $A_1, A_2, \ldots, A_n$ acting between the appropriate Hilbert spaces, then $[A_1, A_2, \ldots, A_n]^{tr}$ denotes the one column operator matrix whose j-th entry is $A_j$.

The next theorem shows that Theorem 5.1 may be obtained as a corollary of Theorem 3.1.

**THEOREM 5.2.** *Let* $Z_j$, $B_{i,j}$ *and* $\tilde{B}_{i,j}$ *for* $i = 0, \ldots, r_j - 1$ *and* $j = 1, \ldots, N$ *be the data in (5.1) for the left tangential Hermite-Fejér interpolation problem. Let* $Z$ *be the operator on* $\mathcal{X} = \ell^2_{r_1}(\mathcal{X}_1) \oplus \cdots \oplus \ell^2_{r_N}(\mathcal{X}_N)$ *and* $B$ *the operator from* $\mathcal{U}$ *to* $\mathcal{X}$ *and* $\tilde{B}$ *the operator from* $\mathcal{Y}$ *to* $\mathcal{X}$ *defined by*

$$Z = \operatorname{diag}[M_1, M_2, \ldots, M_N]$$
$$B = [R_1, R_2, \ldots, R_N]^{tr} \quad \text{and} \quad \tilde{B} = [\tilde{R}_1, \tilde{R}_2, \ldots, \tilde{R}_N]^{tr}. \tag{5.9}$$

*where* $M_j$, $R_j$ *and* $\tilde{R}_j$ *are defined in (5.7) and (5.8). Then* $G$ *in* $H^\infty(\mathcal{Y}, \mathcal{U})$ *is a solution to the left tangential Hermite-Fejér interpolation problem for the data* $Z_j$, $B_{i,j}$, $\tilde{B}_{i,j}$ *with tolerance* $\gamma$ *if and only if* $G$ *is a solution to the standard left Nevanlinna-Pick problem for the data* $Z$, $B$, $\tilde{B}$ *in (5.9) with tolerance* $\gamma$. *In particular, there exists a solution to the left tangential Hermite-Fejér problem if and only if* $\tilde{P} \leq \gamma^2 P$ *where* $P$ *and* $\tilde{P}$ *are the controllability grammians for* $\{Z, B\}$ *and* $\{Z, \tilde{B}\}$, *respectively.*

**PROOF.** Fix $1 \leq j \leq N$ and $G \in H^\infty(\mathcal{Y}, \mathcal{U})$. Let $G_0, G_1, G_2, \ldots$ be the Taylor coefficients of $G$. First notice that the j-th block entry of $\Sigma_{n=0}^{\infty} Z^n B G_n$ is precisely equal to $\Sigma_{n=0}^{\infty} M_j^n R_j G_n$. Thus, in order to prove the theorem, it suffices to show that

$$\sum_{k=0}^{i} \frac{1}{k!} (B_{i-k,j} G)^{(k)} (Z_j)_{\text{left}} = \tilde{B}_{i,j} \qquad (i = 0, \ldots, r_j - 1) \tag{5.10}$$

is equivalent to the equality

$$\sum_{n=0}^{\infty} M_j^n R_j G_n = \tilde{R}_j . \tag{5.11}$$

To prove the equivalence of (5.10) and (5.11), let D on $\ell^2_{r_j}(\mathcal{X}_j)$ be the $r_j \times r_j$ block diagonal operator with $Z_j$ on the main diagonal, and let E on $\ell^2_{r_j}(\mathcal{X}_j)$ be the $r_j \times r_j$ block lower shift, that is, the identity operator on $\mathcal{X}_j$ appears at each spot directly below the main diagonal and all other block entries of E are zero. Obviously, $M_j = D + E$. Using the fact that D and E commute we have

$$M_j^n = (D+E)^n = \sum_{k=0}^{n} \binom{n}{k} E^k D^{n-k} . \tag{5.12}$$

Then using (5.12) along with $E^{r_j} = 0$ we have

$$\sum_{n=0}^{\infty} M_j^n R_j G_n = \sum_{n=0}^{\infty} \sum_{k=0}^{n} \binom{n}{k} E^k D^{n-k} R_j G_n = \sum_{k=0}^{r_j - 1} \sum_{n=k}^{\infty} \binom{n}{k} E^k D^{n-k} R_j G_n . \tag{5.13}$$

Notice that for $0 \le k \le r_j - 1$ we have

$$E^k D^{n-k} R_j G_n = [0, 0, \ldots, 0, Z_j^{n-k} B_{0,j} G_n, Z_j^{n-k} B_{1,j} G_n, \ldots, Z_j^{n-k} B_{r_j-1-k,j} G_n]^{tr}$$

where the first k block entries are zero. So for $i = 0, \ldots, r_j - 1$ the $i+1$ block entry of $E^k D^{n-k} R_j G_n$ is equal to $Z_j^{n-k} B_{i-k,j} G_n$ if $k \le i$ and zero otherwise. Thus, by (5.5) and (5.13), we have

$$\left[\sum_{n=0}^{\infty} M_j^n R_j G_n\right]_{i+1} = \sum_{k=0}^{i} \sum_{n=k}^{\infty} \binom{n}{k} Z_j^{n-k} B_{i-k,j} G_n = \sum_{k=0}^{i} \frac{1}{k!} (B_{i-k,j} G)^{(k)} (Z_j)_{left} .$$

Since the $i+1$ block entry of $\tilde{R}_j$ is equal to $\tilde{B}_{i,j}$ for $i = 0, \ldots, r_j - 1$, the equivalence of (5.10) and (5.11) is proved.

Notice that according to (5.11)

$$(R_j G)(M_j)_{left} = \tilde{R}_j \quad (\text{for } j = 1, 2, \ldots, N) .$$

This shows that the tangential Hermite-Fejér interpolation problem in Theorem 5.1 reduces to the Nevanlinna-Pick interpolation problem with data $M_j$, $R_j$, $\tilde{R}_j$ for $j = 1, \ldots, N$ treated in Section I.3. Therefore Theorem 5.2 can also be inferred from Remark 3.3.

**EXAMPLE 5.3.** Let $B_i$ from $\mathcal{U}$ to $\mathcal{X}_1$ and $\tilde{B}_i$ from $\mathcal{Y}$ to $\mathcal{X}_1$ for $i = 0, 1, ..., r-1$ be the data for the following left tangential Schur interpolation problem: Find (if possible) a function G in $H^\infty(\mathcal{Y}, \mathcal{U})$ satisfying $\|G\|_\infty \leq \gamma$ and

$$\begin{bmatrix} B_0 & 0 & 0 & \cdots & 0 \\ B_1 & B_0 & 0 & \cdots & 0 \\ B_2 & B_1 & B_0 & \cdots & 0 \\ \cdot & \cdot & \cdot & \cdot & \cdot \\ \cdot & \cdot & \cdot & \cdot & \cdot \\ \cdot & \cdot & \cdot & \cdot & \cdot \\ B_{r-1} & B_{r-2} & B_{r-3} & \cdots & B_0 \end{bmatrix} \begin{bmatrix} G_0 \\ G_1 \\ G_2 \\ \cdot \\ \cdot \\ \cdot \\ G_{r-1} \end{bmatrix} = \begin{bmatrix} \tilde{B}_0 \\ \tilde{B}_1 \\ \tilde{B}_2 \\ \cdot \\ \cdot \\ \cdot \\ \tilde{B}_{r-1} \end{bmatrix} \tag{5.14}$$

where $G_k$ for $k = 0, 1, ..., r-1$ are the first Taylor coefficients for G. Notice that this is precisely the left tangential Hermite-Fejér interpolation problem for the case when $N = 1$, $Z_1 = 0$, $B_i = B_{i,1}$ and $\tilde{B}_i = \tilde{B}_{i,1}$ for $i = 0, ..., r-1$. The $r \times r$ block matrix in the left hand side of (5.14) is referred to as the *lower triangular block Toeplitz matrix generated* by $\{B_i\}_0^{r-1}$.

According to Theorem 5.2 the operators Z, B and $\tilde{B}$ in the corresponding standard left Nevanlinna-Pick problem are given by

$$Z = \begin{bmatrix} 0 & & & & \\ I & 0 & & & \\ & I & \ddots & & \\ & & \ddots & \ddots & \\ & & & I & 0 \end{bmatrix}, \quad B = \begin{bmatrix} B_0 \\ B_1 \\ \vdots \\ B_{r-1} \end{bmatrix} \quad \text{and} \quad \tilde{B} = \begin{bmatrix} \tilde{B}_0 \\ \tilde{B}_1 \\ \vdots \\ \tilde{B}_{r-1} \end{bmatrix}, \tag{5.15}$$

where the blank spots in the matrix for Z are zero. Notice that Z is simply a lower shift with the identity below the main diagonal and zeros elsewhere. Using the fact that $Z^r = 0$ we see that the controllability operators for the pairs $\{Z, B\}$ and $\{Z, \tilde{B}\}$ are respectively given by

$$W = [T, 0, 0, 0 \ldots] \quad \text{and} \quad \tilde{W} = [\tilde{T}, 0, 0, 0 \ldots]$$

where T and $\tilde{T}$ are the $r \times r$ lower triangular block Toeplitz matrices generated by $\{B_i\}_0^{r-1}$ and $\{\tilde{B}_i\}_0^{r-1}$, respectively. So the controllability grammians P and $\tilde{P}$ for the pairs $\{Z, B\}$ and $\{Z, \tilde{B}\}$ are given by $P = WW^* = TT^*$ and $\tilde{P} = \tilde{T}\tilde{T}^*$. Therefore there exists a solution to this left tangential Schur interpolation problem if and only if $\gamma^2 TT^* \geq \tilde{T}\tilde{T}^*$.

Notice that with $B_0 = I$ and $B_j = 0$ for $1 \leq j \leq r-1$, and $\tilde{B}_j = A_j$ for $0 \leq j \leq r-1$, the problem (5.14) is precisely the Schur interpolation problem considered at the end of Section I.3.

To complete this section we will present another proof of Theorem 5.2 based on the following lemma which is of independent interest.

**LEMMA 5.4.** *Let* Z *on* $\mathcal{X}$ *and* $\tilde{B}_0, \ldots, \tilde{B}_{r-1}$ *from* $\mathcal{Y}$ *to* $\mathcal{X}$ *be bounded linear operators. and let* $G : \mathbb{D} \to \mathcal{L}(\mathcal{Y}, \mathcal{X})$ *be an analytic operator-valued function. Assume that* $r_{spec}(Z) < 1$. *Then* $G^{(i)}(Z)_{left} = \tilde{B}_i$ *for* $i = 0, \ldots, r-1$ *if and only if*

$$(\lambda I - Z)^{-r}\left\{G(\lambda) - \sum_{\nu=0}^{r-1} \frac{1}{\nu!}(\lambda I - Z)^{\nu}\tilde{B}_{\nu}\right\} \tag{5.16}$$

*extends to an analytic function on* $\mathbb{D}$.

**PROOF.** The proof is split into two parts.

Part (a). First we prove the lemma for $r = 1$. Note that $(\lambda I - Z)^{-1}G(\lambda)$ is analytic on the annulus $\{\lambda \in \mathbb{C} : r_{spec}(Z) < |\lambda| < 1\}$. Consider the Laurent expansion

$$(\lambda I - Z)^{-1}G(\lambda) = \sum_{\nu=-\infty}^{\infty} \lambda^{\nu}F_{\nu} \qquad (r_{spec}(Z) < |\lambda| < 1).$$

Using $(\lambda I - Z)^{-1} = \sum_{1}^{\infty} Z^{n-1}\lambda^{-n}$ for $|\lambda| > r_{spec}(Z)$ we have

$$F_{-\nu} = Z^{\nu-1}G_0 + Z^{\nu}G_1 + Z^{\nu+1}G_2 + \ldots = Z^{\nu-1}G(Z)_{left} \qquad (\nu \geq 1).$$

It follows that

$$\sum_{\nu=-\infty}^{-1} \lambda^{\nu}F_{\nu} = (\lambda I - Z)^{-1}G(Z)_{left} \qquad (\text{for } r_{spec}(Z) < |\lambda| < 1).$$

Hence $(\lambda I - Z)^{-1}[G(\lambda) - \tilde{B}_0]$ extends to an analytic function on $\mathbb{D}$ if and only if $(\lambda I - Z)^{-1}[G(Z)_{left} - \tilde{B}_0]$ extends to an analytic function on $\mathbb{D}$. Since the spectrum of Z is in $\mathbb{D}$, the latter happens if and only if $G(Z)_{left} = \tilde{B}_0$, which proves the lemma for $r = 1$.

Part (b). We proceed by induction. Assume the lemma holds for some $r = k \geq 1$. Take $r = k + 1$. First suppose that

$$(\lambda I - Z)^{-(k+1)}\left[G(\lambda) - \sum_{\nu=0}^{k} \frac{1}{\nu!}(\lambda I - Z)^{\nu}\tilde{B}_{\nu}\right]$$

extends to a function (H say) which is analytic on $\mathbb{D}$. Then

$$G(\lambda) = \sum_{\nu=0}^{k} \frac{1}{\nu!}(\lambda I - Z)^{\nu}\tilde{B}_{\nu} + (\lambda I - Z)^{k+1}H(\lambda) .$$

Thus for $i \le k$ the i-th derivative of G is of the form $G^{(i)}(\lambda) = \tilde{B}_i + (\lambda I - Z)F(\lambda)$, where F is analytic on $\mathbb{D}$. But then we can use the product rule (2.7) to show that

$$G^{(i)}(Z)_{\text{left}} = \tilde{B}_i \qquad (\text{for } i = 0, \ldots, k) . \tag{5.17}$$

To prove the reverse implication, assume that (5.17) holds. Then our induction hypothesis implies that

$$(\lambda I - Z)^{-k} \left[ G(\lambda) - \sum_{\nu=0}^{k-1} \frac{1}{\nu!}(\lambda I - Z)^{\nu}\tilde{B}_{\nu} \right]$$

extends to a function $K(\lambda)$ which is analytic on $\mathbb{D}$. So, in this case,

$$G(\lambda) = \sum_{\nu=0}^{k-1} \frac{1}{\nu!}(\lambda I - Z)^{\nu}\tilde{B}_{\nu} + (\lambda I - Z)^{k}K(\lambda) \qquad (\lambda \in \mathbb{D}) ,$$

and we can use formula (5.17) and the product rule (2.7) to show that

$$\tilde{B}_k = G^{(k)}(Z)_{\text{left}} = k!K(Z)_{\text{left}} .$$

In particular, by the result of Part (a), we see that $(\lambda I - Z)^{-1}[K(\lambda) - (1/k!)\tilde{B}_k]$ extends to a analytic function on $\mathbb{D}$. It follows that

$$\begin{aligned} (\lambda I - Z)^{-(k+1)} \left[ G(\lambda) - \sum_{\nu=0}^{k} \frac{1}{\nu!}(\lambda I - Z)^{\nu}\tilde{B}_{\nu} \right] &= (\lambda I - Z)^{-(k+1)} \left[ (\lambda I - Z)^{k}K(\lambda) - \frac{1}{k!}(\lambda I - Z)^{k}\tilde{B}_k \right] \\ &= (\lambda I - Z)^{-1} \left[ K(\lambda) - \frac{1}{k!}\tilde{B}_k \right] \end{aligned}$$

and the latter extends to an analytic function on $\mathbb{D}$. This completes the proof.

**PROOF OF THEOREM 5.2** (using Lemma 5.4). Let $M_j$, $R_j$ and $\tilde{R}_j$ ($j = 1, \ldots, N$) be given by (5.7) and (5.8). For each $\lambda$ in the resolvent set of $Z_j$ the operator $\lambda I - M_j$ is invertible and $(\lambda I - M_j)^{-1}$ has the following block lower triangular representation:

$$(\lambda I - M_j)^{-1} = \begin{bmatrix} (\lambda I - Z_j)^{-1} & 0 & \cdots & 0 \\ (\lambda I - Z_j)^{-2} & (\lambda I - Z_j)^{-1} & \cdots & 0 \\ \cdot & \cdot & \cdots & \cdot \\ \cdot & \cdot & \cdots & \cdot \\ \cdot & \cdot & \cdots & \cdot \\ (\lambda I - Z_j)^{-r_j} & (\lambda I - Z_j)^{-(r_j-1)} & \cdots & (\lambda I - Z_j)^{-1} \end{bmatrix} \tag{5.18}$$

Obviously, we see that $r_{spec}(M_j) < 1$ for $j = 1, \ldots, N$. Furthermore,

$$(\lambda I - M_j)^{-1} R_j G(\lambda) = \operatorname{col} \left[ \sum_{v=1}^{i} (\lambda I - Z_j)^{-v} B_{i-v,j} G(\lambda) \right]_{i=1}^{r_j}, \tag{5.19}$$

where col denotes the corresponding one column operator matrix, and

$$(\lambda I - M_j)^{-1} \tilde{R}_j = \operatorname{col} \left[ \sum_{v=1}^{i} (\lambda I - Z_j)^{-v} \tilde{B}_{i-v,j} \right]_{i=1}^{r_j}. \tag{5.20}$$

For $i = 0, \ldots, r_j - 1$ and $j = 1, \ldots, N$, let

$$C_{i,j} = \sum_{k=0}^{i} \frac{1}{k!} (B_{i-k,j} G)^{(k)} (Z_j)_{\text{left}} . \tag{5.21}$$

We first show (cf., formula (5.3)) that

$$C_{i,j} = \tilde{B}_{i,j} \quad (i = 0, \ldots, r_j - 1) \ \text{ if and only if } \ (R_j G)(M_j)_{\text{left}} = \tilde{R}_j . \tag{5.22}$$

One computes that

$$\sum_{v=1}^{i} (\lambda I - Z_j)^{-v} C_{i-v,j} = \sum_{v=1}^{i} (\lambda I - Z_j)^{-v} \left\{ \sum_{k=0}^{v-1} \frac{1}{k!} (\lambda I - Z_j)^{k} (B_{i-v,j} G)^{(k)} (Z_j)_{\text{left}} \right\}.$$

Thus, by Lemma 5.4, for $i = 1, \ldots, r_j$ the function

$$\sum_{v=1}^{i} (\lambda I - Z_j)^{-v} [B_{i-v,j} G(\lambda) - C_{i-v,j}] \tag{5.23}$$

extends to an analytic function on $\mathbb{D}$. Hence, if the identities in the left hand side of (5.22) are satisfied, then the function $(\lambda I - M_j)^{-1} \{R_j G(\lambda) - \tilde{R}_j\}$ extends to an analytic function on $\mathbb{D}$. But, according to Lemma 5.4 (with $r = 1$), the latter implies that the equality in the right hand side of (5.22) is fulfilled.

Conversely, assume that the equality in the right hand side of (5.22) is fulfilled. Then, by Lemma 5.4 (with $r = 1$), the function $(\lambda I - M_j)^{-1}\{R_j G(\lambda) - \tilde{R}_j\}$ extends to an analytic function on $\mathbb{D}$. In other words, for $i = 1, \ldots, r_j$ the function

$$\sum_{\nu=1}^{i} (\lambda I - Z_j)^{-\nu}[B_{i-\nu,j}G(\lambda) - \tilde{B}_{i-\nu,j}] \tag{5.24}$$

extends to an analytic function on $\mathbb{D}$. Now subtract the function in (5.23) from the one in (5.24). We conclude that for $i = 1, \ldots, r_j$ the function

$$\sum_{\nu=1}^{i} (\lambda I - Z_j)^{-\nu}[C_{i-\nu,j} - \tilde{B}_{i-\nu,j}]$$

extends to an analytic function on $\mathbb{D}$. But the latter can only happen when the identities in the left hand side of (5.22) are fulfilled.

We conclude that the problem of finding G in $H^\infty(\mathcal{Y}, \mathcal{U})$ such that (5.3) holds is equivalent to the problem of finding G in $H^\infty(\mathcal{Y}, \mathcal{U})$ such that

$$(R_j G)(M_j)_{\text{left}} = \tilde{R}_j \qquad (\text{for } j = 1, \ldots, N) \quad \text{and} \quad \|G\|_\infty \le \gamma. \tag{5.25}$$

By using the operators Z, B and $\tilde{B}$ in (5.9) we readily see that (5.25) holds if and only if there exists a G in $H^\infty(\mathcal{Y}, \mathcal{U})$ satisfying $(BG)(Z)_{\text{left}} = \tilde{B}$ and $\|G\|_\infty \le \gamma$. Therefore the left tangential Hermite-Fejér interpolation problem can be viewed as a special case of the standard left Nevanlinna-Pick problem. This completes the proof.

## I.6. THE NEHARI EXTENSION PROBLEM

Let $K_1, K_2, K_3, \ldots$ be a sequence of operators in $\mathcal{L}(\mathcal{U}, \mathcal{Y})$, where $\mathcal{U}$ and $\mathcal{Y}$ are separable Hilbert spaces. For such a sequence the *Nehari extension problem* is to find an operator valued function G in $L^\infty(\mathcal{U}, \mathcal{Y})$ such that

$$K_n = \frac{1}{2\pi}\int_0^{2\pi} e^{int}G(e^{it})dt \qquad (\text{for } n = 1, 2, 3, \ldots). \tag{6.1}$$

In this case we refer to G as a Nehari *interpolant* of the sequence $K_1, K_2, K_3, \ldots$. So if G is an interpolant of $K_1, K_2, K_3, \ldots$, then $K_n = G_{-n}$ for all $n \ge 1$ where $\{G_n\}$ are the Fourier coefficients of G. We say that G is a Nehari interpolant of $\{K_n\}_1^\infty$ *with tolerance* $\gamma$ if G is a Nehari interpolant of $\{K_n\}_1^\infty$ and $\|G\|_\infty \le \gamma$. The smallest possible $L^\infty$-norm $\|G\|_\infty$ of such an interpolant is of particular interest.

**THEOREM 6.1.** *The sequence of operators* $K_1, K_2, K_3, \ldots$ *in* $\mathcal{L}(\mathcal{U}, \mathcal{Y})$ *has an interpolant* G *in* $L^\infty(\mathcal{U}, \mathcal{Y})$ *if and only if the operator*

$$\Gamma = \begin{bmatrix} K_1 & K_2 & K_3 & \cdots \\ K_2 & K_3 & K_4 & \cdots \\ K_3 & K_4 & K_5 & \cdots \\ \vdots & \vdots & \vdots & \end{bmatrix} : \ell^2_+(\mathcal{U}) \to \ell^2_+(\mathcal{Y}) \tag{6.2}$$

*is well-defined and bounded. In this case*

$$d_\infty = \min \left\{ \|G\|_\infty : G \in L^\infty(\mathcal{U}, \mathcal{Y}) \text{ and } G \text{ is an interpolant of } K_1, K_2, K_3, \ldots \right\} \tag{6.3}$$

*exists and is precisely equal to the operator norm of the operator* $\Gamma$.

The action of the operator $\Gamma$ in (6.2) is defined by

$$\Gamma((x_j)_{j=1}^\infty) = (y_n)_{n=1}^\infty \quad \text{where} \quad y_n = \sum_{j=1}^\infty K_{n+j-1} x_j \,. \tag{6.4}$$

One refers to $\Gamma$ as the *block Hankel operator* associated with $K_1, K_2, K_3, \ldots$.

Let $G \in L^\infty(\mathcal{U}, \mathcal{Y})$ be an interpolant of $K_1, K_2, K_3, \ldots$, and let $L_G$ from $\ell^2(\mathcal{U})$ to $\ell^2(\mathcal{Y})$ be the Laurent operator defined by G; see (1.4). Now let $P_-$ be the orthogonal projection onto $\ell^2_-(\mathcal{Y})$, where $\ell^2_-(\mathcal{Y})$ is the orthogonal complement of $\ell^2_+(\mathcal{Y})$ in $\ell^2(\mathcal{Y})$. By consulting the matrix form for the Laurent operator $L_G$ in (1.4) along with $K_n = G_{-n}$ for $n \geq 1$ we have

$$P_- L_G \,|\, \ell^2_+(\mathcal{U}) = \begin{bmatrix} \cdot & \cdot & \cdot & \cdots \\ \cdot & \cdot & \cdot & \cdots \\ \cdot & \cdot & \cdot & \cdots \\ K_3 & K_4 & K_5 & \cdots \\ K_2 & K_3 & K_4 & \cdots \\ K_1 & K_2 & K_3 & \cdots \end{bmatrix} . \tag{6.5}$$

Here we identify $\ell^2_-(\mathcal{Y})$ with the Hilbert space of all square summable infinite column vectors

$$\begin{bmatrix} \vdots \\ y_{-3} \\ y_{-2} \\ y_{-1} \end{bmatrix} = [\ldots, y_{-3}, y_{-2}, y_{-1}]^{tr} .$$

So by rearranging the rows of $P_- L_G \,|\, \ell^2_+(\mathcal{U})$ this operator can be identified with the block

Hankel operator $\Gamma$. To see this let J be the unitary flip operator from $\ell_-^2(\mathcal{Y})$ onto $\ell_+^2(\mathcal{Y})$ defined by

$$J[\ldots, y_{-3}, y_{-2}, y_{-1}]^{tr} = [y_{-1}, y_{-2}, y_{-3}, \ldots]^{tr}.$$

Now it readily follows that $\Gamma = JP_-L_G \mid \ell_+^2(\mathcal{U})$ when G is an interpolant of $K_1, K_2, K_3, \ldots$. Thus $\|\Gamma\| \le \|L_G\|$. Since $\|L_G\| = \|G\|_\infty$, we see that

$$\|\Gamma\| \le \inf\{\|G\|_\infty : G \in L^\infty(\mathcal{U}, \mathcal{Y}) \text{ and } G \text{ is an interpolant of } K_1, K_2, K_3, \ldots\}. \quad (6.6)$$

Theorem 6.1 tells us that we have equality in (6.6) and that the infimum is attained.

In many applications a slightly different version of the Nehari problem is considered. In this second version one is given a sequence of operators $F_1, F_2, F_3, \ldots$ in $\mathcal{L}(\mathcal{U}, \mathcal{Y})$. For this sequence the Nehari extension problem is to find a function G in $L^\infty(\mathcal{U}, \mathcal{Y})$ satisfying

$$F_n = \frac{1}{2\pi}\int_0^{2\pi} e^{-int} G(e^{it})dt \qquad (n = 1, 2, 3, \ldots) \quad (6.7)$$

Of course this Nehari problem is equivalent to the previous Nehari extension problem, and by a slight abuse of terminology we will also refer to G as an interpolant of $F_1, F_2, F_3, \ldots$ when (6.7) holds. By applying the Nehari extension Theorem 6.1 to $G(e^{-it})$, we see that there exists a G in $L^\infty(\mathcal{U}, \mathcal{Y})$ satisfying (6.7) if and only if the block Hankel operator from $\ell_+^2(\mathcal{U})$ to $\ell_+^2(\mathcal{Y})$ generated by $F_1, F_2, F_3, \ldots$, which will be denoted by $\Gamma_+$, is bounded. Moreover,

$$\|\Gamma_+\| = \inf\{\|G\|_\infty : G \in L^\infty(\mathcal{U}, \mathcal{Y}) \text{ and } G \text{ interpolates } F_1, F_2, F_3 \text{ by (6.7)}\} \quad (6.8)$$

Finally, if G is a function in $L^\infty(\mathcal{U}, \mathcal{Y})$ interpolating $F_1, F_2, F_3, \ldots$, then the Hankel operator $\Gamma_+$ can be identified with

$$P_+L_G \mid \ell_-^2(\mathcal{U}) = \begin{bmatrix} \cdots & F_3 & F_2 & F_1 \\ \cdots & F_4 & F_3 & F_2 \\ \cdots & F_5 & F_4 & F_3 \\ \cdots & \cdot & \cdot & \cdot \\ \cdots & \cdot & \cdot & \cdot \\ \cdots & \cdot & \cdot & \cdot \end{bmatrix}. \quad (6.9)$$

where $P_+$ is the orthogonal projection onto $\ell_+^2(\mathcal{Y})$. To be precise, $\Gamma_+J = P_+L_G \mid \ell_-^2(\mathcal{U})$, where J is now the unitary flip operator from $\ell_-^2(\mathcal{U})$ to $\ell_+^2(\mathcal{U})$. In particular, $\|\Gamma_+\| \le \|G\|_\infty$.

The Nehari extension problem can also be formulated for the case when the underlying Hilbert spaces are not required to be separable. We conclude this section with this nonseparable version. Let $F_{-1}, F_{-2}, \ldots$ be bounded linear operators from $\mathcal{U}$ to $\mathcal{Y}$, where $\mathcal{U}$ and $\mathcal{Y}$ are

Hilbert spaces which may be nonseparable. A block Laurent operator $G = (G_{j-k})_{j,k=-\infty}^{\infty}$ from $\ell^2(\mathcal{U})$ to $\ell^2(\mathcal{Y})$ is said to be a *solution of the Nehari problem for the data* $F_{-1}, F_{-2}, \ldots$ *and tolerance* $\gamma$ if $G_j = F_j$ for $j < 0$ and $\|G\| \leq \gamma$. The next theorem tells us when this extension problem is solvable.

**THEOREM 6.2**. *The Nehari problem for the data* $F_{-1}, F_{-2}, \ldots$ *and tolerance* $\gamma$ *is solvable if and only if the infinite operator matrix*

$$\begin{bmatrix} F_{-1} & F_{-2} & F_{-3} & \cdots \\ F_{-2} & F_{-3} & F_{-4} & \cdots \\ F_{-3} & F_{-4} & F_{-5} & \cdots \\ \vdots & \vdots & \vdots & \end{bmatrix}$$

*defines a bounded linear operator from* $\ell_+^2(\mathcal{U})$ *to* $\ell_+^2(\mathcal{Y})$ *of norm bounded by* $\gamma$.

## I.7. SARASON INTERPOLATION

Throughout this section all Hilbert spaces are separable. Recall that a function $\Theta$ is *inner* if $\Theta \in H^\infty(\mathcal{E}, \mathcal{Y})$ and $\Theta(e^{it})$ is a.e. an isometry. It is well known that a function $\Theta$ in $H^\infty(\mathcal{E}, \mathcal{Y})$ is inner if and only if the multiplication operator $M_\Theta^+$ is an isometry from $H^2(\mathcal{E})$ to $H^2(\mathcal{Y})$. In particular, a function $\Theta$ in $L^\infty(\mathcal{E}, \mathcal{Y})$ is inner if and only if its corresponding block Toeplitz matrix $T_\Theta$ is a lower triangular isometry. For further results on inner functions see Chapter V in Sz.-Nagy-Foias [3] or Chapter IX in Foias-Frazho [4].

Now let F be a function in $H^\infty(\mathcal{U}, \mathcal{Y})$ and $\Theta$ an inner function in $H^\infty(\mathcal{E}, \mathcal{Y})$. Given a tolerance $\gamma \geq 0$, the *Sarason interpolation problem* is to find a $H \in H^\infty(\mathcal{U}, \mathcal{E})$ such that $\|F - \Theta H\| \leq \gamma$. In particular, one is interested in the smallest $\gamma$ for which this interpolation problem is solvable. A non-separable version of the Sarason problem will be discussed in Section I.10.

**THEOREM 7.1**. *Let* F *be in* $H^\infty(\mathcal{U}, \mathcal{Y})$, *and let* $\Theta$ *be an inner function in* $H^\infty(\mathcal{E}, \mathcal{Y})$. *Let* $\mathcal{H}' = H^2(\mathcal{Y}) \ominus \Theta H^2(\mathcal{E})$ *and* $P_{\mathcal{H}'}$ *be the orthogonal projection of* $H^2(\mathcal{Y})$ *onto* $\mathcal{H}'$. *Then there exists* $H \in H^\infty(\mathcal{U}, \mathcal{E})$ *such that* $\|F - \Theta H\|_\infty \leq \gamma$ *if and only if the operator* $P_{\mathcal{H}'} M_F^+$ *is bounded by* $\gamma$, *that is,*

$$\|P_{\mathcal{H}'} M_F^+\| \leq \gamma. \tag{7.1}$$

Let F be in $H^\infty(\mathcal{U}, \mathcal{Y})$, and let $\Theta$ be an inner function in $H^\infty(\mathcal{E}, \mathcal{Y})$. The set of problems described in the paragraph preceding the above theorem will be referred to as the $H^\infty$ optimization problem associated with

$$d_\infty(F, \Theta) = \inf\{\|F - \Theta H\| : H \in H^\infty(\mathcal{U}, \mathcal{E})\}\,. \tag{7.2}$$

The quantity $d_\infty = d_\infty(F, \Theta)$ defined by (7.2) is called the *error* in the Sarason problem. Theorem 7.1 tells us that $d_\infty(F; \Theta)$ is equal to $\|P_{\mathcal{H}'}M_F^+\|$, and the infimum in (7.2) is attained. It is simple to see that

$$\|P_{\mathcal{H}'}M_F^+\| \le d_\infty(F, \Theta)\,. \tag{7.3}$$

Let h be in $H^2(\mathcal{U})$ and H a function in $H^\infty(\mathcal{U}, \mathcal{E})$. Then by $P_{\mathcal{H}'}\Theta Hh = 0$ we have

$$\|P_{\mathcal{H}'}Fh\| = \|P_{\mathcal{H}'}(F - \Theta H)h\| \le \|(F - \Theta H)h\| \le \|F - \Theta H\|_\infty \|h\|\,.$$

Here H in $H^\infty(\mathcal{U}, \mathcal{E})$ is arbitrary, and thus (7.3) holds. The fact that we have equality in (7.3) is a much deeper result.

Recall that an operator valued function $\Theta$ is *two-sided inner* if $\Theta$ is an inner function and $\Theta(e^{it})$ is a.e. unitary. The following result shows that the Sarason problem can be transformed into a Nehari extension problem when $\Theta$ is a two-sided inner function.

**PROPOSITION 7.2.** *Let* F *be in* $H^\infty(\mathcal{U}, \mathcal{Y})$ *and* $\Theta$ *a two-sided inner function in* $H^\infty(\mathcal{E}, \mathcal{Y})$. *Let* $K_1, K_2, K_3, \ldots$ *be the operators in* $\mathcal{L}(\mathcal{U}, \mathcal{E})$ *defined by*

$$K_n = \frac{1}{2\pi}\int_0^{2\pi} e^{int}\Theta(e^{it})^* F(e^{it})dt \qquad (\text{for } n = 1, 2, 3, \ldots)\,. \tag{7.4}$$

*Then* G *is a function in* $L^\infty(\mathcal{U}, \mathcal{E})$ *interpolating* $K_1, K_2, K_3, \ldots$ *(that is* (6.1) *holds) if and only if* $H = \Theta^* F - G$ *is a function in* $H^\infty(\mathcal{U}, \mathcal{E})$. *In this case* G *and* $F - \Theta H$ *have the same* $L^\infty$ *norm. In particular,* H *in* $H^\infty(\mathcal{U}, \mathcal{E})$ *is a solution to the Sarason problem with tolerance* $\gamma$ *associated with* F *and* $\Theta$ *if and only if* $G = \Theta^* F - H$ *is a Nehari interpolant of* $K_1, K_2, K_3, \ldots$ *satisfying* $\|G\|_\infty \le \gamma$. *Finally,*

$$\|\Gamma\| = d_\infty(F, \Theta) = \|P_{\mathcal{H}'}M_F^+\| \tag{7.5}$$

*where* $\Gamma$ *is the block Hankel operator from* $\ell_+^2(\mathcal{U})$ *to* $\ell_+^2(\mathcal{E})$ *generated by* $K_1, K_2, K_3, \ldots$ *and* $\mathcal{H}' = H^2(\mathcal{Y}) \ominus \Theta H^2(\mathcal{E})$.

**PROOF.** Since the Fourier coefficients of $e^{int}$ for $n \le -1$ for any function in $H^\infty$ are zero, it follows that G is an interpolant of $K_1, K_2, K_3, \ldots$ if and only if $H = \Theta^* F - G$ is in $H^\infty(\mathcal{U}, \mathcal{E})$.

Because $\Theta$ is a two-sided inner function $\Theta G = F - \Theta H$. So G and $F - \Theta H$ have the same $L^\infty$ norm. This fact along with Theorem 6.1 and $d_\infty(F, \Theta) = \|P_{\mathcal{H}'} M_F^+\|$ gives (7.5).

One can also prove (7.5) directly. To this end, notice that $H^2(\mathcal{Y}) \ominus \mathcal{H}' := \mathcal{M}$ equals the range of $M_\Theta^+$. Because $M_\Theta^+$ is an isometry, $P_{\mathcal{M}} = M_\Theta^+(M_\Theta^+)^*$ is the orthogonal projection onto $\mathcal{M}$. Using the fact $(M_\Theta^+)^* = P_{H^2} M_{\Theta^*} | H^2(\mathcal{Y})$ we have for h in $H^2(\mathcal{U})$ and $G = \Theta^* F - H$ the following identities:

$$\|P_{\mathcal{H}'} M_F^+ h\|^2 = \|Fh\|^2 - \|P_{\mathcal{M}} Fh\|^2 = \|Fh\|^2 - \|P_{H^2}\Theta^* Fh\|^2 = \|\Theta^* Fh\|^2 - \|P_{H^2}\Theta^* Fh\|^2 =$$

$$\|(I - P_{H^2})\Theta^* Fh\|^2 = \|(I - P_{H^2})(\Theta^* F - H)h\|^2 = \|(I - P_{H^2})Gh\|^2 = \tag{7.6}$$

$$\|P_- L_G \mathcal{F}_{\mathcal{U}}^* h\|^2 = \|\Gamma \mathcal{F}_{\mathcal{U}}^* h\|^2$$

where $P_-$ is the orthogonal projection onto $\ell_-^2(\mathcal{E})$ and $\mathcal{F}_{\mathcal{U}}$ is the Fourier transform (see (1.10)). Since $\mathcal{F}_{\mathcal{U}}$ is unitary, equation (7.6) shows that (7.5) holds, and this completes the proof.

## I.8. NEVANLINNA-PICK INTERPOLATION VIEWED AS A SARASON PROBLEM.

Obviously the left tangential Nevanlinna-Pick interpolation problem is a special case of the left tangential Hermite-Fejér interpolation problem which in turn is a special case of the standard left Nevanlinna-Pick problem. In this section we will show that the standard left Nevanlinna-Pick problem is equivalent to the Sarason problem when the controllability grammian for $\{Z, B\}$ is strictly positive. To this end, we begin with the following useful result.

**THEOREM 8.1**. *Let* W *from* $\ell_+^2(\mathcal{U})$ *to* $\mathcal{X}$ *and* $\tilde{W}$ *from* $\ell_+^2(\mathcal{Y})$ *to* $\mathcal{X}$ *be the controllability operators corresponding to* $\{Z, B\}$ *and* $\{Z, \tilde{B}\}$, *respectively, where* $r_{spec}(Z) < 1$. *Assume that the controllability grammian* $P = WW^*$ *is strictly positive. Let* F *be the function defined by*

$$F(\lambda) = B^*(I - \lambda Z^*)^{-1} P^{-1} \tilde{B} \qquad (|\lambda| < 1). \tag{8.1}$$

*Then* F *is a function in* $H^\infty(\mathcal{Y}, \mathcal{U})$ *satisfying* $WT_F = \tilde{W}$.

**PROOF**. Since $r_{spec}(Z) < 1$, it follows that F is in $H^\infty(\mathcal{Y}, \mathcal{U})$ and F admits a power series expansion of the form

$$F(\lambda) = \sum_0^\infty \lambda^n B^* Z^{*n} P^{-1} \tilde{B}. \tag{8.2}$$

Recall that a function F satisfies $WT_F = \tilde{W}$ if and only if $W[F_0, F_1, F_2, \ldots]^{tr} = \tilde{B}$, where

$F_0, F_1, F_2$ are the Fourier coefficients for F. According to (8.2) we have $F_n = B^* Z^{*n} P^{-1} \tilde{B}$. Thus $[F_0, F_1, F_2, ...]^{tr} = W^* P^{-1} \tilde{B}$ and $W[F_0, F_1, F_2, ...]^{tr} = WW^* P^{-1} \tilde{B} = \tilde{B}$. This completes the proof.

**REMARK 8.2.** For the moment assume that $\mathcal{Y}$ is finite dimensional, and let $H^2(\mathcal{Y}, \mathcal{U})$ be the Hilbert space generated by the set of all analytic functions $F(\lambda)$ in the open unit disc with values in $\mathcal{L}(\mathcal{Y}, \mathcal{U})$ whose norm is defined by

$$\|F\|_2^2 = \sum_{n=0}^{\infty} \text{trace}\,(F_n^* F_n) \tag{8.3}$$

where $F_0, F_1, F_2, ...$ are the Taylor coefficients of F. As in Theorem 8.1, let W and $\tilde{W}$ be the controllability operators for {Z, B} and $\{Z, \tilde{B}\}$, respectively. Now consider the following Wiener or $H^2$ optimization problem associated with:

$$d_2 = \inf \{\|F\|_2 : F \in H^2(\mathcal{Y}, \mathcal{U}) \text{ and } WT_F = \tilde{W}\} . \tag{8.4}$$

If the controllability grammian $P = WW^*$ is strictly positive, then the infimum in (8.4) is attained. Moreover, the function F in $H^2(\mathcal{Y}, \mathcal{U})$ where this infimum is attained is unique and is given by the function F in (8.1). To see this simply notice that the optimization problem in (8.4) is equivalent to

$$d_2^2 = \inf \{ \sum_{n=0}^{\infty} \text{trace}\,(F_n^* F_n) : W[F_0, F_1, F_2, ...]^{tr} = \tilde{B}\} . \tag{8.5}$$

However, this is just a classical least squares optimization problem of the form $d_2 = \inf \{\|x\| : Ax = y\}$ where A is an operator acting between the appropriate Hilbert spaces and y is known. If A is onto, then the unique optimal solution x is given by $x = A^*(AA^*)^{-1} y$ (see Section 6.10 in Luenberger [1]). In our case $A = W$ and $y = \tilde{B}$. So the unique optimal solution is given by

$$[F_0, F_1, F_2, ...]^{tr} = W^*(WW^*)^{-1}\tilde{B} = W^* P^{-1} \tilde{B} .$$

By taking the Fourier transform we arrive at the function F in (8.1). Finally, it is noted that the error $d_2^2 = ((AA^*)^{-1} y, y)$. In our case, $AA^* = P$ and $y = \tilde{B}$. Thus $d_2^2 = \text{trace}\,\tilde{B}^* P^{-1} \tilde{B}$.

**REMARK 8.3.** Let $Z_j$, $B_{i,j}$ and $\tilde{B}_{i,j}$ for $i = 0, ..., r_j - 1$ and $j = 1, ..., N$ be the data for the left tangential Hermite-Fejér interpolation problem. Let Z, B and $\tilde{B}$ be the operators constructed according to equation (5.9) in Theorem 5.2, and assume that the controllability grammian P for {Z, B} is strictly positive. Then, by Theorems 8.1 and 5.2 and the fact that the standard left tangential Nevanlinna-Pick problem is equivalent to (4.5), the function $F(\lambda)$ in (8.1) is a function in $H^\infty(\mathcal{Y}, \mathcal{U})$ satisfying

$$\sum_{k=0}^{i} \frac{1}{k!}(B_{i-k,j}F)^{(k)}(Z_j)_{left} = \tilde{B}_{i,j} \qquad (\text{for } i = 0, \ldots, r_j - 1 \text{ and } j = 1, \ldots, N)\,. \tag{8.6}$$

Moreover, if $\mathcal{Y}$ is finite dimensional, then this function $F(\lambda)$ in (8.1) is the unique function in $H^2(\mathcal{Y}, \mathcal{U})$ with smallest $H^2$ norm satisfying (8.6).

In the remaining part of this section we assume all Hilbert spaces to be separable. Let $S_2$ be the unilateral (forward) shift on $\ell_+^2(\mathcal{U})$. The Beurling-Lax-Halmos Theorem states that $\mathcal{M}$ is an invariant subspace for $S_2$ if and only if $\mathcal{M}$ equals the range of $T_\Theta$ where $\Theta$ is an inner function in $H^\infty(\mathcal{E}, \mathcal{U})$. Moreover, this inner function $\Theta$ is unique up to a unitary constant on the right, that is, if $\Omega$ is another inner function in $H^\infty(\mathcal{E}_1, \mathcal{U})$ satisfying $\mathcal{M} = \operatorname{ran} T_\Omega$, then $\Theta = \Omega U$, where U is a constant unitary operator from $\mathcal{E}$ onto $\mathcal{E}_1$; see Chapter V in Sz.-Nagy-Foias [3], or Chapter IX in Foias-Frazho [4] for further details. Now let W from $\ell_+^2(\mathcal{U})$ to $\mathcal{X}$ be the controllability operator generated by $\{Z, B\}$ where $r_{spec}(Z) < 1$. It is easy to verify that $ZW = WS_2$. Using $ZW = WS_2$ it readily follows that $\mathcal{M} = \ker W$ is an invariant subspace for $S_2$. By the Beurling-Lax-Halmos Theorem, there exists a unique inner function $\Theta$ such that $\ker W = \operatorname{ran} T_\Theta$. (In Chapter III we will provide an explicit construction for this inner function $\Theta$ when the controllability grammian $WW^*$ is invertible.) This sets the stage for the following result which provides an explicit relationship between the standard left Nevanlinna-Pick problem and the Sarason problem.

**THEOREM 8.4.** *Let* W *from* $\ell_+^2(\mathcal{U})$ *to* $\mathcal{X}$ *and* $\tilde{W}$ *from* $\ell_+^2(\mathcal{Y})$ *to* $\mathcal{X}$ *be the controllability operators respectively generated by* $\{Z, B\}$ *and* $\{Z, \tilde{B}\}$, *where* Z, B *and* $\tilde{B}$ *are the data for the standard left Nevanlinna-Pick problem. Let* $\Theta$ *be an inner function in* $H^\infty(\mathcal{E}, \mathcal{U})$ *such that* $\ker W = \operatorname{ran} T_\Theta$. *Finally, assume that there* F *in* $H^\infty(\mathcal{Y}, \mathcal{U})$ *satisfying* $WT_F = \tilde{W}$. *(If the controllability grammian* $P = WW^*$ *is strictly positive, then the function* F *in* (8.1) *is one such* F.) *Then* G *in* $H^\infty(\mathcal{Y}, \mathcal{U})$ *satisfies* $WT_G = \tilde{W}$ *if and only if* $G = F - \Theta H$, *where* H *is a function in* $H^\infty(\mathcal{Y}, \mathcal{E})$. *In particular, a function* G *in* $H^\infty(\mathcal{Y}, \mathcal{U})$ *is a solution to the standard left Nevanlinna-Pick problem with tolerance* $\gamma$ *if and only if* $G = F - \Theta H$ *where* H *in* $H^\infty(\mathcal{Y}, \mathcal{E})$ *is a solution to the Sarason problem with the same tolerance* $\gamma$, *that is,* $\|F - \Theta H\|_\infty \le \gamma$.

**PROOF**. Let H be in $H^\infty(\mathcal{Y}, \mathcal{E})$ and set $G = F - \Theta H$. Then using $WT_\Theta = 0$ and $WT_F = \tilde{W}$ it follows that $WT_G = \tilde{W}$.

On the other hand, assume that $WT_G = \tilde{W}$, where G is in $H^\infty(\mathcal{Y}, \mathcal{U})$. Then we claim that $G = F - \Theta H$ where H is some function in $H^\infty(\mathcal{Y}, \mathcal{E})$. To see this first notice that

$W(T_F - T_G) = 0$. Since the kernel of W equals the range of $T_\Theta$, it follows that the range of $T_F - T_G$ is contained in the range of $T_\Theta$. So given any x in $\ell^2_+(\mathcal{Y})$ there exists a vector $f_x$ in $\ell^2_+(\mathcal{E})$ satisfying $(T_F - T_G)x = T_\Theta f_x$. Because $T_G$, $T_F$ and $T_\Theta$ are all linear operators, the map $f_x$ from $\ell^2_+(\mathcal{Y})$ to $\ell^2_+(\mathcal{E})$ must be linear. Furthermore,

$$\|f_x\| = \|T_\Theta f_x\| = \|(T_F - T_G)x\| \leq (\|T_F\| + \|T_G\|)\|x\| .$$

Hence $f_x$ defines a bounded linear operator Q from $\ell^2_+(\mathcal{Y})$ into $\ell^2_+(\mathcal{E})$ by $Qx = f_x$. Thus $T_F - T_G = T_\Theta Q$. Now let S be the unilateral shift on $\ell^2_+(\mathcal{E})$ and $S_1$ the unilateral shift on $\ell^2_+(\mathcal{Y})$. We claim that $SQ = QS_1$. To see this notice that

$$T_\Theta SQ = S_2 T_\Theta Q = S_2(T_F - T_G) = (T_F - T_G)S_1 = T_\Theta QS_1 .$$

Because $T_\Theta$ is an isometry, $SQ = QS_1$. This implies that there exists a function H in $H^\infty(\mathcal{Y}, \mathcal{E})$ satisfying $Q = T_H$, and thus $T_G = T_F - T_\Theta T_H$. Finally, notice that $T_\Theta T_H = T_{\Theta H}$. (This simply says that convolution in the time domain is multiplication in the frequency domain.) Therefore $T_G = T_F - T_{\Theta H}$ or equivalently, $G = F - \Theta H$, which proves our claim.

So we have shown that if F is any function in $H^\infty(\mathcal{Y}, \mathcal{U})$ satisfying $WT_F = \tilde{W}$, then the set of all functions G in $H^\infty(\mathcal{Y}, \mathcal{U})$ satisfying $WT_G = \tilde{W}$ is given by $G = F - \Theta H$, where H is an arbitrary function $H^\infty(\mathcal{Y}, \mathcal{E})$. Therefore finding a function G in $H^\infty(\mathcal{Y}, \mathcal{U})$ satisfying

$$WT_G = \tilde{W} \quad \text{and} \quad \|G\|_\infty \leq \gamma, \tag{8.7}$$

is equivalent to finding an H in $H^\infty(\mathcal{Y}, \mathcal{E})$ satisfying $\|F - \Theta H\|_\infty \leq \gamma$. In this case $G = F - \Theta H$ satisfies (8.7). Therefore the standard left Nevanlinna-Pick problem can be viewed as a special case of the Sarason problem. This completes the proof.

## I.9 TWO-SIDED NUDELMAN INTERPOLATION

Consider the Hilbert space operators

$$\begin{gathered} Z : \mathcal{X} \to \mathcal{X}, \; B : \mathcal{U} \to \mathcal{X}, \; \tilde{B} : \mathcal{Y} \to \mathcal{X}, \\ \Lambda : \mathcal{F} \to \mathcal{F}, \; C : \mathcal{F} \to \mathcal{Y}, \; \tilde{C} : \mathcal{F} \to \mathcal{U} \quad \text{and} \quad \Gamma : \mathcal{F} \to \mathcal{X}, \end{gathered} \tag{9.1}$$

where the operators Z and $\Lambda$ are assumed to have spectral radii strictly less than one. The *two-sided Nudelman interpolation problem* for the data in (9.1) and tolerance $\gamma$ is to find (if possible) a function G in $H^\infty(\mathcal{Y}, \mathcal{U})$ such that $\|G\|_\infty \leq \gamma$ and

$$(BG)(Z)_{\text{left}} = \tilde{B} \quad \text{and} \quad (GC)(\Lambda)_{\text{right}} = \tilde{C}, \tag{9.2}$$

and the following coupling condition is satisfied

$$\sum_{j,k=0}^{\infty} Z^j B G_{j+k+1} C \Lambda^k = \Gamma , \tag{9.3}$$

where $G_0, G_1, G_1, \ldots$ are the Taylor coefficients of G.

Assume that $G \in H^\infty(\mathcal{Y}, \mathcal{U})$ satisfies (9.2) and (9.3), then

$$Z\Gamma = \sum_{j,k=0}^{\infty} Z^j B G_{j+k} C \Lambda^k - B\tilde{C} \quad \text{and} \quad \Gamma\Lambda = \sum_{j,k=0}^{\infty} Z^j B G_{j+k} C \Lambda^k - \tilde{B}C .$$

Hence for the two-sided Nudelman interpolation problem to be solvable the operator $\Gamma$ must satisfy the following Sylvester equation:

$$Z\Gamma - \Gamma\Lambda = \tilde{B}C - B\tilde{C} . \tag{9.4}$$

This sets the stage for the following existence result.

**THEOREM 9.1**. *Consider the interpolation data in (9.1), and assume that the Sylvester equation (9.4) holds. Let* M *and* N *be the operators defined by*

$$M = \sum_{\nu=0}^{\infty} (\Lambda^\nu)^* [\gamma^2 C^* C - \tilde{C}^* \tilde{C}] \Lambda^\nu \quad \text{and} \quad N = \sum_{\nu=0}^{\infty} Z^\nu [\gamma^2 BB^* - \tilde{B}\tilde{B}^*](Z^\nu)^* . \tag{9.5}$$

*Then the two-sided Nudelman interpolation problem for the data in (9.1) and tolerance $\gamma$ has a solution if and only if the operator*

$$\Xi = \begin{bmatrix} M & \gamma\Gamma^* \\ \gamma\Gamma & N \end{bmatrix} : \mathcal{F} \oplus \mathcal{X} \to \mathcal{F} \oplus \mathcal{X} \tag{9.6}$$

*is positive.*

**REMARK 9.2**. Let Q and $\tilde{Q}$ be the controllability grammians for the pair $\{\Lambda^*, C^*\}$ and $\{\Lambda^*, \tilde{C}^*\}$, respectively. As before, let P and $\tilde{P}$ be the controllability grammians for $\{Z, B\}$ and $\{Z, \tilde{B}\}$, respectively. Then $M = \gamma^2 Q - \tilde{Q}$ and $N = \gamma^2 P - \tilde{P}$. Therefore the two-sided Nudelman problem for the data (9.1) and with tolerance $\gamma$ has a solution if and only if the operator

$$\begin{bmatrix} \gamma^2 Q - \tilde{Q} & \gamma\Gamma^* \\ \gamma\Gamma & \gamma^2 P - \tilde{P} \end{bmatrix} \tag{9.7}$$

is positive.

## I.10 THE TWO-SIDED SARASON PROBLEM

A function $\Omega$ in $H^\infty(\mathcal{U}, \mathcal{E})$ is *inner* if its corresponding Laurent operator $L_\Omega$ is a lower triangular isometry, and $\Omega$ is called *co-inner* if $L_\Omega$ is a lower triangular co-isometry. In the case when the underlying Hilbert spaces are separable (see the beginning of Section I.7), $\Omega \in H^\infty(\mathcal{U}, \mathcal{E})$ is inner (respectively, co-inner) if and only if $\Omega(e^{it})$ is a.e. an isometry (respectively, co-isometry). Notice that a function $\Omega$ in $H^\infty(\mathcal{U}, \mathcal{E})$ is co-inner if and only if $\Omega^{\sim}(\lambda) := \sum_0^\infty \Omega_n^* \lambda^n = \Omega(\bar{\lambda})^*$ is an inner function in $H^\infty(\mathcal{E}, \mathcal{U})$, where $\Omega_n$ for $n \geq 0$ are the Taylor coefficients of $\lambda^n$ for $\Omega$. In other words, a function $\Omega$ is co-inner if and only if the Laurent operator $\tilde{L}$ corresponding to $\Omega^{\sim}$ is a lower triangular isometry.

Now let $\Theta_1$ be a co-inner function in $H^\infty(\mathcal{U}, \mathcal{E})$ with Laurent operator $L_1 = L_{\Theta_1}$ and $\Theta_2$ be an inner function in $H^\infty(\mathcal{F}, \mathcal{Y})$ with Laurent operator $L_2 = L_{\Theta_2}$. Let $L_F$ be a bounded Laurent operator from $\ell^2(\mathcal{U})$ to $\ell^2(\mathcal{Y})$ generated by a bilateral sequence of operator $\{F_n\}_{-\infty}^{\infty}$ in $\mathcal{L}(\mathcal{U}, \mathcal{Y})$. The two-sided Sarason problem with data F, $\Theta_1$ and $\Theta_2$, which has its origins in control theory (see Francis [1] and the last two paragraphs of this section), is to find a H in $H^\infty(\mathcal{E}, \mathcal{F})$ to solve the optimization problem associated with:

$$d_\infty(F, \Theta_1, \Theta_2) = \inf \{\|L_F - L_2 L_H L_1\| : H \in H^\infty(\mathcal{E}, \mathcal{F})\} . \tag{10.1}$$

The following result shows that the infimum in (10.1) is always attained.

**THEOREM 10.1.** *Let $\Theta_1$ in $H^\infty(\mathcal{U}, \mathcal{E})$ be co-inner, $\Theta_2$ in $H^\infty(\mathcal{F}, \mathcal{Y})$ be inner, and the bounded Laurent operator $L_F$ from $\ell^2(\mathcal{U})$ to $\ell^2(\mathcal{Y})$ be the data for the two-sided Sarason problem. Then there exists a function H in $H^\infty(\mathcal{E}, \mathcal{F})$ satisfying*

$$\|L_F - L_2 L_H L_1\| = d_\infty = \inf \{\|L_F - L_2 L_H L_1\| : H \in H^\infty(\mathcal{E}, \mathcal{F})\} \tag{10.2}$$

*where $L_2 = L_{\Theta_2}$ and $L_1 = L_{\Theta_1}$. Moreover, if $\mathcal{H}$ and $\mathcal{H}'$ are the subspaces defined by*

$$\mathcal{H} = \ell^2(\mathcal{U}) \ominus L_1^* \ell_-^2(\mathcal{E}) \quad \text{and} \quad \mathcal{H}' = \ell^2(\mathcal{Y}) \ominus L_2 \ell_+^2(\mathcal{F}) \tag{10.3}$$

*then $d_\infty = \|P_{\mathcal{H}'} L_F \,|\, \mathcal{H}\|$. In particular, there exists a function H in $H^\infty(\mathcal{E}, \mathcal{F})$ satisfying $\|L_F - L_2 L_H L_1\| \leq \gamma$ for some prespecified tolerance $\gamma$ if and only if $\|P_{\mathcal{H}'} L_F \,|\, \mathcal{H}\| \leq \gamma$.*

By taking the Fourier transforms we arrive at the following result in the separable case.

**COROLLARY 10.2.** *Let $\Theta_1$ in $H^\infty(\mathcal{U}, \mathcal{E})$ be co-inner, $\Theta_2$ in $H^\infty(\mathcal{F}, \mathcal{Y})$ be inner and F a function in $L^\infty(\mathcal{U}, \mathcal{Y})$, where $\mathcal{U}$, $\mathcal{Y}$, $\mathcal{E}$ and $\mathcal{F}$ are all separable Hilbert spaces. Then there exists a function H in $H^\infty(\mathcal{E}, \mathcal{F})$ satisfying*

$$\|F - \Theta_2 H \Theta_1\|_\infty = d_\infty = \inf\{\|F - \Theta_2 H \Theta_1\|_\infty : H \in H^\infty(\mathcal{E}, \mathcal{F})\}. \tag{10.4}$$

*Moreover, if* $\mathcal{H}$ *and* $\mathcal{H}'$ *are now the subspaces of* $L^2(\mathcal{U})$ *and* $L^2(\mathcal{Y})$ *defined by*

$$\mathcal{H} = L^2(\mathcal{U}) \ominus \Theta_1^* K^2(\mathcal{E}) \quad \text{and} \quad \mathcal{H}' = L^2(\mathcal{Y}) \ominus \Theta_2 H^2(\mathcal{F}) \tag{10.5}$$

*where* $K^2(\mathcal{E}) = L^2(\mathcal{E}) \ominus H^2(\mathcal{E})$, *then* $d_\infty = \|P_{\mathcal{H}'} M_F \mid \mathcal{H}\|$.

Notice that if $\Theta_1(\lambda) = I$ for all $\lambda \in \mathbb{D}$, and $F \in H^\infty(\mathcal{U}, \mathcal{Y})$, then Corollary 10.2 reduces to Theorem 7.1. It follows that Theorem 10.1 also covers the one-sided Sarason problem.

In control theory one arrives at the following model matching problem, which will be discussed in more detail in Chapter VII. Let F be a function in $L^\infty(\mathcal{U}, \mathcal{Y})$ and $\Omega_1$ be in $H^\infty(\mathcal{U}, \mathcal{E}_o)$ while $\Omega_2$ is in $H^\infty(\mathcal{F}_o, \mathcal{Y})$, with all underlying Hilbert spaces being separable. Then the *model matching problem* corresponding to these data is to solve the $H^\infty$ optimization problem associated with:

$$d'_\infty = \inf\{\|F - \Omega_2 H_o \Omega_1\|_\infty : H_o \in H^\infty(\mathcal{E}_o, \mathcal{F}_o)\}. \tag{10.6}$$

To convert this model matching problem to a two-sided Sarason problem, first recall that a function $\Theta$ in $H^\infty(\mathcal{U}, \mathcal{Y})$ is *outer* if $\Theta H^2(\mathcal{U})$ is dense in $H^2(\mathcal{Y})$. Moreover, $\Theta$ is an *invertible outer* function if $\Theta$ is in $H^\infty(\mathcal{U}, \mathcal{Y})$ and its inverse $\Theta(\lambda)^{-1}$ exists for all $\lambda$ in $\mathbb{D}$ and is also in $H^\infty(\mathcal{Y}, \mathcal{U})$. Obviously an invertible outer function is an outer function. A function $\Theta$ is *co-outer* if $\Theta^\sim(\lambda) := \Theta(\bar{\lambda})^*$ is outer, and $\Theta$ is an *invertible co-outer* function if $\Theta^\sim$ is an invertible outer function.

Now let $\Omega_2 = \Theta_2 \Theta_{2o}$ be an inner-outer factorization of $\Omega_2$, where $\Theta_2$ is an inner function in $H^\infty(\mathcal{F}, \mathcal{Y})$ and $\Theta_{2o}$ is an outer function in $H^\infty(\mathcal{F}_o, \mathcal{F})$. Let $\Omega_1 = \Theta_{1o}\Theta_1$ be the co-outer-co-inner factorization of $\Omega_1$, where $\Theta_{1o}$ is a co-outer function in $H^\infty(\mathcal{E}, \mathcal{E}_o)$ and $\Theta_1$ is a co-inner function in $H^\infty(\mathcal{U}, \mathcal{E})$. It is well known that any function in $H^\infty(\mathcal{U}, \mathcal{Y})$ admits an inner-outer factorization and a co-outer-co-inner factorization; see Chapter V in Sz.-Nagy-Foias [3] or Chapter IX in Foias-Frazho [4]. The results in Appendix A.2 show how one can construct these factorizations for certain rational matrix functions. So using these factorizations it follows that $d'_\infty = d_\infty$, where $d_\infty$ is the quantity appearing in the $H^\infty$ Sarason optimization problem (10.4). In particular, if $\Theta_{2o}$ is an invertible outer function and $\Theta_{1o}$ is an invertible co-outer function, then H in $H^\infty(\mathcal{E}, \mathcal{F})$ is a solution to the Sarason optimization problem (10.4) if and only if $H_o = \Theta_{2o}^{-1} H \Theta_{1o}^{-1}$ is a solution to the model matching optimization problem (10.6). So a solution to the two-sided Sarason problem readily gives a solution to the model matching problem.

## I.11. A FILTERING PROBLEM

In this section we will show how the right-sided Sarason problem naturally occurs in estimation theory. To this end, let us recall some elementary facts concerning weakly stationary processes. First recall that a *random variable* f is a measurable complex-valued function with respect to some $\sigma$-algebra $(\Omega, \mathcal{A}, P)$ where P is a positive measure with area one, that is, $P(\Omega) = 1$. The *expectation*, denoted by E, of the random variable f is defined by

$$Ef = \int f dP . \tag{11.1}$$

A *random process* $x = (x(n))_{-\infty}^{\infty}$ is a sequence of random variables. In other words, a random processes is a sequence x(n) of measurable functions with respect to $(\Omega, \mathcal{A}, P)$. The *covariance matrix* $R_x$ for the random process x is the infinite matrix given by $R_x = E(xx^*)$. In other words, the (n, m)-th entry of $R_x$ is defined by

$$(R_x)_{n,m} = E(x(n)\overline{x(m)}) = \int x(n)\overline{x(m)} dP . \tag{11.2}$$

We say that a matrix R with entries $R_{n,m}$ for n, m in $\mathbb{Z}$ is positive if $(R\alpha, \alpha) \geq 0$ for all sequences $\alpha = (\alpha_n)_{-\infty}^{\infty}$ with compact support. Notice that the covariance matrix $R_x$ is positive. To see this, let $\alpha = (\alpha_n)_{-\infty}^{\infty}$ be a sequence of complex numbers with compact support. Then

$$0 \leq E\left|\sum \alpha_n \overline{x(n)}\right|^2 = \sum_{m,n} \bar{\alpha}_n E(x(n)\overline{x(m)})\alpha_m$$

$$= \sum_{n,m} \bar{\alpha}_n (R_x)_{n,m} \alpha_m = (R_x\alpha, \alpha) .$$

Therefore $(R_x\alpha, \alpha) \geq 0$ for all $\alpha$ and thus $R_x \geq 0$.

*Throughout this section (except for the text after Proposition 11.2) we assume that the covariance matrix* $R_x$ *defines a bounded operator on* $\ell^2$*, that is, we only consider random processes whose covariance matrix* $R_x$ *is a bounded operator on* $\ell^2$.

The random process x(n) is *weakly stationary* if the following two conditions hold, first $Ex(n) = \mu$, the same constant $\mu$ for all n, and secondly its covariance $R_x$ is a Laurent matrix, that is, $E(x(n)\overline{x(m)}) = R_x(n - m)$ is just a function of the difference between n and m for all n and m.

Now let $x = (x(n))_{-\infty}^{\infty}$ and $y = (y(n))_{-\infty}^{\infty}$ be two random processes. The joint covariance matrix $R_{xy}$ is the infinite matrix defined by $R_{xy} = Exy^*$. In particular, the (i, j)- element of $R_{xy}$ is given by $(R_{xy})_{i,j} = E(x(i)\overline{y(j)})$. Notice that $R_{xy}^* = E(yx^*) = R_{yx}$. If we let $z(n) = [x(n), y(n)]^{tr}$ for $-\infty < n < \infty$, then z(n) is a random process with values in $\mathbb{C}^2$, that is, $z = (z(n))_{-\infty}^{\infty}$ is an infinite sequence of measurable functions with values in $\mathbb{C}^2$. Furthermore, if we set $z = [x, y]^{tr}$, then the covariance $R_z$ for z is given by

$$R_z = E\begin{pmatrix} x \\ y \end{pmatrix}[x^*, y^*] = \begin{bmatrix} R_x & R_{xy} \\ R_{yx} & R_y \end{bmatrix}. \tag{11.3}$$

Notice that because $R_x$ and $R_y$ are positive bounded operators, it follows that $R_z$ is a positive bounded operator on $\ell^2 \oplus \ell^2$. In particular, $R_{xy}$ is a bounded operator on $\ell^2$. To see this let $\alpha = (\alpha_n)_{-\infty}^{\infty}$ and $\beta = (\beta_n)_{-\infty}^{\infty}$ be two sequences in $\ell^2$ with compact support. Then using the Cauchy-Schwartz inequality we have

$$|(R_{xy}\beta, \alpha)|^2 = |E(y^*\beta\alpha^*x)|^2 \le E|y^*\beta|^2 \cdot E|x^*\alpha|^2 = (R_x\alpha, \alpha)(R_y\beta, \beta) \le M^2\|\alpha\|^2\|\beta\|^2$$

where the norms of $R_x$ and $R_y$ are bounded by M. Because sequences with compact support are dense in $\ell^2$, this readily shows that $R_{xy}$ is a bounded operator. Since $R_{yx} = R_{xy}^*$, the covariance $R_z$ is bounded on $\ell^2 \oplus \ell^2$. To prove that $R_z$ is positive, notice that the previous inequality yields $|(R_{xy}\beta, \alpha)|^2 \le (R_x\alpha, \alpha)(R_y\beta, \beta)$. Thus

$$(R_z(\alpha \oplus \beta), (\alpha \oplus \beta)) = (R_x\alpha, \alpha) + 2\,\mathrm{Re}\,(R_{xy}\beta, \alpha) + (R_y\beta, \beta) \ge$$

$$(R_x\alpha, \alpha) - 2(R_x\alpha, \alpha)^{1/2}(R_y\beta, \beta)^{1/2} + (R_y\beta, \beta) = ((R_x\alpha, \alpha)^{1/2} - (R_y\beta, \beta)^{1/2})^2 \ge 0\,.$$

Therefore $R_z$ is positive. So $R_z$ is a positive bounded operator on $\ell^2 \oplus \ell^2$.

*Finally, throughout we will also assume that* $R_y$ *is invertible.* Therefore $R_z$ admits a Schur factorization of the form

$$R_z = \begin{bmatrix} R_x & R_{xy} \\ R_{yx} & R_y \end{bmatrix} =$$

$$\begin{bmatrix} I & R_{xy}R_y^{-1} \\ 0 & I \end{bmatrix}\begin{bmatrix} R_x - R_{xy}R_y^{-1}R_{yx} & 0 \\ 0 & R_y \end{bmatrix}\begin{bmatrix} I & 0 \\ R_y^{-1}R_{yx} & I \end{bmatrix}. \tag{11.4}$$

In particular, $R_z$ is an invertible positive operator if and only if its Schur complement $R = R_x - R_{xy}R_y^{-1}R_{yx}$ is invertible.

We say that two random processes, x(n) and y(n) are *jointly weakly stationary* if both x(n) and y(n) are weakly stationary and the joint covariance matrix $R_{xy}$ is a Laurent operator on $\ell^2$, that is, $E(x(n)\overline{y(m)}) = R_{xy}(n - m)$ is just a function of the difference $n - m$ for all n and m. In this case, $R_z$ is a positive 2 by 2 block operator whose entries are Laurent operators. Hence $R_z$ intertwines the bilateral shift $V \oplus V$ on $\ell^2 \oplus \ell^2$ where V is the bilateral shift on $\ell^2$. Therefore $R_z$ is also a Laurent operator when x(n) and y(n) are jointly weakly stationary. Now we are ready state our filtering problem.

**ESTIMATION PROBLEM 11.1.** Let x(n) and y(n) be two jointly weakly stationary random processes, and assume that the joint covariance matrix $R_z$ in (11.3) or (11.4) is a (positive) invertible operator on $\ell^2 \oplus \ell^2$, and thus $R_y$ is invertible. Let $c \in H^\infty$ and $L_c$ the lower triangular Laurent operator generated by c. The *estimate* $\hat{x} = (\hat{x}(n))_{-\infty}^{\infty}$ of the random process x from the process y with respect to c is the weakly stationary random process defined by $\hat{x} = L_c y$. To be precise, $\hat{x} = (\hat{x}(n))_{-\infty}^{\infty}$ is the random process given by

$$\hat{x}(n)(\omega) = \sum_{k=-\infty}^{n} c_{n-k} y(k)(\omega) \qquad (\omega \in \mathcal{A})$$

where $\{c_n\}_0^\infty$ are the Fourier coefficients of $e^{int}$ for $c(e^{it})$. The covariance $R_{\hat{x}}$ for the estimate $\hat{x}$ is given by $R_{\hat{x}} = L_c R_y L_c^*$. Because $L_c$ and $R_y$ are bounded Laurent operators, it follows that $R_{\hat{x}}$ is a positive bounded Laurent operator on $\ell^2$. So the estimate $\hat{x}$ of x is also a weakly stationary process. The function c in $H^\infty$ is called a causal linear time invariant *filter* for the estimate $\hat{x}$ of x. The *error* between the random process x and its estimate $\hat{x}$ is the weakly stationary random process $e = (e(n))_{-\infty}^{\infty}$ defined by $e = x - L_c y$. So our filtering problem is to construct an optimal causal (linear time invariant) filter c in $H^\infty$ which provides the best causal stationary estimate $\hat{x}$ of the process x by solving the following minimization problem

$$d_\infty^2 = \inf \{\|R_e\| : e = x - L_c y \text{ and } c \in H^\infty\} . \tag{11.5}$$

The optimal error in this estimation problem is denoted by $d_\infty$. (The reason we wrote $d_\infty^2$ in (11.5) will become transparent when we present Proposition 11.2 below.) If c is a function in $H^\infty$ and $d_\infty^2 = \|R_e\|$ where $e = x - L_c y$, then c is called an *optimal filter*.

The idea behind this optimization problem is to construct a causal filter c in $H^\infty$ which minimizes the error e(n) between x and $L_c y$ by making the norm of the error covariance $R_e$ as small as possible. One common application of this estimation or filtering problem is the signal plus noise problem. In this problem one is sending a weakly stationary signal $x = (x(n))_{-\infty}^{\infty}$ through a channel which is corrupted by a weakly stationary noise process $N = (N(n))_{-\infty}^{\infty}$, where the signal and noise processes are orthogonal, that is, $R_{xN} = 0$, or equivalently, $Ex(n)\overline{N(m)} = 0$ for all n and m. In the signal plus noise problem, the output y(n) which is received at the end of the channel, is the process $y(n) = x(n) + N(n)$, which is precisely the signal corrupted by the noise processes N(n). The idea is to construct a causal filter c in $H^\infty$ to extract the signal x(n) from the corrupted process $y(n) = x(n) + N(n)$. In almost all practical problems it is not possible to exactly reproduce the signal x(n) from y(n). So we must construct a causal (linear time invariant) filter c in $H^\infty$ to obtain the best possible estimate $\hat{x}(n)$ of x(n) from the corrupted signal y(n). This is done by minimizing the error e between x and $\hat{x}$, that is, by minimizing

$e = x - L_c y$.

To complete our discussion let us show that the processes x and y in the signal plus noise problem are indeed jointly weakly stationary. By definition x is weakly stationary. Using the fact that the processes x and N are orthogonal ($R_{xN} = 0$) we have

$$R_y = E(x+N)(x+N)^* = R_x + R_N .$$

So $R_y$ is a Laurent operator and thus y(n) is weakly stationary. It remains to show that $R_{xy}$ is a Laurent operator. Using once again the fact that x and N are orthogonal processes, we obtain $R_{xy} = E(x(x+N)^*) = R_x$. Thus $R_{xy} = R_x$ is a Laurent operator. Hence x and y are jointly weakly stationary. In the signal plus noise problem we always assume that $R_N$ is invertible and thus $R_y$ is invertible. (In fact in many applications N is a white noise process, that is, $R_N = \gamma I$ for some scalar $\gamma > 0$.) Therefore the signal plus noise problem is a special case of our filtering Problem 11.1.

To solve our filtering problem, let c be a function in $H^\infty$. Then using the fact that the expectation is a linear operator along with $e = x - L_c y$ we obtain

$$R_e = E(ee^*) = E(x - L_c y)(x - L_c y)^* = R_x - L_c R_{yx} - R_{xy} L_c^* + L_c R_y L_c^* . \tag{11.6}$$

Let us assume that $R_z$ in (11.3) is invertible. Then the Laurent operator $R_y$ is strictly positive. Therefore $R_y$ admits a co-outer spectral factorization of the form $R_y = L_\theta L_\theta^*$, where $L_\theta$ is an invertible lower triangular Laurent operator on $\ell^2$ and $L_\theta^{-1}$ is also a lower triangular Laurent operator on $\ell^2$. Here $\theta$ is the symbol of $L_\theta$. Moreover, both $\theta$ and $1/\theta$ are in $H^\infty$, that is, $\theta$ is an invertible outer function. (We say that $L_\theta$ is an invertible *co-outer spectral factor* of a Laurent operator T if $T = L_\theta L_\theta^*$, where $L_\theta$ is a lower triangular invertible Laurent operator on $\ell^2$ and its inverse $L_\theta^{-1}$ is also a lower triangular Laurent operator on $\ell^2$. In this case, the co-outer spectral factorization is unique up to a scalar factor of modulus one, that is, if L is another co-outer spectral factorization of T, then $L = \gamma L_\theta$ where $\gamma$ is a constant of modulus one; see Chapter V in Sz.-Nagy-Foias [3] or Chapter IX in Foias-Frazho [4] for further details.) Now let $L_h$ be the lower triangular Laurent operator defined by $L_h = L_c L_\theta$. Notice that $L_h$ uniquely determines a function h in $H^\infty$. Using this, $R_{xy} = R_{yx}^*$, and $R_y = L_\theta L_\theta^*$ in (11.6), we have

$$\begin{aligned} R_e &= R_x - L_h L_\theta^{-1} R_{yx} - R_{xy} L_\theta^{-*} L_h^* + L_h L_h^* = \\ &R_x - R_{xy} R_y^{-1} R_{yx} + (R_{xy} L_\theta^{-*} - L_h)(R_{xy} L_\theta^{-*} - L_h)^* \end{aligned} \tag{11.7}$$

where $A^{-*}$ denotes is the inverse of $A^*$. Since $R_z$ is invertible, equation (11.4) shows that the Schur complement $R = R_x - R_{xy} R_y^{-1} R_{yx}$ is both positive and invertible. Because $R_x$, $R_{xy}$, $R_{yx}$ and $R_y^{-1}$ are all Laurent operators, it follows that R is an invertible positive Laurent operator. So let $L_o$ be the co-outer invertible spectral factorization of R, that is, $R = L_o L_o^*$ where $L_o$ and $L_o^{-1}$ are invertible lower triangle Laurent operators on $\ell^2$. Now equation (11.7) yields

$$R_e = L_o L_o^* + (R_{xy} L_\theta^{-*} - L_h)(R_{xy} L_\theta^{-*} - L_h)^* . \tag{11.8}$$

Because $\|R_e\| = \sup \{(R_e g, g) : \|g\| = 1\}$ we have

$$\|R_e\| = \left\| \begin{bmatrix} (R_{xy} L_\theta^{-*} - L_h)^* \\ L_o^* \end{bmatrix} \right\|^2 = \|[R_{xy} L_\theta^{-*} - L_h, L_o]\|^2 . \tag{11.9}$$

However, this can be viewed as a special case of the two-sided Sarason problem discussed in Section I.10. To see this, let $L_F$ be the block Laurent operator defined by $L_F = [R_{xy} L_\theta^{-*}, L_o]$, and let $L_1$ be the co-inner Laurent operator defined by $L_1 = [I, 0]$. So if we set $L_2 = I$, then the filtering Problem 11.1 becomes

$$d_\infty^2 = \inf \{\|R_e\| : e = x - L_c y \text{ and } c \in H^\infty\} = \inf \{\|L_F - L_2 L_h L_1\|^2 : h \in H^\infty\} . \tag{11.10}$$

As shown in Section 10 the infimum in (11.10) is attained for some $h_{opt}$ in $H^\infty$. Moreover, if $h_{opt}$ is the solution to this optimization problem, then the optimal filter $L_c = L_{h_{opt}} L_\theta^{-1}$. Summing up our analysis gives the following result.

**PROPOSITION 11.2.** *Let $x = (x(n))_{-\infty}^{\infty}$ and $y = (y(n))_{-\infty}^{\infty}$ be two jointly weakly stationary random processes, and assume that the Laurent covariance matrix $R_z$ in (11.3) is invertible. Let $L_\theta$ and $L_o$ be the co-outer spectral factorizations for*

$$R_y = L_\theta L_\theta^* \quad \text{and} \quad R_x - R_{xy} R_y^{-1} R_{yx} = L_o L_o^* \tag{11.11}$$

*respectively. Finally, let $h_{opt}$ be a function in $H^\infty$ which attains the infimum in the right-hand side Sarason problem*

$$d_\infty = \inf \{\|[R_{xy} L_\theta^{-*} - L_h, L_o]\| : h \in H^\infty\} = \|[R_{xy} L_\theta^{-*} - L_{h_{opt}}, L_o]\| . \tag{11.12}$$

*Then $L_c = L_{h_{opt}} L_\theta^{-1}$, or equivalently, $c = h_{opt}/\theta$ is an optimal filter solving Problem 11.1. Moreover, the norm of the smallest error covariance is $d_\infty^2$.*

To complete this section, let us derive the Wiener filter corresponding to this filtering problem. As before, let $x = (x(n))_{-\infty}^{\infty}$ and $y = (y(n))_{-\infty}^{\infty}$ be jointly weakly stationary processes. In the Wiener problem the estimate $\hat{x} = (\hat{x}(n))_{-\infty}^{\infty}$ of x is given by $\hat{x} = L_c y$, where the filter c is now allowed to be a function in $H^2$. As before, the error process $e = (e(n))_{-\infty}^{\infty}$ in this estimation problem is given by $e = x - L_c y$. This leads to the following Wiener optimization problem

$$d_2^2 = \inf \{(R_e a, a) : e = x - L_c y \text{ and } c \in H^2\}. \tag{11.13}$$

Here a is the vector in $\ell^2$ whose entry in the zero row is one and all other entries are zero, that is, $a = (\delta_{n,0})_{-\infty}^{\infty}$. Obviously, $(R_e a, a) = (R_e)_{0,0}$ the (0, 0)-entry of $R_e$. Finally, because the process e(n) is weakly stationary $(R_e a, a) = E|e(n)|^2$ for all n. Therefore the Wiener filtering problem searches for an optimal filter c which minimizes the variance $E|e(n)|^2$ of the error process e(n).

Notice that in this case $L_c$ is a doubly infinite lower triangular matrix which could be unbounded as an operator on $\ell^2$. Hence in this case the covariance matrix $R_e$ is not assumed to define a bounded operator on $\ell^2$. However, in many practical engineering problems $L_c$ is a bounded operator and thus $R_e$ is bounded. For example, if the spectral density for $R_y$ and the joint spectral density for $R_{xy}$ are both rational, then the function c in $H^2$ which attains the infimum in (11.13) is rational, and thus $L_c$ and $R_e$ are bounded operators; see Caines [1] for further details. (Recall that the spectral density for a Laurent operator is simply the Fourier transform of its zero column, or equivalently, its symbol G in $L^\infty$.)

The idea behind the Wiener optimization problem in (11.13) is to choose a causal (linear time invariant) filter c in $H^2$ to make $(R_e a, a) = (R_e)_{0,0}$ as small as possible. By choosing a filter c to make $(R_e a, a)$ small, this makes the error covariance $R_e$ small. Moreover, if the error covariance $R_e$ is small, then the error process is close to zero, and thus $\hat{x}$ provides a "good" estimate of x. To see this recall that $R_e$ is a positive Laurent matrix. (Since $R_e$ may not be bounded, $R_e$ is positive means that $(R_e\alpha, \alpha) \geq 0$ for all $\alpha = (\alpha_n)_{-\infty}^{\infty}$ with compact support.) Therefore the entries of $R_e$ are of the form $(R_e)_{i,j} = b_{i-j}$ for all i and j in $\mathbb{Z}$ where $\{b_i\}$ are scalars. Since $R_e$ is positive this readily implies that

$$\begin{bmatrix} b_0 & b_{i-j} \\ \overline{b_{i-j}} & b_0 \end{bmatrix} = \begin{bmatrix} (R_e)_{0,0} & (R_e)_{i,j} \\ \overline{(R_e)_{i,j}} & (R_e)_{0,0} \end{bmatrix} \tag{11.14}$$

is a positive operator for all i and j. Because the determinant of a positive matrix is positive, we see that

$$(R_e a, a)^2 \geq |(R_e)_{i,j}|^2$$

for all i and j in $\mathbb{Z}$. So by choosing a filter to make $(R_e a, a)$ as small as possible, we are forcing the entries $(R_e)_{i,j}$ of the error covariance to be small. Theoretically this forces $R_e$ to be small and thus provides a good estimate $\hat{x}$ of x from the process y.

To solve the Wiener optimization problem, recall that the error covariance $R_e$ for $e = x - L_c y$ is given by (11.7) where $L_\theta$ is the co-outer spectral factor for $R_y$ and $L_h = L_c L_\theta$. This readily implies that

$$(R_e a, a) = (R_x a, a) - (R_{xy} R_y^{-1} R_{yx} a, a) + \|(R_{xy} L_\theta^{-*} - L_h)^* a\|^2 . \tag{11.15}$$

Now let $P_+$ be the orthogonal projection onto $\ell_+^2$ and $P_- = I - P_+$ the orthogonal projection onto $\ell^2 \ominus \ell_+^2$. Notice that if L is any Laurent operator, then $\|La\| = \|L^* a\|$. Since $R_{xy} L_\theta^{-*} - L_h$ and $R_y = L_\theta L_\theta^*$ are Laurent operators, (11.15) and $R_y = L_\theta L_\theta^*$ gives

$$\begin{aligned}(R_e a, a) &= (R_x a, a) - \|L_\theta^{-1} R_{yx} a\|^2 + \|(R_{xy} L_\theta^{-*} - L_h) a\|^2 = \\ &(R_x a, a) - \|R_{xy} L_\theta^{-*} a\|^2 + \|P_- R_{xy} L_\theta^{-*} a\|^2 + \|P_+ R_{xy} L_\theta^{-*} a - L_h a\|^2 = \\ &(R_x a, a) - \|P_+ R_{xy} L_\theta^{-*} a\|^2 + \|P_+ R_{xy} L_\theta^{-*} a - L_h a\|^2 .\end{aligned} \tag{11.16}$$

This shows that there exists a unique solution to the Wiener optimization problem by choosing $h(\lambda) = \mathcal{F}_+ P_+ R_{xy} L_\theta^{-*} a$, where $\mathcal{F}_+$ is the Fourier transform from $\ell_+^2$ onto $H^2$. Using $L_c L_\theta = L_h$ we see that the optimal filter is given by $c = h/\theta$, where $\theta$ is the co-outer spectral factor for $R_y$. Summing up this analysis proves the following result.

**PROPOSITION 11.3.** *Let* $x = (x(n))_{-\infty}^{\infty}$ *and* $y = (y(n))_{-\infty}^{\infty}$ *be two jointly weakly stationary processes, and assume that the (Laurent) covariance matrix* $R_y$ *is invertible. Let* $\theta$ *in* $H^\infty$ *be the co-outer spectral factor for* $R_y (= L_\theta L_\theta^*)$. *Then there is a unique solution* c *in* $H^2$ *to the Wiener optimization problem*

$$d_2^2 = \inf \{(R_e a, a) : e = x - L_c y \ \textit{and} \ c \in H^2\} . \tag{11.17}$$

*Moreover, this optimal* c *is given by*

$$c = \frac{\mathcal{F}_+(P_+ R_{xy} L_\theta^{-*} a)}{\theta} \qquad (a = (\delta_{n,0})_{-\infty}^{\infty}) , \tag{11.18}$$

*where* $P_+$ *is the orthogonal projection onto* $\ell_+^2$, *and* $\mathcal{F}_+$ *is the Fourier transform. Finally, the error in estimation* $d_\infty$ *is given by*

$$d_\infty^2 = (R_x a, a) - \|P_+ R_{xy} L_\theta^{-*} a\|^2 . \tag{11.19}$$

In most cases Propositions 11.2 and 11.3 do not give the same filter. Proposition 11.2 provides a filter which minimizes an $L^\infty$ norm, while Proposition 11.3 computes a filter to minimize a $\ell^2$ norm. In Chapter IV we will use the central intertwining lifting in the commutant lifting theorem to construct a trade off between these two optimization problems.

## *Notes to Chapter I:*

The problems discussed in this chapter have their roots deep into classical analysis starting with important contributions of mathematicians like C. Carathéodory, R. Nevanlinna, G. Pick and I. Schur (cf., Fritzsche-Kristein [1]). Here we also mention the Nehari problem which appeared in the fifties (Nehari [1]). Operator techniques to solve interpolation problems were first used by B. Sz.-Nagy and A. Koranyi (Sz.-Nagy-Koranyi [1], [2]). The modern developments, dealing with matrix and operator functions and based on operator theory methods, started with a series of papers by V. M. Adamjan, D. Z. Arov and M. G. Krein (Adamjan-Arov-Krein [1]-[4]), the work of D. Sarason (Sarason [1]) and B. Sz.-Nagy and C. Foias (Sz.-Nagy-Foias [1], [2]). Later developments also include tangential interpolation problems introduced by I. P. Fedcina (Fedcina [1], [2]) and their contour integral version developed by A. A. Nudelman (Nudelman [1], [2]). For the two-sided Sarason problem see also Arov [4] and Kheifets [1]. This subject has been treated and developed further in a number of monographs: Bakonyi-Constantinescu [1], Ball-Gohberg-Rodman [1], Dym [1], Foias-Frazho [4], Helton [3], and Rosenblum-Rovnyak [3]; see also the papers in Gohberg-Sakhnovich [1], the review paper Sakhnovich [1], and the lecture notes Kaashoek [1]. Various aspects of the history of this area can be found in these texts. Interpolation problems have also been considered on Riemann surfaces (see Ball-Vinnikov [1]).

Interpolation problems with norm constraints for operator-valued functions with operator arguments are probably considered here in full generality for the first time (see also Foias-Frazho-Gohberg-Kaashoek [1]). Related problems with operator arguments were treated by M. Rosenblum and J. Rovnyak (see Rosenblum-Rovnyak [1], [2], and Chapter 2 in Rosenblum-Rovnyak [3]).

Finally, the estimation Problem 11.1 is the $H^\infty$ analogue of the standard Wiener filtering problem studied in stochastic processes. For a more in-depth discussion of random processes and Wiener filtering see Caines [1].

## CHAPTER II

# PROOFS USING THE COMMUTANT LIFTING THEOREM

This chapter contains the proofs of the operator-valued interpolation theorems introduced in the previous chapter. The proofs are based on the commutant lifting theorem which is stated for convenience in the first section. In Chapter IV the commutant lifting theorem will be revisited in more detail, and a constructive proof of this theorem will be given there.

## II.1. THE COMMUTANT LIFTING THEOREM

In this section we will present the commutant lifting theorem and use it to solve a basic interpolation problem. Throughout all operators are bounded linear operators acting on Hilbert spaces. If $\mathcal{H}$ is a subspace of $\mathcal{K}$, then $P_{\mathcal{H}}$ is the orthogonal projection of $\mathcal{K}$ onto $\mathcal{H}$. Now let A be an operator from $\mathcal{H}$ to $\mathcal{H}'$. An operator B in $\mathcal{L}(\mathcal{K}, \mathcal{K}')$ is called *lifting* of A if $\mathcal{H}$ and $\mathcal{H}'$ are subspaces of $\mathcal{K}$ and $\mathcal{K}'$, respectively, and

$$P_{\mathcal{H}'}B = AP_{\mathcal{H}} \,. \tag{1.1}$$

Notice that B is a lifting of A if and only if B admits a matrix partitioning of the form

$$B = \begin{bmatrix} A & 0 \\ * & * \end{bmatrix} : \begin{bmatrix} \mathcal{H} \\ \mathcal{H}^{\perp} \end{bmatrix} \to \begin{bmatrix} \mathcal{H}' \\ \mathcal{H}'^{\perp} \end{bmatrix} . \tag{1.2}$$

In particular, B is a lifting of A if and only if $B^*$ maps $\mathcal{H}'$ into $\mathcal{H}$ and $A^* = B^* \mid \mathcal{H}'$.

Clearly an operator U on $\mathcal{K}$ is an isometric lifting of an operator C on $\mathcal{H}$ if U is an isometry satisfying $P_{\mathcal{H}}U = CP_{\mathcal{H}}$. In particular, U on $\mathcal{K}$ is an isometric lifting of C if and only if $\mathcal{H}$ is an invariant subspace for the co-isometry $U^*$ and $C^* = U^* \mid \mathcal{H}$. Obviously, if U is an isometric lifting of C, then C is a contraction, that is, $\|C\| \leq 1$. On the other hand, if C is a contraction, then C admits an isometric lifting. To see this, let D be the positive square root of $I - C^*C$ and $\mathcal{D}$ the closure of the range of D, then following the Sz.-Nagy-Schäffer construction, an isometric lifting U on $\mathcal{K} = \mathcal{H} \oplus \ell_+^2(\mathcal{D})$ of C is given by

$$U = \begin{bmatrix} C & 0 & 0 & 0 & \cdots \\ D & 0 & 0 & 0 & \cdots \\ 0 & I & 0 & 0 & \cdots \\ 0 & 0 & I & 0 & \cdots \\ 0 & 0 & 0 & I & \cdots \\ \cdot & \cdot & \cdot & \cdot & \cdots \\ \cdot & \cdot & \cdot & \cdot & \cdots \\ \cdot & \cdot & \cdot & \cdot & \cdots \end{bmatrix} \quad \text{on } \mathcal{H} \oplus \ell_+^2(\mathcal{D}) . \tag{1.3}$$

This isometry U is called the *Sz.-Nagy-Schäffer isometric lifting* of C. The operator U in (1.3) is a minimal isometric lifting of C, that is, the space $\mathcal{K}$ on which U acts is the closed linear hull of the subspaces $U^n\mathcal{H}$, $n \geq 0$. For further results on isometric lifting see Chapter I in Sz.-Nagy-Foias [3] or Chapter VI in Foias-Frazho [4].

The following theorem is known as the commutant lifting theorem. The theorem is due to Sz.-Nagy-Foias [1]; various proofs of the result may be found in Chapter VII of Foias-Frazho [4]. Finally, recall that an operator A *intertwines* T on $\mathcal{H}$ with T′ on $\mathcal{H}'$ if A is an operator in $\mathcal{L}(\mathcal{H}, \mathcal{H}')$ and AT = T′A.

**THEOREM 1.1.** *Let* T *on* $\mathcal{H}$ *be an isometry,* T′ *a contraction on* $\mathcal{H}'$, *and* U′ *on* $\mathcal{K}'$ *an isometric lifting of* T′. *Assume that* $A \in \mathcal{L}(\mathcal{H}, \mathcal{H}')$ *intertwines* T *with* T′. *Then there exists a lifting* B *in* $\mathcal{L}(\mathcal{H}, \mathcal{K}')$ *of* A *satisfying*

$$BT = U'B \ , \ P_{\mathcal{H}'}B = A \quad \text{and} \quad \|B\| = \|A\| . \tag{1.4}$$

As in the previous theorem assume that A is an operator in $\mathcal{L}(\mathcal{H}, \mathcal{H}')$ intertwining an isometry T with a contraction T′. Let U′ on $\mathcal{K}'$ be an isometric lifting of T′. Then B is called an *intertwining lifting of* A *with respect to* U′, if B is an operator mapping $\mathcal{H}$ into $\mathcal{K}'$ satisfying $U'B = BT$ and $P_{\mathcal{H}'}B = A$. The commutant lifting theorem proves that there exists an intertwining lifting B of A preserving the norm of A, that is, there exists an intertwining lifting B of A such that $\|B\| = \|A\|$.

The block forward (unilateral) shift on the space $\ell_+^2(\mathcal{X})$ will be denoted by $S_{\mathcal{X}}$. In many applications of the commutant lifting theorem the isometrices T and U′ are just block forward shifts. As a first application of the commutant lifting theorem we derive the following abstract interpolation result.

**THEOREM 1.2.** *Consider the operators* W *from* $\ell_+^2(\mathcal{U})$ *to* $\mathcal{X}$ *and* $\tilde{W}$ *from* $\ell_+^2(\mathcal{Y})$ *to* $\mathcal{X}$, *and assume that the following intertwining relations hold:*

$$WS_{\mathcal{U}} = ZW \quad \text{and} \quad \tilde{W}S_{\mathcal{Y}} = Z\tilde{W}\,, \tag{1.5}$$

*for some* Z *in* $\mathcal{L}(\mathcal{X}, \mathcal{X})$. *Then there exists* $G \in H^\infty(\mathcal{Y}, \mathcal{U})$ *such that* $WT_G = \tilde{W}$ *and* $\|G\|_\infty \le \gamma$ *if and only if* $\gamma^2 WW^* - \tilde{W}\tilde{W}^* \ge 0$.

**PROOF**. First, assume that such a G exists. Since the operator norm of the Toeplitz operator $T_G$ generated by G coincides with $\|G\|_\infty$, we have $\|T_G\| \le \gamma$, and hence $T_G T_G^* \le \gamma^2 I$. It follows that $\tilde{W}\tilde{W}^* = WT_G T_G^* W^* \le \gamma^2 WW^*$, and thus $\gamma^2 WW^* - \tilde{W}\tilde{W}^* \ge 0$.

To prove the reverse implication assume that $\gamma^2 WW^* \ge \tilde{W}\tilde{W}^*$. Then there exists an operator $\Lambda$ from $\mathcal{H}'$ into $\ell_+^2(\mathcal{Y})$ satisfying $\Lambda W^* = \tilde{W}^*$ and $\|\Lambda\| \le \gamma$. Here $\mathcal{H}'$ is the closure of the range of $W^*$. Using $S_{\mathcal{U}}^* W^* = W^* Z^*$ it is easy to see that $\mathcal{H}'$ is an invariant subspace for $S_{\mathcal{U}}^*$. So let $T'$ be the contraction on $\mathcal{H}'$ defined by $T'^* = S_{\mathcal{U}}^* | \mathcal{H}'$. Then $U' = S_{\mathcal{U}}$ on $\mathcal{K}' = \ell_+^2(\mathcal{U})$ is an isometric lifting of $T'$. Next let A be the operator from $\ell_+^2(\mathcal{Y})$ into $\mathcal{H}'$ defined by $A = \Lambda^*$. Notice that $WA = \tilde{W}$. This fact and the intertwining relations in (1.5) along with $T = S_{\mathcal{Y}}$ give

$$WAT = \tilde{W}S_{\mathcal{Y}} = Z\tilde{W} = ZWA = WS_{\mathcal{U}}A = WP_{\mathcal{H}'}S_{\mathcal{U}}A = WT'A\,.$$

Because $W | \mathcal{H}'$ is one to one, it follows that $T'A = AT$.

Now we are in a position to apply the commutant lifting theorem. So there exists an intertwining lifting B from $\ell_+^2(\mathcal{Y})$ to $\ell_+^2(\mathcal{U})$ satisfying

$$BS_{\mathcal{Y}} = S_{\mathcal{U}}B\,, \ \|B\| = \|A\| \ \text{and} \ P_{\mathcal{H}'}B = A\,. \tag{1.6}$$

The intertwining condition $BS_{\mathcal{Y}} = S_{\mathcal{U}}B$ implies that $B = T_G$ for some G in $H^\infty(\mathcal{Y}, \mathcal{U})$. Since A is bounded by $\gamma$, the second condition in (1.6) gives $\|G\|_\infty = \|T_G\| = \|B\| = \|A\| \le \gamma$. Finally, the last equation in (1.6) along with $WA = \tilde{W}$ shows that

$$WT_G = WB = WP_{\mathcal{H}'}B = WA = \tilde{W}\,.$$

Therefore $WT_G = \tilde{W}$, where G is a function in $H^\infty(\mathcal{Y}, \mathcal{U})$ satisfying $\|G\|_\infty \le \gamma$. This completes the proof.

Given the operators W and $\tilde{W}$ satisfying (1.5) we may consider the $H^\infty$ optimization problem associated with:

$$d_\infty = d_\infty(W, \tilde{W}) = \inf\{\|F\|_\infty : F \in H^\infty(\mathcal{Y}, \mathcal{U}) \text{ and } WT_F = \tilde{W}\}\,. \tag{1.7}$$

Notice that $d_\infty$ can be infinite. For example, choose $W = [0, 0, \ldots]$ and $\tilde{W} = [I, 0, 0, \ldots]$.

**THEOREM 1.3.** *Let* W *from* $\ell^2_+(\mathcal{U})$ *to* $\mathcal{X}$ *and* $\tilde{W}$ *from* $\ell^2_+(\mathcal{Y})$ *to* $\mathcal{X}$ *be two operators satisfying the intertwining conditions in (1.5), where* Z *is an operator on* $\mathcal{X}$. *Then* $d_\infty(W, \tilde{W})$ *is finite if and only if there exists a bounded linear operator* A *from* $\ell^2_+(\mathcal{Y})$ *to* $\mathcal{H}' := \overline{\text{ran } W^*}$ *such that* $WA = \tilde{W}$. *Moreover, in this case* A *is uniquely determined,* $d_\infty = \|A\|$ *and there exists a* G *in* $H^\infty(\mathcal{Y}, \mathcal{U})$ *solving the* $H^\infty$ *optimization problem in (1.7), that is,* $WT_G = \tilde{W}$ *and*

$$\|G\|_\infty = \|A\| = \inf \{\|F\|_\infty : F \in H^\infty(\mathcal{Y}, \mathcal{U}) \text{ and } WT_F = \tilde{W}\} . \tag{1.8}$$

**PROOF.** Assume that $d_\infty$ is finite. Then there exists an F in $H^\infty(\mathcal{Y}, \mathcal{U})$ satisfying $WT_F = \tilde{W}$. By choosing $A = P_{\mathcal{H}'}T_F$ we have $WA = \tilde{W}$ and $\|A\| \le \|T_F\| = \|F\|_\infty$. Notice that the operator A does not depend on the particular choice of F. Indeed, if $\tilde{F} \in H^\infty(\mathcal{Y}, \mathcal{U})$ and $WT_{\tilde{F}} = \tilde{W}$. Then we have $W(T_{\tilde{F}} - T_F) = 0$, and hence $\text{ran}\,(T_{\tilde{F}} - T_F) \subset (\mathcal{H}')^\perp$. Thus $P_{\mathcal{H}'}T_{\tilde{F}} = P_{\mathcal{H}'}T_F = A$. From the inequality $\|A\| \le \|F\|_\infty$ it now readily follows that $\|A\| \le d_\infty$. On the other hand, assume that $WA = \tilde{W}$ for some bounded linear operator A from $\ell^2_+(\mathcal{Y})$ to $\mathcal{H}'$. Since $W \mid \mathcal{H}'$ is one to one, this is the only operator A from $\ell^2_+(\mathcal{Y})$ to $\mathcal{H}'$ satisfying $WA = \tilde{W}$. According to the proof of the previous theorem there exists a function G in $H^\infty(\mathcal{Y}, \mathcal{U})$ satisfying (1.6) where $B = T_G$. Therefore $WT_G = WA = \tilde{W}$ and $\|G\|_\infty = \|A\| \le d_\infty$. Hence $\|A\| = d_\infty$ and this G solves the optimization problem in (1.8). This completes the proof.

**REMARK 1.4.** As before, let W from $\ell^2_+(\mathcal{U})$ to $\mathcal{X}$ and $\tilde{W}$ from $\ell^2_+(\mathcal{Y})$ to $\mathcal{X}$ be two operators satisfying the intertwining condition (1.5). Now assume that $P := WW^*$ is strictly positive, and let A be the operator from $\ell^2_+(\mathcal{Y})$ to $\mathcal{H}'$ defined by $A = W^*P^{-1}\tilde{W}$. Then A is a bounded operator satisfying $WA = \tilde{W}$ and $\text{ran } A \subset \text{ran } W^*$. Thus, by Theorem 1.3, $d_\infty(W, \tilde{W})$ is finite. Furthermore, if we set $\tilde{P} = \tilde{W}\tilde{W}^*$ and let $P^{1/2}$ be the positive square root of P, then

$$\begin{aligned} d_\infty(W, \tilde{W})^2 &= \|A\|^2 = \|A^*A\| = \|\tilde{W}^*P^{-1}WW^*P^{-1}\tilde{W}\| \\ &= \|(\tilde{W}^*P^{-1}\tilde{W})\| = \|P^{-1/2}\tilde{W}\|^2 = \|\tilde{W}^*P^{-1/2}\|^2 \\ &= \|P^{-1/2}\tilde{W}\tilde{W}^*P^{-1/2}\| = \|P^{-1/2}\tilde{P}P^{-1/2}\| = r_{spec}(\tilde{P}P^{-1}) . \end{aligned} \tag{1.9}$$

The last equality follows from the fact that $P^{-1/2}\tilde{P}P^{-1/2}$ is similar to $\tilde{P}P^{-1}$. To see this compute $P^{-1/2}(\tilde{P}P^{-1})P^{1/2}$. Therefore, if $WW^*$ is strictly positive, then there always exists an G in $H^\infty(\mathcal{Y}, \mathcal{U})$ solving the $H^\infty$-optimization problem in (1.7), that is, $WT_G = \tilde{W}$ and $\|T_G\| = d_\infty(W, \tilde{W})$. Moreover, $d_\infty^2$ is the spectral radius of $\tilde{P}P^{-1}$.

**REMARK 1.5.** As before let W be an operator from $\ell_+^2(\mathcal{U})$ into $\mathcal{X}$ and $\tilde{W}$ an operator from $\ell_+^2(\mathcal{Y})$ into $\mathcal{X}$ satisfying the intertwining conditions in (1.5), where Z is an operator on $\mathcal{X}$. Assume that the relation $WA = \tilde{W}$ holds for a bounded operator A from $\mathcal{H} = \ell_+^2(\mathcal{Y})$ into $\mathcal{H}'$ the closure of the range of $W^*$. Let $U' = S_{\mathcal{U}}$ on $\mathcal{K}' = \ell_+^2(\mathcal{U})$ and $T'$ on $\mathcal{H}'$ be the compression of $U'$ to $\mathcal{H}'$, that is, $T' = P_{\mathcal{H}'}U' | \mathcal{H}'$. Recall that $U'$ is an isometric lifting of $T'$ and $T'A = AT$ where $T = S_{\mathcal{Y}}$. Then B is an intertwining lifting of A with respect to $U'$ if and only if $B = T_F$ where F is a function in $H^\infty(\mathcal{Y}, \mathcal{U})$ and $WT_F = \tilde{W}$.

To prove this first notice that if B is an intertwining lifting of A, then the intertwining condition $S_{\mathcal{U}}B = U'B = BT = BS_{\mathcal{Y}}$ implies that there exists a F in $H^\infty(\mathcal{Y}, \mathcal{U})$ satisfying $B = T_F$. Since $A = P_{\mathcal{H}'}B$ and $WA = \tilde{W}$, it follows that $WT_F = \tilde{W}$. On the other hand, if $WT_F = \tilde{W}$ for some F in $H^\infty(\mathcal{Y}, \mathcal{U})$, then clearly $B = T_F$ satisfies the intertwining condition $U'B = BT$. Moreover, because $W | \mathcal{H}'$ is one to one and $WA = \tilde{W} = WT_F$, we see that $A = P_{\mathcal{H}'}T_F = P_{\mathcal{H}'}B$. Thus B is an intertwining lifting of A with respect to $U'$. This verifies our claim.

Using the fact that there is a one to one correspondence between the set of all F in $H^\infty(\mathcal{Y}, \mathcal{U})$ satisfying $WT_F = \tilde{W}$ and the set of all intertwining liftings B of A, we see that the $H^\infty$ optimization problem in (1.7) is equivalent to the following optimization problem associated with:

$$d_\infty = d_\infty(W, \tilde{W}) = \inf \{\|B\| : \text{B is an intertwining lifting of A}\} . \tag{1.10}$$

Clearly any intertwining lifting B of A satisfies $\|A\| = \|P_{\mathcal{H}'}B\| \leq \|B\|$ and thus $\|A\| \leq d_\infty$. By the commutant lifting theorem, there exists an intertwining lifting B of A preserving the norm of A. In particular, $d_\infty(W, \tilde{W}) = \|A\|$. Moreover, this $B = T_F$ for some F in $H^\infty(\mathcal{Y}, \mathcal{U})$ and $WT_F = \tilde{W}$.

## II.2. PROOF OF THE STANDARD LEFT NEVANLINNA-PICK INTERPOLATION THEOREM

Recall that the left tangential Hermite-Fejér interpolation problem can be viewed as a special case of the standard left Nevanlinna-Pick problem, see Theorem I.5.2. In this section we will use Theorem 1.2 to prove the standard left Nevanlinna-Pick Theorem I.4.1.

To this end, let Z on $\mathcal{X}$, B from $\mathcal{U}$ to $\mathcal{X}$ and $\tilde{B}$ from $\mathcal{Y}$ to $\mathcal{X}$ be the data for the standard left Nevanlinna-Pick problem where $r_{spec}(Z) < 1$. Now let W from $\ell_+^2(\mathcal{U})$ to $\mathcal{X}$ and $\tilde{W}$ from $\ell_+^2(\mathcal{Y})$ to $\mathcal{X}$ be the controllability operators for the pair $\{Z, B\}$ and $\{Z, \tilde{B}\}$, respectively, that is

$$W = [B, ZB, Z^2B, \ldots] \quad \text{and} \quad \tilde{W} = [\tilde{B}, Z\tilde{B}, Z^2\tilde{B}, \ldots] . \tag{2.1}$$

So the standard left Nevanlinna-Pick problem is to find (if possible) a function G in $H^\infty(\mathcal{Y}, \mathcal{U})$ satisfying $WT_G = \tilde{W}$ and $\|G\|_\infty \leq \gamma$. It is easy to show that W and $\tilde{W}$ satisfy the intertwining condition $ZW = WS_{\mathcal{U}}$ and $Z\tilde{W} = \tilde{W}S_{\mathcal{Y}}$. So by Theorem 1.2 there exists a solution G to the standard left Nevanlinna-Pick problem if and only if $\gamma^2 WW^* - \tilde{W}\tilde{W}^*$ is positive, or equivalently $\gamma^2 P \geq \tilde{P}$, where P and $\tilde{P}$ are the controllability grammians for {Z, B} and {Z, $\tilde{B}$}, respectively. This completes the proof of Theorem I.4.1.

The standard left Nevanlinna-Pick problem yields the $H^\infty$ optimization problem associated with:

$$d_\infty = d_\infty(Z, B, \tilde{B}) = \inf \{\|F\|_\infty : F \in H^\infty(\mathcal{Y}, \mathcal{U}) \text{ and } WT_F = \tilde{W}\} . \tag{2.2}$$

According to Theorem 1.3 there exists a solution to this $H^\infty$ optimization problem (i.e., there exists $F \in H^\infty(\mathcal{Y}, \mathcal{U})$ such that $WT_F = \tilde{W}$ and $\|F\|_\infty = d_\infty$) if and only if the relation $WA = \tilde{W}$ holds for a bounded operator from $\ell^2_+(\mathcal{Y})$ to $\mathcal{H}'$ the closure of the range of $W^*$, or equivalently $\gamma^2 P \geq \tilde{P}$ for $\gamma$ sufficiently large. In this case $d_\infty = \|A\|$ and there exists an optimal G in $H^\infty(\mathcal{Y}, \mathcal{U})$ satisfying $WT_G = \tilde{W}$ and $\|G\|_\infty = d_\infty$. Finally, if the controllability grammian P for {Z, B} is strictly positive, then there always exists an optimal solution G to the standard left Nevanlinna-Pick $H^\infty$ optimization problem in (2.2). Moreover, in this case, $A = W^* P^{-1} \tilde{W}$ and $d_\infty^2 = r_{spec}(\tilde{P}P^{-1})$.

## II.3. PROOF OF THE NEHARI EXTENSION THEOREM

In this section we prove the Nehari extension theorem for operator-valued functions (Theorem I.6.1). The proof of the nonseparable version (Theorem I.6.2) follows the same line of reasoning and is therefore omitted. Since the necessity of the Hankel operator condition has already been established in Section I.6, we prove here the sufficiency and the existence of an optimal solution. Recall that an operator L in $\mathcal{L}(\mathcal{K}, \mathcal{F})$ *extends* an operator B in $\mathcal{L}(\mathcal{H}, \mathcal{E})$ if L maps $\mathcal{H}$ into $\mathcal{E}$ and $L | \mathcal{H} = B$. (Of course it is understood that $\mathcal{H} \subseteq \mathcal{K}$ and $\mathcal{E} \subseteq \mathcal{F}$.) Let us begin with the following useful lemma which is of independent interest.

**LEMMA 3.1**. *Let $\mathcal{H}$ be an invariant subspace for a unitary operator* V *on $\mathcal{K}$, and let* T *be the isometry on $\mathcal{H}$ defined by* $T = V | \mathcal{H}$. *Let* V′ *be a unitary operator on $\mathcal{K}'$, and let* B *be an operator from $\mathcal{H}$ to $\mathcal{K}'$ intertwining* T *with* V′, *that is,* $BT = V'B$. *Then there exists an operator* L *from $\mathcal{K}$ to $\mathcal{K}'$ extending* B *and preserving the norm of* B *while also intertwining* V *with* V′, *that is,*

$$L | \mathcal{H} = B, \ \|B\| = \|L\| \ \text{ and } \ LV = V'L . \tag{3.1}$$

*Moreover, if $\mathcal{K} = \bigvee \{V^{*n}\mathcal{H} : n \geq 0\}$, then this* L *is the only operator intertwining* V *with* V′ *and satisfying* $L | \mathcal{H} = B$.

**PROOF**. One can give a direct proof of this result; see Foias-Frazho [4], Corollary VI.2.4. Here we shall present a proof of this result based on the commutant lifting theorem. Clearly $T^*B^* = B^*V'^*$. Since $\mathcal{H}$ is an invariant subspace for V, it follows that $V^*$ is an isometric lifting for $T^*$. Obviously $V'^*$ is an isometric lifting of $V'^*$. By the commutant lifting theorem, there exists an operator C from $\mathcal{K}'$ to $\mathcal{K}$, preserving the norm of B, intertwining $V'^*$ with $V^*$ and satisfying $P_{\mathcal{H}}C = B^*$. Now let $L = C^*$, then L intertwines V with $V'$ and $\|L\| = \|C\| = \|B\|$. Finally, the condition $P_{\mathcal{H}}C = B^*$ implies that $L|\mathcal{H} = B$.

To complete the proof, assume that $\mathcal{K} = \bigvee\{V^{*n}\mathcal{H} : n \geq 0\}$, and let M be any operator intertwining V with $V'$ and satisfying $M|\mathcal{H} = B$. Notice that because both V and $V'$ are unitary both L and M intertwine $V^*$ with $V'^*$. So for h in $\mathcal{H}$ we have

$$LV^{*n}h = V'^{*n}Lh = V'^{*n}Bh = V'^{*n}Mh = MV^{*n}h\,.$$

Since vectors of the form $V^{*n}\mathcal{H}$ for $n \geq 0$ are dense in $\mathcal{K}$, we have $L = M$. This completes the proof.

To see how Lemma 3.1 is used in practice, let S be the forward shift on $\ell_+^2(\mathcal{U})$ and $V'$ the forward bilateral shift on $\ell^2(\mathcal{Y})$. Notice that an operator B from $\ell_+^2(\mathcal{U})$ to $\ell^2(\mathcal{Y})$ intertwines S with $V'$ if and only if B admits a block matrix representation of the form

$$B = \begin{bmatrix} \vdots & \vdots & \vdots & \vdots \\ G_{-1} & G_{-2} & G_{-3} & \cdots \\ \underline{G_0} & G_{-1} & G_{-2} & \cdots \\ G_1 & G_0 & G_{-1} & \cdots \\ G_2 & G_1 & G_0 & \cdots \\ \vdots & \vdots & \vdots & \vdots \end{bmatrix}, \tag{3.2}$$

where $\{G_n\}_{-\infty}^{\infty}$ are operators in $\mathcal{L}(\mathcal{U}, \mathcal{Y})$. In this case Lemma 3.1 implies that there exists a unique operator L from $\ell^2(\mathcal{U})$ to $\ell^2(\mathcal{Y})$ extending B and satisfying $V'L = LV$ where V is the bilateral shift on $\ell^2(\mathcal{U})$. Using $V'L = LV$ and $L|\ell_+^2(\mathcal{U}) = B$ it follows that L is the Laurent operator generated by $\{G_n\}$. Moreover, $\|L\| = \|B\|$. So if B is an operator of the form (3.2), then Lemma 3.1 guarantees that the Laurent operator $L_G$ generated by $\{G_n\}$ (the first column of B) satisfies $L_G|\ell_+^2(\mathcal{U}) = B$ and $\|L_G\| = \|B\|$. Finally, if the spaces $\mathcal{U}$ and $\mathcal{Y}$ are separable, then $\|G\|_\infty = \|L_G\| = \|B\|$ where G is the function in $L^\infty(\mathcal{U}, \mathcal{Y})$ with Fourier series $\Sigma G_n e^{int}$.

To prove the Nehari extension Theorem I.6.1, we have to show that there exists a Nehari interpolant G in $L^\infty(\mathcal{U}, \mathcal{Y})$ for the specified sequence $K_1, K_2, K_3, \ldots$ with the property $\|G\|_\infty = \|\Gamma\|$, where $\Gamma$ is the Hankel operator generated by $K_1, K_2, K_3, \ldots$ given in (I.6.2). Recall that $\mathcal{U}$ and $\mathcal{Y}$ are assumed to be separable. To begin, let A be the operator from $\ell_+^2(\mathcal{U})$ to

$\ell^2_-(\mathcal{Y})$ defined by

$$A = \begin{bmatrix} \cdot & \cdot & \cdot & \cdots \\ \cdot & \cdot & \cdot & \cdots \\ \cdot & \cdot & \cdot & \cdots \\ K_3 & K_4 & K_5 & \cdots \\ K_2 & K_3 & K_4 & \cdots \\ K_1 & K_2 & K_3 & \cdots \end{bmatrix} . \tag{3.3}$$

As shown in Section I.6 the operator A can be identified with $\Gamma$. In fact, $\Gamma = JA$, where J is the unitary flip operator defined in Section I.6, and thus $\|A\| = \|\Gamma\|$.

Now let V and V′ be the bilateral (forward) shifts on $\ell^2(\mathcal{U})$ and $\ell^2(\mathcal{Y})$, respectively. Let T be the isometry on $\mathcal{H} = \ell^2_+(\mathcal{U})$ and T′ the co-isometry on $\mathcal{H}' = \ell^2_-(\mathcal{Y})$ defined by $T = V|\mathcal{H}$ and $T' = P_{\mathcal{H}'}V'|\mathcal{H}'$, respectively. Notice that V′ is an isometric lifting of T′ because $\mathcal{H}'$ is an invariant subspace for $V'^*$. Using the special structure of A it is easy to show that $T'A = AT$. By the commutant lifting (Theorem 1.1), there exists an operator B mapping $\mathcal{H} = \ell^2_+(\mathcal{U})$ into $\mathcal{K}' = \ell^2(\mathcal{Y})$ satisfying

$$\|B\| = \|A\| , \; V'B = BT \quad \text{and} \quad P_{\mathcal{H}'}B = A . \tag{3.4}$$

According to Lemma 3.1 there exists a unique operator L from $\ell^2(\mathcal{U})$ to $\ell^2(\mathcal{Y})$ preserving the norm of A, intertwining V with V′ and $L|\mathcal{H} = B$. Using $V'L = LV$ it follows that there exists a function G in $L^\infty(\mathcal{U}, \mathcal{Y})$ satisfying $L = L_G$. Moreover, $\|G\|_\infty = \|L_G\| = \|B\| = \|\Gamma\|$. Finally, $P_{\mathcal{H}'}B = A$ and $L_G|\mathcal{H} = B$ show that $P_{\mathcal{H}'}L_G|\mathcal{H} = A$. This along with the matrix form for the Laurent operator $L_G$ gives $G_{-n} = K_n$ for $n > 0$, where $\{G_n\}$ are the Fourier coefficients of G. This completes the proof of the Nehari extension theorem.

Obviously, if G is a Nehari interpolant of the data $\{K_n\}_1^\infty$, then $B = L_G|\mathcal{H}$ is an intertwining lifting of A with respect to $U' = V'$. So the above analysis shows that B is an intertwining lifting of A if and only if $B = L_G|\mathcal{H}$ where G is a Nehari interpolant of $\{K_n\}$. In this case $\|B\| = \|G\|_\infty$. So (cf., (I.6.3))

$$\begin{aligned} d_\infty &:= \inf \{\|G\|_\infty : G \text{ is a Nehari interpolant of } \{K_n\}_1^\infty\} = \\ &= \inf \{\|B\| : B \text{ is an intertwining lifting of } A\} . \end{aligned} \tag{3.5}$$

If B is an intertwining lifting of A, then $\|A\| \le \|B\|$. Thus $\|A\| \le d_\infty$. By the commutant lifting theorem there exists an intertwining lifting B of A which preserves the norm of A. Therefore this B uniquely determines a Nehari interpolant G such that $B = L_G|\mathcal{H}$. Moreover, $\|G\|_\infty = \|B\| = \|A\|$ and thus $d_\infty = \|A\|$.

## II.4. PROOF OF THE SARASON THEOREM

In this section we prove the Sarason interpolation theorem for operator-valued functions (Theorem I.7.1). Let $F \in H^\infty(\mathcal{U}, \mathcal{Y})$, and let $\Theta$ be an inner function in $H^\infty(\mathcal{E}, \mathcal{Y})$, where $\mathcal{U}$, $\mathcal{Y}$ and $\mathcal{E}$ are separable Hilbert spaces. Put $\mathcal{H}' = H^2(\mathcal{Y}) \ominus \Theta H^2(\mathcal{E})$. From the discussion in Section I.7 we know that in order to prove Theorem I.7 it suffices to show that there exists a $H \in H^\infty(\mathcal{U}, \mathcal{E})$ such that

$$\|F - \Theta H\|_\infty = \|P_{\mathcal{H}'} M_F^+\| . \tag{4.1}$$

To derive an operator function H as in (4.1) we shall apply Theorem 1.1 with $T = S_{\mathcal{U}}$ and $U' = S_{\mathcal{Y}}$, where $S_{\mathcal{U}}$ and $S_{\mathcal{Y}}$ are now the unilateral forward shifts on $H^2(\mathcal{U}) = \mathcal{H}$ and $H^2(\mathcal{Y}) = \mathcal{K}'$, respectively (that is, $S_{\mathcal{U}}$ and $S_{\mathcal{Y}}$ are multiplications by $\lambda$ on the appropriate $H^2$ spaces). Now let $T'$ on $\mathcal{H}'$ be the compression of $S_{\mathcal{Y}}$ to $\mathcal{H}'$ (that is $T' = P_{\mathcal{H}'} S_{\mathcal{Y}} \mid \mathcal{H}'$). Note that $\Theta H^2(\mathcal{E})$ is invariant under the block forward shift $S_{\mathcal{Y}}$, and hence $U'$ is an isometric lifting of $T'$. Finally, let A be the operator from $\mathcal{H}$ to $\mathcal{H}'$ defined by $A = P_{\mathcal{H}'} M_F^+$. Since $M_F^+ S_{\mathcal{U}} = S_{\mathcal{Y}} M_F^+$, we have

$$AT = P_{\mathcal{H}'} M_F^+ S_{\mathcal{U}} = P_{\mathcal{H}'} S_{\mathcal{Y}} M_F^+ = T' P_{\mathcal{H}'} M_F^+ = T'A ,$$

and thus A intertwines T and $T'$.

We claim that B is an intertwining lifting of A if and only if $B = M_G^+$ where $G = F - \Theta H$ for some H in $H^\infty(\mathcal{U}, \mathcal{E})$. If $B = M_G^+$ where $G = F - \Theta H$ for some H in $H^\infty(\mathcal{U}, \mathcal{E})$, then for f in $H^2(\mathcal{U})$ we have

$$P_{\mathcal{H}'} Bf = P_{\mathcal{H}'}(F - \Theta H)f = P_{\mathcal{H}'} Ff = Af .$$

Hence $P_{\mathcal{H}'} B = A$. Obviously for this B we have $U'B = S_{\mathcal{Y}} B = BT$. Therefore $B = M_G^+$ is an intertwining lifting of A. On the other hand if B is an intertwining lifting of A, then B is an operator from $H^2(\mathcal{U})$ into $H^2(\mathcal{Y})$ satisfying

$$BS_{\mathcal{U}} = S_{\mathcal{Y}} B \quad \text{and} \quad P_{\mathcal{H}'} B = A . \tag{4.2}$$

From the first identity in (4.2) it follows that $B = M_G^+$ for some $G \in H^\infty(\mathcal{U}, \mathcal{Y})$. From the second identity in (4.2) we may conclude that $P_{\mathcal{H}'}(M_F^+ - M_G^+) = 0$, and thus $\operatorname{ran}(M_F^+ - M_G^+) \subset \Theta H^2(\mathcal{E})$. So for each x in $H^2(\mathcal{U})$ there exists a $f_x$ in $H^2(\mathcal{E})$ satisfying $Fx - Gx = \Theta f_x$. Because G, F and $\Theta$ are linear, the map x to $f_x$ is linear. Moreover, the following shows that this map is bounded,

$$\|f_x\| = \|\Theta f_x\| = \|Fx - Gx\| \le (\|F\|_\infty + \|G\|_\infty)\|x\| . \tag{4.3}$$

Therefore the map x to $f_x$ defines a bounded linear operator Q from $H^2(\mathcal{U})$ to $H^2(\mathcal{E})$ by $f_x = Qx$, and thus $Fx - Gx = \Theta Qx$. Notice also that

$$M_\Theta^+ S_\mathcal{E} Q = S_\mathcal{Y} M_\Theta^+ Q = S_\mathcal{Y}(M_F^+ - M_G^+) = (M_F^+ - M_G^+) S_\mathcal{U} = M_\Theta^+ Q S_\mathcal{U} \ . \tag{4.4}$$

Using the fact that $M_\Theta^+$ is an isometry $S_\mathcal{E} Q = Q S_\mathcal{U}$. Hence there exists a H in $H^\infty(\mathcal{U}, \mathcal{E})$ satisfying $Q = M_H^+$. In particular, $F - G = \Theta H$ and $G = F - \Theta H$. So we have shown that there exists a H in $H^\infty(\mathcal{U}, \mathcal{E})$ satisfying $G = F - \Theta H$ and $B = M_G^+$. This verifies our claim concerning the form of intertwining liftings.

Using the fact that the set of all intertwining liftings B of A are given by $B = M_G^+$, where $G = F - \Theta H$ for some H in $H^\infty(\mathcal{U}, \mathcal{E})$ and $\|B\| = \|G\|_\infty$, we see that the Sarason optimization problem is equivalent to the following problem

$$\begin{aligned} d_\infty = d_\infty(F, \Theta) &= \inf\{\|F - \Theta H\|_\infty : H \in H^\infty(\mathcal{U}, \mathcal{E})\} \\ &= \inf\{\|B\| : B \text{ is an intertwining lifting of } A\} \ . \end{aligned} \tag{4.5}$$

Clearly any intertwining lifting B of A satisfies $\|B\| \geq \|A\|$ and thus $\|A\| \leq d_\infty$. By the commutant lifting theorem, there exists an intertwining lifting B of A satisfying $\|B\| = \|A\|$. In particular, $d_\infty(F, \Theta) = \|A\|$. Moreover, because this B is an intertwining lifting of A we have $B = M_G^+$ where $G = F - \Theta H$ for some operator valued function H in $H^\infty(\mathcal{U}, \mathcal{E})$. Therefore $d_\infty = \|A\| = \|B\| = \|F - \Theta H\|_\infty$. This establishes (4.1) and completes the proof.

## II.5. PROOF OF THE TWO-SIDED NUDELMAN THEOREM

To prove the two-sided Nudelman Theorem I.9.1 we need the following useful result whose proof is given in Section IV.2 of Foias-Frazho [4].

**LEMMA 5.1.** *Let* A *be an operator matrix of the form*

$$A = \begin{bmatrix} A_0 & 0 \\ A_{1,0} & A_1 \end{bmatrix} : \begin{bmatrix} \mathcal{H}_0 \\ \mathcal{H}_1 \end{bmatrix} \to \begin{bmatrix} \mathcal{H}'_0 \\ \mathcal{H}'_1 \end{bmatrix} . \tag{5.1}$$

*The norm of the operator* A *is bounded by* $\gamma$ *if and only if* $\|A_0\| \leq \gamma$, $\|A_1\| \leq \gamma$ *and*

$$\gamma A_{1,0} = D_1 \Omega D_0 \ , \tag{5.2}$$

*where* $D_1 = (\gamma^2 I - A_1 A_1^*)^{1/2}$ *and* $D_0 = (\gamma^2 I - A_0^* A_0)^{1/2}$ *and* $\Omega$ *is a contraction from the closure of the range of* $D_0$ *to the closure of the range of* $D_1$.

We will also need the following classical Schur characterization of $2 \times 2$ positive block matrices. A proof of this result is given in Section XVI.1 of Foias-Frazho [4].

**LEMMA 5.2.** *Let $\Xi$ be a $2 \times 2$ operator matrix of the form*

$$\Xi = \begin{bmatrix} M & \Gamma^* \\ \Gamma & N \end{bmatrix} \text{ on } \begin{pmatrix} \mathcal{F} \\ \mathcal{X} \end{pmatrix}. \tag{5.3}$$

*Then $\Xi \geq 0$ if and only if $M \geq 0$ and $N \geq 0$ and $\Gamma = N^{1/2}\Omega_1 M^{1/2}$, where $\Omega_1$ is a contraction from the closure of the range of M to the closure of the range of N.*

Recall that for the two-sided Nudelman interpolation problem the given data are Hilbert space operators:

$$\begin{gathered} Z : \mathcal{X} \to \mathcal{X}, \; B : \mathcal{U} \to \mathcal{X}, \; \tilde{B} : \mathcal{Y} \to \mathcal{X}, \\ \Lambda : \mathcal{F} \to \mathcal{F}, \; C : \mathcal{F} \to \mathcal{Y}, \; \tilde{C} : \mathcal{F} \to \mathcal{U}, \; \Gamma : \mathcal{F} \to \mathcal{X}, \end{gathered} \tag{5.4}$$

where Z and $\Lambda$ have spectral radius strictly less than one. Furthermore, the operator $\Gamma$ is assumed to satisfy the Sylvester equation

$$Z\Gamma - \Gamma\Lambda = \tilde{B}C - B\tilde{C}. \tag{5.5}$$

Given these operators one seeks $G \in H^\infty(\mathcal{Y}, \mathcal{U})$ such that $\|G\|_\infty \leq \gamma$ and

$$(BG)(Z)_{\text{left}} = \tilde{B}, \quad (GC)(\Lambda)_{\text{right}} = \tilde{C} \quad \text{and} \quad \textstyle\sum_{j,k=0}^{\infty} Z^j B G_{j+k+1} C \Lambda^k = \Gamma, \tag{5.6}$$

where $G_0, G_1, G_2, \ldots$ are the Taylor coefficients of G.

To analyze this problem, let $W_c$ from $\ell_+^2(\mathcal{U})$ to $\mathcal{X}$ and $\tilde{W}_c$ from $\ell_+^2(\mathcal{Y})$ to $\mathcal{X}$ be the controllability operators generated by $\{Z, B\}$ and $\{Z, \tilde{B}\}$, respectively, that is,

$$W_c = [B, ZB, Z^2B, \ldots] \quad \text{and} \quad \tilde{W}_c = [\tilde{B}, Z\tilde{B}, Z^2\tilde{B}, \ldots]. \tag{5.7}$$

Let $W_o$ from $\mathcal{F}$ to $\ell_-^2(\mathcal{Y})$ and $\tilde{W}_o$ from $\mathcal{F}$ to $\ell_-^2(\mathcal{U})$ be the operators defined by

$$W_o = [\ldots, C\Lambda^2, C\Lambda, C]^{tr} \quad \text{and} \quad \tilde{W}_o = [\ldots, \tilde{C}\Lambda^2, \tilde{C}\Lambda, \tilde{C}]^{tr} \tag{5.8}$$

where tr denotes the block transpose. Notice that $W_o^*$ and $\tilde{W}_o^*$ can be identified (through the unitary flip operator) with the controllability operators for $\{\Lambda^*, C^*\}$ and $\{\Lambda^*, \tilde{C}^*\}$, respectively. Now let G be any function in $H^\infty(\mathcal{Y}, \mathcal{U})$, and consider the following partitioning of the associated lower triangular block Laurent operator $L_G$:

$$L_G = \begin{bmatrix} T_G^{\#} & 0 \\ H_G & T_G \end{bmatrix} : \ell_-^2(\mathcal{Y}) \oplus \ell_+^2(\mathcal{Y}) \to \ell_-^2(\mathcal{U}) \oplus \ell_+^2(\mathcal{U}) . \tag{5.9}$$

Then the interpolating conditions in (5.6) are respectively equivalent to

$$W_c T_G = \tilde{W}_c,\ T_G^{\#} W_o = \tilde{W}_o\ ,\ \ \Gamma = W_c H_G W_o\ . \tag{5.10}$$

So finding a solution to the two-sided Nudelman interpolation problem is equivalent to finding a function G in $H^\infty(\mathcal{Y},\ \mathcal{U})$ satisfying (5.10) and $\|G\|_\infty \le \gamma$.

To establish some connections to the commutant lifting theorem, let $\mathcal{H}_c$ be the closure of the range of $W_c^*$ and $\mathcal{H}_o$ the closure of the range of $W_o$. Let $\mathcal{H}$ and $\mathcal{H}'$ be the subspaces of $\ell^2(\mathcal{Y})$ and $\ell^2(\mathcal{U})$ defined by $\mathcal{H} = \mathcal{H}_o \oplus \ell_+^2(\mathcal{Y})$ and $\mathcal{H}' = \ell_-^2(\mathcal{U}) \oplus \mathcal{H}_c$. Let V and V′ be the bilateral block forward shifts on $\ell^2(\mathcal{Y})$ and $\ell^2(\mathcal{U})$, respectively. Notice that $\mathcal{H}$ is an invariant subspace for V while $\mathcal{H}'$ is an invariant subspace for $V'^*$. So let T be the isometry on $\mathcal{H}$ and T′ the co-isometry on $\mathcal{H}'$ defined by $T = V|\mathcal{H}$ and $T'^* = V'^*|\mathcal{H}'$. Obviously V′ is an isometric lifting of T′.

We are now ready to start the proof of Theorem I.9.1. It will be convenient to split the proof into two parts. In the first part we prove the necessity of the positivity condition on the operator matrix $\Xi$ in Theorem I.9.1, and in the second part we use the commutant lifting theorem to prove the sufficiency of this condition.

Part (a). Assume that G in $H^\infty(\mathcal{Y},\ \mathcal{U})$ is a solution to the two-sided Nudelman problem, that is, (5.10) holds and $\|G\|_\infty \le \gamma$. (Without loss of generality we assume that $\gamma > 0$.) Let A be the operator from $\mathcal{H}$ to $\mathcal{H}'$ defined by $P_{\mathcal{H}'} L_G | \mathcal{H}$. Clearly $\|A\| \le \|G\|_\infty \le \gamma$. Using $P_{\mathcal{H}'} V' = T' P_{\mathcal{H}'}$ along with the fact that $L_G$ intertwines V with V′ we have $ATh = P_{\mathcal{H}'} V' L_G h = T'Ah$ for all h in $\mathcal{H}$, and thus $AT = T'A$. Because $L_G$ is lower triangular, the operator A admits a matrix representation of the form

$$A = \begin{bmatrix} A_o & 0 \\ A_{2,1} & A_c \end{bmatrix} : \begin{bmatrix} \mathcal{H}_o \\ \ell_+^2(\mathcal{Y}) \end{bmatrix} \to \begin{bmatrix} \ell_-^2(\mathcal{U}) \\ \mathcal{H}_c \end{bmatrix} . \tag{5.11}$$

This and $A = P_{\mathcal{H}'} L_G | \mathcal{H}$ readily implies that

$$\begin{bmatrix} A_o W_o & 0 \\ W_c A_{2,1} W_o & W_c A_c \end{bmatrix} = \begin{bmatrix} I & 0 \\ 0 & W_c \end{bmatrix} A \begin{bmatrix} W_o & 0 \\ 0 & I \end{bmatrix} =$$

$$\begin{bmatrix} T_G^{\#} W_o & 0 \\ W_c H_G W_o & W_c T_G \end{bmatrix} = \begin{bmatrix} \tilde{W}_o & 0 \\ \Gamma & \tilde{W}_c \end{bmatrix} : \begin{bmatrix} \mathcal{F} \\ \ell_+^2(\mathcal{Y}) \end{bmatrix} \to \begin{bmatrix} \ell_-^2(\mathcal{U}) \\ \mathcal{X} \end{bmatrix} .$$

The last equality follows from (5.10). This shows that

$$A_o W_o = \tilde{W}_o\,, \quad W_c A_c = \tilde{W}_c \quad \text{and} \quad \Gamma = W_c A_{2,1} W_o\,. \tag{5.12}$$

Since $\|A_o\| \le \|A\| \le \gamma$, we see that

$$\tilde{Q} := \sum_{n=0}^{\infty} \Lambda^{*n}\tilde{C}^*\tilde{C}\Lambda^n = \tilde{W}_o^*\tilde{W}_o = W_o^* A_o^* A_o W_o \le \gamma^2 W_o^* W_o = \gamma^2 \sum_{n=0}^{\infty} \Lambda^{*n}C^*C\Lambda^n =: \gamma^2 Q\,.$$

So $\gamma^2 Q - \tilde{Q}$ is positive, where $Q = W_o^* W_o$ and $\tilde{Q} = \tilde{W}_o^*\tilde{W}_o$ are the controllability grammians for $\{\Lambda^*, C^*\}$ and $\{\Lambda^*, \tilde{C}^*\}$, respectively. Using $W_c A_c = \tilde{W}_c$, a similar argument shows that $\gamma^2 P - \tilde{P}$ is positive, where $P = W_c W_c^*$ and $\tilde{P} = \tilde{W}_c \tilde{W}_c^*$ are the controllability grammians for $\{Z, B\}$ and $\{Z, \tilde{B}\}$, respectively.

Since the operator norm of A is bounded by $\gamma$, Lemma 5.1 shows that

$$\gamma\Gamma = W_c(\gamma A_{2,1})W_o = W_c D_c \Omega D_o W_o\,, \tag{5.13}$$

where $D_o = (\gamma^2 I - A_o^* A_o)^{1/2}$, $D_c = (\gamma^2 I - A_c A_c^*)^{1/2}$ and $\Omega$ is a contraction. However, according to the classical Schur Lemma 5.2, the matrix

$$\Xi = \begin{bmatrix} \gamma^2 Q - \tilde{Q} & \gamma\Gamma^* \\ \gamma\Gamma & \gamma^2 P - \tilde{P} \end{bmatrix} : \begin{pmatrix} \mathcal{F} \\ \mathcal{X} \end{pmatrix} \to \begin{pmatrix} \mathcal{F} \\ \mathcal{X} \end{pmatrix} \tag{5.14}$$

is positive if and only if

$$\gamma\Gamma = (\gamma^2 P - \tilde{P})^{1/2}\Omega_1(\gamma^2 Q - \tilde{Q})^{1/2} \tag{5.15}$$

where $\Omega_1$ is a contraction from the closure of the range of $\gamma^2 Q - \tilde{Q}$ to the closure of the range of $\gamma^2 P - \tilde{P}$. Using $A_o W_o = \tilde{W}_o$ and x in $\mathcal{F}$ we have

$$\|(\gamma^2 Q - \tilde{Q})^{1/2} x\|^2 = ((\gamma^2 Q - \tilde{Q})x, x) = ((\gamma^2 W_o^* W_o - \tilde{W}_o^*\tilde{W}_o)x, x)$$
$$= (W_o^*(\gamma^2 I - A_o^* A_o)W_o x, x) = \|D_o W_o x\|^2\,.$$

So there exist a unitary operator $\beta_o$ from the closure of the range of $\gamma^2 Q - \tilde{Q}$ to the closure of the range of $D_o$ and a unitary operator $\beta_c$ from the closure of the range of $D_c$ to the closure of the range of $\gamma^2 P - \tilde{P}$ satisfying

$$D_o W_o = \beta_o(\gamma^2 Q - \tilde{Q})^{1/2} \quad \text{and} \quad (\gamma^2 P - \tilde{P})^{1/2}\beta_c = W_c D_c\,. \tag{5.16}$$

The second equation in (5.16) follows by applying a similar argument to $W_c A_c = \tilde{W}_c$ to establish the existence of the unitary operator $\beta_c$. So by combining (5.13) and (5.16) we see that

$\gamma\Gamma$ is given by (5.15) where $\Omega_1$ is now the contraction $\beta_c\Omega\beta_o$. By the Schur Lemma the matrix in (5.14) is positive.

Part (b). To prove the other half let us assume that the operator matrix $\Xi$ in (5.14) is positive. Then obviously $\gamma^2 Q \geq \tilde{Q}$. So for x in $\mathcal{F}$ we have

$$\gamma^2 \|W_o x\|^2 = \gamma^2 (W_o^* W_o x, x) = \gamma^2 (Qx, x) \geq (\tilde{Q}x, x) = (\tilde{W}_o^* \tilde{W}_o x, x) = \|\tilde{W}_o x\|^2 .$$

This implies that there exists an operator $A_o$ from $\mathcal{H}_o$ to $\ell_-^2(\mathcal{U})$ satisfying $A_o W_o = \tilde{W}_o$ and the norm of $A_o$ is bounded by $\gamma$. Now notice that $P_- V W_o = W_o \Lambda$, where $P_-$ is the orthogonal projection of $\ell^2(\mathcal{Y})$ onto $\ell_-^2(\mathcal{Y})$. Therefore $P_- V$ maps $\mathcal{H}_o$ into $\mathcal{H}_o$. Now let $T_o$ be the operator on $\mathcal{H}_o$ defined by $T_o = P_- V | \mathcal{H}_o = P_{\mathcal{H}_o} V | \mathcal{H}_o$. Using this and $A_o W_o = \tilde{W}_o$, we have

$$A_o T_o W_o = A_o P_- V W_o = A_o W_o \Lambda = \tilde{W}_o \Lambda = P'_- V' \tilde{W}_o = P'_- V' A_o W_o .$$

Hence $A_o T_o = P'_- V' A_o$ where $P'_-$ is the orthogonal projection onto $\ell_-^2(\mathcal{U})$ and $A_o$ intertwines $T_o$ with $P'_- V' | \ell_-^2(\mathcal{U})$. A similar argument, involving $\gamma^2 P \geq \tilde{P}$, shows that there exists an operator $A_c$ from $\ell_+^2(\mathcal{Y})$ to $\mathcal{H}_c$ bounded by $\gamma$ satisfying $W_c A_c = \tilde{W}_c$. Moreover, if we let $T_c$ on $\mathcal{H}_c$ be the compression of $V'$ to $\mathcal{H}_c$, then $A_c$ intertwines operator $V | \ell_+^2(\mathcal{Y})$ on $\ell_+^2(\mathcal{Y})$ with $T_c$. Finally, let us notice that this is enough information to establish that the first two identities in (5.12) also hold in this part of the proof.

Now use the fact that the operator matrix $\Xi$ in (5.14) is positive. According to the Schur Lemma (Lemma 5.2) the operator $\gamma\Gamma$ is given by (5.15), where $\Omega_1$ is a contraction. By using (5.16) in (5.15) we see that $\gamma\Gamma$ is given by (5.13), where $\Omega = \beta_c^* \Omega_1 \beta_o^*$. Set $\gamma A_{2,1} = D_c \Omega D_0$ and thus $\Gamma = W_c A_{2,1} W_o$. Notice that because $A_{2,1}$ maps $\mathcal{H}_o = \overline{\operatorname{ran} W_o}$ into $\mathcal{H}_c = \overline{\operatorname{ran} W_c^*}$ the operator $A_{2,1}$ is uniquely determined by $\Gamma$. Now let A be the operator from $\mathcal{H}$ to $\mathcal{H}'$ defined by (5.11). According to Lemma 5.1 this operator A is bounded in norm by $\gamma$.

We claim that A intertwines T with T′. To see this notice that T and T′ admit matrix representations of the form:

$$T = \begin{bmatrix} T_o & 0 \\ P_+ V | \mathcal{H}_o & V | \ell_+^2(\mathcal{Y}) \end{bmatrix} \text{ on } \begin{bmatrix} \mathcal{H}_o \\ \ell_+^2(\mathcal{Y}) \end{bmatrix},$$

$$T' = \begin{bmatrix} P'_- V' | \ell_-^2(\mathcal{U}) & 0 \\ P_{\mathcal{H}_c} V' | \ell_-^2(\mathcal{U}) & T_c \end{bmatrix} \text{ on } \begin{bmatrix} \ell_-^2(\mathcal{U}) \\ \mathcal{H}_c \end{bmatrix}. \tag{5.17}$$

Here $P_+$ is the orthogonal projection onto $\ell_+^2(\mathcal{Y})$. Recall that $A_o$ intertwines $T_o$ with $P'_- V' | \ell_-^2(\mathcal{U})$ and $A_c$ intertwines $V | \ell_+^2(\mathcal{Y})$ with $T_c$. Therefore A in (5.11) intertwines T with T′

if and only if for all h in $\mathcal{H}_o$ we have

$$P_{\mathcal{H}_c} V' A_o h + T_c A_{2,1} h = A_{2,1} T_o h + A_c P_+ V h .$$

Since the range of $W_o$ is dense in $\mathcal{H}_o$, we can set $h = W_o x$ for x in $\mathcal{F}$ and the previous equation holds if and only if

$$P_{\mathcal{H}_c} V' \tilde{W}_o x + T_c A_{2,1} W_o x = A_{2,1} W_o \Lambda x + A_c [Cx, 0, 0, \ldots]^{tr} .$$

Because $W_c \mid \mathcal{H}_c$ is one to one, we can apply $W_c$ to the left, and the previous equation holds if and only if

$$\tilde{B} C x + Z W_c A_{2,1} W_o x = W_c A_{2,1} W_o \Lambda x + \tilde{W}_c [Cx, 0, 0, \ldots]^{tr} .$$

So using $\Gamma = W_c A_{2,1} W_o$ and $\tilde{W}_c [C, 0, 0, \ldots]^{tr} = \tilde{B} C$, we see that $T'A = AT$ if and only if $\Gamma$ satisfies the Sylvester equation (5.5). By assumption $\Gamma$ satisfies (5.5). Therefore A is an operator bounded in norm by $\gamma$ and intertwining T with $T'$.

By the commutant lifting theorem there exists an operator $B_\gamma$ from $\mathcal{H}$ to $\ell^2(\mathcal{U})$, intertwining T with $V'$, preserving the norm of A, and satisfying $P_{\mathcal{H}'} B_\gamma = A$. Since $T = V \mid \mathcal{H}$ Lemma 3.1 shows that there exists a unique operator L from $\ell^2(\mathcal{Y})$ to $\ell^2(\mathcal{U})$, preserving the norm of $B_\gamma$, intertwining V with $V'$, and satisfying $L \mid \mathcal{H} = B_\gamma$. Because $V'L = LV$, the operator L is a block Laurent operator. We have to show that L is block lower triangular. To do this, take $y \in \mathcal{Y}$, and let $x \in \ell^2(\mathcal{Y})$ be the sequence $x = (\delta_{j,0} y)_{-\infty}^{\infty}$. Here $\delta_{j,k}$ is the Kronecker delta. Using (5.11), we see that $Ax \in \mathcal{H}_c \subset \ell_+^2(\mathcal{U})$. Therefore, since $x \in \mathcal{H}$, we have

$$Lx = B_\gamma x = Ax + (I - P_{\mathcal{H}'}) B_\gamma x \in \ell_+^2(\mathcal{U}) .$$

Thus L is block lower triangular, and $L = L_G$ for some $G \in H^\infty(\mathcal{Y}, \mathcal{U})$. Furthermore, $\|G\|_\infty = \|L_G\| = \|B_\gamma\| = \|A\| \le \gamma$. Notice that $A = P_{\mathcal{H}'} L_G \mid \mathcal{H}$. So using $A_o = P'_- L_G \mid \mathcal{H}_o$ and $A_o W_o = \tilde{W}_o$, we have $T_G^\# W_o = P'_- L_G W_o = \tilde{W}_o$. The relation $W_c A_c = \tilde{W}_c$ along with $A_c = P_{\mathcal{H}_c} L_G \mid \ell_+^2(\mathcal{Y})$ gives $W_c T_G = \tilde{W}_c$. Finally, $\Gamma = W_c A_{2,1} W_o = W_c P_{\mathcal{H}_c} L_G W_o = W_c H_G W_o$. Therefore all the Nudelman interpolating conditions in (5.6), or equivalently, (5.10) are satisfied and $\|G\|_\infty = \|A\| \le \gamma$. This completes the proof of the two-sided Nudelman interpolation theorem.

The two-sided Nudelman problem yields the $H^\infty$ optimization problem associated with:

$$d_\infty = \inf \{ \|G\|_\infty : G \in H^\infty(\mathcal{Y}, \mathcal{U}) \text{ and (5.6) holds} \} . \tag{5.18}$$

According to our previous analysis the optimal error $d_\infty$ is finite if and only if there exists a bounded linear operator A of the form (5.11) such that

$$A_o W_o = \tilde{W}_o\ ,\quad W_c A_c = \tilde{W}_c \quad \text{and} \quad \Gamma = W_c A_{2,1} W_o\ . \tag{5.19}$$

In this case A is the only operator of the form (5.11) satisfying (5.19). Moreover, there exists an optimal G in $H^\infty(\mathcal{Y}, \mathcal{U})$ satisfying the interpolating conditions in (5.6), or equivalently, (5.10) and $\|G\|_\infty = d_\infty = \|A\|$.

We say that G is a *Nudelman interpolant*, if G is a function in $H^\infty(\mathcal{Y}, \mathcal{U})$ satisfying the interpolation conditions in (5.6). Notice that there exists a Nudelman interpolant G of the data in (5.4) (which are assumed to satisfy (5.5)) if and only if the relations in (5.19) define a bounded operator A of the form (5.11). So let us assume that there exists a Nudelman interpolant G for the data in (5.4) and let A be a bounded linear operator of the form (5.11) such that the identities in (5.19) are satisfied. Then the proof of the previous theorem shows that $B_o$ is an intertwining lifting of A with respect to $U' = V'$ if and only if $B_o = L_G \mid \mathcal{H}$ where G is a Nudelman interpolant of the data. In this case $B_o$ and G uniquely determine each other. Moreover, the $H^\infty$ optimization problem in (5.18) is equivalent to the following optimization problem

$$d_\infty = \inf\{\|B_o\| : B_o \text{ is an intertwining lifting of } A\}\ . \tag{5.20}$$

Clearly any intertwining lifting $B_o$ of A satisfies $\|A\| \le \|B_o\|$, and thus $\|A\| \le d_\infty$. By the commutant lifting theorem, there exists a intertwining lifting $B_o$ of A which preserves the norm of A. In particular, $\|B_o\| = \|A\|$ and $d_\infty = \|A\|$. Moreover, this $B_o$ uniquely determines a Nudelman interpolant G in $H^\infty(\mathcal{Y}, \mathcal{U})$ by $B_o = L_G \mid \mathcal{H}$ and $d_\infty = \|B_o\| = \|G\|_\infty$.

**REMARK 5.3**. Assume that the controllability grammians $P = W_c W_c^*$ and $Q = W_o^* W_o$ for $\{Z, B\}$ and $\{\Lambda^*, C^*\}$, respectively, are both strictly positive. Clearly the conditions $A_o W_o = \tilde{W}_o$ and $W_c A_c = \tilde{W}_c$ and $\Gamma = W_c A_{2,1} W_o$ uniquely determine $A_o$, $A_c$ and $A_{2,1}$. So by direct substitution, it is easy to show that an explicit formula for the operator A in (5.11) is given by

$$A = \begin{bmatrix} \tilde{W}_o Q^{-1} W_o^* & 0 \\ W_c^* P^{-1} \Gamma Q^{-1} W_o^* & W_c^* P^{-1} \tilde{W}_c \end{bmatrix} : \begin{bmatrix} \mathcal{H}_o \\ \ell_+^2(\mathcal{Y}) \end{bmatrix} \to \begin{bmatrix} \ell_-^2(\mathcal{U}) \\ \mathcal{H}_c \end{bmatrix} . \tag{5.21}$$

Obviously this A is bounded. So if both P and Q are invertible, then there exists a solution G to the $H^\infty$ optimization problem in (5.18), that is, $G \in H^\infty(\mathcal{Y}, \mathcal{U})$, the identities in (5.6) hold and $\|G\|_\infty = d_\infty$.

Let us obtain an explicit formula for the error $d_\infty$ when both P and Q are invertible. To this end, notice that the closure of $W_o \mathcal{F} \oplus \tilde{W}_c^* \mathcal{X}$ contains $(\ker A)^\perp$. For f in $\mathcal{F}$ and x in $\mathcal{X}$ the form of A in (5.21) gives

$$d_\infty^2 = \sup\{\|A(W_o f \oplus \tilde{W}_c^* )x\|^2\} =$$
$$\sup\{\|\tilde{W}_o f\|^2 + \|W_c^* P^{-1}\Gamma f + W_c^* P^{-1}\tilde{P}x\|^2\} = \tag{5.22}$$
$$\sup\{(\tilde{Q}f, f) + (\Gamma^* P^{-1}\Gamma f, f) + 2\,\mathrm{Re}\,(\tilde{P}P^{-1}\Gamma f, x) + (\tilde{P}P^{-1}\tilde{P}x, x)\}$$

where the supremum is taken over all vectors of the form $\|W_o f\|^2 + \|\tilde{W}_c^* x\|^2 = 1$, or equivalently, $(Qf, f) + (\tilde{P}x, x) = 1$. By setting $g = Q^{1/2} f$ and $y = \tilde{P}^{1/2} x$ in (5.22) and using $\|g \oplus y\| = 1$, equation (5.22) now shows that $d_\infty^2 = \|R_o\|$, where $R_o$ is the positive $2 \times 2$ operator matrix defined by

$$R_o = \begin{bmatrix} Q^{-1/2}(\tilde{Q} + \Gamma^* P^{-1}\Gamma)Q^{-1/2} & Q^{-1/2}\Gamma^* P^{-1}\tilde{P}^{1/2} \\ \tilde{P}^{1/2} P^{-1}\Gamma Q^{-1/2} & \tilde{P}^{1/2} P^{-1}\tilde{P}^{1/2} \end{bmatrix} : \begin{bmatrix} \mathcal{F} \\ \mathcal{X} \end{bmatrix} \to \begin{bmatrix} \mathcal{F} \\ \mathcal{X} \end{bmatrix}. \tag{5.23}$$

Finally, assume that $\tilde{P}$ is invertible and set $X = \mathrm{diag}\,[Q^{1/2}, \tilde{P}^{1/2}]$. Then $R_o$ is similar to $R_\infty = X^{-1} R_o X$ which equals

$$R_\infty = \begin{bmatrix} Q^{-1}(\tilde{Q} + \Gamma^* P^{-1}\Gamma) & Q^{-1}\Gamma^* P^{-1}\tilde{P} \\ P^{-1}\Gamma & P^{-1}\tilde{P} \end{bmatrix}. \tag{5.24}$$

So, if P, Q and $\tilde{P}$ are all invertible, then $d_\infty^2 = r_{spec}(R_\infty)$. Notice that the expression for $R_\infty$ in (5.24) does not contain any square roots. However, $R_\infty$ is not a self adjoint operator. So in certain applications it may be advantageous to compute the spectral radius of $R_\infty$ rather than the norm of $R_o$. One can obtain a similar expression for $d_\infty$ when P, Q and $\tilde{Q}$ are all invertible. In this case one simply performs the previous analysis to $AA^*$.

To complete this section let us conclude by solving an abstract Nudelman completion problem. Let U and U′ be isometries of the form:

$$U = \begin{bmatrix} U_{1,1} & 0 & 0 \\ U_{2,1} & U_{2,2} & 0 \\ U_{3,1} & U_{3,2} & U_{3,3} \end{bmatrix} \text{ on } \mathcal{H}_1 \oplus \mathcal{H}_2 \oplus \mathcal{H}_3,$$

$$U' = \begin{bmatrix} U'_{1,1} & 0 & 0 \\ U'_{2,1} & U'_{2,2} & 0 \\ U'_{3,1} & U'_{3,2} & U'_{3,3} \end{bmatrix} \text{ on } \mathcal{H}'_1 \oplus \mathcal{H}'_2 \oplus \mathcal{H}'_3. \tag{5.25}$$

Let $A_o$ from $\mathcal{H}_1$ to $\mathcal{H}'_1$ and $A_c$ from $\mathcal{H}_2$ to $\mathcal{H}'_2$ be operators bounded by $\gamma$ satisfying the following intertwining relations:

$$U'_{1,1}A_o = A_oU_{1,1} \ , \ U'_{2,2}A_c = A_cU_{2,2} \ . \tag{5.26}$$

The *abstract Nudelman completion problem* is to find (if possible) an operator B of the form

$$B = \begin{bmatrix} B_{1,1} & 0 & 0 \\ B_{2,1} & B_{2,2} & 0 \\ B_{3,1} & B_{3,2} & B_{3,3} \end{bmatrix} : \mathcal{H}_1 \oplus \mathcal{H}_2 \oplus \mathcal{H}_3 \to \mathcal{H}'_1 \oplus \mathcal{H}'_2 \oplus \mathcal{H}'_3 \tag{5.27}$$

satisfying the following interpolating conditions:

$$U'B = BU \ , \ B_{1,1} = A_o \ , \ B_{2,2} = A_c \text{ and } \|B\| \le \gamma \tag{5.28}$$

where $\gamma$ is a prespecified bound. To present a solution to this completion problem recall that $D_0$ is the positive square root of $\gamma^2 I - A_o^* A_o$ and $D_c$ is the positive square root of $\gamma^2 I - A_c A_c^*$.

**THEOREM 5.4**. *The abstract Nudelman completion problem with data* U, U′, $A_o$, $A_c$ *and tolerance* $\gamma$ *has a solution* B *of the form (5.27) if and only if the equation*

$$A_{2,1}U_{1,1} - U'_{2,2}A_{2,1} = U'_{2,1}A_o - A_cU_{2,1} \tag{5.29}$$

*has a solution* $A_{2,1}$ *in* $\mathcal{L}(\mathcal{H}_1, \mathcal{H}'_2)$ *where* $\gamma A_{2,1} = D_c \Omega D_0$ *and* $\Omega$ *a contraction from* $\overline{D_0\mathcal{H}_1}$ *to* $\overline{D_c\mathcal{H}'_2}$. *In this case we may take* $B_{2,1}$ *to be this solution* $A_{2,1}$ *of (5.29).*

**PROOF**. The intertwining relations in (5.26) and (5.29) are equivalent to

$$\begin{bmatrix} A_o & 0 \\ A_{2,1} & A_c \end{bmatrix} \begin{bmatrix} U_{1,1} & 0 \\ U_{2,1} & U_{2,2} \end{bmatrix} - \begin{bmatrix} U'_{1,1} & 0 \\ U'_{2,1} & U'_{2,2} \end{bmatrix} \begin{bmatrix} A_o & 0 \\ A_{2,1} & A_c \end{bmatrix} = 0 \ . \tag{5.30}$$

Now, assume that the Nudelman completion problem has a solution B of the form (5.27). Take $A_{2,1} = B_{2,1}$. Then, because of $A_o = B_{1,1}$ and $A_c = B_{2,2}$, we have

$$\begin{bmatrix} A_o & 0 \\ A_{2,1} & A_c \end{bmatrix} = \begin{bmatrix} B_{1,1} & 0 \\ B_{2,1} & B_{2,2} \end{bmatrix} . \tag{5.31}$$

Since B is bounded in norm by $\gamma$, the same holds true for the $2 \times 2$ operator matrix in the right hand side of (5.31). Thus the left hand side of (5.31) is also bounded in norm by $\gamma$. By Lemma 5.1 it follows $\gamma A_{2,1} = D_c \Omega D_0$ for some contraction $\Omega$. The first condition in (5.28) and (5.31) imply that (5.30) holds, and therefore $A_{2,1}$ is a solution of (5.29) of the desired form.

To prove the converse statement, assume that $\gamma A_{2,1} = D_c \Omega D_0$ and $A_{2,1}$ is a solution of (5.29) where $\Omega$ is a contraction. Put $T = U$ and

$$T' = \begin{bmatrix} U'_{1,1} & 0 \\ U'_{2,1} & U'_{2,2} \end{bmatrix}.$$

Now set

$$A = \begin{bmatrix} A_o & 0 & 0 \\ A_{2,1} & A_c & 0 \end{bmatrix} : \mathcal{H}_1 \oplus \mathcal{H}_2 \oplus \mathcal{H}_3 \to \mathcal{H}'_1 \oplus \mathcal{H}'_2 .$$

Our conditions on $A_{2,1}$ imply that $\|A\| \le \gamma$ and A satisfies the intertwining relation $AT = T'A$. Notice that $U'$ is an isometric lifting of $T'$. Thus, by the commutant lifting theorem, there exists an operator B satisfying $U'B = BU$ and $\|B\| \le \gamma$ such that $P_{\mathcal{H}'_1 \oplus \mathcal{H}'_2} B = A$. The latter identity implies that B has the form (5.27) with $B_{1,1} = A_o$ and $B_{2,2} = A_c$. This completes the proof.

## II.6. PROOF OF THE TWO-SIDED SARASON THEOREM

In this section we will prove Theorem I.10.1. To this end, recall that $L_F$ is the bounded Laurent operator generated by the sequence of operators $\{F_n\}_{-\infty}^{\infty}$ in $\mathcal{L}(\mathcal{U}, \mathcal{Y})$ and $L_2 = L_{\Theta_2}$ is the Laurent operator generated by the inner function $\Theta_2$ in $H^\infty(\mathcal{F}, \mathcal{Y})$ while $L_1 = L_{\Theta_1}$ is the Laurent operator generated by the co-inner function $\Theta_1$ in $H^\infty(\mathcal{U}, \mathcal{E})$. Here we will use the commutant lifting theorem to show that there exists an H in $H^\infty(\mathcal{E}, \mathcal{F})$ satisfying

$$\|L_F - L_2 L_H L_1\| = \|P_{\mathcal{H}'} L_F \mid \mathcal{H}\| = \inf\{\|L_F - L_2 L_K L_1\| : K \in H^\infty(\mathcal{E}, \mathcal{F})\} . \tag{6.1}$$

Recall also that $\mathcal{H}$ and $\mathcal{H}'$ are the spaces defined by

$$\mathcal{H} = \ell^2(\mathcal{U}) \ominus L_1^* \ell_-^2(\mathcal{E}) \quad \text{and} \quad \mathcal{H}' = \ell^2(\mathcal{Y}) \ominus L_2 \ell_+^2(\mathcal{F}) . \tag{6.2}$$

According to Lemma 6.1 below if L is any Laurent operator from $\ell^2(\mathcal{U})$ to $\ell^2(\mathcal{Y})$, then $P_{\mathcal{H}'} L \mid \mathcal{H} = 0$ if and only if L admits a factorization of the form $L = L_2 L_H L_1$ where H is in $H^\infty(\mathcal{E}, \mathcal{F})$. Therefore

$$\|P_{\mathcal{H}'} L_F \mid \mathcal{H}\| = \|P_{\mathcal{H}'}(L_F - L_2 L_H L_1) \mid \mathcal{H}\| \le \|L_F - L_2 L_H L_1\| .$$

So this readily implies that

$$\|P_{\mathcal{H}'} L_F \mid \mathcal{H}\| \le \inf\{\|L_F - L_2 L_H L_1\| : H \in H^\infty(\mathcal{E}, \mathcal{F})\} . \tag{6.3}$$

To prove that we have equality in (6.3) let V and $V'$ denote the block bilateral forward shifts on $\ell^2(\mathcal{U})$ and $\ell^2(\mathcal{Y})$, respectively. Since $L_2$ and $L_1$ are (lower triangular) Laurent operators, the space $L_2 \ell_+^2(\mathcal{F})$ is invariant under $V'$ and the space $L_1^* \ell_-^2(\mathcal{E})$ is invariant under $V^*$. It follows that $\mathcal{H}$ is invariant under V and $\mathcal{H}'$ is invariant under $V'^*$. Let T on $\mathcal{H}$ be the isometry defined by $T = V \mid \mathcal{H}$ and $T'$ be the co-isometry on $\mathcal{H}'$ defined by $T' = P_{\mathcal{H}'} V' \mid \mathcal{H}'$. The operator $V'$ is an isometric lifting of $T'$.

Let A from $\mathcal{H}$ into $\mathcal{H}'$ be the operator defined by $A = P_{\mathcal{H}'}L_F|\mathcal{H}$. The operator A intertwines the operators T and $T'$. Indeed, using the lifting property $P_{\mathcal{H}'}V' = T'P_{\mathcal{H}'}$, with $h \in \mathcal{H}$ we have

$$ATh = AVh = P_{\mathcal{H}'}L_FVh = P_{\mathcal{H}'}V'L_Fh = T'P_{\mathcal{H}'}L_Fh = T'Ah.$$

Here we used $L_FV = V'L_F$, which holds because $L_F$ is a Laurent operator.

Next, we apply the commutant lifting theorem (Theorem 1.1) to T, $T'$ and A as above, with $T = V|\mathcal{H}$, $\mathcal{K}' = \ell^2(\mathcal{Y})$ and $U' = V'$. It follows that there exists an operator B from $\mathcal{H}$ to $\ell^2(\mathcal{Y})$ such that

$$BT = V'B, \quad \|B\| = \|A\| \quad \text{and} \quad P_{\mathcal{H}'}B = A. \tag{6.4}$$

Since $\ell^2_+(\mathcal{U}) \subset \mathcal{H}$, the operator V is a minimal unitary extension of $V|\mathcal{H}$, and hence, by Lemma 3.1, there exists a unique operator $B_e$ from $\ell^2(\mathcal{U})$ to $\ell^2(\mathcal{Y})$ preserving the norm of B, intertwining V with $V'$, and satisfying $B_e|\mathcal{H} = B$. Therefore $A = P_{\mathcal{H}'}B_e|\mathcal{H}$ and $\|B_e\| = \|A\|$.

Since $V'B_e = B_eV$, this $B_e$ is also a Laurent operator. Moreover, we have $P_{\mathcal{H}'}(L_F - B_e)|\mathcal{H} = A - P_{\mathcal{H}'}B_e|\mathcal{H} = 0$. Because $L_F - B_e$ is a Laurent operator, Lemma 6.1 below shows that there exists a function H in $H^\infty(\mathcal{E}, \mathcal{F})$ satisfying $L_F - B_e = L_2L_HL_1$, and thus $B_e = L_F - L_2L_HL_1$. Finally,

$$\|L_F - L_2L_HL_1\| = \|B_e\| = \|A\| = \|P_{\mathcal{H}'}L_F|\mathcal{H}\|.$$

So we have equality in (6.3) and (6.1) holds. This completes the proof of Theorem I.10.1.

**LEMMA 6.1**. *Let* L *from* $\ell^2(\mathcal{U})$ *to* $\ell^2(\mathcal{Y})$ *be a Laurent operator,* $\Theta_1$ *a co-inner function in* $H^\infty(\mathcal{U}, \mathcal{E})$ *and* $\Theta_2$ *an inner function in* $H^\infty(\mathcal{F}, \mathcal{Y})$. *Let* $\mathcal{H}$ *and* $\mathcal{H}'$ *be the Hilbert spaces defined in (6.2), where* $L_2 = L_{\Theta_2}$ *and* $L_1 = L_{\Theta_1}$. *Then* $P_{\mathcal{H}'}L|\mathcal{H} = 0$ *if and only if* L *admits a factorization of the form* $L = L_2L_HL_1$, *where* H *is in* $H^\infty(\mathcal{E}, \mathcal{F})$.

**PROOF**. Assume that $P_{\mathcal{H}'}L|\mathcal{H} = 0$. Then $L\mathcal{H} \subseteq L_2\ell^2_+(\mathcal{F})$. Since $\ell^2_+(\mathcal{U}) \subseteq \mathcal{H}$, this implies that $L\ell^2_+(\mathcal{U}) \subseteq L_2\ell^2_+(\mathcal{F})$. So given any x in $\ell^2_+(\mathcal{U})$ there exists a unique $f_x$ in $\ell^2_+(\mathcal{F})$ such that $Lx = L_2f_x$. The uniqueness follows from the fact that $L_2$ is an isometry. Using the fact that L and $L_2$ are linear operators it follows that the map from x to $f_x$ is linear. The following shows that this map is also bounded

$$\|f_x\| = \|L_2f_x\| = \|Lx\| \le \|L\|\,\|x\|.$$

So there exists a bounded linear operator $R_+$ from $\ell^2_+(\mathcal{U})$ to $\ell^2_+(\mathcal{F})$ satisfying $R_+x = f_x$ and

$L = L_2R_+$. Now let $V_{\mathcal{F}}$ be the forward bilateral shift on $\ell^2(\mathcal{F})$, and recall that $S_{\mathcal{G}}$ is the forward unilateral shift on $\ell_+^2(\mathcal{G})$. Then for x in $\ell_+^2(\mathcal{U})$ we have

$$L_2S_{\mathcal{F}}R_+x = L_2V_{\mathcal{F}}R_+x = V'L_2R_+x = V'Lx = LVx = LS_{\mathcal{U}}x = L_2R_+S_{\mathcal{U}}x .$$

Since $L_2$ is one to one, $R_+S_{\mathcal{U}} = S_{\mathcal{F}}R_+$. Hence there exists a function R in $H^\infty(\mathcal{U}, \mathcal{F})$ satisfying $R_+ = T_R$. In particular, $L | \ell_+^2(\mathcal{U}) = L_2T_R = L_2L_R | \ell_+^2(\mathcal{U})$. Because the Laurent operators L and $L_2L_R$ agree on $\ell_+^2(\mathcal{U})$ we have $L = L_2L_R$.

Using $L = L_2L_R$ and $P_{\mathcal{H}'}L | \mathcal{H} = 0$ we obtain

$$L_2L_R(\ell^2(\mathcal{U}) \ominus L_1^*\ell_-^2(\mathcal{E})) = L\mathcal{H} \subseteq L_2\ell_+^2(\mathcal{F}) .$$

Hence $L_R$ maps $\ell^2(\mathcal{U}) \ominus L_1^*\ell_-^2(\mathcal{E})$ into $\ell_+^2(\mathcal{F})$, or equivalently, its adjoint $L_R^*$ maps $\ell_-^2(\mathcal{F})$ into $L_1^*\ell_-^2(\mathcal{E})$, that is, $L_R^*\ell_-^2(\mathcal{F}) \subseteq L_1^*\ell_-^2(\mathcal{E})$. So, by applying our previous argument to $L_R^*$ and $L_1^*$, we see that there exists a function H in $H^\infty(\mathcal{E}, \mathcal{F})$ satisfying $L_R^* = L_1^*L_H^*$ and thus $L_R = L_HL_1$. This and $L = L_2L_R$ show that L admits a factorization of the form $L = L_2L_HL_1$.

On the other hand, assume that $L = L_2L_HL_1$ for some H in $H^\infty(\mathcal{E}, \mathcal{F})$. Notice that $L_1\mathcal{H}$ is orthogonal to $\ell_-^2(\mathcal{E})$, or equivalently, $L_1\mathcal{H}$ is contained in $\ell_+^2(\mathcal{E})$. Thus $L_HL_1\mathcal{H}$ is contained in $\ell_+^2(\mathcal{F})$. Therefore $P_{\mathcal{H}'}L | \mathcal{H} = P_{\mathcal{H}'}L_2L_HL_1 | \mathcal{H}$ is zero. This completes the proof.

If $B_e = L_F - L_2L_HL_1$ where H is in $H^\infty(\mathcal{E}, \mathcal{F})$, then $B = B_e | \mathcal{H}$ is an intertwining lifting of $A = P_{\mathcal{H}'}B_e | \mathcal{H}$ with respect to U'. So by consulting our previous analysis we see that B is an intertwining lifting of A if and only if $B = (L_F - L_2L_HL_1) | \mathcal{H}$ where H is a uniquely determined function in $H^\infty(\mathcal{E}, \mathcal{F})$. In this case $\|B\| = \|L_F - L_2L_HL_1\|$. Hence the two-sided Sarason optimization problem is equivalent to the following optimization problem

$$d_\infty = \inf\{\|L_F - L_2L_HL_1\| : H \in H^\infty(\mathcal{E}, \mathcal{F})\} = \inf\{\|B\| : B \text{ is an intertwining lifting of } A\} . \tag{6.5}$$

If B is an intertwining lifting of A, then $\|A\| \le \|B\|$, and thus $d_\infty \ge \|A\|$. By the commutant lifting theorem there exists an intertwining lifting B of A satisfying $\|B\| = \|A\|$. Hence $d_\infty = \|A\|$. This B admits representation of the form $B = (L_F - L_2L_HL_1) | \mathcal{H}$ where H is a function in $H^\infty(\mathcal{E}, \mathcal{F})$. Therefore there exists a H in $H^\infty(\mathcal{E}, \mathcal{F})$ satisfying $\|L_F - L_2L_HL_1\| = d_\infty$.

## *Notes to Chapter II:*

In 1967 D. Sarason (Sarason [1]) solved scalar interpolation problems of Nevanlinna-Pick type by a new technique which involved lifting of operators to ones commuting with the shift.

B. Sz.-Nagy and C. Foias (Sz.-Nagy-Foias [1]) extended Sarason's result and created a method (nowadays referred to as the commutant lifting theorem) for proving existence of solutions to norm constrained interpolation problems. This commutant lifting approach was systematically used for solving interpolation problems in the monograph Foias-Frazho [4]. The latter book also contains a number of different proofs of the commutant lifting theorem, a synopsis of the history of this method, and its further developments and applications. The use of the commutant lifting theorem in this chapter is more or less standard.

CHAPTER III

# TIME INVARIANT SYSTEMS

Throughout this book the interconnection between interpolation and input-output systems in state space form plays a fundamental role. This chapter introduces discrete time-invariant input-output systems, and various notions connected with them. The chapter also serves as a further motivation for the interpolation problems considered earlier. An introduction to the state space theory is given. Moreover, it is shown that point and operator evaluation naturally occur in linear systems. State space techniques are used to compute the norm of certain Hankel operators, which is precisely the error in the corresponding Nehari interpolation problem. State space techniques are also used to give explicit state space formulas to connect the Nevanlinna-Pick problem to the Sarason problem, and the Nudelman problem to the two-sided Sarason problem. The last section presents some aspects of unitary systems that will be used later.

## III.1 STATE SPACE ANALYSIS

In this section we will introduce some terminology and notation from mathematical system theory. A state space system, denoted by {Z, B, C, D}, is an input-output relation of the form

$$x(n+1) = Zx(n) + Bu(n) \quad \text{and} \quad y(n) = Cx(n) + Du(n) \tag{1.1}$$

where Z is an operator on $\mathcal{X}$ and B maps $\mathcal{U}$ into $\mathcal{X}$ while C maps $\mathcal{X}$ into $\mathcal{Y}$ and D maps $\mathcal{U}$ into $\mathcal{Y}$. Here $\mathcal{U}$, $\mathcal{X}$, and $\mathcal{Y}$ are (possibly infinite dimensional) Hilbert spaces, and the system coefficients Z, B, C, and D are bounded linear operators. The space $\mathcal{X}$ is called the *state space*, the sequence (x(n)) *the state*, the sequence (u(n)) the *input* and (y(n)) is the *output*.

First we assume that the input $u(j) = 0$ for $j < 0$ and the initial condition x(0) is specified. By recursively solving for the state x(n) we obtain

$$x(n) = Z^n x(0) + \sum_{i=0}^{n-1} Z^{(n-i-1)} Bu(i) . \tag{1.2}$$

Substituting this result into the expression for y(n) we have

$$y(n) = CZ^n x(0) + Du(n) + \sum_{i=0}^{n-1} CZ^{(n-i-1)} Bu(i) . \tag{1.3}$$

Writing this out in matrix form yields

$$\begin{bmatrix} y(0) \\ y(1) \\ y(2) \\ y(3) \\ \cdot \\ \cdot \\ \cdot \end{bmatrix} = \begin{bmatrix} C \\ CZ \\ CZ^2 \\ CZ^3 \\ \cdot \\ \cdot \\ \cdot \end{bmatrix} x(0) + \begin{bmatrix} D & 0 & 0 & 0 & \cdots \\ CB & D & 0 & 0 & \cdots \\ CZB & CB & D & 0 & \cdots \\ CZ^2B & CZB & CB & D & \cdots \\ \cdot & \cdot & \cdot & \cdot & \cdots \\ \cdot & \cdot & \cdot & \cdot & \cdots \\ \cdot & \cdot & \cdot & \cdot & \cdots \end{bmatrix} \begin{bmatrix} u(0) \\ u(1) \\ u(2) \\ u(3) \\ \cdot \\ \cdot \\ \cdot \end{bmatrix} \tag{1.4}$$

Obviously the last matrix is a block Toeplitz matrix. The block column matrix

$$W_o = \begin{bmatrix} C \\ CZ \\ CZ^2 \\ CZ^3 \\ \vdots \end{bmatrix} \tag{1.5}$$

appearing in (1.4) is called the *observability operator* for the pair $\{C, Z\}$. (This terminology will be explained in the next section.) We view $W_o$ as a linear map of the state space $\mathcal{X}$ to the linear space consisting of all sequences $[y_0, y_1, y_2, ...]^{tr}$ where $y_j \in \mathcal{Y}$. So the input-output map generated by $\{Z, B, C, D\}$ in (1.4) can be decomposed into an observability map and a block Toeplitz matrix. To be precise, let $\{F_n\}_0^\infty$ be the sequence of operators mapping $\mathcal{U}$ into $\mathcal{Y}$ defined by

$$F_0 = D \quad \text{and} \quad F_n = CZ^{n-1}B \qquad (\text{if } n \geq 1) . \tag{1.6}$$

Then the output $y = (y(n))_0^\infty$ is given by

$$y = W_o x(0) + T_F u \tag{1.7}$$

where the input $u = (u(n))_0^\infty$ and $T_F$ is the block Toeplitz matrix generated by $\{F_n\}_0^\infty$. Notice that the output $y(n)$ is well defined for any input sequence $(u(n))_0^\infty$ and initial condition $x(0)$.

The *transfer function* for the state space system $\{Z, B, C, D\}$ is defined by taking the "Fourier transform" of the sequence $\{F_n\}_0^\infty$ defined in (1.6), that is,

$$F(\lambda) = \sum_0^\infty F_n\lambda^n = D + \lambda C(I - \lambda Z)^{-1}B . \tag{1.8}$$

Because we always assume that Z is a bounded operator, the transfer function $F(\lambda)$ is well defined and analytic in some neighborhood of the origin. Furthermore, if the state space $\mathcal{X}$ is finite dimensional, then $F(\lambda)$ is a rational function, that is, $F(\lambda) = N(\lambda)/d(\lambda)$ where $N(\lambda)$ is an operator valued polynomial and $d(\lambda)$ is a scalar valued polynomial. In fact, we may choose $d(\lambda)$ to be $\det[I - \lambda Z]$, and in this case the degrees of $N(\lambda)$ and $d(\lambda)$ will be at most $\dim \mathcal{X}$. (If R is any operator on a finite dimensional space, then $\det[R]$ is the determinant of any matrix representation of R with respect to its basis for $\mathcal{X}$.) Finally, the terminology transfer function comes from the fact that if the initial condition $x(0) = 0$, then $y = T_F u$, and thus the transfer function F uniquely determines the map $T_F$ which transforms the input sequence $(u(n))_0^\infty$ into the output sequence $(y(n))_0^\infty$ assuming the initial state $x(0) = 0$.

The state space operator Z is said to be *stable* if $r_{spec}(Z) < 1$. In this case we also say that the system $\{Z, B, C, D\}$ is *stable*. Obviously, if the state space $\mathcal{X}$ is finite dimensional, then Z is stable if and only if all the eigenvalues of Z are in the open unit disc. This notion of stability readily leads to the following result.

**PROPOSITION 1.1.** *Let* $F(\lambda)$ *be the transfer function of the stable system* $\{Z, B, C, D\}$. *Then* F *is in* $H^\infty(\mathcal{U}, \mathcal{Y})$, *and thus the block lower triangular Toeplitz operator* $T_F$ *generated by* $\{Z, B, C, D\}$ *is a bounded linear operator from* $\ell_+^2(\mathcal{U})$ *into* $\ell_+^2(\mathcal{Y})$ *with norm* $\|F\|_\infty$. *Furthermore, the observability operator* $W_o$ *is a bounded linear operator from* $\mathcal{X}$ *into* $\ell_+^2(\mathcal{Y})$. *In particular, the system* $\{Z, B, C, D\}$ *in (1.1) defines a bounded linear operator from* $\mathcal{X} \oplus \ell_+^2(\mathcal{U})$ *into* $\ell_+^2(\mathcal{Y})$ *by* $y = W_o x(0) + T_F u$.

In some applications one simply views the state space set up as an input-output map from sequences $u = (u(n))_{-\infty}^\infty$ in $\mathcal{U}$ with finite support (i.e., $u_j$ is nonzero for a finite number of j's only) to sequences $y = (y(n))_{-\infty}^\infty$ in $\mathcal{Y}$. Loosely speaking this corresponds to a zero initial state condition at time $-\infty$. In this case $y = L_F u$, where $L_F$ is the block lower triangular Laurent operator matrix generated by $\{F_n\}_0^\infty$ defined in (1.6). We shall refer to $L_F$ as the *input-output map* of (1.1). If Z is stable, then F is in $H^\infty(\mathcal{U}, \mathcal{Y})$ and $L_F$ is a bounded linear operator from $\ell^2(\mathcal{U})$ into $\ell^2(\mathcal{Y})$. (Recall that $\|L_F\| = \|F\|_\infty$.) Moreover, if $U(\lambda)$ and $Y(\lambda)$ are the Fourier transforms of $(u(n))_{-\infty}^\infty$ and $(y(n))_{-\infty}^\infty$ respectively, then $Y(\lambda) = F(\lambda)U(\lambda)$, so the output $Y(\lambda)$ is obtained by multiplying the input $U(\lambda)$ by the transfer function $F(\lambda)$.

The system $\{Z, B, C, D\}$ is called a *realization* of the sequence $\{F_n\}_0^\infty$ if (1.6) holds. In this case the corresponding block lower triangular Laurent matrix $L_F$ is the input-output map for the system (1.1). On the other hand, given any sequence of operators $\{F_n\}_0^\infty$ in $\mathcal{L}(\mathcal{U}, \mathcal{Y})$, then there exists a realization $\{Z, B, C, D\}$ of $\{F_n\}_0^\infty$, that is, (1.6) holds. Section 4 presents a brief discussion on realization theory. There are many efficient computer algorithms to compute the realization $\{Z, B, C, D\}$ from $\{F_n\}_0^\infty$ when $F(\lambda) = \sum F_n \lambda^n$ is a rational matrix function; see Kalman-Falb-Arbib [1] and Kailath [1] for further details on realization theory. So throughout this monograph we will always assume that one can easily construct a realization $\{Z, B, C, D\}$

for $F(\lambda) = \sum F_n \lambda^n$ in the rational matrix case. Finally, we end this section with the following useful lemma which will be used many times in inverting functions with state space realizations.

**LEMMA 1.2.** *Let* $G(\lambda)$ *be an analytic function with values in* $\mathcal{L}(\mathcal{U}, \mathcal{Y})$ *defined in a neighborhood of zero by*

$$G(\lambda) = D + \lambda C(I - \lambda Z)^{-1} B \tag{1.9}$$

*where* Z *is on* $\mathcal{X}$ *and* B, C, D *are bounded linear operators acting between Hilbert spaces. Assume that* D *is invertible. Then for* $\lambda$ *in a neighborhood of zero, the inverse of* $G(\lambda)$ *is given by*

$$G(\lambda)^{-1} = D^{-1} - \lambda D^{-1} C(I - \lambda(Z - BD^{-1}C))^{-1} BD^{-1} . \tag{1.10}$$

Using $-\lambda BD^{-1}C = (I - \lambda Z) - (I - \lambda(Z - BD^{-1}C))$, the proof of the above lemma is by direct verification of the identities $G(\lambda)H(\lambda) = I_{\mathcal{Y}}$ and $H(\lambda)G(\lambda) = I_{\mathcal{U}}$, where $H(\lambda)$ is the right hand side of (1.10).

## III.2. CONTROLLABILITY AND OBSERVABILITY

In this section we will introduce the concepts of controllability and observability. Since we have already introduced the observability operator $W_o$ in (1.5), let us begin with the definition of observability. First we assume that the state space $\mathcal{X}$ is finite dimensional. The system $\Gamma = \{Z, B, C, D\}$ is called *observable* if given the input $u = (u(n))_0^\infty$ and output $y = (y(n))_0^\infty$ one can uniquely determine the state $x = (x(n))_0^\infty$. Because the state x is uniquely determined by the initial condition x(0) and the input u, the system $\Gamma$ is observable if and only if, given the input u and output y, one can uniquely determine the initial condition x(0). So assume that u and y are known. Recall that $y = W_o x(0) + T_F u$, where $W_o$ is the observability operator defined in (1.5) and $T_F$ is the block lower triangular Toeplitz matrix generated by $\{F_n\}_0^\infty$ in (1.6). Then obviously $y - T_F u = W_o x(0)$. Therefore the system $\Gamma$ is observable if and only if its observable operator $W_o$ is one to one. Moreover, if Z is stable ($r_{spec}(Z) < 1$), then the observability operator $W_o$ is a bounded operator from $\mathcal{X}$ into $\ell_+^2(\mathcal{Y})$. In this case $\Gamma$ is observable if and only if the range of $W_o^*$ equals $\mathcal{X}$. Notice that observability has nothing to do with the operators B and D. So we shall say that the pair {C, Z} is *observable* if and only if $W_o$ is one to one. Finally, if the state space $\mathcal{X}$ has dimension $k < \infty$, then, by the Cayley-Hamilton theorem, each $Z^n$ is a finite linear combination of $Z^j$ for $0 \le j < k$. Therefore, in this case the pair {C, Z} is observable if and only if $CZ^j x(0) = 0$ for $0 \le j < k$ implies that $x(0) = 0$, or equivalently,

$$\mathcal{X} = \bigvee_{j=0}^{k-1} Z^{*j}C^{*}\mathcal{Y} . \tag{2.1}$$

Summing up this discussion readily gives the following result.

**PROPOSITION 2.1.** *Let* {Z, B, C, D} *be a linear system with a finite dimensional state space* $\mathcal{X}$. *Then the following statements are equivalent.*

*(i)* *The system* {Z, B, C, D} *is observable.*

*(ii)* *The observability operator* $W_o$ *in (1.5) generated by* {C, Z} *is one to one.*

*(iii)* *If* $CZ^n x = 0$ *for all* $0 \le n < \dim \mathcal{X}$, *then* $x = 0$.

*(iv)* *We have* $\mathcal{X} = \bigvee \{Z^{*n}C^{*}\mathcal{Y} : 0 \le n < \dim \mathcal{X}\}$.

Now assume that the operator Z is stable with the underlying space $\mathcal{X}$ still finite dimensional. In this case the observability operator $W_o$ generated by the pair {C, Z} is a bounded operator from $\mathcal{X}$ into $\ell_+^2(\mathcal{Y})$. The *observability grammian* for the pair {C, Z} is the operator Q on $\mathcal{X}$ defined by $Q = W_o^* W_o$. Obviously $Q \ge 0$. Moreover, the pair {C, Z} is observable if and only if $Q > 0$. (Since $\mathcal{X}$ is finite dimensional, $Q > 0$ means that $(Qx, x) > 0$ for all nonzero x in $\mathcal{X}$.) Using the definition of $W_o$ along with the fact that Z is stable we have

$$Q = W_o^* W_o = \sum_0^\infty Z^{*n}C^{*}CZ^n . \tag{2.2}$$

By direct substitution it is easy to verify that Q satisfies the Lyapunov equation

$$Q = Z^{*}QZ + C^{*}C . \tag{2.3}$$

We claim that there is only one solution to this Lyapunov equation, that is, $Q = W_o^* W_o$. To see this assume that $Q_1$ is a solution to (2.3). Then by recursively using $Q_1 = C^{*}C + Z^{*}Q_1 Z$ in the right-hand side, we obtain

$$\begin{aligned} Q_1 &= C^{*}C + Z^{*}Q_1 Z = C^{*}C + Z^{*}C^{*}CZ + Z^{*2}Q_1 Z^2 = \\ & C^{*}C + Z^{*}C^{*}CZ + Z^{*2}C^{*}CZ^2 + Z^{*3}Q_1 Z^3 = \cdots = \\ & \sum_{j=0}^{n-1} Z^{*j}C^{*}CZ^j + Z^{*n}Q_1 Z^n . \end{aligned}$$

Because Z is stable, it follows that $Z^{*n}Q_1 Z^n$ converges to zero as n approaches infinity. Thus

$$Q_1 = \sum_0^{\infty} Z^{*n} C^* C Z^n = W_o^* W_o = Q .$$

Therefore the solution to the observability Lyapunov equation in (2.3) is unique.

Finally, in addition to stability, assume that the pair $\{C, Z\}$ is observable. By applying $W_o^*$ to $y - T_F u = W_o x(0)$ we obtain $Qx(0) = W_o^* y - W_o^* T_F u$. Because the pair $\{C, Z\}$ is observable, $Q$ is invertible and thus

$$x(0) = Q^{-1}(W_o^* y - W_o^* T_F u) , \tag{2.4}$$

where $Q^{-1}$ is the inverse of $Q$. The previous equation shows that the observability grammian $Q$ plays a critical role in determining the initial condition $x(0)$ given the input $u$ and output $y$.

Now assume that $\mathcal{X}$ is infinite dimensional. An operator $Z$ on $\mathcal{X}$ is called *pointwise stable* if for each $x \in \mathcal{X}$ the sequence $Z^n x$ approaches zero as $n$ goes to infinity. Notice that for a stable $Z$ we have $\|Z^n\| \to 0$ for $n \to \infty$ and thus such a $Z$ is automatically pointwise stable. The converse statement is not necessarily true. For example, the backward shift on $\ell_+^2$ is pointwise stable but not stable. However, in the finite dimensional case pointwise stability and stability are the same.

In the infinite dimensional case the notion of observability can be defined in different ways. Throughout this book we will work with the strongest version, namely if the state space is infinite dimensional, then we shall say that a pair $\{C, Z\}$ is *observable* if there exist constants $m$ and $M$, with $0 < m \le M$, such that

$$m\|x\|^2 \le \sum_{n=0}^{\infty} \|CZ^n x\|^2 \le M\|x\|^2 \qquad (x \in \mathcal{X}) .$$

The second part of these inequalities is equivalent to the requirement that the observability map $W_o$ given by (1.5) defines a bounded linear operator from $\mathcal{X}$ into $\ell_+^2(\mathcal{Y})$, and hence these inequalities can be restated as $W_o$ is a bounded linear operator and the observability grammian $Q = W_o^* W_o$ is a strictly positive operator on $\mathcal{X}$. By employing the closed graph theorem from functional analysis it can be shown that the pair $\{C, Z\}$ is observable if and only if $W_o$ is a one to one bounded linear operator with closed range. In the literature the notion of observability used here is sometimes referred to as uniform observability.

Let us remark that observability of the pair $\{C, Z\}$ in the infinite dimensional case, implies that $Z$ is pointwise stable. To see this notice that $S^* W_o = W_o Z$, where $S$ is the unilateral shift on $\ell_+^2(\mathcal{Y})$. It follows that $(S^*)^n W_o = W_o Z^n$ for $n \ge 0$. Because $W_o$ has a bounded left inverse, $W_o^+$ say, we have $W_o^+(S^*)^n W_o = Z^n$ for $n \ge 0$. Since $S^*$ is pointwise stable, it follows that $Z$ is pointwise stable. In particular, if the state space is finite dimensional and $W_o$ is bounded, then $Z$ is stable. Finally, it is noted that we apparently have two different notions of observability, one for finite dimensional systems and another one for infinite dimensional systems. However, in all of our applications of systems to either finite dimensional or infinite interpolation problems, the observability operator is always bounded. In this case the concept of observability for finite and infinite dimensional systems coincide. Moreover, whenever ambiguity can occur we will make precise which concept of observability is being used.

Let us note that in the infinite dimensional case, even for a pointwise stable Z, the condition

$$\bigcap_{n=0}^{\infty} \ker CZ^n = \{0\}$$

does not imply observability in our sense. To illustrate this fact we give two examples. In both examples Z is the backward shift on $\mathcal{X} = \ell_+^2$ and $\mathcal{Y} = \mathbb{C}$. First, take $C = [1, 1, 0, 0, ...]$. Then $W_o = I + Z$, and hence $W_o$ is a one to one bounded linear operator. Notice, however, that in this case $W_o$ does not have a bounded left inverse, because 1 is in the boundary of the spectrum of Z. In the second example $C = [1, 1/2, 1/3, ...]$. In this case the observability map $W_o$ is an upper triangular Toeplitz matrix whose symbol is defined by $\phi(e^{it}) = \sum_{n=0}^{\infty} (1/n)e^{-int}$. Since $\phi$ does not belong to $L^\infty$, the operator $W_o$ is not bounded on $\ell_+^2$. Let us notice that even in this case the observability map is one to one (this follows from the fact that the function $\bar{\phi}(e^{it}) = \sum_{n=0}^{\infty} (1/n)e^{int}$ is outer).

Obviously, if Z is stable, then $W_o$ is a bounded operator and its observability grammian $Q = W_o^* W_o$ is the unique solution the Lyapunov equation in (2.3). Moreover, in this case the pair $\{C, Z\}$ is observable if and only if Q is strictly positive.

To discuss controllability let us first assume that the state space $\mathcal{X}$ is finite dimensional and that there is no initial condition. The system $\Gamma = \{Z, B, C, D\}$ is *controllable* if given any z in $\mathcal{X}$, then there exists an input sequence $u = (u(j))_{-\infty}^{n-1}$ of finite support (i.e., $u(j) = 0$ for $-j$ sufficiently large) such that the state at time n is given by z. By recursively solving for x(n) in (1.1) (see (1.2)) we obtain

$$x(n) = \sum_{i=-\infty}^{n-1} Z^{(n-i-1)} Bu(i) . \tag{2.5}$$

Let $W_c$ be the *controllability* map defined by

$$W_c = [B, ZB, Z^2B, Z^3B, ... ] , \tag{2.6}$$

which we view as a linear transformation defined on the space of all sequences $[v_0, v_1, v_2, ...]^{tr}$ of finite support with $v_j \in \mathcal{U}$ for $j \geq 0$ into the state space $\mathcal{X}$. Using (2.5) we have

$$x(n) = W_c[u(n-1), u(n-2), u(n-3), \ldots]^{tr} = W_c u\,. \tag{2.7}$$

By the Cayley-Hamilton theorem the system $\Gamma$ is controllable if and only if $\mathcal{X} = \bigvee \{Z^n B \mathcal{U} : 0 \le n < \dim \mathcal{X}\}$, or equivalently, $B^* Z^{*n} x = 0$ for $0 \le n < \dim \mathcal{X}$ implies that $x = 0$. The controllability of the system does not depend upon the operators C and D, and therefore we shall also speak about controllability of the pair $\{Z, B\}$. Finally, notice that controllability is the dual of observability, that is, the pair $\{Z, B\}$ is controllable if and only if the pair $\{B^*, Z^*\}$ is observable. Summing up we have the following result.

**PROPOSITION 2.2.** *Let* $\{Z, B, C, D\}$ *be a finite dimensional linear system with state space* $\mathcal{X}$. *Then the following results are equivalent*

*(i)* *The system* $\{Z, B, C, D\}$ *is controllable.*

*(ii)* *We have* $\mathcal{X} = \bigvee \{Z^n B \mathcal{U} : 0 \le n < \dim \mathcal{X}\}$.

*(iii)* *If* $B^* Z^{*n} x = 0$ *for all* $0 \le n < \dim \mathcal{X}$, *then* $x = 0$.

*(iv)* *The pair* $\{B^*, Z^*\}$ *is observable.*

Now assume that Z on $\mathcal{X}$ is stable where $\mathcal{X}$ is still finite dimensional. Then the controllability map $W_c$ is a bounded operator from $\ell^2_+(\mathcal{U})$ into $\mathcal{X}$. Therefore, in the stable case, the pair $\{Z, B\}$ is controllable if and only if $W_c$ maps $\ell^2_+(\mathcal{U})$ onto $\mathcal{X}$. The *controllability grammian* P for the pair $\{Z, B\}$ is the operator on $\mathcal{X}$ defined by $P = W_c W_c^*$. Obviously $P \ge 0$. Moreover, P is strictly positive if and only if the pair $\{Z, B\}$ is controllable. By using the definition of $W_c$ we have

$$P = W_c W_c^* = \sum_0^\infty Z^n B B^* Z^{*n}\,. \tag{2.8}$$

Because controllability is the dual of observability, it follows that P is the unique solution of the Lyapunov equation

$$P = ZPZ^* + BB^*\,. \tag{2.9}$$

Assuming that the pair $\{Z, B\}$ is controllable, then

$$u = W_c^* P^{-1} z \tag{2.10}$$

defines a square summable input sequence $u = (u(j))_{-\infty}^{n-1}$ which drives the system to the state z at time n. Therefore the controllability grammian plays a fundamental role in determining the input u such that $z = x(n)$.

Now assume that $\mathcal{X}$ is infinite dimensional. An operator Z on $\mathcal{X}$ is *pointwise $*$-stable*, if for each x in $\mathcal{X}$ the vector $Z^{*n}x$ approaches zero as n approaches infinity. Obviously if Z is stable, then Z is pointwise $*$-stable. We say that the pair {Z, B} is *controllable* if it controllability operator $W_c$ in (2.6) defines a bounded linear operator from $\ell_+^2(\mathcal{U})$ into $\mathcal{X}$ and the range of $W_c$ equals $\mathcal{X}$. So the infinite dimensional pair {Z, B} is controllable if and only if $W_c$ is bounded and its controllability grammian $P = W_c W_c^*$ is strictly positive. Notice that the pair {Z, B} is controllable if and only if the pair $\{B^*, Z^*\}$ is observable, and therefore if {Z, B} is controllable, then Z is pointwise $*$-stable. Also, if Z is stable, then obviously $W_c$ is a bounded linear operator, and its controllability grammian $P = W_c W_c^*$ is the unique solution to the Lyapunov equation in (2.9). With respect to the concept of controllability in the infinite dimensional case remarks similar to the ones made earlier for observability are valid. This is also clear from the duality between controllability and observability. Finally, it is emphasized that in all of our applications of systems to either finite dimensional or infinite dimensional interpolation problems, the controllability operator is always bounded. So in these cases the concept of controllability for finite and infinite dimensional systems coincide.

Recall that $\Gamma = \{Z \text{ on } \mathcal{X}, B, C, D\}$ is a realization for an analytic function $F(\lambda)$ with values in $\mathcal{L}(\mathcal{U}, \mathcal{Y})$ if

$$F(\lambda) = D + \lambda C(I - \lambda Z)^{-1}B \tag{2.11}$$

in some neighborhood of the origin. Notice that this implies that F has to be analytic in a neighborhood of the origin. Recall also that the system $\Gamma$ is *similar* to $\Gamma_1 = \{Z_1 \text{ on } \mathcal{X}_1, B_1, C_1, D_1\}$ if $D = D_1$ and there exists a similarity transformation X from $\mathcal{X}$ onto $\mathcal{X}_1$ (that is, X is a one to one bounded linear operator from $\mathcal{X}$ onto $\mathcal{X}_1$) satisfying

$$XZ = Z_1X, \ XB = B_1 \text{ and } C_1X = C . \tag{2.12}$$

Notice that if $\Gamma$ is similar to $\Gamma_1$, then $\Gamma_1$ is also a realization of $F(\lambda)$. This follows from the following simple calculation

$$D_1 + \lambda C_1(I - \lambda Z_1)^{-1}B_1 = D + \lambda C_1(I - \lambda XZX^{-1})^{-1}B_1 = F(\lambda) .$$

Moreover, similarity preserves controllability and observability. To be precise, if $\Gamma$ is similar to $\Gamma_1$, then $\Gamma$ is controllable (respectively observable) if and only if $\Gamma_1$ is controllable (respectively observability). Furthermore, if $\Gamma$ is similar to $\Gamma_1$, then Z is stable, pointwise stable and pointwise $*$-stable if and only if $Z_1$ is stable, pointwise stable and pointwise $*$-stable, respectively. Finally, we say that $\Gamma$ is *unitarily equivalent* to $\Gamma_1$ if $D = D_1$ and there exists a unitary operator X from $\mathcal{X}$ onto $\mathcal{X}_1$ satisfying (2.12). The following is a classical result in linear systems.

**THEOREM 2.3.** *All controllable and observable realizations for a transfer function* $F(\lambda)$ *with values in* $\mathcal{L}(\mathcal{U}, \mathcal{Y})$ *are similar.*

Notice that the contents of the theorem depends on whether or not the state dimension of the realization is finite. Indeed, in the finite dimensional case the classical definitions of observability and controllability are used, while in the other case we use the uniform versions. Therefore we start the proof for the case when both systems are stable (when there is no difference between the corresponding concept).

**PROOF.** Let $\Gamma = \{Z \text{ on } \mathcal{X}, B, C, D\}$ and $\Gamma_1 = \{Z_1 \text{ on } \mathcal{X}_1, B_1, C_1, D_1\}$ be two stable, controllable and observable realizations of $F(\lambda)$. Since $D = F(0) = D_1$, it follows that $D = D_1$. Now let $W_o$ from $\mathcal{X}$ into $\ell_+^2(\mathcal{Y})$ be the observability operator defined in (1.5) for the pair $\{C, Z\}$ and $\tilde{W}_o$ the corresponding observability operator from $\mathcal{X}_1$ into $\ell_+^2(\mathcal{Y})$ defined by the pair $\{C_1, Z_1\}$, where $C_1$ replaces C and $Z_1$ replaces Z in (1.5). Let $W_c$ be the controllability operator from $\ell_+^2(\mathcal{U})$ onto $\mathcal{X}$ defined in (2.6) for the pair $\{Z, B\}$ and $\tilde{W}_c$ the corresponding controllability operator from $\ell_+^2(\mathcal{U})$ onto $\mathcal{X}_1$ defined by the pair $\{Z_1, B_1\}$. Because $\Gamma$ and $\Gamma_1$ are both realizations of $F(\lambda)$, equations (1.6) and (1.8) show that $CZ^{n-1}B = F_n = C_1Z_1^{n-1}B_1$ for all $n \geq 1$. This readily implies that $W_oW_c = \tilde{W}_o\tilde{W}_c$. So using $P = W_cW_c^*$ we have $W_oP = \tilde{W}_o\tilde{W}_cW_c^*$, or equivalently, $W_o = \tilde{W}_oX$ where $X = \tilde{W}_cW_c^*P^{-1}$. From $W_oW_c = \tilde{W}_o\tilde{W}_c$ and observability it also follows that $\ker W_c = \ker \tilde{W}_c$, and hence $\operatorname{ran} W_c^* = (\ker \tilde{W}_c)^\perp$. By controllability, both $W_c$ and $\tilde{W}_c$ are onto. Thus $\tilde{W}_cW_c^*$ is an invertible operator from $\mathcal{X}$ into $\mathcal{X}_1$, and hence the same is true for $X = \tilde{W}_cW_c^*P^{-1}$.

Using $W_o = \tilde{W}_oX$ we obtain

$$\begin{bmatrix} C \\ W_oZ \end{bmatrix} = W_o = \tilde{W}_oX = \begin{bmatrix} C_1X \\ \tilde{W}_oZ_1X \end{bmatrix} = \begin{bmatrix} C_1X \\ W_oX^{-1}Z_1X \end{bmatrix}.$$

Thus $C = C_1X$ and because $W_o$ is one to one $Z = X^{-1}Z_1X$. To complete the proof it remains to show that $XB = B_1$. To this end, notice that $W_oX^{-1} = \tilde{W}_o$ gives

$$W_oB = [F_1, F_2, F_3, \ldots]^{tr} = \tilde{W}_oB_1 = W_oX^{-1}B_1 .$$

Since $W_o$ is one to one, this implies that $B = X^{-1}B_1$, or equivalently, $XB = B_1$. Therefore (2.12) holds and $\Gamma$ is similar to $\Gamma_1$.

Now assume that $\Gamma$ and $\Gamma_1$ are two arbitrary, controllable and observable realizations of $F(\lambda)$. Choose a nonzero constant $\alpha$ such that both $\alpha Z$ and $\alpha Z_1$ are stable. Then

$$F(\alpha\lambda) = D + \alpha\lambda C(I - \alpha\lambda Z)^{-1}B = D_1 + \alpha\lambda C_1(I - \alpha\lambda Z_1)^{-1}B_1 .$$

Thus both $\Gamma_\alpha = \{\alpha Z, B, \alpha C, D\}$ and $\Gamma_{1,\alpha} = \{\alpha Z_1, B_1, \alpha C_1, D_1\}$ are stable controllable and observable realization for $F(\alpha\lambda)$. So $\Gamma_\alpha$ is similar to $\Gamma_{1,\alpha}$ by the results of the previous paragraph. This readily implies that $\Gamma$ is similar to $\Gamma_1$ and completes the proof.

To complete this section we present the following result which will be used many times throughout this monograph.

**LEMMA 2.4.** *Let* Z *on* $\mathcal{X}$ *and* $\Lambda$ *on* $\mathcal{H}$ *be two stable operators and* M *an operator from* $\mathcal{H}$ *into* $\mathcal{X}$. *Then there exists a unique solution* R *in* $\mathcal{L}(\mathcal{H}, \mathcal{X})$ *solving the following equation*

$$R = ZR\Lambda + M . \tag{2.13}$$

*Moreover, this unique solution* R *is given by*

$$R = \sum_{n=0}^{\infty} Z^n M \Lambda^n . \tag{2.14}$$

An equation of the form (2.13) is called a *Stein equation.* Notice that the Stein equation in (2.13) is a generalization of observability and controllability Lyapunov equations in (2.3) and (2.9), respectively. One refers to the latter equation also as symmetric Stein equations.

**PROOF.** Because both Z and $\Lambda$ are stable, the series in the formula for R in (2.14) converges in the operator topology. Substituting (2.14) into (2.13) shows that R in (2.14) is indeed a solution to the Stein equation in (2.13). If $R_1$ is another solution to the Stein equation in (2.13), then subtracting $R_1 = ZR_1\Lambda + M$ from (2.13) shows that $E = ZE\Lambda$ where $E = R - R_1$. Hence $E = Z^n E \Lambda^n$ for all $n \geq 0$. Because both Z and $\Lambda$ are stable, $E = 0$. Therefore $R = R_1$ is the unique solution to (2.13). This completes the proof.

## III.3. POINT EVALUATION

In this section we will discuss how point and operator evaluation for an F in $H^\infty(\mathcal{U}, \mathcal{Y})$ naturally occurs in linear systems. Actually operator evaluation for the transfer function F of a stable, observable system {Z, B, C, D} already occurred in equation (2.4). Indeed, notice that (2.4) gives

$$Qx(0) = W_o^* y - W_o^* T_F u , \tag{3.1}$$

where $u \in \ell_+^2(\mathcal{U})$ is the input, the vector $x(0) \in \mathcal{X}$ is the initial state at time 0, and $y \in \ell_+^2(\mathcal{Y})$ is the corresponding output. Since $r_{spec}(Z^*) < 1$ by stability, we may consider $(C^*F)(Z^*)_{left}$. In fact, according to the definition (in Section I.2) of left evaluation of an analytic operator valued function at an operator, we have $(C^*F)(Z^*)_{left} = \sum Z^{*n}C^*F_n$, where $\{F_n\}$ are the Taylor coefficients of F. Using this notation we have

$$W_o^*T_F = [(C^*F)(Z^*)_{left}, Z^*(C^*F)(Z^*)_{left}, Z^{*2}(C^*F)(Z^*)_{left}, ...] \tag{3.2}$$

In particular, $W_o^*T_FE_1 = (C^*F)(Z^*)_{left}$, where $E_1 = [I, 0, 0, ...]^*$. So the evaluation of F at operator values naturally occurs in linear systems. Recall also that the term $W_o^*T_F$ appears in the standard left tangential Nevanlinna-Pick interpolation problem, which is a generalization of the classical Nevanlinna-Pick problem; see Section I.4.

To see how (3.1) can be viewed as a linear system of equations involving point evaluations of F, assume that Z is an operator on a m dimensional space. For simplicity of presentation only we also assume that Z has m distinct eigenvalues $\lambda_1, \lambda_2, ..., \lambda_m$ with corresponding eigenvectors $x_1, x_2, ..., x_m$. Now let $Y(\lambda)$ in $H^2(\mathcal{Y})$ and $U(\lambda)$ in $H^2(\mathcal{U})$ be the Fourier transforms of $y = (y(n))_0^\infty$ and $u = (u(n))_0^\infty$, respectively. First notice that

$$(W_o^*y, x_j) = (\sum_0^\infty Z^{*n}C^*y(n), x_j) = \sum_0^\infty (\bar{\lambda}_j^n C^*y(n), x_j) = (C^*Y(\bar{\lambda}_j), x_j)\,. \tag{3.3}$$

Notice that the evaluation of $Y(\lambda)$ at $\bar{\lambda}_j$ makes sense because $Y(\lambda)$ is analytic in the open unit disc. A similar calculation gives

$$\begin{gathered}(W_o^*T_Fu, x_j) = (C^*[I, \bar{\lambda}_jI, \bar{\lambda}_j^2I, \dots]T_Fu, x_j) = \\ (C^*F(\bar{\lambda}_j)[I, \bar{\lambda}_jI, \bar{\lambda}_j^2I, \dots]u, x_j) = (C^*F(\bar{\lambda}_j)U(\bar{\lambda}_j), x_j)\,.\end{gathered} \tag{3.4}$$

So by combining (3.1), (3.3) and (3.4) we see that

$$(Qx(0), x_j) = (C^*Y(\bar{\lambda}_j), x_j) - (C^*F(\bar{\lambda}_j)U(\bar{\lambda}_j), x_j) \qquad (\text{for } j = 1, ..., m)\,. \tag{3.5}$$

Since $x_1, ..., x_m$ forms a basis for $\mathcal{X}$, one can also solve for the initial condition $x(0)$ by solving the linear system of equations in (3.5) involving the point evaluations of $F(\bar{\lambda}_j)$. Indeed, there exists a set of uniquely determined constants $\alpha_1, \alpha_2, ..., \alpha_m$ such that

$$x(0) = \sum_{k=1}^m \alpha_k x_k\,. \tag{3.6}$$

Substituting this into (3.5) yields

$$\sum_{k=1}^{m} \alpha_k (Qx_k, x_j) = (C^* Y(\bar{\lambda}_j), x_j) - (C^* F(\bar{\lambda}_j) U(\bar{\lambda}_j), x_j) =: b_j \qquad (\text{for } j = 1, \ldots, m). \tag{3.7}$$

So if we set $\Lambda_{j,k} = (Qx_k, x_j)$, then (3.7) leads to the matrix equation $\Lambda\alpha = b$. Here $\Lambda$ is the matrix generated by $\{\Lambda_{j,k}\}$ and $\alpha$ is the vector in $\mathbb{C}^m$ determined by $\{\alpha_k\}$ while b is the vector in $\mathbb{C}^m$ determined by $\{b_j\}$. Since $\{x_j\}_1^m$ forms a basis for $\mathcal{X}$ and $Q > 0$, by observability, the matrix $\Lambda$ is invertible. (In fact, $\Lambda$ is the Gram matrix generated by the basis $\{x_j\}_1^m$ with respect to the inner product $(Qf, g)$.) So we can obtain $\alpha = \Lambda^{-1} b$, and then compute the initial condition $x(0)$ by using (3.6). Therefore the point evaluation of F at the eigenvalues of Z are playing a hidden role in the observability problem of finding the initial condition $x(0)$ given the input u and output y. Finally, it is noted that if Z has Jordan blocks of dimension greater than one in its Jordan form, then one can also convert (3.1) to a system of equations similar to those in (3.7). However, in this case the system of equations corresponding to those in (3.7) will contain the appropriate derivatives of F evaluated at $\bar{\lambda}_j$. The details are omitted.

Point evaluation also occurs when one feeds an unbounded input $(u(n))_0^\infty$ into a stable linear system $\{Z, B, C, D\}$ of the form $u(n) = u_o \beta^n$ for $n \geq 0$, where $\beta$ is a scalar satisfying $|\beta| > 1$ and $u_o$ is in $\mathcal{U}$. In this case the identity $y = W_o x(0) + T_F u$, gives

$$y(n) = CZ^n x(0) + (F_n + F_{n-1}\beta + \cdots + F_0 \beta^n) u_o =$$
$$CZ^n x(0) + \beta^n \left\{ F_0 + \frac{F_1}{\beta} + \frac{F_2}{\beta^2} + \cdots + \frac{F_n}{\beta^n} \right\} u_o . \tag{3.8}$$

By using the stability of Z along with the fact that $F_o + F_1/\beta + \cdots + F_n/\beta^n$ converges to $F(1/\beta)$ as n approaches infinity, we obtain the following result.

**PROPOSITION 3.1.** *Let* $\{Z, B, C, D\}$ *be a stable system of the form (1.1) and* $u(n)$ *the unbounded input given by* $u(n) = u_o \beta^n$ *for* $n \geq 0$, *where* $u_o$ *is in* $\mathcal{U}$ *and* $\beta$ *is a scalar satisfying* $|\beta| > 1$. *Then the output* $y(n)$ *in (1.1) converges to* $\beta^n F(1/\beta)$ *when* n *tends to infinity. In particular, the evaluation of* F *at* $1/\beta$ *is given by*

$$\lim_{n \to \infty} \beta^{-n} y(n) = F(1/\beta) . \tag{3.9}$$

For a last example concerning point evaluation, let $F(\lambda)$ be the transfer function for the stable system $\{Z, B, C, D\}$ in (1.1). In this example we assume that there is no initial condition and the input $(u(n))$ in $\ell^2(\mathcal{U})$ is defined by $u(n) = u_o \alpha^{|n|}$ if $n \leq 0$ and $u(n) = 0$ otherwise. Here $\alpha$ is a scalar in the open unit disc. In this case $(y(n))_{-\infty}^{\infty} = y = L_F u$ where $L_F$ is the block lower triangular bounded Laurent operator generated by F. Using $F(\alpha) = \sum F_n \alpha^n$ it follows that $y(n) = \alpha^{|n|} F(\alpha) u_o$ if $n \le 0$. In particular, $y(0)$ is given by the point evaluation $F(\alpha)u_o$.

## III.4 REALIZATION THEORY

In this section we will give a brief introduction to realization theory. Our approach is based on the backward shift realization. For an analysis and history of the backward shift realization see Fuhrmann [1].

Let us begin by establishing a realization theory for rational functions $F(\lambda)$ with values in $\mathcal{L}(\mathcal{U}, \mathcal{Y})$, where $\mathcal{U}$ and $\mathcal{Y}$ are Hilbert spaces. For the moment let us also assume that $\mathcal{U}$ is finite dimensional. At the end of this section we will extend the results to include realizations of $H^\infty$ functions. To begin, recall that $F(\lambda)$ is the transfer function for $\Gamma = \{Z \text{ on } \mathcal{X}, B, C, D\}$ if

$$F(\lambda) = D + \lambda C(I - \lambda Z)^{-1} B\,. \tag{4.1}$$

Here B maps $\mathcal{U}$ into $\mathcal{X}$ and C maps $\mathcal{X}$ into $\mathcal{Y}$. Obviously $F(\lambda)$ is analytic in some neighborhood of the origin and has values in $\mathcal{L}(\mathcal{U}, \mathcal{Y})$. Moreover, $F(\lambda)$ admits a power series expansion of the form $F(\lambda) = \sum_0^\infty F_n \lambda^n$. So $\{Z, B, C, D\}$ is a realization of $F(\lambda)$ if and only if $D = F_0$ and $CZ^{n-1}B = F_n$ for $n \ge 1$. The realization $\{Z, B, C, D\}$ of $F(\lambda)$ is a *minimal realization* if given any other realization $\{Z_1 \text{ on } \mathcal{X}_1, B_1, C_1, D_1\}$ of $F(\lambda)$, then the dimension of the linear space $\mathcal{X}$ is less than or equal to the dimension of the linear space $\mathcal{X}_1$. Clearly the concept of minimal realizations only makes sense when $F(\lambda)$ admits a finite dimensional realization $\{Z \text{ on } \mathcal{X}, B, C, D\}$, that is, (4.1) holds and the state space $\mathcal{X}$ is finite dimensional. Recall that in this case $F(\lambda)$ is both analytic in a neighborhood of the origin and rational; see Section III.1.

The minimal realization problem is the inverse problem, that is, given any rational operator-valued function $F(\lambda)$ analytic in some neighborhood of the origin, find all minimal realizations for $F(\lambda)$. Notice that this problem is equivalent to: Given any rational operator-valued function $F(\lambda)$ whose power series expansion is given by $\sum_0^\infty F_n \lambda^n$, find the set of all minimal realizations $\{Z, B, C, D\}$ satisfying $F_0 = D$ and $F_n = CZ^{n-1}B$ for $n \ge 1$.

To construct a realization for $F(\lambda)$, let $\ell_+(\mathcal{Y})$ be the linear space consisting of all infinite tuples of the form $f = [f_0, f_1, f_2, \ldots]^{tr}$ where $f_j$ is in $\mathcal{Y}$ for all j. Here we do not need a topological structure on $\ell_+(\mathcal{Y})$. Let $S^*$ be the backward shift operator on $\ell_+(\mathcal{Y})$ defined by

$$S^* \begin{bmatrix} f_o \\ f_1 \\ f_2 \\ \vdots \end{bmatrix} = \begin{bmatrix} 0 & 1 & 0 & 0 & \cdots \\ 0 & 0 & 1 & 0 & \cdots \\ 0 & 0 & 0 & 1 & \cdots \\ \vdots & \vdots & \vdots & \vdots & \end{bmatrix} \begin{bmatrix} f_o \\ f_1 \\ f_2 \\ \vdots \end{bmatrix} = \begin{bmatrix} f_1 \\ f_2 \\ f_3 \\ \vdots \end{bmatrix}. \tag{4.2}$$

(Because we did not specify a topology on $\ell_+(\mathcal{Y})$, the $*$ in $S^*$ does not refer to the adjoint of S. We simply denote the linear map in (4.2) by $S^*$.) Let $\Pi_{\mathcal{Y}}$ be the linear map from $\ell_+(\mathcal{Y})$ onto $\mathcal{Y}$ picking out the first component of a vector f in $\ell_+(\mathcal{Y})$, that is

$$\Pi_{\mathcal{Y}} = [I\,,\,0\,,\,0\,,\,0\,,\,...]\,. \tag{4.3}$$

Finally, let $\tilde{B}$ be the linear map from $\mathcal{U}$ into $\ell_+(\mathcal{Y})$ defined by

$$\tilde{B} = [F_1,\, F_2,\, F_3,\, ...]^{tr} \tag{4.4}$$

where $F_n$ for $n \geq 0$ appears in the power series expansion $F(\lambda) = \sum_0^\infty F_n\lambda^n$. It is easy to verify that $\Pi_{\mathcal{Y}} S^{*(n-1)}\tilde{B} = F_n$ for $n \geq 1$. Notice that $\ell_+(\mathcal{Y})$ is not a Hilbert space, and hence according to our definition $\{S^*,\, \tilde{B},\, \Pi_{\mathcal{Y}},\, F_o\}$ cannot be considered to be a realization of $F(\lambda)$. Moreover, $\ell_+(\mathcal{Y})$ is infinite dimensional.

To improve the construction, let $\mathcal{H}$ be the linear submanifold of $\ell_+(\mathcal{Y})$ defined by

$$\begin{aligned}\mathcal{H} &= \operatorname*{span}_{n\geq 0} S^{*n}\tilde{B}\,\mathcal{U} = \text{span}\,\{S^{*n}\tilde{B}u : n = 0,\, 1,\, 2,\, ...,\text{ and } u \in \mathcal{U}\} \\ &= [\tilde{B},\, S^*\tilde{B},\, S^{*2}\tilde{B},\, ...]\ell_{+,c}(\mathcal{U}),\end{aligned} \tag{4.5}$$

where span denotes the linear hull and $\ell_{+,c}(\mathcal{U})$ is the linear submanifold of $\ell_+(\mathcal{U})$ consisting of all sequences of finite support. Clearly, $\mathcal{H}$ is an invariant subspace for $S^*$, that is, $S^*\mathcal{H} \subset \mathcal{H}$. Since $S^{*n}\tilde{B} = [F_{n+1},\, F_{n+2},\, ...]^{tr}$, we see that the space $\mathcal{H}$ is the range of the block Hankel operator H given by

$$H = \begin{bmatrix} F_1 & F_2 & F_3 & \cdots \\ F_2 & F_3 & F_4 & \cdots \\ F_3 & F_4 & F_5 & \cdots \\ \cdot & \cdot & \cdot & \cdots \\ \cdot & \cdot & \cdot & \cdots \\ \cdot & \cdot & \cdot & \cdots \end{bmatrix} : \ell_{+,c}(\mathcal{U}) \to \ell_+(\mathcal{Y})\,. \tag{4.6}$$

We claim that H has finite rank if and only if $F(\lambda)$ is rational. To see this we first write $F(\lambda)$ as $F(\lambda) = N(\lambda)/d(\lambda)$, where $d(\lambda)$ is a scalar polynomial, $d(\lambda) = d_o + d_1\lambda + ... + d_k\lambda^k$, with $d_o \neq 0$, and $N(\lambda)$ is a polynomial with values in $\mathcal{L}(\mathcal{U},\, \mathcal{Y})$ of degree also at most k. So $d(\lambda)F(\lambda) = N(\lambda)$, and therefore

$$d_o F_{n+k} + d_1 F_{n+k-1} + ... + d_k F_n = 0 \qquad (n \geq 1)\,.$$

This readily implies that the m-th block column of H for $m > k$ is a linear combination of the first k block columns of H. It follows that

$$\dim \mathcal{H} = \operatorname{rank} H \le k \cdot \dim \mathcal{U} ,$$

and thus $\mathcal{H}$ is finite dimensional.

Now assume that $\mathcal{H}$ is finite dimensional, and let $\{Z_o \text{ on } \mathcal{H}, B_o, C_o, D_0\}$ be the system defined by

$$Z_o = S^* | \mathcal{H} , \quad B_o = \tilde{B} , \quad C_o = \Pi_{\mathcal{Y}} | \mathcal{H} \text{ and } \quad D_0 = F_o , \tag{4.7}$$

where $B_o$ maps $\mathcal{U}$ into $\mathcal{H}$, and $C_o$ maps $\mathcal{H}$ into $\mathcal{Y}$. Using the fact that $\mathcal{H}$ is an invariant subspace for $S^*$ and the range of $\tilde{B}$ is contained in $\mathcal{H}$, it follows that

$$C_o Z_o^{n-1} B_o = \Pi_{\mathcal{Y}} S^{*(n-1)} \tilde{B} = F_n \qquad (\text{for } n \ge 1) . \tag{4.8}$$

Therefore, $\{Z_o , B_o , C_o , D_0\}$ is a realization for $F(\lambda)$. For obvious reasons this system is called the *restricted backward shift realization* of $F(\lambda)$. In this way we have proved that for a finite dimensional $\mathcal{U}$, the rational operator-valued function $F(\lambda)$ admits a finite dimensional realization, namely the restricted backward shift realization constructed above. This sets the stage for the following result.

**THEOREM 4.1.** *Let* $F(\lambda)$ *be an analytic function in some neighborhood of the origin with values in* $\mathcal{L}(\mathcal{U}, \mathcal{Y})$ *where* $\mathcal{U}$ *is finite dimensional. Then* $F(\lambda)$ *is rational if and only if it admits a finite dimensional realization, i.e., a realization with a finite dimensional state space. In this case the restricted backward shift realization is a minimal realization of* $F(\lambda)$. *Furthermore, all minimal realizations of* $F(\lambda)$ *are similar and their state dimension is the rank of the Hankel operator* H *in (4.6).*

**PROOF**. It remains to prove the statements referring to minimality. Let $F(\lambda)$ be rational, and let $\{Z \text{ on } \mathcal{X}, B , C , D\}$ be any realization of F. Let $W_o$ be the observability operator from $\mathcal{X}$ into $\ell_+(\mathcal{Y})$ defined by $W_o = [C, CZ, CZ^2, \ldots]^{tr}$. In what follows we use the notation introduced in (4.2)-(4.5). Clearly, $S^* W_o = W_o Z$ and $\Pi_{\mathcal{Y}} W_o = C$. Further, using the fact that $F_n = CZ^{n-1}B$, where $\{F_n\}_0^\infty$ are the Taylor coefficients of $F(\lambda)$, we have that $W_o B = \tilde{B}$. This, along with the definition of $\mathcal{H}$, gives

$$\mathcal{H} = \operatorname*{span}_{n\ge0} S^{*n} \tilde{B}\, \mathcal{U} = \operatorname*{span}_{n\ge0} W_o Z^n B\, \mathcal{U} = W_o \operatorname*{span}_{n\ge0} Z^n B\, \mathcal{U} . \tag{4.9}$$

Thus $\mathcal{H} \subseteq \operatorname{ran}(W_o)$. Because $W_o$ is defined on $\mathcal{X}$, it follows that $\dim \mathcal{H} \le \dim \mathcal{X}$. Therefore, the restricted backward shift realization $\{Z_o , B_o , C_o , D_0\}$ is a minimal realization.

Now assume that $\{Z \text{ on } \mathcal{X}, B, C, D\}$ is also a minimal realization of $F(\lambda)$. In this case, $\dim \mathcal{H} = \dim \mathcal{X}$. Since $\mathcal{H} \subseteq \text{ran}(W_0)$, the operator $W_0$ must map $\mathcal{X}$ in a one-to-one way onto $\mathcal{H}$. Let X be the similarity transform from $\mathcal{X}$ onto $\mathcal{H}$ defined by $X = W_0$. Using the definitions of $Z_0$, $B_0$ and $C_0$ in (4.7) we have

$$Z_0X = S^*W_0 = XZ, \quad B_0 = W_0B = XB \quad \text{and} \quad C = \Pi_{\mathcal{Y}}W_0 = C_0X. \tag{4.10}$$

So any minimal realization $\{Z, B, C, D\}$ is similar to the restricted backward shift realization. By transitivity of similar realizations, all minimal realizations are similar. This completes the proof.

As before, assume that $F(\lambda)$ is a rational function analytic in some neighborhood of the origin and $\mathcal{U}$ is finite dimensional. Then the restricted backward shift realization $\{Z_0, B_0, C_0, D_0\}$ is both controllable and observable. The controllability follows from the fact that $Z_0^nB_0 = S^{*n}\tilde{B}$ and $\text{span}_{n\geq 0}S^{*n}\tilde{B}\mathcal{U}$ is all of $\mathcal{H}$. The observability follows from

$$\begin{bmatrix} C_0 \\ C_0Z_0 \\ C_0Z_0^2 \\ \vdots \end{bmatrix} = \begin{bmatrix} \Pi_{\mathcal{Y}} \\ \Pi_{\mathcal{Y}}S^* \\ \Pi_{\mathcal{Y}}S^{*2} \\ \vdots \end{bmatrix} |\mathcal{H} = I|\mathcal{H}, \tag{4.11}$$

where I is the identity on $\ell_+(\mathcal{Y})$. We now are ready to show the equivalence between minimality, controllability and observability for systems with a finite dimensional state space.

**THEOREM 4.2.** *Let* $\Gamma = \{Z, B, C, D\}$ *be a finite dimensional realization of* $F(\lambda)$. *Then* $\Gamma$ *is a minimal realization of* $F(\lambda)$ *if and only if* $\Gamma$ *is both controllable and observable.*

**PROOF.** Suppose that $\Gamma = \{Z, B, C, D\}$ is a minimal realization of $F(\lambda)$, then obviously $\Gamma$ is similar to the restricted backward shift realization. Because the restricted backward shift realization is controllable and observable, it follows that $\Gamma$ is controllable and observable. So minimality implies controllability and observability. Now suppose that $\Gamma$ is controllable and observable. Then $\mathcal{X} = \text{span}_{n\geq 0}Z^nB\mathcal{U}$ and the operator $W_0$ is one-to-one. Furthermore, from equation (4.9), we see that $\mathcal{H} = W_0\text{span}_{n\geq 0}Z^nB\mathcal{U} = W_0\mathcal{X}$. So $\mathcal{H} = \text{ran } W_0$, therefore $W_0$ maps $\mathcal{X}$ in a one-to-one way onto $\mathcal{H}$. In particular, $\mathcal{X}$ and $\mathcal{H}$ have the same dimension. Because the restricted backward shift realization is minimal, so is $\Gamma$. Hence, controllability and observability imply minimality. This completes the proof.

Notice that Theorems 4.1 and 4.2 imply that all finite dimensional, controllable and observable realizations of the same rational transfer function are similar. In this way we recover Theorem 2.3 for the finite dimensional case.

If $F(\lambda)$ is in $H^\infty(\mathcal{U}, \mathcal{Y})$, then the restricted backward shift realization of $F(\lambda)$ can be defined even if $\mathcal{U}$ is infinite dimensional. In this case the restricted backward shift realization $\{Z_o \text{ on } \mathcal{H}, B_o, C_o, D_0\}$ of $F(\lambda)$ is defined by

$$Z_o = S^* | \mathcal{H}, \quad B_o = S^* T_F \Pi_{\mathcal{U}}^*, \quad C_o = \Pi_{\mathcal{Y}} | \mathcal{H} \text{ and } D_0 = F(0), \tag{4.12}$$

where now $\mathcal{H}$ is the closure of $\text{span}_{n\geq 0} S^{*n} B_o \mathcal{U}$ in $\ell_+^2(\mathcal{Y})$. Furthermore, in (4.12) the operator $S^*$ is the backward shift operator on $\ell_+^2(\mathcal{Y})$, the operator $T_F$ is the Toeplitz operator from $\ell_+^2(\mathcal{U})$ into $\ell_+^2(\mathcal{Y})$ generated by F, and $\Pi_{\mathcal{E}}$ stands for the operator from $\ell_+^2(\mathcal{E})$ onto $\mathcal{E}$ defined by $\Pi_{\mathcal{E}} [a_0, a_1, a_2, \ldots]^{tr} = a_0$. Moreover, the subspace $\mathcal{H}$ is now the subspace of $\ell_+^2(\mathcal{Y})$ defined by

$$\mathcal{H} = \bigvee_{n=0}^{\infty} S^{*(n+1)} T_F \Pi_{\mathcal{U}}^* \mathcal{U} . \tag{4.13}$$

Obviously the restricted backward shift realization is a pointwise stable realization of $F(\lambda)$. Moreover, its observability operator given in (4.11) is the isometry $I | \mathcal{H}$ mapping $\mathcal{H}$ into $\ell_+^2(\mathcal{Y})$ where I is now the identity on $\ell_+^2(\mathcal{Y})$. Its controllability operator $W_c$ mapping $\ell_+^2(\mathcal{U})$ into $\mathcal{H}$ is defined by

$$W_c = [B_o, Z_o B_o, Z_o^2 B_o, \ldots] = \begin{bmatrix} F_1 & F_2 & F_3 & \cdots \\ F_2 & F_3 & F_4 & \cdots \\ F_3 & F_4 & F_5 & \cdots \\ \vdots & \vdots & \vdots & \end{bmatrix}, \tag{4.14}$$

where $\{F_n\}_0^\infty$ are the Fourier coefficients of F. In particular, from (4.13) we see that the controllability operator is bounded and has dense range in $\mathcal{H}$. We prove now the following consistency result.

**PROPOSITION 4.3.** *If* $F(\lambda)$ *is a rational function in* $H^\infty(\mathcal{U}, \mathcal{Y})$ *and* $\mathcal{U}$ *is finite dimensional, then the two restricted backward shift realizations defined by (4.7) and (4.12) coincide and are stable.*

**PROOF.** Recall that the restricted backward shift realization $\{Z_o, B_o, C_o, D_0\}$ defined by (4.12) is pointwise stable. Furthermore, the two block Hankel matrices in (4.14) and (4.6) are the same, and hence both have the same finite rank. Therefore, the operator $W_c$ in (4.14) is of finite rank, and hence its range is closed. Since ran $W_c$ is dense in the space $\mathcal{H}$ defined by

(4.13), we conclude that $\mathcal{H} = \text{ran}\, W_c$, and hence the spaces defined by (4.5) and (4.13) are the same. It is now easy to check that the two realizations are the same. Finally, since $Z_0$ is pointwise stable and acts on a finite dimensional space, $Z_0$ is stable. This completes the proof.

A direct consequence of Proposition 4.3 is the following useful fact.

**THEOREM 4.4.** *Let* $\{Z, B, C, D\}$ *be a finite dimensional minimal realization for a rational function* $F(\lambda)$ *with values in* $\mathcal{L}(\mathcal{U}, \mathcal{Y})$ *where* $\mathcal{U}$ *is finite dimensional. Then* $F(\lambda)$ *is in* $H^\infty(\mathcal{U}, \mathcal{Y})$ *if and only if* $Z$ *is stable.*

**PROOF.** Obviously, if $Z$ is stable, then $F(\lambda) = D + \lambda C(I - \lambda Z)^{-1}B$ is in $H^\infty(\mathcal{U}, \mathcal{Y})$. On the other hand, since $F(\lambda)$ is rational and $\mathcal{U}$ is finite dimensional, Proposition 4.3 tells us that the restricted backward shift realization $\{Z_0, B_0, C_0, D\}$ is stable. By Theorem 4.1, the operators $Z$ and $Z_0$ are similar. Hence $Z$ is stable. This completes the proof.

**REMARK 4.5.** Let $Q$ be the observability grammian for a stable, observable realization $\Gamma = \{Z \text{ on } \mathcal{X}, B, C, D\}$ of $F(\lambda)$ in $H^\infty(\mathcal{U}, \mathcal{Y})$ and assume that the space $\mathcal{X} = \bigvee \{Z^n B \mathcal{U} : n \geq 0\}$. Then $Q = I$ if and only if $\Gamma$ is unitarily equivalent to the restricted backward shift realization of $F(\lambda)$.

To prove the above statement, let us first assume $Q = I$. Then the observability operator $W_o$ from $\mathcal{X}$ into $\ell_+^2(\mathcal{Y})$ for $\{C, Z\}$ is an isometry. Since

$$W_o Z^n B = \begin{bmatrix} F_{n+1} \\ F_{n+2} \\ \vdots \end{bmatrix} = S^{*(n+1)} T_F \Pi_{\mathcal{U}}^*,$$

the space $\mathcal{H}$ in (4.13) is the range of $W_o$ restricted to the closure of $\text{span}_{n\geq 0} Z^n B \mathcal{U}$. This space equals $\mathcal{X}$ because $\mathcal{X} = \bigvee \{Z^n B \mathcal{U} : n \geq 0\}$. Thus $X = W_o$ is a unitary operator from $\mathcal{X}$ onto $\mathcal{H}$. Equations similar to (4.10) show that $\Gamma$ is unitarily equivalent to the restricted backward shift realization of $F(\lambda)$. On the other hand, if $\Gamma$ is unitarily equivalent to the restricted backward shift realization of $F(\lambda)$, then $W_o$ is an isometry. This follows from the fact that the observability operator for the restricted backward shift realization is an isometry; see (4.11). Thus $Q = W_o^* W_o = I$. This completes the proof.

To complete the section let us show how one can use the restricted backward shift realization to obtain the classical so called observable canonical realization $\{Z, B, C, D\}$ for a scalar valued rational function $F(\lambda)$ analytic in some neighborhood of the origin. (This may not be the most efficient way to derive this realization. However, it does provide some additional insight into these realizations.) To this end, notice that $F(\lambda)$ is a scalar valued rational analytic function in some neighborhood of the origin if and only if $F(\lambda)$ admits a decomposition of the form $F(\lambda) = f_o + \lambda p(\lambda)/q(\lambda)$ where $f_o$ is a constant, and $p(\lambda)$ and $q(\lambda)$ are polynomials, and $q(0) \neq 0$. Since $q(0) \neq 0$, we can always assume that $q(0) = 1$. So $F(\lambda)$ is a rational analytic function in some neighborhood of the origin if and only if $F(\lambda) = f_o + \lambda p(\lambda)/q(\lambda)$, where p and q are relatively prime polynomials of the form

$$p(\lambda) = \sum_{j=0}^{k-1} p_j \lambda^j \quad \text{and} \quad q(\lambda) = 1 + \sum_{j=1}^{k} q_j \lambda^j , \tag{4.15}$$

where k denotes the maximum of the degrees of $\lambda p(\lambda)$ and $q(\lambda)$. (Two polynomials $n(\lambda)$ and $d(\lambda)$ are *relatively prime* if $n(\lambda)$ and $d(\lambda)$ have no common zeros.) This sets the stage for the following useful result.

**COROLLARY 4.6.** *Let* $F(\lambda) = f_o + \lambda p(\lambda)/q(\lambda)$ *be a scalar valued rational function where* $f_o$ *is a constant and* $p(\lambda)$ *and* $q(\lambda)$ *are relatively prime polynomials of the form (4.15). Let* $\{Z, B, C, D\}$ *be the system given by*

$$Z = \begin{bmatrix} 0 & 0 & 0 & \dots & 0 & -q_k \\ 1 & 0 & 0 & \dots & 0 & -q_{k-1} \\ 0 & 1 & 0 & \dots & 0 & -q_{k-2} \\ \cdot & \cdot & \cdot & \dots & \cdot & \cdot \\ \cdot & \cdot & \cdot & \dots & \cdot & \cdot \\ \cdot & \cdot & \cdot & \dots & \cdot & \cdot \\ 0 & 0 & 0 & \dots & 1 & -q_1 \end{bmatrix}, \quad B = \begin{bmatrix} p_{k-1} \\ p_{k-2} \\ p_{k-3} \\ \cdot \\ \cdot \\ \cdot \\ p_0 \end{bmatrix} \tag{4.16}$$

$$C = [0, 0, 0, \dots, 0, 1] \quad \text{and} \quad D = f_o .$$

*Then* $\{Z, B, C, D\}$ *is a controllable and observable realization of* $F(\lambda)$. *Moreover,* Z *is stable if and only if* $F(\lambda)$ *is in* $H^\infty$, *or equivalently, all the zeros of* $q(\lambda)$ *are in* $|\lambda| > 1$.

**PROOF.** Let $\mathcal{H}_+$ be the linear space consisting of all formal series $g(\lambda)$ of the form

$$g(\lambda) = \sum_{n=0}^{\infty} g_n \lambda^n \qquad (g_n \in \mathbb{C} \text{ for all } n \geq 0) . \tag{4.17}$$

Clearly $\mathcal{H}_+$ contains the set of all rational functions analytic in some neighborhood of the origin. Moreover, $\mathcal{H}_+$ can be identified with $\ell_+ = \ell_+(\mathbb{C})$ through the $\lambda$ transform, that is, $\mathcal{F}_+[g_0, g_1, g_2, \dots]^{tr} = \sum_0^\infty g_n \lambda^n$. Due to this identification we can define the backward shift operator $S^*$ on $\mathcal{H}_+$ and the operator $\Pi_{\mathcal{Y}}$ from $\mathcal{H}_+$ onto $\mathcal{Y}$ by

$$S^*g = \sum_{n=1}^{\infty} g_n\lambda^{n-1} = \frac{g(\lambda)-g(0)}{\lambda} \quad \text{and} \quad \Pi_{\mathcal{Y}}g = g(0) = g_0 \qquad (g \in \mathcal{H}_+). \tag{4.18}$$

The operator $\tilde{B}$ from $\mathcal{U} = \mathbb{C}$ to $\mathcal{H}_+$ is now given by $\tilde{B}u = S^*F(\lambda)u = (p(\lambda)/q(\lambda))u$, where u is in $\mathcal{U}$. Finally, the invariant subspace $\mathcal{H}$ for $S^*$ now becomes

$$\mathcal{H} = \operatorname*{span}_{n\geq 1} S^{*n}F(\lambda) = \operatorname*{span}_{n\geq 0} S^{*n}\left[\frac{p(\lambda)}{q(\lambda)}\right]. \tag{4.19}$$

Notice that if $n = n_0 + n_1\lambda + \cdots + n_{k-1}\lambda^{k-1}$ is any polynomial of degree at most $k-1$, then using $S^*f = (f(\lambda) - f(0))/\lambda$ we have

$$S^*\left[\frac{n}{q}\right] = \frac{n(\lambda) - n(0)q(\lambda)}{\lambda q(\lambda)} =$$

$$\frac{n_1 - n_0q_1 + (n_2 - n_0q_2)\lambda + \cdots + (n_{k-1} - n_0q_{k-1})\lambda^{k-2} - n_0q_k\lambda^{k-1}}{q(\lambda)} = \frac{m(\lambda)}{q(\lambda)}. \tag{4.20}$$

Thus $S^*(n/q) = m/q$, where $m(\lambda)$ is also a polynomial of degree at most $k-1$. Since $p(\lambda)$ is a polynomial of degree at most $k-1$, equations (4.19) and (4.20) show that

$$\mathcal{H} \subseteq \operatorname{span}\{\phi_1, \phi_2, \ldots, \phi_k\}, \tag{4.21}$$

where $\phi_j(\lambda)$ are the functions in $\mathcal{H}_+$ defined by $\phi_1 = \lambda^{k-1}/q(\lambda)$, $\phi_2 = \lambda^{k-2}/q(\lambda)$, ..., $\phi_k = 1/q(\lambda)$. We claim that equality holds in (4.21). If we did not have equality, then $\dim \mathcal{H} < k$. Because the backward shift realization $\{Z_o, B_o, C_o, D_0\}$ is a minimal realization $r(\lambda)/d(\lambda) = C_o(I - \lambda Z_o)^{-1}B_o = p(\lambda)/q(\lambda)$, where $r(\lambda)$ and $d(\lambda)$ are polynomials of degree at most $\dim \mathcal{H} < k$. This contradicts the fact that p and q are relatively prime. So (4.21) holds.

Now let us identify the restricted backward shift realization $\{Z_o, B_o, C_o, D_0\}$ with $\{Z \text{ on } \mathbb{C}^k, B, C, D\}$ through the identification of the basis $\{\phi_j\}_1^k$ for $\mathcal{H}$ with the standard basis $\{e_j\}_1^k$ for $\mathbb{C}^k$. (The j-th entry of $e_j$ is one and all the other entries are zero.) Using $S^*f = (f(\lambda)) - f(0))/\lambda$, we arrive at $Z_o\phi_j = S^*\phi_j = \phi_{j+1}$ for $1 \leq j < k$, and

$$Z_o\phi_k = \frac{1 - q(\lambda)}{\lambda q(\lambda)} = -(q_k\phi_1 + q_{k-1}\phi_2 + \cdots + q_1\phi_k). \tag{4.22}$$

This readily implies that the matrix representation Z for $Z_o$ is given by the Z in (4.16). Clearly $\tilde{B} = p(\lambda)/q(\lambda) = p_{k-1}\phi_1 + \cdots + p_0\phi_k$. So the matrix representation B for $B_o$ is also given by the B in (4.16). Finally, using $\Pi_{\mathcal{Y}}f = f(0)$,

$$Ce_j = \Pi_{\mathcal{Y}}\phi_j = \phi_j(0) = \begin{cases} 0 & \text{for } j = 1, \ldots, k-1, \\ 1 & \text{for } j = k. \end{cases}$$

Hence {Z, B, C, D} given in (4.16) is a minimal realization of $F(\lambda)$.

The statement concerning the stability of Z follows from Theorem 4.4, that is, Z is stable if and only if $F(\lambda)$ is in $H^\infty$. Since p and q are prime and $F = f_0 + \lambda p(\lambda)/q(\lambda)$, it follows that F is in $H^\infty$ if and only if all the roots of $q(\lambda)$ are in $|\lambda| > 1$. This completes the proof.

Finally, it is worth noting that if $F(\lambda)$ is any rational function analytic in some neighborhood of the origin with values in $\mathcal{L}(\mathcal{U}, \mathcal{Y})$, with both $\mathcal{U}$ and $\mathcal{Y}$ finite dimensional, then one can use formulas like (4.20) to develop an algorithm to compute a matrix representation {Z, B, C, D} for the restricted backward shift realization of $F(\lambda)$. The details are omitted.

## III.5 ANTICAUSAL REALIZATIONS

In this section we will introduce anticausal systems which will play a basic role in our solution to the Nehari interpolation problem. To this end, let $L_F$ be the block Laurent matrix generated by the sequence of operators $\{F_n : n \in \mathbb{Z}\}$ with values in $\mathcal{L}(\mathcal{U}, \mathcal{Y})$, that is, the i,j-th block entry of $L_F$ is given by $(L_F)_{i,j} = F_{i-j}$. In systems terminology $L_F$ defines a linear, time-invariant system also denoted by $L_F$. The input u to $L_F$ is a sequence $u = (u(n))_{-\infty}^{\infty}$ with values in $\mathcal{U}$, while the output $y = L_F u = (y(n))_{-\infty}^{\infty}$ is the sequence with values in $\mathcal{Y}$ given by

$$y(n) = \sum_{i=-\infty}^{\infty} F_{n-i}u(i) . \tag{5.1}$$

Obviously, the output sequence y is well defined if {u(n)} has finite support. The system corresponding to $L_F$ is *causal* if the output y(n) at time n depends only on the past inputs u(i) for $i \le n$. Hence $L_F$ is causal if and only if $L_F$ is lower triangular, or equivalently,

$$y(n) = \sum_{i=-\infty}^{n} F_{n-i}u(i) . \tag{5.2}$$

On the other hand, the system defined by $L_F$ is *anticausal,* if the output y(n) at time n depends only on the future inputs u(i) for $i > n$. Therefore $L_F$ is anticausal if and only if $L_F$ is an upper triangular matrix whose main diagonal entries are zero, or equivalently,

$$y(n) = \sum_{i=n+1}^{\infty} F_{n-i}u(i) . \tag{5.3}$$

The symbol $F(\lambda)$ for the Laurent matrix $L_F$ is formally defined by

$$F(\lambda) = \sum_{n=-\infty}^{\infty} F_n \lambda^n . \tag{5.4}$$

In systems terminology $F(\lambda)$ is the *transfer function* for $L_F$. Let $U(\lambda)$ and $Y(\lambda)$ be the $\lambda$-transforms of $u = \{u(n)\}$ and $y = \{y(n)\}$ formally defined by the formal power series

$$U(\lambda) = \sum_{n=-\infty}^{\infty} u(n)\lambda^n \text{ and } Y(\lambda) = \sum_{n=-\infty}^{\infty} y(n)\lambda^n . \tag{5.5}$$

By using (5.1) a formal calculation shows that $Y(\lambda) = F(\lambda)U(\lambda)$. So in systems terminology convolution in the time domain (5.1) corresponds to multiplication $Y(\lambda) = F(\lambda)U(\lambda)$ in the $\lambda$-domain. If $F(\lambda)$ is an analytic function (with values in $\mathcal{L}(\mathcal{U}, \mathcal{Y})$) in some neighborhood of the origin, then $F_n = 0$ for all $n < 0$, and thus $L_F$ is causal. On the other hand, if $\lambda F(\lambda)$ is analytic in some neighborhood of infinity, then $F_n = 0$ for all $n \geq 0$ and therefore $L_F$ is anticausal. Recall that a function $G(\lambda)$ is analytic in some neighborhood of infinity, if $G(1/\lambda)$ is analytic in some neighborhood of the origin.

The input-output map $L_F$ is *stable* if $L_F$ defines a bounded linear operator from $\ell^2(\mathcal{U})$ into $\ell^2(\mathcal{Y})$. Therefore, if $\mathcal{U}$ ad $\mathcal{Y}$ are separable, then $L_F$ is stable if and only if (5.4) defines on the unit circle a function $F$ in $L^\infty(\mathcal{U}, \mathcal{Y})$. In this case $\|L_F\| = \|F\|_\infty$; see Section I.1. Furthermore, if $L_F$ is stable and the input $u = \{u(n)\}$ is in $\ell^2(\mathcal{U})$, then the output $y = L_F u$ is in $\ell^2(\mathcal{Y})$. Moreover, $Y(e^{it}) = F(e^{it})U(e^{it})$ is in $L^2(\mathcal{Y})$. Notice that $L_F$ defines a causal stable system if and only if its transfer function $F$ is in $H^\infty(\mathcal{U}, \mathcal{Y})$. Finally, $L_F$ defines a anticausal stable system if and only if $\lambda^{-1}F(1/\lambda)$ is in $H^\infty(\mathcal{U}, \mathcal{Y})$.

The state space system in (1.1) defines a causal input-output map $L_F$ where $F(\lambda) = D + \lambda C(I - \lambda Z)^{-1}B$. By reversing time the state space set up can also be used to generate an anticausal system. To see this consider the following state space system

$$x(n-1) = Zx(n) + Bu(n) \text{ and } y(n) = Cx(n) \tag{5.6}$$

where the state $x(n)$ is in $\mathcal{X}$, the input $u(n)$ is in $\mathcal{U}$ and the output $y(n)$ is in $\mathcal{Y}$. As before, $Z$ is an operator on $\mathcal{X}$ while $B$ is an operator from $\mathcal{U}$ into $\mathcal{X}$ and $C$ maps $\mathcal{X}$ into $\mathcal{Y}$. By recursively solving for $y(j)$ we obtain

$$\begin{bmatrix} \cdot \\ \cdot \\ \cdot \\ y(n-3) \\ y(n-2) \\ y(n-1) \\ y(n) \end{bmatrix} = \begin{bmatrix} \cdot \\ \cdot \\ \cdot \\ CZ^3 \\ CZ^2 \\ CZ \\ C \end{bmatrix} x(n) + \begin{bmatrix} \cdots & \cdot & \cdot & \cdot & \cdot \\ \cdots & \cdot & \cdot & \cdot & \cdot \\ \cdots & \cdot & \cdot & \cdot & \cdot \\ \cdots & 0 & CB & CZB & CZ^2B \\ \cdots & 0 & 0 & CB & CZB \\ \cdots & 0 & 0 & 0 & CB \\ \cdots & 0 & 0 & 0 & 0 \end{bmatrix} \begin{bmatrix} \cdot \\ \cdot \\ \cdot \\ u(n-3) \\ u(n-2) \\ u(n-1) \\ u(n) \end{bmatrix} . \tag{5.7}$$

So if the initial condition x(n) is set equal to zero at infinity, then the state space system in (5.6) generates an anticausal system $L_F$ whose Laurent operator is given by $F_n = 0$ for all $n \ge 0$ and

$$F_n = CZ^{|1+n|}B \qquad (\text{if } n < 0) . \tag{5.8}$$

Therefore we call the state space system in (5.6) the anticausal system generated by {Z, B, C}. Obviously the concepts of controllability and observability can be extended to the anticausal case. Moreover, the tests for controllability and observability for anticausal systems are the same as those for causal systems. So we say that the anticausal system {Z, B, C} is *controllable*, respectively *observable* if and only if the pair {Z, B} is controllable or respectively the pair {C, Z} is observable.

According to (5.8) the transfer function $F(\lambda)$ for the anticausal system {Z, B, C} is given by

$$F(\lambda) = \sum_{n=1}^{\infty} F_{-n}\lambda^{-n} = \sum_{n=1}^{\infty} CZ^{n-1}B\lambda^{-n} = C(\lambda I - Z)^{-1}B . \tag{5.9}$$

Clearly $F(\lambda)$ is analytic in some neighborhood of infinity. Motivated by this we say that {Z, B, C} is an *anticausal realization* of $F(\lambda)$ if

$$F(\lambda) = C(\lambda I - Z)^{-1}B \tag{5.10}$$

in some neighborhood of infinity, that is, {Z, B, C} is an anticausal realization of $F(\lambda)$ if and only if $F(\lambda)$ is the transfer function for the state space system determined by (5.6). Finally, we say that {Z, B, C} is an anticausal stable realization of $F(\lambda)$ if (5.10) holds and $r_{spec}(Z) < 1$, and in this case the Laurent operator $L_F$ is anticausal and bounded. This follows from the fact that $\lambda^{-1}F(1/\lambda) = C(I - \lambda Z)^{-1}B$ is in $H^\infty(\mathcal{U}, \mathcal{Y})$.

The anticausal realization problem is, given a function $F(\lambda)$ analytic in some neighborhood of infinity satisfying $F(\infty) = 0$, find an anticausal realization of $F(\lambda)$. By replacing $\lambda$ by $1/\lambda$ the anticausal realization problem can be viewed as the causal realization problem discussed in Section 4. Indeed {Z, B, C, 0} is a realization of $F(1/\lambda)$ if and only if {Z, B, C} is an anticausal realization of $F(\lambda)$. Therefore we can apply the results in Section 4 to $F(1/\lambda)$ to compute a minimal anticausal realization {Z, B, C} of F(λ). As expected, we say that {Z, B, C} is a minimal anticausal realization of F(λ) if the state space $\mathcal{X}$ for {Z, B, C} has the smallest possible dimension over the class of all anticausal realizations of F(λ). Moreover, if F(λ) admits a finite dimensional anticausal realization, then {Z, B, C} is a minimal anticausal realization of F(λ) if and only if {Z, B, C} is controllable and observable; see Theorem 4.2. In particular, if {Z, B, C} is a finite dimensional anticausal realization of F(λ), then by inverting the matrix $\lambda I - Z$, we can give to F(λ) the form N(λ)/d(λ) where d(λ) is the scalar valued polynomial $d(\lambda) = \det[\lambda I - Z]$ while N(λ) is a polynomial with values in $\mathcal{L}(\mathcal{U}, \mathcal{Y})$ satisfying $\deg N < \deg d$. Furthermore, by adapting Theorem 4.1 to the anticausal setting we see that if $\mathcal{U}$ is finite dimensional, then F(λ) admits a finite dimensional anticausal realization if and only if F(λ) is a strictly proper rational function, that is, $F(\lambda) = N(\lambda)/d(\lambda)$ where N(λ) is a polynomial with values in $\mathcal{L}(\mathcal{U}, \mathcal{Y})$ and d(λ) is a scalar valued polynomial satisfying $\deg N < \deg d$. It is clear that one can use the restricted backward shift realization techniques in Section 4 to compute a minimal anticausal realization for a strictly proper rational F(λ).

For completeness in this section we will present the classical companion matrix approach to construct a controllable and observable, anticausal realization for a scalar valued strictly proper rational function F(λ). Because F(λ) is a strictly proper rational function, $F(\lambda) = c(\lambda)/d(\lambda)$ where c(λ) and d(λ) are polynomials of the form

$$c(\lambda) = \sum_{i=0}^{k-1} c_i \lambda^i \quad \text{and} \quad d(\lambda) = \lambda^k + \sum_{i=0}^{k-1} d_i \lambda^i . \tag{5.11}$$

Without loss of generality we always assume that c(λ) and d(λ) are *relatively prime* polynomials, that is, c(λ) and d(λ) have no common zeros. We say that Z is the *companion matrix generated by the* polynomial d(λ) if Z is the matrix on $\mathbb{C}^k$ defined by

$$Z = \begin{bmatrix} 0 & 1 & 0 & 0 & \cdots & 0 \\ 0 & 0 & 1 & 0 & \cdots & 0 \\ 0 & 0 & 0 & 1 & \cdots & 0 \\ \vdots & & & & & \\ 0 & 0 & 0 & 0 & \cdots & 1 \\ -d_o & -d_1 & -d_2 & -d_3 & \cdots & -d_{k-1} \end{bmatrix} . \tag{5.12}$$

Now let $v_\alpha$ be the vector on $\mathbb{C}^k$ defined by

$$v_\alpha = [1, \alpha, \alpha^2, \ldots, \alpha^{k-1}]^{tr} . \tag{5.13}$$

This sets the stage for the following elementary result.

**LEMMA 5.1.** *Let* Z *be the companion matrix in (5.12) generated by the polynomial* $d(\lambda)$ *in (5.11). Then the eigenvalues* $\alpha_1, \alpha_2, \ldots, \alpha_m$ *of* Z *are the zeros of* $d(\lambda)$ *and* $\mathbb{C}v_{\alpha_1}, \ldots, \mathbb{C}v_{\alpha_m}$ *are the corresponding eigenspaces. In particular,* Z *is stable if and only if all the zeros of* $d(\lambda)$ *are in the open unit disc.*

**PROOF.** Assume that $Zx = \alpha x$ where $\alpha$ is an eigenvalue and $x = [x_1, x_2, \ldots, x_k]^{tr} \neq 0$. Using the form of Z in (5.12), we obtain

$$\alpha x_1 = x_2,\ \alpha x_2 = x_3,\ \ldots,\ \alpha x_{k-1} = x_k\ . \tag{5.14}$$

If $x_1 = 0$, then $x_j = 0$ for all j and thus $x = 0$. This contradicts the fact x is an eigenvector. So without loss of generality we can assume that $x_1 = 1$. In this case (5.14) shows that $x = v_\alpha$. Now using the structure of Z we have for any $\alpha$

$$(\alpha I - Z)v_\alpha = [0, 0, \ldots, 0, d(\alpha)]^{tr}\ . \tag{5.15}$$

Therefore x is an eigenvector for Z with eigenvalue $\alpha$ if and only if $x = \gamma v_\alpha$ for some nonzero constant $\gamma$ and $d(\alpha) = 0$. This completes the proof.

Now we are ready to present the following classical minimal realization construction.

**PROPOSITION 5.2.** *Let* $F(\lambda) = c(\lambda)/d(\lambda)$ *be a scalar valued rational function where* $c(\lambda)$ *and* $d(\lambda)$ *are the relatively prime polynomials in (5.11). Let* Z *be the companion matrix generated by* $d(\lambda)$ *in (5.12) and* B *and* C *the column and row vectors defined by*

$$B = [0, 0, \ldots, 0, 1]^* \ \text{ and } \ C = [c_0, c_1, \ldots, c_{k-1}]\ . \tag{5.16}$$

*Then* $\{Z, B, C\}$ *is a controllable and observable, anticausal realization of* $F(\lambda)$*. Moreover, all the zeros of* $d(\lambda)$ *are in the open unit disc if and only if* $\{Z, B, C\}$ *is an anticausal, stable, controllable and observable realization of* $F(\lambda)$.

**PROOF.** One can obtain a proof of this result by applying Corollary 4.6 to $F(1/\lambda)$ and then taking the transpose. For completeness let us give a proof of this result by using the companion matrix. To show that $\{Z, B, C\}$ is an anticausal realization of $F(\lambda)$, that is, (5.10) holds first notice that (5.15) gives $(\lambda I - Z)v_\lambda = d(\lambda)B$. By inverting $(\lambda I - Z)$ we readily obtain for $d(\lambda) \neq 0$

$$\frac{1}{d(\lambda)} v_\lambda = (\lambda I - Z)^{-1} B\ .$$

So using C in (5.16) we have $c(\lambda)/d(\lambda) = C(\lambda I - Z)^{-1}B$ for all $\lambda$ large enough. Therefore

$c(\lambda)/d(\lambda)$ is the anticausal transfer function for $\{Z, B, C\}$.

Obviously $\{Z^n B : 0 \le n < k\}$ spans $\mathbb{C}^k$. So $\{Z, B\}$ is controllable. Because $c(\lambda)$ and $d(\lambda)$ have no common zeros the pair $\{C, Z\}$ is observable. To prove this let $\mathcal{M}$ be the invariant subspace for Z defined by ker $W_o$ where $W_o$ is the observability operator in (1.5), that is, $\mathcal{M} = \{x \in \mathbb{C}^k : CZ^n x = 0 \text{ for all } n \ge 0\}$. If $\{C, Z\}$ is not observable, then $\mathcal{M} \neq \{0\}$. Because $\mathcal{M}$ is an invariant subspace for Z, it contains an eigenvector $v_{\alpha_j}$ of Z; see Lemma 5.1. Hence $c(\alpha_j) = Cv_{\alpha_j} = 0$. This contradicts the fact that c and d are relatively prime. Therefore the pair $\{C, Z\}$ is observable. The previous lemma shows that Z is stable if and only if all the zeros of $d(\lambda)$ are in the open unit disc. This completes the proof.

## III.6 COMPUTING THE HANKEL NORM

In this section we will show how one can use state space techniques to compute the norm for certain (block) Hankel operators. To this end, recall that G is a *Nehari interpolant* for a sequence of operators $\{K_n\}_1^\infty$ in $\mathcal{L}(\mathcal{U}, \mathcal{Y})$ if G is a function in $L^\infty(\mathcal{U}, \mathcal{Y})$ satisfying

$$K_n = \frac{1}{2\pi}\int_0^{2\pi} e^{int} G(e^{it})dt \qquad (\text{for } n \ge 1)\,. \tag{6.1}$$

The $H^\infty$ Nehari optimization problem is to find a G in $L^\infty(\mathcal{U}, \mathcal{Y})$ solving the $H^\infty$ optimization problem associated with:

$$d_\infty = \inf\,\{\|G\|_\infty : G \text{ is a Nehari interpolant of } \{K_n\}_1^\infty\}\,. \tag{6.2}$$

The results in Section I.6 show that $d_\infty = \|\Gamma\|$ where $\Gamma$ is the Hankel operator defined by

$$\Gamma = \begin{bmatrix} K_1 & K_2 & K_3 & \cdots \\ K_2 & K_3 & K_4 & \cdots \\ K_3 & K_4 & K_5 & \cdots \\ \vdots & \vdots & \vdots & \end{bmatrix} : \ell_+^2(\mathcal{U}) \to \ell_+^2(\mathcal{Y})\,. \tag{6.3}$$

This $\Gamma$ is called the *Hankel operator associated with* $\{K_n\}$. Finally, the infimum in (6.2) is attained.

Recall that the controllability grammian P and observability grammian Q for a stable system $\{Z, B, C\}$ are obtained by solving the following Lyapunov equations

$$P = ZPZ^* + BB^* \quad \text{and} \quad Q = Z^*QZ + C^*C\,. \tag{6.4}$$

This sets the stage for the following result.

**THEOREM 6.1.** *Let* $\{K_n\}_1^\infty$ *be a sequence of operators in* $\mathcal{L}(\mathcal{U}, \mathcal{Y})$ *and* $\{Z$ *on* $\mathcal{X}$, B, C$\}$ *be a stable, system satisfying*

$$CZ^{n-1}B = K_n \qquad \text{(for all } n \geq 1) . \tag{6.5}$$

*Let* $\Gamma$ *be the Hankel operator associated with* $\{K_n\}_1^\infty$ *in (6.3). Then*

$$\|\Gamma\|^2 = r_{spec}(PQ) = r_{spec}(QP) = r_{spec}(P^{1/2}QP^{1/2}) = r_{spec}(Q^{1/2}PQ^{1/2}) \tag{6.6}$$

*where* P *is the controllability grammian for* $\{Z, B\}$ *and* Q *is the observability grammian for* $\{C, Z\}$ *obtained by solving the Lyapunov equations in (6.4).*

**PROOF.** Let $W_o$ be the observability operator from $\mathcal{X}$ into $\ell_+^2(\mathcal{Y})$ defined in (1.5) and $W_c$ the controllability operator from $\ell_+^2(\mathcal{U})$ into $\mathcal{X}$ defined by (2.6). Then using $CZ^{n-1}B = K_n$ it follows that the Hankel operator $\Gamma$ admits a factorization of the form

$$\Gamma = W_o W_c . \tag{6.7}$$

Recall that $P = W_c W_c^*$ and $Q = W_o^* W_o$. This fact along with $\Gamma = W_o W_c$ gives $\Gamma^*\Gamma = W_c^* W_o^* W_o W_c$. Since $r_{spec}(RS) = r_{spec}(SR)$ for bounded linear operators R from $\mathcal{X}$ into $\mathcal{Y}$ and S from $\mathcal{Y}$ into $\mathcal{X}$ (see formula (3) in Section III.2 of Gohberg-Goldberg-Kaashoek [1]), it follows that

$$r_{spec}(\Gamma^*\Gamma) = r_{spec}(W_o^* W_o W_c W_c^*) = r_{spec}(QP) = r_{spec}(PQ) .$$

Repeating the argument, we have

$$r_{spec}(Q^{1/2}PQ^{1/2}) = r_{spec}(QP) = r_{spec}(P^{1/2}QP^{1/2}) ,$$

which completes the proof.

Assume now additionally that the system $\{Z, B, C\}$ in Theorem 6.1 is controllable and observable. Using $\Gamma = W_o W_c$, we have

$$\Gamma^*\Gamma W_c^* = W_c^* W_o^* W_o W_c W_c^* = W_c^* QP . \tag{6.8}$$

Because $\{Z, B, C\}$ is controllable and observable, the operators $W_o$ and $W_c^*$ are both one to one with closed range. This and $\Gamma = W_o W_c$ implies that $(\ker \Gamma)^\perp$ equals $(\ker W_c)^\perp = \operatorname{ran} W_c^*$. Therefore (6.8) shows that in this case $\Gamma^*\Gamma \,|\, (\ker \Gamma)^\perp$ is similar to QP.

To demonstrate how one can use Theorem 6.1, consider Problem 47 in Halmos [1], that is, find the norm of the Hankel operator $\Gamma$ generated by $K_n = \alpha^{n-1}$ for $n \geq 1$ and some specified $\alpha$ in the open unit disc of $\mathbb{C}$. To compute $\|\Gamma\|$, let $Z = \alpha$ and $B = C = 1$. Then obviously $\{Z, B, C\}$ is stable, controllable and observable realization satisfying (6.5). In this case the Lyapunov equations for P and Q become

$$P = |\alpha|^2 P + 1 \text{ and } Q = |\alpha|^2 Q + 1 .$$

Therefore $P = Q = 1/(1 - |\alpha|^2)$. So the norm of the Hankel operator $\Gamma$ generated by $\{\alpha^{n-1}\}_1^\infty$ is given by $\|\Gamma\| = 1/(1 - |\alpha|^2)$.

In many applications it is convenient to view the data for the Nehari interpolation problem as the transfer function for an anticausal system. To see this, assume that

$$F(e^{it}) = \sum_{n=1}^{\infty} e^{-int} K_n \tag{6.9}$$

defines a function in $L^\infty(\mathcal{U}, \mathcal{Y})$. Then G is a function in $L^\infty(\mathcal{U}, \mathcal{Y})$ satisfying (6.1) if and only if $G = F - H$ for some H in $H^\infty(\mathcal{U}, \mathcal{Y})$. In this case the Nehari optimization problem in (6.2) is equivalent to the following $H^\infty$ optimization problem: Given a function $F(\lambda)$ in $L^\infty(\mathcal{U}, \mathcal{Y})$ of the form (6.9) find an H in $H^\infty(\mathcal{U}, \mathcal{Y})$ to solve the following optimization problem

$$d_\infty = \inf \{\|F - H\|_\infty : H \in H^\infty(\mathcal{U}, \mathcal{Y})\} . \tag{6.10}$$

Here $d_\infty$ is the distance from F to $H^\infty(\mathcal{U}, \mathcal{Y})$ in the $L^\infty$ norm. Moreover, the Nehari theorem shows that $d_\infty = \|\Gamma\|$ and there exists a H in $H^\infty(\mathcal{U}, \mathcal{Y})$ such that $d_\infty = \|F - H\|_\infty$. Now let A be the Hankel operator from $H^2(\mathcal{U})$ into $\mathcal{H}' = L^2(\mathcal{Y}) \ominus H^2(\mathcal{Y})$ defined by $Ah = P_- Fh$ where h is in $H^2(\mathcal{U})$ and $P_-$ is the orthogonal projection onto $\mathcal{H}'$. Let $\mathcal{F}_+$ be the Fourier transform from $\ell_+^2(\mathcal{U})$ onto $H^2(\mathcal{U})$ and $\mathcal{F}_-$ be the Fourier transform from $\ell_-^2(\mathcal{Y})$ onto $\mathcal{H}'$. Then a simple calculation shows that

$$\mathcal{F}_-^* A \mathcal{F}_+ = \begin{bmatrix} \cdot & \cdot & \cdot & \cdots \\ \cdot & \cdot & \cdot & \cdots \\ \cdot & \cdot & \cdot & \cdots \\ K_3 & K_4 & K_5 & \cdots \\ K_2 & K_3 & K_4 & \cdots \\ K_1 & K_2 & K_3 & \cdots \end{bmatrix} : \ell_+^2(\mathcal{U}) \to \ell_-^2(\mathcal{Y}) . \tag{6.11}$$

So by using the unitary flip operator we see that A is unitarily equivalent to $\Gamma$ in (6.3) and thus $d_\infty = \|A\|$.

Let us denote by $K^\infty(\mathcal{U}, \mathcal{Y})$ the subspace of $L^\infty(\mathcal{U}, \mathcal{Y})$ consisting of all functions F in $L^\infty(\mathcal{U}, \mathcal{Y})$ whose Fourier coefficients of $e^{int}$ vanish for all $n \geq 0$, that is, F admits a Fourier series expansion of the form (6.9). Now assume that F admits a stable anticausal realization $\{Z, B, C\}$. Clearly F is in $K^\infty(\mathcal{U}, \mathcal{Y})$. Because A is unitarily equivalent to $\Gamma$ it follows that $d_\infty^2 = \|A\|^2 = r_{spec}(PQ)$. Moreover, by adjusting Theorem 4.1 to the anticausal setting, it follows that, if $\mathcal{U}$ is finite dimensional, then A has finite rank if and only if $F(\lambda)$ is a strictly proper rational function. In Section V.5 we will use the anticausal realization {Z, B, C} of $F(\lambda)$ to compute a state space realization for a function H in $H^\infty(\mathcal{U}, \mathcal{Y})$ satisfying $\|F - H\|_\infty \le \gamma$ for any $\gamma > d_\infty$. This function H will be computed from a special intertwining lifting of A. For now let us summarize the previous analysis as follows.

**COROLLARY 6.2.** *Let* {Z, B, C} *be a stable, anticausal realization for a function* F *in* $K^\infty(\mathcal{U}, \mathcal{Y})$. *Let* A *be the Hankel operator from* $H^2(\mathcal{U})$ *into* $\mathcal{H}' = L^2(\mathcal{Y}) \ominus H^2(\mathcal{Y})$ *defined by* $Ah = P_-Fh$. *Then the error in the* $H^\infty$ *Nehari optimization problem (6.10) is given by*

$$d_\infty^2 = \|A\|^2 = r_{spec}(PQ) = r_{spec}(QP) \tag{6.12}$$

*where* P *is the controllability grammian for* {Z, B} *and* Q *is the observability grammian for* {C, Z}. *Finally, if* F *is any function in* $K^\infty(\mathcal{U}, \mathcal{Y})$ *and* $\mathcal{U}$ *is finite dimensional, then* A *has finite rank if and only if* $F(\lambda)$ *is a strictly proper rational function.*

Computing $d_\infty$ in the scalar rational case is quite simple. To see this assume that $F(\lambda)$ is a rational function in $K^\infty(\mathbb{C}, \mathbb{C})$, or equivalently, $F(\lambda) = c(\lambda)/d(\lambda)$ where $c(\lambda)$ and $d(\lambda)$ are relatively prime scalar valued polynomials of the form (5.11), and all the zeros of $d(\lambda)$ are in the open unit disc. Let Z be the companion matrix given in (5.12), and B and C be the column and row vectors given by (5.16). Then Proposition 5.2 shows that {Z, B, C} is an anticausal stable, controllable and observable realization of $F(\lambda)$. Now compute P and Q by solving the Lyapunov equation in (6.4). Then $d_\infty^2 = \|A\|^2$ is obtained by computing the largest eigenvalue of PQ. (Since PQ is similar to the positive matrix $P^{1/2}QP^{1/2}$, it follows that all the eigenvalues of PQ are positive.) Finally, it is worth noting that in this case, due to the companion form of Z and the structure of B, the controllability grammian P is always a Toeplitz matrix. This fact is left as a simple exercise.

## III.7 COMPUTING THE PROJECTION IN THE SARASON PROBLEM

In this section we return to the Sarason problem for the case when $\Theta$ is a two-sided inner function in $H^\infty(\mathcal{E}, \mathcal{Y})$ given in state variable form. The aim is to compute the orthogonal projection $P_{\mathcal{H}'}$ on the space

$$\mathcal{H}' = \mathcal{H}(\Theta) := H^2(\mathcal{Y}) \ominus \Theta H^2(\mathcal{E})\,. \tag{7.1}$$

This computation is done in the following theorem.

**THEOREM 7.1.** *Let* $\{Z \text{ on } \mathcal{X}, B, C, D\}$ *be a stable, controllable and observable realization of a two-sided inner function* $\Theta$ *in* $H^\infty(\mathcal{E}, \mathcal{Y})$. *Let* $W_o$ *be the observability operator from* $\mathcal{X}$ *into* $H^2(\mathcal{Y})$ *defined by*

$$W_o x = C(I - \lambda Z)^{-1} x \qquad (x \in \mathcal{X}) . \tag{7.2}$$

*Then the range of* $W_o$ *equals the space* $\mathcal{H}' = \mathcal{H}(\Theta)$ *defined in (7.1), and the orthogonal projection* $P_{\mathcal{H}'}$ *onto* $\mathcal{H}'$ *is given by*

$$P_{\mathcal{H}'} = W_o Q^{-1} W_o^* , \tag{7.3}$$

*where* Q *is the observability grammian for the pair* $\{C, Z\}$.

Notice that the operator $W_o$ in (7.2) is the Fourier transform of the observability operator in (1.5). This justifies calling $W_o$ in (7.2) the observability operator for the pair $\{C, Z\}$.

**PROOF OF THEOREM 7.1.** Let S be the unilateral shift on $H^2(\mathcal{Y})$. We claim that

$$\mathcal{H}' = H^2(\mathcal{Y}) \ominus \Theta H^2(\mathcal{E}) = \bigvee_1^\infty S^{*n} \Theta \mathcal{E} . \tag{7.4}$$

To verify this let $\mathcal{H} = \bigvee \{S^{*n} \Theta \mathcal{E} : n \geq 1\}$. Now notice that for all f in $\mathcal{E}$ and h in $H^2(\mathcal{E})$ and $n \geq 1$ we have

$$(S^{*n}\Theta f, \Theta h) = (\Theta f, S^n \Theta h) = (\Theta f, \Theta \lambda^n h) = (f, \lambda^n h) = 0 \qquad (n \geq 1) .$$

Therefore, $\mathcal{H}$ is orthogonal to $\Theta H^2(\mathcal{E})$ and, thus $\mathcal{H} \subset \mathcal{H}'$. To show equality, assume that g is a vector in $\mathcal{H}'$ which is orthogonal to $\mathcal{H}$. Then

$$0 = (g, S^{*n}\Theta f) = (S^n g, \Theta f) = (e^{int} g, \Theta f) = (g, \Theta e^{-int} f) \qquad (n \geq 1) .$$

Hence g is orthogonal to $\Theta e^{-int} \mathcal{E}$ for all $n \geq 1$. Because g is in $\mathcal{H}'$ this g is also orthogonal to $\Theta e^{int} \mathcal{E}$ for all $n \geq 0$. So g is orthogonal to $\Theta L^2(\mathcal{E})$. Since $\Theta$ is two-sided inner $g = 0$. This proves (7.4).

To complete the proof first recall that the adjoint of the unilateral shift is given by

$$(S^* h)(\lambda) = \frac{h(\lambda) - h(0)}{\lambda} \qquad (h \in H^2(\mathcal{Y})) .$$

Using this it is easy to verify that $S^* W_o = W_o Z$. This along with $\Theta = \lambda W_o B + D$ readily shows that $S^{*n}\Theta f = W_o Z^{n-1} B f$ for all $n \geq 1$. Now using the controllability of the pair $\{Z, B\}$ and (7.4) we obtain

$$\mathcal{H}' = \bigvee_{1}^{\infty} S^{*n}\Theta\mathcal{E} = \bigvee_{0}^{\infty} W_o Z^n B\mathcal{E} = W_o\mathcal{X} . \tag{7.5}$$

Therefore $\mathcal{H}'$ is the range of $W_o$. (Recall that the range of $W_o$ is closed because $\{C, Z\}$ is observable.)

To finish the proof it remains to establish the formula for the orthogonal projection $P_{\mathcal{H}'}$ in (7.3). Since $W_o$ is observable, $Q = W_o^* W_o$ is invertible. So the operator $P = W_o(W_o^* W_o)^{-1} W_o^*$ is onto $\mathcal{H}'$. Moreover, it is easy to verify that $P^2 = P^* = P$. Therefore, $P = W_o Q^{-1} W_o^*$ is the orthogonal projection onto $\mathcal{H}'$. Now for x in $\mathcal{X}$ we have

$$(Qx, x) = (W_o x, W_o x) = (\sum_0^\infty \lambda^n CZ^n x, \sum_0^\infty \lambda^n CZ^n x) = \sum_0^\infty \|CZ^n x\|^2 = \left(\sum_0^\infty Z^{*n}C^*CZ^n x, x\right).$$

Therefore $Q = \sum Z^{*n}C^*CZ^n$ is the observability grammian for the pair $\{C, Z\}$. This completes the proof.

Notice that in Theorem 7.1 the space $\mathcal{H}(\Theta)$ is uniquely determined by the stable observable pair $\{C, Z\}$ and does not involve the operators B and D. Moreover, any stable observable pair $\{C, Z\}$ generates in this way a space $\mathcal{H}(\Theta)$ for some two-sided inner function $\Theta$. In fact, the next theorem shows that a stable observable pair $\{C, Z\}$ can be embedded into a stable, controllable and observable system whose transfer functions is two-sided inner.

**THEOREM 7.2.** *Let $\{C, Z$ on $\mathcal{X}\}$ be a stable observable pair, where C maps $\mathcal{X}$ into $\mathcal{Y}$ and whose observability grammian is denoted by Q. Let $R = [Z^*Q^{1/2}, C^*]$ map $\mathcal{X} \oplus \mathcal{Y}$ into $\mathcal{X}$. Then* ker R *and $\mathcal{Y}$ have the same dimension. For any isometry $\Lambda = [B_o^*, D^*]^*$ from $\mathcal{Y}$ into $\mathcal{X} \oplus \mathcal{Y}$ such that* ran $\Lambda$ = ker R, *set $B = Q^{-1/2}B_o$. Then $\{Z, B, C, D\}$ is a stable, controllable and observable realization of the two-sided inner function $\Theta$ in $H^\infty(\mathcal{Y}, \mathcal{Y})$ defined by*

$$\Theta(\lambda) = D + \lambda C(I - \lambda Z)^{-1}B \tag{7.6}$$

*and $Q^{-1}$ is the controllability grammian for $\{Z, B\}$. Moreover, if $\{Z, B_1, C, D_1\}$ is a stable, controllable and observable realization of a two-sided inner function $\Theta_1$ in $H^\infty(\mathcal{U}, \mathcal{Y})$, then there exists a unique unitary operator $\Phi$ from $\mathcal{Y}$ onto $\mathcal{U}$ such that $B = B_1\Phi$, $D = D_1\Phi$, and $\Theta = \Theta_1\Phi$.*

As before, let $W_o$ be the observability operator in (7.2), generated by the stable observable pair $\{C, Z\}$. We say that $\Theta$ is a *two-sided inner function generated by* the pair $\{C, Z\}$ if $\Theta$ is a two-sided inner function in $H^\infty(\mathcal{U}, \mathcal{Y})$ satisfying ran $W_o = \mathcal{H}(\Theta)$. The above theorem gives us a computational procedure for computing a state space realization $\{Z, B, C, D\}$ for a two-sided inner function $\Theta$ generated by $\{C, Z\}$. In this case the observable system $\{C, Z\}$ is embedded in the state space system {Z, B, C, D} and thus Theorem 7.2 is referred to as the *embedding theorem* for {C, Z}. We begin the proof of Theorem 7.2 by establishing the following result.

**LEMMA 7.3.** *Let* $\Theta$ *in* $H^\infty(\mathcal{U}, \mathcal{Y})$ *be the transfer function for the stable controllable and observable system* {Z, B, C, D}. *Let* P *be the controllability grammian for* {Z, B} *and* Q *the observability grammian for* {C, Z}. *Then* $\Theta$ *is an inner function if and only if*

$$D^*D + B^*QB = I \quad \text{and} \quad D^*C + B^*QZ = 0 . \tag{7.7a}$$

*Furthermore,* $\Theta$ *is a two-sided inner function if and only if (7.7a) holds and*

$$DD^* + CPC^* = I \quad \text{and} \quad DB^* + CPZ^* = 0 . \tag{7.7b}$$

**PROOF.** Recall that $\Theta$ admits a Fourier series expansion of the form

$$\Theta(e^{it}) = \sum_{n=0}^{\infty} \Theta_n e^{int} = D + \sum_{n=1}^{\infty} CZ^{n-1}Be^{int} \tag{7.8}$$

where $\Theta_0 = D$ and $\Theta_n = CZ^{n-1}B$ for $n \geq 1$. Moreover, the power series in (7.8) is well defined for all $0 \leq t < 2\pi$ because Z is stable. Now let $F_n$ in $\mathcal{L}(\mathcal{U}, \mathcal{U})$ be the Fourier coefficients of $e^{int}$ in the power series expansion of $\Theta(e^{it})^*\Theta(e^{it})$. Because $\Theta^*\Theta$ is a self adjoint operator, $F_n = F^*_{-n}$ for all n. Moreover, by consulting (7.8) we see that

$$F_0 = \sum_{j=0}^{\infty} \Theta_j^*\Theta_j = D^*D + \sum_{j=0}^{\infty} B^*Z^{*j}C^*CZ^jB$$

$$F_n = \sum_{j=0}^{\infty} \Theta_j^*\Theta_{n+j} = D^*CZ^{n-1}B + \sum_{i=0}^{\infty} B^*Z^{*i}C^*CZ^iZ^nB \quad (n \geq 1) .$$

Recall that $Q = \Sigma_0^\infty Z^{*j}C^*CZ^j$. So using this in the previous equation we arrive at the following equations

$$F_0 = D^*D + B^*QB \quad \text{and} \quad F_n = (D^*C + B^*QZ)Z^{n-1}B \qquad (\text{for } n \geq 1) . \tag{7.9}$$

By definition $\Theta$ is inner if and only if $\Theta(e^{it})^*\Theta(e^{it}) = I$ a.e., or equivalently $F_0 = I$ and $F_n = 0$ for all $n \neq 0$. Equation (7.9) shows that $F_0 = I$ if and only if the first equation in (7.7a) holds. By employing the controllability of {Z, B} equation (7.9) shows that $F_n = 0$ for all $n \neq 0$ if and only if $D^*C + B^*QZ = 0$, or equivalently, the second equation in (7.7a) holds. So the above analysis shows that $\Theta$ is an inner function if and only if (7.7a) holds.

Recall that $\Theta$ is co-inner if $\Theta(\bar{\lambda})^*$ is an inner function. Notice that $\{Z^*, C^*, B^*, D^*\}$ is a realization of $\Theta(\bar{\lambda})^*$. So according to our previous analysis $\Theta(\bar{\lambda})^*$ is inner if and only if (7.7a) holds when we replace {Z, B, C, D} by the system $\{Z^*, C^*, B^*, D^*\}$, respectively. In this case the observability grammian for $\{B^*, Z^*\}$ is P which is precisely the controllability grammian for {Z, B}. So by replacing {Z, B, C, D} and Q in (7.7a) with $\{Z^*, C^*, B^*, D^*\}$ and P, respectively we obtain (7.7b). Therefore $\Theta$ is two-sided inner if and only if (7.7a) and (7.7b) hold. This completes the proof.

We say that {Z, B, C, D} is a *unitary system (respectively an isometric system)* if

$$\Omega = \begin{bmatrix} Z & B \\ C & D \end{bmatrix} : \begin{bmatrix} \mathcal{X} \\ \mathcal{U} \end{bmatrix} \to \begin{bmatrix} \mathcal{X} \\ \mathcal{Y} \end{bmatrix} \tag{7.10}$$

is a unitary operator (respectively an isometric operator). Now let $\Gamma = \{Z, B, C, D\}$ be a stable unitary system. Then $\Omega\Omega^* = I$ implies that $I = ZZ^* + BB^*$. Hence $P = I$ is the controllability grammian for the pair {Z, B}. Since $P = I$ is clearly invertible, it follows that the pair {Z, B} is controllable. Likewise, using $\Omega^*\Omega = I$, it follows that $Q = I$ is the observability grammian for {C, Z} and the pair {C, Z} is observable. So if $\Gamma = \{Z, B, C, D\}$ is a stable unitary system, then $\Gamma$ is automatically controllable and observable. Finally, notice that $\Omega^*\Omega = I$ and $\Omega\Omega^* = I$, along with $P = Q = I$ implies that the relations in (7.7) hold. Therefore the transfer function $\Theta$ for a stable unitary system is inner from both sides. Summing up we obtain the following result.

**LEMMA 7.4.** *Let $\Gamma = \{Z, B, C, D\}$ be a stable unitary system. Then $\Gamma$ is controllable, observable and its transfer function $\Theta$ is two-sided inner.*

**PROOF OF THEOREM 7.2.** Recall that the observability grammian Q for the pair {C, Z} satisfies the Lyapunov equation $Q = Z^*QZ + C^*C$. By applying $Q^{-1/2}$ to both sides of these equation we see that $I = Z_0^*Z_0 + C_0^*C_0$ where

$$Z_0 = Q^{1/2}ZQ^{-1/2} \text{ and } C_0 = CQ^{-1/2} . \tag{7.11}$$

In particular, $\Lambda_0 = [Z_0, C_0]^{tr}$ is an isometry from $\mathcal{X}$ into $\mathcal{X} \oplus \mathcal{Y}$. Furthermore, the space $\mathcal{Y}$ and $\mathcal{R} = (\mathcal{X} \oplus \mathcal{Y}) \ominus \text{ran } \Lambda_0$ have the same dimension. This is obvious when $\mathcal{X}$ is finite dimensional. Although this covers most applications, we want to prove the proceeding result in the general case. To this end, notice that $\phi D_{Z_0} = C_0$ defines an isometry from $\mathcal{D}_{Z_0}$ into $\mathcal{Y}$ such that $\phi\mathcal{D}_{Z_0} = \overline{\text{ran } C_0}$. Here $D_{Z_0} = D_{Z_{0,1}}$ the positive square root of $I - Z_0^*Z_0$ and $\mathcal{D}_{Z_0}$ is the closure of the range of $D_{Z_0}$ while $D_{Z_0^*} = D_{Z_{0,1}^*}$ and $\mathcal{D}_{Z_0^*}$ is the closure of the range of $D_{Z_0}$. Clearly

$$\mathcal{R} = \{x \oplus y : Z_0^*x + D_{Z_0}\phi^*y = 0\} = \{x \oplus \phi f : Z_0^*x + D_{Z_0}f = 0\} \oplus \{\{0\} \oplus \mathcal{Y} \ominus \text{ran } C_0\} .$$

Recall that (see Foias-Frazho [4], Section IV.1)

$$\begin{bmatrix} Z_0 & D_{Z_0^*} \\ D_{Z_0} & -Z_0^* \end{bmatrix} : \begin{bmatrix} \mathcal{X} \\ \mathcal{D}_{Z_0^*} \end{bmatrix} \to \begin{bmatrix} \mathcal{X} \\ \mathcal{D}_{Z_0} \end{bmatrix}$$

is a unitary operator. Thus

$$\{x \oplus \phi f : Z_0^* x + D_{Z_0} f = 0\} = \{D_{Z_0^*} g \oplus -\phi Z_0^* g : g \in \mathcal{D}_{Z_0^*}\},$$

and the dimension of the last space is dim $\mathcal{D}_{Z_0^*}$. According to Lemma 7.12 presented at the end of this section $\mathcal{D}_{Z_0}$ and $\mathcal{D}_{Z_0^*}$ have the same dimension. Therefore

$$\dim \mathcal{R} = \dim \mathcal{D}_{Z_0^*} + \dim (\mathcal{Y} \ominus \operatorname{ran} C_0) = \dim \mathcal{D}_{Z_0} + \dim (\mathcal{Y} \ominus \operatorname{ran} C_0) = \dim \mathcal{Y},$$

which verifies our claim.

Due to the above discussion there exists an isometry $\Lambda$ from $\mathcal{Y}$ onto $\mathcal{R}$. Equation (7.11) shows that $R = [Z^* Q^{1/2}, C^*]$ and $\Lambda_0^*$ have the same kernel. Hence $\Lambda$ is also an isometry from $\mathcal{Y}$ onto ker R. Now let $\Lambda = [B_0, D]^{tr}$ be the matrix representation of $\Lambda$ as a mapping from $\mathcal{Y}$ into $\mathcal{X} \oplus \mathcal{Y}$. By construction

$$[\Lambda_0, \Lambda] = \begin{bmatrix} Z_0 & B_0 \\ C_0 & D \end{bmatrix} : \begin{bmatrix} \mathcal{X} \\ \mathcal{Y} \end{bmatrix} \to \begin{bmatrix} \mathcal{X} \\ \mathcal{Y} \end{bmatrix} \tag{7.12}$$

is a unitary operator. Clearly $Z_0 = Q^{1/2} Z Q^{-1/2}$ is stable. According to Lemma 7.4, $\{Z_0, B_0, C_0, D\}$ is a stable, controllable and observable realization, and its transfer function $\Theta$ is inner from sides. If $B = Q^{-1/2} B_0$, then (7.11) shows that $\{Z_0, B_0, C_0, D\}$ is similar to $\{Z, B, C, D\}$. Therefore $\{Z, B, C, D\}$ is also a controllable and observable realization of the two-sided inner function $\Theta$.

To show that the controllability grammian P for $\{Z, B\}$ equals $Q^{-1}$, first recall that because $\{Z_0, B_0, C_0, D\}$ is a unitary system $I = Z_0 Z_0^* + B_0 B_0^*$. Substituting $Z_0 = Q^{1/2} Z Q^{-1/2}$ and $B_0 = Q^{1/2} B$ into this equation readily shows that $Q^{-1} = Z Q^{-1} Z^* + BB^*$. Hence $P = Q^{-1}$ is the solution to the controllable Lyapunov equations $P = ZPZ^* + BB^*$. Because the solution to this equation is unique $Q^{-1}$ is the controllable grammian for $\{Z, B\}$.

To complete the proof assume that $\{Z, B_1, C, D_1\}$ is a stable, controllable and observable realization of a two-sided inner function $\Theta_1$ in $H^\infty(\mathcal{U}, \mathcal{Y})$. As before, let $\{Z, B, C, D\}$ be the stable, controllable and observable realization for the two-sided inner function $\Theta$ in $H^\infty(\mathcal{Y}, \mathcal{Y})$ constructed by the procedure described above. According to Theorem 7.1, we have $\mathcal{H}(\Theta) = \operatorname{ran} W_o = \mathcal{H}(\Theta_1)$ where $W_o$ from $\mathcal{X}$ into $H^2(\mathcal{Y})$ is the observability operator defined by $W_o x = C(I - \lambda Z)^{-1} x$. By Proposition 7.9 below (the proof of which is independent of Theorem 7.1), there exists a unitary constant $\Phi$ from $\mathcal{Y}$ onto $\mathcal{U}$ satisfying $\Theta(\lambda) = \Theta_1(\lambda)\Phi$. In particular,

$$D+\lambda C(I-\lambda Z)^{-1}B=\Theta(\lambda)=\Theta_1(\lambda)\Phi=D_1\Phi+\lambda C(I-\lambda Z)^{-1}B_1\Phi\,. \tag{7.13}$$

Clearly $D=\Theta(0)=D_1\Phi$. Equation (7.13) also shows that $W_oB=W_oB_1\Phi$. Since the pair $\{C, Z\}$ is observable, $W_o$ is one to one and thus $B=B_1\Phi$. This completes the proof.

The proof of the previous theorem, sets the stage for the following result.

**PROPOSITION 7.5.** *For* $i=1, 2$, *let* $\{Z_i, C_i, B_i, D_i\}$ *be two stable, controllable and observable realizations for the two-sided inner functions* $\Theta_i$ *in* $H^\infty(\mathcal{U}_i, \mathcal{Y})$. *Then* $\mathcal{H}(\Theta_2)=\mathcal{H}(\Theta_1)$ *if and only if there exists a unitary* $\Phi$ *mapping* $\mathcal{U}_2$ *onto* $\mathcal{U}_1$ *such that* $\{Z_2, B_2, C_2, D_2\}$ *is similar to* $\{Z_1, B_1\Phi, C_1, D_1\Phi\}$.

**PROOF.** If $\mathcal{H}(\Theta_2)=\mathcal{H}(\Theta_1)$, then by once again consulting Proposition 7.9 there exists a constant unitary operator $\Phi$ from $\mathcal{U}_2$ onto $\mathcal{U}_1$ satisfying $\Theta_2(\lambda)=\Theta_1(\lambda)\Phi$. Hence $\Gamma_1=\{Z_1, B_1\Phi, C_1, D_1\Phi\}$ and $\Gamma_2=\{Z_2, B_2, C_2, D_2\}$ are two stable, controllable and observable realizations of $\Theta_2(\lambda)$. By virtue of Theorem 2.3, the realization $\Gamma_1$ is similar to $\Gamma_2$. This proves the necessary part of the proposition, and the sufficient part is trivial. This completes the proof.

The following result provides some additional insight into unitary systems, which will be studied in Section 10.

**THEOREM 7.6.** $\Gamma=\{Z, B, C, D\}$ *is a stable, controllable and observable realization for a two-sided inner function* $\Theta$ *if and only if* $\Gamma$ *is similar to a stable, unitary system.*

**PROOF.** If $\Gamma$ is similar to a stable, unitary system, then Lemma 7.4 shows that its transfer function $\Theta$ is two-sided inner. On the other hand, if $\Gamma$ is a stable, controllable and observable realization for a two-sided inner function, then, by Theorem 7.1, ran $W_o=\mathcal{H}(\Theta)$. As before, $W_o$ is the observability operator from $\mathcal{X}$ into $H^2(\mathcal{Y})$ defined by $W_ox=C(I-\lambda Z)^{-1}x$. By consulting the proof of Theorem 7.2, there exists a stable unitary system $\Gamma_u=\{Z_o, B_o, C_o, D_o\}$ whose two-sided inner transfer function $\Theta_1$ also satisfies ran $W_o=\mathcal{H}(\Theta_1)$. According to the previous proposition $\Gamma$ is similar to the stable unitary system $\{Z_o, B_o\Phi, C_o, D_o\Phi\}$ where $\Phi$ is the appropriate unitary operator. This completes the proof.

As an application we now solve a simple algebraic interpolation problem related to the Nevanlinna-Pick problem. To this end, recall that if F is a function in $H^\infty(\mathcal{U}, \mathcal{Y})$, then $T_F$ is the block lower triangular Toeplitz operator from $\ell_+^2(\mathcal{U})$ into $\ell_+^2(\mathcal{Y})$ generated by F. Recall also that $\Theta$ is a two-sided inner function generated by the stable observable pair $\{C, Z\}$, if ran $W_o=\mathcal{H}(\Theta)$; see (7.1).

**THEOREM 7.7.** *Let* $\Theta$ *in* $H^\infty(\mathcal{E}, \mathcal{Y})$ *be a two-sided inner function generated by the stable, observable pair* $\{C, Z$ *on* $\mathcal{X}\}$ *and let* $W = [C^*, Z^*C^*, Z^{*2}C^*, \ldots]$ *be the controllability operator from* $\ell_+^2(\mathcal{Y})$ *into* $\mathcal{X}$ *generated by* $\{Z^*, C^*\}$. *Then* $\ker W = \operatorname{ran} T_\Theta$. *Moreover, if* G *is any function in* $H^\infty(\mathcal{U}, \mathcal{Y})$, *then the following statements are equivalent.*

*(i)* $(C^*G)(Z^*)_{\text{left}} = 0$;

*(ii)* $WT_G = 0$;

*(iii)* $G = \Theta H$ *for some function* H *in* $H^\infty(\mathcal{U}, \mathcal{E})$.

**PROOF.** The equivalence between (i) and (ii) follows from (I.4.4). To complete the proof it remains to show that (ii) is equivalent to (iii). Clearly (ii) is equivalent to $\operatorname{ran} T_G \subseteq \ker W$. Notice that the observability operator $W_o$ equals $\mathcal{F}_+W^*$ where $\mathcal{F}_+$ is the Fourier transform mapping $\ell_+^2(\mathcal{Y})$ onto $H^2(\mathcal{Y})$. Thus $(C^*G)_{\text{left}}(Z^*) = 0$ if and only if $GH^2(\mathcal{U}) \subseteq \mathcal{F}_+ \ker W = \operatorname{ran}(W_o)^\perp$. Because $\Theta$ is generated by $\{C, Z\}$ we have $\operatorname{ran} W_o = \mathcal{H}(\Theta)$ and thus $\operatorname{ran}(W_o)^\perp = \Theta H^2(\mathcal{E})$. Therefore $(C^*G)_{\text{left}}(Z^*) = 0$ if and only if $GH^2(\mathcal{U}) \subseteq \Theta H^2(\mathcal{E})$. Proposition 7.8 below shows that $G = \Theta H$ where H is some function in $H^\infty(\mathcal{U}, \mathcal{E})$. Thus (ii) implies (iii). Finally, because $\operatorname{ran}(W_o)^\perp = \Theta H^2(\mathcal{E})$ and $W_o = \mathcal{F}_+W^*$, it follows that $\ker(W) = \operatorname{ran} T_\Theta$. So if $G = \Theta H$, then $WT_G = WT_\Theta T_H = 0$. Hence (iii) implies (ii). This completes the proof.

The next two useful propositions concerning general inner function have an intrinsic interest and are independent of the preceding results; see Chapter V in Sz.-Nagy-Foias [3] or Chapter IX in Foias-Frazho [4]. For completeness self contained proofs will be given.

**PROPOSITION 7.8.** *Let* G *be a function in* $H^\infty(\mathcal{U}, \mathcal{Y})$ *and* $\Theta$ *an inner function in* $H^\infty(\mathcal{E}, \mathcal{Y})$. *Then* $GH^2(\mathcal{U}) \subseteq \Theta H^2(\mathcal{E})$ *if and only if* G *admits a factorization of the form* $G = \Theta H$ *where* H *is a function in* $H^\infty(\mathcal{U}, \mathcal{E})$.

**PROOF.** If $G = \Theta H$ for some H in $H^\infty(\mathcal{U}, \mathcal{E})$, then

$$GH^2(\mathcal{U}) \subseteq \overline{GH^2(\mathcal{U})} = \overline{\Theta H H^2(\mathcal{U})} \subseteq \Theta H^2(\mathcal{E}) .$$

Now assume that $GH^2(\mathcal{U}) \subseteq \Theta H^2(\mathcal{E})$. If g is any function in $H^2(\mathcal{U})$, then there exists an $f_g$ in $H^2(\mathcal{E})$ satisfying $Gg = \Theta f_g$. Moreover, the mapping from g to $f_g$ is linear in g. Furthermore,

$$\|f_g\| = \|\Theta f_g\| = \|Gg\| \leq \|G\|_\infty \|g\| .$$

Therefore the mapping from g to $f_g$ defines a bounded linear operator T from $H^2(\mathcal{U})$ into $H^2(\mathcal{E})$ by $Tg = f_g$. In particular, $M_G^+ = M_\Theta^+ T$. Now let $S_1$, $S_2$ and S be the unilateral shifts on $H^2(\mathcal{U})$, $H^2(\mathcal{E})$ and $H^2(\mathcal{Y})$, respectively. Then for g in $H^2(\mathcal{U})$ we have

$$M_\Theta^+ TS_1 g = M_G^+ S_1 g = SM_G^+ g = SM_\Theta^+ Tg = M_\Theta^+ S_2 Tg\,.$$

Because $\Theta$ is an inner function, $M_\Theta^+$ is an isometry and the previous equation implies that $S_2T = TS_1$. So there exists an H in $H^\infty(\mathcal{U}, \mathcal{E})$ satisfying $T = M_H^+$. Hence $M_G^+ = M_\Theta^+ M_H^+$, or equivalently, $G(\lambda) = \Theta(\lambda)H(\lambda)$. This completes this proof.

**PROPOSITION 7.9.** *Let $\Theta_1$ in $H^\infty(\mathcal{U}, \mathcal{Y})$ and $\Theta_2$ in $H^\infty(\mathcal{E}, \mathcal{Y})$ be two inner functions. Then $\mathcal{H}(\Theta_2) = \mathcal{H}(\Theta_1)$ if and only if there exists a constant (in $\lambda$) unitary operator $\Phi$ from $\mathcal{E}$ onto $\mathcal{U}$ satisfying $\Theta_2(\lambda) = \Theta_1(\lambda)\Phi$.*

**PROOF.** If $\mathcal{H}(\Theta_2) = \mathcal{H}(\Theta_1)$, then $\Theta_2 H^2(\mathcal{E}) = \Theta_1 H^2(\mathcal{U})$. By the previous proposition, there exists a function $\Phi$ in $H^\infty(\mathcal{E}, \mathcal{U})$ satisfying $\Theta_2(\lambda) = \Theta_1(\lambda)\Phi(\lambda)$. To complete the proof it remains to show that $\Phi$ is a unitary constant (in $\lambda$). Since $\Theta_2$ and $\Theta_1$ are both inner, it follows that $\Phi^*\Phi = \Phi^*\Theta_1^*\Theta_1\Phi = \Theta_2^*\Theta_2 = I$ a.e. on the unit circle. Hence $\Phi$ is is also an inner function. Because $\Theta_1\Phi H^2(\mathcal{E}) = \Theta_2 H^2(\mathcal{E}) = \Theta_1 H^2(\mathcal{U})$ and $\Theta_1$ is inner it follows that $\Phi H^2(\mathcal{E}) = H^2(\mathcal{U})$. In particular, $M_\Phi^+$ is a unitary operator from $H^2(\mathcal{E})$ onto $H^2(\mathcal{U})$. From this it readily follows that $\Phi$ is a unitary constant. To see this simply notice that the Toeplitz operator $T_\Phi$ is a unitary operator from $\ell_+^2(\mathcal{E})$ onto $\ell_+^2(\mathcal{U})$ satisfying $S_1T_\Phi = T_\Phi S_2$ where $S_1$ and $S_2$ are the unilateral shifts on $\ell_+^2(\mathcal{U})$ and $\ell_+^2(\mathcal{E})$, respectively. Since $T_\Phi$ is unitary $S_2T_\Phi^* = T_\Phi^* S_1$ and thus $T_\Phi^*$ is also a lower triangular unitary Toeplitz matrix. So both $T_\Phi$ and $T_\Phi^*$ are lower triangular Toeplitz unitary operators. Therefore $T_\Phi$ is a block diagonal operator with $\Phi(0)$ in the diagonal entries. Hence $\Phi = \Phi(0)$ is a constant unitary operator. On the other hand, if $\Theta_2 = \Theta_1\Phi$ where $\Phi$ is a constant unitary operator, then clearly $\mathcal{H}(\Theta_2) = \mathcal{H}(\Theta_1)$. This completes the proof.

Let us complete this section by obtaining some further results of independent interest concerning isometric systems, unitary systems and the restricted backward shift realization. To this end recall that {Z, B, C, D} is an *isometric system* if the $2 \times 2$ block matrix in (7.10) is an isometry. If $\Gamma = \{Z, B, C, D\}$ is a stable isometric system, then $I = Z^*Z + C^*C$. In particular, the observability grammian Q for $\Gamma$ is $Q = I$. Furthermore (7.7a) holds. By consulting Lemma 7.3, the transfer function $\Theta$ for a stable, isometric controllable system is inner. This proves part of the following result, which is a generalization of Theorem 7.6.

**THEOREM 7.10.** *The system $\Gamma = \{Z, B, C, D\}$ is a stable, controllable and observable realization for an inner function $\Theta$ if and only if $\Gamma$ is similar to a stable controllable and observable isometric system. Moreover, all pointwise stable isometric realizations $\{Z \text{ on } \mathcal{X}, B, C, D\}$ of $\Theta$ in $H^\infty(\mathcal{U}, \mathcal{Y})$ satisfying $\mathcal{X} = \bigvee\{Z^n B\mathcal{U} : n \geq 0\}$ are observable and unitarily equivalent. In particular they are unitarily equivalent to the restricted backward shift realization of $\Theta$.*

**PROOF**. Obviously if $\Gamma$ is similar to a stable, controllable and observable isometric system, then its transfer function $\Theta$ is inner. On the other hand, assume that $\Gamma = \{Z, B, C, D\}$ is a stable, controllable and observable realization for an inner function $\Theta$. Now let us show that the restricted backward shift realization $\Gamma_o = \{Z_o \text{ on } \mathcal{H}, B_o, C_o, D_0\}$ for any inner function $\Theta$ in $H^\infty(\mathcal{U}, \mathcal{Y})$ is an isometric system. To this end, recall that by (4.12) and (4.13) the operators $Z_o, B_o, C_o$ and $D_0$ in the restricted backward shift realization of $\Theta$ are given by

$$\Omega_o = \begin{bmatrix} Z_o & B_o \\ C_o & D_0 \end{bmatrix} = \begin{bmatrix} S^* | \mathcal{H} & S^* T_\Theta \Pi_{\mathcal{U}}^* \\ \Pi_{\mathcal{Y}} | \mathcal{H} & \Theta(0) \end{bmatrix} : \begin{bmatrix} \mathcal{H} \\ \mathcal{U} \end{bmatrix} \to \begin{bmatrix} \mathcal{H} \\ \mathcal{Y} \end{bmatrix} . \tag{7.14}$$

Here $S$ is the unilateral shift on $\ell_+^2(\mathcal{Y})$ and $\mathcal{H}$ is the subspace of $\ell_+^2(\mathcal{Y})$ given by $\mathcal{H} = \bigvee \{S^{*n} T_\Theta \Pi_{\mathcal{U}}^* \mathcal{U} : n \geq 1\}$. By construction $\Gamma_o$ is a pointwise stable realization for $\Theta$. We claim that $\Omega_o$ is an isometry. Clearly the columns $\Phi_1 = [Z_o, C_o]^{tr}$ and $\Phi_2 = [B_o, D_0]^{tr}$ are both isometries. To show that $\Omega_o$ is an isometry, it remains to show that the range of $\Phi_1$ is orthogonal to the range of $\Phi_2$. To this end, let f and g be vectors in $\mathcal{U}$. Then using $I - SS^* = \Pi_{\mathcal{Y}}^* \Pi_{\mathcal{Y}}$ and taking $n \geq 1$, we have

$$(\Phi_1 S^{*n} T_\Theta \Pi_{\mathcal{U}}^* f, \Phi_2 g) = (SS^* S^{*n} T_\Theta \Pi_{\mathcal{U}}^* f, T_\Theta \Pi_{\mathcal{U}}^* g) + (\Pi_{\mathcal{Y}} S^{*n} T_\Theta \Pi_{\mathcal{U}}^* f, \Pi_{\mathcal{Y}} T_\Theta \Pi_{\mathcal{U}}^* g) =$$

$$(S^{*n} T_\Theta \Pi_{\mathcal{U}}^* f, T_\Theta \Pi_{\mathcal{U}}^* g) = (T_\Theta \Pi_{\mathcal{U}}^* f, T_\Theta S_{\mathcal{U}}^n \Pi_{\mathcal{U}}^* g) = 0 .$$

(Here $S_{\mathcal{U}}$ is the unilateral shift on $\ell_+^2(\mathcal{U})$.) The last equality follows from the fact that $T_\Theta$ is an isometry. Since vectors of the form $S^{*n} T_\Theta \Pi_{\mathcal{U}}^* f$ for $n \geq 1$ and f in $\mathcal{U}$ span $\mathcal{H}$, it follows that $\Phi_1$ and $\Phi_2$ have orthogonal ranges. Hence $\Omega_o$ is an isometry.

If $\Gamma = \{Z, B, C, D\}$ is any stable controllable and observable realization of $\Theta$, then its observability operator $W_o = [C, CZ, CZ^2, ...]^{tr}$ from $\mathcal{X}$ into $\ell_+^2(\mathcal{Y})$ is one to one and onto $\mathcal{H}$. In particular, the operator X defined by $X = W_o$ is a similarity transformation from $\mathcal{X}$ onto $\mathcal{H}$. Equation (4.10) shows that $\Gamma$ is similar to the restricted backward shift realization $\Gamma_o$. In particular, because Z is stable, $Z_o$ is stable. Therefore $\Gamma$ is similar to a stable controllable and observable isometric system namely, the restricted backward shift realization.

Let $\Gamma$ be a pointwise stable isometric system. Then $I = Z^*Z + C^*C$ and therefore

$$I = C^*C + Z^*(C^*C + Z^*Z)Z = C^*C + Z^*C^*CZ + Z^{*2}(C^*C + Z^*Z)Z^2 = \cdots$$
$$= \sum_{m=0}^{n-1} Z^{*m}C^*CZ^m + Z^{*n}Z^n . \tag{7.15}$$

Because Z is pointwise stable, $Z^{*n}Z^n$ approaches zero strongly as n approaches infinity. Hence $I = \Sigma_0^\infty Z^{*n}C^*CZ^n = W_o^*W_o$. So if $\Gamma$ is a pointwise stable isometric system, its observability operator $W_o$ is an isometry. Because $\mathcal{X} = \bigvee\{Z^n B\mathcal{U} : n \geq 0\}$ equation (4.9) shows that $W_o$ is an isometry from $\mathcal{X}$ onto $\mathcal{H}$. Hence X from $\mathcal{X}$ onto $\mathcal{H}$ defined by $X = W_o$ is unitary. Equation (4.10) shows that $\Gamma$ is unitary equivalent to the restricted backward shift realization of $\Theta$. This completes the proof.

Equation (7.15) shows that if $\{C, Z\}$ is a pointwise stable pair and $[Z, C]^{tr}$ is an isometry, then $\{C, Z\}$ is observable. Moreover in this case, its observability grammian $Q = I$.

**COROLLARY 7.11.** *The restricted backward shift realization for a two-sided inner function $\Theta$ is a pointwise stable and $*$-stable controllable and observable unitary system. In particular, all pointwise stable and $*$-stable unitary realizations of the same transfer function are controllable, observable and unitarily equivalent.*

**PROOF.** Let $\Gamma_o = \{Z_o \text{ on } \mathcal{H}, B_o, C_o, D_0\}$ the restricted backward shift realization of $\Theta$. Let V be the bilateral shift on $\ell^2(\mathcal{Y})$ and $V_{\mathcal{U}}$ be the bilateral shift on $\ell^2(\mathcal{U})$. Let $P_+$ be the orthogonal projection onto $\ell^2_+(\mathcal{Y})$ and $L_\Theta$ the Laurent operator generated by $\Theta$. Then using the fact that $L_\Theta$ is unitary, the state space $\mathcal{H}$ for the restricted backward shift realization is given by

$$\mathcal{H} = \bigvee_0^\infty S^{*(n+1)} T_\Theta \Pi^*_{\mathcal{U}} \mathcal{U} = \bigvee_0^\infty P_+ V^{*(n+1)} L_\Theta \Pi^*_{\mathcal{U}} \mathcal{U} = \bigvee_1^\infty P_+ L_\Theta V^{*n}_{\mathcal{U}} \Pi^*_{\mathcal{U}} \mathcal{U} = \{P_+ L_\Theta \ell^2_-(\mathcal{U})\}^- =$$

$$\{P_+ L_\Theta(\ell^2(\mathcal{U}) \ominus \ell^2_+(\mathcal{U}))\}^- = \{P_+(\ell^2(\mathcal{Y}) \ominus L_\Theta \ell^2_+(\mathcal{U}))\}^- = \ell^2_+(\mathcal{Y}) \ominus T_\Theta \ell^2_+(\mathcal{U}). \tag{7.16}$$

Hence $P_{\mathcal{H}} = I - T_\Theta T^*_\Theta$. Let $\Omega_o$ be the isometry defined by (7.14). Using $P_{\mathcal{H}} = I - T_\Theta T^*_\Theta$ and $ST_\Theta T^*_\Theta S^* = T_\Theta(I - \Pi^*_{\mathcal{U}} \Pi_{\mathcal{U}})T^*_\Theta$ along with $T^*_\Theta \Pi^*_{\mathcal{Y}} = \Pi^*_{\mathcal{U}} \Theta(0)^*$ we obtain $\Omega_o \Omega^*_o = I$. Hence $\Gamma_o$ is a unitary system.

Obviously $\Gamma_o$ is a pointwise stable, observable system. To show that $\Gamma_o$ is pointwise $*$-stable notice that because $L_\Theta$ is unitary $L_\Theta \ell^2(\mathcal{U}) = \ell^2_+(\mathcal{Y}) \oplus \ell^2_-(\mathcal{Y})$. By (7.16) we have

$$\mathcal{M} := L_\Theta \ell^2_-(\mathcal{U}) = L_\Theta(\ell^2(\mathcal{U}) \ominus \ell^2_+(\mathcal{U})) = \ell^2_-(\mathcal{Y}) \oplus \mathcal{H}. \tag{7.17}$$

Clearly $\mathcal{M}$ is an invariant subspace for $V^*$. So let $S_o$ be the isometry on $\mathcal{M}$ defined by $S_o = V^* | \mathcal{M}$. We claim that $S_o$ is a unilateral shift. The wandering subspace $\mathcal{L}$ for $S_o$ is $\mathcal{L} = \mathcal{M} \ominus S_o \mathcal{M} = L_\Theta \Pi^*_{-1} \mathcal{U}$ where $\Pi_{-1}$ is the operator from $\ell^2(\mathcal{U})$ onto $\mathcal{U}$ defined by $\Pi_{-1}(f_n)^\infty_{-\infty} = f_{-1}$. This readily implies that $\bigoplus_0^\infty S^n_o \mathcal{L} = \mathcal{M}$. Hence $S_o$ is a unilateral shift. Since $\ell^2_-(\mathcal{Y})$ is invariant for $S_o$ and $\mathcal{M} = \ell^2_-(\mathcal{Y}) \oplus \mathcal{H}$ it follows that $\mathcal{H}$ is invariant for $S^*_o$. Moreover, $S^*_o | \mathcal{H} = P_{\mathcal{H}} V | \mathcal{H} = P_{\mathcal{H}} S | \mathcal{H} = Z^*_o$, and thus $Z^*_o$ is pointwise stable. In particular, $\{Z^*_o, C^*_o, B^*_o, D^*_o\}$ is a pointwise stable, controllable unitary system. According to the previous theorem $\{B_o^*, Z_o^*\}$ is observable, or equivalently, $\{Z_o, B_o\}$ is controllable. The previous theorem also shows that all pointwise stable and $*$-stable unitary realizations of the same transfer function are unitarily equivalent. This completes the proof.

To conclude this section we present the operator theoretic result used in the proof of Theorem 7.2. Here we assume that the reader is familiar with elementary spectral theory in Hilbert space.

**LEMMA 7.12.** *Let* $Z_o$ *on* $\mathcal{X}$ *be a stable contraction. Then* $\dim \mathcal{D}_{Z_o} = \dim \mathcal{D}_{Z_o^*}$ *where* $\mathcal{D}_{Z_o}$ *and* $\mathcal{D}_{Z_o^*}$ *are respectively the closures of the ranges of* $D_{Z_o} = (I - Z_o^* Z_o)^{1/2}$ *and* $D_{Z_o^*} = (I - Z_o Z_o^*)^{1/2}$.

**PROOF.** If $\dim \mathcal{D}_{Z_o} \neq \dim \mathcal{D}_{Z_o^*}$, then without loss of generality we can assume that $\dim \mathcal{D}_{Z_o} < \dim \mathcal{D}_{Z_o^*}$. Therefore $\dim \mathcal{D}_{Z_o}$ is finite dimensional and there exists an isometric non-unitary operator $U_o$ from $\mathcal{D}_{Z_o}$ into $\mathcal{D}_{Z_o^*}$. Lemma V.2.1 in Foias-Frazho [4] shows that $U_1 = Z_o | \mathcal{D}_{Z_o}^{\perp}$ is a unitary operator from $\mathcal{D}_{Z_o}^{\perp}$ onto $\mathcal{D}_{Z_o^*}^{\perp}$. Hence $U = U_o \oplus U_1$ is an isometric non-unitary operator on $\mathcal{X}$. It follows that the unit circle is contained in the essential spectrum of U. Clearly $Z_o - U$ is a finite rank operator, and thus the essential spectrum of $Z_o$ and U coincide; see page 191 in Gohberg-Goldberg-Kaashoek [1]. In particular, $r_{spec}(Z_o) = 1$. This contradicts the fact that $Z_o$ is stable, and completes the proof.

## III.8 EXPLICIT CONVERSION FORMULAS

Theorem I.8.4 shows that the standard Nevanlinna-Pick interpolation problem is equivalent to the Sarason interpolation problem where $\Theta$ is a two-sided inner function. Here we will present explicit state space formulas to convert the standard Nevanlinna-Pick problem to the Sarason problem and visa-versa. To this end, recall that the data for the Sarason problem is a specified function F in $H^\infty(\mathcal{U}, \mathcal{Y})$ and an inner function $\Theta$ in $H^\infty(\mathcal{E}, \mathcal{Y})$. We say that $G = F - \Theta H$ is an interpolant for *the Sarason problem with data* F, $\Theta$ *and tolerance* $\gamma$, if $G = F - \Theta H$ for some H in $H^\infty(\mathcal{U}, \mathcal{E})$ and $\|F - \Theta H\|_\infty \leq \gamma$. This sets the stage for the following result whose proof will be given after Theorem 8.2.

**THEOREM 8.1.** *Let* $\{Z_1 \text{ on } \mathcal{X}_1, B_1, C_1, D_1\}$ *and* $\{Z \text{ on } \mathcal{X}, B, C, D\}$ *be two stable controllable and observable realizations for a function* F *in* $H^\infty(\mathcal{U}, \mathcal{Y})$ *and a two-sided inner function* $\Theta$ *in* $H^\infty(\mathcal{E}, \mathcal{Y})$, *respectively. Let* $Q_1$ *be the solution to the Stein equation*

$$Q_1 = Z^* Q_1 Z_1 + C^* C_1 \tag{8.1}$$

*and set*

$$\tilde{B} = C^* D_1 + Z^* Q_1 B_1 \,. \tag{8.2}$$

*Finally, let* W *from* $\ell_+^2(\mathcal{Y})$ *into* $\mathcal{X}$ *and* $\tilde{W}$ *from* $\ell_+^2(\mathcal{U})$ *into* $\mathcal{X}$ *be the controllability operators defined by*

$$W = [C^*, Z^*C^*, Z^{*2}C^*, \ldots] \quad and \quad \tilde{W} = [\tilde{B}, Z^*\tilde{B}, Z^{*2}\tilde{B}, \ldots] \,. \tag{8.3}$$

*Then* $G = F - \Theta H$ *for some* H *in* $H^\infty(\mathcal{U}, \mathcal{E})$ *if and only if* G *is a function in* $H^\infty(\mathcal{U}, \mathcal{Y})$ *satisfying* $WT_G = \tilde{W}$. *Moreover,* $G = F - \Theta H$ *is an interpolant for the Sarason problem with data* F, $\Theta$ *and tolerance* $\gamma$ *if and only if* G *is a solution to the standard Nevanlinna-Pick problem with data* W, $\tilde{W}$ *and tolerance* $\gamma$.

As before, let {C, Z} be a stable observable pair and $W_o$ the observability operator from $\mathcal{X}$ into $H^2(\mathcal{Y})$ defined by $W_o x = C(I - \lambda Z)^{-1} x$. Recall that that $\Theta$ is a *two-sided inner function generated by the pair* {C, Z} if $\Theta$ is a two-sided inner function in $H^\infty(\mathcal{Y}, \mathcal{Y})$ satisfying

$$W_o \mathcal{X} = \mathcal{H}(\Theta) = H^2(\mathcal{Y}) \ominus \Theta H^2(\mathcal{Y}) \,. \tag{8.4}$$

Theorem 7.2 gives an explicit state space procedure to compute a two-sided inner function $\Theta$ satisfying (8.4). Moreover, if {Z, B, C, D} is any stable, controllable and observable realization of a two-sided inner function, then Theorem 7.1 shows that (8.4) holds. So in this case $\Theta$ is precisely the two-sided inner function generated by {C, Z}. This sets the stage for the following result.

**THEOREM 8.2.** *Let* W *from* $\ell_+^2(\mathcal{Y})$ *into* $\mathcal{X}$ *and* $\tilde{W}$ *from* $\ell_+^2(\mathcal{U})$ *into* $\mathcal{X}$ *be the operators in (8.3), where the pair* {C, Z} *is stable and observable. Let* F *be the function in* $H^\infty(\mathcal{U}, \mathcal{Y})$ *defined by*

$$F = C(I - \lambda Z)^{-1} Q^{-1} \tilde{B} \tag{8.5}$$

*where* Q *is the observability grammian for the pair* {C, Z} *and let* $\Theta$ *be the two-sided inner function in* $H^\infty(\mathcal{Y}, \mathcal{Y})$ *generated by* {C, Z}. *Then,* $G = F - \Theta H$ *is an interpolant for the Sarason problem with data,* F, $\Theta$ *and tolerance* $\gamma$ *if and only if* G *is a solution to the standard Nevanlinna-Pick interpolation problem with data* W, $\tilde{W}$ *and tolerance* $\gamma$.

**PROOF OF THEOREM 8.1.** According to Theorem 7.7 we have ran $T_\Theta = \ker W$. So if $G = F - \Theta H$ for some H in $H^\infty(\mathcal{U}, \mathcal{E})$, then $WT_G = \tilde{W}_1$ where $\tilde{W}_1$ is the operator from $\ell_+^2(\mathcal{U})$ into $\mathcal{X}$ defined by $\tilde{W}_1 = WT_F$. Thus G is a solution to the standard Nevanlinna-Pick problem with data W, $\tilde{W}_1$. On the other hand if G is a function in $H^\infty(\mathcal{U}, \mathcal{Y})$ satisfying $WT_G = \tilde{W}_1$, then $W(T_F - T_G) = 0$. By applying Theorem 7.7 once again $F - G = \Theta H$ for some H in $H^\infty(\mathcal{U}, \mathcal{E})$. Hence $G = F - \Theta H$. Therefore G is a function in $H^\infty(\mathcal{U}, \mathcal{Y})$ satisfying $WT_G = \tilde{W}_1$ if and only if $G = G - \Theta H$ for some H in $H^\infty(\mathcal{U}, \mathcal{E})$. In particular, G is a solution to the standard Nevanlinna-Pick problem with tolerance $\gamma$ if and only if $G = F - \Theta H$ and $\|F - \Theta H\|_\infty \le \gamma$.

To complete the proof it remains to show that $\tilde{W}_1 = \tilde{W}$ where $\tilde{W}$ is given by (8.3) and $\tilde{B}$ is computed by (8.2). Since $\{Z_1, B_1, C_1, D_1\}$ is a realization of F, it follows that the Fourier coefficients $\{F_n\}$ of F are given by $F_0 = D_1$ and $F_n = C_1 Z_1^{n-1} B_1$ for all $n \ge 1$. Moreover, the solution to the Stein equation $Q_1$ in (8.1) is given by (see Lemma 2.4)

$$Q_1 = \sum_{n=0}^{\infty} Z^{*n} C^* C_1 Z_1^n .$$

This readily shows that

$$\tilde{W}_1 = WT_F = [C^*, Z^*C^*, Z^{*2}C^*, \dots]\begin{bmatrix} D_1 & 0 & 0 & \cdots \\ C_1B_1 & D_1 & 0 & \cdots \\ C_1Z_1B_1 & C_1B_1 & D_1 & \cdots \\ \cdot & \cdot & \cdot & \cdots \\ \cdot & \cdot & \cdot & \cdots \\ \cdot & \cdot & \cdot & \cdots \end{bmatrix} = [\tilde{B}, Z^*\tilde{B}, Z^{*2}\tilde{B}, \dots]$$

where $\tilde{B}$ is defined in (8.2). This completes the proof.

**PROOF OF THEOREM 8.2.** Now let W and $\tilde{W}$ be the operators in (8.3). Let F be the function defined in (8.5). Then by direct calculation or by consulting Section I.8 (with Z and B replaced by $Z^*$ and $C^*$, respectively) it follows that $WT_F = \tilde{W}$. Moreover, Theorem 7.7 shows that the kernel of W equals the range of $T_\Theta$. So by consulting the proof of Theorem 8.1, it follows that $G = F - \Theta H$ for some H in $H^\infty(\mathcal{U}, \mathcal{Y})$ if and only if G is a function in $H^\infty(\mathcal{U}, \mathcal{Y})$ satisfying $WT_G = \tilde{W}$. This completes the proof.

## III.9 CONNECTING NUDELMAN AND TWO-SIDED SARASON PROBLEMS

In this section we will give explicit state space formulas to show that the Nudelman interpolation problem is equivalent to a special two-sided Sarason or model matching problem. To this end, recall that the data for the two-sided Nudelman problem are the Hilbert space operators

$$Z : \mathcal{X} \to \mathcal{X}, \; B : \mathcal{U} \to \mathcal{X}, \; \tilde{B} : \mathcal{Y} \to \mathcal{X}$$
$$\Lambda : \mathcal{F} \to \mathcal{F}, \; C : \mathcal{F} \to \mathcal{Y}, \; \tilde{C} : \mathcal{F} \to \mathcal{U}, \; \Gamma : \mathcal{F} \to \mathcal{X} \tag{9.1}$$

where both Z and $\Lambda$ have spectral radius strictly less than one. Furthermore, we always assume that the operator $\Gamma$ satisfies the Sylvester equation

$$Z\Gamma - \Gamma\Lambda = \tilde{B}C - B\tilde{C}. \tag{9.2}$$

Recall that $W_c$ from $\ell_+^2(\mathcal{U})$ into $\mathcal{X}$ and $\tilde{W}_c$ from $\ell_+^2(\mathcal{Y})$ into $\mathcal{X}$ are the controllability operators generated by the pairs $\{Z, B\}$ and $\{Z, \tilde{B}\}$ respectively, that is,

$$W_c = [B, ZB, Z^2B, \ldots] \quad \text{and} \quad \tilde{W}_c = [\tilde{B}, Z\tilde{B}, Z^2\tilde{B}, \ldots]. \tag{9.3}$$

Moreover, $W_o$ from $\mathcal{X}$ into $\ell_-^2(\mathcal{Y})$ and $\tilde{W}_o$ from $\mathcal{X}$ into $\ell_-^2(\mathcal{U})$ are the observability operators generated by $\{C, \Lambda\}$ and $\{\tilde{C}, \Lambda\}$, respectively, that is,

$$W_o = [\ldots, C\Lambda^2, \; C\Lambda, C]^{tr} \quad \text{and} \quad \tilde{W}_o = [\ldots, \tilde{C}\Lambda^2, \tilde{C}\Lambda, \tilde{C}]^{tr} \tag{9.4}$$

where tr denotes the transpose. Here for geometrical reasons we chose the ranges of the observability operators in the appropriate $\ell_-^2$ spaces rather than the usual $\ell_+^2$ spaces. If G is a function in $H^\infty(\mathcal{Y}, \mathcal{U})$, then $T_G^\#$, the Toeplitz operator $T_G$ and the Hankel operator $H_G$ are obtained by partitioning the Laurent operator $L_G$ in the following manner

$$L_G = \begin{bmatrix} T_G^\# & 0 \\ H_G & T_G \end{bmatrix} : \ell_-^2(\mathcal{Y}) \oplus \ell_+^2(\mathcal{Y}) \to \ell_-^2(\mathcal{U}) \oplus \ell_+^2(\mathcal{U}). \tag{9.5}$$

We say that a function G is a *Nudelman interpolant for the data* Z, B, $\tilde{B}$, $\Lambda$, C, $\tilde{C}$ and $\Gamma$ if G is a function in $H^\infty(\mathcal{Y}, \mathcal{U})$ satisfying

$$W_cT_G = \tilde{W}_c, \; T_G^\# W_o = \tilde{W}_o \quad \text{and} \quad \Gamma = W_cH_GW_o. \tag{9.6}$$

The following result shows that if $\{Z, B\}$ is controllable and $\{C, \Lambda\}$ is observable, then there exists a Nudelman interpolant for the data Z, B, $\tilde{B}$, $\Lambda$, C, $\tilde{C}$ and $\Gamma$.

**LEMMA 9.1.** *Let* Z, B, $\tilde{B}$, Λ, C, $\tilde{C}$ *and* Γ *be the data for the Nudelman interpolation problem, where* Γ *satisfies the Sylvester equation (9.2). Assume that the pair* {Z, B} *is controllable and* {C, Λ} *is observable. Finally, let* $\{A_2 \text{ on } \mathcal{F} \oplus \mathcal{X}, B_2, C_2\}$ *be the stable system defined by*

$$A_2 = \begin{bmatrix} \Lambda^* & 0 \\ P^{-1}\Gamma Q^{-1} - P^{-1}(Z\Gamma + B\tilde{C})Q^{-1}\Lambda^*, & Z^* \end{bmatrix}$$

$$B_2 = \begin{bmatrix} C^* \\ P^{-1}\tilde{B} - P^{-1}(Z\Gamma + B\tilde{C})Q^{-1}C^* \end{bmatrix} \quad \text{and} \quad C_2 = [\tilde{C}Q^{-1}, \ B^*] \tag{9.7}$$

*where* $P = W_c W_c^*$ *is the controllability grammian for* {Z, B} *and* $Q = W_o^* W_o$ *is the observability grammian for* {C, Λ}. *Then*

$$F(\lambda) = C_2(I - \lambda A_2)^{-1} B_2 \tag{9.8}$$

*is a Nudelman interpolant, that is, (9.6) holds.*

The proof of this lemma follows by directly verifying that the interpolation conditions in (9.6) hold. In Section V.6 we will show that F in (9.8) solves a $H^2$-optimization problem of Nudelman interpolation type, and there it will also be shown where the formulas for $A_2$, $B_2$, and $C_2$ in (9.7) come from. We say that G is a solution to the Nudelman interpolation problem with data Z, B, $\tilde{B}$, Λ, C, $\tilde{C}$ and Γ with tolerance γ if G is a Nudelman interpolant satisfying $\|G\|_\infty \le \gamma$.

Throughout this section we say that F, $\Theta_1$ and $\Theta_2$ is the data for the two-sided Sarason or model matching problem, if F is a function in $H^\infty(\mathcal{Y}, \mathcal{U})$, the functions $\Theta_1$ and $\Theta_2$ are both two-sided inner functions in $H^\infty(\mathcal{Y}, \mathcal{Y})$ and $H^\infty(\mathcal{U}, \mathcal{U})$, respectively. A function G is an *interpolant* for the model matching problem if $G = F - \Theta_2 H \Theta_1$ for some function H in $H^\infty(\mathcal{Y}, \mathcal{U})$. Obviously this G is in $H^\infty(\mathcal{Y}, \mathcal{U})$. Finally, we say that G is *an interpolant for the model matching problem with tolerance* γ, if G is an interpolant for to the model matching problem satisfying $\|G\|_\infty \le \gamma$.

Recall that Θ is a *two-sided inner function generated by a stable observable pair* {C, Z} if Θ is a two-sided inner function in $H^\infty(\mathcal{E}, \mathcal{E})$ satisfying

$$C(I - \lambda Z)^{-1}\mathcal{X} = H^2(\mathcal{E}) \ominus \Theta H^2(\mathcal{E}). \tag{9.9}$$

Theorem 7.2 gives an explicit state space procedure to compute a two-sided inner function satisfying (9.9). Moreover, if {Z, B, C, D} is any stable, controllable and observable realization of a two-sided inner function Θ, then Theorem 7.1 shows that (9.9) holds. So in this case Θ is precisely the two-sided inner function generated by {C, Z}. A function $\Theta_1$ is called the *two-*

*sided inner function $*$-generated by* $\{C, Z\}$ if $\Theta_1$ is the two-sided inner function in $H^\infty(\mathcal{E}, \mathcal{E})$ defined by $\Theta_1(\lambda) = \Theta(\bar{\lambda})^*$ where $\Theta$ is the two-sided inner function generated by $\{C, Z\}$. Notice that if $\{\Lambda^*, C^*, B, D\}$ is a stable, controllable and observable realization of a two-sided inner function $\Theta_1$, then $\Theta_1$ is the two-sided inner function $*$-generated by $\{C, \Lambda\}$. To see this simply apply Theorem 7.1 to $\Theta = \Theta_1(\bar{\lambda})^*$. Finally, it is noted that Theorem 7.2 can be used to compute an explicit state space formula for the two-sided inner function $*$-generated by $\{C, \Lambda\}$, that is, simply use Theorem 7.2 to compute the realization $\{\Lambda, B, C, D\}$ for the inner function $\Theta$ generated by $\{C, \Lambda\}$. Then $\{\Lambda^*, C^*, B^*, D^*\}$ is the realization for the two-sided inner function $\Theta_1$ which is $*$-generated by $\{C, \Lambda\}$.

**THEOREM 9.2.** *Let* $Z$, $B$, $\tilde{B}$, $\Lambda$, $C$, $\tilde{C}$ *and* $\Gamma$ *be the data for the Nudelman interpolation problem, and assume that* $\{Z, B\}$ *is controllable and* $\{C, \Lambda\}$ *is observable. Let* $\Theta_2$ *in* $H^\infty(\mathcal{U}, \mathcal{U})$ *be the two-sided inner function generated by the pair* $\{B^*, Z^*\}$ *and* $\Theta_1$ *the two-sided inner function $*$-generated by the pair* $\{C, \Lambda\}$. *Finally, let* $F$ *be any Nudelman interpolant of the data. (In fact, one such* $F$ *can be computed by Lemma 9.1.) Then* $G$ *is a function in* $H^\infty(\mathcal{Y}, \mathcal{U})$ *satisfying the Nudelman interpolation conditions (9.6) if and only if* $G$ *is a function of the form* $G = F - \Theta_2 H \Theta_1$ *for some* $H$ *in* $H^\infty(\mathcal{Y}, \mathcal{U})$. *Moreover,* $G$ *is a solution to the Nudelman problem with data (9.1) and tolerance* $\gamma$ *if and only if* $G$ *is an interpolant for the two-sided Sarason problem with data* $F$, $\Theta_1$, $\Theta_2$ *and tolerance* $\gamma$.

The previous result gave explicit state space formulas to convert a Nudelman interpolation problem into a two-sided Sarason interpolation problem. The following result gives state space formulas to convert a two-sided Sarason problem into a Nudelman interpolation problem.

**THEOREM 9.3.** *Let* $\{Z_1, B_1, C_1, D_1\}$ *be a stable realization for a function* $F$ *in* $H^\infty(\mathcal{Y}, \mathcal{U})$. *Let* $\{\Lambda^*, C^*, C_o, D_0\}$ *and* $\{Z^*, B_c, B^*, D_c\}$ *be stable, controllable and observable realizations for the two-sided inner functions* $\Theta_1$ *in* $H^\infty(\mathcal{Y}, \mathcal{Y})$ *and* $\Theta_2$ *in* $H^\infty(\mathcal{U}, \mathcal{U})$ *respectively. Let* $P_1$ *and* $Q_1$ *be the solutions to the following Stein equations:*

$$P_1 = ZP_1Z_1 + BC_1 \quad \text{and} \quad Q_1 = Z_1Q_1\Lambda + B_1C \,. \tag{9.10}$$

*Finally, let* $\tilde{B}$ *from* $\mathcal{Y}$ *to* $\mathcal{X}$ *and* $\tilde{C}$ *from* $\mathcal{F}$ *to* $\mathcal{U}$ *and* $\Gamma$ *from* $\mathcal{F}$ *to* $\mathcal{X}$ *be the operators in the Nudelman problem now defined by*

$$\tilde{B} = BD_1 + ZP_1B_1 \,, \ \tilde{C} = D_1C + C_1Q_1\Lambda \quad \text{and} \quad \Gamma = P_1Q_1 \,. \tag{9.11}$$

*Then* $G = F - \Theta_2 H \Theta_1$ *for some* $H$ *in* $H^\infty(\mathcal{Y}, \mathcal{U})$ *if and only if* $G$ *is a Nudelman interpolant in* $H^\infty(\mathcal{Y}, \mathcal{U})$ *for the data* $Z$, $B$, $\tilde{B}$, $\Lambda$, $C$, $\tilde{C}$ *and* $\Gamma$. *Moreover,* $G$ *is an interpolant for the two-sided*

*Sarason problem with data* F, $\Theta_1$, $\Theta_2$ *and tolerance* $\gamma$ *if and only if* G *is a solution to the Nudelman interpolation problem with data (9.1), (9.11) and tolerance* $\gamma$.

In order to prove Theorems 9.1 and 9.2 we need a factorization lemma. Let R be a function in $H^\infty(\mathcal{Y}, \mathcal{U})$. Then recall that $T_R^\#$ is the operator from $\ell_-^2(\mathcal{Y})$ into $\ell_-^2(\mathcal{U})$ defined by $T_R^\# = P_- L_R \mid \ell_-^2(\mathcal{Y})$ where $P_-$ is the orthogonal projection onto $\ell_-^2(\mathcal{U})$ and $L_R$ is the Laurent operator generated by R. Finally, let $P_+$ be the orthogonal projection onto $\ell_+^2(\mathcal{U})$.

**LEMMA 9.4**. *Let* $W_c$ *from* $\ell_+^2(\mathcal{U})$ *into* $\mathcal{X}$ *be the controllability operator in (9.3) generated by the stable, controllable pair* {Z, B} *and* $W_o$ *from* $\mathcal{F}$ *into* $\ell_-^2(\mathcal{Y})$ *be the observability operator in (9.4) generated by the stable observable pair* $\{C, \Lambda\}$. *Let* $\Theta_2$ *in* $H^\infty(\mathcal{U}, \mathcal{U})$ *and* $\Theta_1$ *in* $H^\infty(\mathcal{Y}, \mathcal{Y})$ *be the two-sided inner functions generated by the pair* $\{B^*, Z^*\}$ *and* $*$*-generated by* $\{C, \Lambda\}$, *respectively. Then* R *is a function in* $H^\infty(\mathcal{Y}, \mathcal{U})$ *satisfying*

$$W_c T_R = 0, \; T_R^\# W_o = 0 \quad \textit{and} \quad W_c P_+ L_R W_o = 0 \tag{9.12}$$

*if and only if* R *admits a factorization of the form* $R = \Theta_2 H \Theta_1$ *for some* H *in* $H^\infty(\mathcal{Y}, \mathcal{U})$.

**PROOF**. To verify our claim first notice that by replacing {C, Z} and $\Theta$ in Theorem 7.7 by $\{B^*, Z^*\}$ and $\Theta_2$ respectively, we obtain the first equality in

$$\ker W_c = \operatorname{ran} T_{\Theta_2} \quad \text{and} \quad \operatorname{ran} W_o = \ker T_{\Theta_1}^\# \,. \tag{9.13}$$

To verify the second equality let $\Theta$ be the two-sided inner function generated by the pair $\{C, \Lambda\}$. Then by applying Theorem 7.7 once again

$$\operatorname{ran}\, [C^*, \Lambda^* C^*, \Lambda^{*2} C^*, \ldots]^* = \ell_+^2(\mathcal{Y}) \ominus T_\Theta \ell_+^2(\mathcal{Y}) \,. \tag{9.14}$$

Now let J be the unitary flip operator on $\ell^2(\mathcal{Y})$ which flips and shifts the components of $\ell^2(\mathcal{Y})$ one place, that is,

$$J[\ldots, x_{-2}, x_{-1}, \underline{x_0}, x_1, x_2, \ldots]^{tr} = [\ldots, x_2, x_1, x_0, \underline{x_{-1}}, x_{-2}, \ldots]^{tr} \tag{9.15}$$

where tr denotes the transpose. Notice that $J T_\Theta J \mid \ell_-^2(\mathcal{Y}) = L_{\Theta_1}^* \mid \ell_-^2(\mathcal{Y})$ where $\Theta_1$ is the two-sided inner function given by $\Theta_1(e^{it}) = \Theta(e^{-it})^*$ a.e. So applying J to both sides of (9.14) shows that ran $W_o$ equals $\ell_-^2(\mathcal{Y}) \ominus L_{\Theta_1}^* \mid \ell_-^2(\mathcal{Y})$. Since $(T_{\Theta_1}^\#)^* = L_{\Theta_1}^* \mid \ell_-^2(\mathcal{Y})$ we now obtain the second equality in (9.13).

Assume that R(z) in $H^\infty(\mathcal{Y}, \mathcal{U})$ admits a factorization of the form $R = \Theta_2 H \Theta_1$ for some H in $H^\infty(\mathcal{Y}, \mathcal{U})$. Then (9.13) shows that $W_c T_R = 0$ and $T_R^\# W_o = T_{\Theta_2}^\# T_H^\# T_{\Theta_1}^\# W_o = 0$. To show that the last equality in (9.12) holds notice that

$$W_c P_+ L_R W_o = W_c P_+ L_{\Theta_2} L_H L_{\Theta_1} W_o = W_c T_{\Theta_2} T_H P_+ L_{\Theta_1} W_o + W_c L_{\Theta_2} L_H T_{\Theta_1}^\# W_o = 0\,.$$

Therefore (9.12) holds.

Now assume that (9.12) holds for some R in $H^\infty(\mathcal{Y}, \mathcal{U})$. Since $W_c T_R = 0$ Theorem 7.7 shows that $R = \Theta_2 H_c$ for some $H_c$ in $H^\infty(\mathcal{Y}, \mathcal{U})$. Using the fact that $T_R^\# W_o = 0$ we have for all g in $\ell_-^2(\mathcal{U})$

$$0 = W_o^* (T_R^\#)^* g = W_o^* J T_{R_o} J_{\mathcal{U}}\, g\,,$$

where $J_{\mathcal{U}}$ is the unitary flip operator on $\ell^2(\mathcal{U})$ defined in (9.15) where $\mathcal{U}$ replaces $\mathcal{Y}$, and $T_{R_o} = J L_R^* J_{\mathcal{U}} \mid \ell_+^2(\mathcal{U})$ while $R_o$ is the function in $H^\infty(\mathcal{U}, \mathcal{Y})$ defined by $R_o = R(\bar{\lambda})^*$ for $|\lambda| < 1$. Hence $W_o^* J T_{R_o} = 0$. Recall that $\Theta_1$ is the two-sided inner function $*$-generated by {C, Λ}. By applying Theorem 7.7 with $Z = \Lambda$ and $W = W_o^* J$ and $\Theta(\lambda) = \Theta_1(\bar{\lambda})^*$ we see that $R_o(\lambda) = \Theta_1(\bar{\lambda})^* H_1(\lambda)$ for some $H_1$ in $H^\infty(\mathcal{U}, \mathcal{Y})$. Hence $R(\lambda) = H_o(\lambda)\Theta_1(\lambda)$ where $H_o(\lambda)$ is the function in $H^\infty(\mathcal{Y}, \mathcal{U})$ given by $H_o(\lambda) = H_1(\bar{\lambda})^*$ for $|\lambda| < 1$.

To complete the proof it remains to show that $R = \Theta_2 H \Theta_1$. Notice that (9.12) and (9.13) implies that $\operatorname{ran} P_+ L_R W_o \subseteq \operatorname{ran} T_{\Theta_2}$. Thus

$$L_R W_o = L_{H_o} P_+ L_{\Theta_1} W_o + L_{H_o} T_{\Theta_1}^\# W_o = T_{H_o} P_+ L_{\Theta_1} W_o\,.$$

Hence $\operatorname{ran} L_R W_o$ is contained in $\ell_+^2(\mathcal{U})$ and therefore $\operatorname{ran} L_R W_o \subseteq \operatorname{ran} T_{\Theta_2}$. Since $R = \Theta_2 H_c$ and $\Theta_2$ is inner, it follows that $\operatorname{ran} L_{H_c} W_o \subseteq \ell_+^2(\mathcal{U})$. So $T_{H_c}^\# W_o = P_- L_{H_c} W_o = 0$. The above discussion on this last equation yields $H_c = H\Theta_1$ for some H in $H^\infty(\mathcal{Y}, \mathcal{U})$. This completes the proof.

**PROOF OF THEOREM 9.2.** Let G in $H^\infty(\mathcal{Y}, \mathcal{U})$ be any interpolant of the Nudelman interpolation problem. Obviously the function F in (9.8) is a Nudelman interpolant. Therefore $R = F - G$ is a function in $H^\infty(\mathcal{Y}, \mathcal{U})$ satisfying (9.12). Lemma 9.4 shows that there exists a function H in $H^\infty(\mathcal{Y}, \mathcal{U})$ satisfying $R = \Theta_2 H \Theta_1$. Hence $G = F - \Theta_2 H \Theta_1$. On the other hand, if $G = F - \Theta_2 H \Theta_1$ for some H in $H^\infty(\mathcal{Y}, \mathcal{U})$, then Lemma 9.4 shows that G is an interpolant to the Nudelman problem. Therefore G is a function in $H^\infty(\mathcal{Y}, \mathcal{U})$ of the from $G = F - \Theta_2 H \Theta_1$ for some H in $H^\infty(\mathcal{Y}, \mathcal{U})$ if and only if G is a Nudelman interpolant in $H^\infty(\mathcal{Y}, \mathcal{U})$. This completes the proof.

**PROOF OF THEOREM 9.3.** According to Theorem 7.1, the functions $\Theta_2$ and $\Theta_1$ can be viewed as the two-sided inner functions generated and $*$-generated by $\{B^*, Z^*\}$ and $\{C, \Lambda\}$, respectively. Now let $\tilde{B}$ from $\mathcal{Y}$ to $\mathcal{X}$ and $\tilde{C}$ from $\mathcal{F}$ to $\mathcal{U}$ and $\Gamma$ from $\mathcal{F}$ to $\mathcal{X}$ be the operators defined by

$$\tilde{B} = W_c[F_0, F_1, F_2, \ldots]^{tr} \quad \text{and} \quad \tilde{C} = [\ldots, F_2, F_1, F_0]W_o \quad \text{and} \quad \Gamma = W_c P_+ L_F W_o \tag{9.16}$$

where $\{F_n\}$ are the Fourier coefficients of F. (Recall that the first two equations in (9.16) hold if and only if $\tilde{W}_c = W_c T_F$ and $\tilde{W}_o = T_F^{\#} W_o$.) Then according to Theorem 9.2 it follows that $G = F - \Theta_2 H \Theta_1$ for some H in $H^\infty(\mathcal{Y}, \mathcal{U})$ if and only if G is a Nudelman interpolant for the data Z, B, $\tilde{B}$, $\Lambda$, C, $\tilde{C}$ and $\Gamma$ given by the hypothesis of Theorem 9.3 and (9.16). So to complete the proof it remains to show that $\tilde{B}$, $\tilde{C}$ and $\Gamma$ are given by the formulas in (9.11).

By recursively solving for $P_1$ and $Q_1$ in the Stein equations (9.10) it follows that $P_1$ and $Q_1$ are uniquely given by (see Lemma 2.4)

$$P_1 = \sum_{n=0}^{\infty} Z^n B C_1 Z_1^n \quad \text{and} \quad Q_1 = \sum_{n=0}^{\infty} Z_1^n B_1 C \Lambda^n . \tag{9.17}$$

Because $\{Z_1, B_1, C_1, D_1\}$ is a realization of F it follows that the Fourier coefficients $\{F_n\}$ of F are given by

$$F_0 = D_1 \quad \text{and} \quad F_n = C_1 Z_1^{n-1} B_1 \qquad (\text{if } n \geq 1) . \tag{9.18}$$

So using the expression for $P_1$ in (9.17) along with (9.18) we arrive at

$$\tilde{B} = W_c \begin{bmatrix} F_0 \\ F_1 \\ F_2 \\ \vdots \end{bmatrix} = [B, ZB, Z^2B, \ldots] \begin{bmatrix} D_1 \\ C_1 B_1 \\ C_1 Z_1 B_1 \\ \vdots \end{bmatrix} = BD_1 + ZP_1B_1 .$$

This gives the formula for $\tilde{B}$ in (9.11). A similar calculation involving the second equation in (9.16) and (9.17) establishes the formula for $\tilde{C}$ in (9.11). To compute $\Gamma$ notice that (9.17) and (9.18) give

$$W_cP_+L_FW_o = [B, ZB, Z^2B, \ldots]\begin{bmatrix} \cdots & C_1Z_1^2B_1 & C_1Z_1B_1 & C_1B_1 \\ \cdots & C_1Z_1^3B_1 & C_1Z_1^2B_1 & C_1Z_1B_1 \\ \cdots & C_1Z_1^4B_1 & C_1Z_1^3B_1 & C_1Z_1^2B_1 \\ \cdot & \cdot & \cdot & \cdot \\ \cdot & \cdot & \cdot & \cdot \\ \cdot & \cdot & \cdot & \cdot \end{bmatrix}\begin{bmatrix} \cdot \\ \cdot \\ \cdot \\ C\Lambda^2 \\ C\Lambda \\ C \end{bmatrix} =$$

$$[B, ZB, Z^2B, \ldots][C_1Q_1, C_1Z_1Q_1, C_1Z_1^2Q_1, \ldots]^{tr} = P_1Q_1 .$$

Using $\Gamma = W_cP_+L_FW_o$ shows that $\Gamma = P_1Q_1$. This completes the proof.

## III.10 ISOMETRIC AND UNITARY SYSTEMS

In this section we will present some useful results concerning isometric and unitary systems. To this end, recall that the transfer function $\Theta$ for the system $\{Z \text{ on } \mathcal{X}, B, C, D\}$ is the analytic function with values in $\mathcal{L}(\mathcal{U}, \mathcal{Y})$ defined by

$$\Theta(\lambda) = D + \lambda C(I - \lambda Z)^{-1}B . \tag{10.1}$$

Moreover, the observability operator $W_o$ for the pair $\{C, Z\}$ is the operator from $\mathcal{X}$ into $\ell_+^2(\mathcal{Y})$ defined by

$$W_o = [C, CZ, CZ^2, \ldots]^{tr} . \tag{10.2}$$

We say that $\{Z, B, C, D\}$ is an *isometric system* if the following $2 \times 2$ operator matrix is an isometry

$$\Omega = \begin{bmatrix} Z & B \\ C & D \end{bmatrix} : \begin{bmatrix} \mathcal{X} \\ \mathcal{U} \end{bmatrix} \to \begin{bmatrix} \mathcal{X} \\ \mathcal{Y} \end{bmatrix} . \tag{10.3}$$

Moreover, $\{Z, B, C, D\}$ is a *unitary system* if the matrix $\Omega$ in (10.3) is a unitary operator. Recall that an operator T on $\mathcal{X}$ is *pointwise stable* if $T^n$ approaches zero strongly as n approaches infinity, and T is *pointwise $*$-stable* if $T^*$ is pointwise stable. Finally, recall also that a function $\Theta$ with values in $\mathcal{L}(\mathcal{U}, \mathcal{Y})$ is a *contractive analytic function* if $\Theta$ is in $H^\infty(\mathcal{U}, \mathcal{Y})$ and $\|\Theta\|_\infty \leq 1$. We are now ready to state the following basic result whose proof will be given after the proof of Proposition 10.3.

**THEOREM 10.1.** *Let* $\Theta(\lambda)$ *be the transfer function for the isometric system* $\{Z, B, C, D\}$. *Let* $W_o$ *from* $\mathcal{X}$ *into* $\ell_+^2(\mathcal{Y})$ *be the observability operator generated by* $\{C, Z\}$, *and let* $T_\Theta$ *be the block Toeplitz operator generated by* $\Theta$. *Then the operator*

$$\Phi := [W_o, T_\Theta] : \mathcal{X} \oplus \ell_+^2(\mathcal{U}) \to \ell_+^2(\mathcal{Y}) \tag{10.4}$$

*is a contraction. In particular, $\Theta$ is a contractive analytic function in $H^\infty(\mathcal{U}, \mathcal{Y})$. Moreover, if $Z$ is pointwise stable, then $\Phi$ is an isometry. In this case $\Theta$ is an inner function, $W_o$ is an isometry and the pair $\{C, Z\}$ is observable.*

Notice that $\Theta$ is the transfer function for $\{Z, B, C, D\}$ if and only if $\tilde{\Theta}(\lambda) := \Theta(\bar{\lambda})^*$ is the transfer function for $\{Z^*, C^*, B^*, D^*\}$. Recall that a function $\Theta$ in $H^\infty(\mathcal{U}, \mathcal{Y})$ is *co-inner* if $\Theta(\bar{\lambda})^*$ is an inner function. Finally, we say that $\{Z, B, C, D\}$ is a *co-isometric system* if the matrix $\Omega$ in (10.3) is a co-isometry. By applying the previous theorem to $\{Z^*, C^*, B^*, D^*\}$ we readily obtain the following result.

**COROLLARY 10.2.** *Let $\Theta(\lambda)$ be the transfer function for the co-isometric system $\{Z, B, C, D\}$. Let $W_c = [B, ZB, Z^2B, \ldots]$ be the controllability operator from $\ell_+^2(\mathcal{U})$ into $\mathcal{X}$ generated by $\{Z, B\}$. Then $\Theta$ is a contractive analytic function in $H^\infty(\mathcal{U}, \mathcal{Y})$ and $W_c$ is a contraction. Moreover, if $Z$ is pointwise $*$-stable, then $\Theta$ is co-inner, $W_c$ is a co-isometry and $\{Z, B\}$ is controllable. Finally, if $\{Z, B, C, D\}$ is a unitary system and $Z$ is both pointwise stable and $*$-stable, then $\Theta$ is a two-sided inner function and $\{Z, B, C, D\}$ is both controllable and observable.*

Notice that we need the unitary system $\{Z, B, C, D\}$ to be pointwise stable to guarantee that $\{C, Z\}$ is observable. For example, if $A = I$, $B = 0$, $C = 0$ and $D = I$, then obviously $\{A, B, C, D\}$ is a unitary system. However, this system is not controllable and not observable.

To obtain a proof of the previous theorem we need some elementary facts concerning unitary extensions of isometries. To this end, recall that an operator $U$ on $\mathcal{K}$ is an *extension* of an operator $T$ on $\mathcal{H}$, if $\mathcal{H}$ is an invariant subspace for $U$ and $U|\mathcal{H} = T$. We say that $U$ on $\mathcal{K}$ is *a minimal unitary extension* for an isometry $T$ on $\mathcal{H}$ if $U$ is a unitary operator extending $T$ and

$$\mathcal{K} = \bigvee_{-\infty}^{\infty} U^n \mathcal{H} = \bigvee_{0}^{\infty} U^{*n} \mathcal{H}. \tag{10.5}$$

The last equality follows from the fact that $\mathcal{H}$ is invariant for $U$. We say that two unitary extensions $U$ on $\mathcal{K}$ and $U_1$ on $\mathcal{K}_1$ of $T$ are *isomorphic*, if there exists a unitary operator $W$ from $\mathcal{K}$ onto $\mathcal{K}_1$ satisfying $W|\mathcal{H} = I$ and $WU = U_1W$. This sets the stage for the following result which will be used in proving Theorem 10.1.

**PROPOSITION 10.3.** *Let $T$ be an isometry on $\mathcal{H}$. Then $T$ admits a minimal unitary extension. Moreover, all minimal unitary extensions of $T$ are isomorphic.*

**PROOF.** According to the Wold decomposition (see the end of Section I.1), T admits a unique reducing decomposition of the form $T = S \oplus U_0$ on $\mathcal{M} \oplus \mathcal{H}_0$ where S is a unilateral shift on $\mathcal{M}$ and $U_0$ is a unitary operator on $\mathcal{H}_0$. So without loss of generality, we can assume that S is the unilateral shift on $\ell_+^2(\mathcal{L})$ where $\mathcal{L}$ is the wandering subspace for T defined by $\mathcal{L} = \mathcal{H} \ominus T\mathcal{H}$. Now let V be the bilateral shift on $\ell^2(\mathcal{L})$. Then obviously V is a minimal unitary extension of S. Therefore $U = V \oplus U_0$ on $\mathcal{K} = \ell^2(\mathcal{L}) \oplus \mathcal{H}_0$ is a minimal unitary extension of T.

To complete the proof it remains to show that all minimal unitary extensions are isomorphic. To this end, assume that $U_1$ on $\mathcal{K}_1$ is another minimal unitary extension of T. Then for all sequences $\{h_n\}$ in $\mathcal{H}$ with finite support we have

$$\|\sum_n U^n h_n\|^2 = \sum_{n,m} (U^n h_n, U^m h_m) = \sum_{n \geq m} (U^{n-m} h_n, h_m) + \sum_{m>n} (h_n, U^{m-n} h_m) =$$

$$\sum_{n \geq m} (T^{n-m} h_n, h_m) + \sum_{m>n} (h_n, T^{m-n} h_m) = \|\sum_n U_1^n h_n\|^2 .$$

This along with the minimality condition in (10.5) readily implies that there exists a unitary operator W from $\mathcal{K}$ onto $\mathcal{K}_1$ satisfying $W\sum U^n h_n = \sum U_1^n h_n$. In particular,

$$WU \sum_n U^n h_n = W \sum_n U^{n+1} h_n = \sum_n U_1^{n+1} h_n = U_1 W \sum_n U^n h_n .$$

Since vectors of the form $\{U^n h_n\}$ span a dense set in $\mathcal{K}$, it follows that $WU = U_1 W$. Obviously $W | \mathcal{H} = I$. So U and $U_1$ are isomorphic. This completes the proof.

**PROOF OF THEOREM 10.1.** Because Z is a contraction, $Z^{*n}Z^n$ forms a nonincreasing sequence of positive operators. Let $\Delta^2$ be the strong limit of $Z^{*n}Z^n$ as n approaches infinity, and let $\Delta$ be the positive square root of $\Delta^2$. Obviously for x in $\mathcal{X}$ we have

$$\|\Delta x\|^2 = (\Delta^2 x, x) = \lim_{n \to \infty} \|Z^n x\|^2 = \lim_{n \to \infty} \|Z^n Z x\|^2 = \|\Delta Z x\|^2 .$$

So there exists an isometry T on the closure of the range of $\Delta$, which is denoted by $\mathcal{H}$, satisfying $T\Delta = \Delta Z$. Now let U on $\mathcal{K}$ be any minimal unitary extension of T. Clearly $U\Delta = \Delta Z$.

Let $W_u$ be the controllability operator from $\ell_+^2(\mathcal{U})$ into $\mathcal{K}$ defined by

$$W_u = [U^* \Delta B, U^{*2} \Delta B, U^{*3} \Delta B, U^{*4} \Delta B, \ldots] . \tag{10.6}$$

Let $\Phi_e$ be the operator defined by

$$\Phi_e = \begin{bmatrix} W_o & T_\Theta \\ \Delta & W_u \end{bmatrix} : \begin{bmatrix} \mathcal{X} \\ \ell_+^2(\mathcal{U}) \end{bmatrix} \to \begin{bmatrix} \ell_+^2(\mathcal{Y}) \\ \mathcal{K} \end{bmatrix} . \tag{10.7}$$

Obviously, the operator $\Phi$ in (10.4) is the first row of $\Phi_e$. Moreover, if Z is pointwise stable, $\Delta = 0$ and in this case $\Phi = \Phi_e$. So to complete the proof it is sufficient to show that $\Phi_e$ is an isometry. To prove this first notice that because the $2 \times 2$ operator matrix $\Omega$ in (10.3) is an isometry, we have $I = C^*C + Z^*Z$. Thus

$$I = C^*C + Z^*IZ = C^*C + Z^*(C^*C + Z^*Z)Z = C^*C + Z^*C^*CZ + Z^{*2}IZ^2 .$$

By continuing to replace the identity operator I by $C^*C + Z^*Z$ we arrive at

$$I = \sum_{j=0}^{n} Z^{*j}C^*CZ^j + Z^{*n+1}Z^{n+1} .$$

By passing strong limits we have

$$I = \sum_{n=0}^{\infty} Z^{*n}C^*CZ^n + \Delta^2 = Q + \Delta^2 \tag{10.8}$$

where $Q = \Sigma_0^\infty Z^{*n}C^*CZ^n$. Equation (10.2) implies that Q also equals $W_o^*W_o$. This shows that the observability operator $W_o$ is a well defined bounded operator from $\mathcal{X}$ into $\ell_+^2(\mathcal{Y})$. In fact, $W_o$ is a contraction and its observability grammian $Q = W_o^*W_o$ also equals $I - \Delta^2$. In particular, using $I = Q + \Delta^2$ it follows that first column $[W_o, \Delta]^{tr}$ of $\Phi_e$ is an isometry.

Notice that $\Theta$ in (10.1) is the Fourier transform of $[D, W_oB]^{tr}$. Using this fact it follows that the second column of $\Phi_e$ admits a decomposition of the form

$$\begin{bmatrix} T_\Theta \\ W_u \end{bmatrix} = [Y_0, Y_1, Y_2, \ldots] : \ell_+^2(\mathcal{U}) \to \ell_+^2(\mathcal{Y}) \oplus \mathcal{K} \tag{10.9}$$

where $Y_j$ for $j \geq 0$ are the operators from $\mathcal{U}$ into $\ell_+^2(\mathcal{Y}) \oplus \mathcal{K}$ given by

$$Y_j = S^j \oplus U^{*j} \begin{bmatrix} [D, W_oB]^{tr} \\ U^*\Delta B \end{bmatrix} \tag{10.10}$$

and S is the unilateral shift on $\ell_+^2(\mathcal{Y})$. To prove that $\Phi_e$ is an isometry it remains to show that $\{Y_j\}_0^\infty$ are all isometries with mutually orthogonal ranges, and ran $Y_j$ is orthogonal to ran $[W_o, \Delta]^{tr}$ for all $j \geq 0$. To this end, first notice that by using $D^*D + B^*B = I$, we have

$$Y_j^*Y_j = D^*D + B^*W_o^*W_oB + B^*\Delta^2B = D^*D + B^*(Q + \Delta^2)B = D^*D + B^*B = I .$$

Hence $Y_j$ is an isometry for all $j \geq 0$.

Now let x be in $\mathcal{X}$ and u in $\mathcal{U}$. Then using $W_o^* W_o = Q$ and $Q + \Delta^2 = I$ we have

$$(W_o x \oplus \Delta x,\ Y_j u) = (W_o x \oplus \Delta x,\ S^j [Du,\ W_o Bu]^{tr} \oplus U^{*j+1} \Delta Bu) = \tag{10.11}$$

$$(S^{*j} W_o x \oplus U^{j+1} \Delta x,\ [Du,\ W_o Bu]^{tr} \oplus \Delta Bu) = (W_o Z^j x \oplus \Delta Z Z^j x,\ [Du,\ W_o Bu]^{tr} \oplus \Delta Bu) =$$

$$((D^* C + B^* QZ) Z^j x + B^* \Delta^2 Z Z^j x,\ u) = ((D^* C + B^* Z) Z^j x,\ u) = 0\,.$$

The fourth equality follows from $W_o = [C,\ W_o Z]^{tr}$. The last equality follows from the fact that $B^* Z + D^* C = 0$, because the $2 \times 2$ matrix $\Omega$ in (10.3) is an isometry. Now let V be the isometry on $\ell_+^2(\mathcal{Y}) \oplus \mathcal{K}$ defined by $V = S \oplus U^*$. Equation (10.10) shows that $Y_j = V^j Y_0$ for all $j \geq 0$. Therefore to show that ran $Y_m$ is orthogonal to ran $Y_n$ for all $n \neq m$ it is sufficient to show that ran $Y_0$ is orthogonal to ran $V^n Y_0$ when $n \geq 1$. Hence ran $Y_n$ is orthogonal to ran $Y_m$ for all $n \neq m$ if and only if $V^* Y_0 v$ is orthogonal to $V^j Y_0 u$ for all v, u in $\mathcal{U}$ and $j \geq 0$. So using $V^* Y_0 v = W_o Bv \oplus \Delta Bv$ we have

$$(V^* Y_0 v,\ V^j Y_0 u) = (W_o Bv \oplus \Delta Bv,\ Y_j u) = 0\,.$$

The last equality follows from (10.11) with $x = Bv$. Therefore all the operators $[W_o,\ \Delta]^{tr}$, $Y_0, Y_1, \ldots$ are isometries with mutually orthogonal ranges. So by (10.9), the operator $\Phi_e$ in (10.7) is an isometry. In particular $[W_o,\ T_\Theta]$ is a contraction. Moreover, the Toeplitz operator $T_\Theta$ is contractive, and thus $\Theta$ is a contractive analytic function in $H^\infty(\mathcal{U}, \mathcal{Y})$. Finally, if Z is pointwise stable, then $\Delta = 0$ and $\Phi_e = [W_o,\ T_\Theta]$ is an isometry. In this case $W_o$ is an isometry and $\{C, Z\}$ is observable. Moreover, $T_\Theta$ is an isometry and $\Theta$ is inner. This completes the proof of Theorem 10.1.

We will now make a number of preparations which will give the proof of Theorem 10.4 below. Assume that $\{Z, B, C, D\}$ generates a unitary system, that is, the $2 \times 2$ operator matrix $\Omega$ in (10.3) is unitary. Let $U_*$ on $\mathcal{X} \oplus \ell_+^2(\mathcal{U})$ be the operator defined by

$$U_* = \begin{bmatrix} Z^* & 0 \\ \Pi_o^* B^* & S_{\mathcal{U}} \end{bmatrix} \text{ on } \begin{bmatrix} \mathcal{X} \\ \ell_+^2(\mathcal{U}) \end{bmatrix}. \tag{10.12}$$

where $S_{\mathcal{U}}$ is the unilateral shift on $\ell_+^2(\mathcal{U})$, and $\Pi_o$ is the operator from $\ell_+^2(\mathcal{U})$ onto $\mathcal{U}$ which picks out the first component of $\ell_+^2(\mathcal{U})$, that is, $\Pi_o [u_0, u_1, u_2, \ldots]^{tr} = u_0$. Because $\Omega$ is unitary, $I = ZZ^* + BB^*$. Using this fact it readily follows that $U_*$ is an isometry. As before, let V be the isometry on $\mathcal{V} = \ell_+^2(\mathcal{Y}) \oplus \mathcal{K}$ defined by $V = S \oplus U^*$ and $\Phi_e$ the isometry defined in (10.7). We claim that $\Phi_e$ is a unitary operator satisfying $V\Phi_e = \Phi_e U_*$. To prove this, first notice that for g in $\ell_+^2(\mathcal{U})$ we have

$$\Phi_e U_*(0 \oplus g) = \Phi_e(0 \oplus S_{\mathcal{U}} g) = T_\Theta S_{\mathcal{U}} g \oplus W_u S_{\mathcal{U}} g = S T_\Theta g \oplus U^* W_u g = V\Phi_e(0 \oplus g)\,, \quad (10.13)$$

where $W_u$ is defined in (10.6). Since $\Omega$ is unitary, the identity $\Omega\Omega^* = I$ implies that $CZ^* + DB^* = 0$. This along with $W_o = [C, W_o Z]^{tr}$ and x in $\mathcal{X}$ gives

$$\Phi_e U_* \begin{bmatrix} x \\ 0 \end{bmatrix} = \Phi_e \begin{bmatrix} Z^* x \\ \Pi_0^* B^* x \end{bmatrix} = \begin{bmatrix} W_o Z^* x \\ \Delta Z^* x \end{bmatrix} + \begin{bmatrix} [DB^* x,\ W_o BB^* x]^{tr} \\ U^* \Delta BB^* x \end{bmatrix} =$$

$$\begin{bmatrix} \begin{bmatrix} (CZ^* + DB^*)x \\ W_o(ZZ^* + BB^*)x \end{bmatrix} \\ U^* \Delta(ZZ^* + BB^*)x \end{bmatrix} = \begin{bmatrix} \begin{bmatrix} 0 \\ W_o x \end{bmatrix} \\ U^* \Delta x \end{bmatrix} = \begin{bmatrix} SW_o x \\ U^* \Delta x \end{bmatrix} = V\Phi_e \begin{bmatrix} x \\ 0 \end{bmatrix}. \quad (10.14)$$

The third equality follows from $U\Delta = \Delta Z$ and thus $\Delta = U^* \Delta Z$. The fourth equality follows from $ZZ^* + BB^* = I$. Combining this with (10.13) shows that $V\Phi_e = \Phi_e U_*$.

It remains to show that $\Phi_e$ is unitary. Since $\Phi_e$ is an isometry it is sufficient to show that $\Phi_e$ is onto. Using $S^* W_o = W_o Z$ and $U\Delta = \Delta Z$, it follows that $V^* \Phi_e(x \oplus 0) = \Phi_e(Zx \oplus 0)$ for all x in $\mathcal{X}$. In particular, $V^* \Phi_e(\mathcal{X} \oplus \{0\})$ is contained in the range of $\Phi_e$. By employing $W_o = [C, W_o Z]^{tr}$ once again, we have

$$(I - VV^*)\Phi_e(x \oplus 0) = (I - SS^*)W_o x \oplus 0 = (W_o - SW_o Zx) \oplus 0 = Cx \oplus 0$$

where Cx sits in the first component of $\ell_+^2(\mathcal{Y})$. Since $V\Phi_e = \Phi_e U_*$ and $V^* \Phi_e(\mathcal{X} \oplus \{0\})$ is contained in the range of $\Phi_e$, it follows that $C\mathcal{X} \oplus \{0\})$ is contained in the range of $\Phi_e$. Now let $\mathcal{Y}_c$ be the closed range of C. Then

$$\ell_+^2(\mathcal{Y}_c) = \bigvee_0^\infty S^n(C\mathcal{X}) \subseteq \bigvee_0^\infty V^n \Phi_e \mathcal{K}_* = \bigvee_0^\infty \Phi_e U_*^n \mathcal{K}_* \subseteq \text{ran}\,(\Phi_e) \quad (10.15)$$

where $\mathcal{K}_* = \mathcal{X} \oplus \ell_+^2(\mathcal{U})$. Thus $\ell_+^2(\mathcal{Y}_c) \subseteq \text{ran}\,\Phi_e$.

Now let y be in $\mathcal{Y}_c^\perp = \mathcal{Y} \ominus \mathcal{Y}_c$. Because $\Omega$ is unitary there exists a vector $x \oplus u$ in $\mathcal{X} \oplus \mathcal{U}$ satisfying

$$\begin{bmatrix} Z & B \\ C & D \end{bmatrix} \begin{bmatrix} x \\ u \end{bmatrix} = \begin{bmatrix} 0 \\ y \end{bmatrix} \text{ or equivalently } \begin{bmatrix} x \\ u \end{bmatrix} = \begin{bmatrix} Z^* & C^* \\ B^* & D^* \end{bmatrix} \begin{bmatrix} 0 \\ y \end{bmatrix}. \quad (10.16)$$

Since y is orthogonal to the range of C we have $C^* y = 0$ and thus $x = 0$. In particular, the first equation in (10.16) shows that $Bu = 0$. However, $\Omega^*$ is unitary. So $D^*$ maps $\mathcal{Y}_c^\perp$ unitarily onto a subspace $\mathcal{U}_o$ of $\mathcal{U}$, and $B\mathcal{U}_o = 0$. In other words, the operator $D_0 = D \mid \mathcal{U}_o$ is a unitary operator from $\mathcal{U}_o$ onto $\mathcal{Y}_c^\perp$. By consulting (10.9) and (10.10) we see that $Y_j \mid \mathcal{U}_o = [S^j D_0, 0]^{tr}$. Therefore $\Phi_e \mid \{0\} \oplus \ell_+^2(\mathcal{U}_o) = [T_{D_0}, 0]^{tr}$ where $T_{D_0} = \text{diag}\,[D_0, D_0, D_0, \ldots]$ is the Toeplitz operator from $\ell_+^2(\mathcal{U}_o)$ into $\ell_+^2(\mathcal{Y})$ where the unitary constant $D_o$ appears on the main diagonal and zeros elsewhere. This readily shows that $\ell_+^2(\mathcal{Y}_c^\perp)$ is contained in the range of $\Phi_e$. Combining this with (10.15) shows that $\ell_+^2(\mathcal{Y})$ is contained in the range of $\Phi_e$.

Since $\ell_+^2(\mathcal{Y}) \subseteq \text{ran}\, \Phi_e$ it follows that the range of $\Delta$ is contained in the range of $\Phi_e$. By employing $V^n\Phi_e(x \oplus 0) = \Phi_e U_*^n(x \oplus 0)$ for x in $\mathcal{X}$ and $\ell_+^2(\mathcal{Y}) \subseteq \text{ran}\, \Phi_e$, we see that the subspace $\{0\} \oplus U^{*n}\Delta\mathcal{X}$ is contained in the range of $\Phi_e$ for all $n \geq 0$. Because U is a minimal unitary extension of T, the minimality condition in (10.5) implies that the subspace $\mathcal{K} \subseteq \text{ran}\, \Phi_e$. Combining this with $\ell_+^2(\mathcal{Y}) \subseteq \text{ran}\, \Phi_e$ we see that $\Phi_e$ is onto. Therefore $\Phi_e$ is a unitary operator. Summing up our previous analysis readily proves the following result.

**THEOREM 10.4.** *Let $\Theta$ be the transfer function for a unitary system* {Z, B, C, D}, *and let $\Phi_e$ be the operator defined in (10.7), while $U_*$ is the isometry defined in (10.12). Then $\Phi_e$ is a unitary operator intertwining $U_*$ with $V = S \oplus U^*$. In particular, if Z is pointwise stable, then $\Phi = \Phi_e = [W_o, T_\Theta]$ is a unitary operator from $\mathcal{X} \oplus \ell_+^2(\mathcal{U})$ onto $\ell_+^2(\mathcal{Y})$ satisfying $S\Phi_e = \Phi_e U_*$ where S is the unilateral shift on $\ell_+^2(\mathcal{Y})$. In this case, $\Theta$ is an inner function and*

$$W_o\mathcal{X} = \text{ran}\,(T_\Theta)^\perp = \ell_+^2(\mathcal{Y}) \ominus T_\Theta \ell_+^2(\mathcal{U}) := \mathcal{H}(\Theta)\,. \tag{10.17}$$

Recall that two systems $\{Z \text{ on } \mathcal{X}, B\, C\, D\}$ and $\{Z_1 \text{ on } \mathcal{X}_1, B_1, C_1, D_1\}$ are *unitarily equivalent* if $D = D_1$ and there exists a unitary operator X from $\mathcal{X}$ onto $\mathcal{X}_1$ satisfying $XZ = Z_1X$, $B_1 = XB$ and $C_1X = C$. Notice that {0 on {0}, 0, 0, 1} and {1, 0, 0, 1} are two unitary realizations of the same two-sided inner function $\Theta(\lambda) = 1$. However, these realizations are not unitarily equivalent. Moreover, the realization {1, 0, 0, 1} is not even stable. This sets the stage for the following result.

**THEOREM 10.5.** *A function $\Theta$ is an inner function in $H^\infty(\mathcal{U}, \mathcal{Y})$ if and only if $\Theta$ is the transfer function for some pointwise stable, observable unitary system* {Z, B, C, D}. *Moreover, all pointwise stable unitary realizations for the same inner function $\Theta$ are unitarily equivalent. Finally, $\Theta$ is a two-sided inner function if and only if $\Theta$ is the transfer function for some pointwise stable, $*$-stable, controllable and observable unitary system.*

**PROOF.** The proof of this theorem is similar to the construction of the restricted backward shift realization. If $\Theta$ is an inner function, then obviously $T_\Theta$ is an isometry. Because $ST_\Theta = T_\Theta S_\mathcal{U}$ it follows that ran $T_\Theta$ is an invariant subspace for S. Hence $\mathcal{H} = \ell_+^2(\mathcal{Y}) \ominus \text{ran}\, T_\Theta$ is an invariant subspace for $S^*$. Let $Z_o$ be the pointwise stable contraction on $\mathcal{H}$ defined by $Z_o = S^* | \mathcal{H}$. Let $\Pi_\mathcal{E}$ be the operator from $\ell_+^2(\mathcal{E})$ onto $\mathcal{E}$ defined by $\Pi_\mathcal{E}[f_0, f_1, f_2, \ldots]^{tr} = f_0$.

Let $\{Z_o \text{ on } \mathcal{H}, B_o, C_o, D_0\}$ be the pointwise stable system defined by

$$\Omega = \begin{bmatrix} Z_o & B_o \\ C_o & D_0 \end{bmatrix} = \begin{bmatrix} S^* \mid \mathcal{H} & S^* T_\Theta \Pi_{\mathcal{U}}^* \\ \Pi_{\mathcal{Y}} \mid \mathcal{H} & \Theta(0) \end{bmatrix} : \begin{bmatrix} \mathcal{H} \\ \mathcal{U} \end{bmatrix} \to \begin{bmatrix} \mathcal{H} \\ \mathcal{Y} \end{bmatrix} . \tag{10.18}$$

It is emphasized that $B_o = S^* T_\Theta \Pi_{\mathcal{U}}^*$ does indeed map $\mathcal{U}$ into $\mathcal{H}$. To see this simply observe that ran $T_\Theta$ is orthogonal to ran $S^* T_\Theta \Pi_{\mathcal{U}}^*$, and hence $B_o$ maps $\mathcal{U}$ into $\mathcal{H}$. Using

$$\Pi_{\mathcal{Y}}^* \Pi_{\mathcal{Y}} = I - SS^* , \quad T_\Theta^* \mathcal{H} = \{0\} , \quad P_{\mathcal{H}} = I - T_\Theta T_\Theta^* \text{ and } \Pi_{\mathcal{U}}^* \Pi_{\mathcal{U}} = I - S_{\mathcal{U}} S_{\mathcal{U}}^* ,$$

where $S_{\mathcal{U}}$ is the unilateral shift on $\ell_+^2(\mathcal{U})$, it follows that $\Omega^*\Omega = I$ and $\Omega\Omega^* = I$. Thus $\{Z_o, B_o, C_o, D_0\}$ defines a pointwise stable unitary system. In particular, this system is observable. Now let $\{\Theta_n\}_0^\infty$ be the Fourier coefficients for $\Theta$. Then for u in $\mathcal{U}$ we have

$$C_o Z_o^{n-1} B_o u = \Pi_{\mathcal{Y}} S^{*(n-1)} [\Theta_1 u, \Theta_2 u, \Theta_3 u, \ldots]^{tr} = \Theta_n u \quad (u \in \mathcal{U} \text{ and } n \geq 1) . \tag{10.19}$$

Therefore $\Gamma_o = \{Z_o, B_o, C_o, D_0\}$ is a pointwise stable, observable unitary realization of $\Theta(\lambda)$.

Now assume that $\Gamma = \{Z \text{ on } \mathcal{X}, B, C, D\}$ is another pointwise stable unitary realization of the inner function $\Theta$. We claim that $\Gamma$ is unitarily equivalent to $\Gamma_o$. To this end, recall that Theorem 10.1 and Theorem 10.4 show that $\Gamma$ is an observable system and its observability operator $W_o$ from $\mathcal{X}$ into $\ell_+^2(\mathcal{Y})$ generated by $\{C, Z\}$ is an isometry. Moreover, $\mathcal{H} = \text{ran } W_o$. Hence the operator X mapping $\mathcal{X}$ onto $\mathcal{H}$ defined by $X = W_o$ is unitary. Using $S^* W_o = W_o Z$ it follows that $\mathcal{H}$ is an invariant subspace for $S^*$ and $Z_o X = XZ$. Obviously $C_o X = C_o W_o = C$. Because $\Gamma_o$ and $\Gamma$ are both realizations of $\Theta$, we have $D_0 = \Theta(0) = D$. Furthermore, if $\{\Theta_n : n \geq 0\}$ are the Fourier coefficients of $\Theta(\lambda)$, then using $Z_o = XZX^{-1}$ and $C_o X = C$ we have

$$CZ^{n-1}B = \Theta_n = C_o Z_o^{n-1} B_o = C_o X Z^{n-1} X^{-1} B_o = CZ^{n-1} X^{-1} B_o \qquad (n \geq 1) .$$

Since $\{C, Z\}$ is observable, this implies that $B = X^{-1} B_o$. Therefore $\Gamma_o$ is unitarily equivalent to $\Gamma$. The above analysis shows that any pointwise stable unitary realization of $\Theta$ is unitarily equivalent to $\Gamma_o$. Because the property of unitary equivalence if translative, it follows that all pointwise stable unitary realizations of $\Theta$ are unitarily equivalent. The last statement of this theorem concerning a two-sided inner function $\Theta$ follows from Corollary 7.11. This completes the proof.

**REMARK 10.6**. Let Z be a contraction on $\mathcal{X}$. Let $D_Z = D_{Z,1}$ be the positive square root of $I - Z^*Z$ and $D_{Z^*} = D_{Z^*,1}$ be the positive square root of $I - ZZ^*$. Finally, let $\mathcal{D}_Z$ and $\mathcal{D}_{Z^*}$ be the closed range of $D_Z$ and $D_{Z^*}$, respectively. Then it is easy to verify that

$$\begin{bmatrix} Z^* & D_Z \\ D_{Z^*} & -Z \end{bmatrix} : \begin{bmatrix} \mathcal{X} \\ \mathcal{D}_Z \end{bmatrix} \to \begin{bmatrix} \mathcal{X} \\ \mathcal{D}_{Z^*} \end{bmatrix}$$

is a unitary operator. According to our previous results the function

$$\Theta_Z(\lambda) := \lambda D_{Z^*}(I - \lambda Z^*)^{-1} D_Z \,|\, \mathcal{D}_Z - Z \,|\, \mathcal{D}_Z$$

is a contractive analytic function in $H^\infty(\mathcal{D}_Z, \mathcal{D}_{Z^*})$. Moreover, if $Z^*$ is pointwise stable, then $\Theta_Z(\lambda)$ is an inner function. This function $\Theta_Z$ is called the *characteristic function* for Z and plays an important role in dilation theory. For example, $\Theta_Z$ can be used to determine all the invariant subspaces and the spectrum of Z; see Sz.-Nagy-Foias [3], Foias-Frazho [4] for further details.

## *Notes to Chapter III:*

The interaction between interpolation and system theory plays an important role in recent developments (see the books Francis [1], Ball-Gohberg-Rodman [1], Chui-Chen [1], Foias-Osbay-Tannebaum [1], Green-Limebeer [1], Zhou-Doyle-Glover [1]). On the one hand problems in system theory serve as a motivation for treating various types of matrix function interpolation problems, and vice versa solutions of matrix constrained interpolation problems are used in different control problems. Especially useful is the state space method for input-output systems which originated in Kalman [1], Gilbert [1], Youla [1] and Weiss [1]. At present this state space theory is a well-established subject in engineering with important applications in control and signal processing (see the books Kalman-Falb-Arbib [1], Kailath [1] and Rugh [1]).

The material in the first two sections is standard. Section 3 is based on the paper Ball-Gohberg-Kaashoek [1]. For a more detailed discussion on the shift operator approach to realization theory in Section 4, see Fuhrmann [1]. The connection between the Hankel norm and the controllability and observability grammians in Section 6 is due to Glover [1]. The conversion formulas in Sections 8 and 9 are probably new in this general form. The theory of unitary systems in the tenth section started with the work of M. S. Livšic (Livšic [1], [2]), and was continued by B. Sz.-Nagy and C. Foias (in Sz.-Nagy-Foias [3]) and by M. S. Brodskii (in Brodskii [1], [2]). For further details on this subject see also Chapter 28 in Gohberg-Goldberg-Kaashoek [2], Arocena [1], and Arov [1], [2]. Generalizations of unitary systems, involving several state space operators, may be found in Kravitsky-Livšic-Markus-Vinnikov [1]. Finally, in Chapter VII we shall solve some $H^\infty$-control problems based on interpolation.

## CHAPTER IV

# CENTRAL COMMUTANT LIFTING

This chapter is devoted to the commutant lifting theorem. A distinguished solution, called the central intertwining lifting, is singled out. The special properties of this central solution concern maximum entropy and mixed $H^\infty$ and $L^2$ bounds. Explicit formulas for the central intertwining lifting are given in different settings. As a first application an explicit solution of the operator-valued Schur problem is given.

## IV.1 MINIMAL ISOMETRIC LIFTINGS

This section is devoted to minimal isometric liftings. Recall that an operator B from $\mathcal{K}$ to $\mathcal{K}'$ is a *lifting* of an operator A from $\mathcal{H}$ to $\mathcal{H}'$, if $\mathcal{H} \subseteq \mathcal{K}$ and $\mathcal{H}' \subseteq \mathcal{K}'$ and $P_{\mathcal{H}'}B = AP_{\mathcal{H}}$. Recall that $P_{\mathcal{H}}$ is the orthogonal projection of $\mathcal{K}$ onto $\mathcal{H}$. Notice that B is a lifting of A if and only if B admits a matrix representation of the form

$$B = \begin{bmatrix} A & 0 \\ * & * \end{bmatrix} : \begin{pmatrix} \mathcal{H} \\ \mathcal{H}^\perp \end{pmatrix} \to \begin{pmatrix} \mathcal{H}' \\ \mathcal{H}'^\perp \end{pmatrix} \tag{1.1}$$

where * indicates an unspecified entry. Finally, notice that B is a lifting of A if and only if $B^* \mid \mathcal{H}' = A^*$.

Now let C on $\mathcal{H}$ be a contraction ($\|C\| \leq 1$). An operator U on $\mathcal{K} (\supseteq \mathcal{H})$ is an *isometric lifting* of C if U is an isometry on $\mathcal{K}$ and $P_{\mathcal{H}}U = CP_{\mathcal{H}}$, or equivalently, $U^* \mid \mathcal{H} = C^*$. The operator U is a *minimal isometric lifting* of C if U is an isometric lifting of C and $\mathcal{H}$ is cyclic for U, that is,

$$\mathcal{K} = \bigvee_{0}^{\infty} U^n \mathcal{H} . \tag{1.2}$$

By using the Sz.-Nagy-Schäffer construction it is easy to show that any contraction admits a *minimal isometric lifting*. To see this let U be the operator defined by

$$U = \begin{bmatrix} C & 0 & 0 & 0 & 0 & \cdots \\ D & 0 & 0 & 0 & 0 & \cdots \\ 0 & I & 0 & 0 & 0 & \cdots \\ 0 & 0 & I & 0 & 0 & \cdots \\ 0 & 0 & 0 & I & 0 & \cdots \\ \cdot & \cdot & \cdot & \cdot & \cdot & \cdots \\ \cdot & \cdot & \cdot & \cdot & \cdot & \cdots \\ \cdot & \cdot & \cdot & \cdot & \cdot & \cdots \end{bmatrix} \text{ on } \mathcal{H} \oplus \ell_+^2(\mathcal{D}) . \tag{1.3}$$

where D is the positive square root of $I - C^*C$ and $\mathcal{D}$ is the closure of the range of D. Obviously U is a dilation of C. A simple calculation shows that $U^*U = I$ and $\mathcal{H}$ is cyclic for U. Hence U is a minimal isometric lifting of C. In many applications it may be convenient to express the minimal isometric lifting U in (1.3) in the Fourier domain, that is, let U′ be the minimal isometric lifting of C defined by

$$U' = \begin{bmatrix} C & 0 \\ D & S \end{bmatrix} \text{ on } \mathcal{K}' = \mathcal{H} \oplus H^2(\mathcal{D}) , \tag{1.4}$$

where S is the unilateral shift on $H^2(\mathcal{D})$. Here the operator D maps the space $\mathcal{H}$ into the constant functions of $H^2(\mathcal{D})$, that is,

$$U'(h \oplus g) = Ch \oplus (Dh + \lambda g) \qquad (h \oplus g \in \mathcal{H} \oplus H^2(\mathcal{D})) . \tag{1.5}$$

Finally, we call the minimal isometric lifting U of C in (1.3) or U′ of C in (1.4) *the Sz.-Nagy-Schäffer minimal isometric lifting of* C.

Let C be an operator on $\mathcal{H}$. Let U on $\mathcal{K}$ and U′ on $\mathcal{K}'$ be two liftings of C. Then we say that U and U′ are *isomorphic* if there exists a unitary operator W mapping $\mathcal{K}$ onto $\mathcal{K}'$ satisfying

$$WU = U'W \quad \text{and} \quad W \mid \mathcal{H} = I_{\mathcal{H}} . \tag{1.6}$$

Obviously the isometric liftings U and U′ of C in (1.3) and (1.4), respectively, are isomorphic. The following shows that all minimal isometric lifting are isomorphic.

**THEOREM 1.1.** *Any contraction C on $\mathcal{H}$ admits a minimal isometric lifting. Furthermore, all minimal isometric liftings of C are isomorphic.*

**PROOF.** The Sz.-Nagy-Schäffer construction for U in (1.3) establishes the existence of a minimal isometric lifting of C. To prove uniqueness let U′ on $\mathcal{K}'$ be another minimal isometric lifting of C. If $\{h_n\}_0^\infty$ is a sequence in $\mathcal{H}$ with finite support, then

$$\|\sum U^n h_n\|^2 = \sum (U^n h_n, U^m h_m) = \sum_{n\geq m} (U^{n-m} h_n, h_m) + \sum_{n<m} (h_n, U^{m-n} h_m)$$

$$= \sum_{n\geq m} (C^{n-m} h_n, h_m) + \sum_{n<m} (h_n, C^{m-n} h_m) = \|\sum U'^n h_n\|^2 .$$

The minimality guarantees that the finite sums $\sum U^n h_n$ and $\sum U'^n h_n$ are dense in $\mathcal{K}$ and $\mathcal{K}'$, respectively. Therefore there exists a unitary operator W mapping $\mathcal{K}$ onto $\mathcal{K}'$ such that

$$W \sum U^n h_n = \sum U'^n h_n \quad \text{and thus} \quad WU \sum U^n h_n = U' W \sum U^n h_n$$

for every finite sum $\sum U^n h_n$. From this it easily follows that (1.6) holds. Therefore U is isomorphic to U′. This completes the proof.

Because all minimal isometric liftings U of C are isomorphic we will call any minimal isometric lifting of C *the minimal isometric lifting* of C.

**REMARK 1.2.** As before assume that C is a contraction on $\mathcal{H}$. If $U_1$ on $\mathcal{K}_1$ is an isometric lifting of C on $\mathcal{H}$, then $U_1$ admits a reducing decomposition of the form $U_1 = U_m \oplus U_2$ on $\mathcal{K}_m \oplus \mathcal{K}_2$, where $U_m$ is the minimal isometric lifting of C and

$$\mathcal{K}_m = \bigvee_0^\infty U_1^n \mathcal{H} . \tag{1.7}$$

To see this first notice that $\mathcal{K}_m$ is invariant for $U_1$. Using the lifting property $C^* = U_1^* | \mathcal{H}$ we have

$$U_1^* \mathcal{K}_m = (U_1^* \mathcal{H}) \vee U_1^* \left[ \bigvee_1^\infty U_1^n \mathcal{H} \right] \subseteq \mathcal{H} \vee \left[ \bigvee_0^\infty U_1^n \mathcal{H} \right] = \mathcal{K}_m .$$

Therefore $\mathcal{K}_m$ is also invariant for $U_1^*$. Thus $\mathcal{K}_m$ is a reducing subspace for $U_1$ and $U_m = U_1 | \mathcal{K}_m$ is the minimal isometric lifting for C. This completes the proof.

**PROPOSITION 1.3.** *Assume that* $U_1$ *on* $\mathcal{K}_1$ *is an isometric lifting of a contraction* C *on* $\mathcal{H}$, *and let* U′ *on* $\mathcal{K}' = \mathcal{H} \oplus H^2(\mathcal{D})$ *be the Sz.-Nagy-Schäffer minimal isometric lifting of* C *in (1.4). Then there exists a unique isometry* W *from* $\mathcal{K}'$ *into* $\mathcal{K}_1$ *satisfying*

$$U_1 W = WU' \quad \textit{and} \quad W | \mathcal{H} = I . \tag{1.8}$$

*Moreover, an explicit formula for this isometry is given by*

$$W(h \oplus Df) = h + \sum_{0}^{\infty} U_1^n(U_1 - C)f_n \quad (h \oplus f \in \mathcal{H} \oplus H^2(\mathcal{D})) \tag{1.9}$$

*where* $\{f_n\}$ *are the Taylor coefficients for* f *and* D *is the positive square root of* $I - C^*C$.

**PROOF**. According to the previous remark $U_1 = U_m \oplus U_2$, where $U_m$ is the minimal isometric lifting of C. Since all minimal isometric liftings of C are isomorphic, there exists an isometry W from $\mathcal{K}'$ into $\mathcal{K}_1$ intertwining $U'$ with $U_1$ and satisfying $W | \mathcal{H} = I$. To show that W is unique assume that $W_1$ is another isometry satisfying $U_1 W_1 = W_1 U'$ and $W_1 | \mathcal{H} = I$. Then for all h in $\mathcal{H}$ and $n \geq 0$ we have

$$W_1 U'^n h = U_1^n W_1 h = U_1^n h = U_1^n W h = W U'^n h .$$

Since $\mathcal{H}$ is cyclic for any minimal isometric lifting, this implies that $W = W_1$.

To establish the formula for W in (1.9) notice that for all $f_n$ in $\mathcal{D}$

$$W(0 \oplus \lambda^n D f_n) = W U'^n (0 \oplus D f_n) = W U'^n (U' - C) f_n = U_1^n (U_1 - C) f_n .$$

Using this along with $W | \mathcal{H} = I$, we see that (1.9) holds when $f(\lambda)$ is a polynomial with values in $\mathcal{D}$. Because W is an isometry, it follows that (1.9) holds for all f in $H^2(\mathcal{D})$. This completes the proof.

**REMARK 1.4**. If C is a co-isometry on $\mathcal{H}$, then its minimal isometric lifting U is unitary. Indeed in this case $D = I - C^*C$ is the orthogonal projection onto $\mathcal{D}$. Using this it is easy to verify that $UU^* = I$, where U is the Sz.-Nagy-Schäffer minimal isometric lifting of C in (1.3).

To complete this section let us comment on some terminology used in the literature. An operator U on $\mathcal{K}(\supseteq \mathcal{H})$ is called a *dilation* of an operator C on $\mathcal{H}$ if $C = P_{\mathcal{H}} U | \mathcal{H}$ and

$$P_{\mathcal{H}} U^n | \mathcal{H} = C^n \qquad (\text{for all } n \geq 0) . \tag{1.10}$$

Obviously if U is a lifting of C, then U is a dilation of C. An operator U on $\mathcal{K}(\supseteq \mathcal{H})$ is a *minimal isometric dilation* of C if U is an isometric dilation of C and $\mathcal{H}$ is cyclic for U, that is, (1.2) holds. We claim that U is a minimal isometric dilation of C if and only if U is a minimal isometric lifting of C. Clearly, if U is a minimal isometric lifting of C, then U is a minimal isometric dilation of C. On the other hand, if U is a minimal isometric dilation of C, then for all $n \geq 0$ we have

$$P_{\mathcal{H}} U U^n h = C^{n+1} h = C P_{\mathcal{H}} U^n h \quad (h \in \mathcal{H}) . \tag{1.11}$$

The minimality condition (1.2) implies that $P_{\mathcal{H}} U = C P_{\mathcal{H}}$. Therefore U is a minimal isometric lifting of C, and this proves our claim. Since minimal isometric dilations and minimal isometric liftings are equivalent definitions, we will use this terminology interchangeably. Minimal isometric dilations play a fundamental role in operator theory; see Sz.-Nagy-Foias [3], Foias-Frazho [4] for further details.

## IV.2 THE CENTRAL INTERTWINING LIFTING

In this section we will introduce the central intertwining lifting, and use this intertwining lifting to prove the commutant lifting theorem. We say that an operator C is *bounded by* $\gamma$ if $\|C\| \leq \gamma$ and *strictly bounded by* $\gamma$ if $\|C\| < \gamma$. If C is an operator mapping $\mathcal{H}$ into $\mathcal{H}_1$ bounded by a *specified* $\gamma$, then $D_C$ is the positive square root of $\gamma^2 I - C^* C$ and $\mathcal{D}_C$ is the closure of the range of $D_C$. If we want to emphasize the role of $\gamma$, then we write $D_{C,\gamma}$ or $\mathcal{D}_{C,\gamma}$. Otherwise we simply leave the tolerance $\gamma$ out of the notation. Obviously, if $\|C\| < \gamma$, then $D_C$ is invertible and $\mathcal{D}_C = \mathcal{H}$.

Now, let T be an isometry on $\mathcal{H}$ and $T'$ a contraction on $\mathcal{H}'$. Let A be an operator mapping $\mathcal{H}$ into $\mathcal{H}'$ satisfying $T'A = AT$. Recall that $U_1$ on $\mathcal{K}_1 (\supseteq \mathcal{H}')$ is an isometric lifting of $T'$ if $U_1$ is an isometry satisfying $U_1^* | \mathcal{H}' = T'^*$. An operator B mapping $\mathcal{H}$ into $\mathcal{K}_1$ is called *an intertwining lifting of* A with respect to $U_1$, if

$$U_1 B = BT \quad \text{and} \quad P_{\mathcal{H}'} B = A . \tag{2.1}$$

The commutant lifting theorem states that there exists an intertwining lifting B of A satisfying $\|B\| = \|A\|$. Now let $U'$ on $\mathcal{K}' = \mathcal{H}' \oplus H^2(\mathcal{D}')$ be the Sz.-Nagy-Schäffer minimal isometric dilation of $T'$ given by

$$U' = \begin{bmatrix} T' & 0 \\ D' & S' \end{bmatrix} \quad \text{on} \quad \begin{bmatrix} \mathcal{H}' \\ H^2(\mathcal{D}') \end{bmatrix} \tag{2.2}$$

where $S'$ is the unilateral shift on $H^2(\mathcal{D}')$. Here $D'$ is the positive square root of $I - T'^* T'$, while $\mathcal{D}'$ is the closure of the range of $D'$. Notice that the operator $D'$ in (2.2) maps $\mathcal{H}'$ into the constant functions of $H^2(\mathcal{D}')$. In other words,

$$U'(h \oplus g) = T'h \oplus (D'h + \lambda g) \qquad (h \oplus g \in \mathcal{H}' \oplus H^2(\mathcal{D}')) . \tag{2.3}$$

According to Proposition 1.3 there exists an isometry W from $\mathcal{K}'$ into $\mathcal{K}_1$ intertwining $U'$ with $U_1$ and satisfying $W | \mathcal{H}' = I$. So when proving the commutant lifting theorem, it is sufficient to construct an intertwining lifting $B'$ of A with respect to $U'$ satisfying $\|B'\| = \|A\|$. Indeed, if we set $B = WB'$, then obviously B is an intertwining lifting of A with respect to $U_1$ satisfying $\|B\| = \|B'\| = \|A\|$. In this section without loss of generality we will concentrate on constructing a special intertwining lifting of A with respect to the Sz.-Nagy-Schäffer minimal isometric lifting U′.

Let B be an intertwining lifting of A with respect to the Sz.-Nagy-Schäffer minimal isometric lifting U′. Obviously, B admits a matrix partition of the form $B = [A, X]^{tr}$ mapping $\mathcal{H}$ into $\mathcal{H}' \oplus H^2(\mathcal{D}')$. Notice that $\|B\| \le \gamma$ if and only if for all h in $\mathcal{H}$ we have

$$\|Xh\|^2 \le \gamma^2\|h\|^2 - \|Ah\|^2 = \|D_A h\|^2 \qquad (h \in \mathcal{H}) .$$

So if B is bounded by γ, then there exists a contraction Y mapping $\mathcal{D}_A$ into $H^2(\mathcal{D}')$ satisfying $X = YD_A$. On the other hand, if $X = YD_A$ for some contraction Y, then

$$\|Bh\|^2 = \|Ah\|^2 + \|Xh\|^2 = \|Ah\|^2 + \|YD_A h\|^2 \le \|Ah\|^2 + \|D_A h\|^2 = \gamma^2\|h\|^2 ,$$

and thus $\|B\| \le \gamma$. Therefore B is bounded by γ if and only if there exists a contraction Y mapping $\mathcal{D}_A$ into $H^2(\mathcal{D}')$ satisfying $X = YD_A$. Obviously this contraction Y is uniquely determined by B. Now using the intertwining property $U'B = BT$ along with $X = YD_A$ we have

$$\begin{bmatrix} AT \\ YD_A T \end{bmatrix} = BT = U'B = \begin{bmatrix} T' & 0 \\ D' & S' \end{bmatrix}\begin{bmatrix} A \\ YD_A \end{bmatrix} = \begin{bmatrix} T'A \\ D'A + S'YD_A \end{bmatrix} .$$

This readily shows that $U'B = BT$ implies that $D'A + S'YD_A = YD_A T$. The reverse implication is obtained in a similar way. Summing up this analysis proves the following result.

**LEMMA 2.1** *Let* T *be an isometry on* $\mathcal{H}$ *and* A *be an operator from* $\mathcal{H}$ *into* $\mathcal{H}'$ *bounded by* γ *satisfying* $T'A = AT$, *where* T′ *is a contraction. Let* U′ *on* $\mathcal{H}' \oplus H^2(\mathcal{D}')$ *be the Sz.-Nagy-Schäffer minimal isometric lifting of* T′ *in (2.2). Then* B *is an intertwining lifting of* A *with respect to* U′ *and tolerance* γ *if and only if* B *admits a matrix decomposition of the form*

$$B = \begin{bmatrix} A \\ YD_A \end{bmatrix} : \mathcal{H} \to \begin{bmatrix} \mathcal{H}' \\ H^2(\mathcal{D}') \end{bmatrix} \tag{2.4}$$

*where* Y *is a contraction from* $\mathcal{D}_A$ *into* $H^2(\mathcal{D}')$ *satisfying*

$$D'A + S'YD_A = YD_A T . \tag{2.5}$$

Since below and in later chapters we need to consider bounded linear operators from $\mathcal{U}$ into $H^2(\mathcal{Y})$, where $\mathcal{U}$ and $\mathcal{Y}$ are Hilbert spaces, we make here a short digression concerning such operators. Let us start with an operator-valued function $\Theta(\lambda)$, analytic on the open unit disc $\mathbb{D}$, with values in $\mathcal{L}(\mathcal{U}, \mathcal{Y})$, that is, $\Theta(\lambda) = \sum_{n=0}^{\infty} \lambda^n \Theta_n$, where $\Theta_0, \Theta_1, \ldots$ are bounded linear operators from $\mathcal{U}$ into $\mathcal{Y}$ and the series converges for every $\lambda \in \mathbb{D}$ in the operator norm on $\mathcal{L}(\mathcal{U}, \mathcal{Y})$. If, in addition,

$$\sum_{n=0}^{\infty} \|\Theta_n u\|^2 \le \delta^2 \|u\|^2 \qquad (u \in \mathcal{U}),$$

where $\delta$ is a finite constant, it is clear that the function $\Theta(\lambda)$ can be viewed as an operator $\Theta$ from $\mathcal{U}$ into $H^2(\mathcal{Y})$, namely

$$(\Theta u)(\lambda) = \Theta(\lambda)u, \qquad (u \in \mathcal{U} \text{ and } \lambda \in \mathbb{D}).$$

This operator $\Theta$ is bounded and $\|\Theta\| \le \delta$. Conversely, it is easy to prove that any bounded linear operator $\Theta$ from $\mathcal{U}$ into $H^2(\mathcal{Y})$ arises in this way, and $\delta$ can be taken equal to the norm of $\Theta$. Therefore in the sequel we shall identify the operator $\Theta$ with the corresponding analytic operator-valued function $\Theta(\lambda)$.

Now let us return to the operator B in (2.4). By virtue of the above remark the operator Y from $\mathcal{D}_A$ into $H^2(\mathcal{D}')$ may be identified with an operator-valued function $Y(\cdot)$, analytic on $\mathbb{D}$, whose values are operators from $\mathcal{D}_A$ into $\mathcal{D}'$, via the formula

$$(Yd)(\lambda) = Y(\lambda)d = \sum_{n=0}^{\infty} \lambda^n Y_n d \qquad (d \in \mathcal{D}_A). \tag{2.6}$$

With this identification in mind we shall refer to $\{Y_n\}_0^\infty$ as the *Taylor coefficients* of Y.

By substituting formula (2.6) for Y into (2.5) and matching the coefficients of $\lambda^n$, we obtain the following recursive formulas for $\{Y_n\}$

$$D'A = Y_0 D_A T \quad \text{and} \quad Y_n D_A = Y_{n+1} D_A T \qquad (\text{for } n \ge 0). \tag{2.7}$$

So Y satisfies the intertwining condition in (2.5) if and only if its Taylor coefficients $\{Y_n\}_0^\infty$ satisfy (2.7). Therefore B is an intertwining lifting of A with respect to $U'$ and tolerance $\gamma$ if and only if B admits a matrix representation of the form (2.4) where Y is a contraction from $\mathcal{D}_A$ into $H^2(\mathcal{D}')$ of the form (2.6) satisfying (2.7).

Now let us proceed to construct the set of all operators Y satisfying (2.5), or equivalently, (2.6) and (2.7). To this end, let $\omega$ be the map from $\mathcal{F} := \overline{D_A T \mathcal{H}}$ into $\mathcal{D}' \oplus \mathcal{D}_A$ defined by

$$\omega D_A T h = D'Ah \oplus D_A h \qquad (h \in \mathcal{H}). \tag{2.8}$$

We claim that $\omega$ is an isometry. Using $T'A = AT$ along with h in $\mathcal{H}$ we have

$$\|D_A Th\|^2 = \|Th\|^2 - \|ATh\|^2 = \|h\|^2 - \|Ah\|^2 + \|Ah\|^2 - \|T'Ah\|^2 =$$
$$\|D_A h\|^2 + \|D'Ah\|^2 = \|D'Ah \oplus D_A h\|^2 . \tag{2.9}$$

Therefore $\omega$ is an isometry. Let $\Pi'$ be the operator from $\mathcal{D}' \oplus \mathcal{D}_A$ onto $\mathcal{D}'$ which picks out the first component of $\mathcal{D}' \oplus \mathcal{D}_A$ and $\Pi_A$ the operator from $\mathcal{D}' \oplus \mathcal{D}_A$ onto $\mathcal{D}_A$ which picks out the second component of $\mathcal{D}' \oplus \mathcal{D}_A$. To be precise,

$$\Pi'(d' \oplus d_A) = d' \quad \text{and} \quad \Pi_A(d' \oplus d_A) = d_A \qquad (d' \oplus d_A \in \mathcal{D}' \oplus \mathcal{D}_A) . \tag{2.10}$$

Finally, using (2.8) and (2.10) we obtain the following useful formulas

$$D'A = \Pi' \omega P_{\mathcal{F}} D_A T \quad \text{and} \quad D_A = \Pi_A \omega P_{\mathcal{F}} D_A T . \tag{2.11}$$

Here $P_{\mathcal{F}}$ is the orthogonal projection of $\mathcal{D}_A$ onto $\mathcal{F}$.

Using (2.11) in the recursion formula for $\{Y_n\}$ in (2.7) along with $\mathcal{F} = \overline{D_A T \mathcal{H}}$ we obtain

$$\Pi' \omega P_{\mathcal{F}} D_A T = D'A = Y_0 P_{\mathcal{F}} D_A T \quad \text{and} \quad Y_n \Pi_A \omega P_{\mathcal{F}} D_A T = Y_n D_A = Y_{n+1} P_{\mathcal{F}} D_A T \qquad (n \geq 0) .$$

Now let $\mathcal{G} = \mathcal{D}_A \ominus \mathcal{F}$ and $P_{\mathcal{G}}$ be the orthogonal projection of $\mathcal{D}_A$ onto $\mathcal{G}$. Then the previous equation readily shows that

$$Y_0 = \Pi' \omega P_{\mathcal{F}} + \Gamma_0 P_{\mathcal{G}} \quad \text{and} \quad Y_{n+1} = Y_n \Pi_A \omega P_{\mathcal{F}} + \Gamma_{n+1} P_{\mathcal{G}} \qquad (n \geq 0) \tag{2.12}$$

where $\Gamma_n$ is an operator from $\mathcal{G}$ into $\mathcal{D}'$ for all $n \geq 0$. On the other hand, if $\{Y_n\}_0^\infty$ are given by (2.12) where $\{\Gamma_n\}_0^\infty$ are arbitrary operators from $\mathcal{G}$ into $\mathcal{D}'$, then a simple application of (2.11) shows that (2.7) holds. Thus the set of all $\{Y_n\}$ satisfying (2.7) are given by the recursion formulas in (2.12). Hence $U'B = BT$ if and only if Y satisfies (2.5), or equivalently, the Taylor coefficients $\{Y_n\}$ for Y are given by (2.12). Obviously $\{Y_n\}$ and $\{\Gamma_n\}$ uniquely determine each other. Summing up the previous analysis readily proves the following result.

**LEMMA 2.2.** *Let* T *be an isometry on* $\mathcal{H}$ *and* A *on operator from* $\mathcal{H}$ *into* $\mathcal{H}'$ *bounded by* $\gamma$ *satisfying* $T'A = AT$, *where* $T'$ *is a contraction on* $\mathcal{H}'$. *Let* $U'$ *on* $\mathcal{K}' = \mathcal{H}' \oplus H^2(\mathcal{D}')$ *be the Sz.-Nagy-Schäffer minimal isometric lifting of* $T'$ *in (2.2). Let* B *be an operator from* $\mathcal{H}$ *into* $\mathcal{K}'$. *Then* B *is an intertwining lifting of* A *with tolerance* $\gamma$ *if and only if* B *admits a decomposition of the form* $B = [A, YD_A]^{tr}$, *where* Y *is a contraction mapping* $\mathcal{D}_A$ *into* $H^2(\mathcal{D}')$ *whose Taylor coefficients* $\{Y_n\}$ *(see (2.6)) are recursively determined by (2.12) for some sequence of contraction* $\{\Gamma_n\}$ *from* $\mathcal{G} = D_A \ominus \overline{\mathcal{D}_A T \mathcal{H}}$ *into* $\mathcal{D}'$. *In this case, the sequences of contractions* $\{Y_n\}_0^\infty$ *and* $\{\Gamma_n\}_0^\infty$ *uniquely determine each other.*

The previous lemma shows that B is an intertwining lifting of A with tolerance $\gamma$ if and only if $B = [A, YD_A]^{tr}$, where $Y = \sum \lambda^n Y_n$ is computed according to the recursion formula in (2.12) and $\{\Gamma_n\}_0^\infty$ are chosen such that $\|Y\| \le 1$. So one can construct an intertwining lifting B of A with tolerance $\gamma$ by choosing a sequence of operators $\{\Gamma_n\}_0^\infty$ from $\mathcal{G}$ into $\mathcal{D}'$ such that the operator Y corresponding to the recursion formulas in (2.12) is contractive. Obviously, the simplest choice for $\{\Gamma_n\}$ is $\Gamma_n = 0$ for all n. In this case, the recursion formulas for $\{Y_n\}$ in (2.12) yield

$$Y_n = \Pi'\omega P_{\mathcal{F}}(\Pi_A \omega P_{\mathcal{F}})^n \qquad (n \ge 0) . \tag{2.13}$$

Moreover, the operator Y is given by

$$(Yd)(\lambda) = \sum_{n=0}^{\infty} \lambda^n \Pi'\omega P_{\mathcal{F}}(\Pi_A \omega P_{\mathcal{F}})^n d = \Pi'\omega P_{\mathcal{F}}(I - \lambda \Pi_A \omega P_{\mathcal{F}})^{-1} d , \qquad (d \in \mathcal{D}_A) . \tag{2.14}$$

We shall check later (see the proof of Theorem 2.3 here below) that the expressions in (2.14) are well-defined functions in $H^2(\mathcal{D}')$ and that the corresponding operator Y from $\mathcal{D}_A$ into $H^2(\mathcal{D}')$ is a contraction. Now let $B_\gamma$ be the operator from $\mathcal{H}$ into $\mathcal{K}' = \mathcal{H}' \oplus H^2(\mathcal{D}')$ constructed for this Y, that is,

$$B_\gamma = \begin{bmatrix} A \\ \Pi'\omega P_{\mathcal{F}}(I - \lambda \Pi_A \omega P_{\mathcal{F}})^{-1} D_A \end{bmatrix} : \mathcal{H} \to \begin{bmatrix} \mathcal{H}' \\ H^2(\mathcal{D}') \end{bmatrix} . \tag{2.15}$$

Because the Taylor coefficients $\{Y_n\}$ for this $B_\gamma$ obviously satisfy (2.12) it follows that $U'B_\gamma = B_\gamma T$. The following theorem shows that $B_\gamma$ is indeed an intertwining lifting of A bounded by $\gamma$, that is, $\|B_\gamma\| \le \gamma$. This intertwining lifting $B_\gamma$ of A in (2.15) is called the *central intertwining lifting* of A, because this intertwining lifting is obtained by choosing $\Gamma_n = 0$ for all n. As before, let $U_1$ on $\mathcal{K}_1$ be an isometric lifting of T′. According to Proposition 1.3 there exists a unique isometry W from $\mathcal{K}'$ into $\mathcal{K}_1$ satisfying $U_1 W = WU'$ and $W | \mathcal{H} = I$. Therefore $WB_\gamma$ is an intertwining lifting of A with respect to $U_1$ satisfying $\|WB_\gamma\| \le \gamma$. So without loss of generality we call $WB_\gamma$ *the central intertwining lifting of* A *with respect to* $U_1$ *and tolerance* $\gamma$. Now we are ready to prove the main result of this section.

**THEOREM 2.3.** *Let* T *be an isometry on* $\mathcal{H}$ *and* A *be an operator bounded by* $\gamma$ *mapping* $\mathcal{H}$ *into* $\mathcal{H}'$ *satisfying* $T'A = AT$, *where* T′ *is a contraction on* $\mathcal{H}'$. *Let* $U_1$ *on* $\mathcal{K}_1$ *be an isometric lifting of* T′. *Put* $B = WB_\gamma$, *where* $B_\gamma$ *is given by (2.15) and* W *by (1.9). Then* B *is well-defined and* B *is an intertwining lifting of* A *with respect to* $U_1$ *satisfying* $\|B\| \le \gamma$. *In particular, by choosing* $\gamma = \|A\|$, *then there exists an intertwining lifting* B *of* A *satisfying* $\|B\| = \|A\|$.

**PROOF**. Due to Proposition 1.3, we can without loss of generality, assume that $U_1 = U'$ is the Sz.-Nagy-Schäffer minimal isometric lifting of $T'$ in (2.2) and $B = B_\gamma$ is given by (2.15). Fix $h \in \mathcal{H}$. Since $\|\Pi' f\| = \|f\|^2 - \|\Pi_A f\|^2$ we have

$$\sum_{j=0}^{N} \|\Pi'\omega P_{\mathcal{F}}(\Pi_A \omega P_{\mathcal{F}})^j D_A h\|^2 =$$

$$= \sum_{j=0}^{N} (\|\omega P_{\mathcal{F}}(\Pi_A \omega P_{\mathcal{F}})^j D_A h\|^2 - \|\Pi_A \omega P_{\mathcal{F}}(\Pi_A \omega P_{\mathcal{F}})^j D_A h\|^2) \leq$$

$$\leq \sum_{j=0}^{N} (\|(\Pi_A \omega P_{\mathcal{F}})^j P_{\mathcal{F}} D_A h\|^2 - \|(\Pi_A \omega P_{\mathcal{F}})^{j+1} P_{\mathcal{F}} D_A h\|^2) \tag{2.16}$$

$$= \|P_{\mathcal{F}} D_A h\|^2 - \|(\Pi_A \omega P_{\mathcal{F}})^{N+1} P_{\mathcal{F}} D_A h\|^2 \leq \|P_{\mathcal{F}} D_A h\|^2 .$$

Because this holds for each $N \geq 0$, it follows that $\Pi'\omega P_{\mathcal{F}}(I - \lambda \Pi_A \omega P_{\mathcal{F}})^{-1} D_A h$ belongs to $H^2(\mathcal{D}')$. Thus $B_\gamma h$ is well-defined. Using (2.16) we obtain

$$\|B_\gamma h\|^2 = \|Ah\|^2 + \sum_{j=0}^{\infty} \|\Pi'\omega P_{\mathcal{F}}(\Pi_A \omega P_{\mathcal{F}})^j D_A h\|^2$$

$$\leq \|Ah\|^2 + \|P_{\mathcal{F}} D_A h\|^2 \leq \|Ah\|^2 + \|D_A h\|^2 = \gamma^2 \|h\|^2 . \tag{2.17}$$

Therefore $B_\gamma$ is bounded by $\gamma$.

Obviously we have $P_{\mathcal{H}'} B_\gamma = A$. Now let us directly verify that $U' B_\gamma = B_\gamma T$. By equation (2.11) along with the definition of $B_\gamma$ in (2.15) we have

$$U' B_\gamma = \begin{bmatrix} T'A \\ D'A + \lambda \Pi' \omega P_{\mathcal{F}}(I - \lambda \Pi_A \omega P_{\mathcal{F}})^{-1} D_A \end{bmatrix} =$$

$$\begin{bmatrix} AT \\ \Pi'\omega P_{\mathcal{F}} D_A T + \lambda \Pi' \omega P_{\mathcal{F}}(I - \lambda \Pi_A \omega P_{\mathcal{F}})^{-1} \Pi_A \omega P_{\mathcal{F}} D_A T \end{bmatrix} = \tag{2.18}$$

$$\begin{bmatrix} AT \\ \Pi'\omega P_{\mathcal{F}}(I - \lambda \Pi_A \omega P_{\mathcal{F}})^{-1} D_A T \end{bmatrix} = B_\gamma T .$$

Therefore $B_\gamma$ is an intertwining lifting of A. This completes the proof.

Now let $U'$ on $\mathcal{H}' \oplus \mathcal{M}$ be any minimal isometric lifting of $T'$ and let B from $\mathcal{H}$ into $\mathcal{H}' \oplus \mathcal{M}$ be an intertwining lifting of A with tolerance $\gamma$. The above analysis shows that B uniquely determines a contraction Y from $\mathcal{D}_A$ into $\mathcal{M}$ satisfying $B = [A, YD_A]^{tr}$. If B is the central intertwining lifting of A, then obviously $Y | \mathcal{G} = 0$. The following result shows that the constraint $Y | \mathcal{G} = 0$ uniquely determines the central intertwining lifting of A, when $U'$ is a minimal isometric lifting of T′.

**PROPOSITION 2.4**. *Let* T *be an isometry on* $\mathcal{H}$ *and* A *an operator from* $\mathcal{H}$ *into* $\mathcal{H}'$ *bounded by* $\gamma$ *satisfying* $T'A = AT$, *where* T′ *is a contraction on* $\mathcal{H}'$. *Let* U′ *on* $\mathcal{K}' = \mathcal{H}' \oplus \mathcal{M}$ *be a minimal isometric lifting of* T′. *Finally, let* B *be any intertwining lifting of* A *with tolerance* $\gamma$ *and* Y *be its uniquely determined contraction from* $\mathcal{D}_A$ *into* $\mathcal{M}$, *i.e.*, $B = [A, YD_A]^{tr}$. *Then* B *is the central intertwining lifting of* A *with tolerance* $\gamma$ *if and only if* $Y \mid \mathcal{G} = 0$, *where* $\mathcal{G} = \mathcal{D}_A \ominus \overline{D_A T\mathcal{H}}$. *In particular, if* $\mathcal{G} = \{0\}$, *then there is only one intertwining lifting* B *of* A *satisfying* $\|B\| \le \gamma$, *and this* B *is precisely the central intertwining lifting of* A *with tolerance* $\gamma$.

**PROOF**. Since all minimal isometric liftings are isomorphic, we can assume without loss of generality that U′ is the Sz.-Nagy-Schäffer minimal isometric lifting of A. In this case, $B = [A, YD_A]^{tr}$, where Y is now a contraction from $\mathcal{D}_A$ into $\mathcal{M} = H^2(\mathcal{D}')$. Obviously $Y \mid \mathcal{G} = 0$ if and only if $Y_n \mid \mathcal{G} = 0$ for all $n \ge 0$, where $\{Y_n\}$ are the Taylor coefficients for Y; see (2.6). According to Lemma 2.2, these Taylor coefficients $\{Y_n\}$ must satisfy the recursion formulas in (2.12) for some sequence of contractions $\{\Gamma_n\}$ mapping $\mathcal{G}$ into $\mathcal{D}'$. Moreover, these recursion formulas show that $Y_n \mid \mathcal{G} = 0$ for all n if and only if $\Gamma_n = 0$ for all n. Thus $Y \mid \mathcal{G} = 0$ if and only if $\Gamma_n = 0$ for all n. In this case, $Y_n$ is given by (2.13), or equivalently, Y is given by (2.14), which is precisely the Y component of the central intertwining lifting $B_\gamma$ of A in (2.15). Therefore $Y \mid \mathcal{G} = 0$ if and only if B is the central intertwining lifting of A with tolerance $\gamma$. This completes the proof.

**REMARK 2.5**. If U′ is not a minimal isometric lifting of T′, then the previous proposition is not necessarily true, that is, it may happen that there exists an intertwining lifting B of A, different from the central one, but nevertheless satisfying $Y \mid \mathcal{G} = 0$. For example, let S be the unilateral shift on $H^2$ and T the isometry on $\mathcal{H} = H^2 \oplus \mathbb{C}$ defined by $T = S \oplus 1$. (The notation $C = E \oplus F$ on $\mathcal{H}_1 \oplus \mathcal{H}_2$ means that $C(h \oplus f) = Eh \oplus Ff$.) Let A be the operator from $\mathcal{H}$ into $\mathbb{C}$ defined by $A = [0, 0]$ and T′ the contraction on $\mathcal{H}' = \mathbb{C}$ defined by $T' = 0$. Obviously $T'A = AT$. Moreover, $U' = T$ is an isometric lifting of T′ when $\mathcal{H}'$ is viewed as the subspace of $H^2 \oplus \mathbb{C}$ consisting of the constant functions in $H^2$. However, this U′ is not a minimal isometric lifting of T′. (Notice that S is the minimal isometric lifting of T′.) Now fix $\gamma > 0$. Then $D_A = \gamma I \oplus \gamma$ and $\mathcal{G} = \ker T^* = \{[a, 0, 0, \ldots]^{tr} \oplus \{0\} : a \in \mathbb{C}\}$. Clearly $B_\gamma = 0$ is an intertwining lifting of A with tolerance $\gamma$. In this case $Y = 0$ and $Y \mid \mathcal{G} = \{0\}$. It turns out that this $B_\gamma = 0$ is indeed the central intertwining lifting of A with tolerance $\gamma$. However, this fact is not needed here. All we need to do is establish the existence of another intertwining lifting B of A with tolerance $\gamma$ satisfying $Y \mid \mathcal{G} = 0$. To this end, notice that the diagonal operator $B = 0 \oplus y\gamma$ on

$H^2 \oplus \mathbb{C}$ for any $|y| \le 1$ is also an intertwining lifting of A satisfying $\|B\| \le \gamma$. In this case $Y = 0 \oplus y$ on $H^2 \oplus \mathbb{C}$ and $Y | \mathcal{G} = \{0\}$. So we have infinitely many intertwining lifting B of A satisfying $Y | \mathcal{G} = \{0\}$ and $\|B\| \le \gamma$. Therefore Proposition 2.4 is not necessarily true when U′ is not the minimal isometric lifting of A. In Section 7 we will show that Proposition 2.4 can be extended to the non-minimal isometric dilation case for certain A and T.

In Section VI.2 we shall return to the uniqueness statement appearing in Proposition 2.4. Theorem 2.3 readily yields the following result which is the commutant lifting theorem.

**THEOREM 2.6.** *Let* T *be an isometry on* $\mathcal{H}$ *and* A *an operator mapping* $\mathcal{H}$ *into* $\mathcal{H}'$ *satisfying* $T'A = AT$, *where* T′ *is a contraction on* $\mathcal{H}'$. *Let* U′ *be an isometric lifting of* T′. *Then there exists an intertwining lifting* B *of* A *with respect to* U′ *preserving the norm of* A, *that is,* $\|B\| = \|A\|$.

An operator C from $\mathcal{X}$ into $\mathcal{Y}$ *attains its norm* if there exists a nonzero vector x in $\mathcal{X}$ such that $\|Cx\| = \|C\| \, \|x\|$. In this case we say that x attains the norm of C. Obviously any finite rank operator attains its norm. The following result can be used in computing an intertwining lifting B of A which preserves the norm of A.

**COROLLARY 2.7.** *Let* T *be a unilateral shift on* $H^2$ *and* S *a unilateral shift on* $H^2(\mathcal{E})$. *Let* $\mathcal{H}'$ *be an invariant subspace for* $S^*$ *and* T′ *on* $\mathcal{H}'$ *the contraction obtained by compressing* S *to* $\mathcal{H}'$, *that is,* $T' = P_{\mathcal{H}'} S | \mathcal{H}'$. *Assume that* A *is an operator intertwining* T *with* T′ *which attains its norm. Then* A *admits a unique intertwining lifting* B *satisfying* $\|B\| = \|A\|$. *Moreover, this intertwining lifting* B *is given by* $B = M_g^+$, *where* g *is the function in* $H^\infty(\mathbb{C}, \mathcal{E})$ *defined by*

$$g(\lambda) = \frac{(Ax)(\lambda)}{x(\lambda)}, \tag{2.19}$$

*and* x *is any vector in* $H^2$ *which attains the norm of* A. *Finally,* $g/\|A\|$ *is an inner function, that is,* $g(e^{it})^* g(e^{it}) = \|A\|^2$ *a.e.*

**PROOF.** By the commutant lifting theorem there exists an intertwining lifting B of A satisfying $\|B\| = \|A\|$. Because $BT = SB$ it follows that $B = M_g^+$ for some g in $H^\infty(\mathbb{C}, \mathcal{E})$. Using $P_{\mathcal{H}'}B = A$ we obtain

$$\|A\| \, \|x\| = \|Ax\| = \|P_{\mathcal{H}'}Bx\| \le \|Bx\| \le \|B\| \, \|x\| = \|A\| \, \|x\| .$$

So we have equality. In particular, $P_{\mathcal{H}'}Bx = Bx$ and thus

$$Ax = P_{\mathcal{H}'}Bx = Bx = gx\ . \tag{2.20}$$

Recall that if f is any nonzero function in $H^2$, then $f(e^{it})$ is nonzero a.e.; see Hoffman [1], page 52. So we can divide by x in (2.20) to obtain $g = Ax/x$. Therefore the intertwining lifting B of A satisfying $\|B\| = \|A\|$ is given by $B = M_g^+$, where $g = Ax/x$. Notice that this B must be unique. Because if $B_o$ is another intertwining lifting of A satisfying $\|B_o\| = \|A\|$, then the above analysis shows that $B_o = M_f^+$ for some f in $H^\infty(\mathbb{C}, \mathcal{E})$ and $f = Ax/x$. Therefore $g = f$ and $B = B_o$.

To complete the proof it remains to show that $g/\gamma$ is an inner function when $\gamma = \|A\|$. To this end, first notice that $A^*Ax = \gamma^2 x$. Now let $y = Ax$, then, for all $n \geq 0$ we have

$$\frac{1}{2\pi}\int_0^{2\pi} e^{int}\|y(e^{it})\|^2 dt = (T'^n y,\ y) = (T'^n Ax,\ Ax) =$$

$$(AT^n x,\ Ax) = \gamma^2 (T^n x,\ x) = \frac{1}{2\pi}\int_0^{2\pi} e^{int}\gamma^2 |x(e^{it})|^2 dt\ . \tag{2.21}$$

By taking the complex conjugate in the previous expression (2.21) holds for all n. So $\|y(e^{it})\|_{\mathcal{E}}^2$ and $\gamma^2|x|^2$ have the same Fourier coefficients, and thus $\|y(e^{it})\|_{\mathcal{E}}^2 = \gamma^2|x|^2$ a.e. Since $g = y/x$, it follows that $g^*g = \gamma^2$ a.e. This completes the proof.

By following a similar argument we readily obtain the following result.

**COROLLARY 2.8.** *Let* T *be a unilateral shift on* $H^2$ *and* V *a bilateral shift on* $L^2(\mathcal{E})$. *Let* $\mathcal{H}'$ *be an invariant subspace for* $V^*$ *and* $T'$ *on* $\mathcal{H}'$ *the contraction obtained by compressing* V *to* $\mathcal{H}'$. *Assume that* A *is an operator from* $\mathcal{H}$ *into* $\mathcal{H}'$ *intertwining* T *with* $T'$ *and* x *is a vector in* $H^2$ *which attains the norm of* A. *Then there is only one intertwining lifting* B *of* A *satisfying* $\|B\| = \|A\|$. *Moreover, this intertwining lifting is given by* $B = M_g | H^2$, *where* g *is the function in* $L^\infty(\mathbb{C}, \mathcal{E})$ *defined by*

$$g(e^{it}) = \frac{(Ax)(e^{it})}{x(e^{it})}\ . \tag{2.22}$$

*Finally,* $g(e^{it})^* g(e^{it}) = \|A\|^2$ *a.e.*

One can also use the central intertwining lifting to obtain another proof of Corollary 2.7 and 2.8. To see this notice that the central intertwining lifting B of A with tolerance $\gamma = \|A\|$ is of the form $B = A + YP_{\mathcal{F}}D_A$ where Y is a contraction from $\mathcal{F}$ into $\mathcal{K}' \ominus \mathcal{H}'$. Since $U'B = BT$, where T is a unilateral shift and $U'$ is a unilateral shift (or bilateral shift in the case of Corollary 2.8), we have $B = M_F | H^2$ where F is in $H^\infty(\mathbb{C}, \mathcal{E})$, (or F is in $L^\infty(\mathbb{C}, \mathcal{E})$, in the case of Corollary 2.8). If x attains the norm of A, the $D_A x = 0$ and thus $Fx = Bx = Ax$. Hence $F = Ax/x$.

Moreover, notice that

$$\gamma^2\|x\|^2 - \|Bx\|^2 = \gamma^2\|x\|^2 - \|Ax + YP_{\mathcal{F}}D_Ax\|^2 =$$

$$\gamma^2\|x\|^2 - \|Ax\|^2 - \|YP_{\mathcal{F}}D_Ax\|^2 = \|D_Ax\|^2 - \|YP_{\mathcal{F}}D_Ax\|^2 = 0\,.$$

This implies that

$$0 = \gamma^2\|x\|^2 - \|Bx\|^2 = \gamma^2\|x\|^2 - \|Fx\|^2 = ((\gamma^2 - F^*F)x, x)\,.$$

Because $\|F\|_\infty = \|B\| = \gamma$, we see that $(\gamma^2 - F^*F)x = 0$ and hence $\gamma^2 = F^*F$ a.e. This completes the proof.

**REMARK 2.9**. Let T be a unilateral shift on $H^2$ and A an operator from $H^2$ into $\mathcal{H}'$ satisfying $T'A = AT$ where $T'$ is a contraction on $\mathcal{H}'$. If A attains its norm and $\gamma = \|A\|$, then $\mathcal{G} := \mathcal{D}_A \ominus \overline{D_A T\mathcal{H}}$ is zero. (Here $D_A = D_{A,\gamma}$.) In particular, if $U'$ is a minimal isometric lifting of $T'$, then there is only one intertwining lifting B of A with respect to $U'$ satisfying $\|B\| = \|A\|$; see Proposition 2.4. To see this assume that g is a nonzero vector in $\mathcal{G}$. Then $T^*D_Ag = 0$ and thus $D_Ag = a$ where a is a nonzero constant in $H^2$. Clearly x attains the norm of A if and only if x is a vector in $\ker D_A$. So if x attains the norm of A, then x is orthogonal to $a = D_Ag$ and thus $x(0) = 0$. However, this contradicts the fact that there exists vectors x in $\ker D_A$ such that $x(0) \neq 0$. Therefore $\mathcal{G} = \{0\}$. To prove that there exists an x in $\ker D_A$ such that $x(0) \neq 0$, assume that f attains the norm of A and $f = \lambda^n x$ where $x(0) \neq 0$. Then

$$\|A\|\,\|x\| = \|A\|\,\|f\| = \|Af\| = \|AT^n x\| = \|T'^n Ax\| \leq \|Ax\| \leq \|A\|\,\|x\|\,.$$

So we have equality and x attains the norm of A with $x(0) \neq 0$. This completes the proof.

Let A be an operator satisfying the hypothesis of Remark 2.9. Then it is easy to show that there exists an outer function $x_o$ in $H^2$ attaining the norm of A. To see this let x be any function in $H^2$ attaining the norm of A and $x = x_i x_o$ be its inner outer factorization. Using $T'A = AT$ along with the functional calculus for contractions (Sz.-Nagy-Foias [3]) we have

$$\|A\|\,\|x_o\| = \|A\|\,\|x\| = \|Ax_i x_o\| =$$

$$\|Ax_i(T)x_o\| = \|x_i(T')Ax_o\| \leq \|Ax_o\| \leq \|A\|\,\|x_o\|\,.$$

So we have equality, and the outer function $x_o$ attains the norm of A. In particular, if the kernel of $\|A\|^2 I - A^*A$ is one dimensional, then any vector which attains the norm of A is outer.

Let U on $\mathcal{K}$ be an isometric lifting of $T'$ and assume that $T'A = AT$ where T on $\mathcal{H}$ is an isometry. By applying Proposition 2.3 to (2.15) we obtain the following form for the central intertwining lifting B of A with respect to U and tolerance $\gamma$

$$B = A + \sum_{n=0}^{\infty} U^n \Pi_0' \omega_o P_{\mathcal{F}} (\Pi_A \omega_o P_{\mathcal{F}})^n D_A \,. \tag{2.23}$$

Here $\omega_o$ is the isometry from $\mathcal{D}_A$ into $\mathcal{K} \oplus \mathcal{D}_A$ defined by $\omega_o D_A h = (U - T')Ah \oplus D_A h$ for h in $\mathcal{H}$, and $\Pi_0'$ is the orthogonal projection from $\mathcal{K} \oplus \mathcal{D}_A$ onto $\mathcal{K}$ and $\Pi_A$ is the orthogonal projection onto $\mathcal{D}_A$.

## IV.3 CENTRAL INTERTWINING LIFTING FORMULAS

In this section we will present an explicit formula for the central intertwining lifting $B_\gamma$ when A is strictly bounded by $\gamma$. To this end, we must first establish an operator formula for the orthogonal projection $P_{\mathcal{F}}$. Since $\|A\| < \gamma$, the operator $D_A$ is invertible. Because T is an isometry, the operator $X = D_A T$ on $\mathcal{H}$ is one to one and has closed range. In fact, the range of X is $\mathcal{F}$. Therefore, $X^*X$ is invertible. Moreover, using $T^*T = I$ we have

$$X^*X = T^*(\gamma^2 I - A^*A)T = \gamma^2 I - T^*A^*AT = D_{AT}^2 \,.$$

Notice that the operator $P = X(X^*X)^{-1}X^*$ is onto $\mathcal{F}$ and $P^2 = P = P^*$. Therefore $P = P_{\mathcal{F}}$ is the orthogonal projection of $\mathcal{H}$ onto $\mathcal{F}$. From this we obtain the formula for $P_{\mathcal{F}}$ that we have been looking for, that is,

$$P_{\mathcal{F}} = X(X^*X)^{-1}X^* = D_A T D_{AT}^{-2} T^* D_A \,. \tag{3.1}$$

Finally, assuming that A is strictly bounded by $\gamma$, then $T_A$ is the operator on $\mathcal{H}$ defined by

$$T_A = D_A^2 T D_{AT}^{-2} = (\gamma^2 I - A^*A)T(\gamma^2 I - T^*A^*AT)^{-1} \,. \tag{3.2}$$

We can now state the following useful result.

**PROPOSITION 3.1.** *Let* T *be an isometry on* $\mathcal{H}$ *and* A *an operator strictly bounded by* $\gamma$ *mapping* $\mathcal{H}$ *into* $\mathcal{H}'$ *satisfying* $T'A = AT$, *where* $T'$ *is a contraction on* $\mathcal{H}'$. *Then* $T_A^*$ *is similar to the contraction* $\Pi_A \omega P_{\mathcal{F}}$. *Moreover, the central intertwining lifting* $B_\gamma$ *of* A *with respect to the Sz.-Nagy-Schäffer minimal isometric dilation* $U'$ *can also be computed by*

$$B_\gamma = \begin{bmatrix} A \\ D'AT_A^*(I - \lambda T_A^*)^{-1} \end{bmatrix} \tag{3.3}$$

*where* $D'$ *is the positive square root of* $I - T'^*T'$. *In particular, for all* a *in* $\mathcal{L} = \ker T^*$ *we have*

$$B_\gamma D_A^{-2} a = A D_A^{-2} a \qquad (a \in \ker T^*) \,. \tag{3.4}$$

*Finally, if* T *has no eigenvalue on the unit circle, then* $T_A^*$ *has no eigenvalue on the unit circle.*

**PROOF.** Using the form of the orthogonal projection $P_{\mathcal{F}}$ in (3.1), along with (2.11) we have

$$\Pi_A \omega P_{\mathcal{F}} = \Pi_A \omega D_A T D_{AT}^{-2} T^* D_A = D_A D_{AT}^{-2} T^* D_A = \\ D_A D_{AT}^{-2} T^* D_A^2 D_A^{-1} = D_A T_A^* D_A^{-1} . \tag{3.5}$$

Therefore $\Pi_A \omega P_{\mathcal{F}}$ is similar to $T_A^*$. Substituting

$$D_A T_A^* D_A^{-1} = \Pi_A \omega P_{\mathcal{F}} \tag{3.6}$$

into the second component of the central intertwining lifting $B_\gamma$ of A in (2.15) and using (2.11) we have

$$\Pi' \omega P_{\mathcal{F}} (I - \lambda \Pi_A \omega P_{\mathcal{F}})^{-1} D_A = \Pi' \omega P_{\mathcal{F}} (I - \lambda D_A T_A^* D_A^{-1})^{-1} D_A = \\ \Pi' \omega P_{\mathcal{F}} D_A (I - \lambda T_A^*)^{-1} = \Pi' \omega D_A T D_{AT}^{-2} T^* D_A^2 (I - \lambda T_A^*)^{-1} = \\ D' A T_A^* (I - \lambda T_A^*)^{-1} . \tag{3.7}$$

This yields the form of $B_\gamma$ in (3.3). Now notice that if a is in $\ker T^*$, then $T_A^* D_A^{-2} a = D_{AT}^{-2} T^* a = 0$. This along with the formula for $B_\gamma$ in (3.3) gives (3.4).

Now assume that T has no eigenvalue on the unit circle. To show that $T_A^*$ has no eigenvalue on the unit circle it is sufficient to show that $\Pi_A \omega P_{\mathcal{F}}$ has no eigenvalue of modulus one. If $\Pi_A \omega P_{\mathcal{F}} x = \lambda x$ for some x in $\mathcal{D}_A$ and $|\lambda| = 1$, then $\|x\| = \|\Pi_A \omega P_{\mathcal{F}} x\| \le \|P_{\mathcal{F}} x\| \le \|x\|$. Hence $\|P_{\mathcal{F}} x\| = \|x\|$ and x is in $\mathcal{F}$. Because A is strictly bounded by $\gamma$ we have $x = D_A T h$ for some h in $\mathcal{H}$. It follows that $\lambda D_A T h = \Pi_A \omega P_{\mathcal{F}} D_A T h = D_A h$, due to the second identity in (2.11). Since $D_A$ is invertible, we have $\lambda T h = h$. So $T h = \bar{\lambda} h$. Therefore $h = 0$ and $T_A^*$ has no eigenvalues of modulus one. The proof is now complete.

Notice that the spectral radius of $T_A^*$ can be one. In fact, $T_A^*$ can have eigenvalues on the unit circle. For example, choose $A = 0$, then $T_A^* = T^*$ is a co-isometry. In this case $T_A^*$ and $T^*$ have the same eigenvalues on the unit circle.

Actually one can establish (3.4) directly from the formula for the central intertwining lifting for $B_\gamma$ expressed in equation (2.15), when $\|A\| < \gamma$. To see this first observe that

$$\mathcal{G} := \mathcal{D}_A \ominus \mathcal{F} = D_A^{-1} \mathcal{L} \qquad (\mathcal{L} = \ker T^*) . \tag{3.8}$$

To verify (3.8), notice that a vector g is in $\mathcal{G}$ if and only if g is orthogonal to $D_A T \mathcal{H}$, or equivalently, $D_A g$ is orthogonal to $T\mathcal{H}$. Since $D_A$ is invertible, g is in $\mathcal{G}$ if and only if $g = D_A^{-1} a$ where $D_A g = a$ is orthogonal to $T\mathcal{H}$, or equivalently, $g = D_A^{-1} a$ where a is in $\ker T^*$. Therefore (3.8) holds. Now applying $D_A^{-2} a$ with a in $\ker T^*$ to the central intertwining lifting $B_\gamma$ in (2.15) along with (3.8) gives $B_\gamma D_{\bar{A}}^{-2}a = AD_{\bar{A}}^{-2}a \oplus 0$. This readily establishes (3.4).

The following result provides an explicit formula for the central intertwining lifting which can be used to compute a solution to the standard Nevanlinna-Pick interpolation problem.

**COROLLARY 3.2.** *Let* T *be the unilateral shift on* $\mathcal{H} = H^2(\mathcal{E}_1)$ *and* S *the unilateral shift on* $H^2(\mathcal{E}_2)$. *Let* $\mathcal{H}'$ *be an invariant subspace for* $S^*$ *and* $T'$ *the compression of* S *to* $\mathcal{H}'$, *that is,* $T' = P_{\mathcal{H}'}S \mid \mathcal{H}'$. *Assume that* A *is an operator from* $\mathcal{H} = H^2(\mathcal{E}_1)$ *into* $\mathcal{H}'$ *intertwining* T *with* $T'$ *and* $\|A\| < \gamma$. *Let* $\Psi : \mathbb{D} \to \mathcal{L}(\mathcal{H}, \mathcal{E}_2)$ *be the analytic operator valued function defined by*

$$\Psi(\lambda)h = ((S - T')Ah)(\lambda) \qquad (\lambda \in \mathbb{D} \text{ and } h \in \mathcal{H}), \tag{3.9}$$

*and let* $\Pi_o$ *be the operator from* $H^2(\mathcal{E}_1)$ *onto* $\mathcal{E}_1$ *defined by* $\Pi_o g = g(0)$. *Finally, let* $G_\gamma$ *be the operator valued function defined by*

$$G_\gamma(\lambda) = (A\Pi_o^*)(\lambda) + \Psi(\lambda)T_A^*(I - \lambda T_A^*)^{-1}\Pi_o^* \qquad (\lambda \in \mathbb{D}). \tag{3.10}$$

*Then* $G_\gamma(\lambda)$ *is a well defined function in* $H^\infty(\mathcal{E}_1, \mathcal{E}_2)$ *satisfying* $\|G_\gamma\|_\infty \le \gamma$. *Moreover,* $B = M_{G_\gamma}^+$ *is the central intertwining lifting of* A *with respect to* S *and the tolerance* $\gamma$.

**PROOF.** Obviously S is an isometric lifting of $T'$. By Theorem 2.3 the central intertwining lifting B of A with respect to S is given by $B = WB_\gamma$, where W is the isometry defined in (1.9) for $U_1 = S$. According to (1.9) with $C = T'$ and $D = D'$ we have

$$W\begin{bmatrix} h' \\ D'f \end{bmatrix} = h' + \sum_{n=0}^{\infty} S^n(S - T')f_n .$$

Here $h' \in \mathcal{H}'$ and $f \in H^2(\mathcal{D}')$. Furthermore, $f_n$ is the n-th Taylor coefficient of f. It follows that

$$Bh = W\begin{bmatrix} Ah \\ D'AT_A^*(I - \lambda T_A^*)^{-1}h \end{bmatrix} = Ah + \sum_{n=0}^{\infty} S^n(S - T')A(T_A^*)^{n+1}h \qquad (h \in \mathcal{H}).$$

Now, let $\Psi : \mathbb{D} \to \mathcal{L}(\mathcal{H}, \mathcal{E}_2)$ be given by (3.9). Notice that the k-th Taylor coefficient $\Psi_k$ of $\Psi(\cdot)$ is equal to $\Pi_k'(S - T')A$, where $\Pi_k'$ is the operator from $H^2(\mathcal{E}_2)$ into $\mathcal{E}_2$ which assigns to each function in $H^2(\mathcal{E}_2)$ its k-th Taylor coefficient. It follows that

$$\Pi'_k Bh = \Pi'_k Ah + \sum_{n=0}^{k} \Pi'_{k-n}(S - T')A(T_A^*)^{n+1}h$$

$$= \Pi'_k Ah + \left[\sum_{n=0}^{k} \Psi_{k-n}(T_A^*)^n\right] T_A^* h \qquad (h \in \mathcal{H}).$$

Recall that $\mathcal{H} = H^2(\mathcal{E}_1)$ and T is the forward shift on $H^2(\mathcal{E}_1)$. Since BT = SB, we know that $B = M_G^+$, where G in $\mathcal{H}^\infty(\mathcal{E}_1, \mathcal{E}_2)$ is given by

$$G(\lambda) = \sum_{n=0}^{\infty} \lambda^n \Pi'_n B \Pi_o^* .$$

Thus, by the above calculation, $G = G_\gamma$, where $G_\gamma$ is defined by (3.10). Clearly, $\|G_\gamma\|_\infty = \|G\|_\infty = \|B\| \le \gamma$. This completes the proof.

By mimicking the proof of the previous theorem we readily obtain the following result, which can be used to obtain an explicit formula for the two-sided Sarason interpolation problem.

**COROLLARY 3.3.** *Let* V *and* V′ *be bilateral shifts on* $L^2(\mathcal{E}_1)$ *and* $L^2(\mathcal{E}_2)$, *respectively. Let* $\mathcal{H}$ *be an invariant subspace for* V *containing* $\mathcal{E}_1 (\subseteq \mathcal{H})$ *and* $\mathcal{H}'$ *an invariant subspace for* $V'^*$. *Let* T *on* $\mathcal{H}$ *be the isometry and* T′ *on* $\mathcal{H}'$ *the co-isometry defined by* $T = V | \mathcal{H}$ *and* $T'^* | \mathcal{H}'$. *Assume that* A *is an operator from* $\mathcal{H}$ *into* $\mathcal{H}'$ *intertwining* T *with* T′ *and* $\|A\| < \gamma$. *Let* $\Psi$ *be the operator valued function on the unit circle with values in* $\mathcal{L}(\mathcal{H}, \mathcal{E}_2)$ *defined by*

$$\Psi(e^{it})h = ((V - T')Ah)(e^{it}) \qquad (h \in \mathcal{H}), \tag{3.11}$$

*and let* $\Pi_o$ *be the operator from* $L^2(\mathcal{E}_1)$ *onto* $\mathcal{E}_1$ *which picks out the zero-th Fourier coefficient. Finally, let* $G_\gamma$ *be the operator valued function defined by*

$$G_\gamma(e^{it}) = (A\Pi_o^*)(e^{it}) + \Psi(e^{it})T_A^*(I - e^{it}T_A^*)^{-1}\Pi_o^* . \tag{3.12}$$

*Then* $G_\gamma$ *is a well defined function in* $L^\infty(\mathcal{E}_1, \mathcal{E}_2)$ *satisfying* $\|G_\gamma\|_\infty \le \gamma$. *Moreover,* $B = M_{G_\gamma} | \mathcal{H}$ *is the central intertwining lifting of* A *with respect to* V′ *and tolerance* $\gamma$.

## IV.4 CENTRAL INTERTWINING LIFTING QUOTIENT FORMULAS

In this section we will show that in many cases the central intertwining lifting can be expressed as $PQ^{-1}$, where Q is an outer function. To this end, recall that a subspace $\mathcal{M}$ is cyclic for an operator C on $\mathcal{H}$ if $\mathcal{H}$ equals the closed linear span of $\{C^n \mathcal{M} : n \ge 0\}$. Let $Q(\lambda)$ be an analytic function in the open unit disc whose values are linear operators from $\mathcal{E}_1$ into $\mathcal{E}_2$. Then $Q(\lambda)$ is *outer* if $Q(\lambda)a$ is in $H^2(\mathcal{E}_2)$ for all a in $\mathcal{E}_1$ and $Q\mathcal{E}_1$ is cyclic for the unilateral shift on

$H^2(\mathcal{E}_2)$. Here the symbol Q is also used to denote the mapping from $\mathcal{E}_1$ into $H^2(\mathcal{E}_2)$ defined by $(Qa)(\lambda) = Q(\lambda)a$. The two main results are the following theorems.

**THEOREM 4.1.** *Let* T *be the unilateral shift on* $H^2(\mathcal{E}_1)$ *and* S *the unilateral shift on* $H^2(\mathcal{E}_2)$. *Let* $\mathcal{H}'$ *be an invariant subspace for* $S^*$ *and* $T'$ *on* $\mathcal{H}'$ *the contraction obtained by compressing* S *to* $\mathcal{H}'$. *Assume that* A *is an operator from* $H^2(\mathcal{E}_1)$ *into* $\mathcal{H}'$ *intertwining* T *with* $T'$ *and* $\|A\| < \gamma$. *Let* P *and* Q *be the operator valued analytic functions defined by*

$$P(\lambda)a = (AD_A^{-2}\Pi_o^*a)(\lambda) \quad and \quad Q(\lambda)a = (D_A^{-2}\Pi_o^*a)(\lambda) \qquad (a \in \mathcal{E}_1) \tag{4.1}$$

*where* $\Pi_o$ *is the operator from* $H^2(\mathcal{E}_1)$ *onto* $\mathcal{E}_1$ *defined by* $\Pi_o g = g(0)$. *Then* $Q^{-1}$ *is an outer function in* $H^\infty(\mathcal{E}_1, \mathcal{E}_1)$ *which is given by the following state space formula*

$$Q(\lambda)^{-1} = (\Pi_o D_A^{-2}\Pi_o^*)^{-1}\Pi_o(I - \lambda T_A^*)^{-1}\Pi_o^*, \tag{4.2}$$

*where* $T_A^* := D_{AT}^{-2}T^*D_A^2$. *Furthermore, set*

$$G_\gamma(\lambda) = P(\lambda)Q(\lambda)^{-1}. \tag{4.3}$$

*Then* $G_\gamma$ *is a well defined function in* $H^\infty(\mathcal{E}_1, \mathcal{E}_2)$ *satisfying* $\|G_\gamma\|_\infty \le \gamma$. *Finally,* $B = M_{G_\gamma}^+$ *is the central intertwining lifting of* A *with respect to* S *and tolerance* $\gamma$.

Notice that P and Q in (4.1) are analytic functions defined on $\mathbb{D}$ with values in $\mathcal{L}(\mathcal{E}_1, \mathcal{E}_2)$ and $\mathcal{L}(\mathcal{E}_1, \mathcal{E}_1)$, respectively. Obviously $P(\lambda)a$ and $Q(\lambda)a$ are in $H^2(\mathcal{E}_2)$ and $H^2(\mathcal{E}_1)$, respectively.

**THEOREM 4.2.** *Let* T *be a unilateral shift on* $H^2(\mathcal{E}_1)$ *and* V *a bilateral shift on* $L^2(\mathcal{E}_2)$. *Let* $\mathcal{H}'$ *be an invariant subspace for* $V^*$ *and* $T'$ *on* $\mathcal{H}'$ *the co-isometry obtained by compressing* V *to* $\mathcal{H}'$. *Assume that* A *is an operator from* $H^2(\mathcal{E}_1)$ *into* $\mathcal{H}'$ *intertwining* T *with* $T'$ *and* $\|A\| < \gamma$. *Let* P *and* Q *be the operator valued functions on the unit circle given by*

$$P(e^{it})a = (AD_A^{-2}\Pi_o^*a)(e^{it}) \quad and \quad Q(e^{it})a = (D_A^{-2}\Pi_o^*a)(e^{it}) \qquad (a \in \mathcal{E}_1). \tag{4.4}$$

*Then* $Q^{-1}$ *is an outer function in* $H^\infty(\mathcal{E}_1, \mathcal{E}_1)$ *whose inverse is given by (4.2). Furthermore, set* $G_\gamma = PQ^{-1}$. *Then* $G_\gamma$ *is a well defined function in* $L^\infty(\mathcal{E}_1, \mathcal{E}_2)$ *satisfying* $\|G_\gamma\|_\infty \le \gamma$. *Moreover, the operator* $B = M_{G_\gamma} | H^2(\mathcal{E}_1)$ *from* $H^2(\mathcal{E}_1)$ *into* $L^2(\mathcal{E}_2)$ *is the central intertwining lifting of* A *with respect to* V *and tolerance* $\gamma$.

To prove the previous theorems we need the following two lemmas.

**LEMMA 4.3.** *Let* $T = T_o \oplus T_u$ *on* $\mathcal{H} = \mathcal{H}_o \oplus \mathcal{H}_u$ *be the Wold decomposition of the isometry* T *where* $T_o$ *and* $T_u$ *are respectively the unilateral shift and unitary part of* T. *Moreover, let* $\mathcal{L} = \ker T^*$, *and* $T'$ *be a contraction on* $\mathcal{H}'$. *Let* A *be an operator from* $\mathcal{H}$ *to* $\mathcal{H}'$ *strictly bounded by* $\gamma$ *satisfying* $T'A = AT$. *Then the orthogonal projection of the space* $\mathcal{M} = \bigvee \{T^n D_A^{-2} \mathcal{L} : n \geq 0\}$ *on* $\mathcal{H}_o$ *is onto, that is,* $P_{\mathcal{H}_o} \mathcal{M} = \mathcal{H}_o$. *In particular, if* T *is a unilateral shift, then* $\mathcal{M} = \mathcal{H}$.

**PROOF.** Assume that there exists an h in $\mathcal{H}$ orthogonal to $T^n D_A^{-2} \mathcal{L}$ for all $n \geq 0$. This implies that $D_A^{-2} T^{*n} h$ is orthogonal to $\mathcal{L}$ for all $n \geq 0$. Since $\mathcal{L} = \ker T^*$, we see that $D_A^{-2} T^{*n} h$ is contained in the range of T. Therefore

$$D_A^{-2} T^{*n} h = T h_n \tag{4.5}$$

for some $h_n$ in $\mathcal{H}$. This readily implies that

$$T^{*n+1} h = T^* T^{*n} h = T^* D_A^2 T h_n = D_{AT}^2 h_n .$$

Thus $h_n = D_{AT}^{-2} T^{*n+1} h$. Using this and (4.5) we have

$$T D_{AT}^{-2} T^{*n+1} h = T h_n = D_A^{-2} T^{*n} h . \tag{4.6}$$

By applying $D_A^2$ to both sides of (4.5) and using the definition of $T_A = D_A^2 T D_{AT}^{-2}$, we obtain the formula $T^{*n} h = T_A T^{*n+1} h$ for all $n \geq 0$. Thus

$$h = T_A T^* h = T_A^2 T^{*2} h = T_A^3 T^{*3} h = \cdots = T_A^n T^{*n} h .$$

However, by Proposition 3.1 the operator $T_A$ is similar to a contraction. So $T_A$ is power bounded. In other words, $\|T_A^n\| \leq \beta < \infty$ for some $\beta$ and all $n \geq 0$. In case T is a unilateral shift, $T^{*n} \to 0$ as $n \to \infty$. Thus $h = T_A^n T^{*n} h$ for all $n \geq 0$ implies $h = 0$. Hence $\mathcal{M} = \mathcal{H}$.

For the general case notice that if $h = h_o \oplus h_u$ is in $\mathcal{M}^\perp = \mathcal{H} \ominus \mathcal{M}$, then

$$\|h_o\|^2 + \|h_u\|^2 = \|h\|^2 = \|T_A^n T^{*n} h\|^2 \leq \beta^2 \|T^{*n} h\|^2 \to \beta^2 \|h_u\|^2$$

as n approaches infinity. Clearly $\|h_o\|^2 \leq (\beta^2 - 1)\|h_u\|^2$ and hence $\mathcal{H}_1 = P_{\mathcal{H}_u} \mathcal{M}^\perp$ is a closed subspace. Moreover, there exists a bounded operator X from $\mathcal{H}_1$ into $\mathcal{H}_o$ such that $X h_u = h_o$ for all $h = h_o \oplus h_u$ in $\mathcal{M}^\perp$, and

$$\mathcal{M}^\perp = \{X h_1 \oplus h_1 : h_1 \in \mathcal{H}_1\} .$$

In particular, $\mathcal{M}^\perp = \operatorname{ran} R$ where R is the operator from $\mathcal{H}_1$ into $\mathcal{H}_o \oplus \mathcal{H}_u$ defined by $R h_1 = X h_1 \oplus h_1$. Therefore

$$\mathcal{M} = (\mathcal{M}^{\perp})^{\perp} = \ker R^* = \{h_o \oplus (-X^* h_o \oplus y) : h_o \in \mathcal{H}_o \text{ and } y \in \mathcal{H}_u \ominus \mathcal{H}_1\}.$$

This obviously implies that $P_{\mathcal{H}_o}\mathcal{M} = \mathcal{H}_o$. The proof is now complete.

**LEMMA 4.4.** *Let* T *be a unilateral shift on* $H^2(\mathcal{E}_1)$ *and* A *an operator strictly bounded by* $\gamma$ *satisfying* $T'A = AT$, *where* $T'$ *on* $\mathcal{H}'$ *is a contraction. Let* $Q(\lambda)$ *be the analytic function on* $\mathbb{D}$ *whose values are linear operators on* $\mathcal{E}_1$ *defined by*

$$Q(\lambda)a = (D_A^{-2}\Pi_o^* a)(\lambda) \qquad (a \in \mathcal{E}_1) \tag{4.7}$$

*where* $\Pi_o$ *is the operator from* $H^2(\mathcal{E}_1)$ *into* $\mathcal{E}_1$ *defined by* $\Pi_o g = g(0)$. *Then* $Q(\lambda)$ *is an outer function and*

$$Q(\lambda)^{-1} = (\Pi_o D_A^{-2}\Pi_o^*)^{-1}\Pi_o(I - \lambda T_A^*)^{-1}\Pi_o^* \tag{4.8}$$

*where* $T_A^* = D_{AT}^{-2}T^* D_A^2$.

**PROOF.** Notice that if $h = \Sigma h_n e^{int}$ is in $H^2(\mathcal{E}_1)$, then for $|\lambda| < 1$

$$\Pi_o(I - \lambda T^*)^{-1}h = \sum_{n=0}^{\infty} \lambda^n \Pi_o T^{*n} h = \sum_{n=0}^{\infty} h_n \lambda^n = h(\lambda) \qquad (h \in H^2(\mathcal{E}_1)). \tag{4.9}$$

So using $h(e^{it}) = (D_A^{-2}\Pi_o^* a)(e^{it})$ for a in $\mathcal{E}_1$ in (4.9) we have by the definition of Q in (4.7)

$$Q(\lambda)a = \Pi_o(I - \lambda T^*)^{-1}D_A^{-2}\Pi_o^* a.$$

It follows that

$$Q(\lambda) = \Pi_o(I - \lambda T^*)^{-1}D_A^{-2}\Pi_o^* = \Pi_o D_A^{-2}\Pi_o^* + \lambda\Pi_o(I - \lambda T^*)^{-1}T^* D_A^{-2}\Pi_o^*. \tag{4.10}$$

Notice that $D = \Pi_o D_A^{-2}\Pi_o^*$ is an invertible operator on $\mathcal{E}_1$. Thus by Lemma III.1.2 we obtain

$$Q(\lambda)^{-1} = D^{-1} - \lambda D^{-1}\Pi_o(I - \lambda R)^{-1}T^* D_A^{-2}\Pi_o^* D^{-1}, \tag{4.11}$$

where $R = T^* - T^* D_A^{-2}\Pi_o^* D^{-1}\Pi_o$. Notice formula (4.11) holds in a neighborhood of zero. Using the fact that $T_A^* T = I$ we see that

$$D_{AT}^{-2}T^* D_A^2 = T_A^* = T_A^* T T^* + T_A^* P_{\mathcal{E}_1} = T^* + Y\Pi_o \tag{4.12}$$

where Y is an operator from $\mathcal{E}_1$ to $H^2(\mathcal{E}_1)$ which we are going to determine. By applying $D_A^{-2}\Pi_o^*$ on the right to both sides of (4.12) we have

$$0 = D_{AT}^{-2}T^*\Pi_o^* = T^* D_A^{-2}\Pi_o^* + Y\Pi_o D_A^{-2}\Pi_o^* = T^* D_A^{-2}\Pi_o^* + YD.$$

Thus $Y = -T^*D_A^{-2}\Pi_o^*D^{-1}$. Substituting this into equation (4.12) yields

$$T_A^* = T^* - T^*D_A^{-2}\Pi_o^*D^{-1}\Pi_o = R. \tag{4.13}$$

We see that $R = T_A^*$. Moreover, (4.13) gives

$$T_A^*\Pi_o^* = (T_A^* - T^*)\Pi_o^* = -T^*D_A^{-2}\Pi_o^*D^{-1}\Pi_o\Pi_o^* = -T^*D_A^{-2}\Pi_o^*D^{-1}.$$

Hence for $\lambda$ in a neighborhood of zero we can rewrite (4.11) as follows

$$Q(\lambda)^{-1} = D^{-1} + \lambda D^{-1}\Pi_o(I - \lambda T_A^*)^{-1}T_A^*\Pi_o^* = D^{-1}\Pi_o(I - \lambda T_A^*)^{-1}\Pi_o^*.$$

Since $D = \Pi_o D_A^{-2}\Pi_o^*$, we have established (4.8) for $\lambda$ in a neighborhood of zero. Because $r_{spec}(T_A^*) \le 1$, this implies that (4.11) and thus (4.8) holds for all $|\lambda| < 1$.

Lemma 4.3 shows that $D_A^{-2}\Pi_o^*\mathcal{E}_1 = Q(\lambda)\mathcal{E}_1$ is cyclic for the unilateral shift T. Hence $Q(\lambda)$ is outer. This completes the proof.

**PROOF OF THEOREM 4.1.** Obviously S is an isometric lifting of T′. Proposition 1.3 shows that there exists an isometry W from $\mathcal{K}' = \mathcal{H}' \oplus H^2(\mathcal{D}')$ into $H^2(\mathcal{E}_2)$ satisfying $SW = WU'$ and $W|\mathcal{H}' = I$, where U′ on $\mathcal{K}'$ is the Sz.-Nagy-Schäffer minimal isometric dilation of T′ in (2.2). According to Theorem 2.3 the operator $WB_\gamma$ is the central intertwining lifting of A with respect to S and tolerance $\gamma$. Using the fact that $WB_\gamma$ intertwines the unilateral shift T with S, there exists a function $G_\gamma$ in $H^\infty(\mathcal{E}_1, \mathcal{E}_2)$ such that $WB_\gamma = M_{G_\gamma}^+$. Moreover, $\|G_\gamma\|_\infty = \|B_\gamma\| \le \gamma$. Finally, according to (3.4) and $W|\mathcal{H}' = I$

$$G_\gamma(\lambda)Q(\lambda)a = (WB_\gamma D_A^{-2}\Pi_o^*a)(\lambda) = (AD_A^{-2}\Pi_o^*a)(\lambda) = P(\lambda)a \qquad (a \in \mathcal{E}_1). \tag{4.14}$$

Lemma 4.4 shows that Q is outer and $Q(\lambda)$ is invertible for $\lambda$ the open unit disc. So inverting Q in (4.14) gives $G_\gamma = PQ^{-1}$. Corollary 6.5 or Remark 4.5 below shows that $Q(\lambda)^{-1}$ is in $H^\infty(\mathcal{E}_1, \mathcal{E}_1)$. This completes the proof.

By mimicking the proof of Theorem 4.1 we readily obtain the proof of Theorem 4.2.

**REMARK 4.5.** Theorems 4.1 and 4.2 have the following abstract generalizations which may be useful in certain applications. To be specific, let T on $\mathcal{H}$ be a unilateral shift and V on $\mathcal{K}'$ an isometric lifting of T′ on $\mathcal{H}'$. Assume that A intertwines T with T′ and $\|A\| < \gamma$. Let E be any unitary operator from a Hilbert space $\mathcal{U}$ onto $\mathcal{L} = \ker T^*$. Now let L from $\ell_+^2(\mathcal{U})$ into $\mathcal{K}'$ and R from $\ell_+^2(\mathcal{U})$ into $\mathcal{H}$ be the linear maps defined by

$$L = [AD_A^{-2}E,\ VAD_A^{-2}E,\ V^2AD_A^{-2}E,\ \ldots],$$
$$R = [D_A^{-2}E,\ TD_A^{-2}E,\ T^2D_A^{-2}E,\ \ldots]. \tag{4.15}$$

Clearly, these maps are well defined on a dense set in $\ell_+^2(\mathcal{U})$. Moreover, Lemma 4.3 shows that the range of R is dense in $\mathcal{H}$. We claim that $R^{-1}$ exists as a bounded operator, and $B_\gamma = LR^{-1}$ is precisely the central intertwining lifting of A with tolerance $\gamma$. For a precise interpretation of this formula let $\mathcal{F}$ be the linear space consisting of all sequences $f \in \ell_+^2(\mathcal{U})$ of finite support, and put $\mathcal{M} = R\mathcal{F}$. Then $\mathcal{M}$ is dense in $\mathcal{H}$ (by Lemma 4.3) and there exists a unique bounded linear operator S from $\mathcal{H}$ into $\ell_+^2(\mathcal{U})$ such that

$$RSh = h \quad (h \in \mathcal{M}), \qquad SRf = f \quad (f \in \mathcal{F}).$$

Moreover, we show that $B_\gamma Rf = Lf$ for each $f \in \mathcal{F}$, and hence

$$B_\gamma h = LSh \qquad (h \in \mathcal{M}).$$

So the equality $B_\gamma = LR^{-1}$ has also to be understood in a special way, namely LS is well-defined and bounded on $\mathcal{M}$, and its continuous extension to $\mathcal{H}$ is $B_\gamma$. Let N on $\mathcal{U}$ be the positive square root of $E^*D_A^{-2}E$, and let $\Theta$ be the operator from $\mathcal{H}$ into $\ell_+^2(\mathcal{U})$ defined by

$$\Theta = \mathrm{diag}\,[N, N, N, \ldots]R^{-1}. \tag{4.16}$$

Then $\Theta$ is an operator with dense range satisfying $\gamma^2 I - B_\gamma^* B_\gamma = \Theta^*\Theta$. Using $VB_\gamma = B_\gamma T$ it follows that $T^* D_{B_\gamma}^2 T = D_{B_\gamma}^2$. So up to a Fourier transform $D_{B_\gamma}^2$ is a positive Toeplitz operator and $\Theta$ is its outer spectral factor; see the appendix for the definition of an outer spectral factor.

Let us give a direct proof of some of these results. Recall that if $B_\gamma$ is the central intertwining lifting of A with respect to V, then $B_\gamma D_A^{-2}E = AD_A^{-2}E$. Using $VB_\gamma = B_\gamma T$, it follows that $B_\gamma T^n D_A^{-2}E = V^n AD_A^{-2}E$ for all $n \geq 0$. Because $T^n D_A^{-2}E$ and $V^n AD_A^{-2}E$ are respectively the $n+1^{st}$ column of R and L we see that $B_\gamma R = L$. Lemma 4.3 shows that the range of R is dense in $\mathcal{H}$. Therefore the expression $B_\gamma R = L$ uniquely determines the central intertwining lifting $B_\gamma$. To complete the proof it remains to show that R is invertible and thus $B_\gamma = LR^{-1}$. To see this let $f = (f_n)_0^\infty$ be any sequence in $\ell_+^2(\mathcal{U})$ with finite support. Then

$$\gamma^2\|Rf\|^2 = \gamma^2 \sum_{n=0}^{\infty}\sum_{m=0}^{\infty} (T^n D_A^{-2}Ef_n, T^m D_A^{-2}Ef_m) =$$
$$\gamma^2 \sum_{n\geq m} (T^{n-m} D_A^{-2}Ef_n, D_A^{-2}Ef_m) + \gamma^2 \sum_{m>n} (D_A^{-2}Ef_n, T^{m-n} D_A^{-2}Ef_m). \tag{4.17}$$

Using $P_{\mathcal{H}'}V^k A = T'^k A = AT^k$ for all $k \geq 0$ a similar calculation gives

$$\|Lf\|^2 = \sum_{n\geq m} (V^{n-m}AD_A^{-2}Ef_n, AD_A^{-2}Ef_m) + \sum_{m>n} (AD_A^{-2}Ef_n, V^{m-n}AD_A^{-2}f_m) =$$

$$\sum_{n\geq m} (A^*AT^{n-m}D_A^{-2}Ef_n, D_A^{-2}Ef_m) + \sum_{m>n} (D_A^{-2}Ef_n, A^*AT^{m-n}D_A^{-2}Ef_m)\,. \tag{4.18}$$

By combining (4.17), (4.18) along with the fact that $T^{*k}E = 0$ for all $k > 0$, gives

$$\gamma^2\|Rf\|^2 - \|Lf\|^2 = \sum_{n=0}^{\infty} (D_A^{-2}Ef_n, Ef_n) = \sum_{n=0}^{\infty} \|Nf_n\|^2 \geq \varepsilon^2\|f\| \tag{4.19}$$

for some $\varepsilon > 0$. This readily shows that $(\gamma\varepsilon^{-1})\|Rf\| \geq \|f\|$ for all f in a dense set in $\ell_+^2(\mathcal{U})$. Therefore R admits a bounded inverse $R^{-1}$ on a dense set and by continuity can be extended to a bounded operator on all of $\mathcal{H}$. Moreover, $\|R^{-1}\| \leq \gamma\varepsilon^{-1}$. So using $B_\gamma R = L$, it follows that the central intertwining lifting $B_\gamma$ of A is given by $B_\gamma = LR^{-1}$. (This analysis also shows that the function $Q(\lambda)^{-1}$ in Theorems 4.1 and 4.2 is an outer function in the appropriate $H^\infty$ space.)

To show that $\gamma^2 I - B_\gamma^* B_\gamma = \Theta^*\Theta$, let $g = Rf$ where $f = (f_n)_0^\infty$ is a sequence in $\ell_+^2(\mathcal{U})$ with finite support. By employing $B_\gamma g = B_\gamma Rf = Lf$ along with (4.19) we have

$$((\gamma^2 I - B_\gamma^* B_\gamma)g, g) = \gamma^2\|g\|^2 - \|B_\gamma g\|^2 = \gamma^2\|Rf\|^2 - \|Lf\|^2 =$$

$$\|\operatorname{diag}[N, N, N, \ldots]f\|^2 = \|\Theta g\|^2\,.$$

Since this equality holds on a dense set in $\mathcal{H}$, by continuity, it holds for all g in $\mathcal{H}$. Therefore $((\gamma^2 I - B_\gamma^* B_\gamma)g, g) = (\Theta^*\Theta g, g)$ for all g and by the polarization identity $\gamma^2 I - B_\gamma^* B_\gamma = \Theta^*\Theta$. This completes the proof.

## IV.5 THE CENTRAL SCHUR SOLUTION

In this section we will use our previous results to present an explicit formula for the central solution to the Schur interpolation problem. To recall the Schur interpolation problem, let $F(\lambda)$ be the polynomial in $H^\infty(\mathcal{E}_1, \mathcal{E}_2)$ defined by

$$F(\lambda) = A_0 + A_1\lambda + \cdots + A_{n-1}\lambda^{n-1}\,. \tag{5.1}$$

Here the coefficients $A_i$ are operators mapping $\mathcal{E}_1$ into $\mathcal{E}_2$. Now consider the following $H^\infty$ Schur optimization problem associated with:

$$d_\infty = \inf\{\|F - \lambda^n H\|_\infty : H \in H^\infty(\mathcal{E}_1, \mathcal{E}_2)\} \tag{5.2}$$

where $d_\infty$ is the distance from the polynomial F to $\lambda^n$ $H^\infty(\mathcal{E}_1, \mathcal{E}_2)$ in the $H^\infty$ norm. The $H^\infty$ optimization problem in (5.2) is equivalent to finding a function G in $H^\infty(\mathcal{E}_1, \mathcal{E}_2)$ with the smallest $H^\infty$ norm subject to the constraint that the first n Taylor coefficients of G are $A_0, A_1, \ldots, A_{n-1}$. In practice one does not have to compute the optimal solution to the $H^\infty$ Schur optimization problem in (5.2). Instead one is usually interested in finding a function G in $H^\infty(\mathcal{E}_1, \mathcal{E}_2)$ of the form $G = F - \lambda^n H$ and $\|G\|_\infty \le \gamma$, where $\gamma$ is some prespecified tolerance greater than $d_\infty$. In this case we say that G is a *solution to the Schur interpolation problem with respect to the tolerance* $\gamma$. So G is a solution to the Schur interpolation problem with tolerance $\gamma$ if G is a function in $H^\infty(\mathcal{E}_1, \mathcal{E}_2)$ whose first n Taylor coefficients are $A_0, A_1, \ldots, A_{n-1}$ and $\|G\|_\infty \le \gamma$. If $\gamma = d_\infty$ and $\|G\|_\infty = d_\infty$, then this $G = F - \lambda^n H$ is an *optimal solution* to the Schur minimization problem in (5.2). The following result provides a solution to the suboptimal problem.

**PROPOSITION 5.1.** *Let* $F(\lambda)$ *be a polynomial in* $H^\infty(\mathcal{E}_1, \mathcal{E}_2)$ *of the form (5.1). Let* M *be the lower triangular block Toeplitz matrix defined by*

$$M = \begin{bmatrix} A_0 & 0 & 0 & \cdots & 0 \\ A_1 & A_0 & 0 & \cdots & 0 \\ \vdots & \vdots & \vdots & \vdots & \vdots \\ A_{n-1} & A_{n-2} & A_{n-3} & \cdots & A_0 \end{bmatrix}. \tag{5.3}$$

*Then the optimal error* $d_\infty$ *in the Schur* $H^\infty$ *optimization problem (5.2) is* $d_\infty = \|M\|$. *Now let* $\gamma > d_\infty$, *and let* P *and* Q *be the operator valued polynomials defined by*

$$P(\lambda) = [I, \lambda I, \lambda^2 I, \ldots, \lambda^{n-1} I]MD_M^{-2}\Pi_1^* \quad \textit{and} \quad Q(\lambda) = [I, \lambda I, \lambda^2 I, \ldots, \lambda^{n-1} I]D_M^{-2}\Pi_1^*, \tag{5.4}$$

*where* $\Pi_1 = [I, 0, 0, \ldots, 0]$ *and* $D_M$ *is the square root of* $\gamma^2 I - M^*M$. *Then the central solution* $G_\gamma(\lambda) = P(\lambda)Q(\lambda)^{-1}$ *is a function in* $H^\infty(\mathcal{E}_1, \mathcal{E}_2)$ *solving the Schur interpolation problem with respect to the tolerance* $\gamma$, *that is* $\|G_\gamma\|_\infty \le \gamma$ *and* $G_\gamma = F - \lambda^n H$ *for some* H *in* $H^\infty(\mathcal{E}_1, \mathcal{E}_2)$.

**PROOF.** By choosing $\Theta = \lambda^n$ the $H^\infty$ Schur optimization problem in (5.2) is a special case of the left Sarason interpolation problem discussed in Section I.7. Therefore $d_\infty = \|A\|$, where A is the operator from $H^2(\mathcal{E}_1)$ into $\mathcal{H}' = H^2(\mathcal{E}_2) \ominus \lambda^n H(\mathcal{E}_2)$ defined by $A = P_{\mathcal{H}'} M_F^+$. Clearly $\mathcal{H}'$ is the subspace of $H^2(\mathcal{E}_2)$ consisting of polynomials with values in $\mathcal{E}_2$ of degree at most $n-1$. Now let $\mathcal{F}_1$ the Fourier transform from $\ell_+^2(\mathcal{E}_1)$ onto $H^2(\mathcal{E}_1)$ and $\mathcal{F}_2$ the corresponding Fourier transform from $\ell_+^2(\mathcal{E}_2)$ onto $H^2(\mathcal{E}_2)$. Then using the polynomial F in (5.1) we see that $\mathcal{F}_2[M, 0] = A\mathcal{F}_1$, where M is the block Toeplitz matrix in (5.3). In particular, $d_\infty = \|A\| = \|M\|$.

Let S be the unilateral shift on $H^2(\mathcal{E}_2)$ and $T'$ on $\mathcal{H}'$ the compression of S to $\mathcal{H}'$. Since $\mathcal{H}'$ is an invariant subspace for $S^*$, it follows that S is an isometric lifting of $T'$. Using

$P_{\mathcal{H}'}S = T'P_{\mathcal{H}'}$ it is easy to verify that $T'A = AT$, where T is the unilateral shift on $H^2(\mathcal{E}_1)$. According to Theorem 4.1 the function $G_\gamma = P(\lambda)Q(\lambda)^{-1}$ is in $H^\infty(\mathcal{E}_1, \mathcal{E}_2)$, where P and Q are defined in (4.1). Moreover, $B = M^+_{G_\gamma}$ defines an intertwining lifting of A satisfying $\|G_\gamma\|_\infty \leq \gamma$. We claim that $G_\gamma = F - \lambda^n H$ for some H in $H^\infty(\mathcal{E}_1, \mathcal{E}_2)$. To see this notice that because $B = M^+_{G_\gamma}$ is an intertwining lifting of A we have

$$Fa = Aa = P_{\mathcal{H}'}Ba = P_{\mathcal{H}'}G_\gamma a \qquad (a \in \mathcal{E}_1).$$

Since $\mathcal{H}'$ is the space of polynomial of degree at most $n-1$, it follows that $G_\gamma$ admits a power series expansion of the form

$$G_\gamma(\lambda) = A_0 + \lambda A_1 + \cdots + \lambda^{n-1}A_{n-1} - \lambda^n H$$

where H is in $H^\infty(\mathcal{E}_1, \mathcal{E}_2)$. Therefore $G_\gamma$ is of the form $F - \lambda^n H$, and $\|G_\gamma\|_\infty \leq \gamma$. In other words, $G_\gamma$ is a solution to the Schur interpolation problem with respect to the tolerance $\gamma$.

To compute $G_\gamma = PQ^{-1}$. Notice that $A\mathcal{F}_1 = \mathcal{F}_2[M, 0]$ readily gives

$$P(\lambda)a = (AD_A^{-2}\Pi_0^* a)(\lambda) = \mathcal{F}_2[M, 0]\mathcal{F}_1^{-1}D_A^{-2}\Pi_0^* a = [I, \lambda I, \lambda^2 I, \ldots, \lambda^{n-1}I]MD_M^{-2}\Pi_1^* a, \tag{5.5}$$

where a is in $\mathcal{E}_1$. A similar calculation shows that

$$Q(\lambda)a = (D_A^{-2}\Pi_0^* a)(\lambda) = [I, \lambda I, \ldots, \lambda^{n-1}I]D_M^{-2}\Pi_1^* a. \tag{5.6}$$

This completes the proof.

In many applications it may be difficult to compute the inverse of $Q(\lambda)$ for all $\lambda$ in the open unit disc. The following result eliminates this computation.

**THEOREM 5.2.** *Let* $F(\lambda)$ *in* $H^\infty(\mathcal{E}_1, \mathcal{E}_2)$ *be the polynomial in (5.1) corresponding to the Schur interpolation problem with tolerance* $\gamma > d_\infty$, *where* $\mathcal{E}_1$ *is finite dimensional. Let* M *be the block lower triangular Toeplitz matrix in (5.3) and* $T_n$ *the* $n \times n$ *block lower shift on* $\mathcal{E} = \mathcal{E}_1 \oplus \mathcal{E}_1 \oplus \cdots \oplus \mathcal{E}_1$ *(n times), that is, the identity* I *appears immediately below the main diagonal of* $T_n$ *and zeros elsewhere. Finally, let* $\hat{M}$ *on* $\mathcal{E}$ *be the block matrix given by*

$$\hat{M} = (\gamma^2 I - T_n^* M^* M T_n)^{-1} T_n^* (\gamma^2 I - M^* M). \tag{5.7}$$

*Then* $\hat{M}$ *is stable, i.e.,* $r_{spec}(\hat{M}) < 1$. *Moreover, the central solution* $G_\gamma$ *for the Schur interpolation problem with tolerance* $\gamma$ *is given by*

$$G_\gamma(\lambda) = F(\lambda) + \lambda^n [A_{n-1}, A_{n-2}, \ldots, A_0](I - \lambda\hat{M})^{-1}\hat{M}\Pi_1^* \tag{5.8}$$

*where* $\Pi_1 = [I, 0, 0, \ldots, 0]$.

**PROOF**. According to Corollary 3.2 the central intertwining lifting $G_\gamma$ is given by

$$G_\gamma(\lambda) = (A\Pi_o^*)(\lambda) + \Psi(\lambda)(I - \lambda T_A^*)^{-1} T_A^* \Pi_o^* \tag{5.9}$$

where $A = P_{\mathcal{H}'} M_F^\ddagger$ and $\Psi = (S - T')A$. Using $\mathcal{F}_2[M, 0] = A\mathcal{F}_1$ it follows that $(A\Pi_o^*)(\lambda) = F(\lambda)$. Because $\mathcal{H}'$ is the space of polynomials in $H^2(\mathcal{E}_2)$ of degree at most $n-1$, we have

$$\begin{aligned}\Psi\mathcal{F}_1 &= (S - T')A\mathcal{F}_1 = (I - P_{\mathcal{H}'})SA\mathcal{F}_1 = (I - P_{\mathcal{H}'})\lambda\mathcal{F}_2[M, 0] \\ &= \lambda^n [A_{n-1}, A_{n-2}, \ldots, A_0, 0, 0, \ldots] .\end{aligned} \tag{5.10}$$

Now using $A\mathcal{F}_1 = \mathcal{F}_2[M, 0]$ once again we obtain

$$T_A^* \mathcal{F}_1 = (\gamma^2 I - T^* A^* A T)^{-1} T^* (\gamma^2 I - A^* A)\mathcal{F}_1 = \mathcal{F}_1 \begin{bmatrix} \hat{M} & * \\ 0 & * \end{bmatrix} \tag{5.11}$$

where $\hat{M}$ is defined in (5.7) and $*$ is an unspecified entry. Moreover

$$T_A^* \Pi_o^* = \mathcal{F}_1 [\hat{M}\Pi_1^* , 0]^{tr} . \tag{5.12}$$

Substituting (5.10), (5.11) and (5.12) into (5.9) along with $(A\Pi_o^*)(\lambda) = F(\lambda)$ yields (5.8). To prove the stability of $\hat{M}$ we use Proposition 3.1 to recall that $T_A^*$ is similar to a contraction and has no eigenvalue on the unit circle **T** (because T has no eigenvalue on **T**). Since $\hat{M}$ acts on a finite dimensional space, it follows that the spectrum of $\hat{M}$ is in $\mathbb{D}$, and thus $r_{spec}(\hat{M}) < 1$. This completes the proof.

Now let us present a formula to compute the optimal solution to the $H^\infty$ Schur interpolation problem when $\mathcal{E}_1 = \mathbb{C}$ is one dimensional. To this end, let $f(\lambda)$ be the polynomial with values in $\mathcal{E}_2$ defined by

$$f(\lambda) = a_o + a_1\lambda + a_2\lambda^2 + \ldots + a_{n-1}\lambda^{n-1} . \tag{5.13}$$

In this case the $H^\infty$ Schur optimization problem reduces to

$$d_\infty = \inf \{\|f - \lambda^n h\|_\infty : h \in H^\infty(\mathcal{E}_2)\} . \tag{5.14}$$

Here $d_\infty$ is the distance from f to $\lambda^n H^\infty(\mathcal{E}_2)$ in the $H^\infty$ norm. (By a slight abuse of notation we used $H^\infty(\mathcal{E}_2)$ to denote $H^\infty(\mathbb{C}, \mathcal{E}_2)$.) According to our previous analysis $d_\infty = \|M\|$, where M is now the lower triangular $n \times n$ Toeplitz matrix generated by $\{a_j : 0 \le j < n\}$, that is,

$$M = \begin{bmatrix} a_0 & 0 & \dots & 0 \\ a_1 & a_0 & \dots & 0 \\ \vdots & \vdots & & \vdots \\ a_{n-1} & a_{n-2} & \dots & a_0 \end{bmatrix} . \tag{5.15}$$

Notice that the entries of M in (5.15) are vectors in $\mathcal{E}_2$. The following result allows us to compute the $H^\infty$ optimal solution to this Schur minimization problem.

**PROPOSITION 5.3.** *Let* $f(\lambda)$ *be a polynomial with values in* $\mathcal{E}_2$ *of the form (5.13) and* M *the lower triangular block Toeplitz matrix defined in (5.15). Let* y *be any vector in* $\mathbb{C}^n$ *which attains the norm of* M. *Let* $N(\lambda)$ *be the polynomial with values in* $\mathcal{E}_2$ *and* $d(\lambda)$ *the scalar valued polynomial defined by*

$$N(\lambda) = [I, \lambda I, \lambda^2 I, \dots, \lambda^{n-1} I]My \quad \textit{and} \quad d(\lambda) = [I, \lambda I, \dots, \lambda^{n-1} I]y . \tag{5.16}$$

*Then* $g(\lambda) = N(\lambda)/d(\lambda)$ *is the unique function in* $H^\infty(\mathcal{E}_2)$ *solving the* $H^\infty$ *optimization problem in (5.14), that is,* g *is the only function in* $H^\infty(\mathcal{E}_2)$ *whose first* n *Fourier coefficients are* $a_0, a_1, \dots, a_{n-1}$ *and* $\|g\|_\infty = d_\infty$. *Finally,* $g(e^{it})^* g(e^{it}) = d_\infty^2$ *a.e.*

**PROOF.** Here we follow the notation used in the proof of Proposition 5.1, where now $f = F$ and $\mathbb{C} = \mathcal{E}_1$. Thus $A = P_{\mathcal{H}'} M_f^+$. We claim that there is a one to one correspondence between the set of all intertwining liftings B of A and the set of all function g in $H^\infty(\mathcal{E}_2)$ of the form $g = f - \lambda^n h$ for some h in $H^\infty(\mathcal{E}_2)$. Moreover, in this correspondence $\|B\| = \|g\|$. If B is an intertwining lifting of A, then B intertwines the unilateral shifts T with S. Hence $B = M_g^+$ for some g in $H^\infty(\mathcal{E}_2)$. Furthermore, for any a in $\mathcal{E}_1$ we have

$$fa = Aa = P_{\mathcal{H}'} Ba = P_{\mathcal{H}'} ga . \tag{5.17}$$

Because $\mathcal{H}'$ is the space of polynomial of degree at most $n-1$, it follows that $g = f - \lambda^n h$ for some h in $H^\infty(\mathcal{E}_2)$. Obviously $\|B\| = \|g\|_\infty$. On the other hand, if $g = f - \lambda^n h$, then $B = M_g^+$ intertwines T with S. Moreover $P_{\mathcal{H}'} M_g^+ = P_{\mathcal{H}'} M_f^+ = A$. Therefore B is an intertwining lifting of A.

Due to the one to one correspondence between the set of all intertwining liftings B of A and $H^\infty(\mathcal{E}_2)$ functions of the form $g = f - \lambda^n h$ along with $B = M_g^+$ we see that $d_\infty = \|A\|$. According to Corollary 2.7 there is only one intertwining lifting B of A satisfying $\|B\| = \|A\|$. So there is only one $H^\infty$ optimal solution to the Schur interpolation problem in (5.14). Moreover, this optimal solution is given by $g = Ax/x$ where x is any vector which attains the norm of A. Recall that $\mathcal{F}_2[M, 0] = A\mathcal{F}_1$, where $\mathcal{F}_1$ and $\mathcal{F}_2$ are the appropriate Fourier transforms. So x attains the norm of A if and only if $x = \mathcal{F}_1 y$, where y attains the norm of M. Put, $\hat{y} = [y, 0]^{tr}$. Then

$$g = \frac{Ax}{x} = \frac{\mathcal{F}_2[M, 0]\hat{y}}{\mathcal{F}_1\hat{y}} = \frac{N}{d},$$

where N and d are the polynomials defined in (5.16). This completes the proof.

## IV.6 THE QUASI OUTER FACTOR OF $D^2_{B_\gamma}$

In this section we will present some explicit formulas to compute a quasi outer spectral factor for $\gamma^2 I - B_\gamma^* B_\gamma$, where $B_\gamma$ is the central intertwining lifting of A. The main result is Theorem 6.2 which provides the motivation for the definition of a quasi outer spectral factor. As before, A is an operator bounded by a specified $\gamma$ satisfying $T'A = AT$, where T on $\mathcal{H}$ is an isometry, $T'$ on $\mathcal{H}'$ is a contraction. Recall that the central intertwining lifting $B_\gamma$ of A with respect to the Sz.-Nagy Schäffer minimal isometric dilation $U'$ on $\mathcal{H}' \oplus H^2(\mathcal{D}')$ and tolerance $\gamma$ is given by

$$B_\gamma = \begin{bmatrix} A \\ \Pi'\omega P_{\mathcal{F}}(I - \lambda \Pi_A \omega P_{\mathcal{F}})^{-1} D_A \end{bmatrix} : \mathcal{H} \to \begin{bmatrix} \mathcal{H}' \\ H^2(\mathcal{D}') \end{bmatrix}. \tag{6.1}$$

Here $\mathcal{F} = \overline{D_A T \mathcal{H}}$ and the isometry $\omega$ from $\mathcal{F}$ into $\mathcal{D}' \oplus \mathcal{D}_A$ is given by (2.8) while $\Pi'$ and $\Pi_A$ are defined by (2.10). The following lemma will provide us with some useful information concerning the contraction $\Pi_A \omega P_{\mathcal{F}}$.

**LEMMA 6.1.** *There exists a unique positive operator* R *on* $\mathcal{D}_A$ *such that*

$$\|RD_A h\| = \lim_{n \to \infty} \|(\Pi_A \omega P_{\mathcal{F}})^n D_A h\| \qquad (h \in \mathcal{H}). \tag{6.2}$$

*Furthermore, if* T *is a unilateral shift and* $\|A\| < \gamma$, *then* $R = 0$. *In this case* $T_A^{*n}$ *approaches zero strongly as* n *approaches infinity, where* $T_A^* = D_{AT}^{-2} T^* D_A^2$.

**PROOF.** Obviously $\Pi_A \omega P_{\mathcal{F}} = C$ is a contraction. So $C^{*n}C^n$ is a sequence of decreasing operators. This implies that

$$\text{strong} - \lim_{n \to \infty} (P_{\mathcal{F}} \omega^* \Pi_A^*)^n (\Pi_A \omega P_{\mathcal{F}})^n = R_2 \tag{6.3}$$

defines a positive operator $R_2$ on $\mathcal{D}_A$. Let R be the positive square root of $R_2$. Obviously $R^2 = R_2$. According to (6.3) we have (6.2). Notice that (6.2) determines R uniquely. Now assume that T is a unilateral shift and A is strictly bounded by $\gamma$. Using (2.11) we see that $CD_AT = D_A$. Thus

$$C^n D_A T^n = D_A \qquad \text{(for all } n \geq 0) . \tag{6.4}$$

Let $\mathcal{G}$ be the orthogonal complement of $\mathcal{F}$ in $\mathcal{D}_A$. Recall that $\mathcal{G} = D_A^{-1}\mathcal{L}$, where $\mathcal{L} = \ker T^*$; see (3.8). So equation (6.4) gives

$$C^{n+1} D_A T^n D_A^{-2}\mathcal{L} = CD_A^{-1}\mathcal{L} = \Pi_A \omega P_{\mathcal{F}} D_A^{-1}\mathcal{L} = 0 . \tag{6.5}$$

According to Lemma 4.3, the space $D_A^{-2}\mathcal{L}$ is cyclic for T. Therefore the closed linear span of $D_A T^n D_A^{-2}\mathcal{L}$ for $n \geq 0$ is $\mathcal{H}$. This along with (6.5) shows that the limit in (6.2) is zero. Finally, recall that $T_A^*$ is similar to $C = \Pi_A \omega P_{\mathcal{F}}$; see (3.6). Thus $T_A^{*n} \to 0$ strongly as $n \to \infty$. This completes the proof.

Notice that in general the operator R is not zero. For example, if $\|A\| < \gamma$, then there exists an $\varepsilon > 0$ satisfying $0 < \varepsilon^2 I \leq D_A^2 \leq \gamma^2 I$. If, in addition, T is not a unilateral shift, then there exists a nonzero h in $\mathcal{H}_u = \cap\{T^n\mathcal{H} : n \geq 0\}$. Recall that from the Wold decomposition of an isometry it follows that $T \mid \mathcal{H}_u$ is unitary. So using (2.11) for this h we have

$$\|(\Pi_A \omega P_{\mathcal{F}})^n D_A h\|^2 = \|(\Pi_A \omega P_{\mathcal{F}})^n D_A T^n T^{*n} h\|^2 =$$

$$\|D_A T^{*n} h\|^2 \geq \varepsilon^2 \|T^{*n} h\|^2 = \varepsilon^2 \|h\|^2 \neq 0 .$$

According to (6.2) we see that $\|RD_A h\| \geq \varepsilon \|h\|$. Hence R is nonzero.

Equation (6.2) along with the second equation in (2.11) gives

$$\|RD_A Th\|^2 = \lim_{n\to\infty} \|(\Pi_A \omega P_{\mathcal{F}})^n D_A Th\|^2 = \lim_{n\to\infty} \|(\Pi_A \omega P_{\mathcal{F}})^{n-1} D_A h\|^2 = \|RD_A h\|^2 . \tag{6.6}$$

Let $\mathcal{R}$ be the closure of the range of R. By (6.6) there exists an isometry $W_u$ on $\mathcal{R}$ satisfying $RD_A T = W_u RD_A$. However, because $\mathcal{F} = \overline{D_A T\mathcal{H}}$ and the range of R is contained in $\mathcal{F}$ (see (6.3)), we have that $\{RD_A T\mathcal{H}\}^-$ equals $\overline{R\mathcal{D}_A}$, and so $RD_A T\mathcal{H}$ is dense in $\mathcal{R}$. Thus $W_u$ is onto. In other words, there is a unitary operator $W_u$ on $\mathcal{R}$ satisfying

$$RD_A T = W_u RD_A . \tag{6.7}$$

**THEOREM 6.2.** *Let* T *be an isometry on* $\mathcal{H}$ *and* A *an operator bounded by* $\gamma$ *satisfying* $T'A = AT$, *where* $T'$ *is a contraction on* $\mathcal{H}'$. *Let* $B_\gamma$ *be the central intertwining lifting of* A *defined in (6.1). Let* $\mathcal{G} = \mathcal{D}_A \ominus \mathcal{F}$ *and* $\Xi$ *be the operator from* $\mathcal{H}$ *into* $H^2(\mathcal{G}) \oplus \mathcal{R}$ *defined by*

$$\Xi = \begin{bmatrix} P_{\mathcal{G}}(I - \lambda\Pi_A \omega P_{\mathcal{F}})^{-1} D_A \\ RD_A \end{bmatrix} \tag{6.8}$$

*where* R *is the unique positive operator defined in (6.2) and* $\mathcal{R}$ *is the closure of the range of* R. *Then the range of* $\Xi$ *is dense in* $H^2(\mathcal{G}) \oplus \mathcal{R}$ *and*

$$\Xi^* \Xi = \gamma^2 I - B_\gamma^* B_\gamma . \tag{6.9}$$

*Furthermore, if* S *is the unilateral shift on* $H^2(\mathcal{G})$ *and* $W_u$ *is the unitary operator on* $\mathcal{R}$ *defined in (6.7), then*

$$(S \oplus W_u)\Xi = \Xi T . \tag{6.10}$$

Notice that for every fixed h in $\mathcal{H}$, we see that $P_{\mathcal{G}}(I - \lambda\Pi_A \omega P_{\mathcal{F}})^{-1} D_A h$ is a well defined $\mathcal{G}$-valued function for $|\lambda| < 1$, because $\Pi_A \omega P_{\mathcal{F}}$ is a contraction. Moreover, the previous theorem shows that $P_{\mathcal{G}}(I - \lambda\Pi_A \omega P_{\mathcal{F}})^{-1} D_A h$ is in $H^2(\mathcal{G})$ for all h in $\mathcal{H}$. This theorem also suggests the following terminology: An operator X from the Hilbert space $\mathcal{H}$ into a Hilbert space $\mathcal{K}$ will be called a *quasi outer spectral factor* of $D_{B_\gamma}^2$ if the range of X is dense in $\mathcal{K}$ and $X^* X = D_{B_\gamma}^2$. Notice that formula (6.8) provides an explicit description for such a factor.

To motivate the notion of a quasi outer spectral factor. First notice that $U' B_\gamma = B_\gamma T$ implies that $T^* D_{B_\gamma}^2 T = D_{B_\gamma}^2$, and hence if T is a unilateral shift, then $D_{B_\gamma}^2$ is a block Toeplitz operator. Now let X from $\mathcal{H}$ into $\mathcal{K}$ be a quasi outer spectral factor of $D_{B_\gamma}^2$. Then for each $h \in \mathcal{H}$ we have

$$\|XTh\|^2 = (T^* D_{B_\gamma}^2 Th, h) = (D_{B_\gamma}^2 h, h) = \|Xh\|^2 .$$

So there exist an isometry U on $\mathcal{K}$ such that $UX = XT$. Thus, if T and U are both unilateral shifts, then X can be identified with a lower triangular block Toeplitz operator, and hence, after taking Fourier transforms, X is an outer spectral factor (because its range is dense).

**PROOF OF THEOREM 6.2.** For h in $\mathcal{H}$ we have

$$\|D_{B_\gamma} h\|^2 = \gamma^2 \|h\|^2 - \|B_\gamma h\|^2 = \gamma^2 \|h\|^2 - \|Ah\|^2 - \|\Pi' \omega P_{\mathcal{F}}(I - \lambda\Pi_A \omega P_{\mathcal{F}})^{-1} D_A h\|^2 =$$

$$\|D_A h\|^2 - \lim_{n\to\infty} \sum_{i=0}^{n} \|\Pi' \omega P_{\mathcal{F}}(\Pi_A \omega P_{\mathcal{F}})^i D_A h\|^2 .$$

Thus

$$\|D_{B_\gamma}h\|^2 = \|D_A h\|^2 + \lim_{n\to\infty} \sum_{i=0}^{n} (-\|\omega P_{\mathcal{F}}(\Pi_A \omega P_{\mathcal{F}})^i D_A h\|^2 + \|\Pi_A \omega P_{\mathcal{F}}(\Pi_A \omega P_{\mathcal{F}})^i D_A h\|^2) =$$

$$\|D_A h\|^2 - \|\omega P_{\mathcal{F}} D_A h\|^2 + \lim_{n\to\infty} \left[ \sum_{i=0}^{n} \|(\Pi_A \omega P_{\mathcal{F}})^{i+1} D_A h\|^2 - \sum_{i=1}^{n} \|P_{\mathcal{F}}(\Pi_A \omega P_{\mathcal{F}})^i D_A h\|^2 \right] =$$

$$\|P_{\mathcal{G}} D_A h\|^2 + \lim_{n\to\infty} \left[ \|(\Pi_A \omega P_{\mathcal{F}})^{n+1} D_A h\|^2 + \sum_{i=1}^{n} (\|(\Pi_A \omega P_{\mathcal{F}})^i D_A h\|^2 - \|P_{\mathcal{F}}(\Pi_A \omega P_{\mathcal{F}})^i D_A h\|^2) \right] =$$

$$\|P_{\mathcal{G}} D_A h\|^2 + \sum_{i=1}^{\infty} \|P_{\mathcal{G}}(\Pi_A \omega P_{\mathcal{F}})^i D_A h\|^2 + \lim_{n\to\infty} \|(\Pi_A \omega P_{\mathcal{F}})^n D_A h\|^2 .$$

Therefore we see that

$$\|D_{B_\gamma}h\|^2 = \|P_{\mathcal{G}}(I - \lambda \Pi_A \omega P_{\mathcal{F}})^{-1} D_A h\|^2 + \lim_{n\to\infty} \|(\Pi_A \omega P_{\mathcal{F}})^n D_A h\|^2 , \tag{6.11}$$

where the first norm on the right of the equality is the $H^2(\mathcal{G})$-norm of the $\mathcal{G}-$ valued function $P_{\mathcal{G}}(I - \lambda \Pi_A \omega P_{\mathcal{F}})^{-1} D_A h$ for $|\lambda| < 1$. Equations (6.11) and (6.2) show that $\Xi^* \Xi = \gamma^2 I - B_\gamma^* B_\gamma$. The intertwining property in (6.10) follows from (6.7), (2.11) and the following calculation

$$P_{\mathcal{G}}(I - \lambda \Pi_A \omega P_{\mathcal{F}})^{-1} D_A T = P_{\mathcal{G}} D_A T + \lambda P_{\mathcal{G}}(I - \lambda \Pi_A \omega P_{\mathcal{F}})^{-1} \Pi_A \omega P_{\mathcal{F}} D_A T =$$

$$S P_{\mathcal{G}}(I - \lambda \Pi_A \omega P_{\mathcal{F}})^{-1} D_A .$$

It remains to show that the range of $\Xi$ is dense in $H^2(\mathcal{G}) \oplus \mathcal{R}$. To this end, first notice that equation (6.3) shows that $\mathcal{R} \subseteq \mathcal{F}$. Let $g \oplus f$ be any function in $H^2(\mathcal{G}) \oplus \mathcal{R}$ orthogonal to $\Xi\mathcal{H}$ and let $g = \Sigma g_n z^n$ be the power series expansion of g. According to the definition of $\Xi$ in (6.8) we have for all h in $\mathcal{H}$

$$0 = (g \oplus f, \Xi h) = (f, R D_A h) + \sum_{n=0}^{\infty} (g_n, P_{\mathcal{G}}(\Pi_A \omega P_{\mathcal{F}})^n D_A h) =$$

$$(D_A R f, h) + \sum_{n=0}^{\infty} (D_A (P_{\mathcal{F}} \omega^* \Pi_A^*)^n g_n, h) .$$

This implies that

$$D_A[Rf + g_0 + P_{\mathcal{F}}\omega^*\Pi_A^* g_1 + (P_{\mathcal{F}}\omega^*\Pi_A^*)^2 g_2 + \cdots ] = 0 . \tag{6.12}$$

(The sum in (6.12) is finite because it is obtained by applying the bounded operator $\Xi^*$ to $g \oplus f$.) Since $D_A$ is one to one on $\mathcal{D}_A = \mathcal{F} \oplus \mathcal{G}$ and the range of R is contained in $\mathcal{F}$, we see that $g_0 = 0$. This implies that

$$Rf + P_{\mathcal{F}}\omega^*\Pi_A^* g_1 + (P_{\mathcal{F}}\omega^*\Pi_A^*)^2 g_2 + \cdots = 0 . \tag{6.13}$$

Now notice that

$$\|R\Pi_A\omega P_{\mathcal{F}}x\|^2 = \lim_{n\to\infty} \|(\Pi_A\omega P_{\mathcal{F}})^n \Pi_A\omega P_{\mathcal{F}}x\|^2 = \|Rx\|^2 .$$

This implies that there exists an isometry $W_1$ on $\mathcal{R}$ satisfying

$$W_1 R = R\Pi_A\omega P_{\mathcal{F}} . \tag{6.14}$$

Using (2.11) we have

$$W_1 R D_A T^2 = R\Pi_A\omega P_{\mathcal{F}} D_A T^2 = R D_A T .$$

Since $\mathcal{F} = \overline{D_A T\mathcal{H}}$, this implies that $W_1$ is onto. Therefore $W_1$ is a unitary operator satisfying (6.14). By applying the adjoint of (6.14) in (6.13) we have

$$P_{\mathcal{F}}\omega^*\Pi_A^*[RW_1 f + g_1 + (P_{\mathcal{F}}\omega^*\Pi_A^*)g_2 + (P_{\mathcal{F}}\omega^*\Pi_A^*)^2 g_3 + \cdots ] = 0 . \tag{6.15}$$

Notice that $\Pi_A\omega P_{\mathcal{F}}$ is onto a dense set in $\mathcal{D}_A$. This follows because $\Pi_A\omega P_{\mathcal{F}} D_A T = D_A$. Therefore $P_{\mathcal{F}}\omega^*\Pi_A^*$ is one to one. Since the range of R is contained in $\mathcal{F}$ and $g_n$ is in $\mathcal{G}$, equation (6.15) implies that $g_1 = 0$. By continuing in a similar fashion $g_n = 0$ for all $n \geq 0$. Hence $g = 0$.

To complete the proof it remains to show that $f = 0$. According to equation (6.12) we now have $D_A Rf = 0$. Therefore $Rf = 0$. Since f is in $\mathcal{R}$, we have $f = 0$. This completes the proof.

Now assume that A is strictly bounded by $\gamma$. Then the following result provides us with another outer spectral factor $\Xi_{\mathcal{L}}$ for $D_{B_\gamma}^2$ involving the operators $N_A$ on $\mathcal{L} = \ker T^*$ and $T_A$ on $\mathcal{H}$ defined by

$$N_A = (\Pi_o D_A^{-2}\Pi_o^*)^{-1/2} \quad \text{and} \quad T_A^* = D_{AT}^{-2} T^* D_A^2 . \tag{6.16}$$

As before, $\Pi_o$ is the operator from $\mathcal{H}$ onto $\mathcal{L}$ defined by $\Pi_o = P_{\mathcal{L}}$.

**PROPOSITION 6.3.** *Let* T *be an isometry on* $\mathcal{H}$ *and* $\mathcal{L}$ *be the kernel of* $T^*$. *Let* A *be an operator strictly bounded by* $\gamma$ *satisfying* $T'A = AT$, *where* $T'$ *is a contraction on* $\mathcal{H}'$. *Let* $B_\gamma$ *be the central intertwining lifting of* A *given by (6.1). Let* $\Xi_{\mathcal{L}}$ *be the operator mapping* $\mathcal{H}$ *into*

$H^2(\mathcal{L}) \oplus \mathcal{R}$ *defined by*

$$\Xi_{\mathcal{L}} = \begin{bmatrix} N_A \Pi_o (I - \lambda T_A^*)^{-1} \\ R D_A \end{bmatrix}. \tag{6.17}$$

*Then the range of* $\Xi_{\mathcal{L}}$ *is dense in* $H^2(\mathcal{L}) \oplus \mathcal{R}$ *and*

$$\Xi_{\mathcal{L}}^* \Xi_{\mathcal{L}} = \gamma^2 I - B_\gamma^* B_\gamma . \tag{6.18}$$

*Moreover,* $(S \oplus W_u)\Xi_{\mathcal{L}} = \Xi_{\mathcal{L}} T$, *where* $S$ *is the unilateral shift on* $H^2(\mathcal{L})$.

**PROOF.** According to equation (3.6) we have

$$P_{\mathcal{G}}(I - \lambda \Pi_A \omega P_{\mathcal{F}})^{-1} D_A = P_{\mathcal{G}} D_A (I - \lambda T_A^*)^{-1} . \tag{6.19}$$

By consulting (3.8) we have $\mathcal{G} = D_A^{-1}\mathcal{L}$. Now let X be the operator mapping $\mathcal{L}$ into $\mathcal{H}$ defined by $X = D_A^{-1}\Pi_o^*$. Because $\mathcal{G} = D_A^{-1}\mathcal{L}$, the range of X is $\mathcal{G}$. So the orthogonal projection $P_{\mathcal{G}}$ onto $\mathcal{G}$ is given by

$$P_{\mathcal{G}} = X(X^*X)^{-1}X^* = D_A^{-1}\Pi_o^*(\Pi_o D_A^{-2}\Pi_o^*)^{-1}\Pi_o D_A^{-1} = D_A^{-1}\Pi_o^* N_A^2 \Pi_o D_A^{-1} . \tag{6.20}$$

(It is easy to verify that $P_{\mathcal{G}} = P_{\mathcal{G}}^* = P_{\mathcal{G}}^2$ and thus $P_{\mathcal{G}}$ is indeed the orthogonal projection onto $\mathcal{G}$.) By using (6.20) in (6.19) we obtain

$$P_{\mathcal{G}}(I - \lambda \Pi_A \omega P_{\mathcal{F}})^{-1} D_A = D_A^{-1}\Pi_o^* N_A^2 \Pi_o (I - \lambda T_A^*)^{-1} . \tag{6.21}$$

Notice that

$$\Omega = X(X^*X)^{-1/2} = D_A^{-1}\Pi_o^*(\Pi_o D_A^{-2}\Pi_o^*)^{-1/2} = D_A^{-1}\Pi_o^* N_A \tag{6.22}$$

is a unitary operator mapping $\mathcal{L}$ onto $\mathcal{G}$. Obviously $\Omega$ can be trivially extended to a unitary operator, also denoted by $\Omega$, mapping $H^2(\mathcal{L})$ onto $H^2(\mathcal{G})$ by

$$\Omega f = \Omega \sum_0^\infty f_n \lambda^n = \sum_0^\infty \lambda^n \Omega f_n = \sum_0^\infty \lambda^n D_A^{-1}\Pi_o^* N_A f_n .$$

Moreover, $\Omega$ intertwines with the unilateral shifts on $H^2(\mathcal{L})$ and $H^2(\mathcal{G})$. By using $\Omega$ in (6.21) we have

$$P_{\mathcal{G}}(I - \lambda \Pi_A \omega P_{\mathcal{F}})^{-1} D_A = \Omega N_A \Pi_o (I - \lambda T_A^*)^{-1} .$$

This along with Theorem 6.2 produces the form of $\Xi_{\mathcal{L}}$ in (6.17) and completes the proof.

Let T be the unilateral shift on $\mathcal{H} = H^2(\mathcal{E}_1)$. Recall that an operator function $\Theta$ with values in $\mathcal{L}(\mathcal{E}_1, \mathcal{E}_1)$ is *outer* if $\Theta(\lambda)$ is analytic in the open unit disc, $\Theta(\lambda)a$ is an $H^2(\mathcal{E}_1)$ for all a in $\mathcal{E}_1$ and $\Theta(\cdot)\mathcal{E}_1$ is cyclic for T. If $\Theta$ is in $H^\infty(\mathcal{E}_1, \mathcal{E}_1)$, then $\Theta$ is outer if and only if the range of $M_\Theta^+$ is dense in $H^2(\mathcal{E}_1)$. Recall that an operator $T_o$ on $\mathcal{H}$ is called *Toeplitz* if $T^*T_oT = T_o$. We say that a function $\Theta$ in $H^\infty(\mathcal{E}_1, \mathcal{E}_1)$ is an *outer spectral factor* for the Toeplitz operator $T_o$, if $\Theta$ is an outer function and $M_\Theta^{+*}M_\Theta^+ = T_o$. In this case, by a slight abuse of terminology, we call both the function $\Theta$ and its corresponding operator $M_\Theta^+$ an outer spectral factor for $T_o$. Finally, it is noted that the outer spectral factor $\Theta$ is uniquely determined up to a constant unitary operator on the left. To be precise if $\Psi$ is an outer function in $H^\infty(\mathcal{E}_1, \mathcal{E}_2)$ satisfying $(M_\Psi^+)^*M_\Psi^+ = T_o$, then there exists a unitary constant operator $\Phi$ from $\mathcal{E}_1$ onto $\mathcal{E}_2$ satisfying $\Psi(\lambda) = \Phi\Theta(\lambda)$ for all $|\lambda| < 1$; see the Appendix, Chapter V in Sz.-Nagy-Foias [3] or Chapter IX in Foias-Frazho [4] for further details on outer functions.

As before, let T be the unilateral shift on $H^2(\mathcal{E}_1)$ and A an operator strictly bounded by $\gamma$ satisfying $T'A = AT$, where $T'$ is a contraction on $\mathcal{H}'$. Let B be any central intertwining lifting of A with tolerance $\gamma$. Then $B = WB_\gamma$ for some isometry W, where $B_\gamma$ is the central intertwining lifting of A in (6.1). Using $U'B_\gamma = B_\gamma T$ and $B^*B = B_\gamma^*B_\gamma$, we see that $T^*D_B^2T = D_B^2$, and hence $D_B$ is a Toeplitz operator. Consider the operator $\Xi_\mathcal{L}$ defined in (6.17), where now $\mathcal{L} = \mathcal{E}_1$. Because T is a unilateral shift and $\|A\| < \gamma$, Lemma 6.1 shows that $R = 0$. In particular, $\Xi_\mathcal{L}$ commutes with the unilateral shift T on $H^2(\mathcal{E}_1)$. Therefore $\Xi_\mathcal{L} = M_\Theta^+$ for some function $\Theta$ in $H^\infty(\mathcal{E}_1, \mathcal{E}_1)$. In fact, this function $\Theta$ is given by $\Theta(\lambda)a = (\Xi_\mathcal{L}a)(\lambda)$ for a in $\mathcal{E}_1$. Since $\Xi_\mathcal{L}$ is onto a dense set in $H^2(\mathcal{E}_1)$, it follows that $\Theta$ is an outer function, and thus $\Theta$ is an outer spectral factor for the Toeplitz operator $D_B^2$. Summing up this analysis along with the form of $\Xi_\mathcal{L}$ in (6.17) we obtain the following result.

**THEOREM 6.4.** *Let* T *be the unilateral shift on* $H^2(\mathcal{E}_1)$ *and* A *an operator strictly bounded by* $\gamma$ *satisfying* $T'A = AT$, *where* $T'$ *is a contraction on* $\mathcal{H}'$. *Let* B *be the central intertwining lifting of* A *with tolerance* $\gamma$. *Then the function* $\Theta$ *defined by*

$$\Theta(\lambda) = N_A\Pi_o(I - \lambda T_A^*)^{-1}\Pi_o^* \qquad (\lambda \in \mathbb{D}) \tag{6.23}$$

*is in* $H^\infty(\mathcal{E}_1, \mathcal{E}_1)$ *and is the outer spectral factor for the Toeplitz operator* $D_B^2$. *Here* $\Pi_o$ *is the operator from* $H^2(\mathcal{E}_1)$ *onto* $\mathcal{E}_1$ *defined by* $\Pi_o g = g(0)$ *and* $N_A = (\Pi_o D_A^{-2}\Pi_o^*)^{-1/2}$.

Recall that, when $\|A\| < \gamma$, the analytic function $Q(\lambda) = (D_A^{-2}\Pi_o^*)(\lambda)$ with values in $\mathcal{L}(\mathcal{E}_1, \mathcal{E}_1)$ plays an essential role in describing the central intertwining lifting of A; see Theorem 4.1. By comparing (4.8) with (6.23) we see that $\Theta(\lambda) = N_A^{-1}Q(\lambda)^{-1}$ which gives the following result.

**COROLLARY 6.5.** *Let* T *be the unilateral shift on* $H^2(\mathcal{E}_1)$ *and* A *an operator strictly bounded by* $\gamma$ *satisfying* $T'A = AT$, *where* $T'$ *on* $\mathcal{H}'$ *is a contraction. Let* B *be the central intertwining lifting of* A *with tolerance* $\gamma$ *and* $Q(\lambda)$ *the analytic function whose values are linear operators on* $\mathcal{E}_1$ *defined by*

$$Q(\lambda)a = (D_A^{-2}\Pi_o^* a)(\lambda), \qquad (a \in \mathcal{E}_1). \tag{6.24}$$

*Then* $Q(\lambda)^{-1}$ *is in* $H^\infty(\mathcal{E}_1, \mathcal{E}_1)$ *and* $Q(0)^{1/2}Q(\lambda)^{-1}$ *is precisely the outer spectral factor for the Toeplitz operator* $D_B^2$.

We can obtain another proof of the previous corollary directly from the form of $\Xi$ in (6.8). To see this, let $\Omega$ be the unitary operator from $H^2(\mathcal{L})$ onto $H^2(\mathcal{G})$ defined in (6.22) by $\Omega f = D_A^{-1}\Pi_o^* N_A f$ for f in $H^2(\mathcal{L})$, where $\mathcal{L} = \mathcal{E}_1$. Since $R = 0$ in this case, it follows that

$$\Omega^*\Xi = \Omega^* P_{\mathcal{G}}(I - \lambda\Pi_A\omega P_{\mathcal{F}})^{-1}D_A \tag{6.25}$$

is an outer spectral operator for $I - B^*B$. Since $\Omega^*\Xi$ intertwines two unilateral shifts, it follows that there exists an $\Theta$ in $H^\infty(\mathcal{E}_1, \mathcal{E}_1)$ satisfying $\Omega^*\Xi = M_\Theta^+$. According to (3.8) the space $\mathcal{G} = \mathcal{D}_A \ominus \mathcal{F}$ is given by $\mathcal{G} = D_A^{-1}\mathcal{L}$. So for a in $\mathcal{L} = \mathcal{E}_1$ using the formula for $\Omega$ in (6.22) and $P_{\mathcal{F}}D_A^{-1}a = 0$, we have

$$M_\Theta^+ D_A^{-2}a = \Omega^*\Xi D_A^{-2}a = \Omega^* P_{\mathcal{G}} D_A^{-1}a = N_A^{-1}a.$$

Because $Q(\lambda)a = (D_A^{-2}\Pi_o^* a)(\lambda)$ and $D_A^{-2}\mathcal{L}$ is cyclic for T (see Lemma 4.3) it follows that $\Theta(\lambda) = N_A^{-1}Q(\lambda)^{-1}$. This completes the proof.

By combining the previous corollary along with Theorems 4.1 and 4.2 we readily obtain the following result which summarizes the main results in Sections 4 and 6.

**THEOREM 6.6**. *Let* T *be the unilateral shift on* $H^2(\mathcal{E}_1)$ *and* S *the unilateral shift on* $H^2(\mathcal{E}_2)$ *(respectively* V *the bilateral shift on* $L^2(\mathcal{E}_2)$*). Let* $\mathcal{H}'$ *be an invariant subspace for* $S^*$ *(respectively* $V^*$*) and* $T'$ *on* $\mathcal{H}'$ *the contraction obtained by compressing* S *to* $\mathcal{H}'$ *(respectively compressing* V *to* $\mathcal{H}'$*). Assume that* A *is an operator intertwining* T *with* $T'$ *and* $\|A\| < \gamma$. *Let* P *and* Q *be the functions defined by*

$$P(\lambda)a = (AD_A^{-2}\Pi_o^* a)(\lambda) \quad \textit{and} \quad Q(\lambda)a = (D_A^{-2}\Pi_o^* a)(\lambda) \qquad (|\lambda| = 1 \textit{ and } a \in \mathcal{E}_1), \tag{6.26}$$

*where* $\Pi_o$ *is the operator from* $H^2(\mathcal{E}_1)$ *onto* $\mathcal{E}_1$ *which picks out the zero-th Fourier coefficient. Finally, set*

$$G_\gamma(\lambda) = P(\lambda)Q(\lambda)^{-1}\ . \tag{6.27}$$

*Then* $Q^{-1}$ *is an outer function in* $H^\infty(\mathcal{E}_1, \mathcal{E}_1)$, *and* $G_\gamma$ *is a function in* $H^\infty(\mathcal{E}_1, \mathcal{E}_2)$ *(respectively* $L^\infty(\mathcal{E}_1, \mathcal{E}_2)$*) satisfying* $\|G_\gamma\|_\infty \le \gamma$. *Moreover,* $B = M_{G_\gamma} | H^2(\mathcal{E}_1)$ *is the central intertwining lifting of* A *with tolerance* $\gamma$. *Finally, the outer factor* $\Theta$ *in* $H^\infty(\mathcal{E}_1, \mathcal{E}_1)$ *for* $\gamma^2 I - B^*B$ *is given by*

$$\Theta(\lambda) = N_A \Pi_o (I - \lambda T_A^*)^{-1} \Pi_o^* = Q(0)^{1/2} Q(\lambda)^{-1}\ . \tag{6.28}$$

*In particular,* $\gamma^2 I - G_\gamma(\lambda)^* G_\gamma(\lambda) = \Theta(\lambda)^* \Theta(\lambda)$ for $|\lambda| = 1$.

**COROLLARY 6.7.** *Let* T *be a unilateral shift on* $H^2(\mathcal{E}_1)$ *and* A *an operator on* $H^2(\mathcal{E}_1)$ *strictly bounded by* $\gamma$ *satisfying* $T^*A^*AT \le A^*A$. *Then* $Q(\lambda) = (D_A^{-2}\Pi_o^*)(\lambda)$ *is an outer function and* $Q(\lambda)^{-1}$ *is in* $H^\infty(\mathcal{E}_1, \mathcal{E}_1)$.

**PROOF.** Since $T^*A^*AT \le A^*A$, it follows that there exists a contraction $T'$ on $\mathcal{H}'$, the closure of the range of A, such that $T'A = AT$. By applying Corollary 6.5, the function $Q(\lambda) = (D_A^{-2}\Pi_o^*)(\lambda)$ is outer and $Q(\lambda)^{-1}$ is in $H^\infty(\mathcal{E}_1, \mathcal{E}_1)$. This completes the proof.

The previous corollary leads to another proof for an explicit form for the classical spectral factorization theorem for strictly positive Toeplitz operators; see Clancey-Gohberg [1].

**COROLLARY 6.8.** *Let* $T_o$ *be a strictly positive Toeplitz operator on* $H^2(\mathcal{E}_1)$. *Then the function* $Q_o(\lambda) = (T_o^{-1}\Pi_o^*)(\lambda)$ *is an outer function in* $H^\infty(\mathcal{E}_1, \mathcal{E}_1)$ *and* $Q_o(\lambda)^{-1}$ *is in* $H^\infty(\mathcal{E}_1, \mathcal{E}_1)$. *Moreover, the function* $\Theta_o(\lambda)$ *in* $H^\infty(\mathcal{E}_1, \mathcal{E}_1)$ *defined by*

$$\Theta_o(\lambda) = (\Pi_o T_o^{-1} \Pi_o^*)^{1/2} Q_o(\lambda)^{-1} = Q_o(0)^{1/2} Q_o(\lambda)^{-1} \tag{6.29}$$

*is the outer spectral factor for* $T_o$.

**PROOF.** Let $\gamma > 0$ be any positive scalar such that $\|T_o\| < \gamma^2$. As before, let T be the unilateral shift on $H^2(\mathcal{E}_1)$. Let A be the positive square root of $\gamma^2 I - T_o$. Since $T_o$ is strictly positive, A is strictly bounded by $\gamma$. Using the fact that $T_o$ is Toeplitz we have

$$T^*A^*AT = T^*(\gamma^2 I - T_o)T = \gamma^2 I - T_o = A^*A\ . \tag{6.30}$$

According to the previous corollary the function $(D_A^{-2}\Pi_o^*)(\lambda) = Q(\lambda)$ is outer and $Q(\lambda)^{-1}$ is in $H^\infty(\mathcal{E}_1, \mathcal{E}_1)$. However, $D_A^2 = T_o$. Thus

$$Q(\lambda) = (D_A^{-2}\Pi_o^*)(\lambda) = (T_o^{-1}\Pi_o^*)(\lambda) = Q_o(\lambda)\ . \tag{6.31}$$

By the previous corollary $Q_o(\lambda)$ is an outer function and $Q_o(\lambda)^{-1}$ is in $H^\infty(\mathcal{E}_1, \mathcal{E}_1)$.

To complete the proof it remains to show that the function $\Theta_o(\lambda)$ defined in (6.29) is an outer spectral factor of $T_o$. To see this notice that by (6.30) there exists an isometry $T'$ on $H^2(\mathcal{E}_1)$ satisfying $T'A = AT$. (The range of A equals $H^2(\mathcal{E}_1)$ because $\|T_o\| < \gamma^2$.) Since $T'$ is an isometry the central intertwining lifting B with tolerance $\gamma$ equals A. Thus $D_B^2 = D_A^2 = T_o$. According to Corollary 6.5, and (6.31), the function

$$\Theta_o(\lambda) = (\Pi_o D_A^{-2} \Pi_o^*)^{1/2} Q_o(\lambda)^{-1} = (\Pi_o T_o^{-1} \Pi_o^*)^{1/2} Q_o(\lambda)^{-1}$$

is an outer spectral factor of $T_o$. This readily implies that the function $\Theta_o$ in (6.29) is an outer spectral factor of $T_o$. Since $T_o$ is strictly positive, the operator $M_{\Theta_o}^+$ is invertible. So $\Theta_o(\lambda)^{-1}$ is in $H^\infty(\mathcal{E}_1, \mathcal{E}_1)$. Therefore, $Q_o(\lambda)$ is in $H^\infty(\mathcal{E}_1, \mathcal{E}_1)$. The proof is now complete.

## IV.7 MAXIMUM ENTROPY

In this section we will show that the central intertwining lifting satisfies a certain maximum principle. Then using this maximum principle, we will also show that the central intertwining lifting maximizes a certain entropy function, when T is a unilateral shift of finite multiplicity. To this end, let T be an isometry on $\mathcal{H}$, and B an operator from $\mathcal{H}$ into $\mathcal{K}'$ satisfying $\|B\| \le \gamma$. Let $\Delta(B)$ be the positive operator on $\mathcal{L} = \ker T^*$ defined by the following optimization problem

$$(\Delta(B)a, a) = \inf \{\|D_B(a - Th)\|^2 : h \in \mathcal{H}\} \quad (a \in \mathcal{L}), \tag{7.1}$$

where a is a specified vector in $\mathcal{L}$. One can use the polarization identity to show that $\Delta(B)$ is indeed a well defined positive operator on $\mathcal{L}$. Here we will call $\Delta(B)$ the *Schur complement of* $D_B^2$ (with respect to T). This definition is motivated by the fact that $\Delta(B)$ is indeed the Schur complement of $D_B^2$, when $\|B\| < \gamma$ and $D_B^2$ is partitioned according to its matrix decomposition corresponding to $\mathcal{L} \oplus T\mathcal{H}$; see Lemma 1.2 in Section 1 of the Appendix.

Now let us obtain an explicit formula for $\Delta(B)$. To accomplish this, let $\mathcal{G}(B \cdot T)$ be the subspace of $\mathcal{H}$ defined by $\mathcal{G}(B \cdot T) = \mathcal{D}_B \ominus D_B T\mathcal{H}$. By applying the Projection Theorem to the optimization problem in (7.1) we obtain

$$\inf \{\|D_B(a - Th)\|^2 : h \in \mathcal{H}\} = \inf \{\|D_B a - f\|^2 : f \in D_B T\mathcal{H}\} = \|P_{\mathcal{G}(B\cdot T)} D_B a\|^2 .$$

This along with (7.1) readily shows that

$$\Delta(B) = \Pi_o D_B P_{\mathcal{G}(B\cdot T)} D_B \Pi_o^* , \tag{7.2}$$

where $\Pi_o$ is the operator from $\mathcal{H}$ onto $\mathcal{L}$ defined by $\Pi_o = P_{\mathcal{L}}$.

Obviously, if $T'A = AT$ where A is an operator mapping $\mathcal{H}$ into $\mathcal{H}'$ satisfying $\|A\| \le \gamma$, then $\Delta(A)$ is well defined. In this case, $\mathcal{G}(A\cdot T) = \mathcal{G}$ is precisely the orthogonal complement of $\mathcal{F} = \overline{D_A T \mathcal{H}}$ in $\mathcal{D}_A$ and

$$\Delta(A) = \Pi_o D_A P_{\mathcal{G}} D_A \Pi_o^* \quad \text{where} \quad \mathcal{G} = \mathcal{D}_A \ominus D_A T \mathcal{H} . \tag{7.3}$$

In particular, if $\|A\| < \gamma$, then using the formula for $P_{\mathcal{G}}$ in (6.20) we see that

$$\Delta(A) = (\Pi_o D_A^{-2} \Pi_o^*)^{-1} = N_A^2, \text{ and } \Delta(B) = (\Pi_o D_B^{-2} \Pi_o^*)^{-1} . \tag{7.4}$$

The second equality in (7.4) holds if B is any intertwining lifting of A satisfying $\|B\| < \gamma$. The proof of this fact is identical to the proof of the first equality in (7.4) by replacing $\mathcal{G}$ by $\mathcal{G}(B\cdot T)$ and A by B. The following result involving $\Delta(A)$ shows that the central intertwining lifting satisfies a maximum principle.

**THEOREM 7.1.** *Let* T *be an isometry on* $\mathcal{H}$ *and* A *an operator from* $\mathcal{H}$ *into* $\mathcal{H}'$ *whose norm is bounded by* $\gamma$ *satisfying* $T'A = AT$ *where* $T'$ *is a contraction on* $\mathcal{H}'$. *Let* $U'$ *on* $\mathcal{K}'$ *be an isometric lifting of* $T'$ *and* B *an intertwining lifting of* A *with respect to* $U'$ *and tolerance* $\gamma$. *If* $B_\gamma$ *is the central intertwining lifting of* A *with tolerance* $\gamma$, *then*

$$\Delta(B) \le \Delta(B_\gamma) = \Delta(A) . \tag{7.5}$$

*Moreover,* $\Delta(B) = \Delta(A)$ *if and only if* $Y \mid \mathcal{G} = 0$ *where* Y *the contraction in the matrix representation* $B = [A, YD_A]^{tr}$. *If* $U'$ *is the minimal isometric lifting of* $T'$, *then we have equality in (7.5) if and only if* $B = B_\gamma$.

The proof of this theorem is based on the following known fact.

**LEMMA 7.2.** *Let* A *be an operator mapping* $\mathcal{H}$ *into* $\mathcal{H}'$ *satisfying* $\|A\| \le \gamma$. *Let* B *be an operator mapping* $\mathcal{H}$ *into* $\mathcal{K}' = \mathcal{H}' \oplus \mathcal{M}$ *of the form*

$$B = \begin{bmatrix} A \\ X \end{bmatrix} : \mathcal{H} \to \begin{bmatrix} \mathcal{H}' \\ \mathcal{M} \end{bmatrix} . \tag{7.6}$$

*Then the norm of* B *is bounded by* $\gamma$ *if and only if there exits a contraction* Y *mapping* $\mathcal{D}_A$ *into* $\mathcal{M}$ *satisfying* $X = YD_A$. *In this case the formula*

$$W_Y D_B = D_{Y,1} D_A \tag{7.7}$$

*defines a unitary operator* $W_Y$ *from* $\mathcal{D}_B$ *onto* $\mathcal{D}_{Y,1}$, *where* $D_{Y,1}$ *is the positive square root of* $I - Y^*Y$ *and* $\mathcal{D}_{Y,1}$ *is the closure of the range of* $D_{Y,1}$.

**PROOF.** Assume that Y is a contraction satisfying $X = YD_A$. Then the form of B in (7.6) gives for h in $\mathcal{H}$

$$\gamma^2\|h\|^2 - \|Bh\|^2 = \gamma^2\|h\|^2 - \|Ah\|^2 - \|YD_A h\|^2 \qquad (h \in \mathcal{H})$$

$$= \|D_A h\|^2 - \|YD_A h\|^2 = \|D_{Y,1} D_A h\|^2 \geq 0 .$$

Thus the norm of B is bounded by $\gamma$. Moreover, by comparing the first and last terms $\|D_{B_\gamma} h\| = \|D_{Y,1} D_A h\|$. This shows that there exists a unitary operator $W_Y$ from $\mathcal{D}_B$ onto $\mathcal{D}_{Y,1}$ satisfying (7.7). On the other hand, if the norm of B is bounded by $\gamma$, then

$$\|Xh\|^2 \leq \gamma^2\|h\|^2 - \|Ah\|^2 = \|D_A h\|^2 \qquad (h \in \mathcal{H}) .$$

So there exists a contraction Y from $\mathcal{D}_A$ into $\mathcal{D}_{Y,1}$ such that $X = YD_A$. This completes the proof.

**PROOF OF THEOREM 7.1.** Because B is an intertwining lifting of A, it follows that B admits a matrix representation of the form $B = [A, X]^{tr}$ in (7.6). According to the previous lemma there exists a contraction Y mapping $\mathcal{D}_A$ into $\mathcal{M} = \mathcal{K}' \ominus \mathcal{H}'$ satisfying $X = YD_A$. So using the unitary operator $W_Y$ in (7.7) we have for a in $\mathcal{L} = \ker T^*$

$$(\Delta(B)a, a) = \inf\{\|D_B(a - Th)\|^2 : h \in \mathcal{H}\} = \inf\{\|D_{Y,1} D_A(a - Th)\|^2 : h \in \mathcal{H}\} \leq$$
$$\inf\{\|D_A(a - Th)\|^2 : h \in \mathcal{H}\} = (\Delta(A)a, a) . \tag{7.8}$$

Therefore $\Delta(B) \leq \Delta(A)$.

By construction the central intertwining lifting $B_\gamma$ admits a matrix decomposition of the form $B_\gamma = [A, YD_A]^{tr}$, where Y is a contraction satisfying $Y | \mathcal{G} = 0$ with $\mathcal{G} = \mathcal{D}_A \ominus D_A T\mathcal{H}$; see Proposition 2.4. So assume that B is any intertwining lifting of A with tolerance $\gamma$ such that its corresponding contraction Y satisfies $Y | \mathcal{G} = 0$. In this case we claim that $\Delta(B) = \Delta(A)$. To see this notice that $Y | \mathcal{G} = 0$ gives $(I - Y^*Y)g = g$ for all g in $\mathcal{G}$ and thus $p(D_{Y,1}^2)g = p(I)g$ where $p(\lambda)$ is any polynomial. Because the square root of any positive operator can be obtained by passing strong limits of a sequence of polynomials in that operator (see Halmos [1]), it follows that $D_{Y,1} | \mathcal{G} = I$, the identity on $\mathcal{G}$. Moreover, $D_{Y,1}\mathcal{G} = \mathcal{G}$ and because $D_{Y,1}$ is a self adjoint operator $\mathcal{G}$ is a reducing subspace for $D_{Y,1}$. In particular, $D_{Y,1}\mathcal{F} \subseteq \mathcal{F}$ where $\mathcal{F} = \overline{D_A T\mathcal{H}}$. Using this fact along with the Projection Theorem we obtain

$$(\Delta(B)a, a) = \inf\{\|D_B(a - Th)\|^2 : h \in \mathcal{H}\} = \inf\{\|D_{Y,1}D_A(a - Th)\|^2 : h \in \mathcal{H}\} =$$

$$\inf\{\|D_{Y,1}D_A a - D_{Y,1}f\|^2 : f \in \mathcal{F}\} \geq \inf\{\|D_{Y,1}D_A a - f\|^2 : f \in \mathcal{F}\} = \tag{7.9}$$

$$\|P_{\mathcal{G}} D_{Y,1} D_A a\|^2 = \|D_{Y,1} P_{\mathcal{G}} D_A a\|^2 = \|P_{\mathcal{G}} D_A a\|^2 = (\Delta(A)a, a) .$$

Hence $\Delta(A) \leq \Delta(B)$. Combining this with the fact that $\Delta(B) \leq \Delta(A)$ for any intertwining lifting of A with tolerance $\gamma$, shows that if $Y \mid \mathcal{G} = 0$, then $\Delta(B) = \Delta(A)$. In particular, $\Delta(A) = \Delta(B_\gamma)$. Therefore (7.5) holds.

Now let B be any intertwining lifting of A with tolerance $\gamma$ and Y its corresponding contraction given by $X = YD_A$. Moreover, let $h_n$ be any sequence of vectors in $\mathcal{H}$ such that $D_A(a - Th_n)$ approaches $P_{\mathcal{G}} D_A a$ for a in $\mathcal{L}$. Then we have

$$(\Delta(B)a, a) \leq \|D_B(a - Th_n)\|^2 = \|D_{Y,1}D_A(a - Th_n)\|^2 \to \|D_{Y,1}P_{\mathcal{G}}D_A a\|^2$$

as $n \to \infty$. In particular, this shows that

$$(\Delta(B)a, a) \leq \|D_{Y,1}P_{\mathcal{G}}D_A a\| \leq \|P_{\mathcal{G}}D_A a\|^2 = (\Delta(A)a, a) \qquad (a \in \ker T^*) . \tag{7.10}$$

So if $\Delta(B) = \Delta(A)$, then $D_{Y,1} \mid \mathcal{G}$ is an isometry, or equivalently, $Y \mid \mathcal{G} = 0$. Combining this with our previous analysis, we see that $\Delta(B) = \Delta(A)$ if and only if $Y \mid \mathcal{G} = 0$. Finally, according to Proposition 2.4, if $U'$ is the minimal isometric lifting of $T'$, then $Y \mid \mathcal{G} = 0$ if and only if $B = B_\gamma$. Therefore $\Delta(B) = \Delta(A)$ if and only if B is the central intertwining lifting of A with tolerance $\gamma$. This completes the proof.

As before, let B be an intertwining lifting of A with tolerance $\gamma$ and $B_\gamma$ the central intertwining lifting of A with tolerance $\gamma$. *The maximum principle* for the central intertwining lifting states that $\Delta(B) \leq \Delta(B_\gamma)$. Using $P_{\mathcal{G}(B\cdot T)}D_B T = 0$ for any B lifting A along with (7.2) and (7.3) we see that this *maximum principle is equivalent to the following operator inequalities*

$$D_B P_{\mathcal{G}(B\cdot T)} D_B \leq D_{B_\gamma} P_{\mathcal{G}(B_\gamma\cdot T)} D_{B_\gamma} = D_A P_{\mathcal{G}} D_A . \tag{7.11}$$

*Moreover, we have equality in (7.11) if and only if* $Y \mid \mathcal{G} = 0$ *where* Y *is the contraction corresponding to the matrix representation of* B *given by* $B = [A, YD_A]^{tr}$. *In particular, if* $U'$ *is the minimal isometric lifting of* $T'$, *then we have equality in (7.11) if and only if* $B = B_\gamma$.

Sometimes in the sequel it will be convenient to call an intertwining lifting B of A with tolerance $\gamma$ *maximal* if we have equality in (7.5) or (7.11).

Recall that a Toeplitz operator $T_o$ is an operator on $\mathcal{H} = H^2(\mathcal{E})$ satisfying $T^*T_oT = T_o$ where T is the unilateral shift on $\mathcal{H}$. Moreover, by the famous Brown-Halmos Theorem $T_o$ is a Toeplitz operator if and only if there exists a function N in $L^\infty(\mathcal{E}, \mathcal{E})$ satisfying $T_o = P_+M_N \mid H^2(\mathcal{E})$, where $M_N$ is the multiplication operator on $L^2(\mathcal{E})$ defined by $M_N f = Nf$ if

$f \in L^2(\mathcal{E})$ and $P_+$ is the orthogonal projection onto $H^2(\mathcal{E})$; see Section VIII.4 in Foias-Frazho [4], Chapter XXIII in Gohberg-Goldberg-Kaashoek [2] or Halmos [1]. In this case the function N is unique, and is called the *symbol* for the Toeplitz operator $T_o$. Throughout we will consider Toeplitz operators on both $H^2(\mathcal{E})$ and $\ell_+^2(\mathcal{E})$. Notice that if $\mathcal{H} = \ell_+^2(\mathcal{E})$, then $T_o$ is a Toeplitz operator on $\mathcal{H}$, if $T^*T_oT = T_o$ where T is now the unilateral shift on $\ell_+^2(\mathcal{E})$. In this case the Brown-Halmos Theorem shows that $T_o$ is a Toeplitz operator if and only if $T_o = P_+L_N \mid \ell_+^2(\mathcal{E})$, where N is a function in $L^\infty(\mathcal{E}, \mathcal{E})$ and $P_+$ is the orthogonal projection onto $\ell_+^2(\mathcal{E})$. (Recall that $L_N$ is the Laurent operator on $\ell^2(\mathcal{E})$ generated by N.) As before, N is called the symbol for $T_o$.

Now let $T_o$ be a positive Toeplitz operator on either $\mathcal{H} = H^2(\mathcal{E})$ or $\mathcal{H} = \ell_+^2(\mathcal{E})$, and let T be the unilateral shift on $\mathcal{H}$. Then the operator $\Delta_o$ on $\mathcal{E} = \mathcal{L} = \ker T^*$ is the positive operator defined by the following optimization problem

$$(\Delta_o a, a) = \inf \{(T_o(a - Th), (a - Th)) : h \in \mathcal{H}\} \tag{7.12}$$

where a is a specified vector in $\mathcal{L}$. By using the polarization identity it is easy to show that $\Delta_o$ is indeed positive operator on $\mathcal{L}$. Motivated by the results in the Appendix, $\Delta_o$ is called the *Schur complement for* $T_o$ (with respect to T). This definition is justified by the fact that if $T_o$ is strictly positive, then $\Delta_o$ is indeed the Schur complement of $T_o$ when $T_o$ is decomposed according to the orthogonal decomposition $\mathcal{L} \oplus T\mathcal{H}$ of $\mathcal{H}$; see Lemma 1.2 in the Appendix.

To obtain an explicit formula for $\Delta_o$, let D be the positive square root of $T_o$ and set $\mathcal{G}_o = \overline{D\mathcal{H}} \ominus \overline{DT\mathcal{H}}$. By using the Projection Theorem in the optimization problem (7.12) we obtain

$$\begin{aligned}(\Delta_o a, a) &= \inf \{\|D(a - Th)\|^2 : h \in \mathcal{H}\} \\ &= \inf \{\|Da - f\|^2 : f \in \overline{DT\mathcal{H}}\} = \|P_{\mathcal{G}_o} Da\|^2 \qquad (a \in \ker T^*)\,.\end{aligned} \tag{7.13}$$

This readily implies that

$$\Delta_o = \Pi_o D P_{\mathcal{G}_o} D \Pi_o^*, \text{ where } \mathcal{G}_o = \overline{D\mathcal{H}} \ominus DT\mathcal{H}\,. \tag{7.14}$$

Here $\Pi_o$ is the operator from $\mathcal{H}$ onto $\mathcal{L} = \mathcal{E} = \ker T^*$ defined by $\Pi_o = P_{\mathcal{E}}$. Equation (7.14) shows that $\Delta_o$ is a positive operator on $\mathcal{E}$.

The dimension of a linear space $\mathcal{X}$ is denoted by dim $\mathcal{X}$. If C is an operator on a finite dimensional Hilbert space $\mathcal{E}$, then [C] denotes a matrix representation of C with respect to any basis for $\mathcal{E}$, and det [C] is the determinant of [C]. The following result which is proved in the Appendix (see also Section V.7 in Sz.-Nagy-Foias [3]) is used to obtain our maximal entropy description of the central intertwining lifting $B_\gamma$ of A.

**PROPOSITION 7.3.** *Let* $T_o$ *be a positive Toeplitz operator on* $\mathcal{H} = H^2(\mathcal{E})$ *or* $\mathcal{H} = \ell^2_+(\mathcal{E})$, *where* $\mathcal{E}$ *is finite dimensional, and let* N *in* $L^\infty(\mathcal{E}, \mathcal{E})$ *be the symbol for* $T_o$. *Let* $\Delta_o$ *on* $\mathcal{E}$ *be the Schur complement of* $T_o$ *defined by the optimization problem in (7.12). Then*

$$\frac{1}{2\pi}\int_0^{2\pi} \ln \det [N(e^{it})]dt = \ln \det [\Delta_o] . \tag{7.15}$$

*In particular, the following conditions are equivalent:*

*(i)* $\int_0^{2\pi} \ln \det [N(e^{it})]dt > -\infty$,

*(ii)* $\det [\Delta_o] \neq 0$,

*(iii)* *the Toeplitz operator* $T_o$ *admits an outer spectral factorization of the form* $T_o = (M_\Theta^+)^* M_\Theta^+$ *(or* $T_o = T_\Theta^* T_\Theta$ *if* $\mathcal{H} = \ell^2_+(\mathcal{E})$*) where* $\Theta$ *is an outer function in* $H^\infty(\mathcal{E}, \mathcal{E})$.

*Finally, if any one of the previous conditions hold, then* $\Delta_o = \Theta(0)^*\Theta(0)$, *where* $\Theta$ *is the outer spectral factor for* $T_o$.

Recall that an operator C on $\mathcal{H}$ is *strictly positive* if $C \geq \delta I$ for some $\delta > 0$. Obviously C is strictly positive if and only if C is positive ($\geq 0$) and invertible. This sets the stage for the following useful result.

**COROLLARY 7.4.** *Let* $T_o$ *be a strictly positive Toeplitz operator on* $\mathcal{H} = H^2(\mathcal{E})$ *(respectively* $\mathcal{H} = \ell^2_+(\mathcal{E})$*), where* $\mathcal{E}$ *is finite dimensional, and let* N *in* $L^\infty(\mathcal{E}, \mathcal{E})$ *be the symbol for* $T_o$. *Then* $T_o$ *admits an outer spectral factorization of the form* $T_o = (M_\Theta^+)^* M_\Theta^+$ *(respectively* $T_o = T_\Theta^* T_\Theta$*) where* $\Theta$ *is an outer function in* $H^\infty(\mathcal{E}, \mathcal{E})$. *Moreover, in this case*

$$\frac{1}{2\pi}\int_0^{2\pi} \ln \det [N(e^{it})]dt = -\ln \det [\Pi_o T_o^{-1} \Pi_o^*] \tag{7.16}$$

*where* $\Pi_o$ *is the operator from* $\mathcal{H}$ *onto* $\mathcal{E}$ *defined by* $\Pi_o = P_{\mathcal{E}}$.

**PROOF.** Now let h be in $H^2(\mathcal{E})$. Because $T_o$ is strictly positive there exists a positive constant $\delta > 0$ satisfying

$$(Nh, h) = (T_o h, h) \geq \delta \|h\|^2 .$$

So if V is the bilateral shift on $L^2(\mathcal{E})$, then for all $n \geq 0$

$$(NV^{*n}h, V^{*n}h) = (Nh, h) \geq \delta\|h\|^2 = \delta\|V^{*n}h\| .$$

This implies that $(Nf, f) \geq \delta\|f\|^2$ on a dense set in $L^2(\mathcal{E})$ and thus on all of $L^2(\mathcal{E})$. By a standard measure theoretic argument $N \geq \delta I$ a.e. So Part (i) of Proposition 7.3 holds. The previous proposition also shows that the Toeplitz operator $T_o$ admits an outer spectral factorization of the form $T_o = (M_\Theta^+)^* M_\Theta^+$, where $\Theta$ is an outer function in $H^\infty(\mathcal{E}, \mathcal{E})$.

Now let us establish (7.16). According to (7.14) and (7.15) it suffices to prove the last equality in the following identity

$$\Delta_o = \Pi_o D P_{\mathcal{G}_o} D \Pi_o^* = (\Pi_o T_o^{-1} \Pi_o^*)^{-1} \tag{7.17}$$

where as before D is the positive square root of $T_o$ and $\mathcal{G}_o$ the subspace of $\mathcal{D}(=\mathcal{H})$ defined by $\mathcal{G}_o = \mathcal{D} \ominus \overline{DT\mathcal{H}}$. Notice that g is in $\mathcal{G}_o$ if and only if g is orthogonal to $DT\mathcal{H}$, or equivalently, Dg is orthogonal to $T\mathcal{H}$. So g is in $\mathcal{G}_o$ if and only if Dg is in $\ker T^* = \mathcal{E}$, or equivalently, g is in $D^{-1}\mathcal{E}$. Therefore $\mathcal{G}_o$ equals the range of $D^{-1}\Pi_o^*$. Now let X be an injective operator with closed range $\mathcal{R}$ mapping $\mathcal{X}_1$ into $\mathcal{X}_2$. Then it is well known that the orthogonal projection $P_\mathcal{R}$ onto $\mathcal{R}$ is given by

$$P_\mathcal{R} = X(X^*X)^{-1}X^* . \tag{7.18}$$

To verify this simply notice that the range of $P_\mathcal{R}$ is $\mathcal{R}$ and $P_\mathcal{R} = P_\mathcal{R}^* = P_\mathcal{R}^2$. Now let X be the injective operator defined by $X = D^{-1}\Pi_o^*$. Then $\mathcal{G}_o$ equals the range of X. By using $D^2 = T_o$ in equation (7.18) we obtain

$$P_{\mathcal{G}_o} = D^{-1}\Pi_o^*(\Pi_o T_o^{-1}\Pi_o^*)^{-1}\Pi_o D^{-1} . \tag{7.19}$$

Using this in the second term in (7.17) gives the last equality in (7.17) and completes the proof.

As before, let A be an operator bounded by $\gamma$ mapping $\mathcal{H}$ into $\mathcal{H}'$ satisfying $T'A = AT$ where T is now the unilateral shift on $\mathcal{H} = H^2(\mathcal{E})$ and $\mathcal{E}$ is finite dimensional. Let B be an intertwining lifting of A satisfying $\|B\| \leq \gamma$. Using $U'B = BT$ it follows that $D_B^2$ is a Toeplitz operator on $H^2(\mathcal{E})$. The *entropy* E(B) of B is defined by

$$E(B) = \frac{1}{2\pi}\int_0^{2\pi} \ln \det [N(e^{it})]dt , \tag{7.20}$$

where N in $L^\infty(\mathcal{E}, \mathcal{E})$ is the symbol for the Toeplitz operator $D_B^2$. Throughout this section $N_\gamma$ is the symbol for the Toeplitz operator $D_{B_\gamma}^2$, where $B_\gamma$ is the central intertwining lifting of A. The following result shows that the central intertwining lifting $B_\gamma$ is a maximal entropy intertwining lifting of A, that is, $E(B_\gamma) \geq E(B)$.

**THEOREM 7.5.** *Let* A *be an operator bounded by* $\gamma$ *mapping* $\mathcal{H}$ *into* $\mathcal{H}'$ *satisfying* $T'A = AT$ *where* T *is the unilateral shift on* $\mathcal{H} = H^2(\mathcal{E})$ *(or* $\mathcal{H} = \ell_+^2(\mathcal{E})$*) and* $\mathcal{E}$ *is finite dimensional. Then the central intertwining lifting* $B_\gamma$ *of* A *with tolerance* $\gamma$ *is a maximal entropy intertwining lifting of* A *with tolerance* $\gamma$*, that is,* $E(B_\gamma) \geq E(B)$*, or equivalently,*

$$\frac{1}{2\pi}\int_0^{2\pi} \ln \det [N_\gamma]dt = \ln \det [\Delta(A)] \geq \ln \det [\Delta(B)] = \frac{1}{2\pi}\int_0^{2\pi} \ln \det [N]dt \tag{7.21}$$

*where* B *is any intertwining lifting of* A *satisfying* $\|B\| \leq \gamma$*. Finally, if* $U'$ *on* $\mathcal{K}'$ *is the minimal isometric lifting of* $T'$ *and the entropy* $E(B_\gamma) > -\infty$*, or equivalently,* $\det [\Delta(B_\gamma)]$ *is nonzero, then* $E(B_\gamma) = E(B)$ *if and only if* $B = B_\gamma$ *the central intertwining lifting of* A.

**PROOF.** Let $T_o$ be the Toeplitz operator $T_o = D_B^2$ and $\Delta_o = \Delta(B)$. According to the maximum principle in Theorem 7.1 and (7.15) we have

$$E(B_\gamma) = \ln \det [\Delta(A)] \geq \ln \det [\Delta(B)] = E(B) . \tag{7.22}$$

Therefore (7.21) holds. If $E(B_\gamma) = E(B) > -\infty$, then

$$\det [\Delta(A)] = \det [\Delta(B)] \neq 0 . \tag{7.23}$$

Notice that if C and M are two strictly positive operators on $\mathcal{E}$ satisfying $C \geq M$, then $C = M$ if and only if $\det [C] = \det [M]$. Because $C \geq M$ there exists a contraction $\Gamma$ on $\mathcal{E}$ satisfying $\Gamma C^{1/2} = M^{1/2}$. This implies that $M = C^{1/2}\Gamma^*\Gamma C^{1/2}$. Therefore $\det [M] = \det [C] \det [\Gamma^*\Gamma]$. So if $\det [M] = \det [C]$, then $\det [\Gamma^*\Gamma] = 1$. Since $\Gamma$ is a contraction and $\det [\Gamma^*\Gamma]$ is the product of the eigenvalues of $\Gamma^*\Gamma$, all the eigenvalues of the positive operator $\Gamma^*\Gamma$ are one. This implies that $\Gamma^*\Gamma = I$. Thus $M = C$, which proves our claim. Now using $\Delta(A) \geq \Delta(B)$ along with (7.23) we see that $\Delta(A) = \Delta(B)$. If $U'$ is the minimal isometric lifting of $T'$, it follows that $B = B_\gamma$ the central intertwining lifting of A; see Theorem 7.1. This completes the proof.

**REMARK 7.6.** As before, let A be an operator strictly bounded by $\gamma$ satisfying $T'A = AT$ where T is the unilateral shift on $\mathcal{H} = H^2(\mathcal{E})$ and $\mathcal{E}$ is finite dimensional. Now let B be an intertwining lifting of A satisfying $\|B\| < \gamma$. In this case $D_B^2$ is a strictly positive Toeplitz operator on $H^2(\mathcal{E})$. According to Corollary 7.4 with $T_o = D_B^2$ we have

$$E(B) = \frac{1}{2\pi}\int_0^{2\pi} \ln \det [N(e^{it})]dt = -\ln \det [\Pi_0(\gamma^2 I - B^*B)^{-1}\Pi_0^*] , \tag{7.24}$$

where N in $L^\infty(\mathcal{E}, \mathcal{E})$ is the symbol for the Toeplitz operator $D_B^2$. Here $\Pi_0$ is the operator from $H^2(\mathcal{E})$ onto $\mathcal{E}$ defined by $\Pi_0 = P_{\mathcal{E}}$. Moreover, in this case, the Toeplitz operator $D_B^2$ admits an outer spectral factorization of the form $D_B^2 = (M_\Theta^+)^* M_\Theta^+$, where $\Theta$ is an outer function in

$H^\infty(\mathcal{E}, \mathcal{E})$. In fact, by Corollary 6.8 the outer spectral factor $\Theta$ for $D_B^2$ is given by

$$\Theta(\lambda) = (\Pi_0(\gamma^2 I - B^* B)^{-1}\Pi_0^*)^{1/2} Q(\lambda)^{-1} = Q(0)^{1/2} Q(\lambda)^{-1} . \tag{7.25}$$

Recall that $Q(\lambda)$ is the function in $H^\infty(\mathcal{E}, \mathcal{E})$ given by

$$Q(\lambda) = ((\gamma^2 I - B^* B)^{-1}\Pi_0^*)(\lambda) \qquad (|\lambda| < 1) . \tag{7.26}$$

Since B is strictly contractive, obviously both $Q(\lambda)$ and $Q(\lambda)^{-1}$ are in $H^\infty(\mathcal{E}, \mathcal{E})$.

By consulting (7.4) when $\|A\| < \gamma$ we readily obtain the following result.

**COROLLARY 7.7.** *Let* A *be an operator strictly bounded by* $\gamma$ *mapping* $\mathcal{H}$ *into* $\mathcal{H}'$ *satisfying* $T'A = AT$, *where* T *is the unilateral shift on* $\mathcal{H} = H^2(\mathcal{E})$ *and* $\mathcal{E}$ *is finite dimensional. Let* $B_\gamma$ *be the central intertwining lifting of* A *with tolerance* $\gamma$. *Then the entropy* $E(B_\gamma)$ *of* $B_\gamma$ *is given by*

$$E(B_\gamma) = \frac{1}{2\pi}\int_0^{2\pi} \ln \det [N_\gamma(e^{it})] dt = -\ln \det [\Pi_0(\gamma^2 I - A^* A)^{-1}\Pi_0^*] \tag{7.27}$$

*where* $N_\gamma$ *in* $L^\infty(\mathcal{E}, \mathcal{E})$ *is the symbol for the Toeplitz operator* $D_{B_\gamma}^2$. *Moreover, in this case the Toeplitz operator* $D_{B_\gamma}^2$ *admits an outer spectral factorization of the form* $D_{B_\gamma}^2 = (M_\Theta^+)^* M_\Theta^+$, *where* $\Theta$ *is an outer function in* $H^\infty(\mathcal{E}, \mathcal{E})$.

**REMARK 7.8.** Let A be an operator strictly bounded by $\gamma$ satisfying $T'A = AT$, where T is the unilateral shift on $H^2(\mathcal{E})$. As before, let $B_\gamma$ be the central intertwining lifting of A with tolerance $\gamma$. Then according to Theorem 6.6 the outer spectral factor $\Theta$ in $H^\infty(\mathcal{E}, \mathcal{E})$ for the Toeplitz operator $D_{B_\gamma}^2$ is given by

$$\Theta(\lambda) = (\Pi_0(\gamma^2 I - A^* A)^{-1}\Pi_0^*)^{1/2} Q_\gamma(\lambda)^{-1} = Q_\gamma(0)^{1/2} Q_\gamma(\lambda)^{-1} , \tag{7.28}$$

where $Q_\gamma(\lambda)$ is now the function defined by

$$Q_\gamma(\lambda) = ((\gamma^2 I - A^* A)^{-1}\Pi_0^*)(\lambda) \qquad (|\lambda| < 1) . \tag{7.29}$$

Moreover, $Q_\gamma(\lambda)^{-1}$ is in $H^\infty(\mathcal{E}, \mathcal{E})$. Notice that the formula for the outer spectral factor for the Toeplitz operator $D_{B_\gamma}^2$ in (7.28) is the same as the formula for the outer spectral factor for the Toeplitz operator $D_B^2$ in (7.25) where A replaces B.

As before, let B be an intertwining lifting of A with respect to $U'$ and tolerance $\gamma$. Theorem 7.1 shows that $\Delta(B) \le \Delta(B_\gamma) = \Delta(A)$, where $B_\gamma$ is the central intertwining lifting of A with tolerance $\gamma$. Our previous analysis also shows that $\Delta(B) = \Delta(A)$ if and only if $Y | \mathcal{G} = 0$, where Y is the unique contraction determined by $B = [A, YD_A]^{tr}$. So if $U'$ is the minimal isometric lifting of $T'$, then $Y|\mathcal{G}=0$ if and only if $B=B_\gamma$. In this case we have equality $\Delta(B)=\Delta(A)$ if and only if $B=B_\gamma$. However, according to Remark 2.5 if $U'$ is not the minimal isometric lifting of $T'$, then the condition $Y|\mathcal{G}=0$ does not guarantee that B is the central intertwining lifting of A. In particular, if $U'$ is not the minimal isometric lifting, then $\Delta(B)=\Delta(A)$ does not necessarily imply that $B=B_\gamma$. The following result shows that if $\Pi_A\omega P_{\mathcal{F}}$ is pointwise stable, then the central intertwining lifting $B_\gamma$ is the only maximal intertwining lifting of A with tolerance $\gamma$. To this end, recall that an operator C on $\mathcal{H}$ is *pointwise stable* if $C^n$ approaches zero strongly as n approaches infinity.

**THEOREM 7.9**. *Let* T *be an isometry on* $\mathcal{H}$ *and* A *an operator whose norm is bounded by* $\gamma$ *satisfying* $T'A=AT$, *where* $T'$ *is a contraction on* $\mathcal{H}'$. *Let* $U'$ *be an isometric lifting of* $T'$ *and* B *an intertwining lifting of* A *with tolerance* $\gamma$. *Finally, let* $B_\gamma$ *be the central intertwining lifting of* A *with tolerance* $\gamma$ *and assume that the contraction* $\Pi_A\omega P_{\mathcal{F}}$ *on* $\mathcal{D}_A$ *is pointwise stable. Then*

$$\Delta(B)\le\Delta(B_\gamma)\le\Delta(A)\,, \tag{7.30}$$

*and we have equality in (7.30) if and only if* $B=B_\gamma$. *In particular, if* $\|A\|<\gamma$ *and* T *is a unilateral shift, then* $\Delta(B)=\Delta(A)$ *if and only if* $B=B_\gamma$.

**PROOF**. By Theorem 7.1 it remains to show that if $\Delta(B)=\Delta(A)$, then $B=B_\gamma$. According to Remark 1.2, the isometric lifting $U'$ admits a reducing decomposition of the form $U'=U_m\oplus U$ on $\mathcal{K}_m\oplus\mathcal{K}$, where $U_m$ is the minimal isometric lifting of $T'$. Using this decomposition along with $P_{\mathcal{H}'}B=A$, it follows that B admits a matrix decomposition of the form

$$B=\begin{bmatrix}A\\ YD_A\\ XD_A\end{bmatrix}:\mathcal{H}\to\begin{bmatrix}\mathcal{H}'\\ \mathcal{M}\\ \mathcal{K}\end{bmatrix}, \tag{7.31}$$

where $[Y,X]^{tr}$ is a contraction from $\mathcal{D}_A$ into $\mathcal{M}\oplus\mathcal{K}$ and $\mathcal{M}=\mathcal{K}_m\ominus\mathcal{H}'$. Since $\Delta(B)=\Delta(A)$, we see that both $Y|\mathcal{G}$ and $X|\mathcal{G}$ are zero; see Theorem 7.1. In particular, this implies that $B_c=[A,YD_A]^{tr}$ is the central intertwining lifting of A with respect to $U_m$ and tolerance $\gamma$; see Proposition 2.4. By applying Lemma 7.2 to the contraction $[Y,X]^{tr}$, there exists a contraction Z mapping $\mathcal{D}_{Y,1}$ into $\mathcal{K}$ satisfying $X=ZD_{Y,1}$. (Here $D_{Y,1}$ is the positive square root of $I-Y^*Y$.) Because $Y|\mathcal{G}=0$ we see that $D_{Y,1}P_{\mathcal{G}}=P_{\mathcal{G}}$. This and $X|\mathcal{G}=0$ shows that $Z|\mathcal{G}=0$. Therefore B admits a matrix representation of the form

$$B = [A, YD_A, ZD_{Y,1}D_A]^{tr}, \tag{7.32}$$

where Y and Z are two contractions satisfying $Y | \mathcal{G} = 0$ and $Z | \mathcal{G} = 0$.

Using $U_m B_c = B_c T$ we have for h in $\mathcal{H}$

$$\|D_{B_c} Th\|^2 = \|Th\|^2 - \|B_c Th\|^2 = \|h\|^2 - \|U_m B_c h\|^2 = \|D_{B_c} h\|^2 \qquad (h \in \mathcal{H}).$$

So there exists an isometry $U_c$ on $\mathcal{D}_{B_c}$ satisfying $U_c D_{B_c} = D_{B_c} T$. Since $\Pi_A \omega P_{\mathcal{F}}$ is pointwise stable, Theorem 6.2 shows that $U_c$ is a unilateral shift; see (6.8) and (6.10). (In fact, $U_c$ is unitarily equivalent to the unilateral shift on $H^2(\mathcal{G})$.) By Lemma 7.2 there exists a unitary operator W from $\mathcal{D}_{B_c}$ onto $\mathcal{D}_{Y,1}$ satisfying $WD_{B_c} = D_{Y,1}D_A$. Now let V be the isometry on $\mathcal{D}_{Y,1}$ defined by $V = WU_c W^*$. Obviously V is a unilateral shift. Moreover, the relation $U_c D_{B_c} = D_{B_c} T$ implies that

$$VD_{Y,1}D_A = D_{Y,1}D_A T. \tag{7.33}$$

We claim that $\mathcal{G} = \mathcal{D}_A \ominus D_A T\mathcal{H}$ is the cyclic wandering subspace ker $V^*$ for V. To see this, recall that the cyclic wandering subspace $\mathcal{L}$ in the Wold decomposition for V is given by

$$\mathcal{L} = \mathcal{D}_{Y,1} \ominus V\mathcal{D}_{Y,1} = \mathcal{D}_{Y,1} \ominus VD_{Y,1}D_A\mathcal{H} = \mathcal{D}_{Y,1} \ominus D_{Y,1}D_A T\mathcal{H} = \mathcal{D}_{Y,1} \ominus D_{Y,1}\mathcal{F}.$$

Because $Y | \mathcal{G} = 0$ it follows that $D_{Y,1}g = g$ for all g in $\mathcal{G}$. This along with the fact that $D_{Y,1}$ is a self adjoint operator readily implies that $\mathcal{G}$ is a reducing subspace for $D_{Y,1}$. Thus

$$\mathcal{D}_{Y,1} = \overline{D_{Y,1}\mathcal{G}} \oplus \overline{D_{Y,1}\mathcal{F}} = \mathcal{G} \oplus \overline{D_{Y,1}\mathcal{F}}.$$

Using this in our previous expression for $\mathcal{L}$ shows that $\mathcal{L} = \mathcal{G}$, which proves our claim.

To complete the proof notice that by using the intertwining property $(U_m \oplus U)B = BT$ on the matrix form of B in (7.32) we obtain

$$UZD_{Y,1}D_A = ZD_{Y,1}D_A T = ZVD_{Y,1}D_A. \tag{7.34}$$

Thus $UZ = ZV$. Since $Z | \mathcal{G} = 0$, this implies that $ZV^n g = U^n Zg = 0$ for all g in $\mathcal{G}$ and $n \geq 0$. Because $\mathcal{G}$ is the cyclic wandering subspace for the unilateral shift V, it follows that $Z = 0$. Therefore $B = [B_c, 0]^{tr}$ is the central intertwining lifting of A. The last statement of the theorem follows from the fact that if $\|A\| < \gamma$ and T is a unilateral shift, then $\Pi_A \omega P_{\mathcal{F}}$ is pointwise stable; see Lemma 6.1. This completes the proof.

The following entropy result is an immediate consequence of the previous theorem.

**COROLLARY 7.10.** *Let* T *be a unilateral shift on* $\mathcal{H} = H^2(\mathcal{E})$ *or* $\mathcal{H} = \ell_+^2(\mathcal{E})$, *where* $\mathcal{E}$ *is finite dimensional. Let* A *be an operator whose norm is bounded by* $\gamma$ *satisfying* $T'A = AT$, *where* $T'$ *is a contraction on* $\mathcal{H}'$. *Let* $U'$ *be an isometric lifting of* $T'$ *and* $B$ *an intertwining lifting of* $A$ *with tolerance* $\gamma$. *Finally, let* $B_\gamma$ *be the central intertwining lifting of* $A$ *with tolerance* $\gamma$ *and assume that* $\Pi_A \omega P_{\mathcal{F}}$ *on* $\mathcal{D}_A$ *is pointwise stable. Then the entropy*

$$E(B_\gamma) \geq E(B) . \tag{7.35}$$

*Moreover, if* $E(B_\gamma) > -\infty$, *then we have equality in (7.35) if and only if* $B = B_\gamma$. *In particular, if* $\|A\| < \gamma$, *then* $E(B_\gamma) = E(B)$ *if and only if* $B = B_\gamma$.

**REMARK 7.11.** Let T be an isometry on $\mathcal{H}$ and A an operator from $\mathcal{H}$ into $\mathcal{H}'$ whose norm is bounded by $\gamma$ satisfying $T'A = AT$, where $T'$ is a contraction on $\mathcal{H}'$. Let $U' = U_m \oplus U$ on $\mathcal{K}_m \oplus \mathcal{K}$ be an isometric lifting of $T'$, where $U_m$ is the minimal isometric lifting of $T'$ and U is a unilateral shift. If B is an intertwining lifting of A with tolerance $\gamma$, then $\Delta(B) \leq \Delta(A)$, and $\Delta(B) = \Delta(A)$ if and only if B is the central intertwining lifting of A with tolerance $\gamma$, that is $B = B_\gamma$.

The proof of this fact is a minor modification of the proof of Theorem 7.9. Indeed, if $\Delta(B) = \Delta(A)$, then using the notation in that proof let $V = V_o \oplus V_u$ on $\mathcal{D}_0 \oplus \mathcal{D}_u$ be the Wold decomposition of V, where $V_o$ is a unilateral shift and $V_u$ is the unitary part; see the next to last paragraph of Section I.1. Let $Z = [Z_o, Z_u]$ be the matrix representation of Z mapping $\mathcal{D}_0 \oplus \mathcal{D}_u$ into $\mathcal{K}$. Proceeding as above $Z_o = 0$. Thus $V_u^* Z_u^* = Z_u^* U^*$, and consequently for all h in $\mathcal{K}$ and $n = 0, 1, 2, \ldots$ we have

$$\|Z_u^* h\| = \|V_u^{*n} Z_u^* h\| = \|Z_u^* U^{*n} h\| \leq \|U^{*n} h\| .$$

Since U is a unilateral shift, the last term converges to zero as n approaches infinity. Therefore $Z = 0$ and B is the central intertwining lifting when $\Delta(B) = \Delta(A)$.

## IV.8 SOME MIXED BOUNDS FOR THE CENTRAL INTERTWINING LIFTING

In this section we will present some special bounds for the central intertwining lifting which will be used to solve some mixed $H^\infty$ and $L^2$ interpolation problems. We begin with the main result of this section.

**THEOREM 8.1.** *Let* T *be an isometry on* $\mathcal{H}$ *and* A *be an operator mapping* $\mathcal{H}$ *into* $\mathcal{H}'$ *satisfying* $T'A = AT$ *and* $\|A\| < \gamma$. *Let* $U'$ *be an isometric lifting of* $T'$ *and* B *the central intertwining lifting of* A *with respect to* $U'$ *and tolerance* $\gamma$. *Then the central intertwining lifting* B *of* A *satisfies the following inequality:*

$$\|Ba\|^2 \le \|D_{N_A} a\|^2 \le \frac{\gamma^2\|Aa\|^2}{\gamma^2 - \|A\|^2 + \|Aa\|^2} \qquad (a \in \ker T^* \textit{ and } \|a\| = 1) \tag{8.1}$$

*for all unit vectors a in the kernel of* $T^*$, *where* $N_A$ *is the operator on* $\mathcal{L}$ *defined in (6.16) and* $D_{N_A}$ *is the positive square root of* $\gamma^2 I - N_A^2$.

**PROOF**. Without loss of generality we can assume that $U'$ on $\mathcal{H}' \oplus H^2(\mathcal{D}')$ is the Sz.-Nagy-Schäffer minimal isometry lifting of $T'$ and $B = B_\gamma$ is the corresponding central intertwining lifting in (6.1). Now let a be any unit vector in $\mathcal{L} = \ker T^*$. Using the expression for orthogonal projection for $P_{\mathcal{F}}$ in (3.1) along with (2.17) and $T^* a = 0$ we have

$$\begin{aligned}
&\|B_\gamma a\|^2 \le \|Aa\|^2 + (P_{\mathcal{F}} D_A a,\, D_A a) = \|Aa\|^2 + (D_{AT}^{-2} T^* D_A^2 a,\, T^* D_A^2 a) = \\
&\|Aa\|^2 + (D_{AT}^{-2} T^* A^* Aa,\, T^* A^* Aa) = \\
&(a,\, A^* Aa) + (T(\gamma^2 I - T^* A^* AT)^{-1} T^* A^* Aa,\, A^* Aa) = \\
&(a,\, A^* Aa) + ((\gamma^2 I - TT^* A^* A)^{-1} TT^* A^* Aa,\, A^* Aa) = \\
&\gamma^2 (A^* A(\gamma^2 I - TT^* A^* A)^{-1} a,\, a)\,.
\end{aligned} \tag{8.2}$$

We claim that

$$\gamma^2(\gamma^2 I - TT^* A^* A)^{-1} a = D_A^{-2} \Pi_o^* N_A^2 a \qquad (a \in \ker T^*) \tag{8.3}$$

where $N_A$ on $\mathcal{L}$ is defined in (6.16). To prove this let

$$f = (\gamma^2 I - TT^* A^* A) D_A^{-2} \Pi_o^* N_A^2 a\,. \tag{8.4}$$

Then using $TT^* = I - P_{\mathcal{L}}$ we have

$$\begin{aligned}
(I - P_{\mathcal{L}})f = TT^* f = \gamma^2 TT^* D_A^{-2} \Pi_o^* N_A^2 a - TT^* A^* A D_A^{-2} \Pi_o^* N_A^2 a = \\
TT^*(\gamma^2 I - A^* A) D_A^{-2} \Pi_o^* N_A^2 a = TT^* \Pi_o^* N_A^2 a = 0\,.
\end{aligned}$$

Thus $P_{\mathcal{L}} f = f$. So according to (8.4) we have

$$f = P_{\mathcal{L}} f = P_{\mathcal{L}} (\gamma^2 I - TT^* A^* A) D_A^{-2} \Pi_o^* N_A^2 a = \gamma^2 P_{\mathcal{L}} D_A^{-2} \Pi_o^* N_A^2 a = \gamma^2 a\,.$$

Therefore $f = \gamma^2 a$. So equation (8.3) follows from (8.4). Now using (8.3), we have

$$\begin{aligned}
\gamma^2 (A^* A(\gamma^2 I - TT^* A^* A)^{-1} a,\, a) = (A^* A D_A^{-2} \Pi_o^* N_A^2 a,\, a) = \\
((\gamma^2 D_A^{-2} - I)\Pi_o^* N_A^2 a,\, a) = \gamma^2 (a,\, a) - (N_A^2 a,\, a) = \|D_{N_A} a\|^2\,.
\end{aligned}$$

This along with (8.2) readily yields the first inequality for $\|Ba\|^2$ in (8.1).

To complete the proof notice that

$$\begin{aligned}&\gamma^2 I - N_A^2 = \gamma^2 I - (\Pi_0 D_A^{-2}\Pi_0^*)^{-1} = \\ &\gamma^2 I - \gamma^2(\Pi_0(\gamma^2 I - A^*A + A^*A)D_A^{-2}\Pi_0^*)^{-1} = \\ &\gamma^2 I - \gamma^2(I + \Pi_0 A^*(\gamma^2 I - AA^*)^{-1}A\Pi_0^*)^{-1} .\end{aligned} \tag{8.5}$$

Now let Y be any operator mapping $\mathcal{H}_1$ into $\mathcal{H}_2$. We claim that for any unit vector a in $\mathcal{H}_1$ the following inequality holds:

$$\frac{1}{1+\|Ya\|^2} \le ((I+Y^*Y)^{-1}a, a) \qquad (\|a\|=1) . \tag{8.6}$$

This result follows from the Cauchy-Schwartz inequality, that is,

$$\begin{aligned}&1 = \|a\|^4 = ((I+Y^*Y)^{1/2}a, (I+Y^*Y)^{-1/2}a)^2 \le \\ &\|(I+Y^*Y)^{1/2}a\|^2\|(I+Y^*Y)^{-1/2}a\|^2 = \\ &((I+Y^*Y)a, a)((I+Y^*Y)^{-1}a, a) = (1+\|Ya\|^2)((I+Y^*Y)^{-1}a, a) .\end{aligned} \tag{8.7}$$

Dividing by $1+\|Ya\|^2$ readily produces the inequality in (8.6).

Now let $Y = D_{A^*}^{-1}A\Pi_0^*$ where $D_{A^*}$ is the positive square root of $\gamma^2 I - AA^*$. Using (8.6) in (8.5) along with $\|Ba\|^2 \le \|D_{N_A}a\|^2$ we obtain

$$\begin{aligned}&\|Ba\|^2 \le \|D_{N_A}a\|^2 = ((\gamma^2 I - N_A^2)a, a) = \\ &\gamma^2 - \gamma^2((I+\Pi_0 A^* D_{A^*}^{-2}A\Pi_0^*)^{-1}a, a) = \\ &\gamma^2 - \gamma^2((I+Y^*Y)^{-1}a, a) \le \gamma^2 - \frac{\gamma^2}{1+\|Ya\|^2} = \\ &\frac{\gamma^2\|Ya\|^2}{1+\|Ya\|^2} = \frac{\gamma^2\|D_{A^*}^{-1}Aa\|^2}{1+\|D_{A^*}^{-1}Aa\|^2} .\end{aligned} \tag{8.8}$$

Because $\|A\| < \gamma$, we have $(\gamma^2 - \|A\|^2)I \le D_{A^*}^2$. Recall that if C and D are invertible positive operators satisfying $C \le D$, then $D^{-1} \le C^{-1}$. (Since $C \le D$ there exists a contraction $\Gamma$ such that $C^{1/2} = \Gamma D^{1/2}$. Hence $D^{-1/2} = C^{-1/2}\Gamma$ and $D^{-1/2} = \Gamma^* C^{-1/2}$. So $D^{-1} = D^{-1/2}D^{-1/2} = C^{-1/2}\Gamma\Gamma^* C^{-1/2} \le C^{-1}$.) Thus

$$\|D_{A^*}^{-1}Aa\|^2 = (D_{A^*}^{-2}Aa, Aa) \le \frac{\|Aa\|^2}{\gamma^2 - \|A\|^2} . \tag{8.9}$$

Using (8.9) in the last term of (8.8), along with the fact that the function $\tau(1+\tau)^{-1}$ is increasing for $\tau \geq 0$, equation (8.8) now becomes

$$\|Ba\|^2 \leq \|D_{N_A} a\|^2 \leq \frac{\gamma^2 \|D_{A^*}^{-1} Aa\|^2}{1 + \|D_{A^*}^{-1} Aa\|^2} \leq$$

$$\frac{\gamma^2 \|Aa\|^2 (\gamma^2 - \|A\|^2)^{-1}}{1 + \|Aa\|^2 (\gamma^2 - \|A\|^2)^{-1}} = \frac{\gamma^2 \|Aa\|^2}{\gamma^2 - \|A\|^2 + \|Aa\|^2} .$$

This gives (8.1) and completes the proof.

Let $\mathcal{X}$ and $\mathcal{Y}$ be Hilbert spaces and L be an operator mapping a finite dimensional vector space $\mathcal{X}$ into $\mathcal{Y}$. Recall that the Hilbert-Schmidt norm of L is defined by

$$\|L\|_{HS}^2 = \sum_i \|L\phi_i\|^2 \tag{8.10}$$

where $\{\phi_i\}$ is an orthonormal basis for $\mathcal{X}$. The Hilbert-Schmidt norm of L is independent of the choice for the orthonormal basis; see Section VIII.2 in Gohberg-Goldberg-Kaashoek [1]. The following result, which is essentially folklore, will be useful to solve some mixed $H^\infty$ and $L^2$ interpolation problems.

**LEMMA 8.2.** *Let* L *be a linear operator mapping a n dimensional space* $\mathcal{X}$ *into* $\mathcal{Y}$. *Then*

$$\frac{\|L\|_{HS}^2}{n} = \max \min \|L\phi_i\|^2 \tag{8.11}$$

*where the minimum is taken over a specified orthonormal basis* $\{\phi_i\}_1^n$ *for* $\mathcal{X}$ *and the maximum is taken over all orthonormal basis for* $\mathcal{X}$.

**PROOF.** Let $\{\phi_i\}_1^n$ be any orthonormal basis for $\mathcal{X}$ arranged in the following way $\|L\phi_1\|^2 \geq \|L\phi_2\|^2 \geq \cdots \geq \|L\phi_n\|^2$. According to the definition of the Hilbert-Schmidt norm

$$\|L\|_{HS}^2 = \sum_{i=1}^n \|L\phi_i\|^2 \geq \sum_{i=1}^n \|L\phi_n\|^2 = n\|L\phi_n\|^2 .$$

This readily implies that the mean m of $\{\|L\phi_i\|^2\}$ satisfies the following inequality

$$m = \frac{\|L\|_{HS}^2}{n} \geq \max \min \|L\phi_i\|^2 . \tag{8.12}$$

To complete the proof we must show that there is equality in (8.12). To this end, let r be a scalar in [0,1] and t the positive square root of $1 - r^2$. Let $\psi$ be the unit vector defined by $\psi_1 = r\phi_1 + t\phi_n$, and f(r) the continuous function defined by

$$f(r) = \|L\psi_1\|^2 = \|Lr\phi_1 + Lt\phi_n\|^2 =$$
$$r^2\|L\phi_1\|^2 + 2\,r\,t\,\mathrm{Re}\,(L\phi_1, L\phi_n) + t^2\|L\phi_n\|^2 . \tag{8.13}$$

Notice that because m is the mean of $\{\|L\phi_i\|^2\}$, we have

$$f(0) = \|L\phi_n\|^2 \le m \le \|L\phi_1\|^2 = f(1) .$$

Since f(r) is a positive continuous function for $0 \le r \le 1$, the previous inequality shows that there exists a r such that

$$\|L\psi_1\|^2 = f(r) = m = \frac{\|L\|_{HS}^2}{n} . \tag{8.14}$$

Let $\mathcal{X}_{n-1}$ be the orthogonal complement of $\psi_1$. (In fact, $\{t\phi_1 - r\phi_n, \phi_2, \phi_3, \ldots, \phi_{n-1}\}$ is an orthonormal basis for $\mathcal{X}_{n-1}$.) By using the definition of the Hilbert-Schmidt norm and equation (8.14) we have

$$\|L\|_{HS}^2 = \|L\psi_1\|^2 + \|L\,|\,\mathcal{X}_{n-1}\|_{HS}^2 = \frac{\|L\|_{HS}^2}{n} + \|L\,|\,\mathcal{X}_{n-1}\|_{HS}^2 .$$

This readily implies that

$$\|L\,|\,\mathcal{X}_{n-1}\|_{HS}^2 = \frac{(n-1)\|L\|_{HS}^2}{n} .$$

By applying equation (8.14) to the $n - 1$ dimensional space $\mathcal{X}_{n-1}$, we see that there exists a unit vector $\psi_2$ in $\mathcal{X}_{n-1}$ satisfying

$$\|L\psi_2\|^2 = \frac{\|L\,|\,\mathcal{X}_{n-1}\|_{HS}^2}{n-1} = \frac{\|L\|_{HS}^2}{n} = m .$$

By continuing in this fashion there exist an orthonormal basis $\{\psi_i\}_1^n$ for $\mathcal{X}$ satisfying $\|L\psi_i\|^2 = m$ for all $i = 1, , \ldots, n$. (Moreover, by applying the previous procedure to the orthonormal basis $\{t\phi_1 - r\phi_n, \phi_2, \ldots, \phi_{n-1}\}$ for $\mathcal{X}_{n-1}$ and continuing on, one can easily construct a recursive procedure to compute the orthonormal basis $\{\psi_i\}_1^n$.) Using $\phi_i = \psi_i$ yields equality in (8.12). This completes the proof.

The following result will be used to solve some $H^\infty$ and $L^2$ interpolation problems.

**COROLLARY 8.3.** *Let* T *be an isometry on* $\mathcal{H}$ *and assume that* $\mathcal{L} = \ker T^*$ *is an* n *dimensional vector space. Let* A *be an operator mapping* $\mathcal{H}$ *into* $\mathcal{H}'$ *satisfying* $T'A = AT$ *and* $U'$ *an isometric lifting of* $T'$. *Finally, let* $\delta > 1$ *and* $\gamma$ *the constant defined by* $\gamma = \delta\|A\|$. *Then the central intertwining lifting* B *of* A *with tolerance* $\gamma$ *satisfies the following inequalities*

$$\text{(i)}\quad \|B\| \le \delta\|A\| \tag{8.15}$$

$$\text{(ii)}\quad \|B\,|\,\mathcal{L}\|_{HS}^2 \le \frac{\delta^2\|A\,|\,\mathcal{L}\|_{HS}^2}{\delta^2 - 1 + \dfrac{\|A\,|\,\mathcal{L}\|_{HS}^2}{n\|A\|^2}}\,. \tag{8.16}$$

*In particular, by choosing* $\delta^2 = 2 - \|A\,|\,\mathcal{L}\|_{HS}^2/n\|A\|^2$, *the corresponding central intertwining lifting* B *of* A *with tolerance* $\delta\|A\|$ *satisfies the following bounds*

$$\text{(iii)}\quad \|B\| \le \|A\|\sqrt{2 - \frac{\|A\,|\,\mathcal{L}\|_{HS}^2}{n\|A\|^2}}$$

$$\tag{8.17}$$

$$\text{(iv)}\quad \|B\,|\,\mathcal{L}\|_{HS} \le \|A\,|\,\mathcal{L}\|_{HS}\sqrt{2 - \frac{\|A\,|\,\mathcal{L}\|_{HS}^2}{n\|A\|^2}}\,.$$

**PROOF**. The inequality in part (i) follows from $\|B_\gamma\| \le \gamma$ and $B = WB_\gamma$ where $B_\gamma$ is the central intertwining lifting with tolerance $\gamma$ and W is the appropriate isometry. To obtain part (ii) let $\{\phi_i\}_1^n$ be an orthonormal basis for $\mathcal{L}$. By consulting (8.1) we see that

$$\|B\phi_i\|^2 \le \frac{\delta^2\|A\|^2\|A\phi_i\|^2}{(\delta^2-1)\|A\|^2 + \|A\phi_i\|^2} =$$

$$\tag{8.18}$$

$$\frac{\delta^2\|A\phi_i\|^2}{\delta^2 - 1 + \dfrac{\|A\phi_i\|^2}{\|A\|^2}} \le \frac{\delta^2\|A\phi_i\|^2}{\delta^2 - 1 + \dfrac{\|A\phi_{min}\|^2}{\|A\|^2}}$$

where $\phi_{min}$ is any unit vector from $\{\phi_i\}$ satisfying $\|A\phi_{min}\| \le \|A\phi_i\|$ for all i. According to the proof of the previous lemma we can choose an orthonormal basis $\{\phi_i\}$ such that $\|A\phi_{min}\|^2 = \|A\phi_i\|^2 = \|A\,|\,\mathcal{L}\|_{HS}^2/n$. Notice that we are choosing an orthonormal basis $\{\phi_i\}$ which solves the optimization problem max min $\|A\phi_i\|^2$, because we want to use a basis $\{\phi_i\}$ to make the coefficient $\delta^2(\delta^2 - 1 + \|A\phi_{min}\|^2/\|A\|^2)^{-1}$ of $\|A\phi_i\|^2$ as small as possible. So by choosing a basis such that $\|A\phi_{min}\|^2 = \|A \mid \mathcal{L}\|_{HS}^2/n$, the equation (8.18) yields

$$\|B\phi_i\|^2 \leq \frac{\delta\|A\phi_i\|^2}{\delta^2 - 1 + \dfrac{\|A \mid \mathcal{L}\|_{HS}^2}{n\|A\|^2}} .$$

By using the definition of the Hilbert-Schmidt norm we obtain part (ii). This completes the proof.

## IV.9 A MIXED TWO-SIDED SARASON RESULT

In this section we will use the central intertwining lifting to solve a mixed $H^2$ and $H^\infty$ two-sided Sarason problem. Throughout this section the spaces $\mathcal{U}$ and $\mathcal{Y}$ are finite dimensional. Recall that if C is any operator on $\mathcal{U}$, then

$$\text{trace } (C) = \sum_{i=1}^{n} (C\phi_i, \phi_i) \tag{9.1}$$

where $\{\phi_i\}_1^n$ is an orthonormal basis for $\mathcal{U}$. The trace is independent of the choice of the orthonormal basis. So if C is a square matrix on $\mathbb{C}^n$ and $\{\phi_i\}$ is the standard orthonormal basis (that is, the i-th component of $\phi_i$ is one and all the other entries are zero), then the trace of C is the sum of the diagonal entries of C. Notice that if M is an operator from $\mathcal{U}$ into $\mathcal{Y}$, then

$$\|M\|_{HS}^2 = \sum_1^n \|M\phi_i\|^2 = \sum_1^n (M^*M\phi_i, \phi_i) = \text{trace } (M^*M) . \tag{9.2}$$

In particular, $\|M\|_{HS}^2 = \text{trace } (M^*M)$. The Hilbert space $L^2(\mathcal{U}, \mathcal{Y})$ is the space of all square integrable Lebesgue measurable functions on the unit circle whose values are linear operators from $\mathcal{U}$ into $\mathcal{Y}$. The $L^2(\mathcal{U}, \mathcal{Y})$ norm of a function G is given by

$$\|G\|_2^2 = (G, G)_2 = \frac{1}{2\pi}\int_0^{2\pi} \text{trace } (G(e^{it})^* G(e^{it}))dt = \sum_{-\infty}^{\infty} \text{trace } (G_n^* G_n) , \tag{9.3}$$

where $G = \sum G_n e^{int}$ is the Fourier series expansion of G. Notice that if G is a function in $L^\infty(\mathcal{U}, \mathcal{Y})$, then obviously G is in $L^2(\mathcal{U}, \mathcal{Y})$. Moreover, $\|G\|_2 = \|L_G \mid \mathcal{U}\|_{HS}$, where $L_G$ is the Laurent operator from $L^2(\mathcal{U})$ into $L^2(\mathcal{Y})$ generated by G. Finally, $H^2(\mathcal{U}, \mathcal{Y})$ is the subspace of $L^2(\mathcal{U}, \mathcal{Y})$ consisting of all function G in $L^2$ whose Fourier coefficients $G_n$ of $e^{int}$ are zero for all $n < 0$.

Throughout this section F is a specified function in $L^\infty(\mathcal{U}, \mathcal{Y})$ and $\Theta_1$ is a specified co-inner function in $H^\infty(\mathcal{U}, \mathcal{E})$ and $\Theta_2$ is a specified inner function in $H^\infty(\mathcal{F}, \mathcal{Y})$. Recall that a function $\Theta$ in $H^\infty(\mathcal{E}, \mathcal{F})$ is *inner*, respectively *co-inner*, if $\Theta(e^{it})$ is a.e. an isometry, respectively a co-isometry. Now consider the following $H^\infty$ and respectively $L^2$ optimization problems associated with:

$$d_\infty = d_\infty(F) = \inf\{\|F - \Theta_2 H \Theta_1\|_\infty : H \in H^\infty(\mathcal{E}, \mathcal{F})\},$$
$$d_2 = d_2(F) = \inf\{\|F - \Theta_2 H \Theta_1\|_2 : H \in H^2(\mathcal{E}, \mathcal{F})\}, \tag{9.4}$$

where $d_\infty$ is the distance from F to $\Theta_2 H^\infty(\mathcal{E}, \mathcal{F})\Theta_1$ in the $L^\infty$ norm and $d_2$ is the corresponding distance in the $L^2$ norm. These optimization problems naturally arise in control theory; see Francis [1], Green-Limebeer [1], Zhou-Doyle-Glover [1] and Chapter VII. In most problems the same H does not minimize both $d_\infty$ and $d_2$. In the sequel we will need explicit expressions for $d_\infty$ and $d_2$. Because the $L^2$ norm defines an inner product, we can easily compute $d_2$ by relaying on an argument used in Wiener filtering. To this end, let $[G]_c$ be the causal part of a function G in $L^2(\mathcal{E}, \mathcal{F})$, that is, if $G = \sum_{-\infty}^{\infty} G_n e^{int}$ is the Fourier series expansion of G, then

$$[G]_c = \sum_{n=0}^{\infty} G_n e^{int}. \tag{9.5}$$

We claim that

$$d_2 = d_2(F) = \|F - \Theta_2[\Theta_2^* F \Theta_1^*]_c \Theta_1\|_2. \tag{9.6}$$

By the projection theorem, it is sufficient to show that $F - \Theta_2[\Theta_2^* F \Theta_1^*]_c \Theta_1$ is orthogonal to the linear subspace $\Theta_2 H^2(\mathcal{E}, \mathcal{F})\Theta_1$ with respect to the $L^2$ inner product $(\cdot, \cdot)_2$. To verify (9.6) notice that for any H in $H^2(\mathcal{E}, \mathcal{F})$ we have

$$(F - \Theta_2[\Theta_2^* F \Theta_1^*]_c \Theta_1, \Theta_2 H \Theta_1)_2 = \frac{1}{2\pi}\int_0^{2\pi} \operatorname{trace}\,(\Theta_1^* H^* (\Theta_2^* F - [\Theta_2^* F \Theta_1^*]_c \Theta_1))dt =$$

$$\frac{1}{2\pi}\int_0^{2\pi} \operatorname{trace}\,((\Theta_2^* F - [\Theta_2^* F \Theta_1^*]_c \Theta_1)\Theta_1^* H^*)dt = \frac{1}{2\pi}\int_0^{2\pi} \operatorname{trace}\,((\Theta_2^* F \Theta_1^* - [\Theta_2^* F \Theta_1^*]_c)H^*)dt = 0.$$

The first and third equalities follow from the fact that $\Theta_2$ is inner and $\Theta_1$ is co-inner, respectively. The second equality follows from the fact that trace (BC) = trace (CB) where C maps $\mathcal{U}$ into $\mathcal{F}$ and B maps $\mathcal{F}$ into $\mathcal{U}$. The above analysis shows that $F - \Theta_2[\Theta_2^* F \Theta_1^*]_c \Theta_1$ is orthogonal to $\Theta_2 H^2(\mathcal{E}, \mathcal{F})\Theta_1$. Therefore, by the projection theorem, (9.6) holds.

To establish a formula for $d_\infty$ let $\mathcal{H}$ and $\mathcal{H}'$ be the subspaces defined by

$$\mathcal{H} = L^2(\mathcal{U}) \ominus \Theta_1^* K^2(\mathcal{E}) \quad \text{and} \quad \mathcal{H}' = L^2(\mathcal{Y}) \ominus \Theta_2 H^2(\mathcal{F}), \tag{9.7}$$

where $K^2(\mathcal{E})$ is the orthogonal complement of $H^2(\mathcal{E})$ in $L^2(\mathcal{E})$. Let A(F) be the operator from $\mathcal{H}$ into $\mathcal{H}'$, with symbol F, defined by $A(F)f = P'M_F f$, where f is in $\mathcal{H}$ and $P'$ is the orthogonal projection onto $\mathcal{H}'$. As before, $M_F$ is the multiplication operator from $L^2(\mathcal{U})$ to $L^2(\mathcal{Y})$ defined by $(M_F g)(e^{it}) = F(e^{it})g(e^{it})$ where g is in $L^2(\mathcal{U})$. Now let $V_1$ and $V_2$ be the bilateral shifts (multiplication by $e^{it}$) on $L^2(\mathcal{U})$ and $L^2(\mathcal{Y})$, respectively. Clearly $\mathcal{H}$ is an invariant subspace for $V_1$ and $\mathcal{H}'$ is an invariant subspace for $V_2^*$. Let T on $\mathcal{H}$ be the isometry and $T'$ on $\mathcal{H}'$ be the co-isometry defined by $T = V_1 | \mathcal{H}$ and $T' = P'V_2 | \mathcal{H}'$. Obviously $V_2$ is an isometric lifting of $T'$, that is, $P'V_2 = T'P'$. Using $P'V_2 = T'P'$ and $V_2 M_F = M_F V_1$ it is easy to verify that $T'A(F) = A(F)T$.

According to the results in Section II.6 an operator B is an intertwining lifting of A with respect to $V_2$ if and only if $B = M_G | \mathcal{H}$ where G is a function in $L^\infty(\mathcal{U}, \mathcal{Y})$ of the form $G = F - \Theta_2 H \Theta_1$ for some H in $H^\infty(\mathcal{E}, \mathcal{F})$. Moreover, in this correspondence $\|B\| = \|G\|_\infty$. From this the results in Section II.6 readily show that $d_\infty = \|A\|$ and there exists an optimal $H_{opt}$ in $H^\infty(\mathcal{E}, \mathcal{F})$ satisfying

$$d_\infty = \|A(F)\| = \|F - \Theta_2 H_{opt} \Theta_1\| . \tag{9.8}$$

Moreover, given any $\gamma \geq d_\infty$ the central intertwining lifting B of A with respect to $V_2$ and tolerance $\gamma$ uniquely determines a function $G_\gamma$ in $L^\infty(\mathcal{U}, \mathcal{Y})$ such that $B = M_{G_\gamma} | \mathcal{H}$. By a slight abuse of terminology we also call this function $G_\gamma$ the central intertwining lifting of A with respect to the tolerance $\gamma$. Finally, because the central intertwining lifting B of A is an intertwining lifting, it follows that $G_\gamma$ admits a decomposition of the form $G_\gamma = F - \Theta_2 H \Theta_1$ for some H in $H^\infty(\mathcal{E}, \mathcal{F})$ and $\|G_\gamma\|_\infty \leq \gamma$. Now we are ready to prove the following result.

**THEOREM 9.1.** *Let* F *be a function in* $L^\infty(\mathcal{U}, \mathcal{Y})$ *while* $\Theta_1$ *is co-inner function in* $H^\infty(\mathcal{U}, \mathcal{E})$ *and* $\Theta_2$ *is an inner function in* $H^\infty(\mathcal{F}, \mathcal{Y})$. *Let* $\delta > 1$ *and* $G_\gamma$ *be the central intertwining lifting of* A(F) *with tolerance* $\gamma = \delta d_\infty$. *Then* $G_\gamma$ *admits a decomposition of the form* $G_\gamma = F - \Theta_2 H \Theta_1$ *for some* H *in* $H^\infty(\mathcal{E}, \mathcal{F})$. *Moreover,* $G_\gamma$ *satisfies the following mixed bounds:*

$$\|F - \Theta_2 H \Theta_1\|_\infty \leq \delta d_\infty ,$$

$$\|F - \Theta_2 H \Theta_1\|_2 \leq \delta d_2 \left[\delta^2 - 1 + \frac{d_{21}^2}{n d_\infty^2}\right]^{-1/2} , \tag{9.9}$$

*where* $d_{21} = \|A | \Theta_1^* \mathcal{E}\|_{HS} = \|P_{\mathcal{H}'} F \Theta_1^* | \mathcal{E}\|_{HS}$ *and* n *is the dimension of* $\mathcal{E}$. *In fact, by choosing* $\delta^2 = 2 - d_{21}^2 / n d_\infty^2$ *equation (9.9) becomes*

$$\|F - \Theta_2 H \Theta_1\|_\infty \le d_\infty \sqrt{2 - \frac{d_{21}^2}{n d_2^2}} \quad and \quad \|F - \Theta_2 H \Theta_1\|_2 \le d_2 \sqrt{2 - \frac{d_{21}^2}{n d_\infty^2}} \,. \tag{9.10}$$

*Finally, by choosing* $\delta = \sqrt{2}$ *in (9.9) we get the following mixed bound independent of* $d_{21}$ *and the dimension of* $\mathcal{E}$

$$\|F - \Theta_2 H \Theta_1\|_\infty \le d_\infty \sqrt{2} \quad and \quad \|F - \Theta_2 H \Theta_1\|_2 \le d_2 \sqrt{2} \,. \tag{9.11}$$

**PROOF**. The central intertwining lifting B for A with tolerance of $\gamma$ defines a function $G_\gamma$ in $L^\infty(\mathcal{U}, \mathcal{Y})$ of the form $G_\gamma = F - \Theta_2 H \Theta_1$ satisfying $B = M_{G_\gamma} | \mathcal{H}$. The first inequality in (9.9) follows from the fact that $\|G_\gamma\|_\infty = \|B\| \le \gamma$ when $\gamma = \delta d_\infty$. To obtain the second inequality notice that $\Theta_1^* \mathcal{E} = \ker(T^*) = \mathcal{L}$. Because $\Theta_1^*$ is a.e. an isometry $\|A | \mathcal{L}\|_{HS} = \|A\Theta_1^* | \mathcal{E}\|_{HS}$. According to (8.16) in Corollary 8.3 we have

$$\|(F - \Theta_2 H \Theta_1)\Theta_1^*\|_2^2 = \|G_\gamma \Theta_1^*\|_2^2 = \|B | \mathcal{L}\|_{HS} \le \frac{\delta^2 \|A | \mathcal{L}\|_{HS}^2}{\delta^2 - 1 + \dfrac{d_{21}^2}{n\|A\|^2}} \,. \tag{9.12}$$

A simple application of the Projection Theorem shows that $A(F)\Theta_1^* a = (F\Theta_1^* - \Theta_2[\Theta_2^* F \Theta_1^*]_c a$ for all a in $\mathcal{E}$. This implies that $\|A | \mathcal{L}\|_{HS}^2 = \|G_o \Theta_1^*\|_2^2$ where $G_o = F - \Theta_2[\Theta_2^* F \Theta_1^*]_c \Theta_1$ is the optimal solution to the $L^2$ optimization problem in (9.3), that is, $d_2 = \|G_o\|_2$; see (9.6). Using this along with $d_\infty = \|A\|$ in (9.12) gives

$$\|G_\gamma \Theta_1^*\|_2^2 \le \frac{\delta^2 \|G_o \Theta_1^*\|_2^2}{\delta^2 - 1 + \dfrac{d_{21}^2}{n d_\infty^2}} \,. \tag{9.13}$$

Now let $\Delta_1 = I - \Theta_1^* \Theta_1$. Since $\Theta_1^*$ is isometric, $\Theta_1 \Delta_1 = 0$ and thus $G_\gamma \Delta_1 = F\Delta_1 = G_o \Delta_1$. Notice that $d_{21}^2 = \|A | \mathcal{L}\|_{HS}^2 \le n d_\infty^2$. In particular, $\delta^2 - 1 + d_{21}^2 / n d_\infty^2 \le \delta^2$. This and $G_\gamma \Delta_1 = G_o \Delta_1$ gives

$$\|G_\gamma \Delta_1\|_2^2 \le \frac{\delta^2 \|G_o \Delta_1\|_2^2}{\delta^2 - 1 + \dfrac{d_{21}^2}{n d_\infty^2}} \,. \tag{9.14}$$

Combining this with (9.13) yields

$$\|G_\gamma\Theta_1^*\|_2^2 + \|G_\gamma\Delta_1\|_2^2 \le \frac{\delta^2(\|G_o\Theta_1^*\|_2^2 + \|G_o\Delta_1\|_2^2)}{\delta^2 - 1 + \dfrac{d_{21}^2}{nd_\infty^2}} . \tag{9.15}$$

Because $\Theta_1^*$ is an isometry, $\Delta_1$ is an orthogonal projection. This and the fact that the trace $(M^*M) =$ trace $(MM^*)$ gives

$$\|G_\gamma\Theta_1^*\|_2^2 + \|G_\gamma\Delta_1\|_2^2 = \frac{1}{2\pi}\int_0^{2\pi} (\text{trace}\,(G_\gamma\Theta_1^*\Theta_1 G_\gamma^*) + \text{trace}\,(G_\gamma\Delta_1\Delta_1^* G_\gamma^*))dt = \|G_\gamma\|_2^2 .$$

A similar calculation shows that $\|G_o\Theta_1^*\|_2^2 + \|G_o\Delta_1\|_2^2 = \|G_o\|_2^2$. Therefore (9.15) and $\|G_o\|_2 = d_2$ yields the second equality in (9.9). This completes the proof.

If $\Theta_1 = I$, then the two-sided Sarason problem reduces to the Sarason problem. In this case, the $H^2$ and $H^\infty$ optimization problems in (9.4) simplify to

$$d_\infty = \inf\{\|F - \Theta_2 H\|_\infty : H \in H^\infty(\mathcal{U}, \mathcal{F})\} ,$$
$$d_2 = \inf\{\|F - \Theta_2 H\|_\infty : H \in H^2(\mathcal{U}, \mathcal{F})\} , \tag{9.16}$$

where F is a specified function in $L^\infty(\mathcal{U}, \mathcal{Y})$. In this case $\mathcal{H} = H^2(\mathcal{U})$ and T is the unilateral shift on $H^2(\mathcal{U})$. The operator A simplifies to $A = P'M_F | H^2(\mathcal{U})$ where $P'$ is the orthogonal projection on $\mathcal{H}'$; see (9.7). As before $T'$ on $\mathcal{H}'$ is the compression of the unilateral shift $V_2$ on $L^2(\mathcal{Y})$ to $\mathcal{H}'$. Obviously $d_\infty = \|A\|$. Since $\mathcal{L} = \ker(T^*) = \mathcal{U}$, it now follows that

$$d_2 = \|F - \Theta_2[\Theta_2^* F]_c\|_2 = \|A | \mathcal{U}\|_2 . \tag{9.17}$$

Thus $d_{21} = d_2$. So using this in the previous theorem along with Theorem 6.6, we obtain the following result.

**COROLLARY 9.2.** *Let* F *be a function in* $L^\infty(\mathcal{U}, \mathcal{Y})$ *where* $n = \dim \mathcal{U}$ *is finite and* $\Theta_2$ *is an inner function in* $H^\infty(\mathcal{F}, \mathcal{Y})$. *Let* $\delta > 1$ *and* $G_\gamma$ *in* $L^\infty(\mathcal{U}, \mathcal{Y})$ *be the central intertwining lifting of* $A = P'M_F | H^2(\mathcal{U})$ *with tolerance* $\gamma = \delta d_\infty$. *Then* $G_\gamma = F - \Theta_2 H$ *for some* H *in* $H^\infty(\mathcal{U}, \mathcal{F})$ *and*

$$\|F - \Theta_2 H\|_\infty \le \delta d_\infty ,$$
$$\|F - \Theta_2 H\|_2 \le \frac{\delta d_2}{\sqrt{\delta^2 - 1 + d_2^2/nd_\infty^2}} . \tag{9.18}$$

*If we set* $\Theta_1 = I$, $d_{21} = d_2$, *then (9.10) and (9.11) hold. Moreover, the function* $G_\gamma$ *can be computed by* $G_\gamma(e^{it}) = P(e^{it})Q(e^{it})^{-1}$. *Here* $P(\lambda)$ *and the outer function* $Q(\lambda)$ *are given by*

$$P(e^{it}) = (AD_A^{-2}\Pi_o^*)(e^{it}) \quad and \quad Q(\lambda) = (D_A^{-2}\Pi_o^*)(\lambda)\,, \tag{9.19}$$

*where* $\Pi_o$ *is the operator from* $H^2(\mathcal{U})$ *onto* $\mathcal{U}$ *defined by* $\Pi_o = P_{\mathcal{U}}$. *Furthermore the outer spectral factor* $\Theta(\lambda)$ *for* $\gamma^2 I - G_\gamma^* G_\gamma$ *is given by* $Q(0)^{1/2}Q(\lambda)^{-1}$. *Finally, if* F *is in* $H^\infty(\mathcal{U}, \mathcal{Y})$, *then the central intertwining lifting* $G_\gamma$ *is also in* $H^\infty(\mathcal{U}, \mathcal{Y})$.

Recall that if $F(\lambda) = A_0 + A_1\lambda + \cdots + A_{n-1}\lambda^{n-1}$ and $\Theta_2 = \lambda^n I$, then the Sarason problem reduces to the Schur interpolation problem. In this case $d_\infty = \|M\|$ where M is the lower triangular Toeplitz matrix generated by $\{A_j\}_0^{n-1}$ and

$$d_2^2 = \sum_{j=0}^{n-1} \text{trace}\,(A_j^* A_j)\,. \tag{9.20}$$

Therefore, the results in Section 5 give explicit formulas to compute the central intertwining lifting $G_\gamma$ satisfying the $H^2$ and $H^\infty$ bounds in (9.18).

## *Notes to Chapter IV:*

Except for the material in Section 1, which is standard, this chapter is based on three papers, namely Foias-Frazho [5], Foias-Frazho-Li [1] and Foias-Frazho-Gohberg [1]. Remark 4.5 was taken from Smith-Frazho [1]. Section 7 is related to the maximum entropy principle obtained by the band method technique in Gohberg-Kaashoek-Woerdeman [3] (see also Section XXXIV.4 in Gohberg-Goldberg-Kaashoek [2]). Formulas like (7.21) and (7.27) resemble the maximum entropy formulas appearing in Dym-Gohberg [3], [4], which concern the Nehari case (e.g., when the operator A in (7.27) is a Hankel operator). Earlier maximum entropy results for contractions can be found in Arov-Krein [1]. The first to consider $H^\infty$ and $L^2$ bounds (of the type appearing in Section 8) were V. G. Kaftal, D. R. Larson and G. Weiss (in Kaftal-Larson-Weiss [1]) who dealt with the scalar Nehari case. An improved version of their bounds appears in Section 8 in the general setting of the commutant lifting theorem.

CHAPTER V

# CENTRAL STATE SPACE SOLUTIONS

The main purpose of this chapter is to provide explicit state space formulas for the central solutions of the Nevanlinna-Pick, Sarason, Nehari, Nudelman and two block interpolation problems. The reader mainly interested in direct applications of these formulas can assume that the state space $\mathcal{X}$ is finite dimensional. The formulas are derived by applying the results in the previous chapter for these concrete cases. The corresponding $H^\infty$ and $L^2$ bounds are given. Finally, state space formulas are given for the maximum entropy for some of these problems.

## V.1 THE CENTRAL FORMULA FOR NEVANLINNA-PICK

In this section we will present a state space formula to compute the central intertwining solution to the standard left Nevanlinna-Pick interpolation problem. To this end, let $\{Z$ on $\mathcal{X}$, $B\}$ be a stable, controllable pair and $\{Z, \tilde{B}\}$ a stable pair where B maps $\mathcal{U}$ into $\mathcal{X}$ while $\tilde{B}$ maps $\mathcal{Y}$ into $\mathcal{X}$. As before, let W from $\ell_+^2(\mathcal{U})$ into $\mathcal{X}$ and $\tilde{W}$ from $\ell_+^2(\mathcal{Y})$ into $\mathcal{X}$ be the controllability operators generated by the pair $\{Z, B\}$ and $\{Z, \tilde{B}\}$, respectively, that is,

$$W = [B, ZB, Z^2B, \ldots] \quad \text{and} \quad \tilde{W} = [\tilde{B}, Z\tilde{B}, Z^2\tilde{B}, \ldots]\,. \tag{1.1}$$

Recall that the standard left Nevanlinna-Pick optimization problem is:

$$d_\infty = \inf\{\|G\|_\infty : WT_G = \tilde{W} \quad \text{and} \quad G \in H^\infty(\mathcal{Y}, \mathcal{U})\} \tag{1.2}$$

where $T_G$ is the Toeplitz matrix generated by G. According to the results in Section II.2 it follows that there exists a $G_{opt}$ in $H^\infty(\mathcal{Y}, \mathcal{U})$ solving the $H^\infty$ optimization problem in (1.2), that is, $WT_{G_{opt}} = \tilde{W}$ and $\|G_{opt}\|_\infty = d_\infty$. Moreover, $d_\infty^2 = r_{spec}(\tilde{P}P^{-1})$ where $P = WW^*$ and $\tilde{P} = \tilde{W}\tilde{W}^*$ are the controllably grammians for the pair $\{Z, B\}$ and $\{Z, \tilde{B}\}$, respectively. Recall that, P and $\tilde{P}$ can be obtained as the unique solutions to the following Lyapunov equations

$$P = ZPZ^* + BB^* \quad \text{and} \quad \tilde{P} = Z\tilde{P}Z^* + \tilde{B}\tilde{B}^*\,. \tag{1.3}$$

Finally, $\tilde{P}P^{-1}$ is similar to $P^{-1/2}\tilde{P}P^{-1/2}$. (Compute $P^{-1/2}(\tilde{P}P^{-1})P^{1/2}$.) Notice also that $\tilde{P}P^{-1}$ is similar to $P^{-1}\tilde{P}$. (Compute $P^{-1}(\tilde{P}P^{-1})P$.) Therefore $\tilde{P}P^{-1}$, $P^{-1}\tilde{P}$ and $P^{-1/2}\tilde{P}P^{-1/2}$ all have the same spectral radius $d_\infty^2$.

If $\mathcal{Y}$ is finite dimensional, then associated with the standard left Nevanlinna-Pick problem is the following $H^2$ optimization problem; see Section I.8

$$d_2 = \inf \{\|G\|_2 : WT_G = \tilde{W} \text{ and } G \in H^2(\mathcal{Y}, \mathcal{U})\} . \tag{1.4}$$

According to the results in Section I.8, there exists a unique solution to this $H^2$ optimization problem. Moreover, this optimal $H^2$ solution is given by

$$G_{2*}(\lambda) = B^*(I - \lambda Z^*)^{-1}P^{-1}\tilde{B} \text{ and } d_2^2 = \operatorname{trace} \tilde{B}^* P^{-1}\tilde{B} , \tag{1.5}$$

that is, $WT_{G_{2*}} = \tilde{W}$ and $\|G_{2*}\|_2 = d_2$. Now let A be the linear map from $H^2(\mathcal{Y})$ into $\mathcal{H}' = \overline{\operatorname{ran} W^*}$ defined by $WA = \tilde{W}$. According to the results in Section II.2, the error $d_\infty$ is finite if and only if $WA = \tilde{W}$ defines bounded operator. Moreover, in this case $d_\infty = \|A\|$. So we say that $G_\gamma$ in $H^\infty(\mathcal{Y}, \mathcal{U})$ is the *central interpolant* for the standard Nevanlinna-Pick problem with tolerance $\gamma$ if the Toeplitz operator $T_{G_\gamma}$ is the central intertwining lifting of A with tolerance $\gamma$; see Section IV.2. Clearly this function $G_\gamma$ satisfies $WT_{G_\gamma} = \tilde{W}$ and $\|G_\gamma\|_\infty \le \gamma$. Now we present the state space formulas for the central solution in the next four theorems, which will be subsequently proven.

**THEOREM 1.1.** *Let* $\{Z, B\}$ *a stable, controllable pair and* $\{Z, \tilde{B}\}$ *be the data for the standard left Nevanlinna-Pick interpolation problem. Let* P *and* $\tilde{P}$ *be the controllability grammians for the pair* $\{Z, B\}$ *and* $\{Z, \tilde{B}\}$, *respectively. Let* $\delta > 1$ *and set* $\gamma = \delta d_\infty$. *Then the central interpolant* $G_\gamma$ *with tolerance* $\gamma$ *for the standard left Nevanlinna-Pick problem is given by* $G_\gamma(\lambda) = N(\lambda)D(\lambda)^{-1}$ *where* N *and* D *are the operator valued analytic functions computed by*

$$\begin{aligned} N(\lambda) &= \gamma^2 B^*(I - \lambda Z^*)^{-1}(\gamma^2 P - \tilde{P})^{-1}\tilde{B} \\ D(\lambda) &= I + \tilde{B}^*(I - \lambda Z^*)^{-1}(\gamma^2 P - \tilde{P})^{-1}\tilde{B} . \end{aligned} \tag{1.6}$$

*In particular,* $G_\gamma$ *is a function in* $H^\infty(\mathcal{Y}, \mathcal{U})$ *satisfying* $WT_{G_\gamma} = \tilde{W}$ *such that* $\|G_\gamma\|_\infty < \gamma$ *and if* $\mathcal{Y}$ *is finite dimensional*

$$\|G_\gamma\|_2 \le \delta d_2 \left[\delta^2 - 1 + \frac{d_2^2}{\dim(\mathcal{Y}) d_\infty^2}\right]^{-1/2} . \tag{1.7}$$

In practice it may be hard to compute the inverse of $D(\lambda)$ in the multivariable case. The following result alleviates this problem especially when $\mathcal{X}$ is of low dimension.

**THEOREM 1.2.** *Let* Z, B *and* $\tilde{B}$ *be the data for the standard left Nevanlinna-Pick interpolation problem. Let* P *and* $\tilde{P}$ *be the controllability grammians for the stable, controllable pair* {Z, B} *and the pair* {Z, $\tilde{B}$}, *respectively.*

*(i) Then the central interpolant* $G_\gamma$ *with respect to the tolerance* $\gamma > d_\infty$ *for the standard left Nevanlinna-Pick interpolation problem can also be computed by*

$$G_\gamma(\lambda) = C(I - \lambda M)^{-1}\tilde{B} \tag{1.8}$$

*where* C *from* $\mathcal{X}$ *to* $\mathcal{U}$ *and* M *on* $\mathcal{X}$ *are the operators defined by*

$$C = \gamma^2 B^*(\gamma^2 P - Z\tilde{P}Z^*)^{-1} \quad and \quad M = (\gamma^2 P - \tilde{P})Z^*(\gamma^2 P - Z\tilde{P}Z^*)^{-1} . \tag{1.9}$$

*(ii) The operator* M *in (1.9) is stable, that is,* $r_{spec}(M) < 1$.

Let G in $H^\infty(\mathcal{Y}, \mathcal{U})$ be an interpolant for the standard Nevanlinna-Pick interpolation problem with tolerance $\gamma$, that is, $WT_G = \tilde{W}$ and $\|G\|_\infty \le \gamma$ where $T_G$ the Toeplitz matrix from $\ell_+^2(\mathcal{Y})$ into $\ell_+^2(\mathcal{U})$ generated by G. By a slight abuse of terminology let $\Delta(G)$ be the positive operator on $\mathcal{Y}$ defined according to (IV.7.1), that is,

$$(\Delta(G)a, a) = \inf\{\|D_{T_G}(a \oplus h)\|^2 : h \in \ell_+^2(\mathcal{Y})\}$$

where a is a fixed vector in $\mathcal{Y}$ and $D_{T_G}$ is the positive square root of $\gamma^2 I - T_G^* T_G$. If $\mathcal{Y}$ is finite dimensional, then the entropy of G is defined by

$$E(G) = \frac{1}{2\pi}\int_0^{2\pi} \ln\det[\gamma^2 I - G(e^{it})^* G(e^{it})]dt = \ln\det[\Delta(G)] .$$

The last equality follows from (IV.7.15). We say that F is a *maximal interpolant* for the standard Nevanlinna-Pick problem with tolerance $\gamma$, if F is an interpolant for the standard Nevanlinna-Pick interpolation problem with tolerance $\gamma$ and $\Delta(F) \ge \Delta(G)$ where G is any other interpolant with tolerance $\gamma$. Theorem IV.7.1 shows that the central interpolant $G_\gamma$ is a maximal interpolant for the Nevanlinna-Pick problem with tolerance $\gamma$. If $\mathcal{Y}$ is finite dimensional, then F is a *maximal entropy interpolant*, if $E(F) \ge E(G)$ for all Nevanlinna-Pick interpolants G with tolerance $\gamma$. Moreover, according to the results in Section IV.7, if $E(G_0) > -\infty$ for some interpolant $G_0$, then the central solution is the unique maximal entropy interpolant, that is $E(G_\gamma) \ge E(G)$ and $E(G_\gamma) = E(G)$ if and only if $G_\gamma = G$. This sets the stage for the following result.

**THEOREM 1.3.** *Let* Z, B *and* $\tilde{B}$ *be the data for the standard left Nevanlinna-Pick interpolation problem and assume that* $\gamma > d_\infty$. *Let* P *and* $\tilde{P}$ *be the controllability grammians for* *the stable, controllable pair* $\{Z, B\}$ *and the pair* $\{Z, \tilde{B}\}$*, respectively. Then the central interpolant* $G_\gamma$ *with tolerance* $\gamma$ *is the unique maximal interpolant for the standard left Nevanlinna-Pick problem with tolerance* $\gamma$*, and*

$$\Delta(G_\gamma) = \gamma^2[I + \tilde{B}^*(\gamma^2 P - \tilde{P})^{-1}\tilde{B}]^{-1} .$$

*Moreover, if* $\mathcal{Y}$ *is finite dimensional, then the entropy of* $G_\gamma$ *is given by*

$$E(G_\gamma) = \frac{1}{2\pi}\int_0^{2\pi} \ln\det[\gamma^2 I - G_\gamma^* G_\gamma]dt = -\ln\det[\gamma^{-2}(I + \tilde{B}^*(\gamma^2 P - \tilde{P})^{-1}\tilde{B})] .$$

*In particular, if* $G$ *is any other Nevanlinna-Pick interpolant with tolerance* $\gamma$*, then* $E(G_\gamma) \geq E(G)$ *with equality if and only if* $G_\gamma = G$*, that is,* $G_\gamma$ *is the maximal entropy interpolant.*

Notice that if we let $\gamma = +\infty$, then the central solution in the previous two theorems reduces to the optimal $H^2$ solution $G_{2*}$ in (1.5) to the $H^2$ optimization problem in (1.4). The following result can be used to compute the optimal $H^\infty$ solution when $\mathcal{Y}$ is one dimensional and $\mathcal{X}$ is finite dimensional.

**THEOREM 1.4.** *Let* $Z$ *on* $\mathcal{X}$*,* $B$ *from* $\mathcal{U}$ *into* $\mathcal{X}$ *and* $\tilde{B}$ *from* $\mathbb{C}^1$ *into* $\mathcal{X}$ *be the data for the standard left Nevanlinna-Pick interpolation problem where* $\mathcal{X}$ *is finite dimensional. Let* $P$ *and* $\tilde{P}$ *be the controllability grammians for the stable, controllable pair* $\{Z, B\}$ *and* $\{Z, \tilde{B}\}$*, respectively. Then the unique solution* $g_{opt}$ *to the* $H^\infty$ *Nevanlinna-Pick optimization problem in (1.2) is the function given by*

$$g_{opt}(\lambda) = \frac{d_\infty^2 B^*(I - \lambda Z^*)^{-1}x}{\tilde{B}^*(I - \lambda Z^*)^{-1}x} \tag{1.10}$$

*where* $x$ *is the eigenvector corresponding to the largest eigenvalue* $\lambda_{max}$ *of* $P^{-1}\tilde{P}$*. Moreover, the optimal* $H^\infty$ *error* $d_\infty^2 = \lambda_{max}$ *and* $g_{opt}/d_\infty$ *is an inner function.*

**PROOF OF THEOREM 1.1.** According to the discussion in Section II.2, a function $G$ in $H^\infty(\mathcal{Y}, \mathcal{U})$ satisfies the interpolation constraint $WT_G = \tilde{W}$ if and only if $B' = T_G$ is an intertwining lifting of $A$. Here $A$ is the unique operator from $\ell_+^2(\mathcal{Y})$ into $\mathcal{H}' = W^*\mathcal{X}$ defined by $WA = \tilde{W}$. (Notice that the range of $W^*$ is closed because we have assumed that the pair $\{Z, B\}$ is controllable.) In fact, an explicit formula for $A$ is given by

$$A = W^* P^{-1}\tilde{W} \tag{1.11}$$

where $P = WW^*$ is the controllability grammian for the pair $\{Z, B\}$. Recall also that $T'A = AT$

where T is the unilateral shift on $\ell_+^2(\mathcal{Y})$ and T′ on $\mathcal{H}'$ is the compression of the unilateral shift S on $\ell_+^2(\mathcal{U})$ to $\mathcal{H}'$. Therefore if we choose $B_\gamma$ to be the central intertwining lifting of A with respect to S and the tolerance $\gamma$, then $B_\gamma = T_{G_\gamma}$ for some (unique) $G_\gamma$ in $H^\infty(\mathcal{Y}, \mathcal{U})$, and $WT_{G_\gamma} = \tilde{W}$. Moreover, this $B_\gamma = T_{G_\gamma}$ satisfies the bounds in Theorem IV.8.1.

To find an explicit formula for the central intertwining lifting $G_\gamma$ notice that $T_{G_\gamma}$ is the Fourier transform of $M_{G_\gamma}^+$, that is $\mathcal{F}_{\mathcal{U}} T_{G_\gamma} = M_{G_\gamma}^+ \mathcal{F}_{\mathcal{Y}}$. According to Theorem IV.4.1, an explicit formula for $G_\gamma$ is given by $G_\gamma(\lambda) = N(\lambda)D(\lambda)^{-1}$ where

$$G_\gamma = (\mathcal{F}_{\mathcal{U}} A D_{\tilde{A}}^{-2} \Pi_0^*)(\lambda)[(\mathcal{F}_{\mathcal{Y}} D_{\tilde{A}}^{-2} \Pi_0^*)(\lambda)]^{-1} . \tag{1.12}$$

Here $\Pi_0 = [I, 0, 0, \ldots]$ is the operator from $\ell_+^2(\mathcal{Y})$ onto $\mathcal{Y}$ picking out the first component of a vector in $\ell_+^2(\mathcal{Y})$. Recall also that the inverse of $(\mathcal{F}_{\mathcal{Y}} D_{\tilde{A}}^{-2} \Pi_0^*)(\lambda)$ is an outer function in $H^\infty(\mathcal{Y}, \mathcal{Y})$. So to compute $G_\gamma$ we simply obtain state space formulas for $AD_{\tilde{A}}^{-2}$ and $D_{\tilde{A}}^{-2}$ in (1.12). Using $A = W^* P^{-1} \tilde{W}$ and $P = WW^*$ we obtain

$$D_{\tilde{A}}^2 = (\gamma^2 I - A^* A) = \gamma^2 I - \tilde{W}^* P^{-1} \tilde{W} . \tag{1.12a}$$

Let us search for an inverse of $D_{\tilde{A}}^2$ of the form $\gamma^{-2} I + \tilde{W}^* R \tilde{W}$ where R is an operator on $\mathcal{X}$. Hence

$$I = D_{\tilde{A}}^2(\gamma^{-2} I + \tilde{W}^* R \tilde{W}) = (\gamma^2 I - \tilde{W}^* P^{-1} \tilde{W})(\gamma^{-2} I + \tilde{W}^* R \tilde{W}) =$$

$$I + \tilde{W}^* [(\gamma^2 I - P^{-1} \tilde{P})R - \gamma^{-2} P^{-1}] \tilde{W} .$$

Choosing $R = \gamma^{-2}(\gamma^2 P - \tilde{P})^{-1}$ forces the last term to be zero. A similar calculation shows that for this R we also have that $(\gamma^{-2} I + \tilde{W}^* R \tilde{W}) D_{\tilde{A}}^2$ equals the identity. Therefore the inverse of $D_{\tilde{A}}^2$ is given by

$$D_{\tilde{A}}^{-2} = \gamma^{-2} I + \gamma^{-2} \tilde{W}^* (\gamma^2 P - \tilde{P})^{-1} \tilde{W} . \tag{1.13}$$

Using $A = W^* P^{-1} \tilde{W}$ and $\tilde{P} = \tilde{W}\tilde{W}^*$ in (1.13) we obtain

$$\begin{aligned} AD_{\tilde{A}}^{-2} &= \gamma^{-2} W^* [P^{-1} + P^{-1} \tilde{P}(\gamma^2 P - \tilde{P})^{-1}] \tilde{W} = \\ &\gamma^{-2} W^* [P^{-1}(\gamma^2 P - \tilde{P}) + P^{-1} \tilde{P}](\gamma^2 P - \tilde{P})^{-1} \tilde{W} = W^* (\gamma^2 P - \tilde{P})^{-1} \tilde{W} . \end{aligned} \tag{1.14}$$

Notice that

$$\mathcal{F}_{\mathcal{U}} W^* = \mathcal{F}_{\mathcal{U}} [B, ZB, Z^2 B, \ldots]^* = \sum_{n=0}^{\infty} \lambda^n B^* Z^{*n} = B^*(I - \lambda Z^*) . \tag{1.15}$$

A similar calculation shows that $\mathcal{F}_{\mathcal{Y}}\tilde{W}^* = \tilde{B}^*(I - \lambda Z^*)^{-1}$. Using these Fourier transforms along with $\tilde{W}\Pi_o^* = \tilde{B}$ in (1.12), (1.13) and (1.14), we see that the central intertwining lifting $G_\gamma = ND^{-1}$ where N and D are now defined according to (1.6).

To complete the proof notice that $\mathcal{Y} = \ker T^*$. This readily implies that

$$\mathcal{F}_{\mathcal{U}} A | \mathcal{Y} = \mathcal{F}_{\mathcal{U}}(W^* P^{-1} \tilde{W} \Pi_o^*) = \mathcal{F}_{\mathcal{U}} W^* P^{-1} \tilde{B} = B^*(I - \lambda Z^*)^{-1} P^{-1} \tilde{B} = G_{2*}$$

where $G_{2*}(\lambda)$ is the unique optimal solution to the $H^2$ optimization problem (1.4); see (1.5). In particular, $d_2 = \|G_{2*}\|_2 = \|A | \mathcal{Y}\|_{HS}$. Applying Corollary IV.8.3 we see that the central solution $B_\gamma = M_{G_\gamma}^+$ satisfies the $H^2$ bound in (1.7). Finally, let us notice that (1.13) gives

$$Q(\lambda) := (\mathcal{F}_{\mathcal{Y}} D_{\tilde{A}}^{-2} \Pi_o)(\lambda) = \gamma^{-2} I + \gamma^{-2} \tilde{B}^* (1 - \lambda Z^*)^{-1} (\gamma^2 P - \tilde{P})^{-1} \tilde{B}. \tag{1.16}$$

In particular, the function $Q(\lambda)$ is in $H^\infty(\mathcal{Y}, \mathcal{Y})$. According to Theorem IV.6.5 the function $Q(\lambda)^{-1}$ is also in $H^\infty(\mathcal{Y}, \mathcal{Y})$. So $Q(\lambda)$ is an invertible outer function. Moreover, $\Theta = Q(0)^{1/2} Q(\lambda)^{-1}$ is the outer spectral factor for $\gamma^2 I - G_\gamma^* G_\gamma$. Because both $\Theta$ and $\Theta^{-1}$ are in $H^\infty(\mathcal{Y}, \mathcal{Y})$, this readily implies that $\|G_\gamma\|_\infty < \gamma$. (The fact that $G_\gamma$ is the central intertwining lifting with tolerance $\gamma$ implies only that $\|G_\gamma\|_\infty \le \gamma$.) This completes the proof.

**REMARK 1.5.** Let $D_o$ be the positive square root of $I + \tilde{B}^*(\gamma^2 P - \tilde{P})^{-1} \tilde{B}$ and $G_\gamma$ be the central interwining lifting with tolerance $\gamma$ for the standard Nevanlinna-Pick interpolation problem. Then a state space formula for the outer spectral factor $\Theta$ for the spectral density $\gamma^2 I - G_\gamma^* G_\gamma$ is given by

$$\Theta(\lambda) = \gamma D_o - \gamma D_o \tilde{B}^* (\gamma^2 P - Z\tilde{P}Z^*)^{-1} (I - \lambda M)^{-1} \tilde{B}. \tag{1.17}$$

Using the operator identity $(I + bc^{-1}d)^{-1} = I - b(db + c)^{-1} d$ on $Q(\lambda)$ in (1.16) along with $\gamma^2 P - \tilde{P} + \tilde{B}\tilde{B}^* = \gamma^2 P - Z\tilde{P}Z^*$ we have for all $\lambda$ in $\mathbb{D}$

$$Q(\lambda)^{-1} = \gamma^2 I - \gamma^2 \tilde{B}^* [\tilde{B}\tilde{B}^* + (\gamma^2 P - \tilde{P})(I - \lambda Z^*)]^{-1} \tilde{B} =$$

$$\gamma^2 I - \gamma^2 \tilde{B}^* [\gamma^2 P - \tilde{P} + \tilde{B}\tilde{B}^* - \lambda(\gamma^2 P - \tilde{P}) Z^*]^{-1} \tilde{B} =$$

$$\gamma^2 I - \gamma^2 \tilde{B}^* [(I - \lambda M)(\gamma^2 P - \tilde{P} + \tilde{B}\tilde{B}^*)]^{-1} \tilde{B} =$$

$$\gamma^2 I - \gamma^2 \tilde{B}^* (\gamma^2 P - Z\tilde{P}Z^*)^{-1} (I - \lambda M)^{-1} \tilde{B}.$$

Combining this with $Q(0)^{1/2} = \gamma^{-1} D_o$, shows that the outer spectral factor $\Theta(\lambda) = Q(0)^{1/2} Q(\lambda)^{-1}$ for $\gamma^2 I - G_\gamma^* G_\gamma$ is given by (1.17). Finally, it is noted that since $r_{spec}(M) < 1$, equation (1.17) also shows that $Q(\lambda)^{-1}$ is in $H^\infty(\mathcal{Y}, \mathcal{Y})$.

**PROOF OF THEOREM 1.2 (Part (i)).** Now let R be the operator defined by $R = (\gamma^2 P - \tilde{P})(I - \lambda Z^*)$. According to Theorem 1.1 the central intertwining lifting $G_\gamma$ of A is given by

$$G_\gamma = \gamma^2 B^* R^{-1} \tilde{B}(I + \tilde{B}^* R^{-1} \tilde{B})^{-1} =$$

$$\gamma^2 B^* R^{-1} \tilde{B}[I - (I + \tilde{B}^* R^{-1} \tilde{B})^{-1} \tilde{B}^* R^{-1} \tilde{B}] =$$

$$\gamma^2 B^* [I - R^{-1} \tilde{B}(I + \tilde{B}^* R^{-1} \tilde{B})^{-1} \tilde{B}^*] R^{-1} \tilde{B} =$$

$$\gamma^2 B^* [I - (I + R^{-1} \tilde{B}\tilde{B}^*)^{-1} R^{-1} \tilde{B}\tilde{B}^*] R^{-1} \tilde{B} =$$

$$\gamma^2 B^* (I + R^{-1} \tilde{B}\tilde{B}^*)^{-1} R^{-1} \tilde{B} = \gamma^2 B^* (R + \tilde{B}\tilde{B}^*)^{-1} \tilde{B} =$$

$$\gamma^2 B^* [\gamma^2 P - \tilde{P} + \tilde{B}\tilde{B}^* - \lambda(\gamma^2 P - \tilde{P}) Z^*]^{-1} \tilde{B} =$$

$$\gamma^2 B^* (\gamma^2 P - \tilde{P} + \tilde{B}\tilde{B}^*)^{-1} \left[ I - \lambda(\gamma^2 P - \tilde{P}) Z^* (\gamma^2 P - \tilde{P} + \tilde{B}\tilde{B}^*)^{-1} \right]^{-1} \tilde{B}.$$

Now $\tilde{P} = Z\tilde{P}Z^* + \tilde{B}\tilde{B}^*$ gives the formula for $G_\gamma$ given in (1.8). This completes the proof of Part (i) of Theorem 1.2.

**PROOF OF THEOREM 1.3.** According to the results in Section IV.7, the central intertwining lifting $B_\gamma = T_{G_\gamma}$ is precisely the unique maximal solution for the Nevanlinna-Pick problem with tolerance $\gamma$. Equation (1.13) along with $\Pi_o \tilde{W}^* = \tilde{B}^*$ gives

$$\Pi_o D_{\tilde{A}}^{-2} \Pi_o^* = \gamma^{-2}[I + \tilde{B}^* (\gamma^2 P - \tilde{P})^{-1} \tilde{B}].$$

This along with (IV.7.4), Theorem IV.7.1 and Theorem IV.7.5 yields the formulas for $\Delta(G_\gamma)$ and $E(G_\gamma)$ in Theorem 1.3. The proof is now complete.

**PROOF OF THEOREM 1.4.** Because the state space $\mathcal{X}$ is finite dimensional the operator A has finite rank, and A attains its norm. Let h be a vector in $\ell_+^2$ which attains the norm of A. Corollary IV.2.7 guarantees that there exists a unique vector $g_{opt}$ in $H^\infty(\mathbb{C}, \mathcal{U})$ which solves the $H^\infty$ optimization problem in (1.2), that is, $WT_{g_{opt}} = \tilde{W}$ and $\|g_{opt}\|_\infty = d_\infty$. Moreover, this $g_{opt}$ can be computed by

$$g_{opt}(\lambda) = \frac{(\mathcal{F}_{\mathcal{U}} Ah)(\lambda)}{(\mathcal{F}_+ h)(\lambda)} \tag{1.18}$$

where h is an vector in $\ell_+^2$ which attains the norm of A and $\mathcal{F}_+$ is the Fourier transform from $\ell_+^2$ onto $H^2$. Using $A = W^* P^{-1} \tilde{W}$ we see that $AA^* W^* = W^* P^{-1} \tilde{P}$. Because the pair {Z, B} is controllable, $W^*$ is one to one and onto $\mathcal{H}'$. So the relation $AA^*W^* = W^*P^{-1}\tilde{P}$ implies that $AA^*$ is similar to $P^{-1}\tilde{P}$. Thus $AA^*$ and $P^{-1}\tilde{P}$ have the same eigenvalues. In particular, if $\lambda_{max}$ is the largest eigenvalues of $P^{-1}\tilde{P}$, then $\lambda_{max} = \|AA^*\| = \|A\|^2 = d_\infty^2$.

Now let x be an eigenvector corresponding to the largest eigenvalue $\lambda_{max}$ for $P^{-1}\tilde{P}$. Using $AA^*W^* = W^*P^{-1}\tilde{P}$, it follows that $W^*x$ is the eigenvector corresponding to the largest eigenvalue $\lambda_{max} = d_\infty^2$ of $AA^*$ and thus $W^*x$ attains the norm of $A^*$. We claim that $h = A^*W^*x = \tilde{W}^*x$ attains the norm of A. To prove this it is sufficient to show that h is an eigenvector of $A^*A$ corresponding to $\lambda_{max}$. Moreover h is nonzero, since otherwise $0 = Ah = AA^*W^*x = \lambda_{max}W^*x \neq 0$. Finally,

$$A^*Ah = A^*AA^*W^*x = \lambda_{max}A^*W^*x = \lambda_{max}h\,.$$

Hence $h = \tilde{W}^*x$ attains the norm of A. The Fourier transform $\mathcal{F}_+h$ of $h = \tilde{W}^*x$ is given by $\tilde{B}^*(I - \lambda Z^*)^{-1}x$. The Fourier transform $\mathcal{F}_{\mathcal{U}}Ah$ of $Ah = W^*P^{-1}\tilde{P}x$ is given by $B^*(I - \lambda Z^*)^{-1}P^{-1}\tilde{P}x$. However, $P^{-1}\tilde{P}x = d_\infty^2 x$. This along with (1.18) gives (1.10) and completes the proof.

It remains to prove Part (ii) of Theorem 1.2. Because this kind of stability result is used time and time again we reformulate it as an independent Proposition 1.7 below. We begin by presenting a preliminary result concerning contractions. Recall that an operator C is *contractive* if $\|C\| \leq 1$ and *strictly contractive* if $\|C\| < 1$.

**LEMMA 1.6.** *Let* T *be a contraction on* $\mathcal{H}$ *and* C *a strict contraction from* $\mathcal{H}$ *into* $\mathcal{H}'$ *satisfying* $T^*C^*CT \leq C^*C$. *Let* R *be the operator on* $\mathcal{H}$ *defined by*

$$R = (I - T^*C^*CT)^{-1}T^*(I - C^*C)\,. \tag{1.19}$$

*Then* $r_{spec}(R) \leq 1$. *Moreover, if* T *is a stable operator, then* R *is stable.*

**PROOF.** This proof uses some ideas concerning the operator $\omega$ in Chapter IV. Throughout this proof if $\Gamma$ is a contraction, then $D_\Gamma = D_{\Gamma,1}$ is the positive square root of $I - \Gamma^*\Gamma$. Because C is a strict contraction, the space $\mathcal{H}$ equals the range of $D_C = D_{C,1}$. Notice that $\|D_Ch\|^2 = \|h\|^2 - \|Ch\|^2$. Since $T^*C^*CT \leq C^*C$ there exists a contraction T′ on $\mathcal{H}'$ satisfying $T'C = CT$. Let $\mathcal{F}$ be the subspace of $\mathcal{H} \oplus \mathcal{H}$ defined by $\mathcal{F} = \{D_CTh \oplus D_Th : h \in \mathcal{H}\}^-$. Let $\omega$ be the isometry mapping $\mathcal{F}$ into $\mathcal{H} \oplus \mathcal{D}_{T'}$ defined by

$$\omega\begin{bmatrix} D_CTh \\ D_Th \end{bmatrix} = \begin{bmatrix} D_Ch \\ D_{T'}Ch \end{bmatrix}. \tag{1.20}$$

Observe that $\omega$ is an isometry because using $T'C = CT$ we have

$$\|D_C Th\|^2 + \|D_T h\|^2 = \|Th\|^2 - \|CTh\|^2 + \|h\|^2 - \|Th\|^2 = \|h\|^2 - \|T'Ch\|$$
$$= \|h\|^2 - \|Ch\|^2 + \|Ch\|^2 - \|T'Ch\|^2 = \|D_C h\|^2 + \|D_{T'} Ch\|^2 . \tag{1.21}$$

Let $P_{\mathcal{F}}$ be the orthogonal projection on $\mathcal{H} \oplus \mathcal{H}$ whose range is $\mathcal{F}$. Then, it is easy to verify that

$$P_{\mathcal{F}} = \begin{bmatrix} D_C T \\ D_T \end{bmatrix} D_{CT}^{-2} [T^* D_C, D_T] . \tag{1.22}$$

Indeed, if $\mathcal{F}$ is the range of an operator X such that $X^*X$ is invertible, then $P_{\mathcal{F}} = X(X^*X)^{-1}X^*$; in our case X is the transpose of $[D_C T, D_T]$ and $X^*X = D_{CT}^2$ is invertible because C is strictly contractive. Finally, let $\Pi$ be the operator from $\mathcal{H} \oplus \mathcal{D}_{T'}$ onto $\mathcal{H}$ which picks out the first component, that is,

$$\Pi = \begin{bmatrix} d \\ d' \end{bmatrix} = d . \tag{1.23}$$

Now consider the operator $\Psi = \Pi\omega P_{\mathcal{F}}[I_{\mathcal{H}}, 0]^*$ on $\mathcal{H}$. Obviously $\Psi$ is a contraction. Consulting (1.20), (1.22) and the definition of R in (1.19) we see that

$$\Psi = \Pi\omega \begin{bmatrix} D_C T \\ D_T \end{bmatrix} D_{CT}^{-2} T^* D_C = D_C D_{CT}^{-2} T^* D_C = D_C R D_C^{-1} . \tag{1.24}$$

Therefore R is similar to the contraction $\Psi$ and thus $r_{spec}(R) \le 1$.

To complete the proof assume that T is stable. We have to show that the contraction $\Psi$ has no spectrum on the unit circle. Assuming the contrary, let $\lambda$ be a point on the unit circle in the spectrum of $\Psi$. Then $\lambda$ is in the boundary of the spectrum of $\Psi$ and hence there exists a sequence of unit vectors $\{x_n\}$ in $\mathcal{H}$ such that $\Psi x_n - \lambda x_n \to 0$ as $n \to \infty$; see Problem 78 in Halmos [1]. Notice that $\|\Psi x_n\| \to 1$ as $n \to \infty$. Since $\Psi = \Pi\omega P_{\mathcal{F}}[I, 0]^*$ with $\Pi$ and $\omega$ contractions, $\|P_{\mathcal{F}} x_n\| \to 1$ as $n \to \infty$. Because $x_n$ belongs to $\mathcal{H}$ we may write

$$\begin{bmatrix} x_n \\ 0 \end{bmatrix} = \begin{bmatrix} D_C T h_n \\ D_T h_n \end{bmatrix} + \begin{bmatrix} g_{n,1} \\ g_{n,2} \end{bmatrix} ,$$

where $h_n$ is in $\mathcal{H}$ and $g_n = g_{n,1} \oplus g_{n,2}$ is in $\mathcal{G} = (\mathcal{H} \oplus \mathcal{H}) \ominus \mathcal{F}$. The fact that $\|P_{\mathcal{F}} x_n\| \to 1$ along with $\|x_n\| = 1$ implies that $P_{\mathcal{G}} x_n \to 0$ as $n \to \infty$. In particular, $g_{n,1}$ tends to zero as n goes to infinity. Using the definition of $\Psi$ we have

$$D_C h_n = \Pi\omega \begin{bmatrix} D_C T h_n \\ D_T h_n \end{bmatrix} = \Pi\omega P_{\mathcal{F}} \begin{bmatrix} x_n \\ 0 \end{bmatrix} = \Psi x_n \,.$$

It follows that $D_C h_n - \lambda D_C T h_n = \Psi x_n - \lambda x_n + \lambda g_{n,1}$ tends to zero as n tends to infinity. But $D_C$ is invertible. So $h_n - \lambda T h_n$ approaches zero as n goes to infinity. Using $r_{spec}(T) < 1$ we see that $h_n \to 0$ as $n \to \infty$. Then obviously $x_n$ approaches zero which contradicts the fact that $x_n$ is a unit vector. This completes the proof of the lemma.

**PROPOSITION 1.7.** *Let* P *be the controllability grammian for the stable, controllable pair* $\{Z$ *on* $\mathcal{X}$, $B\}$ *and* $\tilde{P}$ *the controllability grammian for the pair* $\{Z, \tilde{B}\}$. *Let* $\gamma^2 > r_{spec}(\tilde{P}P^{-1})$ *and* M *the operator on* $\mathcal{X}$ *defined by*

$$M = (\gamma^2 P - \tilde{P}) Z^* (\gamma^2 P - Z\tilde{P}Z^*)^{-1} \,. \tag{1.25}$$

*Then* $r_{spec}(M) < 1$.

**PROOF.** Let $T = P^{-1/2} Z P^{1/2}$ where $P^{1/2}$ is the positive square root of P. Because T is similar to Z and Z is stable, T is stable. By applying $P^{-1/2}$ to both sides of $P = ZPZ^* + BB^*$ we have $I = TT^* + P^{-1/2} BB^* P^{-1/2}$. This shows that T is a contraction. Using T in the definition of $M^*$ we obtain

$$P^{1/2} M^* P^{-1/2} = (\gamma^2 I - TQT^*)^{-1} T (\gamma^2 I - Q) \tag{1.26}$$

where $Q = P^{-1/2} \tilde{P} P^{-1/2}$. Notice that $P^{1/2} Q = \tilde{P} P^{-1} P^{1/2}$. Therefore Q and $\tilde{P}P^{-1}$ are similar. In particular, they have the same spectral radius. (As aside, this spectral radius is precisely $d_\infty^2$ the optimal $H^\infty$ error in the $H^\infty$ optimization problem (1.2).) Because $r_{spec}(Q) < \gamma^2$ and Q is self adjoint, $Q < \gamma^2 I$. By multiplying both sides of $\tilde{P} = Z\tilde{P}Z^* + \tilde{B}\tilde{B}^*$ by $P^{-1/2}$ we obtain

$$Q = TQT^* + P^{-1/2} \tilde{B}\tilde{B}^* P^{-1/2} \,.$$

Thus $Q \geq TQT^*$. Now let $C = \gamma^{-1} Q^{1/2}$, then C is a strict contraction and (1.26) becomes

$$P^{1/2} M^* P^{-1/2} = (I - TC^*CT^*)^{-1} T (I - C^*C) = R \,.$$

By replacing T by $T^*$ in Lemma 1.6, the previous equation shows that $M^*$ is similar to R in (1.19) and thus $r_{spec}(M) < 1$. This completes the proof.

**ALTERNATE PROOF OF THEOREM 1.2.** To complete this section, let us use Corollary IV.3.2 to give another proof of Theorem 1.2. This proof establishes the framework which will be used in Section V1.7 to give a complete parameterization for the set of all

Nevanlinna-Pick interpolants with tolerance $\gamma$. According to Corollary IV.3.2 the central intertwining lifting B for A is given by the Toeplitz operator $T_{G_\gamma}$ where $G_\gamma$ is the function in $H^\infty(\mathcal{Y}, \mathcal{U})$ defined by

$$G_\gamma(\lambda) = (\mathcal{F}_{\mathcal{U}} A\Pi_o^*)(\lambda) + \Psi(\lambda)(I - \lambda T_A^*)^{-1} T_A^* \Pi_o^* \quad \text{and}$$
$$(\Psi h)(\lambda) = (\mathcal{F}_{\mathcal{U}}(S - T')Ah)(\lambda) \qquad (h \in \mathcal{H}) . \tag{1.27}$$

Here $\mathcal{F}_{\mathcal{U}}$ is the Fourier transform from $\ell_+^2(\mathcal{U})$ onto $H^2(\mathcal{U})$, the operator $\Pi_o$ from $\ell_+^2(\mathcal{Y})$ onto $\mathcal{Y}$ is defined by $\Pi_o[y_o, y_1, \ldots]^{tr} = y_0$, the unilateral shift on $\ell_+^2(\mathcal{Y})$ is T and S is the unilateral shift on $\ell_+^2(\mathcal{U})$, while $T'$ is the compression of S to $\mathcal{H}'$ the range of $W^*$. Recall also that $A = W^*P^{-1}\tilde{W}$ and that $T_A^* = D_{AT}^{-2} T^* D_A^2$ is defined in (IV.3.2). To obtain the formula for $G_\gamma$ in (1.8), first notice that the orthogonal projection $P'$ onto $\mathcal{H}'$ is given by $P' = W^*P^{-1}W$. This follows by observing that this $P'$ satisfies $P'^2 = P'^* = P'$ and the range of $P'$ is $\mathcal{H}'$. Using $ZW = WS$ along with $P = WW^*$ we have

$$T'W^* = P'SW^* = W^*P^{-1}WSW^* = W^*P^{-1}ZWW^* = W^*P^{-1}ZP .$$

In particular, this shows that

$$T'W^* = W^*P^{-1}ZP . \tag{1.28}$$

To compute $T_A^*\Pi_o^*$ first notice that $T^*\tilde{W}^* = \tilde{W}^*Z^*$ gives

$$T^*(\gamma^2 I - A^*A)\Pi_o^* = -T^*A^*W^*P^{-1}\tilde{W}\Pi_o^* = -T^*\tilde{W}^*P^{-1}WW^*P^{-1}\tilde{B} = -\tilde{W}^*Z^*P^{-1}\tilde{B} . \tag{1.29}$$

Using (1.28) we obtain the following formula involving $D_{AT}^2$

$$(\gamma^2 I - A^*T'^*T'A)\tilde{W}^* = \gamma^2\tilde{W}^* - A^*T'^*T'W^*P^{-1}\tilde{P} =$$
$$\gamma^2\tilde{W}^* - A^*T'^*W^*P^{-1}Z\tilde{P} = \gamma^2\tilde{W}^* - A^*W^*Z^*P^{-1}Z\tilde{P} = \tilde{W}^*(\gamma^2 I - Z^*P^{-1}Z\tilde{P}) . \tag{1.30}$$

We claim that $(\gamma^2 I - Z^*P^{-1}Z\tilde{P})$ is invertible. To verify this notice that $C = P^{1/2}Z^*P^{-1/2}$ is a contraction. This follows by applying $P^{-1/2}$ to both sides of the Lyapunov equation $P = ZPZ^* + BB^*$ and thus $I \geq C^*C$. Now notice that $\gamma^2 I - Z^*P^{-1}Z\tilde{P}$ is similar to

$$P^{1/2}(\gamma^2 I - Z^*P^{-1}ZP^{1/2}P^{-1/2}\tilde{P})P^{-1/2} = \gamma^2 I - CC^*P^{-1/2}\tilde{P}P^{-1/2} .$$

Recall that $d_\infty^2 = r_{spec}(\tilde{P}P^{-1})$. Moreover, $\tilde{P}P^{-1}$ is similar to $P^{-1/2}\tilde{P}P^{-1/2}$ (compute $P^{-1/2}(\tilde{P}P^{-1})P^{1/2}$). In particular, $d_\infty^2 = r_{spec}(P^{-1/2}\tilde{P}P^{-1/2})$ and thus $\gamma^2 I > P^{-1/2}\tilde{P}P^{-1/2}$. Because C is a contraction, it follows that the norm of $CC^*P^{-1/2}\tilde{P}P^{-1/2}$ is strictly less than $\gamma^2$. Hence $\gamma^2 I - CC^*P^{-1/2}\tilde{P}P^{-1/2}$ is invertible. Therefore $\gamma^2 I - Z^* P^{-1} Z\tilde{P}$ is also invertible, which verifies our claim.

By taking inverses in (1.30) along with $AT = T'A$ we arrive at

$$(\gamma^2 I - T^* A^* AT)^{-1} \tilde{W}^* = \tilde{W}^* (\gamma^2 I - Z^* P^{-1} Z\tilde{P})^{-1} . \tag{1.31}$$

Using (1.31) and $T^* \tilde{W}^* = \tilde{W}^* Z^*$ we can now compute an expression for $T_A^*$

$$T_A^* \tilde{W}^* = (\gamma^2 I - T^* A^* AT)^{-1} T^* (\gamma^2 I - A^* A)\tilde{W}^* = (\gamma^2 I - T^* A^* AT)^{-1} T^* \tilde{W}^* (\gamma^2 I - P^{-1}\tilde{P}) =$$
$$(\gamma^2 I - T^* A^* AT)^{-1} \tilde{W}^* Z^* (\gamma^2 I - P^{-1}\tilde{P}) = \tilde{W}^* (\gamma^2 I - Z^* P^{-1} Z\tilde{P})^{-1} Z^* (\gamma^2 I - P^{-1}\tilde{P}) = \tilde{W}^* N , \tag{1.32}$$

where $N = (\gamma^2 I - Z^* P^{-1} Z\tilde{P})^{-1} Z^* (\gamma^2 I - P^{-1}\tilde{P})$. By taking the inverse in (1.32) we obtain

$$(I - \lambda T_A^*)^{-1} \tilde{W}^* = \tilde{W}^* (1 - \lambda N)^{-1} . \tag{1.33}$$

The operator N is similar to M defined in (1.9). Indeed

$$(\gamma^2 P - \tilde{P}) N (\gamma^2 P - \tilde{P})^{-1} = (\gamma^2 P - \tilde{P})(\gamma^2 I - Z^* P^{-1} Z\tilde{P})^{-1} Z^* P^{-1} =$$
$$(\gamma^2 P - \tilde{P}) Z^* (\gamma^2 P - Z\tilde{P}Z^*)^{-1} = M .$$

Because $r_{spec}(M) < 1$, equation (1.33) is well defined for all $|\lambda| < 1$. So by combining (1.29), (1.31) and (1.33) we obtain

$$(I - \lambda T_A^*)^{-1} T_A^* \Pi_o^* = -(I - \lambda T_A^*)^{-1} (\gamma^2 I - A^* T'^* T' A)^{-1} \tilde{W}^* Z^* P^{-1} \tilde{B} =$$
$$-(I - \lambda T_A^*)^{-1} \tilde{W}^* (\gamma^2 I - Z^* P^{-1} Z\tilde{P})^{-1} Z^* P^{-1} \tilde{B} = \tag{1.34}$$
$$-\tilde{W}^* (I - \lambda N)^{-1} (\gamma^2 I - Z^* P^{-1} Z\tilde{P})^{-1} Z^* P^{-1} \tilde{B} = -\tilde{W}^* (I - \lambda N)^{-1} N (\gamma^2 P - \tilde{P})^{-1} \tilde{B} .$$

To complete our expression for the central intertwining lifting $G_\gamma$, we need to find $\Psi(\lambda)$. By using the expression for $T'$ in (1.28) along with the definition of $\Psi(\lambda)$ in (1.27) and $A = W^* P^{-1} \tilde{W}$, we have for x in $\mathcal{X}$

$$(\Psi \tilde{W}^* x)(\lambda) = \mathcal{F}_{\mathcal{U}} (S - T') W^* P^{-1} \tilde{P} x =$$
$$(\mathcal{F}_{\mathcal{U}} S W^* - \mathcal{F}_{\mathcal{U}} W^* P^{-1} Z P) P^{-1} \tilde{P} x = \tag{1.35}$$
$$B^* (I - \lambda Z^*)^{-1} (\lambda I - P^{-1} Z P) P^{-1} \tilde{P} x .$$

So by combining (1.27), (1.34), (1.35) along with $A\Pi_o^* = W^* P^{-1} \tilde{B}$ we see that the central intertwining lifting $G_\gamma$ is given by

$$G_\gamma = B^*(I-\lambda Z^*)^{-1}P^{-1}\tilde{B} - B^*(I-\lambda Z^*)^{-1}(\lambda I - P^{-1}ZP)P^{-1}\tilde{P}(\gamma^2 I - Z^*P^{-1}Z\tilde{P})^{-1}\Omega \quad (1.36)$$

where

$$\Omega = Z^*(\gamma^2 I - P^{-1}\tilde{P})(I-\lambda N)^{-1}(\gamma^2 P - \tilde{P})^{-1}\tilde{B}\,. \quad (1.37)$$

Finally, using once again

$$M = (\gamma^2 P - \tilde{P})Z^*(\gamma^2 P - Z\tilde{P}Z^*)^{-1} = (\gamma^2 P - \tilde{P})N(\gamma^2 P - \tilde{P})^{-1} \quad (1.38)$$

in the previous expression for $G_\gamma$ in (1.36), (1.37), we obtain

$$G_\gamma(\lambda) = B^*(I-\lambda Z^*)^{-1}Q(\lambda)(I-\lambda M)^{-1}\tilde{B} \quad (1.39)$$

where $Q(\lambda)$ is the function given by

$$\begin{aligned} Q(\lambda) &= P^{-1}(I-\lambda M) - P^{-1}(\lambda I - Z)\tilde{P}(\gamma^2 I - Z^*P^{-1}Z\tilde{P})^{-1}Z^*P^{-1} = \\ &P^{-1}(I-\lambda M) - P^{-1}(\lambda I - Z)\tilde{P}Z^*(\gamma^2 P - Z\tilde{P}Z^*)^{-1} = \\ &P^{-1}\left[(I-\lambda M)(\gamma^2 P - Z\tilde{P}Z^*) - (\lambda I - Z)\tilde{P}Z^*\right](\gamma^2 P - Z\tilde{P}Z^*)^{-1} = \\ &P^{-1}(\gamma^2 P - Z\tilde{P}Z^* - \lambda(\gamma^2 P - \tilde{P})Z^* - \lambda\tilde{P}Z^* + Z\tilde{P}Z^*)(\gamma^2 P - Z\tilde{P}Z^*)^{-1} = \\ &\gamma^2(1-\lambda Z^*)(\gamma^2 P - Z\tilde{P}Z^*)^{-1}\,. \end{aligned} \quad (1.40)$$

Substituting $Q(\lambda)$ into our expression for $G_\gamma$ in (1.39), we have

$$G_\gamma(\lambda) = \gamma^2 B^*(\gamma^2 P - Z\tilde{P}Z^*)^{-1}(1-\lambda M)^{-1}\tilde{B}\,. \quad (1.41)$$

This is precisely the central intertwining lifting of A given in Theorem 1.2. The proof is now complete.

## V.2 CENTRAL NEVANLINNA-PICK SOLUTIONS

In this section we will use our previous results to obtain the central solution for the classical tangential Nevanlinna-Pick interpolation problem. To this end, recall that the classical block left Nevanlinna-Pick interpolation problem is: Given a set of distinct complex numbers $z_1, \ldots, z_n$ in the open unit disc along with the operators $B_j$ from $\mathcal{U}$ to $\mathcal{X}_j$ and $\tilde{B}_j$ from $\mathcal{Y}$ to $\mathcal{X}_j$ find (if possible) a function G in $H^\infty(\mathcal{Y}, \mathcal{U})$ satisfying

$$B_j G(z_j) = \tilde{B}_j \quad (\text{for } j = 1, \ldots, n) \quad \text{and} \quad \|G\|_\infty \le \gamma \quad (2.1)$$

where $\gamma$ is a specified tolerance. Associated with this interpolation problem are the following $H^2$

and $H^\infty$ optimization problems:

$$d_2 = \inf\{\|G\|_2 : G \in H^2(\mathcal{Y}, \mathcal{U}) \text{ and } B_j G(z_j) = \tilde{B}_j \text{ for all } j\}$$
$$d_\infty = \inf\{\|G\|_\infty : G \in H^\infty(\mathcal{Y}, \mathcal{U}) \text{ and } B_j G(z_j) = \tilde{B}_j \text{ for all } j\}\,. \tag{2.2}$$

Obviously the $H^2$ optimization problem only makes sense when $\mathcal{U}$ or $\mathcal{Y}$ is finite dimensional.

Remark I.3.3 shows how one can connect the previous Nevanlinna-Pick interpolation to the standard left Nevanlinna-Pick interpolation problem. This is done by setting

$$Z = \begin{bmatrix} z_1 I & & & \\ & z_2 I & & \\ & & \ddots & \\ & & & z_n I \end{bmatrix}, \quad B = \begin{bmatrix} B_1 \\ B_2 \\ \vdots \\ B_n \end{bmatrix}, \quad \tilde{B} = \begin{bmatrix} \tilde{B}_1 \\ \tilde{B}_2 \\ \vdots \\ \tilde{B}_n \end{bmatrix}. \tag{2.3}$$

The blank spots in Z represent zero entries. So Z is a block diagonal matrix whose diagonal entries are $z_j I$. Then the interpolation constraint $B_j G(z_j) = \tilde{B}_j$ for all j is equivalent to $WT_G = \tilde{W}$ where W and $\tilde{W}$ are the controllability operators generated by {Z, B} and {Z, $\tilde{B}$} respectively, given in (1.1). Moreover, the controllability grammians $P = WW^*$ and $\tilde{P} = \tilde{W}\tilde{W}^*$ are given by

$$P = \left[\frac{B_j B_k^*}{1 - z_j \bar{z}_k}\right]_{j,k=1}^{n,n} \quad \text{and} \quad \tilde{P} = \left[\frac{\tilde{B}_j \tilde{B}_k^*}{1 - z_j \bar{z}_k}\right]_{j,k=1}^{n,n} \tag{2.4}$$

In particular, the results in the previous section show that $d_2^2 = \operatorname{trace} \tilde{B}^* P^{-1} \tilde{B}$ and $d_\infty^2 = r_{spec}(P^{-1}\tilde{P})$. So by applying the results in the previous section to our Z, B, $\tilde{B}$, P and $\tilde{P}$ we readily arrive at the following theorem.

**THEOREM 2.1.** *Let $z_j$, $B_j$ and $\tilde{B}_j$ for $j = 1, \ldots, n$ be the data for the classical block left Nevanlinna-Pick interpolation problem. Let Z, B, $\tilde{B}$, P and $\tilde{P}$ be the operators defined in (2.3), (2.4) and assume that {Z, B} is controllable. Let $\delta > 1$ and set $\gamma = \delta d_\infty$. Then the central interpolant $G_\gamma$ for the classical left Nevanlinna-Pick problem is given by $G_\gamma(\lambda) = N(\lambda)D(\lambda)^{-1}$ where N and D are the operator valued analytic functions computed by*

$$N(\lambda) = \gamma^2 B^* (I - \lambda Z^*)^{-1} (\gamma^2 P - \tilde{P})^{-1} \tilde{B}$$
$$D(\lambda) = I + \tilde{B}^* (I - \lambda Z^*)^{-1} (\gamma^2 P - \tilde{P})^{-1} \tilde{B}\,. \tag{2.5}$$

*In particular, $G_\gamma$ is a function in $H^\infty(\mathcal{Y}, \mathcal{U})$ satisfying $B_j G_\gamma(z_j) = \tilde{B}_j$ for all j and $\|G_\gamma\|_\infty < \gamma$ and*

$$\|F\|_2 \le \delta d_2 \left[ \delta^2 - 1 + \frac{d_2^2}{\dim(\mathcal{Y}) d_\infty^2} \right]^{-1/2} . \tag{2.6}$$

*Finally,* $d_2^2 = \text{trace}\, \tilde{B}^* P^{-1} \tilde{B}$ *and* $d_\infty^2 = r_{spec}(\tilde{P}P^{-1})$.

Recall also that if $\gamma > d_\infty$, then the central intertwining interpolant $G_\gamma$ is also the unique maximal interpolant to the Nevanlinna-Pick problem with tolerance $\gamma$. In this case

$$\Delta(G_\gamma) = \gamma^2 [I + \tilde{B}^* (\gamma^2 P - \tilde{P})^{-1} \tilde{B}]^{-1} . \tag{2.7}$$

Moreover, if $\mathcal{Y}$ is finite dimensional, then the entropy of $G_\gamma$ is given by

$$E(G_\gamma) = - \ln \det [\gamma^{-2} (I + \tilde{B}^* (\gamma^2 P - \tilde{P})^{-1} \tilde{B})] . \tag{2.8}$$

Obviously $E(G_\gamma) > -\infty$. So $E(G_\gamma) \ge E(G)$ where G is any other Nevanlinna-Pick interpolant with tolerance $\gamma$. Moreover, $E(G_\gamma) = E(G)$ if and only if $G_\gamma = G$ is the central interpolant with tolerance $\gamma$.

In practice it may be hard to compute the inverse of $D(\lambda)$ in (2.5) in the multivariable case. The following result alleviates this problem.

**THEOREM 2.2.** *Let* $z_j$, $B_j$ *and* $\tilde{B}_j$ *for* $j = 1, \dots, n$ *be the data for the classical block left Nevanlinna-Pick interpolation problem. Let* Z, B, $\tilde{B}$, P *and* $\tilde{P}$ *be the operators defined in (2.3), (2.4), and assume that* {Z, B} *is controllable. Then the central interpolant* $G_\gamma$ *with tolerance* $\gamma > d_\infty$ *for the classical left Nevanlinna-Pick interpolation problem can also be computed by*

$$G_\gamma(\lambda) = C(I - \lambda M)^{-1} \tilde{B} \tag{2.9}$$

*where* C *from* $\mathcal{X}$ *to* $\mathcal{U}$ *and the stable operator* M *on* $\mathcal{X}$ *are defined by*

$$C = \gamma^2 B^* (\gamma^2 P - Z\tilde{P}Z^*)^{-1} \quad \textit{and} \quad M = (\gamma^2 P - \tilde{P}) Z^* (\gamma^2 P - Z\tilde{P}Z^*)^{-1} . \tag{2.10}$$

Notice that if we let $\gamma = +\infty$, then the central solution in the previous two theorems reduces to the optimal $H^2$ solution $G_{2*}$ of the $H^2$ optimization problem in (2.2) given in (1.5). The following result can be used to compute the optimal $H^\infty$ solution when $\mathcal{Y}$ is one dimensional.

**THEOREM 2.3.** *Let* $z_j$, $B_j$ *and* $\tilde{B}_j$ *for* $j = 1, \dots, n$ *be the data for the classical block left Nevanlinna-Pick interpolation problem where* $\mathcal{Y} = \mathbb{C}$. *Let* Z, B, $\tilde{B}$, P *and* $\tilde{P}$ *be the operator* *defined in (2.3), (2.4) and assume that* $\{Z, B\}$ *is controllable. Then the unique solution* $g_{opt}$ *in* $H^\infty(\mathbb{C}, \mathcal{U})$ *to the* $H^\infty$ *Nevanlinna-Pick optimization problem in (2.2) is given by*

$$g_{opt}(\lambda) = \frac{d_\infty^2 B^*(I - \lambda Z^*)^{-1} x}{\tilde{B}^*(I - \lambda Z^*)^{-1} x} \tag{2.11}$$

*where* x *is the eigenvector corresponding to the largest eigenvalue* $\lambda_{max}$ *of* $P^{-1}\tilde{P}$. *Moreover, the optimal* $H^\infty$ *error* $d_\infty^2 = \lambda_{max}$ *and* $g_{opt}/d_\infty$ *is an inner function.*

## V.3 THE CENTRAL HERMITE-FEJER SOLUTION

In this section we will show how one can use the results in Section 1 to compute the central solution for the left tangential Hermite-Fejér interpolation problem. Recall that the data for the left tangential Hermite-Fejér problem for $i = 0, \ldots, r_j - 1$ and $j = 1, \ldots, n$ is given by the bounded linear operators acting between the appropriate Hilbert spaces

$$Z_j : \mathcal{X}_j \to \mathcal{X}_j,\ B_{i,j} : \mathcal{U} \to \mathcal{X}_j \quad \text{and} \quad \tilde{B}_{i,j} : \mathcal{Y} \to \mathcal{X}_j \tag{3.1}$$

and we always assume that

$$r_{spec}(Z_j) < 1 \quad (\text{for } j = 1, \ldots, n)\,. \tag{3.2}$$

The *left tangential Hermite-Fejér interpolation problem* for the data $Z_j$, $B_{i,j}$, $\tilde{B}_{i,j}$ for $i = 0, \ldots, r_j - 1$ and $j = 1, \ldots, n$, with the bound $\gamma$ is to find $G \in H^\infty(\mathcal{Y}, \mathcal{U})$ such that $\|G\|_\infty \le \gamma$ and

$$\sum_{k=0}^{i} \frac{1}{k!}(B_{i-k,j}G)^{(k)}(Z_j)_{left} = \tilde{B}_{i,j} \qquad (i = 0, \ldots, r_j - 1 \text{ and } j = 1, \ldots, n). \tag{3.3}$$

Associated with this interpolation problem are the following $H^2$ and $H^\infty$ optimization problems

$$\begin{aligned} d_2 &= \inf\{\|G\|_2 : G \in H^2(\mathcal{Y}, \mathcal{U}) \text{ and } (3.3) \text{ holds}\} \\ d_\infty &= \inf\{\|G\|_\infty : G \in H^\infty(\mathcal{Y}, \mathcal{U}) \text{ and } (3.3) \text{ holds}\}\,. \end{aligned} \tag{3.4}$$

Theorems I.5.1 and I.5.2 show how to convert the left tangential Hermite-Fejér interpolation problem to the standard left Nevanlinna-Pick interpolation problem. This is done by setting

$$M_j = \begin{bmatrix} Z_j & & & \\ I & Z_j & & \\ & \ddots & \ddots & \\ & & I & Z_j \end{bmatrix} : \text{on } \ell^2_{r_j}(\mathcal{X}_j) \tag{3.5}$$

$$R_j = \begin{bmatrix} B_{0,j} \\ B_{1,j} \\ \vdots \\ B_{r_j-1,j} \end{bmatrix} : \mathcal{U} \to \ell^2_{r_j}(\mathcal{X}_j) \text{ and } \tilde{R}_j = \begin{bmatrix} \tilde{B}_{0,j} \\ \tilde{B}_{1,j} \\ \vdots \\ \tilde{B}_{r_j-1,j} \end{bmatrix} : \mathcal{Y} \to \ell^2_{r_j}(\mathcal{X}_j), \tag{3.6}$$

and the blank spots in the operator matrix representing $M_j$ denote zero entries. Then the operators Z, B and $\tilde{B}$ are given by

$$Z = \text{diag}\,[M_1, M_2, \ldots, M_n]$$
$$B = [R_1, R_2, \ldots, R_n]^{tr} \text{ and } \tilde{B} = [\tilde{R}_1, \tilde{R}_2, \ldots, \tilde{R}_n]^{tr}. \tag{3.7}$$

So according to Theorem I.5.2 the function G in $H^\infty(\mathcal{Y}, \mathcal{U})$ satisfies the interpolation constraint (3.3) if and only if $WT_G = \tilde{W}$ where W and $\tilde{W}$ are the controllability operators in (1.1) generated by {Z,B} and {Z, $\tilde{B}$}, respectively. So applying Theorems 1.1 to 1.3 to the operators in (3.7) we obtain the following central solutions to the left tangential Hermite-Fejér interpolation problem.

**THEOREM 3.1.** *Let* $Z_j$, $B_{i,j}$ *and* $\tilde{B}_{i,j}$ *for* $i = 0, \ldots, r_j - 1$ *and* $j = 1, \ldots, n$ *be the data for the classical block left Tangential Hermite-Fejér problem. Let* Z, B, $\tilde{B}$ *be the operators defined in (3.5), (3.6) and (3.7). Let* P *and* $\tilde{P}$ *be the controllability grammians for the pair* {Z, B} *and* {Z, $\tilde{B}$}, *respectively, and assume that* {Z, B} *is controllable. Let* $\delta > 1$ *and set* $\gamma = \delta d_\infty$. *Then the central interpolant* $G_\gamma$ *for the classical left Tangential Hermite-Fejér problem is given by* $G_\gamma(\lambda) = N(\lambda)D(\lambda)^{-1}$ *where* N *and* D *are the operator valued analytic functions computed by*

$$N(\lambda) = \gamma^2 B^*(I - \lambda Z^*)^{-1}(\gamma^2 P - \tilde{P})^{-1}\tilde{B}$$
$$D(\lambda) = I + \tilde{B}^*(I - \lambda Z^*)^{-1}(\gamma^2 P - \tilde{P})^{-1}\tilde{B}. \tag{3.8}$$

*In particular,* $G_\gamma$ *is a function in* $H^\infty(\mathcal{Y}, \mathcal{U})$ *satisfying (3.3) and* $\|G_\gamma\|_\infty < \gamma$ *and*

$$\|F\|_2 \le \delta d_2 \left[\delta^2 - 1 + \frac{d_2^2}{\dim(\mathcal{Y}) d_\infty^2}\right]^{-1/2} . \tag{3.9}$$

*Finally,* $d_2^2 = \text{trace}(\tilde{B}^* P^{-1} \tilde{B})$ *and* $d_\infty^2 = r_{spec}(\tilde{P}P^{-1})$.

Recall that if $\gamma > d_\infty$, then the central intertwining interpolant $G_\gamma$ is also the unique maximal interpolant to the Hermite-Fejér problem with tolerance $\gamma$. In this case

$$\Delta(G_\gamma) = \gamma^2 [I + \tilde{B}^* (\gamma^2 P - \tilde{P})^{-1} \tilde{B}]^{-1} . \tag{3.10}$$

Moreover, if $\mathcal{Y}$ is finite dimensional, then the entropy of $G_\gamma$ is given by

$$E(G_\gamma) = -\ln \det [\gamma^{-2}(I + \tilde{B}^* (\gamma^2 P - \tilde{P})^{-1} \tilde{B})] . \tag{3.11}$$

Obviously $E(G_\gamma) > -\infty$. So $E(G_\gamma) \ge E(G)$ where G is any other Hermite-Fejér interpolant with tolerance $\gamma$, and $E(G_\gamma) = E(G)$ if and only if $G = G_\gamma$ the central interpolant with tolerance $\gamma$.

In practice it may be hard to compute the inverse of $D(\lambda)$ in the multivariable case. The following result alleviates this problem.

**THEOREM 3.2.** *Let* $Z_j$, $B_{i,j}$ *and* $\tilde{B}_{i,j}$ *for* $i = 0, \ldots, r_j$ *and* $j = 1, \ldots, n$ *be the data for the classical block left Hermite-Fejér interpolation problem. Let* Z, B *and* $\tilde{B}$ *be the operators defined in (3.5), (3.6), (3.7), and assume that* $\{Z, B\}$ *is controllable. Then the central interpolant* $G_\gamma$ *with tolerance* $\gamma > d_\infty$ *for the classical left Nevanlinna-Pick interpolation problem can also be computed by*

$$G_\gamma(\lambda) = C(I - \lambda M)^{-1} \tilde{B} \tag{3.12}$$

*where* C *from* $\mathcal{X}$ *to* $\mathcal{U}$ *and the stable operator* M *on* $\mathcal{X}$ *are defined by*

$$C = \gamma^2 B^* (\gamma^2 P - Z\tilde{P}Z^*)^{-1} \quad \textit{and} \quad M = (\gamma^2 P - \tilde{P}) Z^* (\gamma^2 P - Z\tilde{P}Z^*)^{-1} . \tag{3.13}$$

Notice that if we let $\gamma = +\infty$, then the central solution in the previous two theorems reduces to the optimal $H^2$ solution $G_{2*}$ in (1.5) to the $H^2$ optimization problem in (3.4). The following result can be used to compute the optimal $H^\infty$ solution when $\mathcal{Y}$ is one dimensional.

**THEOREM 3.3.** *Let* $Z_j$, $B_{i,j}$ *and* $\tilde{B}_{i,j}$ *for* $i = 0, \ldots, r_j - 1$ *and* $j = 1, \ldots, n$ *be the data for the classical block left Hermite-Fejér interpolation problem where* $\mathcal{Y} = \mathbb{C}$ *and* $\mathcal{X}_j$ *is finite dimensional for all* j. *Let* Z, B *and* $\tilde{B}$ *the operators defined in (3.5), (3.6), (3.7), and assume that*

$\{Z, B\}$ *is controllable. Then the unique solution* $g_{opt}$ *in* $H^\infty(\mathbb{C}, \mathcal{U})$ *to the* $H^\infty$ *Hermite-Fejér optimization problem in (3.4) is given by*

$$g_{opt}(\lambda) = \frac{d_\infty^2 B^*(I - \lambda Z^*)^{-1} x}{\tilde{B}^*(I - \lambda Z^*)^{-1} x} . \tag{3.14}$$

*where* x *is an eigenvector corresponding to the largest eigenvalue* $\lambda_{max}$ *of* $P^{-1}\tilde{P}$. *Moreover* $d_\infty^2 = \lambda_{max}$ *and* $g_{opt}/d_\infty$ *is an inner function.*

If $Z_j = z_j I$ for $j = 1, ..., n$ where $z_j$ are distinct complex numbers in the open unit disc, then one can give an explicit formula for the controllability grammians P and $\tilde{P}$ in the Hermite-Fejér interpolation problem; see Section X.4 in Foias-Frazho [4]. However, if n is not very large it is probably easier to compute P and $\tilde{P}$ directly by solving the appropriate Lyapunov equations rather than programming the explicit formula for P and $\tilde{P}$. In many applications the computational time saved by the explicit formulas may be minimal at best.

**EXAMPLE 3.4**. Let $B_i$ from $\mathcal{U}$ to $\mathcal{X}$ and $\tilde{B}_i$ from $\mathcal{Y}$ to $\mathcal{X}$ for $i = 0, 1, ..., r-1$ be the data for the following left tangential Schur interpolation problem with tolerance $\gamma$. Find (if possible) a function G in $H^\infty(\mathcal{Y}, \mathcal{U})$ satisfying $\|G\|_\infty \le \gamma$ and

$$\begin{bmatrix} B_0 & 0 & 0 & \dots & 0 \\ B_1 & B_0 & 0 & \dots & 0 \\ B_2 & B_1 & B_0 & \dots & 0 \\ \vdots & \vdots & \vdots & \vdots & \vdots \\ B_{r-1} & B_{r-2} & B_{r-3} & \dots & B_0 \end{bmatrix} \begin{bmatrix} G_0 \\ G_1 \\ G_2 \\ \vdots \\ G_{r-1} \end{bmatrix} = \begin{bmatrix} \tilde{B}_0 \\ \tilde{B}_1 \\ \tilde{B}_2 \\ \vdots \\ \tilde{B}_{r-1} \end{bmatrix} \tag{3.15}$$

where $G_k$ for $k = 0, 1, ..., r-1$ are the first $r-1$ Taylor coefficients for G. Notice that this is precisely the left tangential Hermite-Fejér interpolation problem for the case when $n = 1$, $Z_1 = 0$, $B_i = B_{i,1}$ and $\tilde{B}_i = \tilde{B}_{i,1}$ for $i = 0, ..., r-1$. Associated with this Schur interpolation problem is the following $H^2$ and $H^\infty$ optimization problems:

$$\begin{aligned} d_2 &= \inf \{\|G\|_2 : G \in H^2(\mathcal{Y}, \mathcal{U}) \text{ and } (3.15) \text{ holds}\} \\ d_\infty &= \inf \{\|G\|_\infty : G \in H^\infty(\mathcal{Y}, \mathcal{U}) \text{ and } (3.15) \text{ holds}\} . \end{aligned} \tag{3.16}$$

According to Example I.5.3 this Schur interpolation problem can be viewed as a special case of the corresponding standard left Nevanlinna-Pick problem whose data Z, B and $\tilde{B}$ is given by

$$Z = \begin{bmatrix} 0 & & & & \\ I & 0 & & & \\ & I & \ddots & & \\ & & \ddots & \ddots & \\ & & & I & 0 \end{bmatrix}, \quad B = \begin{bmatrix} B_0 \\ B_1 \\ \vdots \\ B_{r-1} \end{bmatrix} \quad \text{and} \quad \tilde{B} = \begin{bmatrix} \tilde{B}_0 \\ \tilde{B}_1 \\ \vdots \\ \tilde{B}_{r-1} \end{bmatrix} \tag{3.17}$$

where the blank spots in the matrix for Z are zero. Notice that Z is simply a lower shift with the identity below the main diagonal and zeros elsewhere. Now let T be the lower triangular $r \times r$ block Toeplitz matrix generated by $\{B_i\}_0^{r-1}$, that is, T is precisely the $r \times r$ block matrix given on the left hand side of (3.15). Then the controllability grammians P and $\tilde{P}$ for the pair $\{Z, B\}$ and $\{Z, \tilde{B}\}$ are given by $P = TT^*$ and $\tilde{P} = \tilde{T}\tilde{T}^*$. Here $\tilde{T}$ is the lower triangular block $r \times r$ Toeplitz matrix generated by $\{\tilde{B}_i\}_0^{r-1}$. Now assume P is invertible, or equivalently, $\{Z, B\}$ is controllable. Then, our previous analysis along with the results in Example I.5.3 shows that the errors $d_2$ and $d_\infty$ in the Schur optimization problems (3.16) are given by $d_2^2 = \text{trace}\,(\tilde{B}^* P^{-1} \tilde{B})$ and $d_\infty^2 = r_{spec}(\tilde{P}P^{-1})$.

By using Z, B and $\tilde{B}$ in (3.17) along with $P = TT^*$ and $\tilde{P} = \tilde{T}\tilde{T}^*$ in Theorem 3.1 we see that the central solution $G_\gamma$ for the left tangential Schur interpolation problem with tolerance $\gamma > d_\infty$ is given by $G_\gamma(\lambda) = N(\lambda)D(\lambda)^{-1}$. Here $N(\lambda)$ and $D(\lambda)$ are the polynomials given by

$$N(\lambda) = \gamma^2 B^* \Omega(\lambda)(\gamma^2 P - \tilde{P})^{-1}\tilde{B} \quad \text{and} \quad D(\lambda) = I + \tilde{B}^* \Omega(\lambda)(\gamma^2 P - \tilde{P})^{-1}\tilde{B} \tag{3.18}$$

where $\Omega(\lambda) = (I - \lambda Z^*)^{-1}$ now becomes

$$\Omega(\lambda) = \begin{bmatrix} I & \lambda I & \lambda^2 I & \cdots & \lambda^{r-1} I \\ 0 & I & \lambda I & \cdots & \lambda^{r-2} I \\ 0 & 0 & I & \cdots & \lambda^{r-3} I \\ \vdots & \vdots & \vdots & \vdots & \\ 0 & 0 & 0 & \cdots & I \end{bmatrix}. \tag{3.19}$$

Moreover, this central intertwining lifting $G_\gamma$ can also be computed by the state space formulas in (3.12) and (3.13). (It is a simple exercise to show that in the present case the matrix M in (3.13) is a companion matrix.) Finally, if $\mathcal{Y}$ is one dimensional, then there exists a unique solution $g_{opt}$ in $H^\infty(\mathcal{Y}, \mathcal{U})$ to the $H^\infty$ Schur optimization problem in (3.13) and this $\mathcal{Y}$ is given by

$$g_{opt} = \frac{d_\infty^2 B^* \Omega(\lambda)x}{\tilde{B}^* \Omega(\lambda)x} \tag{3.20}$$

where x is the eigenvector corresponding to the largest eigenvalue $\lambda_{max}$ of $P^{-1}\tilde{P}$ and $d_\infty^2 = \lambda_{max}$.

## V.4 THE CENTRAL FORMULA FOR THE SARASON PROBLEM

In this section we will present a state space solution for the following Sarason problem with tolerance $\gamma$: Let F be a function in $H^\infty(\mathcal{U}, \mathcal{Y})$ and $\Theta$ a two-sided inner function in $H^\infty(\mathcal{E}, \mathcal{Y})$, then find a H in $H^\infty(\mathcal{U}, \mathcal{E})$ satisfying $\|F - \Theta H\|_\infty \le \gamma$ for some specified $\gamma$. Associated with this Sarason problem are the following $H^2$ and $H^\infty$ optimization problems

$$\begin{aligned} d_2 &= \inf \{\|F - \Theta H\|_2 : H \in H^2(\mathcal{U}, \mathcal{E})\} \\ d_\infty &= \inf \{\|F - \Theta H\|_\infty : H \in H^\infty(\mathcal{U}, \mathcal{E})\} . \end{aligned} \tag{4.1}$$

Notice that $d_2$ is the distance from F to $\Theta H^2(\mathcal{U}, \mathcal{E})$ in the $H^2$ norm, while $d_\infty$ is the distance from F to $\Theta H^\infty(\mathcal{U}, \mathcal{E})$ in the $H^\infty$ norm. (We always assume that $\mathcal{U}$ is finite dimensional for the $H^2$ optimization problem. If $\mathcal{U}$ is not finite dimensional, then $d_2$ may be infinite.) In Section II.4 we used the commutant lifting theorem to prove that there exists an optimal H in $H^\infty(\mathcal{U}, \mathcal{E})$ satisfying $d_\infty = \|F - \Theta H\|_\infty$.

We say that G is a *Sarason interpolant* for the data F, $\Theta$ with tolerance $\gamma$, if $G = F - \Theta H$ for some H in $H^\infty(\mathcal{U}, \mathcal{E})$ and $\|G\|_\infty \le \gamma$. Obviously, this G is in $H^\infty(\mathcal{U}, \mathcal{Y})$. The function $G_\gamma$ in $H^\infty(\mathcal{U}, \mathcal{Y})$ is the *central interpolant* for the Sarason problem with tolerance $\gamma$ if $M_{G_\gamma}^+$ is the central intertwining lifting for the operator $A = P_{\mathcal{H}'} M_F^+$ used in the commutant lifting theorem to solve the Sarason problem in Section II.4. In order to construct the central interpolant $G_\gamma$ for the Sarason problem with tolerance $\gamma > d_\infty$, recall that $\{Z \text{ on } \mathcal{X}, B, C, D\}$ is realization for some function $\Omega(\lambda)$ in $H^\infty(\mathcal{U}, \mathcal{Y})$ if

$$\Omega(\lambda) = D + \lambda C(I - \lambda Z)^{-1} B \tag{4.2}$$

where Z is an operator on $\mathcal{X}$ and B, C, and D are operator acting between the appropriate spaces. The system {Z, B, C, D} is stable if $r_{spec}(Z) < 1$. Finally, if Z is stable the observability grammian Q for the pair {C, Z} is given by the solution to the following Lyapunov equation

$$Q = Z^* Q Z + C^* C . \tag{4.3}$$

Moreover, the pair {C, Z} is observable if and only if Q is invertible. For further results on realization theory see Chapter III. We are now ready to state the main result of this section.

**THEOREM 4.1.** *Let* $\{Z_1 \text{ on } \mathcal{X}_1, B_1, C_1, D_1\}$ *and* $\{Z \text{ on } \mathcal{X}, B, C, D\}$ *be stable, controllable and observable realizations for* $F$ *in* $H^\infty(\mathcal{U}, \mathcal{Y})$ *and the two-sided inner function* $\Theta$ *in* $H^\infty(\mathcal{E}, \mathcal{Y})$*, respectively. Let* $Q$ *be the observability grammian for the pair* $\{C, Z\}$*. Compute the solution* $Q_1$ *to the following Stein equation*

$$Q_1 = Z^* Q_1 Z_1 + C^* C_1 \quad \textit{and set } \tilde{B} = C^* D_1 + Z^* Q_1 B_1 \tag{4.4}$$

*Finally, let* $\tilde{Q}$ *be the observability grammian for the pair* $\{\tilde{B}^*, Z\}$*. Then the errors* $d_2$ *and* $d_\infty$ *for the* $H^2$ *and* $H^\infty$ *optimization problem in (4.1) are given by*

$$d_2^2 = \text{trace } \tilde{B}^* Q^{-1} \tilde{B} \quad \textit{and} \quad d_\infty^2 = r_{spec}(Q^{-1}\tilde{Q}) . \tag{4.5}$$

*Given any* $\delta > 1$*, then the central intertwining lifting* $M_{G_\gamma}^+$ *or central interpolant* $G_\gamma$ *for the Sarason problem with tolerance* $\gamma = \delta d_\infty$ *is given by* $G_\gamma = F - \Theta H$ *where* $H$ *is the function in* $H^\infty(\mathcal{U}, \mathcal{E})$ *computed by*

$$\begin{aligned} H(\lambda) = {} & B^* \tilde{Q}(\gamma^2 Q - \tilde{Q})^{-1}(1 - \lambda M)^{-1} M \tilde{B} + (D^* D_1 + B^* Q_1 B_1) \\ & + \lambda (D^* C_1 + B^* Q_1 Z_1)(I - \lambda Z_1)^{-1} B_1 \end{aligned} \tag{4.6}$$

*where*

$$M = (\gamma^2 Q - \tilde{Q}) Z (\gamma^2 Q - Z^* \tilde{Q} Z)^{-1} . \tag{4.7}$$

*Moreover,* $M$ *is stable. Finally, this central interpolant satisfies the bounds* $\|G_\gamma\|_\infty < \delta d_\infty$ *and*

$$\|G_\gamma\|_2 \le \delta d_2 \left[ \delta^2 - 1 + \frac{d_2^2}{\dim(\mathcal{U}) d_\infty^2} \right]^{-1/2} . \tag{4.8}$$

Notice that if we let $\delta$ approach infinity, then we obtain the unique optimal solution $H_2$ to the $H^2$ optimization problem in (4.1). To be precise, if we let $\delta$ approach infinity in (4.6) we have

$$H_2 = D^* D_1 + B^* Q_1 B_1 + \lambda (D^* C_1 + B^* Q_1 Z_1)(I - \lambda Z_1)^{-1} B_1 . \tag{4.9}$$

Obviously $H_2$ is in $H^\infty(\mathcal{U}, \mathcal{E})$. Moreover, if we set $G_{2*} = F - \Theta H_2$, then $d_2 = \|G_{2*}\|_2$ and thus $H_2$ is the unique solution to the $H^2$ optimization problem in (4.1). In other words, $G_{2*}a = P_{\mathcal{H}'} Fa$ for all $a$ in $\mathcal{U}$ where $\mathcal{H}'$ is the orthogonal complement of $\Theta H^2(\mathcal{E})$ in $H^2(\mathcal{Y})$. In particular, $\Theta H_2 a$ is the orthogonal projection of $Fa$ onto $\Theta H^2(\mathcal{E})$.

Let $\Delta(G)$ be the positive operator on $\mathcal{U}$ defined according to (IV.7.1), that is,

$$(\Delta(G)a,\, a) = \inf\{\|D_{B'}(a+\lambda h)\|^2 : h \in H^2(\mathcal{U})\} \tag{4.10}$$

where a is a fixed vector in $\mathcal{U}$ and $B' = M_G^+$. Here $D_{B'}$ is the positive square root of $\gamma^2 I - B'^* B'$. If $\mathcal{U}$ is finite dimensional, then the entropy of G is defined by

$$E(G) = \frac{1}{2\pi}\int_0^{2\pi} \ln\det[\gamma^2 I - G(e^{it})^* G(e^{it})]dt = \ln\det[\Delta(G)]\,. \tag{4.11}$$

The last equality follows from (IV.7.15). We say that $G_*$ is a *maximal interpolant* for the Sarason problem with tolerance $\gamma$, if $G_*$ is a Sarason interpolant with tolerance $\gamma$ and $\Delta(G_*) \geq \Delta(G)$ where G is any other Sarason interpolant with tolerance $\gamma$. The function $G_*$ is the *unique maximal interpolant* if $G_*$ is a maximal Sarason interpolant and if $\Delta(G_*) = \Delta(G)$ for any other Sarason interpolant G with tolerance $\gamma$, then $G_* = G$. The results in Section IV.7 show that if $\mathcal{U}$ is finite dimensional, then this maximal interpolant $G_*$ is also the maximal entropy interpolant, that is, $E(G_*) \geq E(G)$ for all Sarason interpolants G with tolerance $\gamma$. Moreover, if $E(G) > -\infty$ for some G, then the maximal interpolant $G_*$ is unique, that is, $E(G_*) = E(G)$ if and only if $G = G_*$ is the maximal entropy interpolant. Finally, the results in Section IV.7 will be used to show that the central interpolant $G_\gamma$ with tolerance $\gamma$, is the unique maximal interpolant with tolerance $\gamma$ for the Sarason problem. This sets the stage for the following result.

**THEOREM 4.2.** *Let* $\{Z_1$ *on* $\mathcal{X}_1$, $B_1$, $C_1$, $D_1\}$ *and* $\{Z$ *on* $\mathcal{X}$, B, C, D$\}$ *be stable, controllable and observable realizations for* F *in* $H^\infty(\mathcal{U}, \mathcal{Y})$ *and the two-sided inner function* $\Theta$ *in* $H^\infty(\mathcal{E}, \mathcal{Y})$, *respectively. Let* Q *be the observability grammian for the pair* {C, Z}. *Compute the solution* $Q_1$ *to the Stein equation in (4.4). Finally, let* $\tilde{Q}$ *be the observability grammian for the pair* $\{\tilde{B}^*, Z\}$ *where* $\tilde{B}$ *is defined in (4.4) and let* $\gamma > d_\infty$. *Then the central interpolant* $G_\gamma = F - \Theta H$ *for the Sarason problem is the unique maximal interpolant for the Sarason problem with tolerance* $\gamma$, *and*

$$\Delta(G_\gamma) = \gamma^2[I + \tilde{B}^*(\gamma^2 Q - \tilde{Q})^{-1}\tilde{B}]^{-1}\,. \tag{4.12}$$

*Moreover, if* $\mathcal{U}$ *is finite dimensional, then the entropy of* $G_\gamma$ *is given by*

$$E(G_\gamma) = \frac{1}{2\pi}\int_0^{2\pi} \ln\det[\gamma^2 I - G_\gamma^* G_\gamma]dt = -\ln\det[\gamma^{-2}(I + \tilde{B}^*(\gamma^2 Q - \tilde{Q})^{-1}\tilde{B})]\,. \tag{4.13}$$

*In particular, if* G *is any other Sarason interpolant with tolerance* $\gamma$, *then* $E(G_\gamma) \geq E(G)$ *with equality if and only if* $G = G_\gamma$.

The following result provides the optimal solution to the $H^\infty$ Sarason optimization problem where $\mathcal{U}$, $\mathcal{E}$ and $\mathcal{Y}$ are all one dimensional, that is, F and $\Theta$ are scalar valued functions in $H^\infty$.

**THEOREM 4.3.** *Let* $\{Z_1 \text{ on } \mathcal{X}_1, B_1, C_1, D_1\}$ *and* $\{Z \text{ on } \mathcal{X}, B, C, D\}$ *be stable, controllable and observable realizations for* F *in* $H^\infty$ *and the scalar valued inner function* $\Theta$ *in* $H^\infty$*, respectively. Let* Q *and* $\tilde{Q}$ *be the respective observability grammians for the pair* $\{C, Z\}$ *and* $\{\tilde{B}^*, Z\}$ *where* $\tilde{B}^*$ *is the operator obtained by solving the Stein equation in (4.4). Then the unique solution* $H_{opt}$ *in* $H^\infty$ *to the* $H^\infty$ *optimization problem in (4.1) is given by*

$$F(\lambda) - \Theta(\lambda)H_{opt}(\lambda) = \frac{d_\infty^2 C(I - \lambda Z)^{-1}x}{\tilde{B}^*(I - \lambda Z)^{-1}x} \tag{4.14}$$

*where* x *is the eigenvector corresponding to the largest eigenvalue* $\lambda_{max}$ *of* $Q^{-1}\tilde{Q}$*. Moreover, the optimal* $H^\infty$ *error* $d_\infty^2 = \lambda_{max}$ *and* $(F - \Theta H_{opt})/d_\infty$ *is an inner function.*

**PROOF OF THEOREM 4.1.** Let S be the unilateral shift on $H^2(\mathcal{Y})$ and $\mathcal{H}'$ the invariant subspace for $S^*$ defined by

$$\mathcal{H}' = \mathcal{H}(\Theta) = H^2(\mathcal{Y}) \ominus \Theta H^2(\mathcal{E}) . \tag{4.15}$$

Recall that for the Sarason problem A is the operator from $H^2(\mathcal{U})$ into $\mathcal{H}'$ defined by $A = P_{\mathcal{H}'}M_F^+$ and T is the unilateral shift on $H^2(\mathcal{U})$, while $T'$ is the comparison of S to $\mathcal{H}'$. Moreover, $B'$ is an intertwining of A with tolerance $\gamma$ if and only if $B' = M_G^+$ where $G = F - \Theta H$ for some H in $H^\infty(\mathcal{U}, \mathcal{E})$ and $\|F - \Theta H\|_\infty \le \gamma$. Here we will use Corollary IV.3.2 to compute the central intertwining lifting $B_\gamma$ of A and thus solve the Sarason problem. First we need a state space formula for A. To this end, let W be the controllability operator from $\ell_+^2(\mathcal{Y})$ into $\mathcal{X}$ generated by the pair $\{Z^*, C^*\}$, that is

$$W = [C^*, Z^*C^*, Z^{*2}C^*, \ldots] . \tag{4.16}$$

Now let $W_o$ be the observability operator from $\mathcal{X}$ into $H^2(\mathcal{Y})$ defined by

$$W_o x = C(I - \lambda Z)^{-1}x \qquad (x \in \mathcal{X}) . \tag{4.17}$$

If $\mathcal{F}_2$ denotes the Fourier transform from $\ell_+^2(\mathcal{Y})$ onto $H^2(\mathcal{Y})$, then (4.16) readily shows that $W_o = \mathcal{F}_2 W^*$. Recall that $P_{\mathcal{H}'} = W_o Q^{-1} W_o^*$; see equation (III.7.3) in Theorem III.7.1. Using this we obtain

$$A = P_{\mathcal{H}'} M_F^+ = \mathcal{F}_2 W^* Q^{-1} W_o^* M_F^+ = \mathcal{F}_2 W^* Q^{-1} W T_F \mathcal{F}_1^* \tag{4.18}$$

where $T_F$ is the lower triangular block Toeplitz matrix from $\ell_+^2(\mathcal{U})$ into $\ell_+^2(\mathcal{Y})$ generated by F and $\mathcal{F}_1$ is the Fourier transform from $\ell_+^2(\mathcal{U})$ onto $H^2(\mathcal{U})$. Now let $Q_1$ be the operator from $\mathcal{X}_1$ into $\mathcal{X}_2$ defined by

$$Q_1 = \sum_0^\infty Z^{*n} C^* C_1 Z_1^n . \tag{4.19}$$

Since Z and $Z_1$ are both stable $Q_1$ is well defined and is also the unique solution to the Stein equation in (4.4); see Lemma III.2.4. By using (4.16), (4.19) and the state space realization $\{Z_1, B_1, C_1, D_1\}$ for F we have

$$\tilde{W} := WT_F = W \begin{bmatrix} D_1 & 0 & 0 & \cdots \\ C_1B_1 & D_1 & 0 & \cdots \\ C_1Z_1B_1 & C_1B_1 & D_1 & \cdots \\ C_1Z_1^2B_1 & C_1Z_1B_1 & C_1B_1 & \cdots \\ \vdots & \vdots & \vdots & \vdots \end{bmatrix} = [\tilde{B}, Z^*\tilde{B}, Z^{*2}\tilde{B}, \ldots] \tag{4.20}$$

where $\tilde{B}$ is defined in (4.4). So the operator A in (4.18) now becomes

$$A = \mathcal{F}_2 W^* Q^{-1} \tilde{W} \mathcal{F}_1^* = W_o Q^{-1} \tilde{W} \mathcal{F}_1^* \tag{4.21}$$

where $\tilde{W}$ is the controllability operator from $\ell_+^2(\mathcal{U})$ into $\mathcal{X}$ generated by $\{Z^*, \tilde{B}\}$ defined in (4.20).

To compute $d_\infty$ notice that

$$AA^*W_o = W_o Q^{-1} \tilde{W}\tilde{W}^* Q^{-1} W_o^* W_o = W_o Q^{-1} \tilde{Q}$$

where $\tilde{Q} = \tilde{W}\tilde{W}^*$ is the observability grammian for the pair $\{\tilde{B}^*, Z\}$. Because the pair {C, Z} is observable, the operator $W_o$ is onto $\mathcal{H}'$; see Lemma III.7.1. So the previous equation shows that $AA^*$ is similar to $Q^{-1}\tilde{Q}$. Therefore $d_\infty^2 = \|A\|^2 = r_{spec}(Q^{-1}\tilde{Q})$. According to the results in Section IV.9, see (IV.9.17), the optimal $H^2$ error is given by

$$d_2^2 = \|A \mid \mathcal{U}\|_2^2 = \|W_o Q^{-1} \tilde{W} \mid \mathcal{U}\|_2^2 =$$

$$\|W_o Q^{-1} \tilde{B}\|_2^2 = \text{trace}\,(\tilde{B}^* Q^{-1} W_o^* W_o Q^{-1} \tilde{B}) = \text{trace}\,(\tilde{B}^* Q^{-1} \tilde{B}) .$$

This establishes the formulas for $d_2$ and $d_\infty$ in (4.5). (This also follows from Theorem III.8.1.)

By Corollary IV.3.2 the central intertwining lifting with tolerance $\gamma$ is given by

$$G_\gamma = (A\Pi_o^*)(\lambda) + \Psi(\lambda)(I - \lambda T_A^*)^{-1} T_A^* \Pi_o^* \tag{4.22}$$

where $\Pi_o$ is the operator from $H^2(\mathcal{U})$ onto $\mathcal{U}$ defined by $\Pi_o = P_{\mathcal{U}}$ and $T_A^* = D_{AT}^{-2} T^* D_A^2$. To compute $\Psi(\lambda)$ first notice that for any h in $\mathcal{H}'$ we have

$$(S - T')h = \Theta\left[\sum_0^\infty \Theta_{n+1}^* h_n\right] \tag{4.23}$$

where $\{\Theta_n\}$ and $\{h_n\}$ are the Fourier coefficients of $\Theta$ and h, respectively. (The sum in (4.23) makes sense because $\Theta_{n+1} = CZ^n B$ for all $n \geq 0$ and Z is stable.) To verify (4.23) notice that $(S - T')h = (I - P_{\mathcal{H}'})\lambda h = P_{\mathcal{M}} \lambda h$ where $\mathcal{M} = \Theta H^2(\mathcal{E})$. Thus $P_{\mathcal{M}} \lambda h = \Theta f$ for some f in $H^2(\mathcal{E})$. By the Projection Theorem $\lambda h - \Theta f$ is orthogonal to $\Theta H^2(\mathcal{E})$, or equivalently, $f = P_+ \Theta^* e^{it} h$ where $P_+$ is the orthogonal projection onto $H^2(\mathcal{E})$. However, h is in $\mathcal{H}'$. So h is orthogonal to $\Theta H^2(\mathcal{E})$, or equivalently, $\Theta^* h$ is in $L^2(\mathcal{E}) \ominus H^2(\mathcal{E})$. Therefore $f = P_+ e^{it} \Theta^* h$ must be in the space $\mathcal{E}$ of constant functions. In other words, f is the Fourier coefficient of $1 = e^{i0t}$ in the power series expansion of $e^{it}\Theta^* h$. Hence $f = \Sigma \Theta_{n+1}^* h_n$. This fact along with $(S - T')h = P_{\mathcal{M}} \lambda h = \Theta f$ establishes (4.23).

Clearly $W_o x$ is in $\mathcal{H}'$ for all x in $\mathcal{X}$; see Theorem III.7.1. So by applying (4.23) to $h = W_o x$ along with the fact that $CZ^{n-1}B = \Theta_n$ for all $n \geq 1$, we have

$$\Psi := (S - T')A = (S - T')W_o Q^{-1} \tilde{W} \mathcal{F}_1^* = \Theta\left[\sum_0^\infty \Theta_{n+1}^* CZ^n Q^{-1} \tilde{W} \mathcal{F}_1^*\right] =$$

$$\Theta B^* \sum_0^\infty Z^{*n} C^* C Z^n Q^{-1} \tilde{W} \mathcal{F}_1^* = \Theta B^* \tilde{W} \mathcal{F}_1^* .$$

Therefore $(\Psi h)(\lambda) = \Theta(\lambda) B^* \tilde{W} \mathcal{F}_1^* h$, where h is in $H^2(\mathcal{U})$.

Now notice that our $A = W_o Q^{-1} \tilde{W} \mathcal{F}_1^*$ is almost identical to the operator A in (1.11) used in the calculation (1.34). The only difference is that our A has Fourier transforms on the right and left; and the role of Z, B, P and $\tilde{P}$ in (1.11) is now played by $Z^*$, $C^*$, Q and $\tilde{Q}$, respectively; see also Theorem III.8.1. So by making this replacement in (1.34) we arrive at

$$(I - \lambda T_A^*)^{-1} T_A^* \Pi_o^* = -\mathcal{F}_1 \tilde{W}^* (1 - \lambda N)^{-1} N (\gamma^2 Q - \tilde{Q})^{-1} \tilde{B} \tag{4.24}$$

where $N = (\gamma^2 I - Z Q^{-1} Z^* \tilde{Q})^{-1} Z (\gamma^2 I - Q^{-1} \tilde{Q})$. As in the computation in (1.38) we define the operator M similar to N by

$$M = (\gamma^2 Q - \tilde{Q}) N (\gamma^2 Q - \tilde{Q})^{-1} = (\gamma^2 Q - \tilde{Q}) Z (\gamma^2 Q - Z^* \tilde{Q} Z)^{-1} . \tag{4.25}$$

But combining (4.22), (4.24), (4.25), with $\Psi = \Theta B^* \tilde{W} \mathcal{F}_1^*$ we see that the central intertwining lifting with tolerance $\gamma$ is given by

$$G_\gamma(\lambda) = P_{\mathcal{H}'} M_F^{+} \Pi_o^* - \Theta B^* \tilde{Q} (\gamma^2 Q - \tilde{Q})^{-1} (1 - \lambda M)^{-1} M \tilde{B} \tag{4.26}$$

where M is given in (4.7). By consulting Theorem 1.2 or Proposition 1.7 with Z, B, P and $\tilde{P}$ replaced by $Z^*$, $C^*$, Q and $\tilde{Q}$, respectively, we see that $r_{spec}(M) < 1$.

So we have shown that the central intertwining lifting $G_\gamma = P_{\mathcal{H}'}F - \Theta H_\gamma$ where $H_\gamma$ is the function in $H^\infty(\mathcal{U}, \mathcal{E})$ immediately to the right of $\Theta$ in (4.26). However, there also exists a function H in $H^\infty(\mathcal{U}, \mathcal{E})$ satisfying $G_\gamma = F - \Theta H$. Thus

$$P_{\mathcal{H}'}F + P_{\mathcal{M}}F - \Theta H = F - \Theta H = G_\gamma = P_{\mathcal{H}'}F - \Theta H_\gamma . \tag{4.27}$$

Hence $\Theta H = \Theta H_\gamma + P_{\mathcal{M}}F$. To complete the proof it remains to show that $P_{\mathcal{M}}F = \Theta H_m$ where $H_m$ is the function given by the the last two terms in (4.6). Because $\Theta$ is an inner function $P_{\mathcal{M}}F = M_\Theta^+ (M_\Theta^+)^* F = M_\Theta^+ H_m$. In particular, $H_m$ is the Fourier transform of $T_\Theta^*\{F_n\}$ where $\{F_n\}$ are the Fourier coefficients of F. So using the realizations {Z, B, C, D} and $\{Z_1, B_1, C_1, D_1\}$ for $\Theta$ and F, respectively along with the formula for $Q_1$ in (4.19) we have

$$T_\Theta^*\{F_n\} = \begin{bmatrix} D^* & B^*C^* & B^*Z^*C^* & \cdots \\ 0 & D^* & B^*C^* & \cdots \\ 0 & 0 & D^* & \cdots \\ \vdots & \vdots & \vdots & \vdots \end{bmatrix} \begin{bmatrix} D_1 \\ C_1B_1 \\ C_1Z_1B_1 \\ \vdots \end{bmatrix} = \begin{bmatrix} D^*D_1 + B^*Q_1B_1 \\ (D^*C_1 + B^*Q_1Z_1)Z_1^0B_1 \\ (D^*C_1 + B^*Q_1Z_1)Z_1B_1 \\ \vdots \end{bmatrix} \tag{4.28}$$

By taking the Fourier transform of the last column vector we arrive at the the last two terms in (4.6). This completes the proof.

**PROOF OF THEOREM 4.2.** Notice that the operator A in (4.21) is precisely the Fourier transform of the operator A in (1.11) used in the Nevanlinna-Pick problem if we replace B, Z, P and $\tilde{P}$ in (1.11) with $C^*$, $Z^*$, Q and $\tilde{Q}$ respectively. In particular, equation (1.13) shows that this replacement gives

$$D_A^{-2} \mathcal{F}_1 = \mathcal{F}_1 \gamma^{-2} [I + \tilde{W}^* (\gamma^2 Q - \tilde{Q})^{-1} \tilde{W}] . \tag{4.29}$$

So using $\Pi_o \mathcal{F}_1 \tilde{W}^* = \tilde{B}^*$ we see that

$$\Pi_o D_A^{-2} \Pi_o^* = \gamma^{-2} [I + \tilde{B}^* (\gamma^2 Q - \tilde{Q})^{-1} \tilde{B}] . \tag{4.30}$$

Now applying Theorems IV.7.1 and Corollary IV.7.7 along with (IV.7.4), we arrive at (4.12) and (4.13). This completes the proof.

Finally, it is noted that (4.29) also gives

$$\Omega(\lambda) := D_{\tilde{A}}^{-2} \mathcal{F}_1 \,|\, \mathcal{U} = \gamma^{-2} I + \gamma^{-2} \tilde{B}^* (I - \lambda Z)^{-1} (\gamma^2 Q - \tilde{Q})^{-1} \tilde{B} . \tag{4.31}$$

Because Z is stable $\Omega$ is in $H^\infty(\mathcal{U}, \mathcal{U})$. According to Corollary IV.6.5 the function $\Omega(0)^{1/2}\Omega(\lambda)^{-1}$ is in $H^\infty(\mathcal{U}, \mathcal{U})$ and is also the outer spectral factor for $\gamma^2 I - G_\gamma^* G_\gamma$. Since both $\Omega$ and $\Omega^{-1}$ are in $H^\infty(\mathcal{U}, \mathcal{U})$ it follows that $\|G_\gamma\|_\infty < \gamma$.

**PROOF OF THEOREM 4.3.** The Proof of Theorem 4.3 is identical to the proof of Theorem 1.4 when one replaces the A, B, P and $\tilde{P}$ in Theorem 1.4 by $W^* Q^{-1} \tilde{W}$, $C^*$, Q and $\tilde{Q}$ respectively. Then formula (1.10) yields (4.14). One can also obtain this result by using Theorem III.8.1 to convert the Sarason problem to a standard Nevanlinna-Pick problem. Then by applying Theorem 1.4 to this Nevanlinna-Pick problem, we obtain the formula for the optimal solution in (4.14). This completes the proof.

## V.5 CENTRAL NEHARI SOLUTIONS

In this section we will use the formulas in Corollary IV.3.3 to derive a state space solution to a certain Nehari interpolation problem. To this end, let $K^\infty(\mathcal{U}, \mathcal{Y})$ be the subspace of $L^\infty(\mathcal{U}, \mathcal{Y})$ consisting of the set of all functions F in $L^\infty(\mathcal{U}, \mathcal{Y})$ whose Fourier coefficients of $e^{int}$ are zero for all $n \geq 0$. Now let F be a function in $K^\infty(\mathcal{U}, \mathcal{Y})$ and consider the following $H^2$ and $H^\infty$ Nehari optimization problems associated with:

$$\begin{aligned} d_2 &= \inf \{ \|F - H\|_2 : H \in H^2(\mathcal{U}, \mathcal{Y}) \} \\ d_\infty &= \inf \{ \|F - H\|_\infty : H \in H^\infty(\mathcal{U}, \mathcal{Y}) \} \end{aligned} \tag{5.1}$$

Here is $d_2$ is the distance from F to $H^2(\mathcal{U}, \mathcal{Y})$ in the $H^2$ norm, while $d_\infty$ is the distance from F to $H^\infty(\mathcal{U}, \mathcal{Y})$ in the $H^\infty$ norm. Obviously, by the Projection Theorem $d_2 = \|F\|_2$. (For the $H^2$ optimization problem we always assume that $\mathcal{U}$ is finite dimensional. If $\mathcal{U}$ is not finite dimensional, then $d_2$ may be infinite.) Moreover, according to the results in Sections I.6, II.3 and III.6 the error $d_\infty = \|A\|$ where A is the Hankel operator from $H^2(\mathcal{U})$ into $K^2(\mathcal{Y}) := L^2(\mathcal{Y}) \ominus H^2(\mathcal{Y})$ defined by $A = P_- M_F \,|\, H^2(\mathcal{U})$. Here $P_-$ is the orthogonal projection onto $K^2(\mathcal{Y})$ and A is the Fourier transform of the operator defined in equation (II.3.3) or (III.6.11). Finally, by a slight abuse of notation we say that $G_\gamma$ is a *central intertwining lifting or central interpolant* for the Nehari problem with tolerance $\gamma$, if $G_\gamma$ is a function in $L^\infty(\mathcal{U}, \mathcal{Y})$ and $M_{G_\gamma} \,|\, H^2(\mathcal{U})$ is the central intertwining lifting of A with tolerance $\gamma$.

Let us note that the optimization problems in (5.1) are also well defined for any F in $L^\infty(\mathcal{U}, \mathcal{Y})$. In this case $d_2 = \|P_- F\|_2$ and $d_\infty$ is still the norm of $A = P_- M_F \,|\, H^2(\mathcal{U})$. Obviously any F in $L^\infty(\mathcal{U}, \mathcal{Y})$ admits a decomposition of the form $F = P_- F + P_+ F$ where $P_+$ is the orthogonal projection onto $H^2(\mathcal{Y})$. In particular, one can replace F in $L^\infty(\mathcal{U}, \mathcal{Y})$ by $P_-F$. However, there is a technical problem. There exists functions F in $L^\infty(\mathcal{U}, \mathcal{Y})$ such that $P_-F$ is not in $K^\infty(\mathcal{U}, \mathcal{Y})$. For example consider the scalar valued function $F(e^{it}) = it$ for $0 \le t < 2\pi$. Fortunately, this does not happen for rational F and many other important applications. So with out much loss of generality we have stated our $H^2$ and $H^\infty$ Nehari optimization problems for F in $K^\infty(\mathcal{U}, \mathcal{Y})$.

Recall that {Z, B, C} is an anticausal realization of a function F in $K^\infty(\mathcal{U}, \mathcal{Y})$ if

$$F(\lambda) = C(\lambda I - Z)^{-1}B \qquad (|\lambda| > 1)\,. \tag{5.2}$$

By taking the power series expansion for sides in (5.2) we see that {Z, B, C} is an anticausal realization for $F(\lambda)$ if and only if $F_n = CZ^{|n+1|}B$ for all $n < 0$ where $\{F_n : n < 0\}$ are the Fourier coefficients of F. Any rational function F in $K^\infty(\mathcal{U}, \mathcal{Y})$ admits a stable, controllable and observable anticausal realization {Z, B, C}; see Section III.5. Finally, as before the controllability grammian P and observability grammian Q for the stable pair {Z, B} and {C, Z} are respectively given by the unique solutions to the Lyapunov equations

$$P = ZPZ^* + BB^* \quad \text{and} \quad Q = Z^*QZ + C^*C\,. \tag{5.3}$$

This sets the stage for the main result of this section.

**THEOREM 5.1**. *Let {Z, B, C} be a stable, controllable and observable anticausal realization for a function* F *in* $K^\infty(\mathcal{U}, \mathcal{Y})$. *Let* P *and* Q *be the controllability and observability grammians for the pairs* {Z, B} *and* {C, Z} *respectively computed by (5.3). Then the errors* $d_2$ *and* $d_\infty$ *for the* $H^2$ *and* $H^\infty$ *optimization problems in (5.1) are given by*

$$d_2 = \operatorname{trace}(B^*QB) \quad \textit{and} \quad d_\infty^2 = r_{spec}(QP)\,. \tag{5.4}$$

*Moreover, given any* $\delta > 1$, *then the central interpolant* $G_\gamma$ *for the* $H^\infty$ *Nehari problem with tolerance* $\gamma = \delta d_\infty$ *is given by* $F - H = G_\gamma(\lambda) = N(\lambda)D(\lambda)^{-1}$ *where* H *is in* $H^\infty(\mathcal{U}, \mathcal{Y})$ *and* $N(\lambda)$ *is in* $K^\infty(\mathcal{U}, \mathcal{Y})$ *and* $D(\lambda)$ *is the invertible outer function in* $H^\infty(\mathcal{U}, \mathcal{U})$ *defined by*

$$N(\lambda) = \gamma^2 C(\lambda I - Z)^{-1}(\gamma^2 I - PQ)^{-1}B \quad \textit{and} \quad D(\lambda) = I + B^*(I - \lambda Z^*)^{-1}(\gamma^2 I - QP)^{-1}QB \tag{5.5}$$

*Finally, the central intertwining lifting* $G_\gamma$ *satisfies the bounds* $\|G_\gamma\|_\infty < \gamma$ *and*

$$\|G_\gamma\|_2 \le \delta d_2 \left[\delta^2 - 1 + \frac{d_2^2}{\dim(\mathcal{U})d_\infty^2}\right]^{-1/2}\,. \tag{5.6}$$

Notice that QP is similar to PQ. (Compute $Q^{-1}(QP)Q$). In particular, QP and PQ have the same spectral radius $d_\infty^2$. In practice it may be hard to invert the outer function $D(\lambda)$ in the multivariable case. The following theorem alleviates this problem and gives a state space formula for the function H in the central intertwining lifting.

**THEOREM 5.2.** *Let* $\{Z, B, C\}$ *be a stable, controllable and observable anticausal realization of a function* F *in* $K^\infty(\mathcal{U}, \mathcal{Y})$. *Let* P *and* Q *be the controllability and observability grammians for the pairs* $\{Z, B\}$ *and* $\{C, Z\}$, *respectively. Then given any* $\delta > 1$, *the central interpolant* $G_\gamma$ *for the* $H^\infty$ *Nehari problem with tolerance* $\gamma = \delta d_\infty$ *is given by* $G_\gamma = F - H$ *where* H *is the function in* $H^\infty(\mathcal{U}, \mathcal{Y})$ *computed by*

$$H = CP(I - \lambda M)^{-1}(\gamma^2 I - Z^* QZP)^{-1} Z^* QB \quad \textit{and}$$

$$M = (\gamma^2 I - Z^* QZP)^{-1} Z^* (\gamma^2 I - QP)\,. \tag{5.7}$$

*Moreover,* $r_{spec}(M) < 1$. *Finally,* $\|F - H\|_\infty < \gamma$ *and* $G_\gamma$ *satisfies the* $H^2$ *bound in (5.6).*

Recall that G is a *Nehari interpolant* for the data F in $K^\infty(\mathcal{U}, \mathcal{Y})$ with tolerance $\gamma$, if $G = F - H$ for some H in $H^\infty(\mathcal{U}, \mathcal{Y})$ and $\|G\|_\infty \le \gamma$. Now let $\Delta(G)$ be the positive operator on $\mathcal{U}$ defined by

$$(\Delta(G)a, a) = \inf\{\|D_{B'}(a + \lambda h)\|^2 : h \in H^2(\mathcal{U})\} \tag{5.8}$$

where a is a fixed vector in $\mathcal{U}$ and $B' = M_G \,|\, H^2(\mathcal{U})$. As before, $D_{B'}$ is the positive square root of $\gamma^2 I - B'^* B'$. If $\mathcal{U}$ is finite dimensional, then the entropy of G is defined by

$$E(G) = \frac{1}{2\pi}\int_0^{2\pi} \ln\det\,[\gamma^2 I - G(e^{it})^* G(e^{it})]dt = \ln\det\,[\Delta(G)]\,. \tag{5.9}$$

The last equality follows from (IV.7.15). We say that $G_*$ is a *maximal interpolant* for the Nehari problem with tolerance $\gamma$, if $G_*$ is a Nehari interpolant with tolerance $\gamma$ and $\Delta(G_*) \ge \Delta(G)$ where G is any other Nehari interpolant with tolerance $\gamma$. The function $G_*$ is the *unique maximal* interpolant if $G_*$ is a maximal entropy interpolant and if $\Delta(G_*) = \Delta(G)$ for any other Nehari interpolant G with tolerance, $\gamma$, then $G_* = G$. The results in Section IV.7 show that if $G_*$ is a maximal interpolant and $\mathcal{U}$ is finite dimensional, then $E(G_*) \ge E(G)$ for all Nehari interpolants G with tolerance $\gamma$. If in addition $E(G) > -\infty$ for some Nehari interpolant G, then the maximal entropy Nehari interpolant $G_*$ is unique and is the maximal interpolant, that is, if $E(G_*) = E(G)$, then $G_* = G$. This sets the stage for the following result.

**THEOREM 5.3**. *Let* {Z, B, C} *be a stable, controllable and observable anticausal realization for a function* F *in* $K^\infty(\mathcal{U}, \mathcal{Y})$. *Let* P *and* Q *be the controllability and observability grammians for the pairs* {Z, B} *and* {C, Z}, *respectively. Then the central intertwining lifting* $G_\gamma = F - H$ *with tolerance* $\gamma$ *is the unique maximal interpolant for the Nehari problem and*

$$\Delta(G_\gamma) = \gamma^2[I + B^*(\gamma^2 I - QP)^{-1}QB]^{-1} . \tag{5.10}$$

*Moreover, if* $\mathcal{U}$ *is finite dimensional, then the entropy of* $G_\gamma$ *if given by*

$$E(G_\gamma) = \frac{1}{2\pi}\int_0^{2\pi} \ln\det[\gamma^2 I - G^*G]dt = -\ln\det[\gamma^{-2}(I + B^*(\gamma^2 I - QP)^{-1}QB)] . \tag{5.11}$$

*In particular, if* G *is any other Nehari interpolant with tolerance* $\gamma$, *then* $E(G_\gamma) \geq E(G)$ *with equality if and only if* $G_\gamma = G$.

The following result can be used to compute the unique optimal H solving the $H^\infty$ optimization problem when $\mathcal{U}$ is one dimensional.

**THEOREM 5.4**. *Let* {Z, B, C} *be a stable, controllable and observable, anticausal realization for a rational function* F *in* $K^\infty(\mathbb{C}, \mathcal{Y})$. *Let* P *and* Q *be the controllability and observability grammians for the pair* {Z, B} *and* {C, Z}, *respectively. Then the unique solution* $H_{opt}$ *in* $H^\infty(\mathbb{C}, \mathcal{Y})$ *to the* $H^\infty$ *optimization problem in (5.1) is computed by*

$$G_{opt} = F - H_{opt} = \frac{C(e^{it}I - Z)^{-1}Px}{B^*(I - e^{it}Z^*)^{-1}x} \tag{5.12}$$

*where* x *is the eigenvector corresponding to the largest eigenvalue* $\lambda_{max}$ *of* QP. *Moreover, the optimal* $H^\infty$ *error* $d_\infty^2 = \lambda_{max}$ *and* $g_{opt}^* g_{opt} = d_\infty^2$ *a.e.*

**PROOF OF THEOREM 5.1**. By choosing $\Theta_2 = I$ and $\Theta_1 = I$, it follows that the $H^\infty$ Nehari optimization problem is a special case of the two-sided Sarason problem discussed in Section I.10 or IV.9. By consulting Sections II.3 or III.6 concerning the $H^\infty$ Nehari problem or Section IV.9 for the two-sided Sarason result, let A be the Hankel operator from $H^2(\mathcal{U})$ into $K^2(\mathcal{Y})$ defined by $A = P_- M_F | H^2(\mathcal{U})$. Let T be the unilateral shift on $H^2(\mathcal{U})$ and V the forward bilateral shift on $L^2(\mathcal{Y})$. Finally, let $T'$ be the co-isometry on $\mathcal{H}' := K^2(\mathcal{Y})$ obtained by compressing V to $\mathcal{H}'$. Since $\mathcal{H}'$ is an invariant subspace for $V^*$, it follows that V is an isometric lifting of $T'$.

By consulting Section IV.9 with $\Theta_2 = I$ and $\Theta_1 = I$ or Section II.3, it follows that $B'$ is an intertwining lifting of A with respect to V if and only if $B' = M_G | H^2(\mathcal{U})$ where $G = F - H$ for some H in $H^\infty(\mathcal{U}, \mathcal{Y})$. In particular, Theorem IV.6.6 shows that the central intertwining lifting with tolerance $\gamma$ is given by

$$G_\gamma(e^{it}) = (AD_{\bar{A}}^{-2}\Pi_o)(e^{it})[(D_{\bar{A}}^{-2}\Pi_o)(e^{it})]^{-1} . \tag{5.13}$$

Here $\Pi_o$ is the operator from $H^2(\mathcal{U})$ onto $\mathcal{U}$ defined by $\Pi_o = P_{\mathcal{U}}$. So to prove Theorem 5.1 it is a simple matter of obtaining state space formulas for $AD_{\bar{A}}^{-2}$ and $D_{\bar{A}}^{-2}$ in (5.13).

To obtain a state space representation for the Hankel operator A, let $W_o$ be the observability operator from $\mathcal{X}$ into $K^2(\mathcal{Y})$ and $W_c$ the controllability operator from $H^2(\mathcal{U})$ into $\mathcal{X}$ defined respectively by

$$W_o x = C(e^{it}I - Z)^{-1}x \quad \text{and} \quad W_c^* x = B^*(I - \lambda Z^*)^{-1}x \qquad (x \in \mathcal{X}) .$$

We claim that $A = W_o W_c$. Before proving this let us establish that

$$T' W_o = W_o Z \quad \text{and} \quad ZW_c = W_c T . \tag{5.14}$$

The following verifies that $T' W_o = W_o Z$

$$T' W_o x = P_- VC(e^{it}I - Z)^{-1}x = P_- e^{it} \sum_{n=1}^{\infty} e^{-int} CZ^{n-1}x = \sum_{n=1}^{\infty} e^{-int} CZ^{n-1}Zx = W_o Zx .$$

Now using $T^*h = (h - h(0))/\lambda$ for h in $H^2(\mathcal{U})$ we have

$$T^* W_c^* x = \frac{B^*(I - \lambda Z^*)^{-1}x - B^*x}{\lambda} = B^*(I - \lambda Z^*)^{-1}Z^*x = W_c^* Z^* x .$$

Therefore $T^* W_c^* = W_c^* Z^*$, or equivalently, $ZW_c = W_c T$, which proves (5.14).

To show that $A = W_o W_c$ first notice that $W_c a = Ba$ for all constant functions a in $\mathcal{U}$. This along with the fact that {Z, B, C} is an anticausal realization of F gives

$$Aa = P_- Fa = P_- C(e^{it}I - Z)^{-1}Ba = W_o Ba = W_o W_c a \qquad (a \in \mathcal{U}) .$$

Hence $Aa = W_o W_c a$. Now using $T'A = AT$ along with (5.14) we have

$$AT^n a = T'^n Aa = T'^n W_o W_c a = W_o Z^n W_c a = W_o W_c T^n a .$$

Because $\mathcal{U}$ is cyclic for T it follows that $A = W_o W_c$.

Now let $P = W_c W_c^*$ and $Q = W_o^* W_o$. We claim that P and Q are the controllability and observability grammians for the pair {Z, B} and {C, Z} respectively. To see this, let x be in $\mathcal{X}$, then

$$(Px, x) = (W_c W_c^* x, x) = (W_c^* x, W_c^* x) = (\sum_0^\infty \lambda^n B^{*n} Z^{*n} x, \sum_0^\infty \lambda^n B^{*n} Z^{*n} x) =$$

$$\sum_0^\infty (B^{*n} Z^{*n} x, B^{*n} Z^{*n} x) = (\sum_0^\infty Z^n BB^* Z^{*n} x, x) .$$

Therefore $P = \Sigma Z^n BB^* Z^{*n}$ is precisely the controllability grammian for the pair $\{Z, B\}$; see Section III.2. A similar calculation shows that $Q = W_0^* W_0$ is the observability grammian for the pair $\{C, Z\}$.

Since F is in $K^\infty(\mathcal{U}, \mathcal{Y})$ it follows that $d_2 = \|F\|_2 = \|A \mid \mathcal{U}\|_{HS}$. So using the fact that $A \mid \mathcal{U} = W_0 W_c \mid \mathcal{U} = W_0 B$ we have

$$d_2^2 = \|W_0 B\|_2^2 = \text{trace}\,(B^* W_0^* W_0 B) = \text{trace}\,(B^* QB) .$$

This establishes the first equality in (5.4). To show that $d_\infty^2 = r_{spec}(QP)$, first recall that $d_\infty = \|A\|$. Now using $A = W_0 W_c$ we have

$$A^* A W_c^* = W_c^* W_0^* W_0 W_c W_c^* = W_c^* QP . \tag{5.15}$$

Because the pair $\{Z, B\}$ is controllable the operator, the operator $W_c^*$ is one to one and onto its range. Moreover, because the pair $\{C, Z\}$ is observable and $A = W_0 W_c$ it follows that $\ker(A)^\perp = H^2(\mathcal{U}) \ominus \ker(A)$ equals the range of $W_c^*$. So equation (5.15) shows that $A^* A \mid \ker(A)^\perp$ is similar to QP. In particular, $A^* A$ and QP have the same spectral radius. Therefore $d_\infty^2 = \|A\|^2 = r_{spec}(QP)$. This completes the proof of equation (5.4).

To complete the proof it remains to establish (5.5) from (5.13). Using $A = W_0 W_c$ and $Q = W_0^* W_0$ we have

$$D_A^2 = \gamma^2 I - A^* A = \gamma^2 I - W_c^* Q W_c .$$

Let us search for an inverse of $D_A^2$ of the form $\gamma^{-2} I + W_c^* R W_c$ where R is an operator on $\mathcal{X}$. Thus

$$I = D_A^2(\gamma^{-2} I + W_c^* R W_c) = (\gamma^2 I - W_c^* Q W_c)(\gamma^{-2} I + W_c^* R W_c) =$$

$$I + W_c^*[(\gamma^2 I - QP)R - \gamma^{-2} Q]W_c .$$

Choosing $R = \gamma^{-2}(\gamma^2 I - QP)^{-1} Q$ forces the last term to be zero. A similar calculation shows that for this choice of R we also have $(\gamma^{-2} I + W_c^* R W_c) D_A^2$ is the identity. Therefore the inverse of $D_A^2$ is given by

$$D_A^{-2} = \gamma^{-2} + \gamma^{-2} W_c^* (\gamma^2 I - QP)^{-1} Q W_c . \tag{5.16}$$

Using $A = W_0 W_c$ and $P = W_c W_c^*$ we obtain

$$AD_A^{-2} = \gamma^{-2}W_o[I + P(\gamma^2 I - QP)^{-1}Q]W_c =$$

$$\gamma^{-2}W_o[I + (\gamma^2 I - PQ)^{-1}PQ]W_c = \tag{5.17}$$

$$\gamma^{-2}W_o(\gamma^2 I - PQ)^{-1}[\gamma^2 I - PQ + PQ]W_c = W_o(\gamma^2 I - PQ)^{-1}W_c \, .$$

Finally, using $B = W_c\Pi_o^*$ in (5.16) and (5.17) we have

$$(AD_A^{-2}\Pi_o)(e^{it}) = W_o(\gamma^2 I - PQ)^{-1}B$$

$$(D_A^{-2}\Pi_o)(\lambda) = \gamma^{-2}I + \gamma^{-2}W_c^*(\gamma^2 I - QP)^{-1}QB \, . \tag{5.18}$$

This along with (5.13) establishes that $G_\gamma = ND^{-1}$ where N and D are given in (5.5). This completes the proof.

**REMARK 5.5.** Let $D_0$ be the positive square root of $I + B^*(\gamma^2 I - QP)^{-1}QB$ and $G_\gamma$ the central intertwining solution to the Nehari interpolation problem with toleration $\gamma$. Then a state space formula for the outer spectral factor $\Theta$ for the spectral density $\gamma^2 I - G_\gamma^* G_\gamma$ is given by

$$\Theta(\lambda) = \gamma D_0 - \gamma D_0 B^*(\gamma^2 I - QZPZ^*)^{-1}(I - \lambda M_o)^{-1}QB \tag{5.19}$$

where $M_o = (\gamma^2 I - QP)Z^*(\gamma^2 I - QZPZ^*)^{-1}$. Notice that $(\gamma^2 I - QP)^{-1}M_o = M(\gamma^2 I - QP)^{-1}$. So $M_o$ is similar to M and thus $M_o$ is stable. Now let $\Omega(\lambda) = (D_A^{-2}\Pi_o^*)(\lambda)$. Then using the operator identity $(I + bc^{-1}d)^{-1} = I - b(db + c)^{-1}d$ along with the Lyapunov equation for P in (5.18) we obtain

$$\Omega(\lambda)^{-1} = \gamma^2 I - \gamma^2 B^*[QBB^* + (\gamma^2 I - QP)(I - \lambda Z^*)]^{-1}QB =$$

$$\gamma^2 I - \gamma^2 B^*[\gamma^2 I - Q(P - BB^*) - \lambda(\gamma^2 I - QP)Z^*]^{-1}QB =$$

$$\gamma^2 I - \gamma^2 B^*[(I - \lambda M_o)(\gamma^2 I - QZPZ^*)]^{-1}QB =$$

$$\gamma^2 I - \gamma^2 B^*(\gamma^2 I - QZPZ^*)^{-1}(I - \lambda M_o)^{-1}QB \, .$$

Since $\Omega(0)^{1/2} = \gamma^{-1}D_0$ it follows that $\Theta(\lambda)$ in (5.19) is given by $\Theta(\lambda) = \Omega(0)^{1/2}\Omega(\lambda)^{-1}$. Therefore, $\Theta(\lambda)$ is the outer spectral factor for $\gamma^2 I - G_\gamma^* G_\gamma$; see Theorem IV.6.6. Because $M_o$ is stable, our analysis also shows that $\Omega(\lambda)^{-1}$ is in $H^\infty(\mathcal{U}, \mathcal{U})$. Clearly $\Omega$ is in $H^\infty(\mathcal{U}, \mathcal{U})$. So both $\Omega$ and $\Omega^{-1}$ are in $H^\infty(\mathcal{U}, \mathcal{U})$. Because $\Omega(0)^{1/2}\Omega(\lambda)^{-1}$ is the outer spectral factor for $\gamma^2 I - G_\gamma^* G_\gamma$ it follows that $\|G_\gamma\|_\infty < \gamma$.

**PROOF OF THEOREM 5.2.** Here we will use Corollary IV.3.3 to compute the central solution $G_\gamma$ with tolerance $\gamma$ for the Nehari interpolation problem. According to this corollary

$$G_\gamma = F - H = A\Pi_0^* + \Psi(\lambda)(I - \lambda T_A^*)^{-1} T_A^* \Pi_0^* \tag{5.20}$$

where $T_A^* = D_{AT}^{-2} T^* D_A^2$ and $\Pi_0$ is the operator from $H^2(\mathcal{U})$ onto $\mathcal{U}$ defined by $\Pi_0 = P_{\mathcal{U}}$. Using $W_c \Pi_0^* = B$ and $A = W_o W_c$ we see that $A\Pi_0^* = W_o B = F$. So by (5.20) the function H in $H^\infty(\mathcal{U}, \mathcal{Y})$ corresponding to the central solution $G_\gamma = F - H$ is given by

$$H = -\Psi(\lambda)(I - \lambda T_A^*)^{-1} T_A^* \Pi_0 . \tag{5.21}$$

Since $T^* \Pi_0^* = 0$ and $A = W_o W_c$, we have

$$T^* D_A^2 \Pi_0^* = -T^* W_c^* W_o^* W_o B = -W_c^* Z^* QB . \tag{5.22}$$

The last equation follows from $T^* W_c^* = W_c^* Z^*$; see (5.14).

Using $T' W_o = W_o Z$ and (5.14) we obtain

$$\begin{aligned} D_{AT}^2 W_c^* &= (\gamma^2 I - A^* T'^* T' A) W_c^* = \\ W_c^*(\gamma^2 I - W_o^* T'^* T' W_o P) &= W_c^*(\gamma^2 I - Z^* QZP) . \end{aligned} \tag{5.23}$$

Because the pair $\{Z, B\}$ is controllable, $W_c^*$ is one to one and has closed range. Obviously $D_{AT}^2$ is invertible. Moreover, equation (5.23) shows that $\mathcal{M} = W_c^* \mathcal{X}$ is an invariant subspace for $D_{AT}^2$. Since $D_{AT}^2$ is a self adjoint operator, $\mathcal{M}$ must be a reducing subspace for $D_{AT}^2$. Therefore $\gamma^2 I - Z^* QZP$ is similar to $D_{AT}^2 | \mathcal{M}$. So $\gamma^2 I - Z^* QZP$ is invertible, and (5.23) gives

$$D_{AT}^{-2} W_c^* = W_c^* (\gamma^2 I - Z^* QZP)^{-1} . \tag{5.24}$$

Using (5.24) in (5.22) along with $T_A^* = D_{AT}^{-2} T^* D_A^2$ we have

$$T_A^* \Pi_0^* = -W_c^* (\gamma^2 I - Z^* QZP)^{-1} Z^* QB . \tag{5.25}$$

Now using (5.24) and $T^* W_c = W_c Z^*$ we have

$$\begin{aligned} T_A^* W_c^* &= D_{AT}^{-2} T^* (\gamma^2 I - W_c^* Q W_c) W_c^* = D_{AT}^{-2} T^* W_c^* (\gamma^2 I - QP) = \\ D_{AT}^{-2} W_c^* Z^* (\gamma^2 I - QP) &= W_c^* (\gamma^2 I - Z^* QZP)^{-1} Z^* (\gamma^2 I - QP) = W_c^* M \end{aligned} \tag{5.26}$$

where M is the operator defined in (5.7). By Lemma IV.6.1 the operator $T_A^{*n} \to 0$ strongly as $n \to \infty$. Because the pair $\{Z, B\}$ is controllable, $W_c^*$ is one to one and has closed range. Thus equation (5.26) shows that $T_A^{*n} W_c^* = W_c^* M^n$ and $M^n \to 0$ strongly as $n \to \infty$. Therefore $r_{spec}(M) \le 1$ and M is stable if $\mathcal{X}$ is finite dimensional. By readjusting the coordinates we can always assume that $P = I$. (To see this replace Z, B and C by $P^{-1/2} Z P^{1/2}$, $P^{-1/2} B$ and $C P^{1/2}$, respectively.) By implementing Proposition 1.7 to $M^*$ with $P = I$ we see that M is stable. Finally using (5.26) once again we obtain

$$(I - \lambda T_A^*)^{-1} W_c^* = W_c^* (I - \lambda M)^{-1} . \tag{5.27}$$

Combining (5.25) and (5.27) yields

$$(I - \lambda T_A^*)^{-1} T_A^* \Pi_o^* = - W_c^* (I - \lambda M)^{-1} (\gamma^2 I - Z^* QZP)^{-1} Z^* QB . \tag{5.28}$$

Notice that for x in $\mathcal{X}$ we have

$$(V - T')W_o x = (I - P_-) e^{it} W_o x = Cx \tag{5.29}$$

where $P_-$ is the orthogonal projection onto $K^2(\mathcal{Y})$. So using the definition of $\Psi(\lambda)$ we obtain

$$(\Psi h)(\lambda) = (V - T')Ah = (V - T')W_o W_c h = CW_c h \qquad (h \in H^2(\mathcal{U})) . \tag{5.30}$$

Combining this with (5.28) and (5.21) we obtain the result that we have been looking for, that is,

$$H(\lambda) = CP(I - \lambda M)^{-1} (\gamma^2 I - Z^* QZP)^{-1} Z^* QB . \tag{5.31}$$

This is precisely the form of H given in Theorem 5.2, which completes the proof.

We will leave to the reader as a tedious exercise to prove Theorem 5.2 as a corollary of Theorem 5.1. In this case one can use Proposition 1.7 to prove the stability of M.

**PROOF OF THEOREM 5.3**. Equation (5.18) along with $W_c \Pi_o^* = B$ yields

$$\Pi_o D_A^{-2} \Pi_o^* = \gamma^{-2} [I + B^* (\gamma^2 I - QP)^{-1} QB] .$$

This along with Theorems IV.7.1 and Corollary IV.7.7 readily establishes (5.10), (5.11) and completes this proof.

**PROOF OF THEOREM 5.4**. Because F is rational, its minimal anticausal realization {Z, B, C} is finite dimensional. Thus $A = W_o W_c$ is a finite rank Hankel operator. In particular, A attains its norm. According to Corollary IV.2.8, there is a unique intertwining lifting $B'$ of A satisfying $\|B'\| = \|A\| = d_\infty$. Moreover, this $B'$ is given by $B' = M_{G_{opt}} | H^2$ where $G_{opt}$ is the function in $L^\infty(\mathbb{C}, \mathcal{Y})$ computed by $G_{opt} = Ah/h$ and h is any vector which attains the norm of A. Since there is a one to one correspondence between the set of all intertwining liftings $B'$ of A and the set of all functions G in $L^\infty(\mathbb{C}, \mathcal{Y})$ of the form $G = F - H$ with H in $H^\infty(\mathbb{C}, \mathcal{Y})$, (that is, $B' = M_G | H^2$) we see that there is a unique solution to the $H^\infty$ Nehari optimization problem in (5.1) and this unique solution is given by $F - H_{opt} = G_{opt} = Ah/h$ for some unique $H_{opt}$ in $H^\infty(\mathbb{C}, \mathcal{Y})$.

Recall that $W_c^*$ maps $\mathcal{X}$ one to one and onto $\ker (A)^\perp$. According to (5.15) the operator $A^* A$ and QP have the same nonzero eigenvalues. So let x be the eigenvector corresponding to the largest eigenvalue $\lambda_{max}$ of QP. Because $\lambda_{max}$ is also the largest eigenvalue of $A^*A$, it follows that $d_\infty^2 = ||A||^2 = \lambda_{max}$. Using (5.15) once again we obtain for $h = W_c^* x$

$$A^* Ah = W_c^* QPx = \lambda_{max} W_c^* x = d_\infty^2 h \,.$$

Thus h attains the norm of A. So by Corollary IV.2.8 the optimal $H^\infty$ solution to the $H^\infty$ Nehari optimization problem is given by

$$G_{opt} = \frac{Ah}{h} = \frac{W_o W_c W_c^* x}{W_c^* x} = \frac{W_o P x}{W_c^* x} \,.$$

This is precisely the form of $G_{opt} = F - H_{opt}$ given in (5.12), which completes the proof.

## V.6 CENTRAL NUDELMAN SOLUTIONS

In this section we will give a state space formula to compute the central intertwining lifting for the two-sided Nudelman problem with tolerance $\gamma$. We will also present a state space formula to compute the $H^\infty$ optimal solution to a scalar valued finite dimensional Nudelman problem.

To begin, recall that the data for the two-sided Nudelman problem are the Hilbert space operators

$$\begin{gathered} Z : \mathcal{X} \to \mathcal{X} \,,\ B : \mathcal{U} \to \mathcal{X} \,,\ \tilde{B} : \mathcal{Y} \to \mathcal{X} \\ \Lambda : \mathcal{F} \to \mathcal{F} \,,\ C : \mathcal{F} \to \mathcal{Y} \,,\ \tilde{C} : \mathcal{F} \to \mathcal{U} \,,\ \Gamma : \mathcal{F} \to \mathcal{X} \end{gathered} \tag{6.1}$$

where both Z and $\Lambda$ have spectral radius strictly less than one. Furthermore, we always assume that the operator $\Gamma$ satisfies the Sylvester equation

$$Z\Gamma - \Gamma\Lambda = \tilde{B}C - B\tilde{C} \,. \tag{6.2}$$

Recall that $W_c$ from $\ell_+^2(\mathcal{U})$ into $\mathcal{X}$ and $\tilde{W}_c$ from $\ell_+^2(\mathcal{Y})$ into $\mathcal{X}$ are the controllability operators generated by the pairs {Z, B} and {Z, $\tilde{B}$} respectively, that is,

$$W_c = [B, ZB, Z^2B, \ldots] \quad \text{and} \quad \tilde{W}_c = [\tilde{B}, Z\tilde{B}, Z^2\tilde{B}, \ldots] \,. \tag{6.3}$$

Moreover, $W_o$ from $\mathcal{X}$ into $\ell_-^2(\mathcal{Y})$ and $\tilde{W}_o$ from $\mathcal{X}$ into $\ell_-^2(\mathcal{U})$ are the observability operators generated by {C, $\Lambda$} and {$\tilde{C}$, $\Lambda$}, respectively, that is,

$$W_o = [\ldots, C\Lambda^2, C\Lambda, C]^{tr} \quad \text{and} \quad \tilde{W}_o[\ldots, \tilde{C}\Lambda^2, \tilde{C}\Lambda, \tilde{C}]^{tr} \tag{6.4}$$

where tr denotes the transpose. Then the two-sided Nudelman problem with tolerance $\gamma$ is to find (if possible) a function G in $H^\infty(\mathcal{Y}, \mathcal{U})$ satisfying $\|G\|_\infty \le \gamma$ and the interpolation constraints

$$W_c T_G = \tilde{W}_c, \; T_G^{\#} W_o = \tilde{W}_o \quad \text{and} \quad \Gamma = W_c H_G W_o \,. \tag{6.5}$$

Recall that if G is any function in $H^\infty(\mathcal{Y}, \mathcal{U})$, then $T_G^{\#}$, the Toeplitz operator $T_G$ and the Hankel operator $H_G$ are obtained by partitioning the Laurent operator $L_G$ in the following manner

$$L_G = \begin{bmatrix} T_G^{\#} & 0 \\ H_G & T_G \end{bmatrix} : \ell_-^2(\mathcal{Y}) \oplus \ell_+^2(\mathcal{Y}) \to \ell_-^2(\mathcal{U}) \oplus \ell_+^2(\mathcal{U}) \,. \tag{6.6}$$

Associated with the Nudelman problem is the following $H^\infty$ optimization problem

$$d_\infty = \inf \{ \|G\|_\infty : G \in H^\infty(\mathcal{Y}, \mathcal{U}) \quad \text{and} \quad (6.5) \text{ holds} \} \,. \tag{6.7}$$

In Section II.5 we used the commutant lifting theorem to show that if {Z, B} is controllable and {C, Λ} is observable, then there exists a G in $H^\infty(\mathcal{Y}, \mathcal{U})$ solving this optimization problem, that is, $\|G\|_\infty = d_\infty$ and (6.5) holds.

Recall that the controllability grammians P and $\tilde{P}$ for the pair {Z, B} and {Z, $\tilde{B}$} are respectively given by $P = W_c W_c^*$ and $\tilde{P} = \tilde{W}_c \tilde{W}_c^*$. Moreover, the observability grammians Q and $\tilde{Q}$ for the pairs {C, Λ} and {$\tilde{C}$, Λ} are respectively given by $Q = W_o^* W_o$ and $\tilde{Q} = \tilde{W}_o^* \tilde{W}_o$. Furthermore, P, $\tilde{P}$, Q and $\tilde{Q}$ can be obtained by solving the following Lyapunov equations

$$\begin{gathered} P = ZPZ^* + BB^* \quad \text{and} \quad \tilde{P} = Z\tilde{P}Z^* + \tilde{B}\tilde{B}^* \\ Q = \Lambda^* Q \Lambda + C^* C \quad \text{and} \quad \tilde{Q} = \Lambda^* \tilde{Q} \Lambda + \tilde{C}^* \tilde{C} \,. \end{gathered} \tag{6.8}$$

Throughout we assume that the pair {Z, B} is controllable and {C, Λ} is observable, or equivalently, both P and Q are invertible. For simplicity of presentation we will also assume that $\tilde{P}$ has closed range. Under these assumptions Section II.5 shows that the error in the $H^\infty$ Nudelman optimization problem is given by $d_\infty^2 = r_{spec}(R_\infty)$ where $R_\infty$ is the block operator on $\mathcal{F} \oplus \mathcal{X}$ defined by

$$R_\infty = \begin{bmatrix} Q^{-1}(\tilde{Q} + \Gamma^* P^{-1} \Gamma) & Q^{-1} \Gamma^* P^{-1} \tilde{P} \\ P^{-1} \Gamma & P^{-1} \tilde{P} \end{bmatrix} . \tag{6.9}$$

The following $2 \times 2$ block operators M and N on $\mathcal{F} \oplus \mathcal{X}$ are used to compute the central *intertwining* lifting for the two-sided Nudelman interpolation problem:

$$M = \begin{bmatrix} \gamma^2 Q - \Lambda^* \tilde{Q} \Lambda - (Z\Gamma + B\tilde{C})^* P^{-1}(Z\Gamma + B\tilde{C}), & -(Z\Gamma + B\tilde{C})^* P^{-1} Z\tilde{P} \\ -Z^* P^{-1}(Z\Gamma + B\tilde{C}), & \gamma^2 I - Z^* P^{-1} Z\tilde{P} \end{bmatrix}$$

$$N = \begin{bmatrix} \Lambda^*(\gamma^2 Q - \tilde{Q}) - (Z\Gamma + B\tilde{C})^* P^{-1}\Gamma, & C^* \tilde{B}^*(\gamma^2 I - P^{-1}\tilde{P}) - \Lambda^* \Gamma^* P^{-1}\tilde{P} \\ -Z^* P^{-1}\Gamma & Z^*(\gamma^2 I - P^{-1}\tilde{P}) \end{bmatrix} \tag{6.10}$$

Now we are ready to state the main results of this section.

**THEOREM 6.1**. *Let* $Z$, $B$, $\tilde{B}$, $\Lambda$, $C$, $\tilde{C}$ *and* $\Gamma$ *be the data for the two-sided Nudelman interpolation problem where* $\Gamma$ *satisfies the Sylvester equation in (6.2). Assume that the pair* $\{Z, B\}$ *is controllable,* $\{C, \Lambda\}$ *is observable and the controllability grammian* $\tilde{P}$ *for the pair* $\{Z, \tilde{B}\}$ *has closed range. Then the optimal error* $d_\infty$ *in the* $H^\infty$ *Nudelman optimization problem (6.7) is given by* $d_\infty^2 = r_{spec}(R_\infty)$ *where* $R_\infty$ *is defined in (6.9). Moreover, given any* $\gamma > d_\infty$ *the central intertwining solution* $G_\gamma$ *in* $H^\infty(\mathcal{Y}, \mathcal{U})$ *with tolerance* $\gamma$ *is given by*

$$G_\gamma = D_n + \lambda C_n (I - \lambda A_n)^{-1} B_n \tag{6.11}$$

*where* $A_n = M^{-1}N$ *and* $B_n$ *from* $\mathcal{Y}$ *to* $\mathcal{F} \oplus \mathcal{X}$ *and* $C_n$ *from* $\mathcal{F} \oplus \mathcal{X}$ *to* $\mathcal{U}$ *and* $D_n$ *from* $\mathcal{Y}$ *to* $\mathcal{U}$ *are the block operators computed by*

$$B_n = M^{-1} \begin{bmatrix} \gamma^2 C^* - (Z\Gamma + B\tilde{C})^* P^{-1}\tilde{B} \\ -Z^* P^{-1}\tilde{B} \end{bmatrix}$$

$$C_n = [\tilde{C} - B^* P^{-1}(Z\Gamma + B\tilde{C}), -B^* P^{-1} Z\tilde{P}] A_n + [B^* P^{-1}\Gamma, -B^*(\gamma^2 I - P^{-1}\tilde{P})] \tag{6.12}$$

$$D_n = B^* P^{-1}\tilde{B} + [\tilde{C} - B^* P^{-1}(Z\Gamma + B\tilde{C}), -B^* P^{-1} Z\tilde{P}] B_n$$

*Finally, the state square operator* $A_n$ *is stable.*

One can also compute the operators $A_n$, $B_n$, $C_n$ and $D_n$ from equations (6.31), (6.32) and (6.39) below, where $R$, $T_1$, $T_2$ are given by (6.19), (6.24), (6.26) and $P_1 = \text{diag}\,[Q, \tilde{P}]$ while $P_2 = \text{diag}\,[\tilde{Q}, P]$. In this form the operators $A_n$ and $B_n$ resemble the corresponding state space operators for the central solution to the Nevanlinna-Pick and Nehari interpolation problems presented in Theorems 1.2 and 5.2.

**REMARK 6.2**. If $\mathcal{Y}$ is finite dimensional, then one readily obtains the following $H^2$ Nudelman optimization problem associated with:

$$d_2 = \inf \{\|G\|_2 : G \in H^2(\mathcal{Y}, \mathcal{U}) \text{ and } (6.5) \text{ holds}\} . \tag{6.13}$$

By letting $\gamma$ approach infinity in the previous theorem, or by using standard $H^2$ optimization techniques we will show that the unique solution $G_{2*}$ to the $H^2$ optimization problem in (6.13) is given by

$$G_{2*} = C_2(I - \lambda A_2)^{-1} B_2 \tag{6.14}$$

where the system $\{A_2 \text{ on } \mathcal{F} \oplus \mathcal{X}, B_2, C_2\}$ is the stable system defined by

$$A_2 = \begin{bmatrix} \Lambda^* & 0 \\ P^{-1}\Gamma Q^{-1} - P^{-1}(Z\Gamma + B\tilde{C})Q^{-1}\Lambda^* , & Z^* \end{bmatrix}$$

$$B_2 = \begin{bmatrix} C^* \\ P^{-1}\tilde{B} - P^{-1}(Z\Gamma + B\tilde{C})Q^{-1}C^* \end{bmatrix} \quad \text{and} \quad C_2 = [\tilde{C}Q^{-1} , \ B^*] \tag{6.15}$$

Clearly $G_{2*}$ is in $H^\infty(\mathcal{Y}, \mathcal{U})$. Moreover, the error $d_2$ to the $H^2$ Nudelman problem is given by

$$d_2^2 = \|G_{2*}\|_2^2 = \text{trace } (B_2^* Q_2 B_2) \tag{6.16}$$

where $Q_2$ is the observability grammian for the pair $\{C_2, A_2\}$. Finally, if $\delta > 1$ and $\gamma = \delta d_\infty$, then the central intertwining lifting $G_\gamma$ computed by (6.11) satisfies $\|G_\gamma\| \le \delta d_\infty$ and the following $H^2$ bound

$$\|G_\gamma\|_2 \le \delta d_2 \left[ \delta^2 - 1 + \frac{d_2^2}{\dim(\mathcal{Y}) d_\infty^2} \right]^{-1/2} . \tag{6.17}$$

The following result solves the $H^\infty$ Nudelman optimization problem when $\mathcal{Y}$ is one dimensional.

**THEOREM 6.3.** *Let* Z, B, $\tilde{B}$, Λ, C, $\tilde{C}$ *and* Γ *be the data for the two-sided Nudelman interpolation problem where* $\mathcal{Y} = \mathbb{C}$, *and both* $\mathcal{F}$ *and* $\mathcal{X}$ *are finite dimensional. Assume that the pair* $\{Z \text{ on } \mathcal{X}, B\}$ *is controllable and* $\{C, \Lambda \text{ on } \mathcal{F}\}$ *is observable. Then there is a unique solution* $g_{opt}$ *to the* $H^\infty$ *Nudelman optimization problem in (6.7) and this* $g_{opt}$ *is the function given by*

$$g_{opt} = \frac{\tilde{C}(\lambda I - \Lambda)^{-1} f + B^*(I - \lambda Z^*)^{-1} P^{-1}(\Gamma f + \tilde{P}x)}{C(\lambda I - \Lambda)^{-1} f + \tilde{B}^*(I - \lambda Z^*)^{-1} x} \tag{6.18}$$

*where* $f \oplus x$ *in* $\mathcal{F} \oplus \mathcal{X}$ *is the eigenvector corresponding to the largest eigenvalue* $\lambda_{max}$ *of the* $2 \times 2$ *block matrix* $R_\infty$ *defined in (6.9). Moreover, the optimal* $H^\infty$ *error* $d_\infty^2 = \lambda_{max}$ *and* $g_{opt}/d_\infty$

*is an inner function.*

**PROOF OF THEOREM 6.1.** Here we will use Corollary IV.3.3 along with our notation established in Section II.5 to construct the central intertwining lifting $G_\gamma$ of A. To this end, let us first establish some notation. Let R be the $2 \times 2$ block matrix on $\mathcal{F} \oplus \mathcal{X}$ defined by

$$R = \begin{bmatrix} Q^{-1} & 0 \\ P^{-1}\Gamma Q^{-1} & P^{-1} \end{bmatrix}. \tag{6.19}$$

Recall that $\mathcal{H}_o$ is the range of $W_o$ while $\mathcal{H}_c$ is the range of $W_c^*$. (The range of both $W_o$ and $W_c^*$ are closed because Q and P are invertible.) Let $W_1$ from $\mathcal{H} = \mathcal{H}_o \oplus \ell_+^2(\mathcal{Y})$ into $\mathcal{F} \oplus \mathcal{X}$ and $W_2$ from $\mathcal{F} \oplus \mathcal{X}$ into $\ell_-^2(\mathcal{U}) \oplus \mathcal{H}_c = \mathcal{H}'$ be the operators defined by

$$W_1 = \begin{bmatrix} W_o^* & 0 \\ 0 & \tilde{W}_c \end{bmatrix} \quad \text{and} \quad W_2 = \begin{bmatrix} \tilde{W}_o & 0 \\ 0 & W_c^* \end{bmatrix}. \tag{6.20}$$

Now let A be the operator from $\mathcal{H}$ into $\mathcal{H}'$ defined by $A = W_2 R W_1$. Let T on $\mathcal{H}$ be the isometry and T′ on $\mathcal{H}'$ the co-isometry defined by $T = V \mid \mathcal{H}$ and $T' = P_{\mathcal{H}'} V' \mid \mathcal{H}'$ where V and V′ are the bilateral shifts on $\ell^2(\mathcal{Y})$ and $\ell^2(\mathcal{U})$, respectively. Then according to the results in Section II.5 we have $T'A = AT$. Moreover, a function G in $H^\infty(\mathcal{Y}, \mathcal{U})$ is a solution to the Nudelman interpolation problem if and only if $B' = L_G \mid \mathcal{H}$ is an intertwining lifting of A with respect to V′ and the tolerance γ. Notice that this function G is given by the Fourier transform of $B' \mid \mathcal{Y}$. By Corollary IV.3.3 along with the fact that an intertwining lifting for the Nudelman problem defines a function in $H^\infty(\mathcal{Y}, \mathcal{U})$, the central intertwining lifting with tolerance γ is given by $B_\gamma = L_{G_\gamma} \mid \mathcal{H}$ where

$$G_\gamma = \mathcal{F}_{\mathcal{U}} A \Pi_o^* + \Psi(1 - \lambda T_A^*)^{-1} T_A^* \Pi_o^*. \tag{6.21}$$

Here $\mathcal{F}_{\mathcal{U}}$ is the Fourier transform from $\ell^2(\mathcal{U})$ onto $L^2(\mathcal{U})$ and $(\Psi h)(\lambda) = \mathcal{F}_{\mathcal{U}}(V' - T')Ah$ for all h in $\mathcal{H}$. Moreover, $T_A^* = D_{AT}^{-2} T^* D_A^2$ and $\Pi_o$ is the operator from $\mathcal{H}$ onto $\mathcal{Y}$ defined by $\Pi_o = P_{\mathcal{Y}}$. To complete the proof, we convert the $G_\gamma$ in (6.21) to to the state space formula in Theorem 6.1.

Let f be in $\mathcal{F}$ and g in $\ell_+^2(\mathcal{Y})$. Then using $T = V \mid \mathcal{H}$ we see that

$$T(W_o f \oplus g) = W_o A f \oplus (Cf + Sg) \quad (f \in \mathcal{F} \text{ and } g \in \ell_+^2(\mathcal{Y})) \tag{6.22}$$

where S is the unilateral shift on $\ell_+^2(\mathcal{Y})$. Using (6.22) or the fact that $P_{\mathcal{H}_o} = W_o Q^{-1} W_o^*$ and $P_{\mathcal{H}} V^* = T^* P_{\mathcal{H}}$ we will show that

$$T^*\begin{bmatrix} W_o f \\ g \end{bmatrix} = \begin{bmatrix} W_o Q^{-1}(\Lambda^* Qf + C^*\Pi_o g) \\ S^* g \end{bmatrix}. \tag{6.23}$$

To verify (6.23) let $\Pi_{-1}$ be the operator from $\ell_-^2(\mathcal{Y})$ onto $\mathcal{Y}$ defined by $\Pi_{-1}[\ldots, y_{-2}, y_{-1}] = y_{-1}$. Then

$$T^*(W_o f \oplus g) = P_{\mathcal{H}} V^*(W_o f \oplus g) = P_{\mathcal{H}_o}(V^* W_o f + \Pi_{-1}^* \Pi_o g) \oplus S^* g =$$

$$W_o Q^{-1} W_o^*(V^* W_o f + \Pi_{-1}^* \Pi_o g) \oplus S^* g = W_o Q^{-1}(\Lambda^* Qf + C^* \Pi_o g) \oplus S^* g$$

which proves (6.23). In particular, using $S^* \tilde{W}_c^* = \tilde{W}_c^* Z^*$ we see that $T^* W_1^* = W_1^* T_1^*$ where $T_1$ is the $2 \times 2$ block stable matrix on $\mathcal{F} \oplus \mathcal{X}$ defined by

$$T_1 = \begin{bmatrix} Q\Lambda Q^{-1} & 0 \\ \tilde{B} C Q^{-1} & Z \end{bmatrix}. \tag{6.24}$$

Now let $k = [\ldots, k_{-3}, k_{-2}, k_{-1}]^{tr}$ be in $\ell_-^2(\mathcal{U})$ where tr denotes the transpose and x be in $\mathcal{X}$, then we claim that

$$T'\begin{bmatrix} k \\ W_c^* x \end{bmatrix} = \begin{bmatrix} P_- V' k \\ W_c^*(P^{-1} B k_{-1} + P^{-1} Z P x) \end{bmatrix} \qquad (k \in \ell_-^2(\mathcal{U}) \text{ and } x \in \mathcal{X}). \tag{6.25}$$

This follows from the fact that $P_{\mathcal{H}'} V' = T' P'_{\mathcal{H}'}$ and $P_{\mathcal{H}_c} = W_c^* P^{-1} W_c$, that is,

$$T'(k \oplus W_c^* x) = P_{\mathcal{H}'} V'(k \oplus W_c^* x) = P_- V' k \oplus P_{\mathcal{H}_c}(k_{-1} + V' W_c^* x) =$$

$$P_- V' k \oplus W_c^* P^{-1} W_c (k_{-1} + V' W_c^* x) = P_- V' k \oplus W_c^* P^{-1}(B k_{-1} + Z P x).$$

This readily gives (6.25). Now using (6.25) we readily see that $T' W_2 = W_2 T_2$ where $T_2$ is the stable $2 \times 2$ block matrix on $\mathcal{F} \oplus \mathcal{X}$ defined by

$$T_2 = \begin{bmatrix} \Lambda & 0 \\ P^{-1} B \tilde{C} & P^{-1} Z P \end{bmatrix}. \tag{6.26}$$

Finally, by using the Sylvester equation (6.2) it is easy to verify that $T_2 R = R T_1$.

Now let $P_1 = W_1 W_1^* = \operatorname{diag}[Q, \tilde{P}]$ and $P_2 = W_2^* W_2 = \operatorname{diag}[\tilde{Q}, P]$. Using $A = W_2 R W_1$ it follows that $A^* A = W_1^* R^* P_2 R W_1$. Recall that if M from $\mathcal{M}$ into $\mathcal{N}$ and N from $\mathcal{N}$ into $\mathcal{M}$ are two bounded linear operators, then $r_{spec}(MN) = r_{spec}(NM)$; see formula (3) in Section III.2 of Gohberg-Goldberg-Kaashoek [1]. So using this and $W_1 W_1^* = P_1$ we have

$$\|A\|^2 = r_{spec}(A^* A) = r_{spec}(W_1^* R^* P_2 R W_1) = r_{spec}(R^* P_2 R W_1 W_1^*) = r_{spec}(R^* P_2 R P_1).$$

Therefore the error $d_\infty$ in the Nudelman optimization problem is given by

$$d_\infty^2 = ||A||^2 = r_{spec}(R^* P_2 R P_1) .$$

A simple calculation shows that $R_\infty = R^* P_2 R P_1$ which proves part of Theorem 6.1. Finally, let us notice that $T'A = W_2 T_2 R W_1$. In particular, $A^* T'^* T' A^* = W_1^* R^* T_2^* P_2 T_2 R W_1$. By repeating the above argument we see that $||T'A||^2$ equals the spectral radius of $R^* T_2^* P_2 T_2 R P_1$. Since $||T'A|| \leq ||A|| = d_\infty$, this readily shows that if $\gamma > d_\infty$, then $\gamma^2 I - R^* T_2^* P_2 T_2 R P_1$ is invertible. This fact will be used in constructing $A_n$ and $B_n$.

Now using $D_A^2 = \gamma^2 I - W_1^* R^* P_2 R W_1$ and $T^* W_1^* = W_1^* T_1^*$ along with the expression for $T^*$ in (6.23) we arrive at

$$T^* D_A^2 \Pi_o^* = W_1^* \begin{bmatrix} \gamma^2 Q^{-1} C^* \\ 0 \end{bmatrix} - W_1^* T_1^* R^* P_2 R \begin{bmatrix} 0 \\ \tilde{B} \end{bmatrix} . \tag{6.27}$$

To compute $T_A^* = D_{AT}^{-2} T^* D_A^2$ we need an expression of $D_{AT}^{-2}$. Since $AT = T'A = W_2 T_2 R W_1$ we have

$$D_{AT}^2 = \gamma^2 I - W_1^* R^* T_2^* P_2 T_2 R W_1 . \tag{6.28}$$

So we look for an inverse of $D_{AT}^2$ of the form $\gamma^{-2} I + W_1^* L W_1$. In this case

$$I = (\gamma^2 I - W_1^* R^* T_2^* P_2 T_2 R W_1)(\gamma^{-2} I + W_1^* L W_1) =$$

$$I + W_1^* [-\gamma^{-2} R^* T_2^* P_2 T_2 R + (\gamma^2 I - R^* T_2^* P_2 T_2 R P_1) L] W_1 .$$

By choosing L to make the term in the brackets [ ] zero we see that the inverse of $D_{AT}^2$ is given by

$$D_{AT}^{-2} = \gamma^{-2} I + \gamma^{-2} W_1^* (\gamma^2 I - R^* T_2^* P_2 T_2 R P_1)^{-1} R^* T_2^* P_2 T_2 R W_1 . \tag{6.29}$$

Our previous analysis shows that the inverse in (6.29) is well defined as long as $\gamma > d_\infty$. A straight forward calculation using (6.29) shows that

$$D_{AT}^{-2} W_1^* = W_1^* (\gamma^2 I - R^* T_2^* P_2 T_2 R P_1)^{-1} . \tag{6.30}$$

This and (6.27) shows that $T_A^* \Pi_o^* = D_{AT}^{-2} T^* D_A^2 \Pi_o^* = W_1^* B_n$ where $B_n$ is defined by

$$B_n = (\gamma^2 I - R^* T_2^* P_2 T_2 R P_1)^{-1} \begin{bmatrix} \gamma^2 Q^{-1} C^* \\ 0 \end{bmatrix} - (\gamma^2 I - R^* T_2^* P_2 T_2 R P_1)^{-1} T_1^* R^* \begin{bmatrix} 0 \\ \tilde{B} \end{bmatrix} . \tag{6.31}$$

Using (6.30) once again we have $T_A^* W_1^* = W_1^* A_n$ where $A_n$ is the operator on $\mathcal{F} \oplus \mathcal{X}$ defined by

$$A_n = (\gamma^2 I - R^* T_2^* P_2 T_2 R P_1)^{-1} T_1^* (\gamma^2 I - R^* P_2 R P_1) . \tag{6.32}$$

We claim that T is a unilateral shift on $\mathcal{H}$. To see this notice that (6.23) along with the stability of $\Lambda$ shows $T^{*k}h$ approaches zero as k approaches infinity for all h of the form $W_o f \oplus g$ where g has compact support in $\ell_+^2(\mathcal{Y})$. Because these h's form a dense set in $\mathcal{H}$ we see that $T^{*k}$ strongly approaches zero. So T is a unilateral shift. By consulting Lemma IV.6.1 the operator $T_A^*$ is pointwise stable. (Recall that an operator C is pointwise stable, if $C^n$ approaches zero pointwise as n approaches infinity.)

Now we claim that $A_n$ is pointwise stable. To see this first notice that $\ker W_1^*$ is an invariant subspace for $A_n$. If h is in $\ker W_1^*$, then $W_1^* A_n h = T_A^* W_1^* h = 0$, and thus $A_n h$ is also in $\ker W_1^*$. The space $\ker W_1^*$ is also invariant for $T_1^*$. (Use $T^* W_1^* = W_1^* T_1^*$.) Since $P_1$ and $W_1^*$ have the same kernel equation (6.32) shows that $A_n \mid \ker W_1^* = T_1^* \mid \ker W_1^*$. Because $T_1^*$ is stable it follows that $A_n \mid \ker W_1^*$ is also stable. To show that $A_n$ is pointwise stable assume $\mathcal{M} = \ker (W_1^*)^{\perp}$ and recall that by assumption $\tilde{P}$ has closed range. So $W_1^* \mid \mathcal{M}$ is a bounded invertible transformation from $\mathcal{M}$ onto $\operatorname{ran} W_1^*$. Since $\mathcal{M}^{\perp}$ is invariant for $A_n$ we have $P_{\mathcal{M}} A_n = P_{\mathcal{M}} A_n P_{\mathcal{M}}$. Because $T_A^* W_1^* = W_1^* A_n$ the operator $T_A^* \mid \operatorname{ran} W_1^*$ is similar to $P_{\mathcal{M}} A_n \mid \mathcal{M}$. According to Lemma IV.6.1, $T_A^*$ is pointwise stable and thus $P_{\mathcal{M}} A_n \mid \mathcal{M}$ is pointwise stable. Now by using the decomposition of $A_n$ on $\mathcal{M}^{\perp} \oplus \mathcal{M}$ along with the fact that $A_n \mid \mathcal{M}^{\perp}$ is stable, it follows that $A_n$ is pointwise stable. By using Proposition 1.7 one can also show that $A_n$ is stable.

Using $T_A^* W_1^* = W_1^* A_n$ it follows that $(I - \lambda T_A^*)^{-1} W_1^* = W_1^* (I - \lambda A_n)^{-1}$ holds for all $|\lambda| < 1$. Combining this with $W_1^* B_n = T_A^* \Pi_o$ and $A = W_2 R W_1$ we have

$$A(I - \lambda T_A)^{-1} T_A^* \Pi_o^* = W_2 R P_1 (I - \lambda A_n)^{-1} B_n . \tag{6.33}$$

Now let $\hat{W}_c^* = \mathcal{F}_{\mathcal{U}} W_c^* = B^*(I - \lambda Z^*)^{-1}$, where $\mathcal{F}_{\mathcal{U}}$ is the Fourier transform. Notice that $\hat{W}_c^*$ maps $\mathcal{X}$ into $H^2(\mathcal{U})$. Using $T' W_2 = W_2 T_2$ we have

$$\mathcal{F}_{\mathcal{U}} (V' - T') W_2 = \mathcal{F}_{\mathcal{U}} \begin{bmatrix} \tilde{W}_o \Lambda & 0 \\ \tilde{C} & V' W_c^* \end{bmatrix} - \mathcal{F}_{\mathcal{U}} W_2 T_2 =$$

$$[\tilde{C}, 0] + [0, \hat{W}_c^*] \left[ \begin{bmatrix} \Lambda & 0 \\ 0 & \lambda I \end{bmatrix} - T_2 \right] . \tag{6.34}$$

Combining (6.21), (6.33), (6.34) along with $\mathcal{F}_{\mathcal{U}} A \Pi_o^* = [0, \hat{W}_c^*] R [0, \tilde{B}]^{tr}$ we see that the central intertwining lifting $G_\gamma$ of A is given by

$$G_\gamma = [0, \hat{W}_c^*]R[0, \tilde{B}]^{tr} - [0, \hat{W}_c^*]T_2RP_1(I - \lambda A_n)^{-1}B_n$$
$$\lambda[0, \hat{W}_c^*]RP_1(I - \lambda A_n)^{-1}B_n + [\tilde{C}, 0]RP_1(I - \lambda A_n)^{-1}B_n\ . \tag{6.35}$$

In order to simplify the previous expression for $G_\gamma$ we will factor $(I - \lambda A_n)^{-1}B_n$ out on the right. To this end, notice that the first term in $G_\gamma$ becomes

$$[0, \hat{W}_c^*]R\begin{bmatrix}0\\ \tilde{B}\end{bmatrix} = B^*P^{-1}\tilde{B} + \lambda[0, \hat{W}_c^*]T_1^*R^*\begin{bmatrix}0\\ \tilde{B}\end{bmatrix} =$$
$$B^*P^{-1}\tilde{B} - \lambda[0, \hat{W}_c^*](\gamma^2 I - R^*T_2^*P_2T_2RP_1)B_n = \tag{6.36}$$
$$B^*P^{-1}\tilde{B} - \lambda[0, \hat{W}_c^*](\gamma^2 I - R^*T_2^*P_2T_2RP_1)(I - \lambda A_n)(I - \lambda A_n)^{-1}B_n\ .$$

So the first term for $G_\gamma$ in (6.35) can be replaced by the last two terms in (6.36). Now by collecting all the terms in (6.35) and ( 6.36) containing $(I - \lambda A_n)^{-1}B_n$ on the right and $[0, \hat{W}_c^*]$ on the left, and equating like coefficients of $\lambda^0$, $\lambda$, and $\lambda^2$ we obtain

$$-[0, \hat{W}_c^*]T_2RP_1 + \lambda[0, \hat{W}_c^*]RP_1 - \lambda[0, \hat{W}_c^*](\gamma^2 I - T_1^*R^*P_2T_2RP_1) +$$
$$\lambda^2[0, \hat{W}_c^*]T_1^*(\gamma^2 I - R^*P_2RP_1) =$$
$$-[0, \hat{W}_c^*](I - \lambda T_1^*)T_2RP_1 + \lambda[0, \hat{W}_c^*](I - \lambda T_1^*)RP_1 - \lambda\gamma^2[0, \hat{W}_c^*](I - \lambda T_1^*) = \tag{6.37}$$
$$-[0, B^*]T_2RP_1 + \lambda[0, B^*]RP_1 - \lambda\gamma^2[0, B^*]\ .$$

In moving from the first equality to the second notice that the term $R^*P_2$ disappears due to the upper triangular structure of $T_1^*$ and $[0, \hat{W}_c^*]$. Substituting (6.36) into (6.35) and using the identity in (6.37) we now arrive at

$$G_\gamma = B^*P^{-1}\tilde{B} - [0, B^*]T_2RP_1(I - \lambda A_n)^{-1}B_n$$
$$-\lambda[0, B^*](\gamma^2 I - RP_1)(I - \lambda A_n)^{-1}B_n + [\tilde{C}, 0]RP_1(I - \lambda A_n)^{-1}B_n\ .$$

Now using $(I - \lambda A_n)^{-1} = I + \lambda A_n(I - \lambda A_n)^{-1}$ we readily see that

$$G_\gamma = \lambda C_n(I - \lambda A_n)^{-1}B_n + D_n \tag{6.38}$$

when $C_n$ and $D_n$ are defined by

$$C_n = [\tilde{C}, 0]A_n - [0, B^*]T_2RP_1A_n - [0, B^*](\gamma^2 I - RP_1)$$
$$D_n = B^*P^{-1}\tilde{B} + [\tilde{C}, 0]B_n - [0, B^*]T_2RP_1B_n \ . \tag{6.39}$$

Finally, using $A_n$, $B_n$, $C_n$ and $D_n$ in (6.32), (6.31), (6.39) and the Sylvester equation (6.2) along with

$$M = \text{diag}\,[Q, I](\gamma^2 I - R^*T_2^*P_2T_2RP_1) \quad \text{and} \quad N = \text{diag}\,[Q, I]T_1^*(\gamma^2 I - R^*P_2RP_1)$$

we readily arrive at $A_n = M^{-1}N$ and the formula for $B_n$, $C_n$ and $D_n$ in (6.12). This completes the proof.

**VERIFICATION OF REMARK 6.2.** Showing that the central solution $G_\gamma$ satisfies the mixed $H^2$ and $H^\infty$ bounds in (6.17) is slightly more complicated, then the Nehari or Nevanlinna-Pick case. The main reason for this is because in the Nudelman problem $\mathcal{L} = \ker T^*$ does not necessarily equal the space $\mathcal{Y}$. Let $\Omega$ be the operator from $\mathcal{F} \oplus \ell_+^2(\mathcal{Y})$ onto $\mathcal{H}$ defined by

$$\Omega(f \oplus g) = W_o Q^{-1/2} f \oplus g \qquad (f \in \mathcal{F} \text{ and } g \in \ell_+^2(\mathcal{Y}))\ .$$

Using $Q = W_o^* W_o$ it follows that $\Omega$ is indeed a unitary operator. By consulting (6.23) we see that $\mathcal{L} = \ker(T^*) = \Omega(f \oplus g_o)$ where $f \oplus g_o$ in $\mathcal{F} \oplus \mathcal{Y}$ is any vector in $\ker X$ where X is the operator from $\mathcal{F} \oplus \mathcal{Y}$ into $\mathcal{F}$ defined by $X = [\Lambda^* Q^{1/2}, C^*]$. Notice that $Q = XX^*$; see the Lyapunov equation for Q in (6.8). Since Q is invertible, this shows that X is onto $\mathcal{F}$, and thus $\ker X$ and $\mathcal{Y}$ have the same dimension. Now let $\Phi'$ be any isometry from $\mathcal{Y}$ onto $\ker X$ and partition $\Phi'$ in the following manner $\Phi' = [\phi_1^*, \phi_2^*]^*$ maps $\mathcal{Y}$ into $\mathcal{F} \oplus \mathcal{Y}$. Then according to our previous discussion the operator

$$\Phi := \Omega\Phi' = \begin{bmatrix} W_o Q^{-1/2}\phi_1 \\ \phi_2 \end{bmatrix} \tag{6.40}$$

is an isometry from $\mathcal{Y}$ onto $\mathcal{L}$.

Recall that $T = V\,|\,\mathcal{H}$ is a unilateral shift and thus $\mathcal{L}$ is a cyclic wandering subspace for T, that is, $\mathcal{H} = \oplus\{T^n\mathcal{L} : n \geq 0\}$. Since $\ell_+^2(\mathcal{Y}) \subseteq \mathcal{H}$ it follows that $\mathcal{L}$ is also a wandering subspace for the bilateral shift V, that is, $\ell^2(\mathcal{Y}) = \oplus\{V^n\mathcal{L} : n \in \mathbb{Z}\}$. From this and $\mathcal{L} = \Phi\mathcal{Y}$ it follows that $VL = LV$ where L is the operator on $\ell^2(\mathcal{Y})$ defined by

$$L = [\ldots,\ V^{*2}\Phi,\ V^*\Phi,\ \Phi,\ V\Phi,\ V^2\Phi \ldots]\ .$$

The isometry $\Phi$ appears in the zero-th column. Since $V^n\Phi$ is orthogonal to $V^m\Phi$ for all $n \neq m$

and $\Phi$ is an isometry, the operator L is unitary. Because $VL = LV$ the operator L is Laurent and there exists a function $\Theta_o$ in $L^\infty(\mathcal{Y}, \mathcal{Y})$ satisfying $L = L_{\Theta_o}$. Furthermore, using the fact that L is unitary it follows that $\Theta_o(e^{it})$ is a.e. a unitary operator on $\mathcal{Y}$. Finally, this function $\Theta_o$ can be obtained by taking the Fourier transform of the zeroth column of L, that is, by (6.40)

$$\Theta_o = \mathcal{F}_{\mathcal{Y}}\Phi = C(e^{it}I - \Lambda)^{-1}Q^{-1/2}\phi_1 + \phi_2\,, \tag{6.41}$$

where $\mathcal{F}_{\mathcal{Y}}$ is the Fourier transform from $\ell^2(\mathcal{Y})$ onto $L^2(\mathcal{Y})$.

Recall that $B'$ is an intertwining lifting of A if and only if $B' = L_G \mid \mathcal{H}$ where G is a function in $H^\infty(\mathcal{Y}, \mathcal{U})$ satisfying the Nudelman interpolation conditions (6.5). Moreover, since $\Theta_o\mathcal{Y}$ is the Fourier transform of $\mathcal{L}$ we have $\|B' \mid \mathcal{L}\|_{HS} = \|G\Theta_o\|_2 = \|G\|_2$. (The last equality follows from the fact that trace (EF) = trace (FE) and $\Theta_o$ is a.e. unitary.) This and $A = P_{\mathcal{H}'}B'$ shows that $\|A \mid \mathcal{L}\|_{HS} \leq \|B' \mid \mathcal{L}\|_{HS} = \|G\|_2$. Therefore $\|A \mid \mathcal{L}\|_{HS} \leq d_2$ the optimal error in the $H^2$ Nudelman optimization problem (6.13). On the other hand, if $B_\gamma = L_{G_\gamma} \mid \mathcal{H}$ is the central intertwining lifting of A, then Corollary IV.8.3 implies that

$$d_2 \leq \|G_\gamma\|_2 = \|B_\gamma \mid \mathcal{L}\|_{HS} \leq \frac{\delta\|A \mid \mathcal{L}\|_{HS}}{\sqrt{\delta^2 - 1}} \tag{6.42}$$

where $\gamma = \delta d_\infty$. In particular, letting $\delta$ approach infinity we see that $d_2 \leq \|A \mid \mathcal{L}\|_{HS}$ and thus $d_2 = \|A \mid \mathcal{L}\|_{HS}$. By Corollary IV.8.3 we readily obtain the $H^2$ bound for $G_\gamma$ in (6.17).

To obtain the formula for the optimal $H^2$ solution $G_{2*}$ in (6.14) and (6.15) one can either use the projection theorem or let $\gamma$ approach infinity in the previous theorem. We will leave it to the reader as a simple exercise to obtain the optimal $G_{2*}$ by projection techniques. Now let us obtain $G_{2*}$ by letting $\gamma$ approach infinity. To this end, it is probably easiest to use the formula for $G_\gamma$ in (6.35). In this case as $\gamma$ approaches infinity we see that $A_n$ approaches $T_1^*$ and $B_n$ approaches $[CQ^{-1}, 0]^*$; see (6.31) and (6.32). So using this in (6.35) and the upper triangular structure of $T_1^*$ we obtain

$$G_{2*} = \hat{W}_c^* P^{-1}\tilde{B} - \hat{W}_c^* P^{-1}(Z\Gamma + B\tilde{C})(I - \lambda Q^{-1}\Lambda^* Q)^{-1}Q^{-1}C^* +$$

$$\lambda\hat{W}_c^* P^{-1}\Gamma(I - \lambda Q^{-1}\Lambda^* Q)^{-1}Q^{-1}C^* + \tilde{C}(I - \lambda Q^{-1}\Lambda^* Q)^{-1}Q^{-1}C^* =$$

$$\hat{W}_c^* P^{-1}\tilde{B} + \hat{W}_c^* P^{-1}(\lambda\Gamma - Z\Gamma - B\tilde{C})Q^{-1}(I - \lambda\Lambda^*)^{-1}C^* + \tilde{C}Q^{-1}(I - \lambda\Lambda^*)^{-1}C^* =$$

$$[\tilde{C}Q^{-1}, B^*]\begin{bmatrix} (I - \lambda\Lambda^*) & 0 \\ -\lambda P^{-1}\Gamma Q^{-1} + P^{-1}(Z\Gamma + B\tilde{C})Q^{-1} & (I - \lambda Z^*) \end{bmatrix}^{-1}\begin{bmatrix} C^* \\ P^{-1}\tilde{B} \end{bmatrix}.$$

So if we let M and $A_1$ be the matrices defined by

$$M = \begin{bmatrix} I & 0 \\ P^{-1}(Z\Gamma + B\tilde{C})Q^{-1} & I \end{bmatrix} \quad \text{and} \quad A_1 = \begin{bmatrix} \Lambda^* & 0 \\ P^{-1}\Gamma Q^{-1} & Z^* \end{bmatrix}$$

then the above expression becomes

$$G_{2*} = [\tilde{C}Q^{-1}, B^*](M - \lambda A_1)^{-1}B_1 =$$

$$[\tilde{C}Q^{-1}, B^*](I - \lambda M^{-1}A_1)^{-1}M^{-1}B_1 = C_2(I - \lambda A_2)^{-1}B_2$$

where $B_1 = [C, \tilde{B}^* P^{-1}]^*$. Finally, setting $A_2 = M^{-1}A_1$ and $B_2 = M^{-1}B_1$ we arrive at the formula for the optimal $H^2$ solution $G_{2*}$ in (6.14) and (6.15).

To compute the error $d_2$ recall that the Fourier coefficients $G_j$ of $G_{2*}$ are given by $G_j = C_2A_2^jB_2$ for $j \geq 0$. Thus

$$d_2^2 = \|G_{2*}\|_2^2 = \sum_0^\infty \text{trace}\,(G_j^*G_j) = \text{trace}\,(B_2 \sum_0^\infty A_2^{*j}C_2^*C_2A_2^jB_2) = \text{trace}\,(B_2^*Q_2B_2)$$

where $Q_2$ is the observability grammian for the pair $\{C_2, A_2\}$. This completes the proof.

**PROOF OF THEOREM 6.3.** Using $A = W_2RW_1$ we have $A^*A = W_1^*R^*P_2RW_1$. Applying $W_1^*$ to the left we obtain $A^*AW_1^* = W_1^*R^*P_2RP_1$. Recall that $\|A\|^2 = d_\infty^2$ equals the largest eigenvalues $\lambda_{max}$ for $R^*P_2RP_1$. Now let $f \oplus x$ in $\mathcal{F} \oplus \mathcal{X}$ be the eigenvector for $R^*P_2RP_1$ corresponding to the eigenvalue $\lambda_{max}$, and set $y = W_1^*(f \oplus x)$. Notice that y is nonzero. If $y = 0$, then $W_1y = P_1(f \oplus x)$ is zero and $f \oplus x$ is in the kernel of $P_1$. Thus $R^*P_2RP_1(f \oplus x)$ is zero and $d_\infty^2 = \lambda_{max} = 0$. So without loss of generality we can assume that $d_\infty$, or equivalently, A is nonzero. Then y is nonzero. Moreover,

$$A^*Ay = A^*AW_1^*(f \oplus x) = W_1^*R^*P_2RP_1(f \oplus x) = \lambda_{max}W_1^*(f \oplus x) = d_\infty^2 y\,.$$

Therefore y is a vector in $\mathcal{H}$ which attains the norm of A. Notice the Corollary IV.2.8 and its proof can be extended to our setting where T is now the operator $T = V | \mathcal{H}$ and $\mathcal{H}$ is an invariant subspace for the bilateral shift V on $L^2$. Recall also that the intertwining lifting $B'$ of A in the Nudelman problem uniquely determine a function G in $H^\infty(\mathcal{Y}, \mathcal{U})$ such that $B' = L_G | \mathcal{H}$. Since $\mathcal{Y} = \mathbb{C}$ there is a unique function $g_{opt}$ in $H^\infty(\mathbb{C}, \mathcal{U})$ solving the $H^\infty$ optimization problem in (6.7) and this $g_{opt}$ is the function given by $g_{opt} = \mathcal{F}_\mathcal{U} Ay / \mathcal{F}_1 y$ where $\mathcal{F}_\mathcal{U}$ and $\mathcal{F}_1$ are the Fourier transforms from $\ell^2(\mathcal{U})$ onto $L^2(\mathcal{U})$ and $\ell^2$ onto $L^2$, respectively. Using $A = W_2RW_1$ and $y = W_1^*(f \oplus x)$ we have

$$\mathcal{F}_{\mathcal{U}} Ay = \mathcal{F}_{\mathcal{U}} W_2 R W_1 W_1^*(f \oplus x) = [\tilde{C}(\lambda I - \Lambda)^{-1}, B^*(I - \lambda Z^*)^{-1}] R(Qf \oplus \tilde{P}x)$$

$$\mathcal{F}_1 y = \mathcal{F}_1 W_1^*(f \oplus x) = C(\lambda I - \Lambda)^{-1} f + \tilde{B}^*(I - \lambda Z^*)^{-1} x .$$

Using the definition of R in (6.19) we arrive at the form of $g_{opt} = \mathcal{F}_{\mathcal{U}} Ay/\mathcal{F}_1 y$ in (6.18). This completes the proof.

## V.7 THE CENTRAL TWO BLOCK SOLUTION

In this section we will use state space techniques to compute the central intertwining lifting for a special two block interpolation problem. To this end, recall that $K^\infty(\mathcal{U}, \mathcal{Y})$ is the subspace of $L^\infty(\mathcal{U}, \mathcal{Y})$ consisting of all functions F in $L^\infty(\mathcal{U}, \mathcal{Y})$ whose Fourier coefficients of $e^{int}$ are zero for all $n \geq 0$. Now let $F_1$ and $F_2$ be functions in $K^\infty(\mathcal{U}, \mathcal{Y}_1)$ and $K^\infty(\mathcal{U}, \mathcal{Y}_2)$, respectively, and consider the following $H^2$ and $H^\infty$ optimization problems

$$d_j = \inf \left\{ \left\| \begin{bmatrix} F_1 - H \\ F_2 \end{bmatrix} \right\|_j : H \in H^j(\mathcal{U}, \mathcal{Y}_1) \right\} \qquad (\text{for } j = 1, \infty) . \tag{7.1}$$

Here $d_2$ is the distance from $[F_1, F_2]^{tr}$ to $[H^2(\mathcal{U}, \mathcal{Y}_1), \{0\}]^{tr}$, in the $L^2$ norm, while $d_\infty$ is the distance from $[F_1, F_2]^{tr}$ to $[H^\infty(\mathcal{U}, \mathcal{Y}_1), \{0\}]^{tr}$ in the $L^\infty$ norm. Obviously the $L^2$ optimization problem in (7.1) only makes sense when $\mathcal{U}$ is finite dimensional. So throughout this section we always assume that $\mathcal{U}$ is finite dimensional. In this case the Projection Theorem shows that $d_2 = \|[F_1, F_2]^{tr}\|_2$, and the optimal $L^2$ solution is given by $H = 0$. Finally, it is noted that one can also state the $H^2$ and $H^\infty$ optimization problem in (7.1) for any $F_1$ and $F_2$ in the appropriate $L^\infty$ spaces. However, in many applications (see Section VII.4 or Green-Limebeer [1]) the functions $F_1$ and $F_2$ are in the appropriate $K^\infty$ spaces. Moreover, it is this $K^\infty$ assumption on $F_1$ and $F_2$ which leads to some elegant state space formulas for computing the central intertwining lifting.

The $L^\infty$ optimization problem in (7.1) is a special case of the Sarason problem discussed in Section I.7 or I.10. To see this let $\Theta$ be the inner function in $H^\infty(\mathcal{Y}_1, \mathcal{Y}_1 \oplus \mathcal{Y}_2)$ defined by $\Theta = [I, 0]^*$ and F the function in $K^\infty(\mathcal{U}, \mathcal{Y}_1 \oplus \mathcal{Y}_2)$ defined by $F = [F_1, F_2]^{tr}$. Then the $L^\infty$ optimization problem in (7.1) is equivalent to the following Sarason optimization problem

$$d_\infty = \inf \{ \|F - \Theta H\|_\infty : H \in H^\infty(\mathcal{U}, \mathcal{Y}_1) \} . \tag{7.2}$$

According to the results in Section I.7 or I.10 the optimal error $d_\infty = \|A\|$, where A is the operator from $H^2(\mathcal{U})$ into $\mathcal{H}' = L^2(\mathcal{Y}_1 \oplus \mathcal{Y}_2) \ominus \Theta H^2(\mathcal{U})$ defined by $A = P_{\mathcal{H}'} M_F | H^2(\mathcal{U})$. Moreover, using the form of $\Theta = [I, 0]^*$ it follows that

$$\mathcal{H}' = K^2(\mathcal{Y}_1) \oplus L^2(\mathcal{Y}_2) \quad \text{and} \quad A = P_{\mathcal{H}'} M_F \,|\, H^2(\mathcal{U})\,. \tag{7.3}$$

We say that H is a *solution for the two block interpolation problem with tolerance* $\gamma$, if H is a function in $H^\infty(\mathcal{U}, \mathcal{Y}_1)$ satisfying

$$\left\| \begin{bmatrix} F_1 - H \\ F_1 \end{bmatrix} \right\|_\infty \leq \gamma. \tag{7.4}$$

If (7.4) holds, then obviously $\gamma \geq d_\infty$. In this section we will obtain a state space solution H for the two block interpolation problem by using the formula for the central intertwining lifting B of A in Corollary IV.3.3.

Now assume that $F = [F_1, F_2]$ admits an anticausal, stable, controllable and observable realization $\{Z, B, [C_1, C_2]^{tr}\}$. By an anticausal realization (see Section III.5) we mean that

$$F(\lambda) = \begin{bmatrix} F_1 \\ F_2 \end{bmatrix} = \begin{bmatrix} C_1 \\ C_2 \end{bmatrix} (\lambda I - Z)^{-1} B \qquad (|\lambda| > 1)\,. \tag{7.5}$$

Let $Q_2$ be the observability grammian for the pair $\{C_2, Z\}$. Consider the following Riccati difference equation

$$P_{n+1} = ZP_nZ^* + (B + ZP_nZ^*Q_2B)(\gamma^2 I - B^*Q_2B - B^*Q_2ZP_nZ^*Q_2B)^{-1}(B + ZP_nZ^*Q_2B)^* \tag{7.6}$$

subject to the initial condition $P_0 = 0$. As before, we say that P is a *positive steady state solution* to the Riccati difference equation, if $\{P_n\}$ converge to a positive operator P as n approaches infinity and $\gamma^2 I - B^*Q_2B - B^*Q_2ZPZ^*Q_2B$ is strictly positive. If $\{P_n\}$ does not converge to a positive operator or $\gamma^2 I - B^*Q_2B - B^*Q_2ZPZ^*Q_2B$ is not strictly positive, then in our terminology there is no positive steady state solution. If P is a steady state solution to the Riccati difference equation, then P is also a solution to the corresponding algebraic Riccati equation

$$P = ZPZ^* + (B + ZPZ^*Q_2B)(\gamma^2 I - B^*Q_2B - B^*Q_2ZPZ^*Q_2B)^{-1}(B + ZPZ^*Q_2B)^*\,. \tag{7.7}$$

Recall that P is a *positive solution to the algebraic Riccati equation* in (7.7), if P is a positive operator satisfying (7.7) and $\gamma^2 I - B^*Q_2B - B^*Q_2ZPZ^*Q_2B$ is strictly positive. Moreover, P is a *minimal solution* to the algebraic Riccati equation (7.7) if P is a positive solution to (7.7) and $P \leq P_1$ where $P_1$ is any positive solution to (7.7). In the proof of Theorem 7.1 we will see that the Riccati difference equation (7.6) admits a steady state solution if and only if there exists a positive solution to the algebraic Riccati equation (7.7). In this case the algebraic Riccati equation admits a unique minimal solution P. Moreover, this minimal P is precisely the steady state solution to the Riccati difference equation (7.6). Finally, we say that $P = \mathrm{RIC}(\gamma)$ if P is the positive steady state solution to the Riccati difference equation (7.6), or equivalently, P is the minimal solution to the algebraic Riccati equation (7.7). Notice that for a specified $\gamma$, the Riccati difference equation (7.6) may not have a steady state solution, or the algebraic Riccati equation may not admit a positive solution. In this case for convenience we set $P = +\infty$. This sets the stage for the following iterative procedure for computing $\|A\| = d_\infty$.

**THEOREM 7.1.** *Let* $\{Z, B, [C_1, C_2]^{tr}\}$ *be an anticausal, stable, controllable and observable realization for a function* $F = [F_1, F_2]^{tr}$ *in* $K^\infty(\mathcal{U}, \mathcal{Y}_1 \oplus \mathcal{Y}_2)$. *Let* $Q_1$ *and* $Q_2$ *be the observability grammians for the pair* $\{C_1, Z\}$ *and* $\{C_2, Z\}$, *respectively. Finally, let* $A$ *be the operator from* $H^2(\mathcal{U})$ *into* $\mathcal{H}' = K^2(\mathcal{Y}_1) \oplus L^2(\mathcal{Y}_2)$ *defined by* $A = P_{\mathcal{H}'} M_F \,|\, H^2(\mathcal{U})$. *Then the optimal errors* $d_2$ *and* $d_\infty$ *for the* $L^2$ *and* $L^\infty$ *optimization problems in (7.1) are given by*

$$\begin{aligned} &\|A \,|\, \mathcal{U}\|_2 = d_2^2 = \operatorname{trace} B^*(Q_1 + Q_2)B \quad \textit{and} \\ &\|A\| = d_\infty = \inf\{\gamma > 0 : P = \mathrm{RIC}(\gamma) \ \textit{and} \ r_{spec}(PQ_1) < 1\}\,. \end{aligned} \tag{7.8}$$

Obviously the operator $A$ in (7.3) has infinite rank. Computing the norm of an infinite rank operator can in general be a difficult problem. However, one can use an iterative bisection method to compute the infimum $d_\infty$ in (7.8), and thus $\|A\| = d_\infty$. To be more precise, assume that $\gamma_1$ and $\gamma_2$ are two positive scalars such that $P = \mathrm{RIC}(\gamma_2)$ is finite and $r_{spec}(PQ_1) < 1$, while $P = \mathrm{RIC}(\gamma_1)$ is infinite or $r_{spec}(PQ_1) > 1$. Then according to the previous theorem $\gamma_1 \le d_\infty \le \gamma_2$. Now set $\gamma = (\gamma_1 + \gamma_2)/2$. If $P = \mathrm{RIC}(\gamma)$ is finite and $r_{spec}(PQ_1) < 1$, then $\gamma_1 \le d_\infty \le \gamma$, otherwise $\gamma \le d_\infty \le \gamma_2$. If $\gamma_1 \le d_\infty \le \gamma$, then redefine $\gamma_2$ by setting $\gamma_2 = \gamma$, otherwise redefine $\gamma_1$ by $\gamma_1 = \gamma$. Now $d_\infty$ is in the smaller interval $[\gamma_1, \gamma_2]$ half the size of the original interval. So by continuing in this fashion the interval $[\gamma_1, \gamma_2]$ will shrink to the error $d_\infty = \|A\|$. This bisection method to compute the norm of $A$ is known as $\gamma$-iteration; see Francis [1], Green-Limebeer [1], Zhou-Doyle-Glover [1] for further details.

We say that $H$ is the *central solution* for the two block interpolation problem with tolerance $\gamma$ if $H$ is the unique function in $H^\infty(\mathcal{U}, \mathcal{Y}_1)$ given by $B_\gamma = M_G \,|\, H^2(\mathcal{U})$, where $B_\gamma$ is the central intertwining lifting of $A$ with tolerance $\gamma$ and $G = [F_1 - H, F_2]^{tr}$. The following result gives us a state space formula for computing the central solution $H$.

**THEOREM 7.2.** *Let* $\{Z \text{ on } \mathcal{X}, B, [C_1, C_2]^{tr}\}$ *be an anticausal, stable, controllable and observable realization for a function* $F = [F_1, F_2]$ *in* $K^\infty(\mathcal{U}, \mathcal{Y}_1 \oplus \mathcal{Y}_2)$. *Let* $Q_1$ *and* $Q_2$ *be the observability grammian for the pair* $\{C_1, Z\}$ *and* $\{C_2, Z\}$, *respectively and assume that the range of* $Q_1$ *is closed. Finally, for* $\delta > 1$ *set* $\gamma = \delta d_\infty$ *and let* $P = \mathrm{RIC}(\gamma)$. *Then the solution* $H$ *in* $H^\infty(\mathcal{U}, \mathcal{Y}_1)$ *for the two block interpolation problem with tolerance* $\gamma$ *corresponding to the* *central intertwining lifting is given by*

$$H = C_1P(I - \lambda M)^{-1}(I - Z^*Q_1ZP)^{-1}Z^*(Q_1 + Q_2)B \text{ where}$$
$$M = (I - Z^*Q_1ZP)^{-1}Z^*(I - Q_1P)\,. \tag{7.9}$$

*Moreover, M is stable and the interpolant* H *satisfies the following* $L^2$ *bound*

$$\left\| \begin{bmatrix} F_1 - H \\ F_2 \end{bmatrix} \right\|_2 \le \delta d_2 \left[ \delta^2 - 1 + \frac{d_2^2}{\dim(\mathcal{U})d_\infty^2} \right]^{-1/2} . \tag{7.10}$$

We say that G is a *two block interpolant* for the data $[F_1, F_2]^{tr}$ in $K^\infty(\mathcal{U}, \mathcal{Y}_1 \oplus \mathcal{Y}_2)$ with tolerance $\gamma$, if $G = [F_1 - H, F_2]^{tr}$ for some H in $H^\infty(\mathcal{U}, \mathcal{Y}_1)$ and $\|G\|_\infty \le \gamma$. Obviously this G is in $L^\infty(\mathcal{U}, \mathcal{Y}_1 \oplus \mathcal{Y}_2)$. Now let $\Delta(G)$ be the positive operator on $\mathcal{U}$ defined by

$$(\Delta(G)a, a) = \inf\{\|D_{B'}(a + \lambda h)\|^2 : h \in H^2(\mathcal{U}_1)\} \tag{7.11}$$

where a is a fixed vector in $\mathcal{U}$ and $B'$ is the intertwining of A defined by $B' = M_G \mid H^2(\mathcal{U})$. As before, $D_{B'}$ is the positive square root of $\gamma^2 I - (B')^* B'$. If $\mathcal{U}$ is finite dimensional, then the *entropy* of G is defined by

$$E(G) = \frac{1}{2\pi}\int_0^{2\pi} \ln\det[\gamma^2 I - G(e^{it})^* G(e^{it})]dt = \ln\det[\Delta(G)]\,. \tag{7.12}$$

The last equality follows from (IV.7.15). We say that $G_\gamma$ is a *maximal interpolant* for the two block problem with tolerance $\gamma$, if $G_\gamma$ is a two block interpolant with tolerance $\gamma$ and $\Delta(G_\gamma) \ge \Delta(G)$ where G is any other two block interpolant with tolerance $\gamma$. The function $G_\gamma$ is the *unique maximal* interpolant if $G_\gamma$ is a maximal entropy interpolant and if $\Delta(G_\gamma) = \Delta(G)$ for any other two block interpolant G with tolerance, $\gamma$, then $G_\gamma = G$. The results in Section IV.7 show that if $G_\gamma$ is a maximal interpolant and $\mathcal{U}$ is finite dimensional, then $E(G_\gamma) \ge E(G)$ for all two block interpolants G with tolerance $\gamma$. If in addition $E(G) > -\infty$ for some interpolant G, then the maximal entropy interpolant $G_\gamma$ is unique, that is, if $E(G_\gamma) = E(G)$, then $G_\gamma = G$. This sets the stage for the following result.

**THEOREM 7.3.** *Let* $\{Z, B, [C_1, C_2]^{tr}\}$ *be a stable, controllable and observable anticausal realization for a function* $F = [F_1, F_2]^{tr}$ *in* $K^\infty(\mathcal{U}, \mathcal{Y}_1 \oplus \mathcal{Y}_2)$. *Let* H *be the function in* $H^\infty(\mathcal{U}, \mathcal{Y}_1)$ *defined in (7.9) for some* $\gamma > d_\infty$. *Then the central solution* $G_\gamma = [F_1 - H, F_2]^{tr}$ *is the unique maximal interpolant for the two block problem with tolerance* $\gamma$. *Moreover, if* $\mathcal{U}$ *is finite dimensional, then the entropy of* G *is given by*

$$E(G_\gamma) = \frac{1}{2\pi}\int_0^{2\pi} \ln\det[\gamma^2 I - G^* G]dt = -\ln\det[N^2] \tag{7.13}$$

*where* $N^2$ *is the positive operator on* $\mathcal{U}$ *defined by*

$$N^2 = D^{-1} + D^{-1}B^*(I + Q_2 ZPZ^*)Q_1(I - PQ_1)^{-1}(I + ZPZ^*Q_2)BD^{-1}$$

$$D = \gamma^2 I - B^*Q_2 B - B^*Q_2 ZPZ^*Q_2 B\,. \tag{7.14}$$

*In particular, if* G *is any other two block interpolant with tolerance* $\gamma$*, then* $E(G_\gamma) \geq E(G)$ *with equality if and only if* $G_\gamma = G$.

The positive square root N of the operator $N^2$ in (7.14) will play an important part in the parameterization of all solutions for the two block problem in Section VI.9.

**PROOF OF THEOREM 7.1**. Because $F = [F_1, F_2]^{tr}$ is in $K^\infty(\mathcal{U}, \mathcal{Y}_1 \oplus \mathcal{Y}_2)$ the Projection Theorem shows that $d_2 = \|F\|_2$. As before, let A be the operator from $H^2(\mathcal{U})$ into $\mathcal{H}' = K^2(\mathcal{Y}_1) \oplus L^2(\mathcal{Y}_2)$ defined by (7.3). Then obviously $d_2 = \|F\|_2 = \|A \mid \mathcal{U}\|_2$. To establish the state space formula for $d_2$ in (7.8), let $\{F_{n,j} : j \leq -1\}$ be the Fourier coefficients for $F_n$ where $n = 1, 2$. Then using $F_{n,j} = C_n Z^{|j+1|}B$ for all $j < 0$ we have

$$d_2^2 = \|F\|_2^2 = \sum_{n=1,j=-1}^{2,-\infty} \operatorname{trace}(F_{n,j}^* F_{n,j}) =$$

$$\operatorname{trace} B^* \sum_{j=0}^{\infty} Z^{*j}C_1^*C_1 Z^j B + \operatorname{trace} B^* \sum_{j=0}^{\infty} Z^{*j}C_2^*C_2 Z^j B = \operatorname{trace} B^*(Q_1 + Q_2)B\,.$$

This establishes the first equation in (7.8).

As noted earlier, one can choose $\Theta_2 = [I, 0]^*$ and $\Theta_1 = I$ in Theorem I.10.1 to show that $d_\infty = \|A\|$. For completeness let us give a direct proof of this fact. To this end, let T be the unilateral shift on $H^2(\mathcal{U})$ and V the bilateral shift on $L^2(\mathcal{Y}_1 \oplus \mathcal{Y}_2)$. Let $T'$ be the co-isometry on $\mathcal{H}'$ obtained by compressing V to $\mathcal{H}'$, that is, $T' = P_{\mathcal{H}'}V \mid \mathcal{H}'$. Since $\mathcal{H}'$ is an invariant subspace for $V^*$, it follows that V is the minimal isometric lifting of $T'$. Moreover, using $P_{\mathcal{H}'}V = T'P_{\mathcal{H}'}$ we have $T'A = AT$.

We claim that B is an intertwining lifting of A with respect to V if and only if $B = M_G \mid H^2(\mathcal{U})$ where G is a function in $L^\infty(\mathcal{U}, \mathcal{Y}_1 \oplus \mathcal{Y}_2)$ of the form

$$G = \begin{bmatrix} G_1 \\ G_2 \end{bmatrix} = \begin{bmatrix} F_1 - H \\ F_2 \end{bmatrix} \tag{7.15}$$

for some H in $H^\infty(\mathcal{U}, \mathcal{Y}_1)$. Obviously if G is given by (7.15), then $B = M_G \mid H^2 \mid \mathcal{U})$ is an intertwining lifting of A, that is, $VB = BT$ and $P_{\mathcal{H}'}B = A$. On the other hand, if B is an intertwining lifting of A, then the intertwining property $VB = BT$ implies that there exists a unique $G = [G_1, G_2]^{tr}$ in $L^\infty(\mathcal{U}, \mathcal{Y}_1 \oplus \mathcal{Y}_2)$ such that $B = M_G \mid H^2(\mathcal{U})$. Now for a in $\mathcal{U}$ the lifting property $P_{\mathcal{H}'}B = A$ gives

$$\begin{bmatrix} P_-G_1 a \\ G_2 a \end{bmatrix} = P_{\mathcal{H}'} \begin{bmatrix} G_1 a \\ G_2 a \end{bmatrix} = P_{\mathcal{H}'}Ba = Aa = \begin{bmatrix} F_1 a \\ F_2 a \end{bmatrix} \tag{7.16}$$

where $P_-$ is the orthogonal projection onto $K^2(\mathcal{Y}_1)$ and thus $P_{\mathcal{H}'} = P_- \oplus I$. Equation (7.16) shows that $G_2 = F_2$. Because $P_-G_1 a = F_1 a$ is in $K^2(\mathcal{Y}_1)$ for all a in $\mathcal{U}$, it follows that $F_1$ and $G_1$ have the same Fourier coefficients of $e^{int}$ for all $n < 0$. So $F_1 - G_1 = H$ for some analytic function H in the open unit disc with values in $\mathcal{L}(\mathcal{U}, \mathcal{Y}_1)$. Because both $F_1$ and $G_1$ are in $L^\infty(\mathcal{U}, \mathcal{Y}_1)$, this implies that H is in $H^\infty(\mathcal{U}, \mathcal{Y}_1)$. Therefore $G_1 = F_1 - H$ for some H in $H^\infty(\mathcal{U}, \mathcal{Y}_1)$ which proves our claim.

Recall that if $B = M_G \mid H^2(\mathcal{U})$, then $\|B\| = \|G\|_\infty$. Since there is a one to one correspondence between the set of all intertwining liftings B of A and the set of all function G of the form $[F_1 - H, F_2]$ for H in $H^\infty(\mathcal{U}, \mathcal{Y}_1)$, the $L^\infty$ optimization problem in (7.1) is equivalent to

$$d_\infty = \inf \{\|B\| : \text{B is an intertwining lifting of A}\}\ .$$

Because $P_{\mathcal{H}'}B = A$ for any intertwining lifting of A we obtain $\|B\| \geq \|A\|$ and thus $d_\infty \geq \|A\|$. On the other, hand by the commutant lifting theorem, there exists an intertwining lifting B of A such that $\|B\| = \|A\|$. Therefore $d_\infty = \|A\|$.

To complete the proof it remains to show that $\|A\|$ is given by the infimum in (7.8). To this end, let $\Gamma_c$ be the Hankel operator from $H^2(\mathcal{U})$ into $K(\mathcal{Y}_1)$ defined by $\Gamma_c = P_-M_{F_1} \mid H^2(\mathcal{U})$. Let $W_o$ from $\mathcal{X}$ into $K^2(\mathcal{Y}_1)$ and $W_c$ from $H^2(\mathcal{U})$ into $\mathcal{X}$ be the observability and controllability operators respectively defined by

$$W_o x = C_1(e^{it}I - Z)^{-1}x \quad \text{and} \quad W_c^* x = B^*(I - e^{it}Z^*)^{-1}x \qquad (x \in \mathcal{X})\ . \tag{7.17}$$

Notice that $W_o^* W_o = Q_1$ the observability grammian for $\{C_1, Z\}$. By replacing C by $C_1$ in the proof of Theorem 5.1 we see that $\Gamma_c$ admits a factorization of the form $\Gamma_c = W_o W_c$.

Clearly the operator A admits Hankel-Toeplitz decomposition of the form

$$A = \begin{bmatrix} \Gamma_c \\ M_{F_2} \mid H^2(\mathcal{U}) \end{bmatrix} = \begin{bmatrix} W_o W_c \\ M_{F_2} \mid H^2(\mathcal{U}) \end{bmatrix}\ . \tag{7.18}$$

So for $\gamma > 0$ we have

$$\gamma^2 I - A^*A = \gamma^2 I - P_+ M_{F_2}^* M_{F_2} \mid H^2(\mathcal{U}) - \Gamma_c^*\Gamma_c = \hat{T}_{R,\gamma} - \Gamma_c^*\Gamma_c \tag{7.19}$$

where $P_+$ is the orthogonal projection onto $H^2(\mathcal{U})$ and $\hat{T}_R$ is the Toeplitz operator on $H^2(\mathcal{U})$ defined by $\hat{T}_{R,\gamma} = \gamma^2 I - P_+ M_{F_2}^* M_{F_2} \mid H^2(\mathcal{U})$. Notice that $\hat{T}_{R,\gamma}$ is precisely the Fourier transform of the Toeplitz operator $T_{R_\gamma}$ on $\ell_+^2(\mathcal{U})$ generated by the function $R_\gamma = \gamma^2 I - F_2^* F_2$ in $L^\infty(\mathcal{U}, \mathcal{U})$. Moreover, if $\gamma > \|A\| = d_\infty$, then (7.19) shows that the Toeplitz operator $\hat{T}_{R,\gamma}$ is strictly positive. In this case $\hat{T}_{R,\gamma}$ admits an invertible outer spectral factor $\Theta$, that is, there exists an outer function $\Theta$ in $H^\infty(\mathcal{U}, \mathcal{U})$ such that $\Theta^{-1}$ is in $H^\infty(\mathcal{U}, \mathcal{U})$ and $\hat{T}_{R,\gamma} = (M_\Theta^+)^* M_\Theta^+$. So if $\gamma > d_\infty$, then $\hat{T}_{R,\gamma}$ admits an invertible outer spectral factor. The converse of this statement is: If $\hat{T}_{R,\gamma}$ does not admit an invertible outer spectral factor, then $\gamma \le d_\infty$. (Recall that we can use the Riccati techniques in Section 2 of the Appendix to determine whether or not $\hat{T}_{R,\gamma}$ admits a square outer spectral factor, that is, an outer spectral factor in $H^\infty(\mathcal{U}, \mathcal{U})$.) So for the moment assume that $\hat{T}_{R,\gamma}$ admits an invertible outer spectral factor $\Theta$. Then (7.19) gives

$$\gamma^2 I - A^*A = (M_\Theta^+)^* M_\Theta^+ - \Gamma_c^*\Gamma_c = (M_\Theta^+)^* \left[ I - (M_\Theta^+)^{-*}\Gamma_c^*\Gamma_c (M_\Theta^+)^{-1} \right] M_\Theta^+ \tag{7.20}$$

where $N^{-*}$ denotes the inverse of $N^*$. Therefore, if $\hat{T}_{R,\gamma}$ admits an invertible outer spectral factor $\Theta$, then $\|A\| \le \gamma$ if and only if $\Gamma_c (M_\Theta^+)^{-1}$ is contractive. Summing up the above analysis readily shows that

$$d_\infty = \inf\{\gamma > 0 : \hat{T}_{R,\gamma} \text{ admits an invertible outer spectral factor } \Theta \text{ and } \|\Gamma_c (M_\Theta^+)^{-1}\| \le 1\} \tag{7.21}$$

Lemma 7.4 below shows that $\hat{T}_{R,\gamma}$ admits an invertible outer spectral factor $\Theta$, that is, an outer spectral factor in $H^\infty(\mathcal{U}, \mathcal{U})$ if and only if $P = \mathrm{RIC}(\gamma)$ exists. So to obtain the infimum for $d_\infty$ in (7.8) it remains to show that $\Gamma = \Gamma_c (M_\Theta^+)^{-1}$ is contractive if and only if $r_{spec}(PQ_1) \le 1$. To this end, assume that $\gamma > d_\infty$ and $\Theta$ is the invertible outer spectral factor for $\hat{T}_{R,\gamma}$. Then Lemma 7.4 below shows that $\mathrm{RIC}(\gamma) = P = W_c \hat{T}_{R,\gamma}^{-1} W_c^*$. So using $\Gamma_c = W_o W_c$ and $(M_\Theta^+)^* M_\Theta^+ = \hat{T}_{R,\gamma}$ we have

$$\Gamma\Gamma^* = W_o W_c (M_\Theta^+)^{-1} (M_\Theta^+)^{-*} W_c^* W_o^* = W_o W_c \hat{T}_{R,\gamma}^{-1} W_c^* W_o^* = W_o P W_o^* . \tag{7.22}$$

Recall that if M from $\mathcal{M}$ into $\mathcal{N}$ and N from $\mathcal{N}$ into $\mathcal{M}$ are two bounded linear operators, then $r_{spec}(MN) = r_{spec}(NM)$; see formula (3) in Section III.2 of Gohberg-Goldberg-Kaashoek [1]. Recall also that $Q_1 = W_o^* W_o$ is the observability grammian for the pair $\{C_1, Z\}$. Using this in (7.22) we have

$$\|\Gamma\|^2 = r_{spec}(\Gamma\Gamma^*) = r_{spec}(W_o P W_o^*) = r_{spec}(P W_o^* W_o) = r_{spec}(PQ_1) .$$

Therefore $\Gamma$ is contractive if and only if $r_{spec}(PQ_1) \le 1$. Using this in (7.21) along with the fact that $\hat{T}_{R,\gamma}$ admits a square outer spectral factorization if and only if $P = RIC(\gamma)$ is finite gives (7.8). This completes the proof.

The following lemma which was used in the proof of the previous theorem is of independent interest.

**LEMMA 7.4.** *Let* $\{Z$ *on* $\mathcal{X}$, $B$, $C_2\}$ *be an anticausal, stable, and controllable realization for a function* $F_2$ *in* $K^\infty(\mathcal{U}, \mathcal{Y}_2)$ *and* $Q_2$ *the observability grammian for* $\{C_2, Z\}$. *For a specified* $\gamma > 0$ *let* $R$ *be the function in* $L^\infty(\mathcal{U}, \mathcal{U})$ *defined by* $R = \gamma^2 I - F_2^* F_2$, *and* $T_R$ *the Toeplitz operator on* $\ell_+^2(\mathcal{U})$ *generated by* $R$. *Then* $T_R$ *admits a square outer spectral factor* $\Theta$ *if and only if* $P = RIC(\gamma)$ *is exists. In this case the outer spectral factor for* $T_R$ *is given by*

$$\Theta = D_0 - \lambda C_o(I - \lambda Z^*)^{-1} Z^* Q_2 B$$

$$D_0 = (\gamma^2 I - B^* Q_2 B - B^* Q_2 Z P Z^* Q_2 B)^{1/2} \quad \textit{and} \quad C_o = D_0^{-1} B^* (I + Q_2 Z P Z^*) . \tag{7.23}$$

*Finally, if* $T_R$ *is strictly positive, then* $P = RIC(\gamma)$ *is given by*

$$P = W T_R^{-1} W^* \quad \textit{where} \quad W = [B, ZB, Z^2B, \ldots] \tag{7.24}$$

*is the controllability operator from* $\ell_+^2(\mathcal{U})$ *into* $\mathcal{X}$ *generated by* $\{Z, B\}$.

**PROOF.** To prove this lemma we simply apply the Positive Real Lemma A2.2 and Remark A2.3 to the Toeplitz operator $T_R$ generated by R. (Here A2 refers to Section 2 in the Appendix.) To this end, recall that the Fourier coefficients of $F_2$ are given by $F_{2,j} = C_2 Z^{|j+1|} B$ for $j \leq -1$. Now let a, b be vectors in $\mathcal{U}$ and $\{R_n\}_{-\infty}^{\infty}$ be the Fourier coefficients of R. Then using $Q_2 = \sum Z^{*j} C_2^* C_2 Z^j$ we see that the zero-th Fourier coefficient $R_0$ of R is given by

$$(R_0 a, b) = (Ra, b) = \gamma^2 (a, b) - (F_2 a, F_2 b) = \gamma^2 (a, b) - \sum_{j=-1}^{-\infty} (F_{2,j} a, F_{2,j} b) =$$

$$\gamma^2 (a, b) - \sum_{0}^{\infty} (C_2 Z^j B a, C_2 Z^j B b) = \gamma^2 (a, b) - (B^* Q_2 B a, b) .$$

Thus $R_0 = \gamma^2 I - B^* Q_2 B$. For $n \geq 1$ we have

$$(R_n a, b) = (Ra, e^{int} b) = -(F_2 a, e^{int} F_2 b) =$$

$$-\sum_{j=-1}^{-\infty} (F_{2,j} a, F_{2,j-n} b) = -\sum_{j=0}^{\infty} (C_2 Z^j B a, C_2 Z^j Z^n B b) = -(B^* Z^{*n} Q_2 B a, b) .$$

By combining the two previous equation we arrive at

$$R_0 = \gamma^2 I - B^* Q_2 B \quad \text{and} \quad R_n = -B^* Z^{*n-1} Z^* Q_2 B \qquad (\text{for } n \geq 1)\,. \tag{7.25}$$

Now we can use (7.25) in the Positive Real Lemma A2.2 and the discussion in Remark A2.3 to determine whether or not $T_R$ admits a square outer spectral factor. So by replacing C by $B^*$, Z by $Z^*$ and B by $-Z^*Q_2B$ and $R_0$ by $\gamma^2 - B^*Q_2B$ in the Riccati difference equation (A2.1) or the algebraic Riccati equation (A2.4) we arrive at the Riccati equations in (7.6) and (7.7). In particular, $T_R$ admits a square outer spectral factor $\Theta$ if and only if $P = \mathrm{RIC}(\gamma)$ exists. Moreover, if $P = \mathrm{RIC}(\gamma)$ exists, then by replacing C by $B^*$, Z by $Z^*$, and B by $-Z^*Q_2B$ and using $R_0 = \gamma^2 I - B^*Q_2B$ in (A2.2) and (A2.3) we arrive at the outer spectral factor $\Theta$ for $T_R$ given in (7.23). Finally, if $T_R$ is strictly positive, then Remark A2.3 with C replaced by $B^*$ and Z by $Z^*$ shows that $P = WT_R^{-1}W^*$. This completes the proof.

**PROOF OF THEOREM 7.2.** Here we will obtain a proof of Theorem 7.2 based on the state space formula for the central intertwining lifting $B_\gamma$ of A given by Corollary IV.3.3. Because $\gamma > d_\infty$, the Toeplitz operator $\hat{T}_{R,\gamma}$ admits an invertible outer spectral factor $\Theta$ which is given by (7.23). Now let $W_\theta$ be the observability operator from $\mathcal{X}$ into $H^2(\mathcal{U})$ defined by

$$W_\theta x = C_o(I - \lambda Z^*)^{-1}x \qquad (x \in \mathcal{X})\,. \tag{7.26}$$

We claim that $W_c(M_\Theta^+)^{-1} = W_\theta^*$. To prove this first notice that by substituting the formula for $C_o$ into the algebraic Riccati equation (7.7) we see that $P = ZPZ^* + C_o^*C_o$. In particular, P is the controllability grammian for the pair $\{Z, C_o^*\}$, that is, $P = W_\theta^* W_\theta$. Using the fact the $P = \sum Z^n C_o^* C_o Z^{*n}$ we obtain

$$B = C_o^* D_0 - ZPZ^*Q_2B = [C_o^*, ZC_o^*, Z^2C_o^*, \ldots]\begin{bmatrix} D_0 \\ -C_oZ^*Q_2B \\ -C_oZ^*Z^*Q_2B \\ \vdots \end{bmatrix}. \tag{7.27}$$

Since the last column of (7.27) contains the Fourier coefficients of $\Theta$, equation (7.27) implies that

$$[B, ZB, Z^2B, \ldots] = [C_o^*, ZC_o^*, Z^2C_o^*, \ldots]T_\Theta \tag{7.28}$$

where $T_\Theta$ is the Toeplitz operator on $\ell_+^2(\mathcal{U})$ generated by $\Theta$. By taking the Fourier transform of (7.28) we arrive at $W_c = W_\theta^* M_\Theta^+$. Therefore $W_c(M_\Theta^+)^{-1} = W_\theta^*$ which proves our claim. Finally, using this result we obtain

$$\Gamma := \Gamma_c(M_\Theta^+)^{-1} = W_o W_c (M_\Theta^+)^{-1} = W_o W_\theta^* . \tag{7.29}$$

Now let $B_\gamma$ be the central intertwining lifting of A with respect to the bilateral shift V and tolerance $\gamma$. Using $VB = BT$ where T is the unilateral shift, it follows that $B = M_G \,|\, H^2(\mathcal{U})$ for some G in $H^\infty(\mathcal{U}, \mathcal{Y}_1 \oplus \mathcal{Y}_2)$. Moreover, this $G = [F_1 - H, F_2]^{tr}$ for some H in $H^\infty(\mathcal{U}, \mathcal{Y}_1)$; see (7.15). According to Corollary IV.3.3 this G is given by the state space formula

$$G = A\Pi_o^* + \Psi(\lambda)(I - \lambda T_A^*)^{-1} T_A^* \Pi_o^* \quad \text{where} \quad T_A^* = D_{AT}^{-2} T^* D_A^2 \tag{7.30}$$

and $\Pi_o$ is the operator from $H^2(\mathcal{U})$ onto $\mathcal{U}$ defined by $\Pi_o g = g(0)$. Using the form of A in (7.18) along with the definition of $\Psi(\lambda)$ we obtain

$$\Psi(\lambda)h = (V - T')Ah = (V - T')\begin{bmatrix} W_o W_c h \\ F_2 h \end{bmatrix} = \begin{bmatrix} C_1 W_c h \\ 0 \end{bmatrix} \qquad (h \in H^2(\mathcal{U})) . \tag{7.31}$$

Substituting this into (7.30) along with $A\Pi_o^* = [F_1, F_2]^{tr}$ and $G = [F_1 - H, F_2]^{tr}$ we see that the central interpolant

$$H = -\, C_1 W_c (I - \lambda T_A^*)^{-1} T_A^* \Pi_o^* . \tag{7.32}$$

To complete the proof it is a simple matter to convert the state space formula in (7.32) to the state space formula for H in (7.9). By virtue of (7.20) and $TM_\Theta^+ = M_\Theta^+ T$ we have

$$D_A^2 = (M_\Theta^+)^* (I - \Gamma^* \Gamma) M_\Theta^+ \quad \text{and} \quad D_{AT}^2 = (M_\Theta^+)^* (I - T^* \Gamma^* \Gamma T) M_\Theta^+ . \tag{7.33}$$

This readily implies that

$$T_A^* = (M_\Theta^+)^{-1} T_\Gamma^* M_\Theta^+ \quad \text{where} \quad T_\Gamma^* := (I - T^* \Gamma^* \Gamma T)^{-1} T^* (I - \Gamma^* \Gamma) . \tag{7.34}$$

Using this in (7.32) along with $W_c (M_\Theta^+)^{-1} = W_\theta^*$ we obtain the following expression for the central interpolant H

$$H = -\, C_1 W_\theta^* (I - \lambda T_\Gamma^*)^{-1} T_\Gamma^* M_\Theta^+ \Pi_o^* . \tag{7.35}$$

Using $\Gamma = W_o W_\theta^* = W_o W_c (M_\Theta^+)^{-1}$ and $Q_1 = W_o^* W_o$ we obtain

$$(I - \Gamma^* \Gamma) M_\Theta^+ \Pi_o^* = M_\Theta^+ \Pi_o^* - \Gamma^* W_o W_c \Pi_o^* = M_\Theta^+ \Pi_o^* - W_\theta Q_1 B . \tag{7.36}$$

Applying $T^*$ to both sides of the previous expression along with $T^* W_\theta = W_\theta Z^*$ and $T^* \Theta a = -\, W_\theta Z^* Q_2 B$ (see (7.23) and (7.26)) we obtain

$$T^* (I - \Gamma^* \Gamma) M_\Theta^+ \Pi_o^* = -\, W_\theta Z^* (Q_1 + Q_2) B . \tag{7.37}$$

Using $T^* W_\theta = W_\theta Z^*$ and $W_\theta^* T = Z W_\theta^*$ with $W_\theta^* W_\theta = P$ we have

$$(I - T^* \Gamma^* \Gamma T) W_\theta = W_\theta - T^* \Gamma^* W_o W_\theta^* T W_\theta = W_\theta - T^* W_\theta W_o^* W_o Z P = W_\theta (I - Z^* Q_1 Z P) \,. \quad (7.38)$$

Since $\gamma > d_\infty$ the Toeplitz operator $\hat{T}_{R,\gamma}$ is strictly positive and thus $P = W_c \hat{T}_{R,\gamma}^{-1} W_c^*$. Because $\{Z, B\}$ is controllable P is strictly positive. By virtue of $P = W_\theta^* W_\theta$ we see that $W_\theta$ is one to one and has closed range. In particular, (7.38) shows that $\mathcal{H}_\theta = \operatorname{ran} W_\theta$ is an invariant subspace for $I - T^* \Gamma^* \Gamma T$. Moreover, $(I - T^* \Gamma^* \Gamma T) | \mathcal{H}_\theta$ is similar to $I - Z^* Q_1 Z P$. Using $\gamma > d_\infty$ once again we see that $\Gamma$ is strictly contractive; see (7.20). Hence $\|\Gamma T\| = \|T^* \Gamma^*\| < 1$. So $\Gamma T$ is strictly contractive and $I - T^* \Gamma^* \Gamma T$ is invertible. This readily implies that $I - Z^* Q_1 Z P$ is also invertible. By taking the inverses in (7.38) we obtain the result we have been looking for

$$(I - T^* \Gamma^* \Gamma T)^{-1} W_\theta = W_\theta (I - Z^* Q_1 Z P)^{-1} \,. \quad (7.39)$$

Substituting this in (7.37) gives

$$T_\Gamma^* M_\Theta^+ \Pi_o^* = - W_\theta (I - Z^* Q_1 Z P)^{-1} Z^* (Q_1 + Q_2) B \,. \quad (7.40)$$

Now we need an expression for $T_\Gamma^* W_\theta$ in terms of M in (7.9). To this end, notice that $\Gamma = W_o W_\theta^*$ gives

$$T^* (I - \Gamma^* \Gamma) W_\theta = T^* W_\theta - T^* W_\theta W_o^* W_o W_\theta^* W_\theta = W_\theta^* Z^* - W_\theta Z^* Q_1 P = W_\theta Z^* (I - Q_1 P) \,. \quad (7.41)$$

Using this along with (7.39) we obtain

$$T_\Gamma^* W_\theta = W_\theta (I - Z^* Q_1 Z P)^{-1} Z^* (I - Q_1 P) = W_\theta M \,. \quad (7.42)$$

The previous equation shows that $\mathcal{H}_\theta = \operatorname{ran} W_\theta$ is invariant to $T_\Gamma^*$. Recall that $W_\theta$ is one to one and has closed range. So $T_\Gamma^* | \mathcal{H}_\theta$ is similar to M. Because $T_\Gamma^*$ is similar to $T_A^*$ and $T_A^*$ strongly approaches zero as n approaches infinity (see Lemma IV.6.1), it follows that $M^n$ tends to zero strongly as n tends to infinity. In fact, by using Lemma 1.6 one can show that $r_{spec}(M) < 1$. (Since the proof is similar to Proposition 1.7, the details are omitted.) This proves the stability results for M in Theorem 7.2. In particular, for all $|\lambda| < 1$ equation (7.42) gives

$$(I - \lambda T_\Gamma^*)^{-1} W_\theta = W_\theta (I - \lambda M)^{-1} \,. \quad (7.43)$$

Substituting (7.43) and (7.40) into our expression for H in (7.35) yields

$$H = C_1 W_\theta^* W_\theta (I - \lambda M)^{-1} (I - Z^* Q_1 Z P)^{-1} Z^* (Q_1 + Q_2) B \,. \quad (7.44)$$

Finally, using $P = W_\theta^* W_\theta$ we obtain the state space formula for H in (7.9). The $L^2$ bound for G in (7.10) is a simple consequence of Theorem 7.1 with the $H^2$ and $H^\infty$ bound for the central intertwining lifting given in Corollary IV.8.3. This completes the proof.

**PROOF OF THEOREM 7.3.** First let us show that $N^2$ in (7.14) is given by

$$N^2 = \Pi_o D_A^{-2} \Pi_o^* . \tag{7.45}$$

Then Theorem 7.3 follows from Corollary IV.7.7. Using $Q_l = W_o^* W_o$ and $\Gamma = W_o W_\theta^*$ we have

$$I - \Gamma^* \Gamma = I - W_\theta Q_l W_\theta^* . \tag{7.46}$$

Let us search for an inverse of $(I - \Gamma^* \Gamma)$ of the form $I + W_\theta R W_\theta^*$ for some R on $\mathcal{X}$. In this case (7.46) and $P = W_\theta^* W_\theta$ show that

$$(I + W_\theta R W_\theta^*)(I - \Gamma^* \Gamma) = (I + W_\theta R W_\theta^*)(I - W_\theta Q_l W_\theta^*) = I + W_\theta [R - RPQ_l - Q_l] W_\theta^* = I .$$

So if we set $R = Q_l (I - PQ_l)^{-1}$, then we obtain the last equality. Because $r_{spec}(PQ_l) < 1$ the inverse of $I - PQ_l$ is well defined. A similar calculation shows that for this R we have $(I - \Gamma^* \Gamma)(I + W_\theta R W_\theta^*) = I$. Thus

$$(I - \Gamma^* \Gamma)^{-1} = I + W_\theta Q_l (I - PQ_l)^{-1} W_\theta^* . \tag{7.47}$$

Now using (7.33) and $\Pi_o W_\theta = C_o$ we obtain

$$\begin{aligned} &\Pi_o D_A^{-2} \Pi_o^* = \Pi_o (M_\Theta^+)^{-1} (I - \Gamma^* \Gamma)^{-1} (M_\Theta^+)^{-*} \Pi_o^* = \\ &\Theta(0)^{-1} \Pi_o (I + W_\theta Q_l (I - PQ_l)^{-1} W_\theta^*) \Pi_o^* \Theta(0)^{-*} = \\ &D_o^{-1} (I + C_o Q_l (I - PQ_l)^{-1} C_o^*) D_o^{-1} = N^2 . \end{aligned} \tag{7.48}$$

The second equality follows from the fact that for any invertible outer function $\Theta$ in $H^\infty(\mathcal{U}, \mathcal{U})$ we have $\Pi_o (M_\Theta^+)^{-1} = \Theta(0)^{-1} \Pi_o$. (To see this simply notice that $\Theta(0)\Pi_o = \Pi_o M_\Theta^+$ and recall that $\Theta(0)$ is invertible for any invertible outer function.) By substituting the definitions of $C_o$ and $D_o$ in (7.23) into (7.48) we arrive at the formula for $N^2$ in (7.14). Finally, an application of Corollary IV.7.7 completes the proof.

## V.8 THE FOUR BLOCK PROBLEM

In this section we will use Corollary IV.8.3 to solve a mixed $H^2$ and $H^\infty$ four block Nehari type interpolation problem. To this end, let $\mathcal{U}$ be finite dimensional, and recall that $L^2(\mathcal{U}, \mathcal{Y})$ is the Hilbert space of all square integrable Lebesgue measurable functions on $[0, 2\pi)$ whose values are bounded linear operators mapping $\mathcal{U}$ into $\mathcal{Y}$. The norm of a function $\Psi$ in $L^2(\mathcal{U}, \mathcal{Y})$ is defined by

$$\|\Psi\|_2^2 = \frac{1}{2\pi}\int_0^{2\pi} \operatorname{trace}\,(\Psi(t)^*\Psi(t))dt = \|M_\Psi \mid \mathcal{U}\|_{HS}^2 . \tag{8.1}$$

Notice that the $L^2$ norm of $\Psi$ equals the Hilbert-Schmidt norm of the multiplication operator $M_\Psi$ restricted to the subspace $\mathcal{U}$. To present our $H^2$ and $H^\infty$ Nehari result, let G be the function defined by

$$G = \begin{bmatrix} F_{11} & F_{12} \\ F_{21} & F_{22} \end{bmatrix} \text{ in } L^\infty(\mathcal{U}_1 \oplus \mathcal{U}_2, \mathcal{Y}_1 \oplus \mathcal{Y}_2) \tag{8.2}$$

where $\mathcal{U}_1 \oplus \mathcal{U}_2$ is finite dimensional. Now consider the following $H^2$ and $H^\infty$ Nehari type optimization problems

$$d_i = \inf \left\{ \left\| \begin{bmatrix} F_{11} - H & F_{12} \\ F_{21} & F_{22} \end{bmatrix} \right\|_i : H \in H^i(\mathcal{U}_1, \mathcal{Y}_1) \right\} \qquad (\text{for } i = 2, \infty) \tag{8.3}$$

where $\|\cdot\|_2$ is $L^2$ norm and $\|\cdot\|_\infty$ is the $L^\infty$ norm. Notice that $d_2$ is the distance from G to $H^2(\mathcal{U}_1, \mathcal{Y}_1) \oplus \{0\}$ in the $L^2$ norm, while $d_\infty$ is the distance from G to $H^\infty(\mathcal{U}_1, \mathcal{Y}_1) \oplus \{0\}$ in the $L^\infty$ norm. The $H^\infty$ optimization problem in (8.3) is known as the four block optimization problem. This optimization problem plays an important role in control theory; see Francis [1], Green-Limebeer [1].

By using the projection theorem it is easy to solve the $H^2$ optimization problem in (8.3). To this end, let $P_+$ be the orthogonal projection from $L^2(\cdot)$ onto $H^2(\cdot)$ and let $P_- = I - P_+$. Then for any H in $H^2(\mathcal{U}_1, \mathcal{Y}_1)$ it follows that

$$\left\| \begin{bmatrix} F_{11} - H & F_{12} \\ F_{21} & F_{22} \end{bmatrix} \right\|_2^2 = \|F_{11} - H\|_2^2 + \|F_{12}\|_2^2 + \|F_{21}\|_2^2 + \|F_{22}\|_2^2 \geq$$

$$\|F_{11} - P_+F_{11}\|_2^2 + \|F_{12}\|_2^2 + \|F_{21}\|_2^2 + \|F_{22}\|_2^2 .$$

Therefore the optimal H in $H^2(\mathcal{U}_1, \mathcal{Y}_1)$ solving the $H^2$ optimization problem in (8.3) is given by $H = P_+F_{11}$. Moreover, the $H^2$ error $d_2$ is given by

$$d_2^2 = \|P_-F_{11}\|_2^2 + \|F_{12}\|_2^2 + \|F_{21}\|_2^2 + \|F_{22}\|_2^2 . \tag{8.4}$$

Notice that the $H^\infty$ four block optimization problem in (8.3) is a special case of the two-sided Sarason problem discussed in Sections I.10, II.6 and IV.9. To see this simple set $\Theta_2 = [I, 0]^*$ and $\Theta_1 = [I, 0]$. (One can also show that the two-sided Sarason problem can be converted into a four block type optimization problem; see Chu-Doyle-Lee [1], Francis [1], Green-Limebeer [1]. Section VIII.7 in Foias-Frazho [4] or Frazho [1] shows that the commutant lifting theorem can also be viewed as a generalized four block problem.) So we can use the results in those sections to solve the four problem. Because the four bock problem plays an important role in control theory we will present a solution of this problem directly, without using the two-sided Sarason results. In order to obtain a solution to the $H^\infty$ four block optimization problem in (8.3), let $A(G)$ be the operator, with symbol $G$, mapping $H^2(\mathcal{U}_1)\oplus L^2(\mathcal{U}_2)$ into $K^2(\mathcal{Y}_1)\oplus L^2(\mathcal{Y}_2)$ defined by $A(G)=(P_-\oplus I)M_G\,|\,(H^2(\mathcal{U}_1)\oplus L^2(\mathcal{U}_2))$ where $K^2(\mathcal{Y}_1)=L^2(\mathcal{Y}_1)\ominus H^2(\mathcal{Y}_1)$. To be precise,

$$A(G)\begin{bmatrix} x \\ y \end{bmatrix} = \begin{bmatrix} P_-(F_{11}x+F_{12}y) \\ F_{21}x+F_{22}y \end{bmatrix} \qquad (x\oplus y \text{ in } H^2(\mathcal{U}_1)\oplus L^2(\mathcal{U}_2))\,. \tag{8.5}$$

By using the commutant lifting theorem, the proof of the following theorem shows that $d_\infty = \|A(G)\|$. Moreover, there exists an $H$ in $H^\infty(\mathcal{U}_1,\mathcal{Y}_1)$ satisfying

$$d_\infty = \left\| \begin{vmatrix} F_{11}-H & F_{12} \\ F_{21} & F_{22} \end{vmatrix} \right\|_\infty . \tag{8.6}$$

In other words, there exists an optimal solution to the $H^\infty$ four block Nehari type optimization problem in (8.3).

**THEOREM 8.1**. *Let* $G$ *in* $L^\infty(\mathcal{U}_1\oplus\mathcal{U}_2,\mathcal{Y}_1\oplus\mathcal{Y}_2)$ *be the function defined in (8.2) where* $n=\dim(\mathcal{U}_1)$. *Then* $d_\infty=\|A(G)\|$. *Moreover, given any* $\delta>1$, *there exists an* $H$ *in* $H^\infty(\mathcal{U}_1,\mathcal{Y}_1)$ *satisfying the following mixed* $H^2$ *and* $H^\infty$ *bounds:*

$$\begin{gathered} \left\| \begin{bmatrix} F_{11}-H & F_{12} \\ F_{21} & F_{22} \end{bmatrix} \right\|_\infty \le \delta d_\infty\,, \\ \left\| \begin{bmatrix} F_{11}-H & F_{12} \\ F_{21} & F_{22} \end{bmatrix} \right\|_2 \le \delta d_2 \left[\delta^2-1+\frac{d_{21}^2}{n d_\infty^2}\right]^{-1/2}, \end{gathered} \tag{8.7}$$

*where* $d_{21}^2=\|P_-F_{11}\|_2^2+\|F_{21}\|_2^2$. *In fact, by choosing* $\delta^2=2-d_{21}^2/nd_\infty^2$ *equation (8.7) becomes*

$$\left\| \begin{bmatrix} F_{11}-H & F_{12} \\ F_{21} & F_{22} \end{bmatrix} \right\|_\infty \le d_\infty \sqrt{2-\frac{d_{21}^2}{n d_\infty^2}}\,, \tag{8.8}$$

$$\left\|\begin{bmatrix} F_{11} - H & F_{12} \\ F_{21} & F_{22} \end{bmatrix}\right\|_2 \le d_2 \sqrt{2 - \frac{d_{21}^2}{n d_\infty^2}} .$$

**PROOF**. First notice that the operator $A(\Psi) = 0$ for any symbol $\Psi$ in the space $L^\infty(\mathcal{U}_1 \oplus \mathcal{U}_2, \mathcal{Y}_1 \oplus \mathcal{Y}_2)$ if and only if

$$\Psi \in \begin{bmatrix} H^\infty(\mathcal{U}_1, \mathcal{Y}_1) & 0 \\ 0 & 0 \end{bmatrix} . \tag{8.9}$$

Obviously, if $\Psi$ satisfies (8.9), then $A(\Psi) = 0$. To verify the other half, assume that $\Psi$ is given by

$$\Psi = \begin{bmatrix} \psi_{11} & \psi_{12} \\ \psi_{21} & \psi_{22} \end{bmatrix}$$

and $A(\Psi) = 0$. Clearly this implies that both $\psi_{21}$ and $\psi_{22}$ are zero. Since $A(\Psi)(0 \oplus y) = 0$ for all y on $L^2(\mathcal{U}_2)$, we see that $P_- \psi_{12} e^{-int} \mathcal{U}_2 = 0$ for all $n \ge 0$. Hence $\psi_{12} = 0$. So the only possible nonzero term in $\Psi$ is $\psi_{11}$. Finally, using $0 = A(\Psi)(a \oplus 0) = P_- \psi_{11} a$ for all a in $\mathcal{U}_1$, we see that $\psi_{11}$ is in $H^\infty(\mathcal{U}_1, \mathcal{Y}_1)$. Therefore $A(\Psi) = 0$ if and only if $\Psi$ satisfies (8.9).

Now let S be the unilateral shift on $H^2(\mathcal{U}_1)$ and $V_2$ the bilateral shift (multiplication by $e^{it}$) on $L^2(\mathcal{U}_2)$. Let T be the isometry on $\mathcal{H} = H^2(\mathcal{U}_1) \oplus L^2(\mathcal{U}_2)$ defined by $T = S \oplus V_2$. Let $V_1'$ and $V_2'$ be the bilateral shifts on $L^2(\mathcal{Y}_1)$ and $L^2(\mathcal{Y}_2)$, respectively. Let $T'$ on $\mathcal{H}' = K^2(\mathcal{Y}_1) \oplus L^2(\mathcal{Y}_2)$ be the co-isometry defined by $T' = (P_- V_1' \mid K^2(\mathcal{Y}_1)) \oplus V_2'$. Obviously, $V' = V_1' \oplus V_2'$ is the minimal isometric dilation of $T'$. Using $P_{\mathcal{H}'} V' = T' P_{\mathcal{H}'}$ it follows that $T'A(G) = A(G)T$. According to Corollary IV.8.3, the central intertwining lifting $B_\gamma$ mapping $\mathcal{H} = H^2(\mathcal{U}_1) \oplus L^2(\mathcal{U}_2)$ into $\mathcal{K}' = L^2(\mathcal{Y}_1) \oplus L^2(\mathcal{Y}_2)$ of A satisfies the bounds in (IV.8.15) and (IV.8.16) where $\gamma = \delta\|A\|$. Thus $\|B_\gamma\| \le \delta\|A\|$.

Using $V'B_\gamma = B_\gamma T$ it follows that $B_\gamma = M_{G_\gamma} \mid \mathcal{H}$ where $G_\gamma$ is a uniquely determined function in $L^\infty(\mathcal{U}_1 \oplus \mathcal{U}_2, \mathcal{Y}_1 \oplus \mathcal{Y}_2)$ and $\|B_\gamma\| = \|G_\gamma\|_\infty$. To see this notice that the bilateral shift V on $L^2(\mathcal{U}_1 \oplus \mathcal{U}_2)$ is a minimal unitary extension of T. According to Lemma II.3.1, there exists a unique operator $\hat{B}_\gamma$ from $L^2(\mathcal{U}_1 \oplus \mathcal{U}_2)$ into $L^2(\mathcal{Y}_1 \oplus \mathcal{Y}_2)$ satisfying

$$\hat{B}_\gamma \mid \mathcal{H} = B_\gamma , \; V'\hat{B}_\gamma = \hat{B}_\gamma V \;\text{ and }\; \|\hat{B}_\gamma\| = \|B_\gamma\| .$$

Since $\hat{B}_\gamma$ intertwines two bilateral shifts, there exists a function $G_\gamma$ in $L^\infty(\mathcal{U}_1 \oplus \mathcal{U}_2, \mathcal{Y}_1 \oplus \mathcal{Y}_2)$ such that $\hat{B}_\gamma = M_{G_\gamma}$. Hence $\|B_\gamma\| = \|\hat{B}_\gamma\| = \|G_\gamma\|_\infty$. Finally, employing $\hat{B}_\gamma \mid \mathcal{H} = B_\gamma$ it follows that $B_\gamma = M_{G_\gamma} \mid \mathcal{H}$ which proves our claim.

Using $B_\gamma = M_{G_\gamma} | \mathcal{H}$ along with the fact that $B_\gamma$ is an intertwining lifting of A yields

$$A(G_\gamma) = P_{\mathcal{H}'} M_{G_\gamma} | \mathcal{H} = P_{\mathcal{H}'} B_\gamma = A(G) .$$

Therefore $A(G - G_\gamma) = 0$. In other words, $\Psi = G - G_\gamma$ satisfies (8.9). So there exists an H in $H^\infty(\mathcal{U}_1, \mathcal{Y}_1)$ such that

$$G_\gamma = \begin{bmatrix} F_{11} - H & F_{12} \\ F_{21} & F_{22} \end{bmatrix} . \tag{8.10}$$

Moreover, $\|G_\gamma\|_\infty = \|B_\gamma\| \le \delta\|A\|$. However, if G′ is any function given by the right hand side of (8.10) where H is any function in $H^\infty(\mathcal{U}_1, \mathcal{Y}_1)$, then $A(G) = A = A(G')$. This implies that $\|A\| \le \|G'\|_\infty$. By taking the infimum of $\|G'\|_\infty$ over all H in $H^\infty(\mathcal{U}_1, \mathcal{Y}_1)$ we see that $\|A\| \le d_\infty$. Therefore we have

$$\|A\| \le d_\infty \le \|G_\gamma\|_\infty = \|B_\gamma\| \le \delta\|A\| .$$

For $\delta = 1$, we obtain equality in the previous formula and thus $d_\infty = \|A\|$. Furthermore, if we choose $\delta = 1$, then $\|G_\gamma\| = d_\infty$ and there exists a H in $H^\infty(\mathcal{U}_1, \mathcal{Y}_1)$ satisfying (8.6).

Notice that $\mathcal{U}_1 = \ker(T^*) = \mathcal{L}$. Moreover,

$$d_{21}^2 = \|P_- F_{11}\|_2^2 + \|F_{21}\|_2^2 = \|A | \mathcal{U}_1\|_{HS}^2 . \tag{8.11}$$

So the inequality in equation (IV.8.16) of Corollary IV.8.3 becomes

$$\|B_\gamma | \mathcal{U}_1\|_{HS}^2 \le \frac{\delta^2 \|A | \mathcal{U}_1\|_2^2}{\delta^2 - 1 + \dfrac{d_{21}^2}{n d_\infty^2}} . \tag{8.12}$$

According to (8.10) we see that

$$\|B_\gamma | \mathcal{U}_2\|_{HS}^2 = \|M_{G_\gamma} | \mathcal{U}_2\|_{HS}^2 = \|M_G | \mathcal{U}_2\|_{HS}^2 = \|F_{12}\|_2^2 + \|F_{22}\|_2^2 .$$

Using this equation (8.12) along with (8.11), and the fact that $\delta^2(\delta^2 - 1 + d_{21}^2/nd_\infty^2)^{-1}$ is greater than or equal to one (because $d_{21}^2 \le nd_\infty^2$), we have

$$\|G_\gamma\|_2^2 = \|M_{G_\gamma} | (\mathcal{U}_1 \oplus \mathcal{U}_2)\|_{HS}^2 = \|B_\gamma | \mathcal{U}_1\|_{HS}^2 + \|B_\gamma | \mathcal{U}_2\|_{HS}^2 \leq$$

$$\frac{\delta^2 \|A | \mathcal{U}_1\|^2}{\delta^2 - 1 + \dfrac{d_{21}^2}{nd_\infty^2}} + \|F_{12}\|_2^2 + \|F_{22}\|_2^2 \leq$$

$$\frac{\delta(\|P_-F_{11}\|_2^2 + \|F_{21}\|_2^2 + \|F_{12}\|_2^2 + \|F_{22}\|_2^2)}{\delta^2 - 1 + \dfrac{d_{21}^2}{nd_\infty^2}} = \frac{\delta^2 d_2^2}{\delta^2 - 1 + \dfrac{d_{21}^2}{nd_\infty^2}} .$$

Obviously $\|G_\gamma\|_\infty = \|B_\gamma\| \leq \delta d_\infty$. This along with the form of $G_\gamma$ in (8.10) completes the proof.

By choosing $\mathcal{U}_2 = \mathcal{Y}_2 = \{0\}$ with $\mathcal{U} = \mathcal{U}_1$ and $\mathcal{Y} = \mathcal{Y}_1$, we arrive at the following $H^2$ and $H^\infty$ Nehari optimization problems associated with:

$$d_i = \inf \{\|F - H\|_i : H \in H^\infty(\mathcal{U}, \mathcal{Y})\} \qquad (\text{for } i = 2, \infty) \tag{8.13}$$

where F is a specified function in $L^\infty(\mathcal{U}, \mathcal{Y})$. In this case the optimal $H^2$ solution is $d_2 = \|P_-F\|_2$ and $H = P_+F$. Section I.6 shows that the error $d_\infty$ to the $H^\infty$ Nehari optimization problem in (8.13) is $d_\infty = \|A\|$ where A is the Hankel operator from $H^2(\mathcal{U})$ into $K^2(\mathcal{Y})$ defined by $A = P_-M_F | H^2(\mathcal{U})$. Moreover, there exists a function in $H^\infty(\mathcal{U}, \mathcal{Y})$ satisfying $d_\infty = \|F - H\|_\infty$. The previous theorem immediately gives the following result, which is also a special case of Corollary IV.9.2.

**COROLLARY 8.2.** *Let* F *be a function in* $L^\infty(\mathcal{U}, \mathcal{Y})$ *and* $n = \dim(\mathcal{U})$. *Let* $d_2$ *and* $d_\infty$ *be the error in the optimization problems in (8.13). Then* $d_2 = \|P_-F\|_2$ *and* $d_\infty = \|A\|$ *where* $A = P_-M_F | H^2(\mathcal{U})$. *Moreover, given any* $\delta > 1$, *there exists a* H *in* $H^\infty$ *satisfying the following mixed* $H^2$ *and* $H^\infty$ *bounds*

$$\|F - H\|_\infty \leq \delta d_\infty \quad \textit{and} \quad \|F - H\|_2 \leq \frac{\delta d_2}{\sqrt{\delta^2 - 1 + d_2^2/nd_\infty^2}} . \tag{8.14}$$

*In fact by choosing* $\delta^2 = 2 - d_2^2/nd_\infty^2$ *equation (8.14) becomes*

$$\|F - H\|_\infty \leq d_\infty \sqrt{2 - \frac{d_2^2}{nd_\infty^2}} \quad \textit{and} \quad \|F - H\|_2 \leq d_2 \sqrt{2 - \frac{d_2^2}{nd_\infty^2}} . \tag{8.15}$$

*Finally, if* $B_\gamma$ *is the central intertwining lifting of* A *with tolerance* $\gamma = \delta d_\infty$, *then* $B_\gamma = M_{G_\gamma} | H^2(\mathcal{U})$ *where* $G_\gamma = F - H$ *for some* H *in* $H^\infty(\mathcal{U}, \mathcal{Y})$ *and this* G – H *satisfies (8.14).*

To complete this section we will show how Theorem IV.6.6 can be used to give an explicit formula to solve a general two block $H^2$ and $H^\infty$ Nehari problem. To this end, let $G = [F_{11}, F_{21}]^{tr}$ be a fixed function in $L^\infty(\mathcal{U}, \mathcal{Y}_1 \oplus \mathcal{Y}_2)$. In this case, $\mathcal{U}_2 = \{0\}$, and the $H^2$ and $H^\infty$ optimization problems in (8.3) reduces to

$$d_i = \inf\left\{ \left\| \begin{bmatrix} F_{11} - H \\ F_{21} \end{bmatrix} \right\|_i : H \in H^\infty(\mathcal{U}, \mathcal{Y}_1) \right\} \qquad (\text{for } i = 2, \infty)\,. \tag{8.16}$$

The two block problem considered in Section 7 is a special case of the two block problem in (8.16), because here the functions $F_{11}$ and $F_{21}$ are not required to be in the appropriate $K^\infty$ spaces. In this setting, the operator $A = A(G)$ mapping $H^2(\mathcal{U})$ into $K^2(\mathcal{Y}_1) \oplus L^2(\mathcal{Y}_2)$ is defined by

$$Au = \begin{bmatrix} P_- F_{11} u \\ F_{21} u \end{bmatrix} \qquad (u \in H^2(\mathcal{U}))\,.$$

As before, it is easy to show that $T'A = AT$ where $T = S$ is now the unilateral shift on $\mathcal{H} = H^2(\mathcal{U})$ and $T'$ is the compression of $V_1' \oplus V_2'$ to $\mathcal{H}' = K^2(\mathcal{Y}_1) \oplus L^2(\mathcal{Y}_2)$. Obviously, $d_\infty = \|A\|$ and

$$d_2^2 = \|P_- F_{11}\|_2^2 + \|F_{21}\|_2^2 = \|A \mid \mathcal{U}\|_{HS}^2 = d_{21}^2\,. \tag{8.17}$$

Finally, notice that in this case $\mathcal{L} = \ker(T^*) = \mathcal{U}$ and $\Pi_o$ is the operator from $H^2(\mathcal{U})$ onto $\mathcal{U}$ defined by $\Pi_o = P_{\mathcal{U}}$, which picks out the zero degree component of a function in $H^2(\mathcal{U})$. Recall that $D_A^2 = \gamma^2 I - A^*A$. The following result provides us with a simple formula to compute the central intertwining lifting $B_\gamma$ for the two block problem corresponding to (8.16).

**THEOREM 8.3.** *Let* $G = [F_{11}, F_{21}]^{tr}$ *be a function in* $L^\infty(\mathcal{U}, \mathcal{Y}_1 \oplus \mathcal{Y}_2)$ *and* $A = A(G)$ *its corresponding linear operator, where* $n = \dim(\mathcal{U})$ *is finite. Let* $\delta > 1$ *and* $\gamma = \delta d_\infty$. *Let* $P(e^{it})$ *and* $Q(\lambda)$ *be the functions defined by*

$$P(e^{it}) = (AD_A^{-2}\Pi_o^*)(e^{it}) \quad \textit{and} \quad Q(\lambda) = (D_A^{-2}\Pi_o^*)(\lambda)\,. \tag{8.18}$$

*Then the central intertwining lifting* $B_\gamma$ *of* $A$ *with tolerance* $\gamma$ *is given by* $B_\gamma = M_{G_\gamma} \mid H^2(\mathcal{U})$ *where* $G_\gamma$ *is the function in* $L^\infty(\mathcal{U}, \mathcal{Y}_1 \oplus \mathcal{Y}_2)$ *defined by*

$$G_\gamma(e^{it}) = P(e^{it})Q(e^{it})^{-1}\,. \tag{8.19}$$

*In particular, this* $G_\gamma$ *admits a decomposition of the form*

$$G_\gamma = \begin{bmatrix} F_{11} - H \\ F_{21} \end{bmatrix} \tag{8.20}$$

*where* H *is a function in* $H^\infty(\mathcal{U}, \mathcal{Y}_1)$. *Furthermore, the central interpolant* $G_\gamma$ *satisfies the following* $H^2$ *and* $H^\infty$ *bounds*

$$\left\| \begin{bmatrix} F_{11} - H \\ F_{21} \end{bmatrix} \right\|_\infty \leq \delta d_\infty \quad and \quad \left\| \begin{bmatrix} F_{11} - H \\ F_{21} \end{bmatrix} \right\|_2 \leq \delta d_2 \left( \delta^2 - 1 + \frac{d_2^2}{n d_\infty^2} \right)^{-1/2} . \tag{8.21}$$

*Finally, if we choose* $\delta^2 = 2 - d_2^2/nd_\infty^2$ *we obtain*

$$\left\| \begin{bmatrix} F_{11} - H \\ F_{21} \end{bmatrix} \right\|_\infty \leq d_\infty \sqrt{2 - \frac{d_2^2}{n d_\infty^2}} \quad and \quad \left\| \begin{bmatrix} F_{11} - H \\ F_{21} \end{bmatrix} \right\|_2 \leq d_2 \sqrt{2 - \frac{d_2^2}{n d_\infty^2}} . \tag{8.22}$$

Before proving this result let us first notice that according to Theorem IV.6.6 the function $Q(\lambda)$ is an outer function and $Q(\lambda)^{-1}$ is in $H^\infty(\mathcal{U}, \mathcal{U})$. Moreover, the function $Q(0)^{1/2}Q(\lambda)^{-1}$ is an outer spectral factor for the Toeplitz operator $\gamma^2 I - B_\gamma^* B_\gamma$, or equivalently, $Q(0)^{1/2}Q(\lambda)^{-1}$ is an outer spectral factor for $\gamma^2 I - G_\gamma^* G_\gamma$.

**PROOF OF THEOREM 8.3**. Obviously $V' = V_1' \oplus V_2'$ is a minimal isometric dilation of $T'$. According to Theorem IV.6.6 the central intertwining lifting $B_\gamma = M_{G_\gamma} \mid H^2(\mathcal{U})$ where $G_\gamma$ is the function in $L^\infty(\mathcal{U}, \mathcal{Y}_1 \oplus \mathcal{Y}_2)$ defined in (8.19). The fact that $G_\gamma$ admits a decomposition of the form (8.20) and satisfies the $H^2$ and $H^\infty$ bounds in (8.21) readily follows from (8.17) and the proof of Theorem 8.1. This completes the proof.

**REMARK 8.4**. In many applications it is hard to invert the infinite dimensional operator $D_A^{-2}$ in (8.18). (As before we assume that $\|A\| < \gamma$.) To complete this section we will follow some ideas in Chu-Doyle-Lee [1], and Francis [1] to help alleviate this problem. Let $T_o$ be the Toeplitz operator on $H^2(\mathcal{U})$ defined by

$$T_o = P_+(\gamma^2 I - M_{F_{21}}^* M_{F_{21}}) \mid H^2(\mathcal{U}) .$$

We claim that the Toeplitz operator $T_o$ is strictly positive. This follows because $\|A\| < \gamma$. So for some $\varepsilon > 0$ we have for all h in $\mathcal{H} = H^2(\mathcal{U})$

$$\varepsilon\|h\|^2 \leq \gamma^2\|h\|^2 - \|A(G)h\|^2 = \gamma^2\|h\|^2 - \|F_{21}h\|^2 - \|P_- F_{11} h\|^2 \leq (T_o h, h) .$$

Hence $T_o$ is strictly positive. Let $\Theta$ in $H^\infty(\mathcal{U}, \mathcal{U})$ be the outer spectral factor for $T_o$, that is, $\Theta$

is an outer function and $(M_{\Theta}^{+})^{*}M_{\Theta}^{+} = T_o$. Since $T_o$ is strictly positive, $\Theta(\lambda)^{-1}$ is also in $H^{\infty}(\mathcal{U}, \mathcal{U})$. Let $A_{11}$ be the Hankel operator from $H^2(\mathcal{U})$ to $K^2(\mathcal{Y}_1)$ defined by $A_{11} = P_{-}M_{F_{11}} | H^2(\mathcal{U})$. Then

$$D_A^{-2}\Pi_o^{*} = (T_o - A_{11}^{*}A_{11})^{-1}\Pi_o^{*} = ((M_{\Theta}^{+})^{*}M_{\Theta}^{+} - A_{11}^{*}A_{11})^{-1}\Pi_o^{*} =$$

$$(M_{\Theta}^{+})^{-1}(I - (M_{\Theta}^{+})^{-*}A_{11}^{*}A_{11}(M_{\Theta}^{+})^{-1})^{-1}(M_{\Theta}^{+})^{-*}\Pi_o^{*} = (M_{\Theta}^{+})^{-1}(I - A_{new}^{*}A_{new})^{-1}\Pi_o^{*}\Theta(0)^{-*}$$

where $A_{new}$ is the strictly contractive Hankel operator from $H^2(\mathcal{U})$ to $K^2(\mathcal{Y}_1)$ defined by $A_{new} = A_{11}(M_{\Theta}^{+})^{-1}$. (Recall that $N^{-*}$ is the inverse of $N^{*}$.) Notice that $A_{new}$ is strictly contractive because $(M_{\Theta}^{+})^{*}(I - A_{new}^{*}A_{new})M_{\Theta}^{+} = D_A^2$ is strictly positive. By canceling out $\Theta(0)$, the functions P and Q in (8.18) can be replaced by

$$P_n(e^{it}) = (A(M_{\Theta}^{+})^{-1}D_{new}^{-2}\Pi_o^{*})(e^{it}) \quad \text{and} \quad Q_n(\lambda) = \Theta(\lambda)^{-1}(D_{new}^{-2}\Pi_o^{*})(\lambda)\,, \tag{8.23}$$

where $D_{new}^2 = I - A_{new}^{*}A_{new}$. As before, the central intertwining lifting $B_{\gamma}$ of A is given by $M_{G_{\gamma}} | H^2(\mathcal{U})$ where $G_{\gamma} = P_n Q_n^{-1}$. If we let $P_{new}$ and $Q_{new}$ be the functions defined by

$$P_{new} = (A_{new}D_{new}^{-2}\Pi_o^{*})(e^{it}) \quad \text{and} \quad Q_{new} = (D_{new}^{-2}\Pi_o^{*})(\lambda)\,, \tag{8.24}$$

then we see that the function $G_{\gamma}$ determined by the central intertwining lifting $B_{\gamma} = M_{G_{\gamma}} | H^2(\mathcal{U})$ of A is given by (8.20) where

$$F_{11} - H = P_{new}Q_{new}^{-1}\Theta\,. \tag{8.25}$$

The formulas for $P_{new}$ and $Q_{new}$ in (8.24) are easier to use than the formulas for P and Q in (8.19) in the rational case. Because if both $F_{11}$ and $F_{21}$ are rational functions, then $A_{11}$ is a finite rank Hankel operator and the outer function $\Theta$ is rational. This implies that the Hankel operator $A_{new}$ is finite rank. Therefore one can use standard state space techniques to compute $\Theta$ and $D_{new}^{-2}\Pi_o^{*}$. (The outer factor $\Theta$ is usually computed by solving a Riccati equation, and $d_{\infty}$ is computed by an iteration on $\gamma$; see Section 7 and the Appendix.) So in the rational setting one can compute state space formulas for $P_{new}$ and $Q_{new}$ in (8.24) and thus find the central intertwining lifting $B_{\gamma}$. The details are omitted. For further results on converting two and four block problems to standard Nehari problems see Chu-Doyle-Lee [1], Green-Limebeer [1]; and for its corresponding central solution in the setting of the commutant lifting theorem see Frazho [1].

Finally, let us notice that in the Nehari or one block setting, that is, when $F_{21} = 0$ or $\mathcal{Y}_2 = \{0\}$, our formula for $G_{\gamma} = PQ^{-1}$ is precisely the central solution for the Nehari problem.

*Notes to Chapter V:*

The state space formulas for the central solution in this chapter are derived by specifying the results of the previous chapter for different concrete interpolation problems. Some of these formulas appeared before (in Foias-Frazho [5] and Frazho-Kherat [1]), and others (for example the ones for the Nudelman, the Sarason and the two block problems) are new. The state space formulas for the central solution in the Nehari problem were first obtained probably in Gohberg-Kaashoek-Van Schagen [1] (see also Section XXXV.7 in Gohberg-Goldberg-Kaashoek [2]).

## CHAPTER VI

# PARAMETERIZATION OF INTERTWINING LIFTINGS AND ITS APPLICATIONS

In this chapter it is shown that the set of all intertwining liftings with tolerance $\gamma$ in the commutant lifting theorem is parameterized by a natural set of contractive analytic operator-valued functions. A Redheffer scattering interpolation of this parameterization is given. In the final three sections, state space formulas are given for the Redheffer scattering matrix parameterizing the set of all solutions for the standard left Nevanlinna-Pick, Nehari and two-block interpolation problems.

## VI.1 THE MÖBIUS TRANSFORMATION

In this section we will introduce the Möbius transform for a contraction and present some elementary facts concerning the Möbius transform and its minimal isometric dilation. In the following sections the Möbius transformation will be used to prove that our Schur parameterization for the commutant lifting theorem yields all intertwining liftings with tolerance $\gamma$.

Let $b(\lambda)$ and $a(\lambda)$ be the Möbius transforms defined by

$$b(\lambda) = \frac{\lambda - \alpha}{1 - \bar{\alpha}\lambda} \quad \text{and} \quad a(\lambda) = \frac{\lambda + \alpha}{1 + \bar{\alpha}\lambda} \tag{1.1}$$

where $\alpha$ is a fixed complex number in the open unit disc. Recall also that b (or a) is a conformal mapping which maps the open disc in a one to one way onto its self. Clearly one obtains a by replacing $\alpha$ by $-\alpha$ in b. Moreover, it is easy to verify that $\lambda = b(a(\lambda))$ and $\lambda = a(b(\lambda))$.

Now let X be a contraction on $\mathcal{H}$. Then the Möbius transform of X is the operator on $\mathcal{H}$ defined by

$$X_\alpha = b(X) = (X - \alpha I)(I - \bar{\alpha}X)^{-1}\,. \tag{1.2}$$

Obviously $X_{-\alpha} = a(X)$. Moreover, it is easy to show that $X = b(a(X))$. In particular, $\mathcal{H}_1$ is an invariant subspace for X if and only if $\mathcal{H}_1$ is an invariant subspace for $X_\alpha$.

We claim that $X_\alpha$ is also a contraction. To see this, let h be in $\mathcal{H}$. Then using $g = (I - \bar{\alpha}X)^{-1}h$ we have

$$\|h\|^2 - \|X_\alpha h\|^2 = \|(I - \bar{\alpha}X)g\|^2 - \|(X - \alpha I)g\|^2 =$$

$$\|g\|^2 - 2\,\mathrm{Re}\,(g, \bar{\alpha}Xg) + |\alpha|^2\|Xg\|^2 - \|Xg\|^2 + 2\,\mathrm{Re}\,(Xg, \alpha g) - |\alpha|^2\|g\|^2 =$$

$$(1 - |\alpha|^2)\|D_X g\|^2 = d_\alpha^2\|D_X(I - \bar{\alpha}X)^{-1}h\|^2 \geq 0$$

where $D_X = D_{X,1}$ is now the positive square root of $I - X^*X$ and $d_\alpha = (1 - |\alpha|^2)^{1/2}$. Therefore $X_\alpha$ is a contraction. From the previous equation we see that there exists a unitary operator $Z_\alpha$ from $\mathcal{D}_{X_\alpha}$ onto $\mathcal{D}_X$ satisfying

$$Z_\alpha D_{X_\alpha} = d_\alpha D_X(I - \bar{\alpha}X)^{-1}\ . \tag{1.3}$$

Here $D_{X_\alpha}$ is the positive square root of $I - X_\alpha^* X_\alpha$. Finally, equation (1.3) shows that X is an isometry if and only if $X_\alpha$ is an isometry. In particular, by replacing X by $X^*$, it follows that X is unitary if and only if $X_\alpha$ is unitary.

Clearly the unilateral shift S is unitarily equivalent to its corresponding Möbius transform $S_\alpha$. This fact is also established in the next lemma where the unitary equivalence is explicitly given.

**LEMMA 1.1.** *Let* $b(\lambda)$ *and* $a(\lambda)$ *be the Möbius transforms defined in (1.1) for some fixed* $\alpha$ *in the open disc. Let* S *be the unilateral shift on* $H^2(\mathcal{D})$ *(where* $\mathcal{D}$ *is any Hilbert space) and* $\Phi_{\mathcal{D}}$ *the operator defined on* $H^2(\mathcal{D})$ *by*

$$(\Phi_{\mathcal{D}} f)(\lambda) = \frac{d_\alpha f(b(\lambda))}{1 - \bar{\alpha}\lambda} \qquad (|\lambda| < 1 \ \textit{and} \ f \in H^2(\mathcal{D}))\ . \tag{1.4}$$

*Then* $\Phi_{\mathcal{D}}$ *is a unitary operator on* $H^2(\mathcal{D})$ *satisfying* $\Phi_{\mathcal{D}} S = S_\alpha \Phi_{\mathcal{D}}$. *Moreover, the inverse of* $\Phi_{\mathcal{D}}$ *is given by*

$$(\Phi_{\mathcal{D}}^* g)(\lambda) = \frac{d_\alpha g(a(\lambda))}{1 + \bar{\alpha}\lambda} \qquad (g \in H^2(\mathcal{D}))\ . \tag{1.5}$$

**PROOF.** Using $e^{it} = b(e^{i\sigma})$ a simple calculation shows that

$$dt = \frac{d_\alpha^2 d\sigma}{|1 - \bar{\alpha}e^{i\sigma}|^2}\ . \tag{1.6}$$

In particular, if f is in $H^2(\mathcal{D})$, then

$$\|f\|^2 = \frac{1}{2\pi}\int_0^{2\pi} \|f(e^{it})\|^2 dt = \frac{1}{2\pi}\int_0^{2\pi} \frac{d_\alpha^2 \|f(b(e^{i\sigma}))\|^2}{|1-\bar{\alpha}e^{i\sigma}|^2} d\sigma = \|g\|^2 , \tag{1.7}$$

where g is the analytic function defined by

$$g(\lambda) = \frac{d_\alpha f(b(\lambda))}{1-\bar{\alpha}\lambda} = (\Phi_{\mathcal{D}} f)(\lambda) \qquad (|\lambda| < 1) . \tag{1.8}$$

Equation (1.7) shows that g is in $H^2(\mathcal{D})$ and $\Phi_{\mathcal{D}}$ is an isometry.

To obtain f from g notice that by using $b(a(\lambda)) = \lambda$ in (1.8) we obtain

$$g(a(\lambda)) = \frac{d_\alpha f(b(a(\lambda)))}{1-\bar{\alpha}a(\lambda)} = \frac{(1+\bar{\alpha}\lambda)f(\lambda)}{d_\alpha} .$$

So the inverse of $\Phi_{\mathcal{D}}$ is given by

$$\Phi_{\mathcal{D}}^{-1} g = f(\lambda) = \frac{d_\alpha g(a(\lambda))}{1+\bar{\alpha}\lambda} . \tag{1.9}$$

Now let g be any function in $H^2(\mathcal{D})$. Then equation (1.7) with f, b and $\alpha$ replaced by g, a and $-\alpha$ respectively shows that

$$\|g\|^2 = \frac{1}{2\pi}\int_0^{2\pi} \frac{d_\alpha^2 \|g(a(e^{it}))\|^2 dt}{|1+\bar{\alpha}e^{it}|^2}$$

is finite. In particular, the function f defined by (1.9) is in $H^2(\mathcal{D})$. Moreover, $\Phi_{\mathcal{D}} f = \Phi_{\mathcal{D}} \Phi_{\mathcal{D}}^{-1} g = g$. Because g is now an arbitrary vector in $H^2(\mathcal{D})$ it follows that $\Phi_{\mathcal{D}}$ is onto. Therefore $\Phi_{\mathcal{D}}$ is a unitary operator on $H^2(\mathcal{D})$.

Using $S_\alpha = b(S)$ along with the definition of $\Phi_{\mathcal{D}}$ in (1.4) we have

$$\Phi_{\mathcal{D}} Sf = \frac{d_\alpha b(\lambda) f(b(\lambda))}{1-\bar{\alpha}\lambda} = \frac{b(S_\alpha) d_\alpha f(b(\lambda))}{1-\bar{\alpha}\lambda} = S_\alpha \Phi_{\mathcal{D}} f .$$

So $\Phi_{\mathcal{D}}$ intertwines S with $S_\alpha$. This completes the proof.

As before, let $T'$ be a contraction on $\mathcal{H}'$ and $U'$ on $\mathcal{K}' = \mathcal{H}' \oplus H^2(\mathcal{D}')$ the Sz.-Nagy-Schäffer minimal isometric dilation of $T'$, that is,

$$U' = \begin{bmatrix} T' & 0 \\ D' & S \end{bmatrix} \text{ on } \begin{bmatrix} \mathcal{H}' \\ H^2(\mathcal{D}') \end{bmatrix} \tag{1.10}$$

where $D'$ is the positive square root of $I - T'^*T'$ while $\mathcal{D}'$ is the closure of the range of $D'$ and S is the unilateral shift on $H^2(\mathcal{D}')$. Recall that in the definition of the Sz.-Nagy-Schäffer minimal isometric dilation $D'$ maps $\mathcal{H}'$ into the constant functions in $H^2(\mathcal{D}')$; see Section IV.1 for further details. It is easy to show that $U'_\alpha = b(U')$ is a minimal isometric dilation of $T'_\alpha = b(T')$. Indeed, $U'_\alpha$ is an isometry and $\mathcal{H}'$ is an invariant subspace for $U'^*_\alpha$ satisfying $U'^*_\alpha \mid \mathcal{H}' = T'^*_\alpha$. The minimality condition $\mathcal{K}' = \bigvee \{U'^n_\alpha \mathcal{H}' : n \geq 0\}$ follows from the fact that $U' = a(U'_\alpha)$. Thus $\{U'^n \mathcal{H}' : n \geq 0\}$ and $\{U'^n_\alpha \mathcal{H}' : n \geq 0\}$ span the same space. Therefore, $U'_\alpha$ is the minimal isometric dilation of $T'_\alpha$. Finally, a straightforward calculation shows that

$$U'_\alpha = \begin{bmatrix} T'_\alpha & 0 \\ \dfrac{d^2_\alpha D'(1-\bar{\alpha}T')^{-1}}{1-\bar{\alpha}\lambda} & S_\alpha \end{bmatrix} \text{ on } \begin{bmatrix} \mathcal{H}' \\ H^2(\mathcal{D}') \end{bmatrix} \tag{1.11}$$

where $S_\alpha = b(S)$.

Let $U_\alpha$ be the Sz.-Nagy-Schäffer minimal isometric dilation of $T'_\alpha$, that is, $U_\alpha$ is the isometry on $\mathcal{K}_\alpha = \mathcal{H}' \oplus H^2(\mathcal{D}'_\alpha)$ defined by

$$U_\alpha = \begin{bmatrix} T'_\alpha & 0 \\ D'_\alpha & S_1 \end{bmatrix} \text{ on } \mathcal{H}' \oplus H^2(\mathcal{D}'_\alpha) \tag{1.12}$$

where $S_1$ is the unilateral shift on $H^2(\mathcal{D}'_\alpha)$. Here $D'_\alpha$ is the positive square root of $I - T'^*_\alpha T'_\alpha$ while $\mathcal{D}'_\alpha$ is the closure of the range of $D'_\alpha$. Since all minimal isometric dilations of $T'_\alpha$ are isomorphic (see Theorem IV.1.1), there exists a unique unitary operator $\Phi$ from $\mathcal{K}_\alpha$ onto $\mathcal{K}'$ satisfying

$$\Phi U_\alpha = U'_\alpha \Phi \quad \text{and} \quad \Phi \mid \mathcal{H}' = I. \tag{1.13}$$

To complete this section, let us present an explicit formula for $\Phi$. For this purpose let $Z_\alpha$ be the unitary operator from $\mathcal{D}'_\alpha$ onto $\mathcal{D}'$ defined according to (1.3) where $X = T'$, that is,

$$Z_\alpha D'_\alpha = d_\alpha D'(I - \bar{\alpha}T')^{-1}. \tag{1.14}$$

Now let $Z$ be the unitary operator from $H^2(\mathcal{D}'_\alpha)$ onto $H^2(\mathcal{D}')$ defined by $(Zf)(\lambda) = Z_\alpha f(\lambda)$ for $f$ in $H^2(\mathcal{D}'_\alpha)$. Obviously, $Z$ intertwines the canonical unilateral shift $S_1$ on $H^2(\mathcal{D}'_\alpha)$ with $S$. We claim that the unitary operator $\Phi$ from $\mathcal{K}'_\alpha$ onto $\mathcal{K}'$ satisfying (1.13) is given by

$$\Phi = \begin{bmatrix} I & 0 \\ 0 & \Phi_{\mathcal{D}'} Z \end{bmatrix}. \tag{1.15}$$

Here $\Phi_{\mathcal{D}'}$ is the unitary operator on $H^2(\mathcal{D}')$ defined by (1.4) in Lemma 1.1. Obviously $\Phi$ is a unitary operator satisfying $\Phi \mid \mathcal{H}' = I$. Using (1.14) along with $\Phi_{\mathcal{D}'} S = S_\alpha \Phi_{\mathcal{D}'}$ we have

$$\Phi U_\alpha = \begin{bmatrix} I & 0 \\ 0 & \Phi_{\mathcal{D}'}Z \end{bmatrix} \begin{bmatrix} T'_\alpha & 0 \\ D'_\alpha & S_1 \end{bmatrix} = \begin{bmatrix} T'_\alpha & 0 \\ \dfrac{d_\alpha^2 D'(1-\bar{\alpha}T')^{-1}}{1-\bar{\alpha}\lambda} & \Phi_{\mathcal{D}'}SZ \end{bmatrix} = U'_\alpha \Phi .$$

Therefore $\Phi U_\alpha = U'_\alpha \Phi$ which proves our claim. Summing up our previous analysis we obtain the following result.

**PROPOSITION 1.2.** *Let* $U'$ *on* $\mathcal{K}' = \mathcal{H}' \oplus H^2(\mathcal{D}')$ *be the Sz.-Nagy-Schäffer minimal isometric dilation of a contraction* $T'$ *on* $\mathcal{H}'$ *and* $\alpha$ *be a fixed complex number in the open unit disc. Then* $U'_\alpha = b(U')$ *is a minimal isometric dilation of* $T'_\alpha$. *Moreover, if* $U_\alpha$ *on* $\mathcal{K}_\alpha = \mathcal{H}' \oplus H^2(\mathcal{D}'_\alpha)$ *is the Sz.-Nagy-Schäffer minimal isometric dilation of* $T'_\alpha$, *then* $\Phi U_\alpha = U'_\alpha \Phi$ *where* $\Phi$ *is the unitary operator from* $\mathcal{K}_\alpha$ *onto* $\mathcal{K}'$ *given by (1.15).*

## VI.2 THE SCHUR PARAMETERIZATION

In this section we will give a Schur type parameterization for the set of all intertwining liftings of A with tolerance $\gamma$. Throughout this section, T on $\mathcal{H}$ is an isometry, $T'$ is a contraction on $\mathcal{H}'$ and $U'$ on $\mathcal{K}' = \mathcal{H}' \oplus H^2(\mathcal{D}')$ is the Sz.-Nagy-Schäffer minimal isometric dilation of $T'$ in (1.10). Moreover, A is an operator from $\mathcal{H}$ into $\mathcal{H}'$ intertwining T with $T'$ and the norm of A is bounded by $\gamma$. Recall that an operator B from $\mathcal{H}$ into $\mathcal{K}'$ is called an *intertwining lifting* of A with tolerance $\gamma$ if $P_{\mathcal{H}'}B = A$ and $U'B = BT$ and the norm of B is bounded by $\gamma$. The commutant lifting theorem states that there exists an intertwining lifting B of A with tolerance $\gamma$. In this chapter we will provide a simple proof, based on a Möbius transform method which shows that the set of all intertwining liftings B of A with tolerance $\gamma$ are parameterized by the closed ball of contraction-valued analytic functions in $H^\infty(\mathcal{G}, \mathcal{G}')$.

Recall that $D_A$ is the positive square root of $\gamma^2 I - A^* A$ and $\mathcal{D}_A$ is the closure of the range of $D_A$. To obtain a specific formula for the set of all intertwining liftings B of A with tolerance $\gamma$, let $\Pi'$ be the operator from $\mathcal{D}' \oplus \mathcal{D}_A$ onto $\mathcal{D}'$ which picks out the $\mathcal{D}'$ component, and $\Pi_A$ the operator from $\mathcal{D}' \oplus \mathcal{D}_A$ onto $\mathcal{D}_A$ which picks out the $\mathcal{D}_A$ component, that is,

$$\Pi'(x \oplus y) = x \quad \text{and} \quad \Pi_A(x \oplus y) = y \qquad (x \oplus y \in \mathcal{D}' \oplus \mathcal{D}_A) . \tag{2.1}$$

As before, let $\mathcal{F}$ be the subspace of $\mathcal{D}_A$ defined by $\mathcal{F} = (D_A T \mathcal{H})^-$. Let $\omega$ be isometry from $\mathcal{F}$ into $\mathcal{D}' \oplus \mathcal{D}_A$ defined by

$$\omega D_A Th = D'Ah \oplus D_A h \qquad (h \in \mathcal{H}) . \tag{2.2}$$

The results in Section IV.2 show that $\omega$ is indeed an isometry; see (IV.2.9).

Recall that $\Omega(\lambda)$ is a *contractive analytic function* if $\Omega(\lambda)$ is an analytic function in the open unit disc with values in $\mathcal{L}(\mathcal{U}, \mathcal{Y})$ and $\|\Omega(\lambda)\| \leq 1$ for all $|\lambda| < 1$. We say that $W(\lambda)$ is a *Schur contraction* if $W(\lambda)$ is a contractive analytic function in $H^\infty(\mathcal{D}_A, \mathcal{D}' \oplus \mathcal{D}_A)$ and $W(\lambda) | \mathcal{F} = \omega$ for all $|\lambda| < 1$. Notice that $W(\lambda)$ is a Schur contraction if and only if $W(\lambda)$ is a contractive analytic function in $H^\infty(\mathcal{D}_A, \mathcal{D}' \oplus \mathcal{D}_A)$ and $W(0) | \mathcal{F} = \omega$. To verify this assume that $W(\lambda)$ is a contractive analytic function satisfying $W(0) | \mathcal{F} = \omega$. Now let f be in $\mathcal{F}$ and $\{W_n\}_0^\infty$ the Fourier coefficient of $W(\lambda)$. Then

$$\|f\|^2 \geq \|Wf\|^2 = \sum_{n=0}^{\infty} \|W_n f\|^2 = \|\omega f\|^2 + \sum_{1}^{\infty} \|W_n f\|^2 = \|f\|^2 + \sum_{1}^{\infty} \|W_n f\|^2 . \tag{2.3}$$

Thus $W_n f = 0$ for all $n \geq 1$. This readily implies that $W(\lambda) | \mathcal{F} = \omega$ and W is a Schur contraction, which proves our claim.

Now let $\mathcal{G}$ be the orthogonal complement of $\mathcal{F}$ in $\mathcal{D}_A$, the subspace $\mathcal{F}'$ be the range of $\omega$ and let $\mathcal{G}'$ be the orthogonal complement of $\mathcal{F}'$, that is,

$$\mathcal{G} = \mathcal{D}_A \ominus \overline{D_A T \mathcal{H}}, \ \mathcal{F}' = \{D'Ah \oplus D_A h : h \in \mathcal{H}\}^- \text{ and } \mathcal{G}' = (\mathcal{D}' \oplus \mathcal{D}_A) \ominus \mathcal{F}' . \tag{2.4}$$

We claim that $W(\lambda)$ is a Schur contraction if and only if $W(\lambda)$ admits a decomposition of the form $W(\lambda) = \omega P_{\mathcal{F}} + R(\lambda) P_{\mathcal{G}}$ where R is a contractive analytic function in $H^\infty(\mathcal{G}, \mathcal{G}')$. Obviously, if R is a contractive analytic function in $H^\infty(\mathcal{G}, \mathcal{G}')$, then $W = \omega P_{\mathcal{F}} + RP_{\mathcal{G}}$ is a Schur contraction. On the other hand, if $W(\lambda)$ is a Schur contraction, then $\tilde{W}(\lambda) := W(\bar{\lambda})^*$ is a contractive analytic function in $H^\infty(\mathcal{D}' \oplus \mathcal{D}_A, \mathcal{D}_A)$ satisfying $\tilde{W}(0) | \mathcal{F}' = \omega^*$. Moreover, $\{W_n^*\}_0^\infty$ are the Fourier coefficient of $\tilde{W}(\lambda)$. (Recall that $\{W_n\}_0^\infty$ are the Fourier coefficients of W.) In particular, for any f in $\mathcal{F}'$

$$\|f\|^2 \geq \|\tilde{W}f\|^2 = \|\omega^* f\|^2 + \sum_{1}^{\infty} \|W_n^* f\|^2 = \|f\|^2 + \sum_{1}^{\infty} \|W_n^* f\|^2 .$$

This readily shows that $W_n^* | \mathcal{F}' = 0$ for all $n \geq 1$. According to (2.3) we also have that $W_n | \mathcal{F} = 0$ for all $n \geq 1$. Hence $W_n$ maps $\mathcal{G}$ into $\mathcal{G}'$ for all $n \geq 1$. Because $\omega$ can be viewed as a unitary operator from $\mathcal{F}$ onto $\mathcal{F}'$ and $W_0$ is a contraction in $\mathcal{L}(\mathcal{D}_A, \mathcal{D}' \oplus \mathcal{D}_A)$, it follows that $W_0$ admits a decomposition of the form $W_0 = \omega P_{\mathcal{F}} + R_0 P_{\mathcal{G}}$ where $R_0$ is a contraction in $\mathcal{L}(\mathcal{G}, \mathcal{G}')$. So if we set $R_n = W_n | \mathcal{G}$ for all $n \geq 0$, then

$$W(\lambda) = \sum_{0}^{\infty} \lambda^n W_n = \omega P_{\mathcal{F}} + \sum_{n=0}^{\infty} \lambda^n R_n P_{\mathcal{G}} = \omega P_{\mathcal{F}} + R(\lambda) P_{\mathcal{G}}$$

where R is the analytic function with values in $\mathcal{L}(\mathcal{G}, \mathcal{G}')$ whose Fourier coefficients are $W_n | \mathcal{G}$ for all $n \geq 0$. Because $W(\lambda)$ is a contractive analytic function, it follows that $R(\lambda)$ is a contractive analytic function in $H^\infty(\mathcal{G}, \mathcal{G}')$. This verifies our claim.

We are now ready to state the main result of this section which shows that the set of all intertwining lifting with tolerance $\gamma$ are parameterized by the Schur contractions.

**THEOREM 2.1.** *Let* T *on* $\mathcal{H}$ *be an isometry,* T′ *be a contraction on* $\mathcal{H}'$ *and* U′ *on* $\mathcal{K}' = \mathcal{H}' \oplus H^2(\mathcal{D}')$ *be the Sz.-Nagy-Schäffer minimal isometric dilation in (1.10) of* T′. *Let* A *be an operator from* $\mathcal{H}$ *into* $\mathcal{H}'$ *intertwining* T *with* T′ *and assume that* $\|A\| \le \gamma$. *Then the set of all intertwining liftings* B *of* A *with respect to* U′ *and tolerance* $\gamma$ *are given by*

$$B = \begin{bmatrix} A \\ \Pi' W(\lambda)(I - \lambda \Pi_A W(\lambda))^{-1} D_A \end{bmatrix} : \mathcal{H} \to \begin{bmatrix} \mathcal{H}' \\ H^2(\mathcal{D}') \end{bmatrix}, \tag{2.5}$$

*where* $W(\lambda)$ *is a Schur contraction. Precisely, (2.5) provides a one-to-one correspondence between the set of all intertwining liftings* B *of* A *with tolerance* $\gamma$ *and the set of all Schur contractions.*

Notice that Theorem 2.1 also tells us when the intertwining lifting is unique. In fact, there is precisely one intertwining lifting of B with respect to a minimal isometric dilation U′ and tolerance $\gamma$ if and only if at least one of the spaces $\mathcal{G}$ or $\mathcal{G}'$ in (2.4) is equal to zero.

Recall that any Schur contraction W admits a decomposition of the form $W = W_R = \omega P_{\mathcal{F}} + R P_{\mathcal{G}}$ where R is a contractive analytic function in $H^\infty(\mathcal{G}, \mathcal{G}')$. Therefore the previous theorem shows that there is a one to one correspondence between the set of all intertwining liftings B of A with tolerance $\gamma$ and the set of all contractive analytic functions R in $H^\infty(\mathcal{G}, \mathcal{G}')$. In particular, if we choose $R = 0$, then the Schur contraction $W = \omega P_{\mathcal{F}}$. In this case the intertwining lifting B in (2.5) is precisely the central intertwining lifting $B_\gamma$ studied in Chapter IV. Because $R = 0$ is the center of the closed unit ball in $H^\infty(\mathcal{G}, \mathcal{G}')$, the intertwining lifting $B = B_\gamma$ obtained by choosing $R = 0$, or equivalently, $W = \omega P_{\mathcal{F}}$ is called the *central intertwining lifting* of A.

The proof of Theorem 2.1 will be given over the next several sections. First let us verify that B in (2.5) satisfies $\|B\| \le \gamma$ when W is a Schur contraction. To this end, let h be in $\mathcal{H}$ and $f(\lambda)$ be the analytic function in $|\lambda| < 1$ defined by

$$f(\lambda) = (I - \lambda \Pi_A W(\lambda))^{-1} D_A h, \text{ or equivalently, } (I - \lambda \Pi_A W(\lambda)) f(\lambda) = D_A h. \tag{2.6}$$

Let $f = \sum f_n \lambda^n$ and $W = \sum W_n \lambda^n$ be the power series expansion for f and W, respectively. By matching like coefficients of $\lambda^n$ in (2.6) along with the fact that $W_n$ maps $\mathcal{D}_A$ into $\mathcal{D}' \oplus \mathcal{D}_A$ we obtain

$$\begin{bmatrix} W_0 & 0 & \dots & 0 \\ W_1 & W_0 & \dots & 0 \\ \vdots & \vdots & \dots & \vdots \\ W_{n-1} & W_{n-2} & \dots & W_0 \end{bmatrix} \begin{bmatrix} f_0 \\ f_1 \\ \vdots \\ f_{n-1} \end{bmatrix} = \begin{bmatrix} d_0 \oplus f_1 \\ d_1 \oplus f_2 \\ \vdots \\ d_{n-1} \oplus f_n \end{bmatrix} \tag{2.7}$$

where $f_0 = D_A h$ and $d_j$ are vectors in $\mathcal{D}'$ for all j. Because $W(\lambda)$ is a Schur contraction, the $n \times n$ lower triangular Toeplitz matrix in (2.7) is a contraction for all n. This and $f_0 = D_A h$ gives

$$\|D_A h\|^2 + \sum_{j=1}^{n-1} \|f_j\|^2 \geq \sum_{j=0}^{n-1} \|d_j \oplus f_{j+1}\|^2 = \sum_0^{n-1} \|d_j\|^2 + \sum_1^n \|f_j\|^2 .$$

Using the fact that $\|f_j\|^2$ for $j = 1, \dots, n-1$ appears on both sides of the inequality we obtain

$$\|D_A h\|^2 \geq \sum_0^{n-1} \|d_j\|^2 \qquad \text{(for all n)} .$$

So the function $d(\lambda) = \sum_0^\infty d_j \lambda^j$ is in $H^2(\mathcal{D}')$ and $\|d\| \leq \|D_A h\|$. However, according to the definition of $f(\lambda)$ in (2.6) along with (2.7) we have

$$d(\lambda) = \Pi' W(\lambda) f(\lambda) = \Pi' W(\lambda)(I - \lambda \Pi_A W(\lambda))^{-1} D_A h . \tag{2.8}$$

Combining this with our definition of B in (2.5), we obtain

$$\gamma^2 \|h\|^2 - \|Bh\|^2 = \gamma^2 \|h\|^2 - \|Ah\|^2 - \|d\|^2 = \|D_A h\|^2 - \|d\|^2 \geq 0 .$$

Therefore B is bounded by $\gamma$.

Obviously $P_{\mathcal{H}'} B = A$. So to complete the sufficiency part of the proof of Theorem 2.1, it remains to show that B intertwines T with U′. To this end, we need the following useful identities which follow directly from the definition of $\omega$ in (2.2); see also (IV.2.11)

$$\Pi' \omega D_A T = D' A \quad \text{and} \quad \Pi_A \omega D_A T = D_A . \tag{2.9}$$

Using these identities, along with the definition of B in (2.5) and $W \mid \mathcal{F} = \omega$, we have

$$BT = \begin{bmatrix} AT \\ \Pi' W (I - \lambda \Pi_A W)^{-1} D_A T \end{bmatrix} = \begin{bmatrix} AT \\ \Pi' W D_A T + \lambda \Pi' W (I - \lambda \Pi_A W)^{-1} \Pi_A W D_A T \end{bmatrix} =$$

$$= \begin{bmatrix} T' A \\ D' A + \lambda \Pi' W (I - \lambda \Pi_A W)^{-1} D_A \end{bmatrix} = U' B .$$

Here U′ is the Sz.-Nagy-Schäffer minimal isometric dilation of T′ in (1.10). Therefore $U'B = BT$ and B is a intertwining lifting of A satisfying $\|B\| \leq \gamma$.

**REMARK 2.2**. Now let $U' = U_m \oplus S$ where $U_m$ is the Sz.-Nagy-Schäffer minimal isometric lifting of $T'$ and S is a unilateral shift on $H^2(\mathcal{E})$. Then Theorem 2.1 also holds when $U'$ is replaced by $U_m \oplus S$ and $\mathcal{G}'$ is replaced by $\mathcal{G}' \oplus \mathcal{E}$. The proof of this fact is essentially the same as the proof of Theorem 2.1. This fact also follows by noting that this $U'$ can be identified with the minimal isometric dilation of $T' \oplus 0$ on $\mathcal{H}' \oplus \mathcal{E}$, and then applying Theorem 2.1 with $T'$ replaced by $T' \oplus 0$ and A replaced by $[A, 0]^{tr}$.

## VI.3 RECOVERING THE SCHUR CONTRACTION

In this section we will show how one can recover the Schur contraction $W(\lambda)$ directly from the contractive intertwining lifting B in (2.5) of A.

Let us recall that an operator B from $\mathcal{H}$ into $\mathcal{H}' \oplus \mathcal{M}$ of the form $B = [A^*, X^*]^*$ satisfies $\|B\| \le \gamma$ if and only if X admits a factorization of the form $X = YD_A$ where Y is a contraction from $\mathcal{D}_A$ into $\mathcal{M}$. (As always we assume that A is an operator from $\mathcal{H}$ into $\mathcal{H}'$ satisfying $\|A\| \le \gamma$.) To see this, notice that if B is bounded by $\gamma$, then $\|Ah\|^2 + \|Xh\|^2 = \|Bh\|^2 \le \gamma^2\|h\|^2$ implies that $\|Xh\|^2 \le \gamma^2\|h\|^2 - \|Ah\|^2 = \|D_A h\|^2$, for all $h \in \mathcal{H}$. So there exists a contraction Y from $\mathcal{D}_A$ into $\mathcal{M}$ satisfying $X = YD_A$. (In our case $\mathcal{M} = H^2(\mathcal{D}')$.) On the other hand, if $B = [A^*, D_A Y^*]^*$ for some contraction Y, then

$$\|Bh\|^2 = \|Ah\|^2 + \|YD_A h\|^2 \le \|Ah\|^2 + \gamma^2\|h\|^2 - \|Ah\|^2 = \gamma^2\|h\|^2 .$$

Hence B is bounded by $\gamma$.

From our above discussion we see that the intertwining lifting B of A in (2.5) admits a decomposition of the form

$$B = \begin{bmatrix} A \\ Y(\lambda)D_A \end{bmatrix} = \begin{bmatrix} A \\ \Pi' W(\lambda)(I - \lambda \Pi_A W(\lambda))^{-1} D_A \end{bmatrix} \tag{3.1}$$

where Y is the contraction from $\mathcal{D}_A$ into $H^2(\mathcal{D}')$ uniquely defined by B. From (3.1) we infer that this Y is given by the formula

$$Y(\lambda) = \Pi' W(\lambda)(I - \lambda \Pi_A W(\lambda))^{-1} . \tag{3.2}$$

We are now ready to state the main result of this section.

**THEOREM 3.1**. *Let* B *be the intertwining lifting of* A *defined in (2.5) where* $W(\lambda)$ *is a Schur contraction. Let* Y *be the unique contraction from* $\mathcal{D}_A$ *into* $H^2(\mathcal{D}')$ *be defined by* $YD_A = (I - P_{\mathcal{H}'})B$. *Let* $E(\lambda)$ *and* $F(\lambda)$ *be the analytic functions in* $|\lambda| < 1$ *defined by*

$$E(\lambda) = (I - \lambda T^*)^{-1}(D_A - \lambda A^* D'Y(\lambda))$$

$$F(\lambda) = \frac{E(\lambda) - D_A}{\lambda} = (I - \lambda T^*)^{-1}(T^* D_A - A^* D'Y(\lambda)) . \tag{3.3}$$

*Then the following identities hold*

$$\begin{bmatrix} \Pi' W(\lambda) Q(\lambda) \\ D_A \Pi_A W(\lambda) Q(\lambda) \end{bmatrix} = \begin{bmatrix} Y(\lambda) \\ F(\lambda) \end{bmatrix} \tag{3.4}$$

*where* $D_A Q(\lambda) = E(\lambda)$ *and* $Q(\lambda)$ *is the analytic function in* $|\lambda| < 1$ *whose values are linear operators on* $\mathcal{D}_A$ *defined by*

$$Q(\lambda) = (I - \lambda \Pi_A W(\lambda))^{-1} . \tag{3.5}$$

Notice, that if $||A|| < \gamma$, then $D_A$ is invertible. In particular the previous theorem shows the Schur contraction $W(\lambda)$ is given by

$$W(\lambda) = \begin{bmatrix} Y(\lambda) E(\lambda)^{-1} D_A \\ D_A^{-1} F(\lambda) E(\lambda)^{-1} D_A \end{bmatrix} . \tag{3.6}$$

Also note that the intertwining lifting B of A in (2.5) uniquely determines the functions E and F in (3.3); these functions in turn uniquely determine the Schur contraction $W(\lambda)$ by (3.4). This observation proves the following result.

**COROLLARY 3.2.** *The intertwining lifting* B *of* A *given by (2.5) uniquely determines its Schur contraction* $W(\lambda)$.

**PROOF OF THEOREM 3.1.** Recall that $\omega$ is the isometry from $\mathcal{F} = \overline{D_A T \mathcal{H}}$ into $\mathcal{D}' \oplus \mathcal{D}_A$ defined by (2.2). Thus $T^* D_A \omega^* = [A^* D', D_A]$. So using $P_{\mathcal{F}} = \omega^* \omega P_{\mathcal{F}}$ we have

$$T^* D_A = T^* D_A P_{\mathcal{F}} = [A^* D', D_A] \omega P_{\mathcal{F}} . \tag{3.7}$$

Now recall that the Schur contraction W satisfies $W(\lambda) | \mathcal{F} \equiv \omega$. Moreover, the range of $\omega$ is precisely $\mathcal{F}' = \{D'Ah \oplus D_A h : h \in \mathcal{H}\}^-$ which is the orthogonal complement of the kernel of $[A^* D', D_A]$. Therefore we can replace $\omega$ in (3.7) by W. Moreover using $P_{\omega\mathcal{F}} W(\lambda) = W(\lambda) P_{\mathcal{F}}$ we obtain

$$
\begin{aligned}
T^* D_A &= [A^* D', D_A] W P_{\mathcal{F}} = [A^* D', D_A] P_{\omega \mathcal{F}} W = [A^* D', D_A] W = \\
&= A^* D' \Pi' W + D_A \Pi_A W .
\end{aligned} \tag{3.8}
$$

The relation (3.8) and the definition (3.5) for $Q(\lambda)$ yields

$$
\begin{aligned}
(I - \lambda T^*) D_A Q(\lambda) &= D_A Q(\lambda) - \lambda A^* D' \Pi' W(\lambda) Q(\lambda) - \lambda D_A \Pi_A W(\lambda) Q(\lambda) = \\
&D_A (I - \lambda \Pi_A W(\lambda)) Q(\lambda) - \lambda A^* D' Y(\lambda) = D_A - \lambda A^* D' Y(\lambda) .
\end{aligned}
$$

So by taking the inverse, we see that $D_A Q(\lambda) = E(\lambda)$ where $E(\lambda)$ is defined in (3.3). Now notice that

$$
F(\lambda) = \frac{E(\lambda) - D_A}{\lambda} = \frac{D_A Q(\lambda) - D_A}{\lambda} = D_A \Pi_A W(\lambda) Q(\lambda) . \tag{3.9}
$$

A simple calculation shows that $F(\lambda)$ is also given by the last equation in (3.3). Finally, using $Y = \Pi' W Q$ and $D_A \Pi_A W Q = F$ we obtain (3.4) which completes the proof.

## VI.4 CONSTRUCTING THE SCHUR CONTRACTION

In this section we will present some preliminary notation and results which will be used to show that any intertwining lifting B of A with tolerance $\gamma$ admits a representation of the form (2.5) where W is a Schur contraction. To this end, let B be any intertwining lifting of A with respect to $U'$ and tolerance $\gamma$. According to Lemma IV.2.1 this B admits a matrix decomposition of the form

$$
B = \begin{bmatrix} A \\ Y D_A \end{bmatrix} : \mathcal{H} \to \begin{bmatrix} \mathcal{H}' \\ H^2(\mathcal{D}') \end{bmatrix} \tag{4.1}
$$

where Y is a uniquely determined contraction from $\mathcal{D}_A$ into $H^2(\mathcal{D}')$. Notice that $Y = Y(\lambda)$ can be viewed as an analytic function for $|\lambda| < 1$, that is, $Y(\lambda)$ admits a power series expansion of the form

$$
Yh = \sum_0^\infty \lambda^n Y_n h \quad \text{and} \quad \|Yh\|^2 = \sum_0^\infty \|Y_n h\|^2 \le \|h\|^2 \qquad (h \in \mathcal{D}_A) . \tag{4.2}
$$

Finally, by consulting Lemma IV.2.1 once again or using the form of B in (4.1) along with $U'B = BT$ where $U'$ is the Sz.-Nagy-Schäffer minimal isometric dilation of $T'$ in (1.10), we obtain

$$Y(\lambda)D_A T = D'A + \lambda Y(\lambda)D_A \qquad (|\lambda| < 1). \tag{4.3}$$

The following result shows how to construct the Schur contraction $W(\lambda)$ from an arbitrary intertwining lifting B of A satisfying $\|B\| \le \gamma$.

**THEOREM 4.1.** *Let* B *be an intertwining lifting of* A *with tolerance* $\gamma$ *and* $Y(\lambda)$ *the contraction from* $\mathcal{D}_A$ *into* $H^2(\mathcal{D}')$ *obtained by the decomposition of* B *in (4.1). Let* $E(\lambda)$ *and* $F(\lambda)$ *be the functions defined (for* $|\lambda| < 1$*) by*

$$\begin{aligned} E(\lambda) &= (I - \lambda T^*)^{-1}(D_A - \lambda A^* D' Y(\lambda)) \\ F(\lambda) &= (I - \lambda T^*)^{-1}(T^* D_A - A^* D' Y(\lambda)). \end{aligned} \tag{4.4}$$

*Then* $\lambda F(\lambda) = E(\lambda) - D_A$ *and the functions* E *and* F *admit factorizations of the form*

$$E(\lambda) = D_A Q(\lambda) \text{ and } F(\lambda) = D_A G(\lambda) \qquad (\text{for } |\lambda| < 1) \tag{4.5}$$

*where* $Q(0) = I$ *and* $Q(\lambda)^{-1}$ *exists for all* $|\lambda| < 1$. *Moreover, the function* $W(\lambda)$ *defined by*

$$W(\lambda)d = \begin{bmatrix} Y(\lambda)Q(\lambda)^{-1}d \\ G(\lambda)Q(\lambda)^{-1}d \end{bmatrix} \qquad (d \in \mathcal{D}_A) \tag{4.6}$$

*is a Schur contraction. Finally, the intertwining lifting* B *is given by the corresponding Schur representation in (2.5).*

The rest of this section will be devoted to proving Theorem 4.1. This along with Theorem 3.1 completes the proof of Theorem 2.1. As we will show later by using a Möbius transformation, it is sufficient to verify that (4.5) holds for $\lambda = 0$ and that $W(0)$ defined by (4.6) is a contraction satisfying $W(0) | \mathcal{F} = \omega$. To this end, we establish the following result.

**LEMMA 4.2.** *Let* B *be an intertwining lifting of* A *satisfying* $\|B\| \le \gamma$. *Let* $E(\lambda)$ *and* $F(\lambda)$ *be the functions defined by (4.4). Then* $E(0) = D_A$ *and* $F(0) = D_A \Gamma_0$ *where* $\Gamma_0$ *is a contraction on* $\mathcal{D}_A$. *Moreover,*

$$W(0) = \begin{bmatrix} Y(0) \\ \Gamma_0 \end{bmatrix} : \mathcal{D}_A \to \begin{bmatrix} \mathcal{D}' \\ \mathcal{D}_A \end{bmatrix} \tag{4.7}$$

*is a contraction satisfying* $W(0) | \mathcal{F} = \omega$.

**PROOF**. Clearly $E(0) = D_A$ and thus $Q(0) = I$. By to (4.4), we have for all h in $\mathcal{H}$

$$\|F(0)^*h\|^2 = \|D_A Th - Y(0)^* D'Ah\|^2 =$$

$$= \|\omega D_A Th\|^2 - 2 \operatorname{Re} (Y(0)D_A Th, D'Ah) + \|Y(0)^* D'Ah\|^2 = \tag{4.8}$$

$$= \|D_A h\|^2 + \|D'Ah\|^2 - 2\|D'Ah\|^2 + \|Y(0)^* D'Ah\|^2 \leq \|D_A h\|^2 .$$

The third equality follows from the definition of $\omega$ in (2.2) along with the fact that $Y(0)D_A T = D'A$; see (4.3). The inequality follows from the fact that $Y(0)$ is contractive. According to (4.8) there exists a contraction $\Gamma$ on $\mathcal{D}_A$ satisfying $F(0)^* = \Gamma D_A$. Therefore $F(0) = D_A \Gamma_0$ where $\Gamma_0$ is the contraction on $\mathcal{D}_A$ defined by $\Gamma_0 = \Gamma^*$.

Using $F(0)^* = \Gamma_0^* D_A$ equation (4.8) also gives the following

$$\|D_{\Gamma_0^*} D_A h\|^2 = \|D_A h\|^2 - \|\Gamma_0^* D_A h\|^2 = \|D_A h\|^2 - \|F(0)^* h\|^2 = \|D_{Y(0)^*} D'Ah\|^2 . \tag{4.9}$$

Here $D_{\Gamma_0^*}$ is the positive square root of $I - \Gamma_0 \Gamma_0^*$ while $D_{Y(0)^*}$ is the positive square root of $I - Y(0)Y(0)^*$.

To show that $W(0)$ in (4.7) is a contraction, let h be in $\mathcal{H}$ and d in $\mathcal{D}'$. Then using $Y(0)D_A T = D'A$ and $\|D_{\Gamma_0^*} D_A h\| = \|D_{Y(0)^*} D'Ah\|$; (see (4.9)), we have

$$\|d \oplus D_A h\|^2 - \|W(0)^* (d \oplus D_A h)\|^2 = \|d\|^2 + \|D_A h\|^2 - \|Y(0)^* d + \Gamma_0^* D_A h\|^2 =$$

$$= \|D_{Y(0)^*} d\|^2 + \|D_A h\|^2 - 2 \operatorname{Re} (Y(0)^* d, \Gamma_0^* D_A h) - \|\Gamma_0^* D_A h\|^2 =$$

$$= \|D_{Y(0)^*} d\|^2 + \|D_{\Gamma_0^*} D_A h\|^2 - 2 \operatorname{Re} (Y(0)^* d, F(0)^* h) =$$

$$= \|D_{Y(0)^*} d\|^2 + \|D_{Y(0)^*} D'Ah\|^2 - 2 \operatorname{Re} (Y(0)^* d, D_A Th - Y(0)^* D'Ah) =$$

$$= \|D_{Y(0)^*} d\|^2 + \|D_{Y(0)^*} D'Ah\|^2 - 2 \operatorname{Re} (d, (I - Y(0)Y(0)^*)D'Ah) =$$

$$= \|D_{Y(0)^*} (d - D'Ah)\|^2 \geq 0 .$$

Therefore $W(0)^*$ is a contraction.

To complete the proof it remains to show that $W(0) \mid \mathcal{F} = \omega$. To this end, first notice that $T'A = AT$ and $Y(0)D_A T = D'A$ gives

$$D_A \Gamma_0 D_A T = F(0)D_A T = (T^* D_A - A^* D'Y(0))D_A T = \gamma^2 I - T^* A^* AT - A^* D'^2 A = D_A^2 .$$

So using $D_A \Gamma_0 D_A T = D_A^2$ we have

$$\begin{bmatrix} I & 0 \\ 0 & D_A \end{bmatrix} W(0)D_A T = \begin{bmatrix} Y(0)D_A T \\ D_A \Gamma_0 D_A T \end{bmatrix} = \begin{bmatrix} D'A \\ D_A^2 \end{bmatrix} = \begin{bmatrix} I & 0 \\ 0 & D_A \end{bmatrix} \omega D_A T .$$

Therefore $W(0)D_A T = \omega D_A T$ or equivalently $W(0) \mid \mathcal{F} = \omega$. This completes the proof.

Now we will use some of our previous results on the Möbius transformation in Section 1 to prove Theorem 4.1. As before, let B be an intertwining lifting of A of the form (4.1), and $\alpha$ a fixed complex number in the open unit disc and $b(\lambda)$ the möbius transformation in (1.1). Obviously $T'_\alpha A = AT_\alpha$ where $T_\alpha = b(T)$ and $T'_\alpha = b(T')$. Recall that $U'_\alpha = b(U')$ is a minimal isometric dilation of $T'_\alpha$. From this fact it readily follows that B is also an intertwining lifting of A with respect to $T_\alpha$ and $U'_\alpha$, and tolerance $\gamma$, that is, B is an operator whose norm is bounded by $\gamma$ satisfying $P_{\mathcal{H}'}B = A$ and $U'_\alpha B = BT_\alpha$. By Proposition 1.2 we have $\Phi U_\alpha = U'_\alpha \Phi$ where $U_\alpha$ is the Sz.-Nagy-Schäffer minimal isometric dilation of $T'_\alpha$. Therefore $B_\alpha = \Phi^* B$ is an intertwining of A with respect to $U_\alpha$ and $T_\alpha$ and tolerance $\gamma$. By applying our previous results to $B_\alpha$ instead of B we see that $B_\alpha$ admits a decomposition of the form

$$B_\alpha = \begin{bmatrix} A \\ Y_\alpha(\lambda)D_A \end{bmatrix} = \begin{bmatrix} A \\ Z^* \Phi^*_{\mathcal{D}'} Y(\lambda) D_A \end{bmatrix} = \Phi^* B . \tag{4.10}$$

Here $Y_\alpha(\lambda)$ is a contraction mapping $\mathcal{D}_A$ onto $H^2(\mathcal{D}'_\alpha)$. Equation (4.3) applied to $Y_\alpha$ and $T_\alpha$ shows that

$$Y_\alpha(\lambda)D_A T_\alpha = D'_\alpha A + \lambda Y_\alpha(\lambda) D_A \qquad (\lambda, \alpha \in \mathbb{D}) \tag{4.11}$$

where $\mathbb{D}$ is the open unit disc. Finally, the functions $E(\lambda)$ and $F(\lambda)$ for the intertwining lifting $B_\alpha$ of A become

$$\begin{aligned} E_\alpha(\lambda) &= (I - \lambda T^*_\alpha)^{-1}(D_A - \lambda A^* D'_\alpha Y_\alpha(\lambda)) \\ F_\alpha(\lambda) &= (I - \lambda T^*_\alpha)^{-1}(T^*_\alpha D_A - A^* D'_\alpha Y_\alpha(\lambda)) . \end{aligned} \tag{4.12}$$

Now let us establish some connection between Y, E, F and $Y_\alpha$, $E_\alpha$, $F_\alpha$, respectively. To this end, first notice that (4.10) along with the formula for $\Phi^*_{\mathcal{D}}$ in (1.5) gives $Y_\alpha(0) = d_\alpha Z^*_\alpha Y(a(0))$, or equivalently,

$$Z_\alpha Y_\alpha(0) = d_\alpha Y(\alpha) . \tag{4.13}$$

By applying Lemma 4.2 to $B_\alpha$ and $F_\alpha$ we see that $F_\alpha(0) = D_A \Gamma_\alpha$ where $\Gamma_\alpha$ is a contraction of $\mathcal{D}_A$. We claim that

$$d^2_\alpha E(\alpha) = D_A(\alpha \Gamma_\alpha + I) \text{ and } d^2_\alpha F(\alpha) = D_A(\Gamma_\alpha + \bar{\alpha} I) . \tag{4.14}$$

To verify the first equation notice that (4.13), (1.14) and $T'A = AT$ gives

$$F_\alpha(0) = T_\alpha^* D_A - A^* D'_\alpha Y_\alpha(0) = T_\alpha^* D_A - A^* D'_\alpha Z_\alpha^* d_\alpha Y(\alpha) =$$

$$= T_\alpha^* D_A - d_\alpha^2 A^*(I - \alpha T'^*)^{-1} D' Y(\alpha) = (I - \alpha T^*)^{-1}[(T^* - \bar{\alpha} I)D_A - d_\alpha^2 A^* D' Y(\alpha)] =$$

$$= (I - \alpha T^*)^{-1}[(T^* D_A - A^* D' Y(\alpha)) - \bar{\alpha}(D_A - \alpha A^* D' Y(\alpha))] = F(\alpha) - \bar{\alpha} E(\alpha) .$$

This yields the following useful expression for $F_\alpha(0)$,

$$D_A \Gamma_\alpha = F_\alpha(0) = F(\alpha) - \bar{\alpha} E(\alpha) . \tag{4.15}$$

Combining this with $E(\alpha) = D_A + \alpha F(\alpha)$ (see (3.3)) we obtain

$$\begin{bmatrix} I & -\alpha I \\ -\bar{\alpha} I & I \end{bmatrix} \begin{bmatrix} E(\alpha) \\ F(\alpha) \end{bmatrix} = \begin{bmatrix} D_A \\ D_A \Gamma_\alpha \end{bmatrix} . \tag{4.16}$$

Solving (4.16) for $E(\alpha)$ and $F(\alpha)$ readily gives (4.14). Because $\Gamma_\alpha$ is a contraction $\alpha\Gamma_\alpha + I$ is invertible for all $|\alpha| < 1$. So if we set $Q(\lambda) = d_{\bar{\lambda}}^{-2}(\lambda\Gamma_\lambda + I)$, then $Q(\lambda)$ is invertible for all $|\lambda| < 1$ and $Q(0) = I$. This shows that $E(\lambda)$ admits a factorization of the form $E(\lambda) = D_A Q(\lambda)$ with $Q(\lambda)$ invertible and $Q(0) = I$. Likewise by setting $G(\lambda) = d_{\bar{\lambda}}^{-2}(\Gamma_\lambda + \bar{\lambda} I)$ we see that $F(\lambda)$ admits a factorization of the form $D_A G(\lambda)$. Since both $E(\lambda)$ and $F(\lambda)$ are analytic in $|\lambda| < 1$, it follows that both $Q(\lambda)$ and $G(\lambda)$ are analytic in $|\lambda| < 1$; see Lemma 4.3 below. Moreover, because $Q(\lambda)$ is invertible, $Q(\lambda)^{-1}$ is also analytic in $|\lambda| < 1$. Therefore the function $W(\lambda)$ defined in (4.6) is analytic in the open unit disc. This proves part of Theorem 4.1.

Now let us show that $W(\alpha)$ is a contraction for all $|\alpha| < 1$. Using $E(\alpha) = D_A Q(\alpha)$ and $F(\alpha) = D_A G(\alpha)$ in (4.14) we obtain

$$\begin{bmatrix} Q(\alpha) \\ G(\alpha) \end{bmatrix} = \frac{1}{d_\alpha^2} \begin{bmatrix} I & \alpha I \\ \bar{\alpha} I & I \end{bmatrix} \begin{bmatrix} I \\ \Gamma_\alpha \end{bmatrix} . \tag{4.17}$$

Recall that $W(\alpha)$ is the transpose of $[Y(\alpha)Q(\alpha)^{-1}, G(\alpha)Q(\alpha)^{-1}]$. By using (4.17) along with $Z_\alpha Y_\alpha(0) = d_\alpha Y(\alpha)$ and $f$ in $\mathcal{D}_A$ we have

$$\begin{aligned} &\|Q(\alpha)f\|^2 - \|W(\alpha)Q(\alpha)f\|^2 = \|Q(\alpha)f\|^2 - \|G(\alpha)f\|^2 - \|Y(\alpha)f\|^2 = \\ &= d_{\bar{\alpha}}^{-4}\|f + \alpha\Gamma_\alpha f\|^2 - d_{\bar{\alpha}}^{-4}\|\bar{\alpha} f + \Gamma_\alpha f\|^2 - \|Y(\alpha)f\|^2 = \\ &= d_{\bar{\alpha}}^{-2}(\|f\|^2 - \|\Gamma_\alpha f\|^2 - \|Z_\alpha Y_\alpha(0)f\|^2) = \\ &= d_{\bar{\alpha}}^{-2}(\|f\| - \|Y_\alpha(0)f \oplus \Gamma_\alpha f\|^2) = d_{\bar{\alpha}}^{-2}(\|f\| - \|W_\alpha(0)f\|^2) \geq 0 . \end{aligned} \tag{4.18}$$

The last two relations are obtained by applying Lemma 4.2 to $B_\alpha$, $E_\alpha$ and $F_\alpha$. In this setting

$W_\alpha(0)$ is the contraction defined by the transpose of $[Y_\alpha(0), \Gamma_\alpha]$. Because $Q(\alpha)$ is invertible, equation (4.18) shows that $W(\alpha)$ is a contraction for all $|\alpha| < 1$. Therefore $W(\lambda)$ is a contractive analytic function, as desired. Moreover, by Lemma 4.2 we also have $W(0) | \mathcal{F} = \omega$ where $\omega$ is the isometry defined in (2.2). So by the maximum principle $W(\lambda) | \mathcal{F} = \omega$ for all $|\lambda| < 1$. Therefore $W(\lambda)$ is a Schur contraction. This proves yet another part of Theorem 4.1.

To complete the proof, it remains to show that B is given by the Schur representation in (2.5). According to (4.6) and (4.17) we have

$$\begin{aligned} I - \lambda\Pi_A W(\lambda) &= I - \lambda G(\lambda)Q(\lambda)^{-1} = (Q(\lambda) - \lambda G(\lambda))Q(\lambda)^{-1} = \\ &= d_{\bar\lambda}^{-2}(I + \lambda\Gamma_\lambda - |\lambda|^2 I - \lambda\Gamma_\lambda)Q(\lambda)^{-1} = Q(\lambda)^{-1} . \end{aligned} \tag{4.19}$$

Therefore $Q(\lambda)$ is the inverse of $I - \lambda\Pi_A W(\lambda)$. Using this fact in (4.6) we obtain

$$Y(\lambda) = \Pi' W(\lambda)Q(\lambda) = \Pi' W(\lambda)(I - \lambda\Pi_A W(\lambda))^{-1} . \tag{4.20}$$

So according to the form of B in (4.1)

$$B = \begin{bmatrix} A \\ \Pi' W(\lambda)(I - \lambda\Pi_A W(\lambda))^{-1} D_A \end{bmatrix} . \tag{4.21}$$

This is precisely the Schur representation of B in Theorem 2.1. This completes the proof of both Theorem 4.1 and Theorem 2.1.

**LEMMA 4.3.** *Let* $H(\lambda)$ *be an operator-valued analytic function in* $\mathbb{D}$ *and let* $K(\lambda)$ *be an operator-valued locally bounded function in* $\mathbb{D}$*. Let moreover* X *be a one-to-one operator such that*

$$H(\lambda) = XK(\lambda) \qquad (\lambda \in \mathbb{D}) .$$

*Then* $K(\lambda)$ *is also analytic in* $\mathbb{D}$.

**PROOF**. Let $\mathcal{X}$, $\mathcal{Y}$, $\mathcal{Z}$ denote the Hilbert spaces that $H(\lambda)$ maps $\mathcal{Y}$ into $\mathcal{Z}$, $K(\lambda)$ maps $\mathcal{Y}$ into $\mathcal{X}$ (for all $\lambda \in \mathbb{D}$), while X maps $\mathcal{X}$ into $\mathcal{Z}$. Let moreover

$$H(\lambda) = \sum_{n=0}^{\infty} \lambda^n H_n$$

be the Taylor expansion of $H(\lambda)$ at the orgin and let $r \in (0, 1)$. Then for any $n \geq 0$ and $z \in \mathcal{Z}$ we have

$$H_n^* z = \frac{1}{2\pi r^n} \int_0^{2\pi} e^{in\theta} H(re^{i\theta})^* z d\theta . \tag{4.22}$$

Notice that

$$\|H(re^{i\theta})^* z\| = \|K(re^{i\theta})^* X^* z\| \leq \left(\max_{|\lambda|\leq r} \|K(\lambda)\|\right) \|X^* z\| .$$

Using this (4.22) yields

$$r^n \|H_n^* z\| \leq \left(\max_{|\lambda|\leq r} \|K(\lambda)\|\right) \|X^* z\| .$$

Therefore there exists a unique operator $K_{*n}$ from $\mathcal{X}$ into $\mathcal{Y}$ (the uniqueness of $K_{*n}$ comes from the fact that the range of $X^*$ is dense) such that

$$H_n^* = K_{*n} X^* \quad \text{and} \quad \|K_{*n}\| \leq \frac{\max_{|\lambda|\leq r} \|K(\lambda)\|}{r^n} .$$

Set $K_n = K_{*n}^*$ $(n = 0, 1, \ldots)$. Then $H_n = XK_n$. Finally, let

$$M(\lambda) = \sum_{n=0}^{\infty} \lambda^n K_n .$$

This operator-valued power series converges for $|\lambda| < r$. Since $r \in (0, 1)$ is arbitrary and $K_n$ are independent of r, $M(\lambda)$ is well defined and analytic in $\mathbb{D}$. Moreover it is clear that

$$XM(\lambda) = \sum_{n=0}^{\infty} \lambda^n XK_n = \sum_{n=0}^{\infty} \lambda^n H_n = H(\lambda) \qquad (\lambda \in \mathbb{D}) .$$

It follows $X(M(\lambda) - K(\lambda)) = 0$ for $\lambda \in \mathbb{D}$. Since X is one-to-one, we conclude that $K(\lambda) \equiv M(\lambda)$ is analytic in $\mathbb{D}$.

**REMARK 4.4**. It is worth noting that (see R. Teodorescu [1]) that the operator $(I - \lambda T^*)^{-1}$ occurring in Theorems 3.1 and 4.1 can be eliminated. To see this notice that (3.2) can be written as

$$\Pi' W(\lambda) = Y(\lambda)(I - \lambda \Pi_A W(\lambda)) . \tag{4.23}$$

Using (4.23) and (3.8) we have

$$(D_A - \lambda A^* D' Y(\lambda)) \Pi_A W(\lambda) = T^* D_A - A^* D' Y(\lambda) \quad (\lambda \in \mathbb{D}) . \tag{4.24}$$

Since $Q(\lambda) = d_{\bar{\lambda}}^{-2}(I + \lambda \Gamma_\lambda)$ where $\Gamma_\lambda$ is the contraction on $\mathcal{D}_A$ in the proof of Theorem 4.1, we obtain

$$\Omega(\lambda) := D_A - \lambda A^* D' Y(\lambda) = (I - \lambda T^*)E(\lambda) =$$

$$(I - \lambda T^*)D_A Q(\lambda) = (I - \lambda T^*)D_A(I + \lambda\Gamma_\lambda)d_{\bar\lambda}^{-2} \quad (\lambda \in \mathbb{D}) .$$

From this representation of $\Omega(\lambda)$ it follows that $\Omega(\lambda)$ is one to one from $\mathcal{D}_A$ into $\mathcal{H}$. Therefore (4.24) uniquely determines $\Pi_A W(\lambda)$ for $\lambda \in \mathbb{D}$ and consequently by (4.23), it also uniquely determines $W(\lambda)$.

## VI.5 THE REDHEFFER SCATTERING PARAMETERIZATION

In this section we will give a more condensed representation for the set of all intertwining lifting B of A with tolerance $\gamma$. Then we will briefly present a Redheffer scattering interpretation of this parameterization of all intertwining liftings.

As before, A is a contraction mapping $\mathcal{H}$ into $\mathcal{H}'$ intertwining the isometry T with T' satisfying $\|A\| \le \gamma$. Throughout this section $\Pi'$ and $\Pi_A$ are the operators defined in (2.1) and $\omega$ is the isometry defined in (2.2), while $\mathcal{F} = \overline{D_A T\mathcal{H}}$ and $\mathcal{F}'$, $\mathcal{G}$, $\mathcal{G}'$ are the subspaces defined in (2.4). Now let $\Phi_{1,1}$, $\Phi_{1,2}$, $\Phi_{2,1}$ and $\Phi_{2,2}$ be the operator valued analytic functions in $|\lambda| < 1$ defined by

$$\begin{aligned}
\Phi_{1,1}(\lambda) &= \lambda P_{\mathcal{G}}(I - \lambda\Pi_A\omega P_{\mathcal{F}})^{-1}\Pi_A \mid \mathcal{G}' \\
\Phi_{1,2}(\lambda) &= P_{\mathcal{G}}(I - \lambda\Pi_A\omega P_{\mathcal{F}})^{-1} \mid \mathcal{D}_A \\
\Phi_{2,1}(\lambda) &= \lambda\Pi'\omega P_{\mathcal{F}}(I - \lambda\Pi_A\omega P_{\mathcal{F}})^{-1}\Pi_A \mid \mathcal{G}' + \Pi' \mid \mathcal{G}' \\
\Phi_{2,2}(\lambda) &= \Pi'\omega P_{\mathcal{F}}(I - \lambda\Pi_A\omega P_{\mathcal{F}})^{-1} \mid \mathcal{D}_A .
\end{aligned} \tag{5.1}$$

Notice that $\Phi_{1,1}(\lambda)$ has values in $\mathcal{L}(\mathcal{G}', \mathcal{G})$ and $\Phi_{1,2}(\lambda)$ has values in $\mathcal{L}(\mathcal{D}_A, \mathcal{G})$, while $\Phi_{2,1}(\lambda)$ and $\Phi_{2,2}(\lambda)$ live in $\mathcal{L}(\mathcal{G}', \mathcal{D}')$ and $\mathcal{L}(\mathcal{D}_A, \mathcal{D}')$, respectively. The following result uses these analytic functions to give a complete description of the set of all intertwining lifting B with tolerance $\gamma$.

**THEOREM 5.1**. *Let* A *be an operator from* $\mathcal{H}$ *into* $\mathcal{H}'$ *intertwining an isometry* T *with a contraction* T' *satisfying* $\|A\| \le \gamma$. *Let* U' *on* $\mathcal{H}' \oplus H^2(\mathcal{D}')$ *be the Sz.-Nagy-Schäffer minimal isometric lifting of* T' *given in (1.10). Then the set of all intertwining liftings* B *of* A *with respect to* U' *and tolerance* $\gamma$ *is given by*

$$B = B(R) = \begin{bmatrix} A \\ \Phi_{2,2}D_A + \Phi_{2,1}R(I - \Phi_{1,1}R)^{-1}\Phi_{1,2}D_A \end{bmatrix} \tag{5.2}$$

*where* R *is a contractive analytic function in* $H^\infty(\mathcal{G}, \mathcal{G}')$. *Moreover, the mapping from* R *to* B(R) *in (5.2) is a bijection from the closed unit ball in* $H^\infty(\mathcal{G}, \mathcal{G}')$ *onto the set of all intertwining liftings* B *of* A *with tolerance* $\gamma$.

Notice that by choosing $R = 0$ in the previous theorem, we obtain $B = B_\gamma = [A, \Phi_{2,2}D_A]^{tr}$ which is precisely the central intertwining lifting of A with with respect to U′ and tolerance $\gamma$; see (IV.2.15). In fact (5.2), is precisely the explicit form of formula (2.5) for $W = W_R = \omega P_{\mathcal{F}} + RP_{\mathcal{G}}$ in which the explicit dependence on R is explicated.

**PROOF**. Recall that a Schur contraction W admits a decomposition of the form $W = \omega P_{\mathcal{F}} \oplus RP_{\mathcal{G}}$ where R is a contractive analytic function in $H^\infty(\mathcal{G}, \mathcal{G}')$. Using this representation of W in the second component for the intertwining lifting B of A in (2.5) of Theorem 2.1, along with $X(\lambda) = (I - \lambda\Pi_A\omega P_{\mathcal{F}})^{-1}$ on $\mathcal{D}_A$ we obtain

$$\Pi'W(\lambda)(I - \lambda\Pi_A W(\lambda))^{-1} = \Pi'(\omega P_{\mathcal{F}} + RP_{\mathcal{G}})[I - \lambda\Pi_A\omega P_{\mathcal{F}} - \lambda\Pi_A RP_{\mathcal{G}}]^{-1} =$$

$$\Pi'(\omega P_{\mathcal{F}} + RP_{\mathcal{G}})[(I - \lambda\Pi_A\omega P_{\mathcal{F}})(I - \lambda X\Pi_A RP_{\mathcal{G}})]^{-1} = \Pi'(\omega P_{\mathcal{F}} + RP_{\mathcal{G}})(I - \lambda X\Pi_A RP_{\mathcal{G}})^{-1}X =$$

$$\Pi'\omega P_{\mathcal{F}}(I - \lambda X\Pi_A RP_{\mathcal{G}})^{-1}X + \Pi'RP_{\mathcal{G}}(I - \lambda X\Pi_A RP_{\mathcal{G}})^{-1}X =$$

$$\Pi'\omega P_{\mathcal{F}}X + \lambda\Pi'\omega P_{\mathcal{F}}X\Pi_A RP_{\mathcal{G}}(I - \lambda X\Pi_A RP_{\mathcal{G}})^{-1}X + \Pi'RP_{\mathcal{G}}(I - \lambda X\Pi_A RP_{\mathcal{G}})^{-1}X =$$

$$\Phi_{2,2}(\lambda) + (\lambda\Pi'\omega P_{\mathcal{F}}X\Pi_A + \Pi')R(I - \Phi_{1,1}R)^{-1}P_{\mathcal{G}}X = \Phi_{2,2}(\lambda) + \Phi_{2,1}(\lambda)R(I - \Phi_{1,1}(\lambda)R)^{-1}\Phi_{1,2}$$

Substituting this into the form of B in (2.5) readily yields B(R) in (5.2). This completes the proof.

To complete this section we will develop some simple connections between Redheffer scattering and the characterization of all intertwining liftings B(R) of A in (5.2). To this end, let $\Phi$ be the operator defined by

$$\Phi := \begin{bmatrix} \Phi_{1,1} & \Phi_{1,2} \\ \Phi_{2,1} & \Phi_{2,2} \end{bmatrix} : \begin{bmatrix} H^2(\mathcal{G}') \\ \mathcal{D}_A \end{bmatrix} \to \begin{bmatrix} H^2(\mathcal{G}) \\ H^2(\mathcal{D}') \end{bmatrix}. \tag{5.3}$$

Proposition 5.2 below shows that $\Phi$ is a contraction. Notice that the first column $[\Phi_{1,1}, \Phi_{2,1}]^{tr}$ of $\Phi$ is a multiplication operator from $H^2(\mathcal{G}')$ into $H^2(\mathcal{G})$, while the second column $[\Phi_{1,2}, \Phi_{2,2}]^{tr}$ of $\Phi$ is an "observability operator" which maps $\mathcal{D}_A$ into $H^2(\mathcal{D}')$. Now consider the following Redheffer scattering system

$$\begin{bmatrix} g \\ y \end{bmatrix} = \begin{bmatrix} \Phi_{1,1} & \Phi_{1,2} \\ \Phi_{2,1} & \Phi_{2,2} \end{bmatrix} \begin{bmatrix} x \\ u \end{bmatrix} \tag{5.4}$$

$$\text{subject to} \quad x = Rg$$

where R is a contractive analytic function in $H^\infty(\mathcal{G}, \mathcal{G}')$. Here u is in $\mathcal{D}_A$ and x is in $H^2(\mathcal{G}')$ while g is in $H^2(\mathcal{G})$ and y is in $H^2(\mathcal{D}')$. By solving for (5.4) we see that $y = Y(R)u$ where Y(F) is the operator from $\mathcal{D}_A$ into $H^2(\mathcal{D}')$ defined by

$$Y(R) = \Phi_{2,2} + \Phi_{2,1}R(I - \Phi_{1,1}R)^{-1}\Phi_{1,2} . \tag{5.5}$$

Notice that the intertwining lifting B(R) of A given by (5.2) is precisely $B(R) = [A, Y(R)D_A]^{tr}$. Therefore the set of all intertwining lifting B of A with tolerance $\gamma$ is given by $B = [A, Y(R)D_A]$ where Y(R) is the solution to the Redheffer scattering system in (5.4) and R is an arbitrary contractive analytic function in $H^\infty(\mathcal{G}, \mathcal{G}')$. (For some further results on Redheffer scattering see Chapter XIV in Foias-Frazho [4].) The following result shows that the scattering matrix $\Phi$ is contractive. To this end, recall that an operator Z on $\mathcal{X}$ is pointwise stable if $Z^n$ approaches zero strongly as n approaches infinity.

**PROPOSITION 5.2.** *Let* A *be an operator from* $\mathcal{H}$ *into* $\mathcal{H}'$ *intertwining the isometry* T *with the contraction* T′ *satisfying* $\|A\| \le \gamma$. *Let* $\Phi$ *be the Redheffer scattering matrix in (5.3) generated by the operator valued analytic functions in (5.1). Then* $\Phi$ *is a contraction whose range is onto. In particular,* $\Phi_{1,1}$ *and* $\Phi_{2,1}$ *are contractive analytic functions in* $H^\infty(\mathcal{G}', \mathcal{G})$ *and* $H^\infty(\mathcal{G}', \mathcal{D}')$, *respectively. Moreover, if* $\Pi_A \omega P_{\mathcal{F}}$ *is pointwise stable, then* $\Phi$ *is unitary. In this case,* $[\Phi_{1,1}(\lambda), \Phi_{2,1}(\lambda)]^{tr}$ *defines an inner function in* $H^\infty(\mathcal{G}', \mathcal{G} \oplus \mathcal{D}')$.

**PROOF.** Recall that the transfer function $\Theta$ for the system $\{Z \text{ on } \mathcal{X}, B, C, D\}$ is the analytic function defined by

$$\Theta(\lambda) = D + \lambda C(I - \lambda Z)^{-1}B . \tag{5.6}$$

Moreover, the observability grammian $W_o$ for the pair $\{C, Z\}$ is the operator from $\mathcal{X}$ into $\ell_+^2(\mathcal{Y})$ defined by

$$W_o = [C, CZ, CZ^2, \ldots]^{tr} . \tag{5.7}$$

Finally, recall that $\{Z, B, C, D\}$ is a *unitary system* if the following $2 \times 2$ block matrix is unitary

$$\begin{bmatrix} Z & B \\ C & D \end{bmatrix} : \begin{bmatrix} \mathcal{X} \\ \mathcal{U} \end{bmatrix} \to \begin{bmatrix} \mathcal{X} \\ \mathcal{Y} \end{bmatrix} . \tag{5.8}$$

By choosing

$$Z = \Pi_A \omega P_{\mathcal{F}},\ B = \Pi_A \mid \mathcal{G}',\ C = \begin{bmatrix} P_{\mathcal{G}} \\ \Pi' \omega P_{\mathcal{F}} \end{bmatrix} \text{ and } D = \begin{bmatrix} 0 \\ \Pi' \mid \mathcal{G}' \end{bmatrix} \tag{5.9}$$

we see that {Z, B, C, D} is a unitary system. Here $\mathcal{X} = \mathcal{D}_A$ and $\mathcal{U} = \mathcal{G}'$ while $\mathcal{Y} = \mathcal{G} \oplus \mathcal{D}'$. In this case, the transfer function $\Theta$ for {Z, B, C, D} is given by $\Theta(\lambda) = [\Phi_{1,1}(\lambda), \Phi_{2,1}(\lambda)]^{tr}$. Moreover, $[\Phi_{1,1}, \Phi_{2,1}]$ is now the Fourier transform of the Toeplitz operator $T_\Theta$ while $[\Phi_{1,2}, \Phi_{2,2}]$ is the Fourier transform of the observability operator $W_o$. By applying Theorems III.10.1 and III.10.4, we readily see that the operator $\Phi$ in (5.3) is a contraction and unitary if $\Pi_A \omega P_{\mathcal{F}}$ is pointwise stable. This completes the proof.

The following result is a simple consequence of the previous proposition and Lemma IV.6.1.

**COROLLARY 5.3.** *Let* A *be an operator from* $\mathcal{H}$ *into* $\mathcal{H}'$ *intertwining a unilateral shift* T *with a contraction* T′ *and assume that* $\|A\| < \gamma$. *Then the Redheffer scattering matrix* $\Phi$ *in (5.3) is unitary. In particular,* $[\Phi_{1,1}(\lambda), \Phi_{2,1}(\lambda)]^{tr}$ *is an inner function in* $H^\infty(\mathcal{G}', \mathcal{G} \oplus \mathcal{D}')$.

Now let $\Phi_A$ be the Redheffer scattering matrix given by

$$\Phi_A := \begin{bmatrix} \Phi_{1,1} & \Phi_{1,2} D_A \\ \begin{bmatrix} 0 \\ \Phi_{2,1} \end{bmatrix} & \begin{bmatrix} A \\ \Phi_{2,2} D_A \end{bmatrix} \end{bmatrix} : \begin{bmatrix} H^2(\mathcal{G}') \\ \mathcal{H} \end{bmatrix} \to \begin{bmatrix} H^2(\mathcal{G}) \\ \begin{bmatrix} \mathcal{H}' \\ H^2(\mathcal{D}') \end{bmatrix} \end{bmatrix} . \tag{5.10}$$

Since $\gamma^{-1}[A, D_A]^{tr}$ is an isometry and $\Phi$ is a contraction, it follows that $\Phi_A(I \oplus \gamma^{-1} I)$ is a contraction. Moreover, if $\Pi_A \omega P_{\mathcal{F}}$ is pointwise stable, then $\Phi_A(I \oplus \gamma^{-1} I)$ is an isometry. Now consider the Redheffer scattering system

$$\begin{bmatrix} g \\ y \end{bmatrix} = \begin{bmatrix} \Phi_{1,1} & \Phi_{1,2} D_A \\ \begin{bmatrix} 0 \\ \Phi_{2,1} \end{bmatrix} & \begin{bmatrix} A \\ \Phi_{2,2} D_A \end{bmatrix} \end{bmatrix} \begin{bmatrix} x \\ u \end{bmatrix} \tag{5.11}$$

subject to $x = Rg$

where R is a contractive analytic function in $H^\infty(\mathcal{G}, \mathcal{G}')$. Here u is a vector in $\mathcal{H}$ and x is in $H^2(\mathcal{G}')$, while g is in $H^2(\mathcal{G})$ and y is in $\mathcal{K}' = \mathcal{H}' \oplus H^2(\mathcal{D}')$. By solving the Redheffer system in (5.11) we see that $y = B(R)u$ where $B(R)$ is the intertwining lifting of A with respect to U′ and tolerance $\gamma$ given by (5.2). So the set of all intertwining lifting with tolerance $\gamma$ can be obtained by solving the Redheffer scattering system in (5.11).

Now let U on $\mathcal{K}$ be any minimal isometric lifting of T′. According to Proposition IV.1.3 there exists a unitary operator $\Psi$ from $\mathcal{K}'$ onto $\mathcal{K}$ satisfying $\Psi U' = U\Psi$ and $\Psi | \mathcal{H}' = I$. In fact, a formula for this $\Psi$ is given by

$$\Psi(h \oplus D'f) = h + \sum_{n=0}^{\infty} U^n(U - T')f_n \qquad (h \oplus f \in \mathcal{H}' \oplus H^2(\mathcal{D}')) \tag{5.12}$$

where $\{f_n\}$ are the Fourier coefficients of f. In particular, C is an intertwining lifting of A with respect to U and tolerance $\gamma$ if and only if $C = \Psi B$ where B is an intertwining lifting of A with respect to U′ and tolerance $\gamma$. Therefore the set of all intertwining liftings C of A with respect to U and tolerance $\gamma$ are given by $C = \Psi B(R)$ where R is a contractive analytic function in $H^\infty(\mathcal{G}, \mathcal{G}')$ and $B(R)$ is the intertwining lifting of A with respect to U′ and tolerance $\gamma$ given in (5.2). In this case the Redheffer scattering matrix $\Psi_A$ becomes $\Psi_A = (I \oplus \Psi)\Phi_A$, that is,

$$\Psi_A = \begin{bmatrix} \Phi_{1,1} & \Phi_{1,2}D_A \\ \Psi\Phi_{2,1} & A + \Psi\Phi_{2,2}D_A \end{bmatrix} : \begin{bmatrix} H^2(\mathcal{G}') \\ \mathcal{H} \end{bmatrix} \to \begin{bmatrix} H^2(\mathcal{G}) \\ \mathcal{K} \end{bmatrix}. \tag{5.13}$$

Obviously $\Psi_A(I \oplus \gamma^{-1}I)$ is a contraction, and if $\Pi_A \omega P_{\mathcal{F}}$ is pointwise stable, then $\Psi_A(I \oplus \gamma^{-1}I)$ is an isometry. In this setting the Redheffer scattering system in (5.11) becomes

$$\begin{bmatrix} g \\ y \end{bmatrix} = \begin{bmatrix} \Phi_{1,1} & \Phi_{1,2}D_A \\ \Psi\Phi_{2,1} & A + \Psi\Phi_{2,2}D_A \end{bmatrix} \begin{bmatrix} x \\ u \end{bmatrix} \tag{5.14}$$

$$\text{subject to } x = Rg$$

where R is a contractive analytic function in $H^\infty(\mathcal{G}, \mathcal{G}')$. Here u is in $\mathcal{H}$ and x is in $H^2(\mathcal{G}')$ while g is in $H^2(\mathcal{G})$ and y is in $\mathcal{K}$. As before, by solving the Redheffer system in (5.14) we see that $y = C(R)u$ where $C(R)$ is an intertwining lifting of A with respect to U and tolerance $\gamma$.

We claim that

$$\Psi_A(S' \oplus T) = (S_{\mathcal{G}} \oplus U)\Psi_A \tag{5.15}$$

where S′ is the unilateral shift on $H^2(\mathcal{G}')$ and $S_{\mathcal{G}}$ is the unilateral shift on $H^2(\mathcal{G})$. Because $\Phi_{1,1}$ is a multiplication operator from $H^2(\mathcal{G}')$ into $H^2(\mathcal{G})$, it is clear that $\Phi_{1,1}S' = S_{\mathcal{G}}\Phi_{1,1}$. Moreover, since $B_\gamma = A + \Psi\Phi_{2,2}D_A$ is precisely the central intertwining lifting of A with respect to U and tolerance $\gamma$, it follows that $UB_\gamma = B_\gamma T$. So to verify (5.15) it remains to show that $\Phi_{1,2}D_A T = S_{\mathcal{G}}\Phi_{1,2}D_A T$ and $\Psi\Phi_{2,1}S' = U\Psi\Phi_{2,1}$. Because $\Phi_{2,1}$ is a multiplication operator

$S\Phi_{2,1} = \Phi_{2,1}S'$. (Recall that S is the unilateral shift on $H^2(\mathcal{D}')$.) However, $\Psi S = \Psi U' | H^2(\mathcal{D}') = U\Psi | H^2(\mathcal{D}')$. Thus $\Psi\Phi_{2,1}S' = U\Psi\Phi_{2,1}$. Finally, using the definition of $\Phi_{1,2}$ in (5.1) along with (2.9) we have

$$\Phi_{1,2}D_AT = P_{\mathcal{G}}D_AT + \lambda P_{\mathcal{G}}(I - \lambda\Pi_A\omega P_{\mathcal{F}})^{-1}\Pi_A\omega P_{\mathcal{F}}D_AT =$$

$$\lambda P_{\mathcal{G}}(I - \lambda\Pi_A\omega P_{\mathcal{F}})^{-1}D_A = S_{\mathcal{G}}\Phi_{1,2}D_A .$$

Therefore $\Phi_{1,2}D_AT = S_{\mathcal{G}}\Phi_{1,2}D_A$ and (5.15) holds.

Now assume that T is a unilateral shift on $\mathcal{H} = H^2(\mathcal{E}_1)$ and U is a unilateral shift on $\mathcal{K} = H^2(\mathcal{E}_2)$. Since $\Psi_A$ intertwines the unilateral shift $S' \oplus T$ with the unilateral shift $S_{\mathcal{G}} \oplus U$, it follows that $\Psi_A = M^+_{\Psi_o}$ for some function $\Psi_o$ in the appropriate $H^\infty$ space. In fact this function $\Psi_o$ is given by

$$\Psi_o(\lambda) = \begin{bmatrix} \Psi_{1,1}(\lambda) & \Psi_{1,2}(\lambda) \\ \Psi_{2,1}(\lambda) & \Psi_{2,2}(\lambda) \end{bmatrix} \text{ in } H^\infty(\mathcal{G}' \oplus \mathcal{E}_1, \ \mathcal{G} \oplus \mathcal{E}_2) . \tag{5.16}$$

Here $\Psi_{i,j}(\lambda)$ for i, j = 1, 2 are given by

$$\Psi_{1,1}(\lambda) = \Phi_{1,1}(\lambda), \ \Psi_{1,2}(\lambda) = (\Phi_{1,2}D_A\Pi_o^*)(\lambda) ,$$
$$\tag{5.17}$$
$$\Psi_{2,1}(\lambda) = \Psi(\lambda)\Phi_{2,1}(\lambda) \text{ and } \Psi_{2,2}(\lambda) = (A\Pi_o^*)(\lambda) + (\Psi\Phi_{2,2}D_A\Pi_o^*)(\lambda) .$$

Here $\Pi_o$ is the operator from $H^2(\mathcal{E}_1)$ into $\mathcal{E}_1$ defined $\Pi_o h = h(0)$. Notice that the $\Psi_{2,1}(\lambda)$ term is given by $\Psi(\lambda)\Phi_{2,1}(\lambda)$ because $\Psi S = U\Psi | H^2(\mathcal{D}')$. So $\Psi | H^2(\mathcal{D}')$ intertwines two unilateral shifts. Hence $\Psi | H^2(\mathcal{D}')$ can be viewed as a multiplication operator $\Psi(\lambda)$. Since $\Psi_A(I \oplus \gamma^{-1}I)$ is a contraction it follows that $\Psi_o(\lambda)(I \oplus \gamma^{-1}I)$ is a contractive analytic function in $H^\infty(\mathcal{G}' \oplus \mathcal{E}_1, \ \mathcal{G} \oplus \mathcal{E}_2)$. In particular, if $\Pi_A\omega P_{\mathcal{F}}$ is pointwise stable, then $\Psi_o$ is inner.

On the other hand, if U is the bilateral shift on $L^2(\mathcal{E}_2)$, then $\Psi_A = M_{\Psi_o} | H^2(\mathcal{G}') \oplus H^2(\mathcal{E}_1)$ where $\Psi_o$ is the function in $L^\infty(\mathcal{G}' \oplus \mathcal{E}_1, \ \mathcal{G} \oplus \mathcal{E}_2)$ given by (5.16) where the $\Psi_{i,j}$ terms are computed for $\lambda = e^{it}$ in (5.17). (Notice that in this case $\Psi_{1,1}(\lambda)$ is still a function in the appropriate $H^\infty$ space.) Obviously $\Psi_o(I \oplus \gamma^{-1}I)$ is a.e. contractive, and if $\Pi_A\omega P_{\mathcal{F}}$ is pointwise stable, then $\Psi_o(I \oplus \gamma^{-1}I)$ is a.e. an isometry. Summing up our previous analysis readily leads to the following result.

**THEOREM 5.4.** *Let* A *be an operator from* $\mathcal{H}$ *into* $\mathcal{H}'$ *intertwining an isometry* T *with a contraction* T′ *satisfying* $\|A\| \le \gamma$. *Let* U *be a minimal isometric lifting of* T′ *and* Ψ *the unitary operator intertwining the Sz.-Nagy-Schäffer minimal isometric lifting* U′ *on* $\mathcal{K}' = \mathcal{H}' \oplus H^2(\mathcal{D}')$ *of* T′ *with* U, *that is,* $\Psi U' = U\Psi$ *and* $\Psi | \mathcal{H}' = I$. *Then the set of all intertwining lifting* B *of* A *with respect to* U *and tolerance* $\gamma$ *are given by*

$$B = A + \Psi\Phi_{2,2}D_A + \Psi\Phi_{2,1}R(I - \Phi_{1,1}R)^{-1}\Phi_{1,2}D_A \tag{5.18}$$

*where R is a contractive analytic function in* $H^\infty(\mathcal{G}, \mathcal{G}')$. *Moreover, the operator* $\Psi_A(I \oplus \gamma^{-1}I)$ *is a contraction and isometric if* $\Pi_A \omega P_{\mathcal{F}}$ *is pointwise stable. Finally, the following holds*

*(i)* *If* T *is a unilateral shift on* $H^2(\mathcal{E}_1)$ *and* U *is a unilateral shift on* $H^2(\mathcal{E}_2)$, *then* $\Psi_o(I \oplus \gamma^{-1}I)$ *is a contractive analytic function in* $H^\infty(\mathcal{G}' \oplus \mathcal{E}_1, \mathcal{G} \oplus \mathcal{E}_2)$. *Moreover, if* $\Pi_A \omega P_{\mathcal{F}}$ *is pointwise stable or* $\|A\| < \gamma$, *then* $\Psi_o(I \oplus \gamma^{-1}I)$ *is inner.*

*(ii)* *If* T *is a unilateral shift on* $H^2(\mathcal{E}_1)$ *and* U *is a bilateral shift on* $L^2(\mathcal{E}_2)$ *then* $\Psi_o(I \oplus \gamma^{-1}I)$ *is a.e. a contraction in* $L^\infty(\mathcal{G}' \oplus \mathcal{E}_1, \mathcal{G} \oplus \mathcal{E}_2)$. *Moreover, if* $\Pi_A \omega P_{\mathcal{F}}$ *is pointwise stable or* $\|A\| < \gamma$, *then* $\Psi_o(I \oplus \gamma^{-1}I)$ *is a.e. an isometry.*

We now proceed to a harmonic-type maximal principle for the Schur parameterization presented in Section VI.2. We will need a few additional properties of the Schur parameterization which we establish in the lemmas below. Finally, let $H_1^\infty(\mathcal{G}, \mathcal{G}')$, respectively $H_{1,0}^\infty(\mathcal{G}, \mathcal{G}')$ be the closed, respectively open unit balls in $H^\infty(\mathcal{G}, \mathcal{G}')$.

**LEMMA 5.5.** *Let* $W = \omega P_{\mathcal{F}} + R(\lambda)P_{\mathcal{G}}$ *where* $R \in H_{1,0}^\infty(\mathcal{G}, \mathcal{G}')$ *and let* $Q(\lambda)$ *be defined by* $Q(\lambda) = (I - \lambda\Pi_A W(\lambda))^{-1}$; *see (3.5). Then*

$$\|P_{\mathcal{G}} Qh\|_{H^2} \le \frac{\|h\|}{\sqrt{1 - \|R\|_\infty^2}} \qquad (h \in \mathcal{D}_A).$$

**PROOF.** The $H^2$ norm $\|P_{\mathcal{G}} Qh\|_{H^2}$ is the $H^2(\mathcal{G})$ norm of the $\mathcal{G}$ valued analytic function $P_{\mathcal{G}} Q(\lambda)h$. Throughout $P_n$ denotes the orthogonal projection of any $H^2$ space onto its subspace of polynomials of degree at most n. Using this notation we have

$$P_n \frac{Q(\lambda) - I}{\lambda} h = P_n \Pi_A W P_n Q(\lambda)h \qquad (h \in \mathcal{D}_A).$$

Therefore, for $h \in \mathcal{D}_A$, we obtain

$$\|P_n \frac{Q(\lambda) - I}{\lambda} h\|_{H^2}^2 \le \|(\omega P_{\mathcal{F}} \oplus R(\lambda)P_{\mathcal{G}})P_n Q(\lambda)h\|_{H^2}^2$$

$$= \|\omega P_{\mathcal{F}} P_n Q(\lambda)h\|_{H^2}^2 + \|R(\lambda)P_{\mathcal{G}} P_n Q(\lambda)h\|_{H^2}^2 \le \|P_{\mathcal{F}} P_n Q(\lambda)h\|_{H^2}^2 + \|R\|_\infty^2 \|P_{\mathcal{G}} P_n Q(\lambda)h\|_{H^2}^2 .$$

Let $Q_n$ for n = 0, 1, 2 ... be the Taylor coefficient of $\lambda^n$ in $Q(\lambda)$. Since $Q(0) = Q_0 = I$, we have

$$\sum_{i=1}^{n+1} \|Q_i h\|^2 \le \|P_{\mathcal{F}} h\|^2 + \|R\|_\infty^2 \|P_{\mathcal{G}} h\|^2 + \sum_{i=1}^{n} (\|P_{\mathcal{F}} Q_i h\|^2 + \|R\|_\infty^2 \|P_{\mathcal{G}} Q_i h\|^2) .$$

This readily implies that

$$(1 - \|R\|_\infty^2) \sum_{i=1}^{n} \|P_{\mathcal{G}} Q_i h\|^2 + \|Q_{n+1} h\|^2 \le \|P_{\mathcal{F}} h\|^2 + \|R\|_\infty^2 \|P_{\mathcal{G}} h\|^2 .$$

Now, recalling that $Q_0 = I$, we obtain

$$(1 - \|R\|_\infty^2) \sum_{i=0}^{n} \|P_{\mathcal{G}} Q_i h\|^2 \le \|P_{\mathcal{F}} h\|^2 + \|R\|_\infty^2 \|P_{\mathcal{G}} h\|^2 + (1 - \|R\|_\infty^2) \|P_{\mathcal{G}} h\|^2 = \|h\|^2 .$$

The lemma follows by letting n tend to infinity. This completes the proof.

**LEMMA 5.6.** *Let* $B(R)$ *be the intertwining lifting of* A *corresponding to the Schur contraction* $W = W_R = \omega P_{\mathcal{G}} + RP_{\mathcal{G}}$ *with* $R \in H_{1,0}^\infty(\mathcal{G}, \mathcal{G}')$. *Let* $B_\gamma$ *the central lifting with tolerance* $\gamma$, *that is, the intertwining lifting associated to* $R = 0$. *Then*

$$\|(B_\gamma - B(R))h\| \le \|R\|_\infty \|P_{\mathcal{G}} (1 - \lambda \Pi_A W)^{-1} D_A h\|_{H^2} \qquad (h \in \mathcal{H}) . \tag{5.19}$$

**PROOF**. Using (2.5) we have

$$(B_\gamma - B(R))h = \begin{bmatrix} 0 \\ \Pi' \omega P_{\mathcal{F}} (I - \lambda \Pi_A \omega P_{\mathcal{F}})^{-1} D_A h - \Pi' W (I - \lambda \Pi_A W)^{-1} D_A h \end{bmatrix} .$$

Notice that

$$\begin{aligned} &\Pi' \omega P_{\mathcal{F}} (I - \lambda \Pi_A \omega P_{\mathcal{F}})^{-1} - \Pi' W (I - \lambda \Pi_A W)^{-1} \\ &\quad = \Pi' \omega P_{\mathcal{F}} (I - \lambda \Pi_A \omega P_{\mathcal{F}})^{-1} - \Pi' \omega P_{\mathcal{F}} (I - \lambda \Pi_A W)^{-1} - \Pi' R(\lambda) P_{\mathcal{G}} (I - \lambda \Pi_A W)^{-1} \\ &\quad = \Pi' \omega P_{\mathcal{F}} (I - \lambda \Pi_A \omega P_{\mathcal{F}})^{-1} (- \lambda \Pi_A R(\lambda)) P_{\mathcal{G}} (I - \lambda \Pi_A W)^{-1} - \Pi' R(\lambda) P_{\mathcal{G}} (I - \lambda \Pi_A W)^{-1} \\ &\quad = - [\lambda \Pi' \omega P_{\mathcal{F}} (I - \lambda \Pi_A \omega P_{\mathcal{F}})^{-1} \Pi_A + \Pi'] R(\lambda) P_{\mathcal{G}} (I - \lambda \Pi_A W)^{-1} \\ &\quad = - \Phi_{2,1}(\lambda) R(\lambda) P_{\mathcal{G}} (I - \lambda \Pi_A W)^{-1} \end{aligned}$$

where $\Phi_{2,1}$ is as in (5.1). Clearly, from Proposition 5.2 it follows that $\Phi_{2,1}$ is a contractive, analytic function and hence we have the relation (5.19). This completes the proof.

**LEMMA 5.7.** *Let* $\lambda$ *be in the open unit disc,* $R \in H_{1,0}^\infty(\mathcal{G}, \mathcal{G}')$, *set* $W = W_R$ *and let* $\{h_j\}$ *be a sequence of vectors in* $\mathcal{H}$.

*i) If* $\|P_{\mathcal{G}}(I-\lambda\Pi_A W)^{-1}D_A h_j\|_{H^2} \to 0$, *then* $\|(B_\gamma - B(R))h_j\| \to 0$.

*ii) If* $\|P_{\mathcal{G}}(I-\lambda\Pi_A \omega P_{\mathcal{F}})^{-1}D_A h_j\|_{H^2} \to 0$, *then* $\|P_{\mathcal{G}}(I-\lambda\Pi_A W)^{-1}D_A h_j\|_{H^2} \to 0$.

**PROOF**. The proof of (i) is immediate from (5.19). To verify (ii), notice that

$$\begin{aligned} P_{\mathcal{G}}(I-\lambda\Pi_A W)^{-1} &= P_{\mathcal{G}}(I-\lambda\Pi_A\omega P_{\mathcal{F}} - \lambda\Pi_A R(\lambda)P_{\mathcal{G}})^{-1} \\ &= P_{\mathcal{G}}[I-\lambda(I-\lambda\Pi_A\omega P_{\mathcal{F}})^{-1}\Pi_A R(\lambda)P_{\mathcal{G}}]^{-1}(1-\lambda\Pi_A\omega P_{\mathcal{F}})^{-1} \\ &= [1-\lambda P_{\mathcal{G}}(I-\lambda\Pi_A\omega P_{\mathcal{F}})^{-1}\Pi_A R(\lambda)P_{\mathcal{G}}]^{-1}P_{\mathcal{G}}(I-\lambda\Pi_A\omega P_{\mathcal{F}})^{-1} \\ &= [I-\Phi_{1,1}(\lambda)R(\lambda)P_{\mathcal{G}}]^{-1}P_{\mathcal{G}}(I-\lambda\Pi_A\omega P_{\mathcal{F}})^{-1} \end{aligned}$$

where $\Phi_{1,1}$ is as in (5.1). From Proposition 5.2 we have that $\Phi_{1,1}$ is a contractive analytic function and hence for all $\lambda \in \mathbb{D}$, we have $\|\Phi_{1,1}(\lambda)R(\lambda)P_{\mathcal{G}}\| \le \|R\|_\infty < 1$. Consequently, $[I-\Phi_{1,1}(\lambda)R(\lambda)P_{\mathcal{G}}]^{-1}$ is in $H^\infty(\mathcal{G}, \mathcal{G})$. Now, the statement (ii) is obvious. This completes the proof.

We are ready to prove the following harmonic maximal principle.

**THEOREM 5.8**. *Let* $\|A\| < \gamma$ *and let* $B_\gamma$ *be the central intertwining lifting of* A *with tolerance* $\gamma$. *Then* $\|B_\gamma\| = \gamma$ *implies* $\|B(R)\| = \gamma$ *for all* $R \in H^\infty_{1,0}(\mathcal{G}, \mathcal{G}')$. *Conversely, if there exists a* $R_0$ *in* $H^\infty_{1,0}(\mathcal{G}, \mathcal{G}')$ *such that* $\|B(R_0)\| = \gamma$, *where* $W = W_{R_0}$, *then* $\|B_\gamma\| = \gamma$.

**PROOF**. Suppose $\|B_\gamma\| = \gamma$, then there exists a sequence of unit vectors $\{h_j\}$ in $\mathcal{H}$ such that $\|B_\gamma h_j\| \to \gamma$ as j tends to infinity. This implies that $D_{B_\gamma}h_j$ approach zero. By virtue of (IV.6.11) we have

$$\|D_{B_\gamma}h_j\| \ge \|P_{\mathcal{G}}(I-\lambda\Pi_A\omega P_{\mathcal{F}})^{-1}D_A h_j\|^2_{H^2} .$$

Therefore, $P_{\mathcal{G}}(I-\lambda\Pi_A\omega P_{\mathcal{F}})^{-1}D_A h_j$ approaches zero in the $H^2$ norm. By part (ii) of Lemma 5.7 the sequence $P_{\mathcal{G}}(I-\lambda\Pi_A W)^{-1}D_A h_j$ also approaches zero in the $H^2$ norm. This in turn, by virtue of Lemma 5.7 (i), implies that $\|(B(R)-B_\gamma)h_j\|)$ approaches zero. Hence $\|B(R)\| = \gamma$.

Conversely, suppose now that $\|B(R_0)\| = \gamma$ for some $R_0 \in H^\infty_{1,0}(\mathcal{G}, \mathcal{G}')$. Then there exists a sequence of unit vectors $\{h_j\}$ in $\mathcal{H}$ such that $\|D_{B(R_0)}h_j\|$ approaches zero as j tends to infinity. Let $W = W_{R_0} = \omega P_{\mathcal{F}} + R_0 P_{\mathcal{G}}$, and recall that

$$B(R_0) = \begin{bmatrix} A \\ \Pi' W Q D_A \end{bmatrix},$$

where $Q(\lambda) = (I - \lambda \Pi_A W)^{-1}$. Then for all f in $\mathcal{H}$ we have

$$\|D_{B(R_0)}f\|^2 = \|D_A f\|^2 - \sum_{n=0}^{\infty} \|\Pi'(WQ)_n D_A f\|^2$$

$$= \|D_A f\|^2 - \sum_{n=0}^{\infty} (\|(WQ)_n D_A f\|^2 - \|\Pi_A (WQ)_n D_A f\|^2), \tag{5.20}$$

where for any F in a $H^\infty$ space we denote by $(F)_n$ the n-th Taylor coefficient of F. Now, since $\lambda \Pi_A W(\lambda) Q(\lambda) = Q(\lambda) - I$, we have by comparing Taylor coefficients

$$Q_0 = I \text{ and } Q_n = \Pi_A (WQ)_{n-1} \qquad (\text{for } n \geq 1). \tag{5.21}$$

Then by (5.20) and (5.21) we have

$$\|D_{B(R_0)}f\|^2 = \|D_A f\|^2 - \lim_{n\to\infty} \left[ \sum_{i=0}^{n} \|(WQ)_i D_A f\|^2 - \sum_{i=1}^{n+1} \|Q_i D_A f\|^2 \right]$$

$$\geq \|D_A f\|^2 - \|(WQ)_0 D_A f\|^2 + \lim_{n\to\infty} \sum_{i=1}^{n} (\|Q_i D_A f\|^2 - \|(WQ)_i D_A f\|^2)$$

$$= \lim_{n\to\infty} (\|P_n Q D_A f\|_{H^2}^2 - \|P_n W P_n Q D_A f\|_{H^2}^2)$$

$$= \lim_{n\to\infty} (\|P_n Q D_A f\|_{H^2}^2 - \|P_n \omega P_{\mathcal{F}} P_n Q D_A f\|_{H^2}^2 - \|P_n R_0(\lambda) P_{\mathcal{G}} P_n Q D_A f\|_{H^2}^2)$$

$$= \lim_{n\to\infty} (\|P_n P_{\mathcal{G}} Q D_A f\|_{H^2}^2 - \|P_n R_0(\lambda) P_{\mathcal{G}} Q D_A f\|_{H^2}^2).$$

Note that $\|P_{\mathcal{G}} Q D_A f\|_{H^2}^2 < \infty$ by Lemma 5.5. Hence,

$$\lim_{n\to\infty} \|P_n P_{\mathcal{G}} Q D_A f\|_{H^2}^2 = \|P_{\mathcal{G}} Q D_A f\|_{H^2}^2$$

and

$$\lim_{n\to\infty} \|P_n R_0(\lambda) P_{\mathcal{G}} Q D_A f\|_{H^2}^2 = \|R_0(\lambda) P_{\mathcal{G}} Q D_A f\|_{H^2}^2 .$$

Therefore, it follows that

$$\|D_{B(R_0)}f\|^2 \geq \|P_{\mathcal{G}} Q D_A f\|^2 - \|R_0\|_\infty^2 \|P_{\mathcal{G}} Q D_A f\|_{H^2}^2 = (I - \|R_0\|_\infty^2) \|P_{\mathcal{G}} Q D_A f\|_{H^2}^2$$

Now, with $f = h_j$ and letting j go to infinity we see that $\|P_{\mathcal{G}} Q D_A h_j\|_{H^2}^2$ approaches zero. Part (i) of Lemma 5.7 (i), implies that $\|(B(R_0) - B_\gamma) h_j\|$ approaches zero as j tends to infinity. Consequently $\|B_\gamma\| = \gamma$. The proof of the theorem is now complete.

**REMARK 5.9.** Note that Theorem 5.8 also states that if $\|B(R_0)\| = \gamma$ for some $R_0$ in $H^\infty_{1,0}(\mathcal{G}, \mathcal{G}')$, then $\|B(R)\| = \gamma$ for all R in $H^\infty_{1,0}(\mathcal{G}, \mathcal{G}')$.

## VI.6 THE PARAMETERIZATION FOR $\|A\| < \gamma$

Assume that A is an operator from $\mathcal{H}$ into $\mathcal{H}'$ intertwining the isometry T with a contraction T′ and $\|A\| < \gamma$. Recall that $T_A$ is the operator on $\mathcal{H}$ defined by (see (IV.3.2) and (IV.3.6))

$$T_A = (\gamma^2 I - A^* A)T(\gamma^2 I - T^* A^* A T)^{-1} \text{ and } \Pi_A \omega P_{\mathcal{F}} = D_A T_A^* D_A^{-1} . \tag{6.1}$$

Now let $\mathcal{L}$ be the kernel of $T^*$ and N on $\mathcal{L}$ and $N_1$ on $\mathcal{D}'$ be the positive operators defined by the positive square roots of

$$N^2 = \Pi_o D_A^{-2} \Pi_o \text{ and } N_1^2 = (I + D'AD_A^{-2}A^*D') \mid \mathcal{D}' \tag{6.2}$$

where $\Pi_o$ is the operator from $\mathcal{H}$ onto $\mathcal{L}$ defined by $\Pi_o = P_{\mathcal{L}}$. As before, D′ is the positive square root of $I - T'^*T'$ while $\mathcal{D}'$ is the closure of the range of D′. Finally, let $Y_{1,1}$, $Y_{1,2}$, $Y_{2,1}$ and $Y_{2,2}$ be the operator valued analytic function defined by

$$\begin{aligned}
Y_{1,1}(\lambda) &= -\lambda N^{-1}\Pi_o(I - \lambda T_A^*)^{-1} D_A^{-2} A^* D' N_1^{-1} \\
Y_{1,2}(\lambda) &= N^{-1}\Pi_o(I - \lambda T_A^*)^{-1} \\
Y_{2,1}(\lambda) &= N_1 - D'A(I - \lambda T_A^*)^{-1} D_A^{-2} A^* D' N_1^{-1} \\
Y_{2,2}(\lambda) &= D'AT_A^*(I - \lambda T_A^*)^{-1} .
\end{aligned} \tag{6.3}$$

It turns out that $Y_{1,1}$ and $Y_{2,1}$ are contractive analytic functions in $H^\infty(\mathcal{D}', \mathcal{L})$ and $H^\infty(\mathcal{D}', \mathcal{D}')$. So $[Y_{1,1}, Y_{2,1}]^{tr}$ can be viewed as a multiplication operator from $H^2(\mathcal{D}')$ into $H^2(\mathcal{L}) \oplus H^2(\mathcal{D}')$. Moreover, $[Y_{1,2}, Y_{2,2}]^{tr}$ can be viewed as an operator from $\mathcal{H}$ into $H^2(\mathcal{L}) \oplus H^2(\mathcal{D}')$, respectively. This sets the stage for the following result.

**THEOREM 6.1.** *Let* A *be an operator from* $\mathcal{H}$ *into* $\mathcal{H}'$ *intertwining an isometry* T *with a contraction* T′ *and assume that* $\|A\| < \gamma$. *Then the set of all intertwining liftings* B *of* A *with respect to the Sz.-Nagy-Schäffer isometric lifting* U′ *in (1.10) and tolerance* $\gamma$ *is given by*

$$B = B(R) = \begin{bmatrix} A \\ Y_{2,2} + Y_{2,1}R(I - Y_{1,1}R)^{-1}Y_{1,2} \end{bmatrix} \tag{6.4}$$

*where* R *is a contractive analytic function in* $H^\infty(\mathcal{L}, \mathcal{D}')$. *Moreover, the mapping from* R *to*

B(R) *is a bijection from the closed unit ball in* $H^\infty(\mathcal{L}, \mathcal{D}')$ *onto the set of all intertwining lifting with* B *of* A *with respect to* U′ *and tolerance* $\gamma$.

Notice that if $R = 0$, then $B(0) = [A, Y_{2,2}]^{tr}$ is precisely the central intertwining lifting of A with respect to U′ and tolerance $\gamma$; see Proposition IV.3.1.

**PROOF**. The proof follows by connecting the $\Phi_{i,j}$ for i, j = 1,2 in (5.1) to the $Y_{i,j}$ in (6.3) and then implementing Theorem 5.1. To this end, first notice that equation (IV.3.7) shows that $Y_{2,2}(\lambda) = \Phi_{2,2}(\lambda)D_A$. According to (IV.3.8) the space $\mathcal{G} = \operatorname{ran} D_A^{-1}\Pi_o^*$. Notice that if $X = D_A^{-1}\Pi_o^*$, then $X(X^*X)^{-1/2}$ is an isometry from $\mathcal{L}$ into $\mathcal{D}_A$ whose range equals $\mathcal{G}$. Since $X^*X = \Pi_o D_A^{-2}\Pi_o^* = N^2$, it follows that

$$E = D_A^{-1}\Pi_o^* N^{-1} \tag{6.5}$$

is an isometry from $\mathcal{L}$ into $\mathcal{D}_A$ whose range is $\mathcal{G}$. Clearly $P_{\mathcal{G}} = EE^*$ and thus $E^*P_{\mathcal{G}} = E^*$. Observe that $f \oplus g$ in $\mathcal{D}' \oplus \mathcal{D}_A$ is in $\mathcal{G}'$ if and only if $f \oplus g$ is orthogonal to $\{D'Ah \oplus D_A h : h \in \mathcal{H}\}$, or equivalently, $A^*D'f + D_A g = 0$. Hence $f \oplus g$ is in $\mathcal{G}'$ if and only if $f \oplus g$ is in $\mathcal{D}' \oplus \mathcal{D}_A$ and $g = -D_A^{-1}A^*D'f$. In other words, $\mathcal{G}'$ is the range of the operator Y from $\mathcal{D}'$ into $\mathcal{D}' \oplus \mathcal{D}_A$ defined by

$$Y = \begin{bmatrix} I \\ -D_A^{-1}A^*D' \end{bmatrix} : \mathcal{D}' \to \begin{bmatrix} \mathcal{D}' \\ \mathcal{D}_A \end{bmatrix}.$$

Obviously $E_1 = Y(Y^*Y)^{-1/2}$ is an isometry from $\mathcal{D}'$ into $\mathcal{D}' \oplus \mathcal{D}_A$ whose range is $\mathcal{G}'$. Since $Y^*Y = N_1^2$ it follows that

$$E_1 = \begin{bmatrix} I \\ -D_A^{-1}A^*D' \end{bmatrix} N_1^{-1} \tag{6.6}$$

is an isometry from $\mathcal{D}'$ into $\mathcal{D}' \oplus \mathcal{D}_A$ whose range is $\mathcal{G}'$.

Now using (6.1) along with (6.5), (6.6) and $E^*P_{\mathcal{G}} = E^*$ we have

$$\begin{aligned} &E^*\Phi_{1,1}(\lambda)E_1 = -\lambda E^*P_{\mathcal{G}}(I - \lambda\Pi_A\omega P_{\mathcal{F}})^{-1}D_A^{-1}A^*D'N_1^{-1} = \\ &-\lambda N^{-1}\Pi_o D_A^{-1}(I - \lambda D_A T_A^* D_A^{-1})^{-1}D_A^{-1}A^*D'N_1^{-1} = \\ &-\lambda N^{-1}\Pi_o(I - \lambda T_A^*)^{-1}D_A^{-2}A^*D'N_1^{-1} = Y_{1,1}(\lambda). \end{aligned} \tag{6.7}$$

Equation (IV.3.1) shows that $P_{\mathcal{F}} = D_A T D_{AT}^{-2} T^* D_A$. Now a similar calculation on $\Phi_{1,2}$ in (5.1) gives

$$E^*\Phi_{1,2}(\lambda)D_A = N^{-1}\Pi_o D_A^{-1}(I-\lambda\Pi_A\omega P_{\mathcal{F}})^{-1}D_A =$$
$$N^{-1}\Pi_o D_A^{-1}(I-\lambda D_A T_A^* D_A^{-1})^{-1}D_A = Y_{1,2}(\lambda)\,. \tag{6.8}$$

Finally, computing $\Phi_{2,1}E_1$ gives

$$\Phi_{2,1}(\lambda)E_1 = -\lambda\Pi'\omega P_{\mathcal{F}}(I-\lambda D_A T_A^* D_A^{-1})^{-1}D_A^{-1}A^*D'N_1^{-1} + N_1^{-1} =$$
$$-\lambda\Pi'\omega P_{\mathcal{F}}D_A(I-\lambda T_A^*)^{-1}D_A^{-2}A^*D'N_1^{-1} + N_1^{-1}$$
$$-\lambda\Pi'\omega D_A T D_{AT}^{-2}T^*D_A^2(I-\lambda T_A^*)^{-1}D_A^{-2}A^*D'N_1^{-1} + N_1^{-1} =$$
$$-\lambda D'A(I-\lambda T_A^*)^{-1}T_A^*D_A^{-2}A^*D'N_1^{-1} + N_1^{-1} = \tag{6.9}$$
$$D'A[I-(I-\lambda T_A^*)^{-1}]D_A^{-2}A^*D'N_1^{-1} + N_1^{-1} =$$
$$(I+D'AD_A^{-2}A^*D')N_1^{-1} - D'A(I-\lambda T_A^*)^{-1}D_A^{-2}A^*D'N_1^{-1} = Y_{2,1}(\lambda)\,.$$

By substituting (6.7) to (6.9) along with $Y_{2,2} = \Phi_{2,2}D_A$ in (5.2) of Theorem 5.1 we readily obtain the form of B(R) in (6.4) where R is replaced by $E_1^*RE$. Because E and $E_1$ are isometrics whose ranges are $\mathcal{G}$ and $\mathcal{G}'$, respectively, we can choose R to be in $H^\infty(\mathcal{L}, \mathcal{D}')$. This completes the proof.

According to Proposition 5.2 and (6.7), (6.9) we see that $Y_{1,1}(\lambda)$ and $Y_{2,1}(\lambda)$ are contractive analytic functions in $H^\infty(\mathcal{D}', \mathcal{L})$ and $H^\infty(\mathcal{D}', \mathcal{D}')$ respectively. In fact, if $\Pi_A\omega P_{\mathcal{F}}$ is pointwise stable, then $[Y_{1,1}, Y_{2,1}]^{tr}$ is an inner function. Moreover, for h in $\mathcal{H}$ the vector $Y_{1,2}(\lambda)h$ is in $H^2(\mathcal{L})$, while $Y_{2,2}(\lambda)h$ is in $H^2(\mathcal{D}')$.

To be precise let $Y_A$ be the operator defined by

$$Y_A = \begin{bmatrix} Y_{1,1} & Y_{1,2} \\ [0,\ Y_{2,1}]^{tr} & [A,\ Y_{2,2}]^{tr} \end{bmatrix} : \begin{bmatrix} H^2(\mathcal{D}') \\ \mathcal{H} \end{bmatrix} \to \begin{bmatrix} H^2(\mathcal{L}) \\ \mathcal{H}'\oplus H^2(\mathcal{D}') \end{bmatrix}. \tag{6.10}$$

Then $Y_A(I\oplus\gamma^{-1}I)$ is a contraction. Furthermore, if $\Pi_A\omega P_{\mathcal{F}}$ is pointwise stable, then $Y_A(I\oplus\gamma^{-1}I)$ is an isometry.

**REMARK 6.2.** As before, let A be an operator from $\mathcal{H}$ into $\mathcal{H}'$ intertwining an isometry T on $\mathcal{H}$ with a contraction T′ on $\mathcal{H}'$ and assume that $\|A\| < \gamma$. Let U on $\mathcal{K}$ be a minimal isometric lifting of T′, and U′ on $\mathcal{K}' = \mathcal{H}'\oplus H^2(\mathcal{D}')$ be the Sz.-Nagy-Schäffer minimal isometric lifting of T′. Finally, let $\Psi$ be the unique unitary operator from $\mathcal{K}'$ onto $\mathcal{K}$ satisfying $\Psi U = U'\Psi$ and $\Psi | \mathcal{H}' = I$. Then the set of all intertwining lifting B of A with respect to U and tolerance $\gamma$ is given by

$$B = B(R) = A + \Psi Y_{2,2} + \Psi Y_{2,1} R(I - \lambda Y_{1,1} R)^{-1} Y_{1,2} \tag{6.11}$$

where R is a contractive analytic function in $H^\infty(\mathcal{L}, \mathcal{D}')$. Moreover, the map from R to B(R) is a bijection from the set of all contractive analytic functions R in $H^\infty(\mathcal{L}, \mathcal{D}')$ onto the set of all intertwining lifting B of A with respect to U and tolerance $\gamma$. Furthermore, this $B = B(R)$ can be obtained as the solution to the Redheffer scattering system

$$\begin{bmatrix} g \\ y \end{bmatrix} = \begin{bmatrix} Y_{1,1} & Y_{1,2} \\ \Psi Y_{2,1} & A + \Psi Y_{2,2} \end{bmatrix} \begin{bmatrix} x \\ u \end{bmatrix} \tag{6.12}$$

$$\text{subject to } x = Rg$$

Here u is a vector in $\mathcal{H}$ and x is in $H^2(\mathcal{D}')$ while g is in $H^2(\mathcal{L})$ and y is in $\mathcal{K}$. In this setting the Redheffer scattering matrix $\Psi_A$ is defined by

$$\Psi_A = \begin{bmatrix} Y_{1,1} & Y_{1,2} \\ \Psi Y_{2,1} & A + \Psi Y_{2,2} \end{bmatrix} : \begin{bmatrix} H^2(\mathcal{D}') \\ \mathcal{H} \end{bmatrix} \to \begin{bmatrix} H^2(\mathcal{L}) \\ \mathcal{K} \end{bmatrix} . \tag{6.13}$$

As before $\Psi_A(I \oplus \gamma^{-1} I)$ is a contraction. If $\Pi_A \omega P_{\mathcal{F}}$ is pointwise stable, then $\Psi_A(I \oplus \gamma^{-1} I)$ is inner.

If T is a unilateral shift on $H^2(\mathcal{E}_1)$ and U is a unilateral shift on $H^2(\mathcal{E}_2)$, then $\Psi_A = M^+_{Y_o}$ where $Y_o$ is the function in $H^\infty(\mathcal{D}' \oplus \mathcal{E}_1, \mathcal{L} \oplus \mathcal{E}_2)$ defined by

$$Y_o(\lambda) = \begin{bmatrix} Y_{1,1}(\lambda) & (Y_{1,2}\Pi_o^*)(\lambda) \\ \Psi(\lambda) Y_{2,1}(\lambda) & (A\Pi_o^*)(\lambda) + (\Psi Y_{2,2}\Pi_o^*)(\lambda) \end{bmatrix} . \tag{6.14}$$

Here $\Pi_o$ is the operator from $H^2(\mathcal{E}_1)$ onto $\mathcal{E}_1$ defined by $\Pi_o h = h(0)$. (Recall that $\Psi S = \Psi U' | H^2(\mathcal{D}') = U\Psi | H^2(\mathcal{D}')$. So $\Psi | H^2(\mathcal{D}')$ intertwines two unilateral shifts and $\Psi(\lambda)$ denotes the symbol for the multiplication operator $\Psi | H^2(\mathcal{D}')$. Moreover, $\Psi(\lambda) Y_{2,1}(\lambda) = (\Psi Y_{2,1} | \mathcal{D}')(\lambda)$.) In particular, $Y_o(I \oplus \gamma^{-1} I)$ is a contractive analytic function. Finally, if $\Pi_A \omega P_{\mathcal{F}}$ is pointwise stable or $\|A\| < \gamma$, then $Y_o(I \oplus \gamma^{-1} I)$ is an inner function.

On the other hand, if T is a unilateral shift on $H^2(\mathcal{E}_1)$ and U is the bilateral on $L^2(\mathcal{E}_2)$, then $\Psi_A = M_{Y_o} | (H^2(\mathcal{D}') \oplus H^2(\mathcal{E}_1))$ where $Y_o$ is the function in $L^\infty(\mathcal{D}' \oplus \mathcal{E}_1, \mathcal{L} \oplus \mathcal{E}_2)$ defined according to (6.14) with $\lambda = e^{it}$. In particular, $Y_o(I \oplus \gamma^{-1} I)$ is a.e. a contraction. Finally, if $\Pi_A \omega P_{\mathcal{F}}$ is pointwise stable or $\|A\| < \gamma$, then $Y_o(I \oplus \gamma^{-1} I)$ is a.e. an isometry.

## VI.7 THE NEVANLINNA-PICK PARAMETERIZATION

Here we will use the Schur parameterization for the commutant lifting theorem, to obtain the set of all solutions to the standard left Nevanlinna-Pick interpolation problem with tolerance $\gamma$. Recall that Z on $\mathcal{X}$, B and $\tilde{B}$ is the data for the standard left Nevanlinna-Pick problem, where B maps $\mathcal{U}$ into $\mathcal{X}$ and $\tilde{B}$ maps $\mathcal{Y}$ into $\mathcal{X}$. Throughout it is always assumed that $\{Z, B\}$ is a stable, controllable pair. As before W from $\ell_+^2(\mathcal{U})$ into $\mathcal{X}$ and $\tilde{W}$ from $\ell_+^2(\mathcal{Y})$ into $\mathcal{X}$ are the controllability operators generated by the pair $\{Z, B\}$ and $\{Z, \tilde{B}\}$, i.e.,

$$W = [B, ZB, Z^2B, \ldots] \text{ and } \tilde{W} = [\tilde{B}, Z\tilde{B}, Z^2\tilde{B}, \ldots] . \tag{7.1}$$

Recall also that G is a *Nevanlinna-Pick interpolant* if G is a function in $H^\infty(\mathcal{Y}, \mathcal{U})$ satisfying $WT_G = \tilde{W}$. So the left standard Nevanlinna-Pick interpolation problem with tolerance $\gamma$ is to find the set of all function G in $H^\infty(\mathcal{Y}, \mathcal{U})$ satisfying $WT_G = \tilde{W}$ and $\|G\|_\infty \leq \gamma$. Now let $P = WW^*$ and $\tilde{P} = \tilde{W}\tilde{W}^*$ be the controllability grammians for the pair $\{Z, B\}$ and $\{Z, \tilde{B}\}$, respectively. In particular, P and $\tilde{P}$ can be obtained as the unique solutions of

$$P = ZPZ^* + BB^* \text{ and } \tilde{P} = Z\tilde{P}Z^* + \tilde{B}\tilde{B}^* . \tag{7.2}$$

According to the results in Section II.2 there exists a solution to the Nevanlinna-Pick problem with tolerance $\gamma$ if and only if $\gamma \geq d_\infty$ where $d_\infty^2 = r_{spec}(\tilde{P}P^{-1})$.

To obtain the set of all solutions to the standard Nevanlinna-Pick problem with tolerance $\gamma > d_\infty$, let $\Theta$ be the two-sided inner function in $H^\infty(\mathcal{U}, \mathcal{U})$ generated by $\{Z^*, B^*\}$, computed according to Proposition III.7.2. To be precise, let C from $\mathcal{X}$ to $\mathcal{U}$ and E on $\mathcal{U}$ be the operators computed by Proposition III.7.2 and $\Theta$ the transfer function for $\{Z^*, C^*, B^*, E^*\}$ given by

$$\Theta(\lambda) = E^* + \lambda B^*(I - \lambda Z^*)^{-1}C^* . \tag{7.3}$$

By construction $\{C, Z\}$ is observable and

$$\mathcal{F}_+ W^* \mathcal{X} = \{B^*(I - \lambda Z^*)^{-1}x : x \in \mathcal{X}\} = H^2(\mathcal{U}) \ominus \Theta H^2(\mathcal{U}) \tag{7.4}$$

where $\mathcal{F}_+$ is the Fourier transform from $\ell_+^2(\mathcal{U})$ onto $H^2(\mathcal{U})$. Now let N on $\mathcal{Y}$ and $N_*$ on $\mathcal{U}$ be the positive operators defined by

$$N = \gamma^{-1}[I + \tilde{B}^*(\gamma^2 P - \tilde{P})^{-1}\tilde{B}]^{1/2} \text{ and } N_* = [I + C\tilde{P}(\gamma^2 P - \tilde{P})^{-1}PC^*]^{1/2} . \tag{7.5}$$

As in Section V.1 equation (V.1.9), let M be the stable operator on $\mathcal{X}$ defined by

$$M = (\gamma^2 P - \tilde{P})Z^*(\gamma^2 P - Z\tilde{P}Z^*)^{-1} . \tag{7.5a}$$

Now let $\Phi_{1,1}$, $\Phi_{1,2}$, $\Phi_{2,1}$ and $\Phi_{2,2}$ be the operator valued analytic function defined by

$$\begin{aligned}
\Phi_{1,1}(\lambda) &= -\lambda N^{-1}\tilde{B}^*(\gamma^2 P - \tilde{P})^{-1}(I - \lambda M)^{-1} P C^* N_*^{-1} \\
\Phi_{1,2}(\lambda) &= N^{-1} - \lambda N^{-1}\tilde{B}^*(\gamma^2 P - \tilde{P})^{-1}(I - \lambda M)^{-1} M\tilde{B} \\
\Phi_{2,1}(\lambda) &= \Theta(\lambda)N_* - \Theta(\lambda)C\tilde{P}(\gamma^2 P - \tilde{P})^{-1}(I - \lambda M)^{-1} P C^* N_*^{-1} \\
\Phi_{2,2}(\lambda) &= \gamma^2 B^*(\gamma^2 P - Z\tilde{P}Z^*)^{-1}(I - \lambda M)^{-1}\tilde{B} .
\end{aligned} \tag{7.6}$$

It turns out that the $2 \times 2$ block matrix generated by $\{\Phi_{i,j}(\lambda) : i, j = 1, 2\}$ is a function in $H^\infty(\mathcal{U} \oplus \mathcal{Y}, \mathcal{Y} \oplus \mathcal{U})$; see Remark 7.2 below.

**THEOREM 7.1.** *Let Z, B and $\tilde{B}$ be the data for the standard left Nevanlinna-Pick problem with tolerance $\gamma > d_\infty$. Assume that $\{Z, B\}$ is a stable controllable pair. Let $\Phi_{i,j}(\lambda)$ for $i, j = 1, 2$ be the operator valued analytic functions given by (7.6). Then the set of all Nevanlinna-Pick interpolants G with tolerance $\gamma$ is given by*

$$G = \Phi_{2,2} + \Phi_{2,1} R (I - \Phi_{1,1} R)^{-1} \Phi_{1,2} \tag{7.7}$$

*where R is a contractive analytic function in $H^\infty(\mathcal{Y}, \mathcal{U})$. Moreover, the mapping from R into G given by (7.7) is a bijection from the set of contractive analytic functions R in $H^\infty(\mathcal{Y}, \mathcal{U})$ onto the set of Nevanlinna-Pick interpolants G with tolerance $\gamma$.*

**REMARK 7.2.** Equation (7.7) shows that the set of all Nevanlinna-Pick interpolants G with tolerance $\gamma$ is given by solving the following Redheffer scattering system

$$\begin{bmatrix} g \\ u \end{bmatrix} = \begin{bmatrix} \Phi_{1,1}(\lambda) & \Phi_{1,2}(\lambda) \\ \Phi_{2,1}(\lambda) & \Phi_{2,2}(\lambda) \end{bmatrix} \begin{bmatrix} x \\ y \end{bmatrix} \tag{7.8}$$

$$\text{subject to } x = R(\lambda) g$$

where R is a contractive analytic function in $H^\infty(\mathcal{Y}, \mathcal{U})$. Here $x \oplus y$ is in $\mathcal{U} \oplus \mathcal{Y}$ while $g \oplus u$ is in $\mathcal{Y} \oplus \mathcal{U}$. In other words, if y is in $\mathcal{Y}$, then the solution $u = G(\lambda)y$ to the Redheffer scattering system in (7.8) is precisely the Nevanlinna-Pick interpolant G with tolerance $\gamma$. Moreover, if $\Phi(\lambda)$ is the $2 \times 2$ block matrix in (7.8), then $\Phi(\lambda)$ is in $H^\infty(\mathcal{U} \oplus \mathcal{Y}, \mathcal{Y} \oplus \mathcal{U})$. Furthermore, $\Phi(I \oplus \gamma^{-1} I)$ is inner. In particular, if both $\mathcal{U}$ and $\mathcal{Y}$ are finite dimensional, then $\Phi(I \oplus \gamma^{-1} I)$ is inner from both sides. Finally, recall that the central interpolant $G_\gamma = \Phi_{2,2}$ is obtained by setting

$R = 0$, and $\|G_\gamma\|_\infty < \gamma$; see Theorem V.1.1. So if $\|R\|_\infty < 1$, then Theorem 5.8 shows that $\|G\|_\infty < \gamma$.

**PROOF OF THEOREM 7.1.** Recall that the operators T, A and T′ in the commutant lifting theorem for the standard Nevanlinna-Pick problem are given by $A = W^* P^{-1} \tilde{W}$ mapping $\mathcal{H} = \ell_+^2(\mathcal{Y})$ into $\mathcal{H}' = \operatorname{ran} W^*$ while T is the unilateral shift on $\ell_+^2(\mathcal{Y})$ and T′ is the compression of the unilateral shift S on $\ell_+^2(\mathcal{U})$ to $\mathcal{H}'$, that is, $T' = P_{\mathcal{H}'} S | \mathcal{H}'$. Later, we will need the following useful identifies collected from (V.1.13) and (V.1.14)

$$D_A^{-2} = \gamma^{-2}[I + \tilde{W}^* (\gamma^2 P - \tilde{P})^{-1} \tilde{W}] \text{ and } D_A^{-2} A^* = \tilde{W}^* (\gamma^2 P - \tilde{P})^{-1} W . \tag{7.9}$$

For the moment assume that the kernel of B is zero. We claim that S is the minimal isometric lifting of T′. Since $\mathcal{H}'$ is an invariant subspace for $S^*$ satisfying $S^* | \mathcal{H}' = T'^*$, it follows that S is an isometric lifting of T′. To verify that S is the minimal isometric lifting it remains to show that $\mathcal{H}'$ is cyclic for S. To this end, recall that $T'^* W^* = S^* W^* = W^* Z^*$. Using this we readily obtain for x in $\mathcal{X}$

$$(I - ST'^*)W^* x = W^* x - SW^* Z^* x = B^* x \oplus 0 .$$

Notice that $B^* x$ sits in the first position of $\ell_+^2(\mathcal{U})$. Because the range of $B^*$ is dense in $\mathcal{U}$ we see that $\mathcal{U} \subseteq (I - ST'^*)\mathcal{H}' \subseteq \mathcal{H}' \vee S\mathcal{H}'$. This readily shows that

$$\ell_+^2(\mathcal{U}) \subseteq \bigvee_0^\infty S^n \mathcal{H}' \subseteq \ell_+^2(\mathcal{U}) .$$

Therefore $\mathcal{H}'$ is cyclic for S and S is a minimal isometric dilation of T′.

Let Ψ be the operator from $\mathcal{H}' \oplus H^2(\mathcal{D}')$ into $\ell_+^2(\mathcal{U})$ defined by

$$\Psi(h \oplus D'f) = h + \sum_0^\infty S^n (S - T') f_n \qquad (h \oplus f \in \mathcal{H}' \oplus H^2(\mathcal{D}')) . \tag{7.10}$$

where $\{f_n\}_0^\infty$ are the Fourier coefficients of f. Proposition IV.1.3 shows that Ψ is a unitary operator intertwining the Sz.-Nagy-Schäffer minimal isometric dilation U′ with S and $\Psi | \mathcal{H}' = I$. Recall that B′ is an intertwining lifting of A with respect to S satisfying $\|B'\| \le \gamma$ if and only if $B' = T_G$ where G is a Nevanlinna-Pick interpolant with tolerance γ. Since $S\Psi = \Psi U'$ it follows that $B_o$ is an intertwining lifting of A with respect to U′ and tolerance γ if and only if $B' = \Psi B_o$ is an intertwining lifting of A with respect to S and tolerance γ. Moreover, the corresponding Nevanlinna-Pick interpolant G with tolerance γ satisfying $B' = T_G = \Psi B_o$ is given by $G = \mathcal{F}_+ \Psi B_o \Pi_o^*$ where $\Pi_o$ is the operator from $\ell_+^2(\mathcal{Y})$ onto $\mathcal{Y}$ given by picking out the first component of $\ell_+^2(\mathcal{Y})$, that is, $\Pi_o = P_{\mathcal{Y}}$. So according to Theorem 6.1 or Remark 6.2, the set of all Nevanlinna-Pick interpolants G with tolerance $\gamma$ is given by

$$G = \mathcal{F}_+ A\Pi_o^* + \mathcal{F}_+ \Psi Y_{2,2}(\lambda)\Pi_o^* + \mathcal{F}_+ \Psi Y_{2,1} R(I - Y_{1,1}R)^{-1} Y_{1,2}\Pi_o^* \tag{7.11}$$

where $Y_{i,j}$ is defined in (6.3) and R is a contractive analytic function in $H^\infty(\mathcal{Y}, \mathcal{D}')$. To complete the proof we identify $\mathcal{D}'$ with $\mathcal{U}$ and convert (7.11) to the state space formulation given by (7.6).

Now let us obtain an identification between $\mathcal{D}'$ and $\mathcal{U}$. To this end, first notice that

$$\mathcal{F}_+ \Psi D'W^* = \Theta CP\,. \tag{7.12}$$

Here $D'W^*x$ is viewed as an element in $\{0\} \oplus \mathcal{D}' \subseteq \mathcal{H}' \oplus H^2(\mathcal{D}')$. To establish (7.12) let x be in $\mathcal{X}$ and observe that $S^* \mid \mathcal{H}' = T'^*$ implies that $(S - T')W^*$ is orthogonal to $\mathcal{H}' = \operatorname{ran} W^*$. However, according to Theorem III.7.1, the space $\ell_+^2(\mathcal{U}) = \mathcal{H}' \oplus \operatorname{ran} T_\Theta$. Thus $(S - T')W^*x = T_\Theta h$ for some h in $\ell_+^2(\mathcal{U})$. We claim that $(S - T')W^*x$ is also orthogonal to $\operatorname{ran}(T_\Theta S)$. For any f in $\ell_+^2(\mathcal{U})$ we have

$$((S - T')W^*x,\ T_\Theta Sf) = (W^*x,\ S^*T_\Theta Sf) - 0 = 0\,.$$

Hence $(S - T')W^*x$ is orthogonal to both $\mathcal{H}'$ and $\operatorname{ran} T_\Theta S$. Therefore $(S - T')W^*x = T_\Theta(u \oplus 0)$ for some vector u in $\mathcal{U}$ the first component of $\ell_+^2(\mathcal{U})$. To obtain this u notice that $(S - T')W^*x - T_\Theta(u \oplus 0)$ is orthogonal to $T_\Theta(v \oplus 0)$ for all v in $\mathcal{U}$. This along with $T_\Theta(v \oplus 0) = E^*v \oplus W^*C^*v$ gives

$$0 = ((S - T')W^*x - T_\Theta(u \oplus 0),\ T_\Theta(v \oplus 0)) = (W^*x,\ W^*C^*v) - (u,\ v) = (CPx - u,\ v)\,.$$

Since v is arbitrary in $\mathcal{U}$ we have $u = CPx$. Finally, using the definition of $\Psi$ we obtain (7.12) by the following calculation

$$\mathcal{F}_+ \Psi D'W^*x = \mathcal{F}_+(S - T')W^*x = \mathcal{F}_+ T_\Theta(CP \oplus 0) = \Theta(\lambda)CP\,.$$

Because $\Psi$ is a unitary operator mapping $\mathcal{H}' \oplus H^2(\mathcal{D}')$ onto $\ell_+^2(\mathcal{U}) = \mathcal{H}' \oplus \operatorname{ran} T_\Theta$ and $\Psi \mid \mathcal{H}' = I$, it follows that $\mathcal{F}_+ \Psi H^2(\mathcal{D}') = \Theta H^2(\mathcal{U})$. However, equation (7.12) along with the fact that $\Psi \mid H^2(\mathcal{D}')$ intertwines two unilateral shifts, show that $\mathcal{F}_+ \Psi H^2(\mathcal{D}') = \Theta H^2(\overline{\operatorname{ran} C})$. Hence the range of C is dense in $\mathcal{U}$. Therefore $\mathcal{D}'$ can be identified with $\mathcal{U}$. To obtain a more explicit formula for this identification notice that $W^*P^{-1/2}$ is an isometry from $\mathcal{X}$ into $\ell_+^2(\mathcal{U})$ whose range is $\mathcal{H}'$. This readily implies that $D'W^*P^{-1/2} = W^*P^{-1/2}D$ for some positive operator D on $\mathcal{X}$. In fact, using $T'W^* = W^*P^{-1}ZP$ (see (V.1.28)) it follows that D is the positive root of $I - P^{1/2}Z^*P^{-1}ZP^{1/2}$. Using (7.12) we have for x in $\mathcal{X}$

$$\|Dx\|^2 = \|W^*P^{-1/2}Dx\|^2 = \|D'W^*P^{-1/2}x\|^2 = \|\Theta CP^{1/2}x\|^2 = \|CP^{1/2}x\|^2 .$$

So there exists an isometry $\phi$ mapping $\mathcal{U} = \overline{\operatorname{ran} C}$ into $\mathcal{X}$ satisfying $D = \phi CP^{1/2}$. Now let $\phi_1$ be the isometry from $\mathcal{U}$ into $\mathcal{H}'$ whose range is $\mathcal{D}'$ defined by

$$\phi_1 = W^*P^{-1/2}\phi \text{ where } D = \phi CP^{1/2} . \tag{7.13}$$

Finally, notice that $\mathcal{F}_+\Psi D'W^* = \Theta CP$ gives

$$\mathcal{F}_+\Psi\phi_1 CP^{1/2} = \mathcal{F}_+\Psi W^*P^{-1/2}D = \mathcal{F}_+\Psi D'W^*P^{-1/2} = \Theta CP^{1/2} .$$

This gives the following useful identity $\mathcal{F}_+\Psi\phi_1 = \Theta$.

By using $\Pi_o\tilde{W}^* = \tilde{B}^*$ in the expression for $D_A^{-2}$ in (7.9) on $N^2 = \Pi_o D_A^{-2}\Pi_o^*$ we arrive at the expression for N in (7.5). Now let us show that $N_1\phi_1 = \phi_1 N_*$ where $N_1$ is defined in (6.2). To this end, notice that $\phi^*D = CP^{1/2}$ and by taking adjoints $D\phi = P^{1/2}C^*$. By employing (7.13) we have $D'\phi_1 = W^*P^{-1/2}D\phi = W^*C^*$. So using $D'\phi_1 = W^*C^*$ and (7.9) we have

$$N_1^2\phi_1 = \phi_1 + D'AD_A^{-2}A^*W^*C^* = \phi_1 + D'A\tilde{W}^*(\gamma^2P - \tilde{P})^{-1}WW^*C^* =$$

$$\phi_1 + D'W^*P^{-1/2}P^{-1/2}\tilde{W}\tilde{W}^*(\gamma^2P - \tilde{P})^{-1}PC^* = \phi_1 + W^*P^{-1/2}\phi\phi^*DP^{-1/2}\tilde{P}(\gamma^2P - \tilde{P})^{-1}PC^* =$$

$$\phi_1 + \phi_1 C\tilde{P}(\gamma^2P - \tilde{P})^{-1}PC^* = \phi_1 N_*^2$$

Thus $N_1^2\phi_1 = \phi_1 N_*^2$, and $N_1\phi_1 = \phi_1 N_*$ which proves our claim. To convert (6.3) to the operator valued analytic function $\Phi_{i,j}$ in (7.6) recall from (V.1.32), (V.1.33), (V.1.34) and (V.1.38) that

$$\begin{gathered} T_A^*\tilde{W}^*(\gamma^2P - \tilde{P})^{-1} = \tilde{W}^*(\gamma^2P - \tilde{P})^{-1}M, \\ (I - \lambda T_A^*)\tilde{W}^*(\gamma^2P - \tilde{P})^{-1} = \tilde{W}^*(\gamma^2P - \tilde{P})^{-1}(I - \lambda M)^{-1} , \\ (I - \lambda T_A^*)^{-1}T_A^*\Pi_o^* = -\tilde{W}^*(\gamma^2P - \tilde{P})^{-1}(I - \lambda M)^{-1}M\tilde{B} . \end{gathered} \tag{7.14}$$

Using this and (7.9) on the $Y_{1,1}$ term in (6.3) along with $D'\phi_1 = W^*C^*$ we obtain

$$\begin{gathered} \Pi_o(I - \lambda T_A^*)^{-1}D_A^{-2}A^*D'N_1^{-1}\phi_1 = \Pi_o(I - \lambda T_A^*)^{-1}D_A^{-2}A^*W^*C^*N_*^{-1} = \\ \Pi_o(I - \lambda T_A^*)^{-1}\tilde{W}^*(\gamma^2P - \tilde{P})^{-1}WW^*C^*N_*^{-1} = \tilde{B}^*(\gamma^2P - \tilde{P})^{-1}(I - \lambda M)^{-1}PC^*N_*^{-1} . \end{gathered} \tag{7.15}$$

From this it readily implies that $\Phi_{1,1}(\lambda) = Y_{1,1}\phi_1$.

Now using the last equation in (7.14) on the $Y_{1,2}$ term in (6.3) we have

$$Y_{1,2}(\lambda)\Pi_o^* = N^{-1}\Pi_o(I-\lambda T_A^*)^{-1}\Pi_o^* = N^{-1} + \lambda N^{-1}\Pi_o(I-\lambda T_A^*)^{-1}T_A^*\Pi_o^* =$$

$$N^{-1} - \lambda N^{-1}\tilde{B}^*(\gamma^2 P - \tilde{P})^{-1}(I-\lambda M)^{-1}M\tilde{B} = \Phi_{1,2}(\lambda)\,.$$

By following the computation in (7.15) along with $\mathcal{F}_+\Psi\phi_1 u = \Theta(\lambda)u$ for u in $\mathcal{U}$ and (7.12), (7.9), (7.14), we obtain

$$\mathcal{F}_+\Psi Y_{2,1}\phi_1 = \Theta N_* - \mathcal{F}_+\Psi D'A(I-\lambda T_A^*)^{-1}D_A^{-2}A^*D'\phi_1 N_*^{-1} =$$

$$\Theta N_* - \mathcal{F}_+\Psi D'A\tilde{W}^*(\gamma^2 P - \tilde{P})^{-1}(I-\lambda M)^{-1}PC^*N_*^{-1} =$$

$$\Theta N_* - \Theta CPP^{-1}\tilde{W}\tilde{W}^*(\gamma^2 P - \tilde{P})^{-1}(I-\lambda M)^{-1}PC^*N_*^{-1} = \Phi_{2,1}(\lambda)$$

The last term $\Phi_{2,2}(\lambda)$ is simply the central solution $\mathcal{F}_+A\Pi_o^* + \mathcal{F}_+\Psi Y_{2,2}(\lambda)\Pi_o^*$ computed in Theorem V.1.2. So by replacing R by $\phi_1^* R$ in (7.11) we obtain the Redheffer scattering formula in (7.7).

If $\ker B \neq \{0\}$, then let $\mathcal{H}' = \operatorname{ran} W^* \oplus \ker B$ and as before $T' = P_{\mathcal{H}'}S\,|\,\mathcal{H}'$. Here $\ker B$ is viewed as a subspace of $\mathcal{U}$ and thus $T'\,|\,\ker B = \{0\}$. Moreover, S is a minimal isometric lifting of $T'$. So $D'$ admits a reducing decomposition of the form $D' = D_0' \oplus I$ on $\operatorname{ran} W^* \oplus \ker B$. Notice that $E\,|\ker(B)$ is a unitary operator from $\ker B$ onto $\operatorname{ran}(C)^{\perp}$. Because $\Theta$ is two-sided inner $\{Z, B, C, E\}$ is similar to a unitary system; see Theorem III.7.6. So without loss of generality we can assume that $\{Z, B, C, E\}$ is unitary, that is,

$$\begin{bmatrix} Z & B \\ C & E \end{bmatrix} : \begin{bmatrix} \mathcal{X} \\ \mathcal{U} \end{bmatrix} \to \begin{bmatrix} \mathcal{X} \\ \mathcal{Y} \end{bmatrix}$$

is unitary. Using this it follows that $E\,|\,\ker B$ is a unitary map from $\ker B$ onto $\operatorname{ran}(C)^{\perp}$. Now by performing minor modifications of the above argument there exists a unitary operator $\phi_1$ from $\mathcal{U}$ onto $\mathcal{D}'$ satisfying $\mathcal{F}_+\Psi\phi_1 = \Theta$ and (7.13) holds. Finally, using this in the above argument, we obtain the parameterization of all solutions in (7.7). This fact also follows from Remark 2.2. Remark 7.2 follows from Remark 6.2. This completes the proof.

## VI.8 THE NEHARI PARAMETERIZATION

In this section we will use the Schur representation for the commutant lifting theorem, to obtain a state space parameterization for the set of all solutions for the Nehari interpolation problem with tolerance $\gamma > d_\infty$. To this end, let $\{Z, \text{ on } \mathcal{X}, B, C\}$ be a stable, controllable and observable, anticausal realization for a function F in $K^\infty(\mathcal{U}, \mathcal{Y})$, that is,

$$F(\lambda) = C(\lambda I - Z)^{-1}B \qquad (|\lambda| > 1)\,. \tag{8.1}$$

Let P be the controllability grammian for {Z, B} and Q the observability grammian for {C, Z}. In particular, P and Q can be obtained by solving the following Lyapunov equations

$$P = ZPZ^* + BB^* \quad \text{and} \quad Q = Z^*QZ + C^*C\,. \tag{8.2}$$

Recall that H is a *solution* to the Nehari interpolation problem with tolerance $\gamma$, if H is the function in $H^\infty(\mathcal{U}, \mathcal{Y})$ satisfying $\|F - H\|_\infty \leq \gamma$. Moreover, there exists a solution to the Nehari interpolation problem with tolerance $\gamma$ if and only if $\gamma \geq d_\infty = \|A\|$. Here A is the Hankel operator from $H^2(\mathcal{U})$ into $K^2(\mathcal{Y}) := L^2(\mathcal{Y}) \ominus H^2(\mathcal{Y})$ defined by $A = P_- M_F \,|\, H^2(\mathcal{U})$ where $P_-$ is the orthogonal projection onto $K^2(\mathcal{Y})$.

To obtain a state space parameterization for the set of all solutions to the Nehari problem with tolerance $\gamma > d_\infty$ let N on $\mathcal{U}$ and $N_*$ on $\mathcal{Y}$ be the positive operator defined by

$$N = \gamma^{-1}[I + B^*(\gamma^2 I - QP)^{-1}QB]^{1/2} \quad \text{and} \quad N_* = [I + C(\gamma^2 I - PQ)^{-1}PC^*]^{1/2}\,. \tag{8.3}$$

As in Section V.5 equation (V.5.7), let M be the stable operator on $\mathcal{X}$ defined by

$$M = (\gamma^2 I - Z^*QZP)^{-1}Z^*(\gamma^2 I - QP)\,. \tag{8.4}$$

Finally, let $\Phi_{i,j}(\lambda)$ for i, j = 1, 2 be the operator valued analytic function in $|\lambda| < 1$ defined by

$$\Phi_{1,1}(\lambda) = -\lambda N^{-1}B^*(I - \lambda M)^{-1}(\gamma^2 I - QP)^{-1}C^*N_*^{-1} \tag{8.5}$$

$$\Phi_{1,2}(\lambda) = N^{-1} - \lambda N^{-1}B^*(I - \lambda M)^{-1}(\gamma^2 I - Z^*QZP)^{-1}Z^*QB$$

$$\Phi_{2,1}(\lambda) = CP(I - \lambda M)^{-1}(\gamma^2 I - QP)^{-1}C^*N_*^{-1} - N_*$$

$$\Phi_{2,2}(\lambda) = CP(I - \lambda M)^{-1}(\gamma^2 I - Z^*QZP)^{-1}Z^*QB\,.$$

The function $\Phi_{1,1}$ is in $H^\infty(\mathcal{Y}, \mathcal{U})$ and $\Phi_{2,1}$ is in $H^\infty(\mathcal{Y}, \mathcal{Y})$, while $\Phi_{1,2}$ is in $H^\infty(\mathcal{U}, \mathcal{U})$ and $\Phi_{2,2}$ is in $H^\infty(\mathcal{U}, \mathcal{Y})$. This sets the stage for the main result of this section.

**THEOREM 8.1.** *Let* {Z, B, C} *be a stable, controllable and observable anticausal realization for a function* F *in* $K^\infty(\mathcal{U}, \mathcal{Y})$. *Let* $\Phi_{i,j}(\lambda)$ *for* i, j = 1, 2 *be the operator valued analytic functions defined in (8.3), (8.4) and (8.5) and assume that* $\gamma > d_\infty$. *Then the set of all solutions* H *to the Nehari interpolation problem with data* F *and tolerance* $\gamma$ *is given by*

$$H(\lambda) = \Phi_{2,2}(\lambda) + \Phi_{2,1}(\lambda)R(\lambda)(I - \Phi_{1,1}(\lambda)R(\lambda))^{-1}\Phi_{1,2}(\lambda) \tag{8.6}$$

*where* R *is a contractive analytic function in* $H^\infty(\mathcal{U}, \mathcal{Y})$. *Moreover, the mapping from* R *into* H *given by (8.6) is a bijection from the set of all contractive analytic functions* R *in* $H^\infty(\mathcal{U}, \mathcal{Y})$ *onto the set of all functions* H *in* $H^\infty(\mathcal{U}, \mathcal{Y})$ *satisfying* $\|F - H\|_\infty \leq \gamma$.

Notice that if $R = 0$ in the previous theorem, then $H = \Phi_{2,2}$ is precisely the central solution to the Nehari optimization problem with tolerance $\gamma$; see Theorem V.5.2.

**REMARK 8.2.** Equation (8.6) shows that the set of all Nehari interpolants $G = F - H$ with tolerance $\gamma$ is given by solving the following Redheffer scattering system:

$$\begin{bmatrix} g \\ y \end{bmatrix} = \begin{bmatrix} \Phi_{1,1} & \Phi_{1,2} \\ -\Phi_{2,1} & F - \Phi_{2,2} \end{bmatrix} \begin{bmatrix} x \\ u \end{bmatrix} \tag{8.7}$$

$$\text{subject to } x = Rg$$

where R is a contractive analytic function in $H^\infty(\mathcal{U}, \mathcal{Y})$. Here u is in $\mathcal{U}$, while x, g and y are functions with values in $\mathcal{Y}$, $\mathcal{U}$ and $\mathcal{Y}$, respectively. Then the corresponding Nehari interpolant $G = F - H$ with tolerance $\gamma$ is given by $y = Gu$. Moreover, if $\Phi$ is the $2 \times 2$ block matrix in (8.7), then $\Phi$ is a function in $L^\infty(\mathcal{Y} \oplus \mathcal{U}, \mathcal{U} \oplus \mathcal{Y})$ and $\Phi(I \oplus \gamma^{-1} I)$ is a.e. an isometry. The matrix formed by $\{\Phi_{i,j}(\lambda)\}$ is a function in $H^\infty(\mathcal{Y} \oplus \mathcal{U}, \mathcal{U} \oplus \mathcal{Y})$, and $[\Phi_{1,1}, \Phi_{2,1}]$ is an inner function. Finally, if both $\mathcal{U}$ and $\mathcal{Y}$ are finite dimensional, then $\Phi(I \oplus \gamma^{-1} I)$ is a unitary operator a.e. Finally, recall that the central interpolant $G_\gamma = F - \Phi_{2,2}$ is obtained by setting $R = 0$ and $\|G_\gamma\|_\infty < \gamma$; see Theorem V.5.1. So if $\|R\|_\infty < 1$, then Theorem 5.8 shows that $\|F - H\|_\infty < \gamma$.

**PROOF OF THEOREM 8.1.** Recall that the operators T, A and T′ in the commutant lifting set up for the solution to the Nehari problem are given by letting T be the unilateral shift on $\mathcal{H} = H^2(\mathcal{U})$, while T′ is the co-isometry on $\mathcal{H}' = K^2(\mathcal{Y})$ obtained by compressing the bilateral shift V on $L^2(\mathcal{Y})$ to $K^2(\mathcal{Y})$, that is, $T' = P_- V | K^2(\mathcal{Y})$, and finally $A = P_- M_F | H^2(\mathcal{U})$. Since $V^* | \mathcal{H}' = T'^*$ and $\mathcal{H}'$ is cyclic for V, it follows that V is the minimal isometric lifting of T′. Moreover, $D' = e^{-it}\Pi_{-1}$ where $\Pi_{-1}$ is the operator from $K^2(\mathcal{Y})$ onto $\mathcal{Y}$ which picks out the Fourier coefficient of $e^{-it}$, that is,

$$\Pi_{-1} f = \frac{1}{2\pi} \int_0^{2\pi} e^{it} f(e^{it}) dt \qquad (f \in K^2(\mathcal{U})) . \tag{8.8}$$

To verify that $D' = e^{-it}\Pi_{-1}$ simply notice that $D'^2$ is an orthogonal projection and thus $D' = D'^2$. Since $I - T'^* T' = e^{-it}\Pi_{-1}$ we have $D' = e^{-it}\Pi_{-1}$.

Since $\mathcal{D}' = e^{-it}\mathcal{Y}$, it follows that d is in $H^2(\mathcal{D}')$ if and only if $d(\lambda) = e^{-it} f(\lambda)$ where $f(\lambda)$ is a function in $H^2(\mathcal{Y})$. Because V is a minimal isometric lifting of T′ there exists a unique unitary operator $\Psi$ from $\mathcal{K}' = \mathcal{H}' \oplus H^2(\mathcal{D}')$ onto $L^2(\mathcal{Y})$ satisfying $V\Phi = \Phi U'$ and $\Phi | \mathcal{H}' = I$ where U′ is the Sz.-Nagy-Schäffer minimal isometric lifting of T′ in (1.10). By consulting Proposition IV.1.3 we see that

$$\Psi D'f = (V - T')f = \Pi_{-1} f \qquad (f \in K^2(\mathcal{Y})) . \tag{8.9}$$

Here D′f is viewed as a vector in $\{0\} \oplus \mathcal{D}' \subseteq \mathcal{H}' \oplus H^2(\mathcal{D}')$. Moreover, $(\Psi e^{-it} f)(\lambda) = f(\lambda)$ where $e^{-it} f(\lambda)$ is in $H^2(\mathcal{D}')$. To see this let $\{f_n\}$ be the Taylor coefficients of $\lambda^n$ for $f(\lambda)$. Then using (IV.1.9) we have

$$\Psi e^{-it} f = \sum_{n=0}^{\infty} V^n (V - T') e^{-it} f_n = \sum_{n=0}^{\infty} V^n f_n = f(e^{it}) .$$

This readily shows that $(\Psi e^{-it} f)(\lambda) = f(\lambda)$. Now let $\phi_1$ be the unitary operator from $\mathcal{Y}$ onto $\mathcal{D}'$ defined by $\phi_1 a = e^{-it} a = \Pi_{-1}^* a$ where a is in $\mathcal{Y}$. Obviously $\Psi \phi_1 a = a$ for all a in $\mathcal{Y}$.

Recall that B′ is an intertwining lifting of A with respect to V and tolerance $\gamma$ if and only if $B' = M_G \mid H^2(\mathcal{U})$ where $G = F - H$ for some H in $H^\infty(\mathcal{U}, \mathcal{Y})$ and $\|G\|_\infty \le \gamma$; see Section V.5. In other words, H is a solution to the Nehari interpolation problem with tolerance $\gamma$ if and only if $B' = M_G \mid H^2(\mathcal{U})$ is an intertwining lifting of A with respect to V and tolerance $\gamma$, where $G = F - H$. Moreover, this function G can be recovered from B′ by $G(\lambda) = (B' \Pi_o^*)(\lambda)$ where $\Pi_o$ is the operator from $H^2(\mathcal{U})$ onto $\mathcal{U}$ defined by $\Pi_o g = g(0)$. In particular, $H(\lambda) = -(P_+ B' \Pi_o^*)(\lambda)$ where $P_+$ is the orthogonal projection onto $H^2(\mathcal{Y})$. Since $\Psi U' = V \Psi$ it follows that B′ is an intertwining lifting of A with respect to V and tolerance $\gamma$ if and only if $B' = \Psi B_o$ where $B_o$ is an intertwining lifting of A with respect to U′ and tolerance $\gamma$. Therefore the set of all solutions H to the Nehari interpolation problem with tolerance $\gamma$ are given by $G - H = (\Psi B_o \Pi_o^*)(\lambda)$ where $B_o$ is an intertwining lifting of A with respect to U′ and tolerance $\gamma$. Using $\Psi \mid \mathcal{H}' = I$ it follows that $F(\lambda) = (\Psi A \Pi_o^*)(\lambda)$. Combining this along with Theorem 6.1 we see that the set of all solutions H to the Nehari interpolation problem with tolerance $\gamma$ are given by

$$H(\lambda) = -(\Psi Y_{2,2} \Pi_o^*)(\lambda) - (\Psi Y_{2,1} R (I - Y_{1,1} R)^{-1} Y_{1,2} \Pi_o^*)(\lambda) \tag{8.10}$$

where R is a contractive analytic function in $H^\infty(\mathcal{U}, \mathcal{D}')$ and $Y_{i,j}$ for i, j = 1, 2 and defined in (6.3). (Notice that $\mathcal{L} = \ker T = \mathcal{U}$ in this setting.)

Notice that $-(\Psi Y_{2,2} \Pi_o^*)(\lambda)$ is precisely the central solution $\Phi_{2,2}$ with tolerance $\gamma$ discussed in detail in Section V.5. So to complete the proof it remains to find state space expressions for N, $N_1$, $Y_{1,1}$, $Y_{2,1}$ and $Y_{1,2}$. To this end, let us recall that from Section V.5 we have $A = W_o W_c$ where $W_o$ is the observability operator from $\mathcal{X}$ into $K^2(\mathcal{Y})$ and $W_c$ is the controllability operator from $H^2(\mathcal{U})$ into $\mathcal{X}$ defined by

$$W_o x = C(e^{it} I - Z)^{-1} x \text{ and } W_c^* x = B^* (I - \lambda Z^*)^{-1} x \qquad (x \in \mathcal{X}) . \tag{8.11}$$

Recall that the controllability grammian $P = W_c W_c^*$ while the observability grammian $Q = W_o^* W_o$. Throughout the rest of the proof we will use the following formulas taken from (5.16), (5.17), (5.26), (5.27) and (5.28) in Chapter V

$$D_{A^*}^{-2} = \gamma^{-2}[I + W_c^*(\gamma^2 I - QP)^{-1} QW_c] \text{ and } D_{A^*}^{-2} A^* = W_c^*(\gamma^2 I - QP)^{-1} W_o^*$$

$$T_A^* W_c^* = W_c^* M \text{ and } (I - \lambda T_A^*)^{-1} W_c^* = W_c^*(I - \lambda M)^{-1} \tag{8.12}$$

$$(I - \lambda T_A^*)^{-1} T_A^* \Pi_o^* = - W_c^*(I - \lambda M)^{-1}(\gamma^2 I - Z^* QZP)^{-1} Z^* QB .$$

So using $\Pi_o W_c^* = B^*$ along with our previous expression for $D_{A^*}^{-2}$ we see that the N in (6.2) becomes

$$N^2 = \Pi_o D_{A^*}^{-2} \Pi_o^* = \gamma^{-2}[I + B^*(\gamma^2 I - QP)^{-1} QB] . \tag{8.13}$$

This readily yields the expression for N in (8.3). Next let us show that $N_1 \phi_1 = \phi_1 N_*$ where $N_1$ is defined in (6.2). This follows from $D' \phi_1 = \Pi_{-1}^*$ and $D' W_o = e^{-it} \Pi_{-1} W_o = e^{-it} C$ along with our formulas for $D_{A^*}^{-2} A^*$ in (8.12) and $\Pi_{-1} W_o = C$, that is,

$$N_1^2 \phi_1 = \phi_1 + D' A D_{A^*}^{-2} A^* D' \phi_1 = \phi_1 + D' A W_c^*(\gamma^2 I - QP)^{-1} W_o^* \Pi_{-1}^* =$$

$$\phi_1 + e^{-it} \Pi_{-1} W_o W_c W_c^*(\gamma^2 I - QP)^{-1} C^* = \phi_1 + \phi_1 CP(\gamma^2 I - QP)^{-1} C^* = \phi_1 N_*^2 . \tag{8.14}$$

Hence $N_1 \phi_1 = \phi_1 N_*$ and $N_*$ is given by the state space formula in (8.3). Notice that $\Pi_o W_c^* = B^*$. Now using (8.12) and $D' \phi_1 = \Pi_{-1}^*$ along with $W_o^* \Pi_{-1}^* = C^*$ once again we obtain the following expression for $Y_{1,1}$

$$Y_{1,1}(\lambda) \phi_1 = - \lambda N^{-1} \Pi_o (I - \lambda T_A^*)^{-1} D_{A^*}^{-2} A^* \Pi_{-1}^* N_*^{-1} =$$

$$- \lambda N^{-1} \Pi_o (I - \lambda T_A^*)^{-1} W_c^*(\gamma^2 I - QP)^{-1} C^* N_*^{-1} = \tag{8.15}$$

$$- \lambda N^{-1} B^*(I - \lambda M)^{-1}(\gamma^2 I - QP)^{-1} C^* N_*^{-1} = \Phi_{1,1}(\lambda) .$$

Using (8.12) and $\Pi_o W_c^* = B^*$ once again we obtain

$$(Y_{1,2} \Pi_o^*)(\lambda) = N^{-1} \Pi_o (I - \lambda T_A^*)^{-1} \Pi_o^* = N^{-1} + \lambda N^{-1} \Pi_o (I - \lambda T_A^*)^{-1} T_A^* \Pi_o^* =$$

$$N^{-1} - \lambda N^{-1} B^*(I - \lambda M)^{-1}(\gamma^2 I - Z^* QZP)^{-1} Z^* QB = \Phi_{1,2}(\lambda) . \tag{8.16}$$

Finally, using $N_1 \phi_1 = \phi_1 N_*$ a similar computation involving (8.12) yields

$$Y_{2,1}(\lambda) \phi_1 = \phi_1 N_* - D' A (I - \lambda T_A^*)^{-1} D_{A^*}^{-2} A^* D' \phi_1 N_*^{-1} =$$

$$\phi_1 N_* - D' A (I - \lambda T_A^*)^{-1} W_c^*(\gamma^2 I - QP)^{-1} C^* N_*^{-1} = \tag{8.17}$$

$$\phi_1 N_* - D' W_o W_c W_c^*(I - \lambda M)^{-1}(\gamma^2 I - QP)^{-1} C^* N_*^{-1} = - \phi_1 \Phi_{2,1}(\lambda) .$$

The previous calculations with $\Psi \phi_1 = I$ shows that

$$\Phi_{1,1}(\lambda) = Y_{1,1}(\lambda)\phi_1 \text{ and } (Y_{1,2}\Pi_o^*)(\lambda) = \Phi_{1,2}(\lambda) \text{ and}$$
$$\Psi Y_{2,1}(\lambda)\phi_1 = -\Phi_{2,1}(\lambda) \text{ and } (\Psi Y_{2,2}\Pi_o^*)(\lambda) = -\Phi_{2,2}(\lambda)\,. \tag{8.18}$$

By replacing the contractive analytic function R in $H^\infty(\mathcal{U}, \mathcal{D}')$ in (8.10) by the contractive analytic function $\phi_1 R$ where R is now in $H^\infty(\mathcal{U}, \mathcal{Y})$ and using (8.18) we arrive at our expression for the set of all Nehari interpolants in (8.6). The comments in Remark 8.2 follows from Remark 6.2. This completes the proof.

## VI.9 THE TWO BLOCK PARAMETERIZATION

In this section we will use the Schur representation for the commutant lifting theorem to obtain a state space parameterization for the set of all solutions for the two block interpolation problem (discussed in Section V.7) with tolerance $\gamma > d_\infty$. To this end, let $\{Z \text{ on } \mathcal{X}, B, [C_1, C_2]^{tr}\}$ be a stable, anticausal, controllable and observable realization for a function $[F_1, F_2]^{tr}$ in $K^\infty(\mathcal{U}, \mathcal{Y}_1 \oplus \mathcal{Y}_2)$, that is,

$$\begin{bmatrix} F_1(\lambda) \\ F_2(\lambda) \end{bmatrix} = \begin{bmatrix} C_1 \\ C_2 \end{bmatrix} (\lambda I - Z)^{-1} B \qquad (|\lambda| > 1)\,. \tag{9.1}$$

Let $Q_1$ and $Q_2$ be the observability grammians for the pair $\{C_1, Z\}$ and $\{C_2, Z\}$, respectively. In particular, $Q_1$ and $Q_2$ can be obtained by solving the following Lyapunov equations

$$Q_1 = Z^* Q_1 Z + C_1^* C_1 \text{ and } Q_2 = Z^* Q_2 Z + C_2^* C_2\,. \tag{9.2}$$

Recall that H is a solution to the two block interpolation problem with tolerance $\gamma$, if H is a function in $H^\infty(\mathcal{U}, \mathcal{Y}_1)$ and $\|G\|_\infty \le \gamma$ where $G = [F_1 - H, F_2]^{tr}$. Moreover, there exists a solution to the two block interpolation with tolerance $\gamma$, if and only if $\gamma \ge d_\infty = \|A\|$. Here A is the operator from $H^2(\mathcal{U})$ into $\mathcal{H}' = K^2(\mathcal{Y}_1) \oplus L^2(\mathcal{Y}_2)$ defined by $Ah = P_{\mathcal{H}'}(F_1 h \oplus F_2 h)$ for h in $H^2(\mathcal{U})$; see Section V.7.

Now assume that $\gamma > d_\infty$ and let $P_n$ be the solution to the following Riccati difference equation

$$P_{n+1} = ZP_nZ^* + (B + ZP_nZ^*Q_2B)(\gamma^2 I - B^*Q_2B - B^*Q_2ZP_nZ^*Q_2B)^{-1}(B + ZP_nZ^*Q_2B)^* \tag{9.3}$$

subject to the initial condition $P_0 = 0$. Since $\gamma > d_\infty$, the results in Section V.7, show that $P_n$ converges strongly to the positive operator $P := \mathrm{RIC}(\gamma)$. This P is also the minimal positive solution to the following algebraic Riccati equation

$$P = ZPZ^* + (B + ZPZ^*Q_2B)(\gamma^2 I - B^*Q_2B - B^*Q_2ZPZ^*Q_2B)^{-1}(B + ZPZ^*Q_2B)^* . \quad (9.4)$$

Finally, the results in Section V.7 show that $r_{spec}(PQ_1) < 1$ when $\gamma > d_\infty$. Because P and $Q_1$ are self adjoint operators and $r_{spec}(R) = r_{spec}(R^*)$ for any operator R on $\mathcal{X}$ we also have $r_{spec}(Q_1 P) < 1$. In particular, both $I - PQ_1$ and $I - Q_1 P$ are invertible operators.

To obtain a state space parameterization for the set of all solutions to the two block problem with tolerance $\gamma > d_\infty$, let N on $\mathcal{U}$ and $N_*$ on $\mathcal{Y}_1$ be the positive operators defined by taking the positive square roots of the following operators

$$N_*^2 = I + C_1 P(I - Q_1 P)^{-1} C_1^*$$

$$N^2 = D^{-1} + D^{-1}B^*(I + Q_2ZPZ^*)Q_1(I - PQ_1)^{-1}(I + ZPZ^*Q_2)BD^{-1} \quad (9.5)$$

$$D = \gamma^2 I - B^*Q_2B - B^*Q_2ZPZ^*Q_2B$$

as in equation (V.7.9) let M be the stable operator on $\mathcal{X}$ defined by

$$M = (I - Z^*Q_1ZP)^{-1}Z^*(I - Q_1P) . \quad (9.6)$$

Finally, let $\Phi_{i,j}(\lambda)$ for $i = 1, 2$ be the operator valued analytic functions in the open unit disc defined by

$$\Phi_{1,1}(\lambda) = -\lambda N^{-1}D^{-1}B^*(I + Q_2ZPZ^*)(I - \lambda M)^{-1}(I - Q_1P)^{-1}C_1^*N_*^{-1}$$

$$\Phi_{1,2}(\lambda) = N^{-1} - \lambda N^{-1}D^{-1}B^*(I + Q_2ZPZ^*)(I - \lambda M)^{-1}(I - Z^*Q_1ZP)^{-1}Z^*(Q_1 + Q_2)B$$

$$\Phi_{2,1}(\lambda) = C_1P(I - \lambda M)^{-1}(I - Q_1P)^{-1}C_1^*N_*^{-1} - N_* \quad (9.7)$$

$$\Phi_{2,2}(\lambda) = C_1P(I - \lambda M)^{-1}(I - Z^*Q_1ZP)^{-1}Z^*(Q_1 + Q_2)B .$$

The function $\Phi_{1,1}$ is in $H^\infty(\mathcal{Y}_1, \mathcal{U})$ and $\Phi_{2,1}$ is in $H^\infty(\mathcal{Y}_1, \mathcal{Y}_1)$ while $\Phi_{1,2}$ is in $H^\infty(\mathcal{U}, \mathcal{U})$ and $\Phi_{2,2}$ is in $H^\infty(\mathcal{U}, \mathcal{Y}_1)$. This sets the stage for the main result of this section.

**THEOREM 9.1.** *Let* $\{Z, B, [C_1, C_2]^{tr}\}$ *be a stable, anticausal controllable and observable realization for a function* $F = [F_1, F_2]^{tr}$ *in* $K^\infty(\mathcal{U}, \mathcal{Y}_1 \oplus \mathcal{Y}_2)$. *Let* $\Phi_{i,j}(\lambda)$ *for* $i, j = 1, 2$ *be the operator valued analytic functions defined in (9.5), (9.6) and (9.7) and assume that* $\gamma > d_\infty$. *Then the set of all solutions* H *to the two block interpolation problem with data* $F = [F_1, F_2]^{tr}$ *and tolerance* $\gamma$ *is given by*

$$H(\lambda) = \Phi_{2,2}(\lambda) + \Phi_{2,1}(\lambda)R(\lambda)(I - \Phi_{1,1}(\lambda)R(\lambda))^{-1}\Phi_{1,2}(\lambda) \quad (9.8)$$

*where* R *is a contractive analytic function in* $H^\infty(\mathcal{U}, \mathcal{Y}_1)$. *Moreover, the map from* R *into* H *given by (9.8) is a bijection from the set of all contractive analytic functions* R *in* $H^\infty(\mathcal{U}, \mathcal{Y}_1)$ *onto the set of all functions* H *in* $H^\infty(\mathcal{U}, \mathcal{Y}_1)$ *satisfying* $\|F - [H, 0]^{tr}\|_\infty \leq \gamma$.

Notice that if $R = 0$ in the previous theorem, then $H = \Phi_{2,2}$ is precisely the central solution to the two block problem with tolerance $\gamma$; see Theorem V.7.2. Finally, it is noted that by choosing $C_2 = 0$, Theorem 8.1 can be obtained as a corollary of Theorem 9.1. (In this case $Q_2 = 0$ and if P solves (9.4), then $\gamma^2 P$ solves (8.2).)

**REMARK 9.2.** Equation (9.8) shows that the set of all two block interpolants with tolerance $\gamma$ are given by $G = [F_1 - H, F_2]^{tr}$ where G is the solution to the following Redheffer scattering system:

$$\begin{bmatrix} g \\ y \end{bmatrix} = \begin{bmatrix} \Phi_{1,1} & \Phi_{1,2} \\ \begin{bmatrix} -\Phi_{2,1} \\ 0 \end{bmatrix} & \begin{bmatrix} F_1 - \Phi_{2,2} \\ F_2 \end{bmatrix} \end{bmatrix} \begin{bmatrix} x \\ u \end{bmatrix} \tag{9.9}$$

$$\text{subject to } x = Rg$$

where R is a contractive analytic function in $H^\infty(\mathcal{U}, \mathcal{Y}_1)$. Here u is in $\mathcal{U}$ while x, g and y are vectors with values in $\mathcal{Y}_1$, $\mathcal{U}$ and $\mathcal{Y}_1 \oplus \mathcal{Y}_2$, respectively. Then the corresponding two block interpolant G with tolerance $\gamma$ is given by $y = Gu$. Moreover, if $\Phi$ is the $2 \times 2$ block matrix in (9.9), then $\Phi$ is a function in $L^\infty(\mathcal{Y}_1 \oplus \mathcal{U}, \mathcal{U} \oplus (\mathcal{Y}_1 \oplus \mathcal{Y}_2)$ and $\Phi(I \oplus \gamma^{-1} I)$ is a.e. an isometry. The $2 \times 2$ matrix formed by $\{\Phi_{k,j}(\lambda)\}$ is a function in $H^\infty(\mathcal{Y}_1 \oplus \mathcal{U}, \mathcal{U} \oplus \mathcal{Y}_1)$, and $[\Phi_{1,1}, \Phi_{2,1}]^{tr}$ is an inner function.

**PROOF OF THEOREM 9.1.** Here we will follow the notation established in Section V.7. Recall that the operators T, A and T′ in the commutant lifting set up for the solution to the two block problem are given by letting T be the unilateral shift on $\mathcal{H} = H^2(\mathcal{U})$, while T′ is the co-isometry on $\mathcal{H}' = K^2(\mathcal{Y}_1) \oplus L^2(\mathcal{Y}_2)$ obtained by compressing the bilateral shift V on $L^2(\mathcal{Y}_1 \oplus \mathcal{Y}_2)$ to $\mathcal{H}'$, that is $T' = P_{\mathcal{H}'} V | \mathcal{H}'$, and finally $A = P_{\mathcal{H}'} M_F | H^2(\mathcal{U})$ where $F = [F_1, F_2]^{tr}$. Since $V^* | \mathcal{H}' = T'^*$ and $\mathcal{H}'$ is cyclic for V, it follows that V is the minimal isometric lifting of T′. Let $\Pi_{-1}$ be the operator from $\mathcal{H}'$ onto $\mathcal{Y}_1$ defined by

$$\Pi_{-1} h = \frac{1}{2\pi} \int_0^{2\pi} e^{it} h_1(e^{it}) dt \qquad (h = h_1 \oplus h_2 \in K^2(\mathcal{Y}_1) \oplus L^2(\mathcal{Y}_2)). \tag{9.10}$$

We claim that $D' = e^{-it} \Pi_{-1} \oplus 0$. To verify this notice that $D'^2 = I - T'^* T'$ is the orthogonal projection onto $e^{-it} \mathcal{Y}_1 \oplus \{0\}$. Because $D'^2 = e^{-it} \Pi_{-1} \oplus 0$ it follows that $D' = D'^2 = e^{-it} \Pi_{-1} \oplus 0$

which proves our claim.

Since $\mathcal{D}' = e^{-it}\mathcal{Y}_1 \oplus \{0\}$, it follows that $d(\lambda)$ is in $H^2(\mathcal{D}')$ if and only if $d(\lambda) = e^{-it}f(\lambda) \oplus 0$ where $f(\lambda)$ is in $H^2(\mathcal{Y}_1)$. Because V is the minimal isomeric lifting of T′, there exists a unique operator $\Psi$ from $\mathcal{K}' = \mathcal{H}' \oplus H^2(\mathcal{D}')$ onto $L^2(\mathcal{Y}_1 \oplus \mathcal{Y}_2)$ satisfying $V\Psi = \Psi U'$ and $\Psi | \mathcal{H}' = I$ where $U'$ is the Sz.-Nagy-Schäffer minimal isometric lifting of $T'$ in (1.10). We claim that $\Psi$ maps $H^2(\mathcal{D}')$ onto $H^2(\mathcal{Y}_1) \oplus \{0\}$ in the following way

$$(\Psi(e^{-it}f \oplus 0))(\lambda) = f(\lambda) \oplus 0 \qquad (f \in H^2(\mathcal{Y}_1)) . \tag{9.11}$$

To verify this let $\{f_n\}$ be the Taylor coefficients of $\lambda^n$ for $f(\lambda)$. Then using equation (IV.1.9) in Proposition IV.1.3 we have

$$\Psi(e^{-it}f \oplus 0) = \sum_{n=0}^{\infty} V^n(V - T')e^{-it}f_n \oplus 0 = \sum_{n=0}^{\infty} V^n(f_n \oplus 0) = f(e^{it}) \oplus 0 . \tag{9.12}$$

This yields (9.11). Finally, let $\phi_1$ be the unitary operator from $\mathcal{Y}_1$ onto $\mathcal{D}'$ defined by $\phi_1 a = e^{-it}a \oplus 0 = \Pi_{-1}^* a$ where a is in $\mathcal{Y}_1$. Obviously $\Psi\phi_1 a = a \oplus 0$.

Recall that $B'$ is an intertwining lifting of A with respect to V and tolerance $\gamma$ if and only if $B' = M_G | H^2(\mathcal{U})$ where $G = [F_1 - H, F_2]^{tr}$ for some H in $H^\infty(\mathcal{U}, \mathcal{Y}_1)$ and $\|G\|_\infty \le \gamma$, see Section V.7. In other words, H is a solution to the two block interpolation problem with tolerance $\gamma$ if and only if $B' = M_G | H^2(\mathcal{U})$ is an intertwining lifting of A with respect to V and tolerance $\gamma$, where $G = [F_1 - H, F_2]^{tr}$. Moreover, this function G can be recovered from $B'$ by $G(\lambda) = (B'\Pi_o^*)(\lambda)$ where $\Pi_o$ is the operator from $H^2(\mathcal{U})$ onto $\mathcal{U}$ defined by $\Pi_o g = g(0)$. In particular, $H(\lambda) = -(P_+ B'\Pi_o^*)(\lambda)$ where $P_+$ is the orthogonal projection from $L^2(\mathcal{Y}_1 \oplus \mathcal{Y}_2)$ onto $H^2(\mathcal{Y}_1)$. Since $\Psi U' = V\Psi$ it follows that $B'$ is an intertwining lifting of A with respect to V and tolerance $\gamma$ if and only if $B' = \Psi B_o$ where $B_o$ is an intertwining lifting of A with respect to $U'$ and tolerance $\gamma$. Therefore the set of all solutions H to the two block interpolation problem with tolerance $\gamma$ are given by $[F_1 - H, F_2]^{tr} = (\Psi B_o \Pi_o^*)$ where $B_o$ is an intertwining lifting of A with respect to $U'$ and tolerance $\gamma$. Using $\Psi | \mathcal{H}' = I$ it follows that $[F_1(\lambda), F_2(\lambda)]^{tr} = (\Psi A \Pi_o^*)(\lambda)$. Combining this along with Theorem 6.1 we see that the set of all solutions H to the Nehari interpolation problem with tolerance $\gamma$ are given by

$$H(\lambda) = -(\Psi Y_{2,2}\Pi_o^*)(\lambda) - (\Psi Y_{2,1}R(I - Y_{1,1}R)^{-1}Y_{1,2}\Pi_o^*)(\lambda) \tag{9.13}$$

where R is a contractive analytic function in $H^\infty(\mathcal{U}, \mathcal{D}')$ and $Y_{i,j}$ for i, j = 1, 2 are defined in (6.3). (Notice that $\mathcal{L} = \ker T = \mathcal{U}$ in this setting.)

Notice that $-(\Psi Y_{2,2}\Pi_o^*)(\lambda) = \Phi_{2,2}(\lambda)$ is precisely the central solution with tolerance $\gamma$ discussed in detail in Section V.7. Moreover, according to (V.7.14) and (V.7.45) the operator $N^2$ in (6.2) is given by (9.5). So to complete the proof it remains to find state space expressions for $N_1$, $Y_{1,1}$, $Y_{1,2}$ and $Y_{2,1}$ in (6.2) and (6.3). To accomplish this we follow all the notation developed in Section V.7. To begin, recall that $A = [W_o W_c, M_{F_2} | H^2(\mathcal{U})]^{tr}$ where $W_o$ is the observability operator from $\mathcal{X}$ into $K^2(\mathcal{Y}_1)$ and $W_c$ is the controllability operator from $H^2(\mathcal{U})$ onto $\mathcal{X}$ defined by

$$W_o x = C_1(e^{it}I - Z)^{-1} x \text{ and } W_c^* x = B^*(I - \lambda Z^*)^{-1} x \qquad (x \in \mathcal{X}). \tag{9.14}$$

Moreover, $W_\theta$ is the observability operator from $\mathcal{X}$ into $H^2(\mathcal{U})$ defined by

$$W_\theta x = C_o(I - \lambda Z^*)^{-1} x \text{ and } W_c(M_\Theta^+)^{-1} = W_\theta^* \tag{9.15}$$

where $\Theta$ is the invertible outer function defined in (V.7.23). Recall also that $Q_1 = W_o^* W_o$ is the observability grammian for the pair $\{C_1, Z\}$ and $P = RIC(\gamma) = W_\theta^* W_\theta$ is the controllability grammian for the pair $\{Z, C_o^*\}$. Throughout the rest of the proof we will also need the following formulas taken from (7.29), (7.34), (7.40), (7.42), (7.43) and (7.47) in Chapter V

$$\begin{gathered}\Gamma = W_o W_\theta^* \text{ and } (I - \Gamma^*\Gamma)^{-1} = I + W_\theta Q_1 (I - PQ_1)^{-1} W_\theta^* \\ M_\Theta^+ T_A^* = T_\Gamma^* M_\Theta^+,\ T_\Gamma^* W_\theta = W_\theta M \text{ and } (I - \lambda T_\Gamma^*)^{-1} W_\theta = W_\theta (I - \lambda M)^{-1} \\ T_\Gamma^* M_\Theta^+ \Pi_o^* = -W_\theta (I - Z^* Q_1 Z P)^{-1} Z^* (Q_1 + Q_2) B .\end{gathered} \tag{9.16}$$

To compute $N_1$ we need the following useful formula

$$D_A^{-2} A^* D' \phi_1 = (M_\Theta^+)^{-1} W_\theta (I - Q_1 P)^{-1} C_1^* . \tag{9.17}$$

To verify this recall that $D_A^2 = (M_\Theta^+)^* (I - \Gamma^*\Gamma) M_\Theta^+$; see (V.7.33). This identity along with (9.15), (9.16) and $P = W_\theta^* W_\theta$ gives

$$\begin{gathered}D_A^{-2} A^* D' \phi_1 = D_A^{-2} A^* \Pi_{-1}^* = (M_\Theta^+)^{-1} (I - \Gamma^*\Gamma)^{-1} (M_\Theta^+)^{-*} W_c^* W_o^* \Pi_{-1}^* = \\ (M_\Theta^+)^{-1} (I - \Gamma^*\Gamma)^{-1} W_\theta C_1^* = (M_\Theta^+)^{-1} W_\theta (I + Q_1 (I - PQ_1)^{-1} P) C_1^* = \\ (M_\Theta^+)^{-1} W_\theta (I - Q_1 P)^{-1} C_1^* .\end{gathered} \tag{9.18}$$

The previous equation yields (9.17). Equation (9.17) along with (9.15) shows that

$$\begin{gathered}N_1^2 \phi_1 = \phi_1 + D'AD_A^{-2} A^* D' \phi_1 = \phi_1 + D'A(M_\Theta^+)^{-1} W_\theta (I - Q_1 P)^{-1} C_1^* = \\ \phi_1 + D' W_o W_\theta^* W_\theta (I - Q_1 P)^{-1} C_1^* = \phi_1 + e^{-it} C_1 P (I - Q_1 P)^{-1} C_1^* = \phi_1 N_*^2 .\end{gathered} \tag{9.19}$$

This readily implies that $N_1^2 \phi_1 = \phi_1 N_*^2$ where $N_*$ is the operator on $\mathcal{Y}_1$ defined in (9.5). In particular, $p(N_1^2)\phi_1 = \phi_1 p(N_*^2)$ where $p(\lambda)$ is any polynomial. So by passing strong limits on the polynomials to take the square roots we obtain $N_1 \phi_1 = \phi_1 N_*$.

Notice that for any invertible outer function $\Theta$ we have $\Pi_o(M_\Theta^+)^{-1} = \Theta(0)^{-1}\Pi_o$. Using this fact along with (9.15), (9.16) and (9.17) we obtain the following expression for $Y_{1,1}$ in (6.3)

$$\begin{aligned} Y_{1,1}(\lambda)\phi_1 &= -\lambda N^{-1}\Pi_o(I-\lambda T_A^*)^{-1}D_A^{-2}A^*D'\phi_1 N_*^{-1} = \\ &-\lambda N^{-1}\Pi_o(I-\lambda T_A^*)^{-1}(M_\Theta^+)^{-1}W_\theta(I-Q_1P)^{-1}C_1^*N_*^{-1} = \\ &-\lambda N^{-1}\Pi_o(M_\Theta^+)^{-1}(I-\lambda T_\Gamma^*)^{-1}W_\theta(I-Q_1P)^{-1}C_1^*N_*^{-1} = \\ &-\lambda N^{-1}\Theta(0)^{-1}C_o(I-\lambda M)^{-1}(I-Q_1P)^{-1}C_1^*N_*^{-1} = \Phi_{1,1}(\lambda)\,. \end{aligned} \tag{9.20}$$

The second from the last equality follows from $\Pi_o W_\theta = C_o$. Finally, using the formulas for $\Theta(0) = D_0$ and $C_o$ in (V.7.23) we arrive at the state space expression for $\Phi_{1,1}(\lambda)$ in (9.7).

Using (9.16) and $\Pi_o(M_\Theta^+)^{-1} = D_0^{-1}\Pi_o$ we arrive at the following state space formula for $Y_{1,2}$ in (6.3)

$$\begin{aligned} (Y_{1,2}\Pi_o^*)(\lambda) &= N^{-1}\Pi_o(I-\lambda T_A^*)^{-1}\Pi_o^* = N^{-1} + \lambda N^{-1}\Pi_o(I-\lambda T_A^*)^{-1}T_A^*\Pi_o^* = \\ &N^{-1} + \lambda N^{-1}\Pi_o(M_\Theta^+)^{-1}(I-\lambda T_\Gamma^*)^{-1}T_\Gamma^* M_\Theta^+\Pi_o^* = \\ &N^{-1} - \lambda N^{-1}D_0^{-1}\Pi_o(I-\lambda T_\Gamma^*)^{-1}W_\theta(I-Z^*Q_1ZP)^{-1}Z^*(Q_1+Q_2)B = \\ &N^{-1} - \lambda N^{-1}D_0^{-1}C_o(I-\lambda M)^{-1}(I-Z^*Q_1ZP)^{-1}Z^*(Q_1+Q_2)B = \Phi_{1,2}\,. \end{aligned} \tag{9.21}$$

By virtue of the formulas for $D_0$ and $C_o$ in (V.7.23) we arrive at the expression for $\Phi_{1,2}(\lambda)$ in (9.7).

By employing (9.15), (9.16) and (9.17) we obtain the following state space formula for $Y_{2,1}$ in (6.3)

$$\begin{aligned} Y_{2,1}\phi_1 &= \phi_1 N_* - D'A(I-\lambda T_A^*)^{-1}D_A^{-2}A^*D'\phi_1 N_*^{-1} = \\ &\phi_1 N_* - D'A(I-\lambda T_A^*)^{-1}(M_\Theta^+)^{-1}W_\theta(I-Q_1P)^{-1}C_1^*N_*^{-1} = \\ &\phi_1 N_* - D'A(M_\Theta^+)^{-1}(I-\lambda T_\Gamma^*)^{-1}W_\theta(I-Q_1P)^{-1}C_1^*N_*^{-1} = \\ &\phi_1 N_* - D'W_oW_\theta^*W_\theta(I-\lambda M)^{-1}(I-Q_1P)^{-1}C_1^*N_*^{-1} = \\ &\phi_1 N_* - \phi_1 C_1P(I-\lambda M)^{-1}(I-Q_1P)^{-1}C_1^*N_*^{-1} = -\phi_1\Phi_{2,1}(\lambda)\,. \end{aligned} \tag{9.22}$$

Therefore $-Y_{2,1}\phi_1 = \phi_1\Phi_{2,1}(\lambda)$ where $\Phi_{2,1}$ is given by the state space formula in (9.7).

The previous calculations along with $\Psi\phi_1 = I$ show that

$$Y_{1,1}(\lambda)\phi_1 = \Phi_{1,1}(\lambda) \text{ and } (Y_{1,2}\Pi_o^*)(\lambda) = \Phi_{1,2}(\lambda)$$
$$(\Psi Y_{2,1}\phi_1)(\lambda) = -\Phi_{2,1} \text{ and } (\Psi Y_{2,2}\Pi_o^*)(\lambda) = -\Phi_{2,2}(\lambda)\,. \tag{9.23}$$

By replacing the contractive analytic function R in $H^\infty(\mathcal{U}, \mathcal{D}')$ in (9.13) by the contractive analytic function $\phi_1 R$ where R is now in $H^\infty(\mathcal{U}, \mathcal{Y}_1)$ and using (9.23) we arrive at our expression for the set of all two block interpolants in (9.8). The comments in Remark 9.2 follows from Remark 6.2. This completes the proof.

## *Notes to Chapter VI:*

The material in Sections 1-4 of this chapter is based on the paper Foias-Frazho [6]. Formula (2.5), which describes the set of all intertwining liftings, first appeared in Arsene-Ceausescu-Foias [1], where the proof is based on choice sequences. The proof of Theorem 2.1 given here is more direct. Sections 5 and 6 present an improved version of some of the material in Chapter XIV of Foias-Frazho [4]. However, Theorem 5.8 and its proof are due to Biswas [1]. The results in Sections 7-9 are derived by specifying the formula describing the set of all intertwining liftings for the concrete interpolation problems considered in these sections. The first description of all solutions of the Nehari problem in state space form is due to Glover [1]. Let us also mention that the band method yields another way to derive the parameterization of all solutions for different interpolation problems; in particular, for the Nehari problem the band method leads to the same formulas as the ones appearing in (8.6) (see Section XXXV.7 in Gohberg-Goldberg-Kaashoek [2]). State space formulas for problems related to the one of Nehari appear in Ball-Ran [1], [2], [3], where these formulas are based on the Grasmannian approach in Ball-Helton [1]. The monograph Ball-Gohberg-Rodman [1] is entirely devoted to state space formulas for rational interpolants; it also contains remarks on the history of state space formulas for interpolation problems. For some other approaches to parameterizing the set of all solution to the Nehari problem see Chui-Chen [1] and Green-Limebeer [1]. Results related to unitary systems and parametrization of solutions of interpolation problems can be found in Arocena [1].

## CHAPTER VII

# APPLICATIONS TO CONTROL SYSTEMS

The aim of this chapter is to show how norm constrained interpolation is used to solve problems in systems and control. We concentrate on scalar input and scalar output systems, which already gives a flavor of the subject. In the first section a basic feedback control problem is presented. Then for this problem the Youla parameterization of all stabilizing controllers is given (in Section 2). The third section employs some of the interpolation results of the previous chapters to solve a $H^\infty$ and $H^2$ error tracking problem. Section four presents a special two block problem as a natural problem in control. The final section presents a brief introduction to multivariable case and introduces the model matching problem. Finally, some numerical examples are also given.

## VII.1. FEEDBACK CONTROL

In this section we present a basic feedback control problem and introduce the concept of internal stability. For simplicity of presentation, throughout the first four sections all transfer functions, input and output signals are scalar valued.

Consider the following state space system

$$x_{n+1} = Zx_n + Bu_n \text{ and } y_n = Cx_n + Du_n \tag{1.1}$$

where Z is an operator on $\mathcal{X}$, and the input sequence $(u_n)_0^\infty$ and output sequence $(y_n)_0^\infty$ both have values in $\mathbb{C}$, while the state $x_n$ has values in $\mathcal{X}$. In this case the transfer function g for $\{Z, B, C, D\}$ is given by

$$g(\lambda) = D + \lambda C(I - \lambda Z)^{-1}B . \tag{1.2}$$

Obviously, g is analytic in some neighborhood of the origin. Moreover, assuming that the initial state $x_o = 0$, the output sequence $(y_n)_0^\infty$ is computed by $(y_n)_0^\infty = T_g(u_n)_0^\infty$ where $T_g$ is the Toeplitz matrix generated by g. That is

$$\begin{bmatrix} y_0 \\ y_1 \\ y_2 \\ \vdots \end{bmatrix} = \begin{bmatrix} g_0 & 0 & 0 & \cdots \\ g_1 & g_0 & 0 & \cdots \\ g_2 & g_1 & g_0 & \\ \vdots & \vdots & & \ddots \end{bmatrix} \begin{bmatrix} u_0 \\ u_1 \\ u_2 \\ \vdots \end{bmatrix} = T_g (u_n)_0^\infty , \tag{1.3}$$

where $g_n$ is the coefficient of $\lambda^n$ for $n \geq 0$ in the power series expansion of g at the origin. Furthermore, if Z is stable, then g is in $H^\infty$, the Toeplitz operator $T_g$ defines a bounded linear operator on $\ell_+^2$ and $\|T_g\| = \|g\|_\infty$.

We call the function g in (1.2) the *transfer function* of the causal linear time invariant system (1.1). Notice that in this case $g_0 = D$ and $g_n = CZ^{n-1}B$ for $n \geq 1$. A transfer function is always analytic in a neighborhood of the origin, and conversely any function analytic in some neighborhood of zero is a transfer function of some causal system of the form (1.1). The latter statement follows from the realization theory in Section III.4. As a consequence the product of two transfer functions is again a transfer function, and if g is transfer function with $g(0) \neq 0$, then $1/g$ is also a transfer function. We say that the transfer function g is *stable* if its Toeplitz matrix $T_g$ defines a bounded linear operator on $\ell_+^2$. Therefore g is stable if and only if g is in $H^\infty$. In this case $\|T_g\| = \|g\|_\infty$.

The block diagram associated with the transfer function g is given by

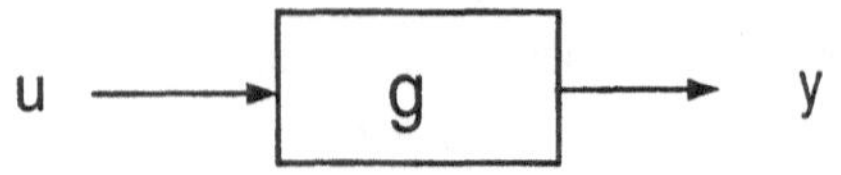


Figure 1

Let $u(\lambda)$ and $y(\lambda)$ be the functions formally defined by

$$u(\lambda) = \sum_{n=0}^{\infty} u_n \lambda^n \quad \text{and} \quad y(\lambda) = \sum_{n=0}^{\infty} y_n \lambda^n . \tag{1.4}$$

Now assume that $u(\lambda)$ is analytic in some neighborhood of the origin. By consulting (1.3), it follows that $y(\lambda) = g(\lambda)u(\lambda)$ and thus $y(\lambda)$ is also analytic in some neighborhood of the origin. So the convolution formula $(y_n)_0^\infty = T_g(u_n)_0^\infty$ in the time domain corresponds to the multiplication formula $y(\lambda) = g(\lambda)u(\lambda)$ in the $\lambda$ or frequency domain. Obviously, if $(u_n)_0^\infty$ is in $\ell_+^2$, then $u(\lambda)$ is in $H^2$. Moreover, if g is also stable ($g \in H^\infty$), then the output $y(\lambda)$ is in $H^2$ and the multiplication formula $y(\lambda) = g(\lambda)u(\lambda)$ makes sense for all $|\lambda| < 1$.

One of the main objectives of feedback control theory is to design a compensator c, that is, a transfer function c to meet a specified design criteria or minimize some objective function. A typical block diagram for accomplishing this is given by the block diagram in Figure 2 below.

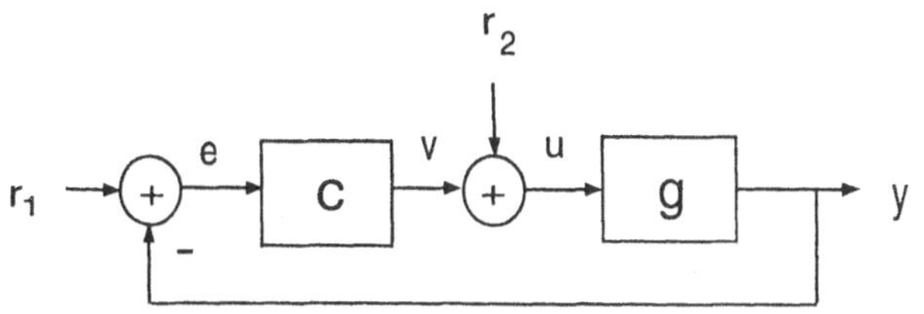


Figure 2

Here the open loop transfer function g in Figure 1 is specified and g may or may not be stable. The closed loop system for g is given by the block diagram in Figure 2, and c is called the *compensator or controller*. The compensator $c = c(\cdot)$ is always understood to be an analytic function in some neighborhood of the origin. This guarantees that the lower triangular Toeplitz matrix $T_c$ corresponding to c defines a map from the error signal e to v by $(v_n)_0^\infty = T_c(u_n)_0^\infty$, or equivalently, in the frequency domain $v(\lambda) = c(\lambda)e(\lambda)$. The signal $e = r_1 - y$ is the error between the input $r_1$ and the output y. Moreover, $u = v + r_2$ is the input to the open loop transfer function g.

A basic control problem is to design a compensator c such that the closed loop system from the input $r_1 \oplus r_2$ to the signal $e \oplus u$ is stable, and at the same time minimize some performance criteria. The equations corresponding to the block diagram in Figure 2 in the $\lambda$ or frequency domain, are given by $e = r_1 - gu$ and $u = r_2 + ce$. By simultaneously solving these equations we arrive at

$$\begin{bmatrix} e \\ u \end{bmatrix} = \begin{bmatrix} (1+gc)^{-1} & -g(1+cg)^{-1} \\ c(1+gc)^{-1} & (1+cg)^{-1} \end{bmatrix} \begin{bmatrix} r_1 \\ r_2 \end{bmatrix}. \tag{1.5}$$

The closed loop system in Figure 2 is *internally stable* if all the entries in the previous $2 \times 2$ matrix

$$\frac{1}{1+gc}, \quad \frac{c}{1+gc} \quad \text{and} \quad \frac{g}{1+cg} \tag{1.6}$$

are in $H^\infty$. Internal stability is important because it guarantees that the internal signals e and u do not blow up when the inputs $r_1$, $r_2$ are well defined. Finally, let us note that if the system is internally stable, then the transfer function from $r_1 \oplus r_2$ to the output y is also stable. To see this notice that (1.5) gives

$$y = gu = \frac{gc}{1+gc} r_1 + \frac{g}{1+cg} r_2 . \tag{1.7}$$

Because the system is internally stable, $g/(1+cg)$ is in $H^\infty$; see (1.6). Moreover,

$$\frac{gc}{1+gc} = 1 - \frac{1}{1+gc}$$

is also in $H^\infty$. Therefore if the feedback system in Figure 2 is internally stable, then the transfer function from $r_1 \oplus r_2$ to the output y is also stable.

We say that c is a *stabilizing controller* for g if c is a transfer function, and the three functions in (1.6) are all in $H^\infty$. If g is rational and c is a rational controller, then all the transfer functions in (1.6) are rational. In this case c is a rational stabilizing controller for g if and only if all the poles of the functions in (1.6) are in $\{\lambda : |\lambda| > 1\}$. So if g is rational and c is a stabilizing controller for g, then c moves all the unstable poles of g (that is, all the poles of g in the closed unit disc) to the corresponding poles of the transfer functions in (1.6) in the stable region $\{\lambda : |\lambda| > 1\}$. For example, if $g(\lambda) = (\lambda - 2)/(\lambda - .2)$, then the constant function $c(\lambda) = 1$ is a stabilizing controller, which moves the unstable pole $\lambda = .2$ of $g(\lambda)$ to the stable pole $\lambda = 1.1$ for the transfer functions in (1.6).

We are now ready to present the following basic feedback control problem: Given an open loop transfer function g find a stabilizing controller c for g to meet some specified design criteria. For example, one design criteria could be to solve the following optimization

$$\inf \{\|(1 + gc)^{-1}\| : \text{c is a stabilizing controller}\} . \tag{1.8}$$

In this problem one searches for a stabilizing controller c to make the transfer function $(1 + gc)^{-1}$ as small as possible with respect to some norm $\|\cdot\|$. This problem naturally occurs in tracking when one is trying to design a compensator c to force the output y to follow the input $r_1$ as close as possible. According to (1.5) the transfer function from $r_1$ to e is given by $(1 + gc)^{-1}$. So if the signal $r_2 = 0$, then $e = (1 + gc)^{-1} r_1 = r_1 - y$. In particular, if $(1 + gc)^{-1}$ is small, then $r_1 \approx y$ and the output y is tracking the input. Obviously there are some practical constraints to this problem. For instance assume that the constant function $c(\lambda) \equiv \alpha$ is a stabilizing controller for all large $\alpha > \alpha_0$. For a specific example consider the previous transfer function $g(\lambda) = (\lambda - 2)/(\lambda - .2)$. In this case one can choose a constant stabilizing controller $c(\lambda) \equiv \alpha$ for $\alpha$ large enough such that $\|(1 + \alpha g)^{-1}\| \approx 0$ and thus $r_1 \approx y$. Moreover, in this case the infimum in (1.8) is now zero. However, this controller $c(\lambda) \equiv \alpha$ for very large $\alpha$ cannot be implemented in practice. Finally, it is noted that tracking problems frequency occurs in guidance control, autopilots, and cruise control systems.

## VII.2 THE YOULA PARAMETERIZATION

In this section we will derive the Youla parameterization of all stabilizing controllers, which shows in particular that the set of all stabilizing controllers for a rational function g is parameterized by $H^\infty$. Then this parameterization will be used to develop some $H^2$ and $H^\infty$ optimal control problems.

We say that a transfer function g admits a *coprime factorization*, if there exists four functions n, d, p and q in $H^\infty$ satisfying

$$g = n/d \text{ and } np + dq = 1 . \tag{2.1}$$

Since g is a transfer function, by definition $g(\lambda)$ is analytic in some neighborhood of the origin, and thus $g(\lambda) = n(\lambda)/d(\lambda)$ holds in some neighborhood of the origin. The Bezout identity $n(\lambda)p(\lambda) + d(\lambda)q(\lambda) = 1$ holds for all $|\lambda| < 1$ and a.e. on the unit circle. Moreover, because $g(\lambda) = n(\lambda)/d(\lambda)$ is a transfer function $d(0) \neq 0$ and, thus $1/d(\lambda)$ is also analytic in some neighborhood of the origin. If $d(0) = 0$, then the Bezout identity shows that $n(0) \neq 0$. This contradicts the fact that $g = n/d$ is analytic in some neighborhood of the origin. Hence $d(0) \neq 0$. Furthermore, if g admits a coprime factorization, then $g(e^{it})$ is Lebesgue measurable. However, g may not be in $L^\infty$. For example, $g(\lambda) = 1/(\lambda - 1)$ admits a coprime factorization and g is not in $L^\infty$. If g is in $H^\infty$, then one can choose $n = g$, $d = 1$, $p = 0$ and $q = 1$. So all functions in $H^\infty$ admit a coprime factorization. In many engineering problems g is rational. All rational transfer functions g admit a coprime factorization. If g is rational, then $g = n/d$, where n and d are polynomials with no common zeros. By implementing the Euclidean algorithm one can compute polynomials p and q satisfying (2.1). (For completeness the Euclidean algorithm is presented at the end of this section.) Moreover, one can also compute n and d in the rational case by using state space techniques; see Francis [1], Green-Limebeer [1], Zhou-Doyle-Glover [1]. If g admits a coprime factorization, then it makes sense that one should only consider coprime compensators c. We say that c is a *coprime compensator* if c is a transfer function and c admits a coprime factorization. The following famous result (see Youla-Jabr-Bongiorno [1]) provides a complete characterization of all coprime stabilizing controllers.

**THEOREM 2.1.** *Assume that the transfer function* g *admits a coprime factorization of the form* $g = n/d$ *where* n, d, p *and* q *are all functions in* $H^\infty$ *satisfying (2.1). Then the set of all coprime stabilizing compensators* c *for* g *is given by*

$$c = \frac{p + dh}{q - nh} , \tag{2.2}$$

*where* h *is a function in* $H^\infty$ *and* $q(0) \neq n(0)h(0)$. *In particular,* $p + dh$ *and* $q - nh$ *have no common zeros in* $|\lambda| < 1$.

**PROOF.** Let $c = x/y$ where x and y are functions in $H^\infty$. By employing $g = n/d$ we obtain

$$\frac{1}{1+gc} = \frac{dy}{nx+dy}, \quad \frac{c}{1+gc} = \frac{dx}{nx+dy} \quad \text{and} \quad \frac{g}{1+cg} = \frac{ny}{nx+dy}. \tag{2.3}$$

However, if $x = p + dh$ and $y = q - nh$ for some h in $H^\infty$, then using $np + dq = 1$ we see that $nx + dy = 1$. Therefore all the functions in (2.3), or equivalently, in (1.6) are in $H^\infty$. Moreover, if $y(0) = q(0) - n(0)h(0)$ is not zero, it follows that $c(\lambda) = x(\lambda)/y(\lambda)$ is analytic in some neighborhood of the origin. So if c is a controller of the form (2.2), then c is a stabilizing controller. Since $nx + dy = 1$, it also follows that x/y is a coprime factorization of c and thus $p + dh$ and $q - nh$ have no common zeros in $|\lambda| < 1$.

On the other hand, assume that c is a stabilizing controller and x/y is a coprime factorization of c where $ax + by = 1$ and x, y, a, b are all in $H^\infty$. Since $c = x/y$ is a stabilizing controller, we have

$$1 - \frac{1}{1+gc} = \frac{gc}{1+gc} = \frac{nx}{nx+dy} \tag{2.4}$$

is a function in $H^\infty$. Now using the fact that all the functions in (2.3) and (2.4) are in $H^\infty$ along with $ax + by = 1$ and (2.1) we obtain

$$[p,\ q]\begin{bmatrix} nx(nx+dy)^{-1} & ny(nx+dy)^{-1} \\ dx(nx+dy)^{-1} & dy(nx+dy)^{-1} \end{bmatrix}\begin{bmatrix} a \\ b \end{bmatrix} = \frac{1}{nx+dy}. \tag{2.5}$$

Therefore both $nx + dy$ and $1/(nx + dy)$ are in $H^\infty$. In particular, $nx + dy = \phi$ where $\phi$ is an invertible outer function in $H^\infty$, that is, both $\phi$ and $1/\phi$ are in $H^\infty$. Hence

$$n(x/\phi) + d(y/\phi) = 1 \quad (\text{where } \phi,\ x/\phi \text{ and } y/\phi \text{ are in } H^\infty). \tag{2.6}$$

Notice that equation (2.1) gives

$$\begin{bmatrix} p & d \\ q & -n \end{bmatrix}\begin{bmatrix} n & d \\ q & -p \end{bmatrix} = I \tag{2.7}$$

where I is the identity matrix on $\mathbb{C}^2$. Now let $\mathcal{N}$ be the subspace of $H^\infty(\mathbb{C}^2)$ defined by

$$\mathcal{N} = \{[h_1,\ h_2]^{tr} : nh_1 + dh_2 = 1 \text{ and } h_1,\ h_2 \in H^\infty\}. \tag{2.8}$$

Equation (2.7) is used to show that $\mathcal{N} = \mathcal{N}_1$, where $\mathcal{N}_1$ is defined by

$$\mathcal{N}_1 := \begin{bmatrix} p \\ q \end{bmatrix} + \begin{bmatrix} d \\ -n \end{bmatrix} H^\infty . \tag{2.9}$$

Clearly $\mathcal{N}_1 \subseteq \mathcal{N}$. If $nh_1 + dh_2 = 1$ for some $h_1$ and $h_2$ in $H^\infty$, then (2.7) gives

$$\begin{bmatrix} n & d \\ q & -p \end{bmatrix} \begin{bmatrix} h_1 \\ h_2 \end{bmatrix} = \begin{bmatrix} 1 \\ h \end{bmatrix} \text{ and } \begin{bmatrix} h_1 \\ h_2 \end{bmatrix} = \begin{bmatrix} p \\ q \end{bmatrix} + \begin{bmatrix} d \\ -n \end{bmatrix} h \tag{2.10}$$

where $h = qh_1 - ph_2$ clearly is in $H^\infty$. Hence $\mathcal{N} \subseteq \mathcal{N}_1$ and $\mathcal{N} = \mathcal{N}_1$. Equation (2.6) implies that $nh_1 + dh_2 = 1$, where $h_1 = x/\phi$ and $h_2 = y/\phi$. Equations (2.8), (2.9) and $\mathcal{N} = \mathcal{N}_1$ readily show that $x = \phi(p + dh)$ and $y = \phi(q - nh)$, where h is a function in $H^\infty$. Therefore the compensator $c = x/y$ is given by the formula in (2.2). By definition all compensators are analytic in some neighborhood of the origin. Because c admits a coprime factorization of the form $c(\lambda) = x(\lambda)/y(\lambda)$, it readily follows that $y(0)$ is nonzero. Hence $q(0) \neq n(0)h(0)$. This completes the proof.

In many applications g is a rational function in $L^\infty$. In general it is quite difficult to implement nonrational compensators, because they require infinite dimensional realizations. So in practice if g is rational, then one only considers rational controllers. The following is an immediate consequence of Theorem 2.1.

**COROLLARY 2.2.** *Assume* g *is a rational transfer function* $g = n/d$, *where* n, d, p *and* q *are all rational functions in* $H^\infty$ *satisfying (2.1). Then the set of all rational stabilizing compensators* c *for* g *are given by (2.2), where* h *is a rational function in* $H^\infty$ *and* $q(0) \neq n(0)h(0)$. *In particular,* $p + dh$ *and* $q - nh$ *have no common zeros in* $|\lambda| < 1$.

Because rational scalar valued transfer functions play an important role in applications, we conclude this section by presenting the Euclidean algorithm to compute polynomials $p(\lambda)$ and $q(\lambda)$ satisfying $a(\lambda)p(\lambda) + b(\lambda)q(\lambda) = 1$ when $a(\lambda)$ and $b(\lambda)$ are relatively prime polynomials, that is, $a(\lambda)$ and $b(\lambda)$ have no common zeros. To this end, let $a(\lambda)$ and $b(\lambda)$ be two polynomials and without loss of generality assume that $\deg a \leq \deg b$. (The degree of a polynomial is denoted by deg.) By performing division there exists two unique polynomials $r(\lambda)$ and $f(\lambda)$ satisfying $b(\lambda) = r(\lambda)a(\lambda) + f(\lambda)$ where $\deg r = \deg b - \deg a$ and $\deg f < \deg a$. The polynomial $r(\lambda)$ is the *remainder* polynomial obtained by dividing $b(\lambda)/a(\lambda)$. Moreover, one can compute the unique polynomials $r(\lambda)$ and $f(\lambda)$ by matching in decreasing order like coefficients of $\lambda^m, \lambda^{m-1}, \ldots, \lambda^0$ in $b(\lambda) = r(\lambda)a(\lambda) + f(\lambda)$ to first obtain $r(\lambda)$ and then $f(\lambda)$. (Here $m = \deg b$.) Clearly

$$[a(\lambda),\ b(\lambda)]\begin{bmatrix} -r(\lambda) & 1 \\ 1 & 0 \end{bmatrix} = [f(\lambda),\ a(\lambda)] := [a_1(\lambda),\ b_1(\lambda)]\,, \tag{2.11}$$

where $a_1(\lambda)$ and $b_1(\lambda)$ are polynomials such that $\deg a_1 < \deg b_1$. So we can use division once again to find the unique polynomials $r_1(\lambda)$ and $f_1(\lambda)$ satisfying $b_1(\lambda) = r_1(\lambda)a_1(\lambda) + f_1(\lambda)$ where $\deg r_1 = \deg b_1 - \deg a_1$ and $\deg f_1 < \deg a_1$. Continuing in this fashion we arrive at the following difference equation

$$[a_j(\lambda),\ b_j(\lambda)]\begin{bmatrix} -r_j(\lambda) & 1 \\ 1 & 0 \end{bmatrix} = [a_{j+1}(\lambda),\ b_{j+1}(\lambda)] \tag{2.12}$$

with the initial condition $a_0(\lambda) = a(\lambda)$ and $b_0(\lambda) = b(\lambda)$. Here $r_j(\lambda)$ is the unique remainder polynomial determined by $b_j(\lambda) = r_j(\lambda)a_j(\lambda) + f_j(\lambda)$ where $\deg r_j = \deg b_j - \deg a_j$ and $\deg f_j < \deg a_j$. Notice that $\deg a_j < \deg b_j$ for all $j \geq 1$. So the difference equation in (2.12) must eventually end in a finite number of steps with $a_{k+1}(\lambda) = 0$ for some $j = k$. This observation readily leads to the following Euclidean algorithm for determining greatest polynomial divisor of $a(\lambda)$ and $b(\lambda)$. By definition $c(\lambda)$ is the *greatest common divisor of* $a(\lambda)$ *and* $b(\lambda)$ if $c(\lambda)$ is a polynomial satisfying $a(\lambda) = c(\lambda)\alpha(\lambda)$ and $b(\lambda) = c(\lambda)\beta(\lambda)$ where $\alpha(\lambda)$ and $\beta(\lambda)$ are relatively prime polynomials. Obviously the greatest common divisor is unique up to a constant factor.

**THEOREM 2.3.** *Let* $a(\lambda)$ *and* $b(\lambda)$ *be polynomials satisfying* $\deg a \leq \deg b$, *and set* $a_0(\lambda) = a(\lambda)$ *and* $b_0(\lambda) = b(\lambda)$. *By using (2.12) recursively compute* $a_{j+1}(\lambda)$, $b_{j+1}(\lambda)$ *and the remainder polynomial* $r_j(\lambda)$ *for* $b_j(\lambda)/a_j(\lambda)$ *until* $a_{k+1}(\lambda) = 0$ *for some* $k = j$. *Then* $b_{k+1}(\lambda)$ *is the greatest common polynomial divisor of* $a(\lambda)$ *and* $b(\lambda)$. *In particular,* $a(\lambda)$ *and* $b(\lambda)$ *are relatively prime if and only if* $b_{k+1}(\lambda)$ *is a constant. In this case*

$$\begin{bmatrix} p(\lambda) \\ q(\lambda) \end{bmatrix} = \begin{bmatrix} -r_0(\lambda) & 1 \\ 1 & 0 \end{bmatrix}\begin{bmatrix} -r_1(\lambda) & 1 \\ 1 & 0 \end{bmatrix}\cdots\begin{bmatrix} -r_k(\lambda) & 1 \\ 1 & 0 \end{bmatrix}\begin{bmatrix} 0 \\ 1/b_{k+1} \end{bmatrix} \tag{2.13}$$

*are two polynomials of degree at most* $\deg b$ *satisfying* $a(\lambda)p(\lambda) + b(\lambda)q(\lambda) = 1$.

**PROOF.** Let $R_j(\lambda)$ be the $2 \times 2$ polynomial matrix in (2.12). Then

$$[a(\lambda),\ b(\lambda)]R_0(\lambda)R_1(\lambda)\ \ldots\ R_k(\lambda) = [0,\ b_{k+1}(\lambda)]\,. \tag{2.14}$$

To prove that $b_{k+1}(\lambda)$ is the greatest common divisor of $a(\lambda)$ and $b(\lambda)$, first notice that $[a(\lambda),\ b(\lambda)] = [0,\ b_{k+1}(\lambda)]R_k(\lambda)^{-1}\ \ldots\ R_0(\lambda)^{-1}$. Since $R_j(\lambda)^{-1}$ is also a polynomial matrix, that is, a matrix whose entries are polynomials, $b_{k+1}(\lambda)$ divides both $a(\lambda)$ and $b(\lambda)$. Now let $\alpha(\lambda)$ and $\beta(\lambda)$ be the polynomials given by $a(\lambda) = b_{k+1}(\lambda)\alpha(\lambda)$ and $b(\lambda) = b_{k+1}(\lambda)\beta(\lambda)$. So dividing (2.14) by $b_{k+1}(\lambda)$ we obtain $[\alpha(\lambda), \beta(\lambda)]R_0(\lambda) \ldots R_k(\lambda) = [0, 1]$. Let $p(\lambda)$ and $q(\lambda)$ be the polynomials defined by $[p(\lambda), q(\lambda)]^{tr} = R_0(\lambda) \ldots R_k(\lambda)[0, 1]^{tr}$. Then equation (2.14) shows that $\alpha(\lambda)p(\lambda) + \beta(\lambda)q(\lambda) = 1$. Thus $\alpha(\lambda)$ and $\beta(\lambda)$ are relatively prime. Therefore $b_{k+1}(\lambda)$ is the greatest polynomial divisor of $a(\lambda)$ and $b(\lambda)$.

If $a(\lambda)$ and $b(\lambda)$ are relatively prime, then $b_{k+1}$ is a constant. In this case $p(\lambda)$ and $q(\lambda)$ given by (2.13) satisfies $a(\lambda)p(\lambda) + b(\lambda)q(\lambda) = 1$. Moreover, since $b_j(\lambda) = a_{j-1}(\lambda)$ for all $j \geq 1$, it follows that $\deg r_j = \deg a_{j-1} - \deg a_j$ for all $j \geq 1$. In particular, if $b_{k+1}$ is a constant, or equivalently, $a(\lambda)$ and $b(\lambda)$ are relatively prime, then the degree of the polynomials $p(\lambda)$ and $q(\lambda)$ in (2.13) is less than or equal to

$$\sum_{j=0}^{k} \deg r_j \leq \deg b - \deg a_o + \deg a_o - \deg a_1 + \cdots + \deg a_k - \deg a_{k+1} = \deg b\,.$$

This completes the proof.

Let $a(\lambda)$, $b(\lambda)$, $p(\lambda)$, $q(\lambda)$ be polynomials of degree at most m and $\gamma(\lambda)$ a polynomial of degree at most 2m. Let $T_a$ and $T_b$ be the lower triangular matrices in $\mathcal{L}(\mathbb{C}^{m+1}, \mathbb{C}^{2m+1})$ generated by the polynomials $a(\lambda)$ and $b(\lambda)$, respectively. (The entries of $T_a$ and $T_b$ are given by $(T_a)_{i,j} = a_{i-j}$ and $(T_b)_{i,j} = b_{i-j}$ where $a_j$ and $b_j$ are the coefficients of $\lambda^j$ for $a(\lambda)$ and $b(\lambda)$, respectively.) Let $\vec{p}$, $\vec{q}$ and $\vec{\gamma}$ be the vectors given by

$$\vec{p} = [p_0, p_1, \ldots, p_m]^{tr}, \vec{q} = [q_0, q_1, \ldots, q_m]^{tr} \text{ and } \vec{\gamma} = [\gamma_0, \gamma_1, \ldots, \gamma_{2m}]^{tr}\,, \tag{2.15}$$

where $p_j$, $q_j$ and $\gamma_j$ are the coefficients of $\lambda^j$ for $p(\lambda)$, $q(\lambda)$ and $\gamma(\lambda)$, respectively. Then $\gamma(\lambda) = a(\lambda)p(\lambda) + b(\lambda)q(\lambda)$ if and only if $\vec{\gamma} = T_a\vec{p} + T_b\vec{q}$. In particular, $a(\lambda)p(\lambda) + b(\lambda)q(\lambda) = 1$ if and only if $e = T_a\vec{p} + T_b\vec{q}$ where $e = [1, 0, 0, \ldots, 0]^{tr}$. According to the previous theorem $a(\lambda)$ and $b(\lambda)$ are relatively prime if and only if there exists polynomials $p(\lambda)$ and $q(\lambda)$ of degree at most m satisfying $a(\lambda)p(\lambda) + b(\lambda)q(\lambda) = 1$. Therefore $a(\lambda)$ and $b(\lambda)$ are relatively prime if and only if e is in the range of $[T_a, T_b]$ in $\mathcal{L}(\mathbb{C}^{2m+2}, \mathbb{C}^{2m+1})$. So a rather naive method to find polynomials $p(\lambda)$ and $q(\lambda)$ satisfying $a(\lambda)p(\lambda) + b(\lambda)q(\lambda) = 1$, is to first check to see if e is in the range of $[T_a, T_b]$. If this is true, compute the vectors $\vec{p}$ and $\vec{q}$ satisfying $[T_a, T_b]\vec{p} \oplus \vec{q} = e$. Then the polynomials $p(\lambda)$ and $q(\lambda)$ corresponding to $\vec{p}$ and $\vec{q}$ satisfy $a(\lambda)p(\lambda) + b(\lambda)q(\lambda) = 1$. This approach for computing the polynomials $p(\lambda)$ and $q(\lambda)$ involves inverting a matrix $[T_a, T_b]$ in $\mathcal{L}(\mathbb{C}^{2m+2}, \mathbb{C}^{2m+1})$. Even though this matrix is loaded with zeros this method is probably not as efficient as the Euclidean approach in Theorem 2.3. However, it is easy to program and appears to work well for polynomials of degree less than 150 or so.

## VII.3 MIXED $H^\infty$ and $H^2$ CONTROL PROBLEMS

In this section we will present a $H^2$ and $H^\infty$ control problem. As before, let g be the open loop transfer function for the block diagram in Figure 1, and c the corresponding compensator for the feedback control system in Figure 2. According to (1.5) the transfer function from $r_1$ to e corresponding to the tracking error is given by $1/(1+gc)$, that is, in the $\lambda$-domain $e(\lambda) = r_1(\lambda)/(1 + g(\lambda)c(\lambda))$. The *weighted tracking error* is defined by $we = w(r_1 - y)$ where w is a weighting function in $H^\infty$ chosen by the designer to emphasize a certain frequency range. The weighted tracking error transfer function is defined by $w(1+gc)^{-1}$ which is precisely the transfer function from $r_1(\lambda)$ to the weighted tracking error $w(\lambda)e(\lambda)$. A basic control problem is to design a stabilizing coprime controller to minimize the weighted tracking error over the class of all stabilizing controller. In mathematical terminology this corresponds to the following optimization problem

$$d_j = \inf \{\|w(1+gc)^{-1}\|_j : c \text{ is a coprime stabilizing controller for } g\} . \tag{3.1}$$

Notice that one can choose the weight $w = 1$. In this case the optimization problem in (3.1) attempts to minimize the tracking error over the class of all coprime stabilizing controllers. The weight w is usually chosen by the engineers to minimize the tracking error e over a critical range of frequencies. For example, if one wants to design a control system to force the output y to track the input $r_1$ for low frequencies, then the designer would choose w to be a low pass filter, that is, w would pass through low frequencies and reject high frequencies.

Obviously, the solution to the minimization problem in (3.1) depends upon the norm $\|\cdot\|_j$ one is using. In practice, one uses the $H^2$ norm $\|\cdot\|_2$ or $H^\infty$ norm $\|\cdot\|_\infty$. So in (3.1) we take $j = 2$ or $\infty$. If one chooses the $H^2$ norm, then the optimal solution to the minimization problem in (3.1) is a coprime stabilizing controller c which minimizes the expected value of the weighted error $|(L_w(e_n)_{-\infty}^{\infty})_0|^2$ at time 0 when the input $r_1$ is white noise and $L_w$ is the Laurent operator generated by w. In this case, c is a stabilizing controller which minimizes the area under the spectral density $S_e(e^{it}) = |w(e^{it})(1 + g(e^{it})c(e^{it}))^{-1}|^2$ of the weighted error over the class of all coprime stabilizing controllers; see Section I.11 and Caines [1] for further results on stochastic processes and Wiener optimization problems. If one uses the $H^\infty$ norm, then an optimal solution to the minimization problem in (3.1) is a coprime stabilizing controller c, which minimizes the maximum value of $\|we\|$ over the class of all inputs $r_1$ in the closed unit ball of $H^2$.

To convert the optimization problem in (3.1) to a standard $H^2$ or $H^\infty$ optimization problem let $g = n/d$ be a coprime factorization of g where n, d, p and q are all functions in $H^\infty$ satisfying (2.1). According to the Youla parameterization in Theorem 2.1, the set of all coprime stabilizing controllers c is given by (2.2). So using this c in (2.2) along with $g = n/d$ we see that the tracking error transfer function is given by

$$\frac{1}{1+gc} = \frac{d(q-nh)}{qd-ndh+pn+ndh} = dq-ndh\,, \tag{3.2}$$

where h is in $H^\infty$. Therefore the optimal control problem in (3.1) yields the following $H^2$ or $H^\infty$ Sarason interpolation problem

$$d_j = \inf\{\|wd(q-nh)\|_j : h \in H^\infty\} \qquad (\text{for } j = 2, \infty)\,. \tag{3.3}$$

Notice that we have ignored the constraint $q(0) \neq n(0)h(0)$ in the Youla parameterization when converting the optimization problem from (3.1) to (3.3). Removing this constraint does not effect the error $d_j$. Moreover, in practice if $d_j \neq 0$, then the optimal h will almost always satisfy $q(0) \neq n(0)h(0)$. (In Example 3.1 below we will present a case where $d_\infty \neq 0$ and the optimal h satisfies $q(0) = n(0)h(0)$.) If $j = 2$, then (3.3) is an $H^2$ optimization problem. On the other hand, if $j = \infty$, then (3.3) is an $H^\infty$ Sarason optimization problem. Now let $\theta_i\theta_o$ be the inner-outer factorization of wnd, where $\theta_i$ is inner and $\theta_o$ is outer. Using $\theta_i\theta_o = wnd$ in (3.3), we readily see that (3.3) is precisely the Sarason problem discussed in detail in Section I.7.1. Moreover, if h is the solution to one of these optimization problems, and $q(0) \neq n(0)h(0)$, then the corresponding compensator $c = (p+dh)/(q-nh)$. If $q(0) = n(0)h(0)$, then a minor perturbation of h will give a stabilizing controller via (2.2) and $d_j \approx \|w(1+gc)^{-1}\|_j$. In the rational case we can readily apply the state space techniques in Section V.4 to find an optimal rational h in $H^\infty$ which solves the Sarason problem in (3.3) for $j = 2$ or $\infty$. (If g is rational, then it is understood that n, d, p and q are all taken to be rational functions.) For certain nonrational g one can implement the skew Toeplitz techniques in Foias-Osbay-Tannenbaum [1], to compute the optimal controller.

In the next three paragraphs we will discuss the case when $d_\infty$ in (3.3) is zero. According to the Sarason Theorem I.7.1, the optimal error $d_\infty = \|A\|$, where A is the operator from $H^2$ into $\mathcal{H} := H^2 \ominus \theta_i H^2$ defined by $A = P_{\mathcal{H}} M_{wdq} | H^2$. In particular, $d_\infty = 0$ if and only if $\|A\| = 0$. However, A is zero if and only if $wdqH^2 \subseteq \theta_i H^2$. Hence $d_\infty = 0$ if and only if $\phi_i q_i H^2 \subseteq \theta_i H^2$, where $\phi_i$ and $q_i$ is the inner part of wd and q, respectively. By employing Lemma III.7.8 we see that $d_\infty = 0$ if and only if $\phi_i q_i = \theta_i m$, where m is an inner function in $H^\infty$. Since $\theta_i$ is the inner part of (wd)n, the error $d_\infty = 0$ if and only if $q_i = n_i m$, where $n_i$ is the inner part of n. The Bezout identity $np + dq = 1$ for all $|\lambda| < 1$, implies that n and q have no common inner part. Thus $d_\infty = 0$ if and only if n is outer. According to Section IV.9, the optimal $H^2$ error $d_2 = \|A | \mathcal{L}\|$, where $\mathcal{L} = \ker S^*$ is now the set of all constant functions in $H^2$ and S is the unilateral shift on $H^2$. So $d_2 = 0$ if and only if $A | \mathcal{L} = 0$. Since $T'^n Aa = AS^n a$ for all a in $\mathcal{L}$ and $n \geq 0$, it follows that $d_2 = 0$ if and only if $A = 0$. Therefore $d_2 = 0$ if and only if $d_\infty = 0$, or equivalently, n is outer.

If $d_\infty = 0$, then the optimal compensator c is may not be in $L^\infty$. If $d_\infty = 0$, then n is outer and thus $0 = \inf\{\|q - nh\|_\infty : h \in H^\infty\}$. So there exists a sequence of functions $h_k$ in $H^\infty$ such that $q - nh_k$ approaches zero in the $H^\infty$ norm as k approaches infinity. This readily implies that $\|wdq - wndh_k\|_\infty$ approaches zero as k approaches infinity. Hence the corresponding stabilizing controllers $(p + dh_k)/(q - nh_k)$ may not converge to an $L^\infty$ function as k approaches infinity. So if $d_\infty = 0$, then one usually searches for stabilizing controllers c in $L^\infty$ such that $\|w(1 + gc)^{-1}\|_\infty \le \gamma$, where $\gamma$ is some specified tolerance.

To provide some additional insight, let us assume that n is an invertible outer function. (Recall that f is an *invertible outer function* if both f and 1/f are in $H^\infty$.) In this case the unique optimal solution to (3.3) is given by $h = q/n$, and $d_\infty = 0$. Furthermore, $q - nh = 0$ and the controller $c = (p + dh)/(q - nh)$ is not well defined. Notice that, if n is an invertible outer function, then the constant function $c(\lambda) \equiv \alpha$ for any $|\alpha| > \alpha_0 := 1/\|g\|_\infty$ is a stabilizing controller. To see this first notice that $1/g = d/n$ is in $H^\infty$. So if $|\alpha| > \alpha_0$, then $\|1/\alpha g\|_\infty < 1$ and thus $(1 + 1/\alpha g)$ is an invertible outer function. Since $1/\alpha g$ is in $H^\infty$,

$$\frac{1}{1+gc} = \frac{1/\alpha g}{1 + 1/\alpha g}$$

is also in $H^\infty$ for all $|\alpha| > \alpha_0$. This readily implies that all the functions in (1.6) are in $H^\infty$. Therefore $c(\lambda) \equiv \alpha$ is a stabilizing controller for all $|\alpha| > \alpha_0$. Clearly $\|w(1 + \alpha g)^{-1}\|_j$ approaches $d_\infty = 0$ as $\alpha$ approaches infinity. Obviously one cannot implement the controller $c(\lambda) \equiv +\infty$. However, one can find a constant controller $c(\lambda) \equiv \alpha$ for some large $\alpha$ such that $\|w(1 + gc)^{-1}\|_j \le \gamma$, where $\gamma$ is some specified bound. Finally, it is noted that one can use classical root locus techniques to show that if $g = n/d$, where n and d are polynomials and all the zeros of $n(\lambda)$ are in $|\lambda| > 1$, then $c(\lambda) \equiv \alpha$ is a stabilizing controller for all large $\alpha$; see Ogata [1].

If $n(0) \ne 0$, then obviously one can choose a h in $H^\infty$ such that $q(0) = n(0)h(0)$, and thus the corresponding controller c is not a transfer function, or equivalently, c is not analytic in some neighborhood of the origin. Although the (anticausal) condition $q(0) = n(0)h(0)$ rarely occurs at the optimal solution h of (3.3), when $d_\infty \ne 0$ it can happen.

**EXAMPLE 3.1**. To construct an example of a transfer function whose optimal $H^\infty$ solution of (3.3) violates the (causality) condition $q(0) \ne n(0)h(0)$, let $\theta$ be a rational inner function and f a rational invertible outer function. Consider the following $H^\infty$ optimization problem $d_\infty = \inf\{\|f - f\theta h\|_\infty : h \in H^\infty\}$. Because f and $\theta$ are both rational, the optimal solution to this Sarason problem is given by $f - f\theta h_1 = d_\infty b$, where $h_1 \in H^\infty$ and $b(\lambda)$ is a Blaschke product; see Corollary IV.2.7 or Theorem V.4.3. Now choose any invertible rational outer function f and a Blaschke product $\theta$ such that the corresponding Blaschke product $b(\lambda)$ is not a unitary constant. (For example, $f = 1/(2-\lambda)$ and $\theta = \lambda^2$.) Then $b(\alpha) = 0$ for some $|\alpha| < 1$. Let m be the Möbius transformation defined by $m(\lambda) = (\lambda+\alpha)/(1+\bar{\alpha}\lambda)$. Let $w(\lambda) = f(m(\lambda))$ and $n(\lambda) = \theta(m(\lambda))$. Obviously $w(\lambda)$ is a rational invertible outer function and $n(\lambda)$ is a Blaschke product. Now let $d = 1$, $p = 0$ and $q = 1$. Then $g = n/d$ admits a coprime factorization. Moreover, the $H^\infty$ optimization problem in (3.3) now reduces to $d_\infty = \inf\{\|w - wnh\| : h \in H^\infty\}$. Since the Möbius transformation does not affect the $H^\infty$ norm, it follows that the unique solution to this $H^\infty$ optimization problem is given by $w(\lambda) - w(\lambda)n(\lambda)h(\lambda) = d_\infty b(m(\lambda))$ where $h(\lambda) = h_1(m(\lambda))$. Using $w(0) \neq 0$ and $b(m(0)) = 0$, we have $q(0) = 1 = n(0)h(0)$. In this case $d_\infty \neq 0$ and the $H^\infty$ optimal solution yields a controller c which is not a transfer function.

The previous example can be expanded to demonstrate some of the practical problems associated with $H^\infty$ optimal controllers. To see this let $g = n/d$ be a coprime rational transfer function, where n, d, p and q are rational transfer functions satisfying (2.1), and let w be a rational weighting function in $H^\infty$. If $\phi_i\phi_o$ is the inner-outer factorization of wd, then the $H^\infty$ optimization problem in (3.3) is equivalent to the following $H^\infty$ optimization problem $d_\infty = \inf\{\|\phi_o(q - nh)\|_\infty : h \in H^\infty\}$. If $\phi_o$ and the outer part $n_o$ of n are both invertible outer functions, then Corollary IV.2.7 or Theorem V.4.3 shows that the unique optimal solution h in $H^\infty$ to this optimization problem is given by $\phi_o(q - nh) = d_\infty b$, where b is a Blaschke product. Because $\phi_o$ is outer, $q - nh$ and b have the same zeros in $\mathbb{D}$. Since the inner part $\phi_i$ of wd does not affect the $H^\infty$ optimal solution, this h is also the unique solution to the $H^\infty$ optimization problem in (3.3), and thus $c = (p + dh)/(q - nh)$ is the corresponding $H^\infty$ optimal stabilizing controller. Recall that $p + dh$ and $q - nh$ have no common zeros in $\mathbb{D}$. So the unstable poles of c, or equivalently, the poles of c in $\mathbb{D}$, are precisely the zeros of $q - nh$ in $\mathbb{D}$. In particular, c is in $H^\infty$ if and only if $b(\lambda) \equiv b(0)$ is a constant of modulus one. Now let A be the operator from $H^2$ into $\mathcal{H} = H^2 \ominus n_i H^2$ associated with the previous Sarason $H^\infty$ optimization problem, that is, $A = P_{\mathcal{H}} M_{\phi_o q} | H^2$ where $n_i$ is the inner part of n. Theorem III.7.1 shows that the dimension $N_i$ of $\mathcal{H}$ equals the number of zeros of $n_i$ including multiplicities. In particular, A has finite rank. According to Corollary IV.2.7 we have $d_\infty b = Ax/x$ where x attains the norm of A. (Since $An_iH^2 = \{0\}$, this x must be in $\mathcal{H}$.) So if $N_i = 1$, then b is a constant function of modulus one. However, if $N_i > 1$, then the expression $d_\infty b = Ax/x$ shows that b is not a constant of modulus one in most cases. (One can always contrive a special situation where $N_i > 1$ and $b(\lambda) \equiv b(0)$. For instance, $w = 1$ and $g = n_i$ with $d = 1$, $p = 0$ and $q = 1$.) Therefore, in almost all practical problems, if $N_i > 1$, then the optimal stabilizing controller c is not in $H^\infty$. This means that the transfer function associated with c is unstable and its corresponding Toeplitz operator $T_c$ is not a bounded operator on $\ell^2_+$. If c is not in $H^\infty$, then the fact that $T_c$ is unbounded makes it extremely difficult to implement this controller c in some engineering applications, even though c is a stabilizing controller. For instance, if one of the sensors or actuators fails and $c \notin H^\infty$, then we are essentially left with an unstable open loop system which cannot be used in practice. So if $N_i > 1$, then in general the $H^\infty$ optimal stabilizing controller cannot be used to solve these problems. In many cases the optimal $H^\infty$ controller is not in $H^\infty$ but there exists stabilizing controllers which are in $H^\infty$. For example, if $w = 1/(2-\lambda)$ and $g = \lambda^2$ with $n = \lambda^2$, $d = 1$, $p = 0$ and $q = 1$, then the optimal $H^\infty$ stabilizing controller is not in $H^\infty$, and yet $c(\lambda) = \alpha$ for any constant $\alpha$ in $\mathbb{D}$ is a stabilizing controller. Therefore an important problem in control is to find a h in $H^\infty$ such that $q - nh$ is outer. Then $c = (p + dh)/(q - nh)$ is a stabilizing controller in $H^\infty$. For some further results and open problems concerning interpolation with outer functions, see Doyle-Francis-Tannenbaum [1] and Vidyasagar [1].

Now let us convert the Sarason problem in (3.3) to the form of the Nehari problem studied in Section V.5. Then one can also apply the state space techniques in Section V.5 to compute a solution h to the Sarason problem in (3.3) and the corresponding compensator $c = (p + dh)/(q - nh)$. To this end, let x be any function in $L^2$. Then $P_+x$ is the orthogonal projection of x onto $H^2$ and $P_-x$ is the orthogonal projection of x onto $L^2 \ominus H^2$. Clearly $x = P_-x + P_+x$. As before, let $\theta_i\theta_o = wnd$ be the inner-outer factorization of wnd where $\theta_i$ is an inner function and $\theta_o$ is an outer function. Without much loss of generality, let us also assume that $\theta_o$ is an invertible outer function, that is, $1/\theta_o$ is also in $H^\infty$. Notice that for any h in $H^\infty$ the functions $P_-\bar{\theta}_i wdq + P_+\bar{\theta}_i wdq - \theta_o h$ and $wdq - wndh$ both have the same $H^2$ and $H^\infty$ norms. (To eliminate some technical problems we always assume that $P_-\bar{\theta}_i wdq$ is in $L^\infty$. This always happens in the rational case.) Moreover, because $\theta_o$ is an invertible outer function $\theta_o H^2 = H^2$. Hence the Sarason optimization problems in (3.3) is equivalent to the following Nehari optimization problems:

$$d_j = \inf \{\|P_-(\bar{\theta}_i wdq) - h_o\|_j : h_o \in H^\infty\} \qquad (j = 2 \text{ or } \infty)\,. \tag{3.4}$$

Furthermore, if $h_o$ is a solution to one of the previous Nehari optimization problems, then $h_o = \theta_o h - P_+(\bar{\theta}_i wdq)$, and thus $h = (h_o + P_+(\bar{\theta}_i wdq))/\theta_o$ is the solution to the corresponding Sarason problem in (3.3). In particular, if $h_o$ solves (3.4), then the corresponding compensator c is given by

$$c = \frac{\theta_o p + dh_o + dP_+(\bar{\theta}_i wdq)}{\theta_o q - nh_o - nP_+(\bar{\theta}_i wdq)}\,. \tag{3.5}$$

As always we assume that $q(0) \neq n(0)h(0)$, or equivalently, the denominator of c in (3.5) is not zero at the origin.

Clearly $h_o = 0$ is the solution to the $H^2$ Nehari optimization problem in (3.4) and $d_2 = \|P_-\bar{\theta}_i wdq\|_2$. Moreover, the $H^2$ optimal compensator is given by $h_o = 0$ in (3.5), that is,

$$c = \frac{\theta_o p + dP_+(\bar{\theta}_i wdq)}{\theta_o q - nP_+(\bar{\theta}_i wdq)} . \tag{3.6}$$

Furthermore, the optimal $H^\infty$ error is given by $d_\infty = \|H\|$, where H is the Hankel operator from $H^2$ into $L^2 \ominus H^2$ defined by $Hf = P_-(P_-\bar{\theta}_i wdq)f = P_-\bar{\theta}_i wdqf$, where f is in $H^2$. In the rational case one can use the state space techniques in Section V.5 to compute the optimal $h_o$ solving the Nehari optimization problem in (3.4). Then the corresponding compensator c is given by (3.5). Moreover, one can use the central Nehari solution in Section V.5 to compute a rational $h_o$ in $H^\infty$ satisfying the mixed $H^2$ and $H^\infty$ bounds in Theorem V.5.1, and then the mixed $H^2$ and $H^\infty$ compensator c is also given by (3.5) for this $h_o$. These facts will be demonstrated on a simple rational function g in Example 3.2 below. Summing up the analysis of this and the previous paragraph we arrive at the following result.

**THEOREM 3.2.** *Let* $g = n/d$, *where* n, d, p *and* q *are functions in* $H^\infty$ *satisfying* $np + dq = 1$, *and let* w *be a specified weight function in* $H^\infty$. *Moreover, let* $\theta_i\theta_o$ *be the inner outer factorization of* wnd, *where* $\theta_i$ *is inner and* $\theta_o$ *is an invertible outer function. Then the* $H^2$ *and* $H^\infty$ *tracking optimization problem in (3.1) is equivalent to the corresponding* $H^2$ *and* $H^\infty$ *Nehari optimization problem in (3.4).*

*(i)* *The solution to the* $H^2$ *Nehari optimization problem is* $h_o = 0$, *and the* $H^2$ *tracking error* $d_2 = \|P_-\bar{\theta}_i wdq\|_2$. *If* $\theta_o q - nP_+\bar{\theta}_i wdq$ *is nonzero at* $\lambda = 0$, *then the optimal* $H^2$ *tracking controller* c *is given by (3.6).*

*(ii)* *The optimal* $H^\infty$ *tracking error* $d_\infty$ *is the norm of the Hankel operator* $P_-M_{\bar{\theta}_i wdq} | H^2$. *If* $h_o$ *is the optimal solution to the* $H^\infty$ *Nehari optimization problem in (3.4), and* $\theta_o q - nh_o - nP_+\bar{\theta}_i wdq$ *is nonzero at* $\lambda = 0$, *then the optimal* $H^\infty$ *tracking compensator* c *is given by (3.5).*

*(iii)* *The tracking error* $d_\infty = 0$, *or equivalently,* $d_2 = 0$ *if and only if* n *is an outer function.*

The following example shows how one can use the results in Section V.5 to numerically solve a tracking problem.

**EXAMPLE 3.3.** Consider the unstable open loop transfer function $g(\lambda) = n(\lambda)/d(\lambda)$ where $n(\lambda) = -.31 + .85\lambda$ and $d(\lambda) = .63 + .92\lambda + \lambda^2$ and the weighting function $w = 1$. The roots of $d(\lambda)$ are $-.46 \pm .65i$. So $g(\lambda)$ is clearly unstable. Two polynomials $p(\lambda)$ and $q(\lambda)$ satisfying the Bezout identity are given by $p(\lambda) = -1.38 - 1.07\lambda$ and $q(\lambda) = .91$. (Here we have only expressed p ad q up to two decimal places. So $np + dq \approx 1$. Throughout we will only give numbers up to two decimal places. However, our computer simulation will include accuracy beyond two decimal places.)

On this example we first used fast Fourier transform techniques to compute the inner-outer spectral factorization $\theta_i\theta_o$ of wnd. (Recall that the outer spectral factor $\theta$ for any b in $H^\infty$ is given by $\theta(\lambda) = \exp(a(\lambda))$ where $a(\lambda) = a_o + 2a_1\lambda + 2a_2\lambda^2 + \cdots$ and $\{a_n\}_{-\infty}^{\infty}$ are the Fourier coefficients of $\log |b|$; see Hoffman [1]. So using fast Fourier transforms we can compute both $\theta_i$ and $\theta_o$.) One can also obtain $\theta_i$ and $\theta_o$ by computing the poles and zeros of wnd, or by the state space techniques in the Appendix. All of these methods have their advantages and disadvantages. (Even though the weight $w = 1$ in our specific example we have included it here to present a discussion of the general case.) Next we used fast Fourier transform techniques to compute the Fourier coefficients of $P_-\bar{\theta}_i wdq$. Then we applied a standard Kalman-Ho algorithm (see Kalman-Falb-Arbib [1]) to compute a suitable anticausal realization $\{Z, B, C\}$ satisfying $C(\lambda I - Z)^{-1}B = P_-\bar{\theta}_i wdq$. Now for various values of $\delta$ we used the state space techniques in Section V.5 to compute the function $h_o$ in $H^\infty$ for the central solution to the Nehari problem corresponding to (3.4). Then using this $h_o$ with fast Fourier transform techniques we arrived at the stabilizing controller c in (3.5). Finally, implementing a Kalman-Ho algorithm once again we arrived at the polynomials $x(\lambda)$ and $y(\lambda)$ satisfying $c(\lambda) = x(\lambda)/y(\lambda)$. By combining the above into a Matlab program we first derived that $d_\infty = 1.29$ and $d_2 = 1.20$.

Now recall that for any $\delta > 1$, the central solution $h_o$ for the Nehari problem in (3.4) satisfies the bounds (see Section IV.9)

$$\|P_-\bar{\theta}_i wdq - h_o\|_\infty \le \delta d_\infty \quad \text{and} \quad \|P_-\bar{\theta}_i wdq - h_o\|_2 \le \delta d_2(\delta^2 - 1 + d_2^2/d_\infty^2)^{-1/2} . \tag{3.7}$$

This readily implies that for an arbitrary $\delta > 1$, the corresponding stabilizing controller c in (3.5) yields a weighted tracking error transfer function $w(1 + gc)^{-1}$ which satisfies the following mixed $H^2$ and $H^\infty$ bounds

$$\|w(1 + gc)^{-1}\|_\infty \le \delta d_\infty \quad \text{and} \quad \|w(1 + gc)^{-1}\|_2 \le \delta d_2(\delta^2 - 1 + d_2^2/d_\infty^2)^{-1/2} . \tag{3.8}$$

Moreover, if $\delta = 1$, then the corresponding stabilizing controller c solves the $H^\infty$ optimization problem in (3.1). In this case $w(1 + gc)^{-1}$ is in $H^\infty$ and $|w(1 + gc)^{-1}| = d_\infty$ a.e. If $\delta = +\infty$, then the corresponding stabilizing controller c solves the $H^2$ optimization problem in (3.1), that is, $w(1 + gc)^{-1}$ is in $H^2$ and $\|w(1 + gc)^{-1}\|_2 = d_2$. Equation (3.8) shows that any choice of $\delta$ in

$(1, \infty)$ provides a trade off between the optimal $H^\infty$ stabilizing controller and the optimal $H^2$ stabilizing controller. In particular, if we choose $\delta^2 = 2 - d_2^2/d_\infty^2$, then the corresponding controller c yields a transfer function $(1+gc)^{-1}$ which satisfies the same bounds, that is,

$$\|w(1+gc)^{-1}\|_\infty \le d_\infty\sqrt{2-d_2^2/d_\infty^2} \text{ and } \|w(1+gc)^{-1}\|_\infty \le d_2\sqrt{2-d_2^2/d_\infty^2}\,. \tag{3.9}$$

Finally, if we choose $\delta = \sqrt{2}$, then (3.8) shows that the corresponding stabilizing controller c yields a transfer function which satisfies the following bounds

$$\|w(1+gc)^{-1}\|_\infty \le d_\infty\sqrt{2} \text{ and } \|w(1+gc)^{-1}\|_2 \le d_2\sqrt{2}\,. \tag{3.10}$$

Obviously, one can obtain the bounds in (3.10) by choosing $\delta^2 = 2 - d_2^2/d_\infty^2$ and thus (3.9) satisfies the bounds in (3.10). However, in practice the choice of $\delta = \sqrt{2}$ can have certain advantages over $\delta^2 = 2 - d_2^2/d_\infty^2$ and vise versa, as shown by computer simulation.

By running the previous algorithm on Matlab with $\delta = 1$ we found that the optimal $H^\infty$ stabilizing controller $c(\lambda) = x(\lambda)/y(\lambda)$, where $x(\lambda)$ and $y(\lambda)$ are the polynomials given by $x(\lambda) = -.46 - .6\lambda$ and $y(\lambda) = 1$. (Because n has only one zero in $\mathbb{D}$, this controller c is in $H^\infty$.) Next by choosing $\delta = (2 - d_2^2/d_\infty^2)^{1/2} = 1.06$ we found that the corresponding stabilizing controller $c = x/y$ is now given by $x(\lambda) = -.51 - .64\lambda - .03\lambda^2$ and $y(\lambda) = 1$. For $\delta = \sqrt{2}$ the corresponding stabilizing compensator $c = x/y$ is given by $x = -.66 - .78\lambda - .12\lambda^2$ and $y(\lambda) = 1$. Finally, for $\delta = +\infty$, the optimal $H^2$ stabilizing controller $c = x/y$ is given by $x(\lambda) = -.85 - .96\lambda - .24\lambda^2$ and $y(\lambda) = 1$. Notice that for $\delta = 1$, the polynomial $x(\lambda)$ is of order 1 while for $\delta > 1$, the polynomial $x(\lambda)$ is of order 2. This explains the small coefficient of $-.03\lambda^2$ of $x(\lambda)$ for $\delta = 1.06$.

The graph in Figure 3 plots $|(1+g(e^{it})c(e^{it}))^{-1}|$ on the y axis versus $0 \le t \le 2\pi$ on the x axis for $\delta = 1$, 1.06, $\sqrt{2}$ and $+\infty$, respectively. The flat line corresponds to $\delta = 1$, the optimal $H^\infty$ controller. The next line below $d_\infty = 1.29$ at $t = \pi$ corresponds to $\delta^2 = 2 - d_2^2/d_\infty^2 = 1.06$. The next graph beneath that one at $\pi$ corresponds to $\delta = \sqrt{2}$. Finally the bottom graph at $\pi$ is the optimal $H^2$ controller with $\delta = +\infty$. (The graphs are symmetric about $\pi$ because g is a rational function with real coefficients.) Notice that the optimal $H^\infty$ controller ($\delta = 1$) has a smaller error $|(1+gc)^{-1}|$ at low frequencies $0 < t < 1$ while the optimal $H^2$ filter ($\delta = +\infty$) performs better at high frequencies $1 < t < \pi$. By choosing $\delta = 1.06$ or $\sqrt{2}$ we see that the corresponding stabilizing compensator yields a transfer function whose error $|(1+gc)^{-1}|$ is smaller than the error for the optimal $H^2$ controller at low frequencies $0 < t < 1$ and at the same time is smaller than the optimal error for the $H^\infty$ controller at high frequencies $1 < t < \pi$. So by varying $\delta$ we can find a stabilizing controller which provides a trade off between the optimal $H^2$ and $H^\infty$ controllers and performs well over all frequency ranges. Therefore the control engineer can vary $\delta$ to find the appropriate controller for a specific design objective. Finally, it is noted that if we vary $\delta$ from one to infinity, then the graphs in Figure 3 move continuously from the optimal $H^\infty$ controller to the corresponding optimal $H^2$ controller.

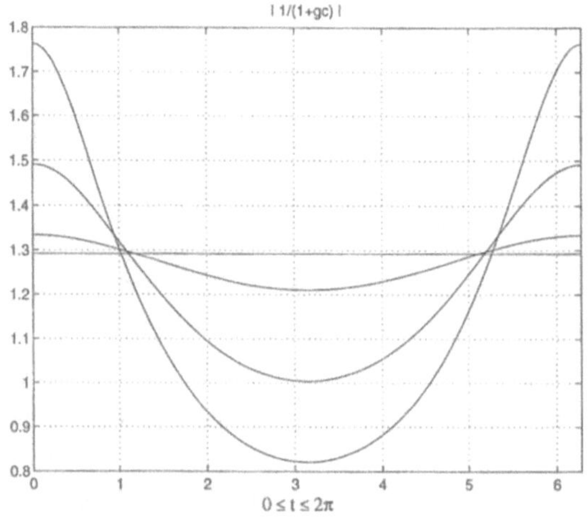


Figure 3

The previous example shows that the optimal $H^2$ controller performs better than the optimal $H^\infty$ controller over high frequencies. The theory states that the optimal $H^\infty$ controller will perform better than any other stabilizing controller over some frequency range. However, computer simulations show that this frequency range can be small and does not necessarily occur at high or low frequency ranges. The following provides an example where the optimal $H^\infty$ controller performs better at high frequencies.

**EXAMPLE 3.4**. Consider the unstable open loop transfer function $g(\lambda) = n(\lambda)/d(\lambda)$ given by $n(\lambda) = -.65 - .8\lambda + \lambda^2$ and $d(\lambda) = 1.5 + -3.5\lambda + \lambda^2$ and the weight $w = 1$. The roots of $d(\lambda)$ are .5 and 3. So clearly $g(\lambda)$ is unstable. Here the $p(\lambda)$ and $q(\lambda)$ we used for the Bezout identity $np + dq = 1$ are $p(\lambda) = -1.29 + .16\lambda^2$ and $q(\lambda) = .11 - .44\lambda - .16\lambda^2$. By applying our previous Matlab program we found that $d_\infty = 1.25$ and $d_2 = 1.08$. If $\delta = 1$, then the optimal $H^\infty$ stabilizing controller $c = x/y$ is given by $x(\lambda) = -1.38 + .46\lambda$ and $y(\lambda) = 1 - .77\lambda$. If $\delta = (2 - d_2^2/d_\infty^2)^{1/2} = 1.12$, then the corresponding stabilizing compensator $c = x/y$ is given by $x(\lambda) = -16.31 + 6.67\lambda - .41\lambda^2$ and $y(\lambda) = 10 - 7.69\lambda$. For $\delta = \sqrt{2}$, the corresponding stabilizing compensator $c = x/y$ is given by $x(\lambda) = -20 + 9.75\lambda - 1.03\lambda^2$ and $y(\lambda) = 10 - 7.69\lambda$. Finally, for $\delta = +\infty$ the optimal $H^2$ controller $c = x/y$ is given by $x(\lambda) = -26.15 + 14.87\lambda - 2.05\lambda^2$ and $y(\lambda) = 10 - 7.69\lambda$.

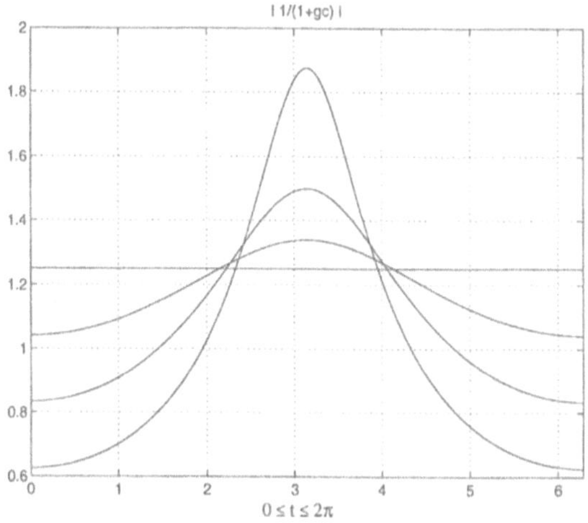


Figure 4

We ran our previous Matlab program on this example for $\delta = 1$, $1.12$, $\sqrt{2}$ and $\infty$. Figure 4 displays the tracking error $|(1 + g(e^{it})c(e^{it}))^{-1}|$ for these four different controllers where $0 \le t \le 2\pi$. In this example the flat line $d_\infty = 1.25$ corresponds to the optimal $H^\infty$ controller $\delta = 1$. The curve immediately above the constant function $d_\infty = 1.25$ at $t = \pi$ corresponds to the controller c with $\delta = (2 - d_2^2/d_\infty^2)^{1/2} = 1.12$. Immediately above this line is $|(1 + gc)^{-1}|$ for the controller c with $\delta = \sqrt{2}$. Finally the highest curve at $t = \pi$ corresponds to the optimal $H^2$ controller with $\delta = +\infty$. Notice that in this example the optimal $H^2$ controller yields less error in $|(1 + g(e^{it})c(e^{it})|^{-1}$ for low frequencies $0 < t < 2$ while the optimal $H^\infty$ controller performs better at high frequencies $t > 2$. As expected for $\delta = 1.12$ and $\sqrt{2}$ we obtain controllers which provide a trade off between the optimal $H^\infty$ and $H^2$ controllers. In particular, the controllers c for $\delta = 1.12$ or $\sqrt{2}$ yield less error in $|(1 + gc)^{-1}|$ at lower frequencies than the optimal $H^\infty$ controller and perform better at high frequencies, than the optimal $H^2$ controllers.

## VII.4 A TWO BLOCK CONTROL PROBLEM

In this section we will show how the two block problem discussed in Section V.7 naturally occurs in feedback control. In addition to the tracking problem, there are many other control problems associated with the feedback control system in Figure 2. For example, consider the feedback control system given in Figure 5. Here $r_1$ is an input signal, and $r_3$ is additive disturbance corrupting the feedback loop. The error signal e is now given by $e = r_1 - (y + r_3)$. So using $y = gce$ we have $e = (1 + gc)^{-1}r_1 - (1 + gc)^{-1}r_3$. Substituting this back into $y = gce$ we obtain

$$\begin{bmatrix} e \\ y \end{bmatrix} = \begin{bmatrix} (1+gc)^{-1} & -(1+gc)^{-1} \\ gc(1+gc)^{-1} & -gc(1+gc)^{-1} \end{bmatrix} \begin{bmatrix} r_1 \\ r_3 \end{bmatrix} . \tag{4.1}$$

One control problem is to construct a stabilizing controller for g (that is, the functions in (1.6) are in $H^\infty$) to minimize the effect of the disturbance $r_3$ on the output y. The transfer function from the disturbance $r_3$ to the output y is given by $-gc(1+gc)^{-1}$.

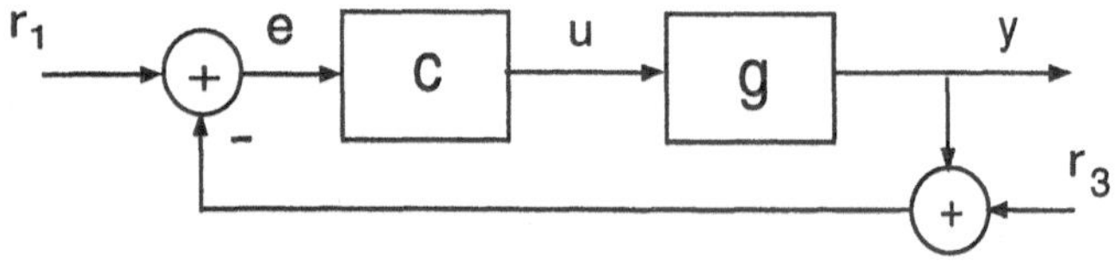


Figure 5

This readily leads to the following optimization problem

$$d_j = \inf \{\|w_o gc/(1+gc)\|_j : c \text{ is a stabilizing coprime compensator for } g\} \qquad (j = 2 \text{ or } \infty) . \tag{4.2}$$

Here $w_o$ in $H^\infty$ is a weighting function chosen by the engineer to minimize the effect of the disturbance $r_3$ on the output y over certain critical frequency ranges. One can also choose $w_o = 1$.

As before assume that g admits a coprime factorization of the form $g = n/d$ and p, q are functions in $H^\infty$ such that $np + dq = 1$. Moreover, let c be the coprime stabilizing controller given by the Youla parameterization in (2.2). Then

$$\frac{w_o gc}{1+gc} = \frac{w_o(np+ndh)}{qd - ndh + pn + ndh} = w_o np + w_o ndh .$$

So using this set up and ignoring the (causality) condition $q(0) \neq n(0)h(0)$, the optimal control problem in (4.2) yields

$$d_j = \inf \{\|w_o np + w_o ndh\|_j : h \in H\} \qquad (j = 2 \text{ or } \infty) . \tag{4.3}$$

This control problem also leads to $H^2$ and $H^\infty$ Sarason optimization problem of the type discussed in Sections I.7 and V.4. In fact, if $\theta_i\theta_o$ is the inner-outer factorization of $w_o nd$, then the $H^2$ or $H^\infty$ optimization problem in (4.2) is precisely the Sarason problem discussed in Section V.4. So in the rational case one can use the state space techniques of Section V.4 to compute a h which solves the $H^2$ or $H^\infty$ optimization problem in (4.3). If $q(0) \neq n(0)h(0)$, then the corresponding stabilizing compensator c is given by $c = (p+dh)/(q-nh)$. Moreover, if we assume that all the functions are rational, then the optimal $H^\infty$ solution h to (4.3) is given by $w_o n(p+dh) = d_\infty b$, where b is a Blaschke product. Clearly the zeros of $p+dh$ play no role in determining whether or not $c=(p+dh)/(q-nh)$ is in $H^\infty$. So unlike the tracking problem in (3.3), the $H^\infty$ optimal solutions to (4.3) do not generally yield stabilizing controllers $c \notin H^\infty$. By following our previous analysis $d_\infty = 0$, or equivalently, $d_2 = 0$ if and only if d is a outer function. In this case $0 = \inf\{\|p+dh\|_\infty : h \in H^\infty\}$. So there exists a h in $H^\infty$ such that $0 = p + dh$ or $\|p+dh\|_\infty \approx 0$. According to (2.2) the corresponding compensator $c = 0$ or $c \approx 0$. In particular, if d is an invertible outer function, then g is in $H^\infty$ and the obvious choice of $c=0$ is a stabilizing controller which minimizes the norm of the transfer function $w_o gc(1+gc)^{-1}$. Therefore if $d_\infty = 0$ the control problem in (4.2) is not very interesting. Finally, it is noted that one can follow the analysis in Section 3 to convert the control problem in (4.3) to a $H^2$ or $H^\infty$ Nehari optimization problem of the type given in (3.4). The details are omitted.

In applications one may want to obtain a stabilizing controller to minimize the effect of the disturbance $r_3$ on both the error signal e and the output y. According to (4.1) the transfer function from $r_3$ to $e \oplus y$ is given by $-[(1+gc)^{-1}, gc(1+gc)^{-1}]^{tr}$. This readily leads to the following $H^2$ and $H^\infty$ optimization problems

$$d_j = \inf\left\{\left\|\begin{bmatrix} w(1+gc)^{-1} \\ w_o gc(1+gc)^{-1}\end{bmatrix}\right\|_j : c \text{ is a stabilizing coprime controller for } g\right\} = \tag{4.4}$$

$$\inf\left\{\left\|\begin{bmatrix} wdq \\ w_o np\end{bmatrix} - \begin{bmatrix} wnd \\ -w_o nd\end{bmatrix} h\right\|_j : h \in H^\infty\right\} \quad (j = 2 \text{ or } \infty).$$

Here w in $H^\infty$ is a weight to be chosen by the engineer to minimize the effect of the disturbance $r_3$ on the tracking error e over certain critical frequency ranges. As before w can be one. If one choses $w_o = 0$, respectively $w = 0$, then the optimization problem in (4.4) reduces to the corresponding optimization problems in (3.3), respectively (4.3). Notice that the first transfer function $(1+gc)^{-1}$ in (4.4) is also the transfer function from $r_1$ to the tracking error e. So the optimization problem in (4.4) attempts to construct a stabilizing controller for g to minimize the effect of the disturbance $r_3$ on $e \oplus y$ and at the same time it also tries to force the output y to track the input $r_1$. So let us call this $H^2$ or $H^\infty$ optimization problem the *disturbance reduction problem.* Using (4.1) we see that the transfer function from $r_1$ to $e \oplus y$ is given by $[(1+gc)^{-1}, gc(1+gc)^{-1}]^{tr}$. Notice that this is precisely the transfer functions appearing before the weights w and $w_o$ in the optimization problem (4.4). So if $r_3 = 0$, then the optimization problem in (4.4) attempts to minimize the effect of the reference signal $r_1$ on $e \oplus y$. Of course for nonzero $r_1$ we cannot drive both $e = r_1 - y$ and y to zero. So if $r_3 = 0$, then the optimization problem in (4.4) attempts to find a trade off between minimizing the tracking error e and minimizing the effect of $r_1$ on the output y.

The disturbance reduction problem in (4.4) reduces to a $H^2$ or $H^\infty$ Sarason problem discussed in Section I.7. Moreover, if h is a solution to this Sarason problem, and $q(0) \neq n(0)h(0)$, then the optimal compensator c which solves the corresponding disturbance reduction control problem is given by $c = (p + dh)/(q - nh)$. (If $q(0) = n(0)h(0)$, then a slight perturbation of h will yield a stabilizing controller which approximately solves the optimization problem in (4.4).) Notice that if w and $w_o$ are both nonzero, then $d_\infty \neq 0$. If $d_\infty = 0$, then the first equation in (4.4) shows that both $(1 + gc)^{-1} \approx 0$ and $gc(1 + gc)^{-1} \approx 0$ for some controller c. These two equations cannot occur simultaneously, thus $d_\infty \neq 0$.

Now let us convert the disturbance reduction optimization problem in (4.4) to a $H^2$ or $H^\infty$ two block optimization problem of the type discussed in Section V.7. To this end, let

$$\begin{bmatrix} \theta_1 \\ \theta_2 \end{bmatrix} \theta_o = \begin{bmatrix} wnd \\ -w_o nd \end{bmatrix} \tag{4.5}$$

be the inner-outer factorization of $[wnd, -w_o nd]^{tr}$, where $[\theta_1, \theta_2]^{tr}$ is an inner function in $H^\infty(\mathbb{C}^2)$ and $\theta_o$ is an outer function. As before, let us assume that $\theta_o$ is an invertible outer function in $H^\infty$, that is, $1/\theta_o$ is also in $H^\infty$. Now let $\Theta$ be the function in $L^\infty(\mathbb{C}^2, \mathbb{C}^2)$ defined by

$$\Theta = \begin{bmatrix} \theta_1 & \bar{\theta}_2 \\ \theta_2 & -\bar{\theta}_1 \end{bmatrix} . \tag{4.6}$$

Because $[\theta_1, \theta_2]^{tr}$ is an inner function, it follows that $\Theta(e^{it})$ is a.e. unitary. Now using (4.5) and (4.6) we have for h in $H^\infty$

$$\Theta^* \begin{bmatrix} wdq \\ w_o np \end{bmatrix} - \Theta^* \begin{bmatrix} wnd \\ -w_o nd \end{bmatrix} h = \begin{bmatrix} a \\ b \end{bmatrix} - \begin{bmatrix} \theta_o h \\ 0 \end{bmatrix} . \tag{4.7}$$

Here a is the function in $L^\infty$ and b is the function in $H^\infty$ defined by

$$a = \bar{\theta}_1 wdq + \bar{\theta}_2 w_o np \text{ and } b = \theta_2 wdq - \theta_1 w_o np \qquad (\text{a.e. on } [0, 2\pi)) . \tag{4.8}$$

Since $\Theta$ is a.e. unitary, equation (4.7) shows that the Sarason optimization problem in (4.4) is equivalent to the following $H^2$ or $H^\infty$ optimization problem

$$d_j = \inf\left\{ \left\| \begin{bmatrix} a - \theta_o h \\ b \end{bmatrix} \right\|_j : h \in H^\infty \right\} \qquad (j = 2 \text{ or } \infty) . \tag{4.9}$$

Recall that $K^\infty$ is the subspace of $L^\infty$ consisting of all functions in $L^\infty$ whose Fourier coefficients of $e^{int}$ for $n \geq 0$ are all zero. Now let $f_1 = P_- a$, where $P_-$ is the orthogonal projection onto $L^2 \ominus H^2$. Let $h_o = \theta_o h - P_+ a$ and observe that $f_1 - h_o = a - \theta_o h$. Moreover, let $f_2$ be the function in $K^\infty$ defined by $f_2(e^{it}) = e^{-it}\overline{b(e^{it})}$ a.e. Notice that $|f_2(e^{it})| = |b(e^{it})|$ a.e. Finally, because $\theta_o$ is outer and $f_1 - h_o = f_1 - (\theta_o h - P_+ a) = a - \theta_o h$, it now follows that the $H^2$ or $H^\infty$ optimization problem in (4.9) is equivalent to the following $H^2$ or $H^\infty$ optimization problem

$$d_j = \inf\left\{ \left\| \begin{bmatrix} f_1 - h_o \\ f_2 \end{bmatrix} \right\|_j : h_o \in H^\infty \right\} \qquad (j = 2 \text{ or } \infty) . \tag{4.10}$$

This is precisely the two block optimization problems discussed in Section V.7. Furthermore, $h_o$ is a solution to one of two block optimization problems in (4.10) if and only if $h_o = \theta_o h - P_+ a$, where h is the optimal solution to the corresponding disturbance reduction problem in (4.4). In particular, if $h_o$ is an optimal solution to one of the two block optimization problems in (4.10), then the corresponding optimal disturbance reducing controller

$$c = \frac{\theta_o p + dP_+ a + dh_o}{\theta_o q - nP_+ a - nh_o} . \tag{4.11}$$

This result is obtained by simply substituting $h = (h_o + P_+ a)/\theta_o$ into the form for c in the Youla parameterization (2.2).

By consulting Section V.7, it follows that $h_o = 0$ is the optimal $H^2$ solution to the $H^2$ two block optimization problem in (4.10). In this case the $H^2$ error is $d_2 = \|f_1 \oplus f_2\|_2$. Moreover, the optimal $H^2$ disturbance reducing compensator is given by setting $h_o = 0$ in (4.11), that is,

$$c = \frac{\theta_o p + dP_+ a}{\theta_o q - nP_+ a} . \tag{4.12}$$

Finally, the optimal $H^\infty$ tracking error $d_\infty = \|A\|$, where A is the operator from $H^2$ into $K^2 \oplus L^2$ defined by $Ax = P_- f_1 x \oplus f_2 x$ for x in $H^2$. Summing up our previous analysis readily yields the following result.

**THEOREM 4.1.** *Assume that* $g = n/d$ *and* $np + dq = 1$, *where* n, d, p *and* q *are functions in* $H^\infty$. *Let* $[\theta_1, \theta_2]^{tr}\theta_o$ *be the inner-outer factorization for* $[wnd, -w_o nd]^{tr}$, *where* w *and* $w_o$

*are weight functions in* $H^\infty$ *and* $\theta_o$ *is an invertible outer function. Let a in* $L^\infty$ *and* $f_1$, $f_2$ *be the functions in* $K^\infty$ *defined by*

$$a = (\bar{\theta}_1 wdq + \bar{\theta}_2 w_o np),\ f_1 = P_- a \ \ \textit{and} \ \ f_2(e^{it}) = e^{-it}[(\theta_2 wdq)(e^{it}) - (\theta_1 w_o np)(e^{it})]^* \ . \quad (4.13)$$

*Then the* $H^2$ *or* $H^\infty$ *disturbance reduction problem in (4.4) is equivalent to the corresponding* $H^2$ *or* $H^\infty$ *two block optimization problem in (4.10). Moreover, the following results hold.*

*(i)* *The optimal solution to the* $H^2$ *two block optimization problem in (4.10) is* $h_o = 0$, *the* $H^2$ *disturbance reduction error is* $d_2 = \|f_1 \oplus f_2\|_2$. *If* $(\theta_o q)(0) \neq (nP_+a)(0)$, *then the* $H^2$ *optimal disturbance reducing compensator* c *is given by (4.12).*

*(ii)* *The* $H^\infty$ *disturbance reduction error is* $d_\infty = \|A\|$, *where* A *is the operator from* $H^2$ *into* $K^2 \oplus L^2$ *defined by* $Ax = P_- f_1 x \oplus f_2 x$ *for* x *in* $H^2$. *If* $h_o$ *is the optimal solution to the* $H^\infty$ *two block optimization in (4.10) and* $(\theta_o q - nP_+a)(0) \neq (nh_o)(0)$, *then the optimal disturbance reducing controller* c *is given by (4.11).*

One can use the state space formulas in Section V.7 to compute the optimal controllers in Theorem 4.1 in the rational case. Moreover, one can also use Theorem V.7.2, to obtain some mixed $H^2$ and $H^\infty$ controllers corresponding to the optimization problems in (4.4) and (4.10). Computer simulations for this two block problem generate graphs similar to those in Figures 3 and 4. Finally, it is noted that associated with the block diagram in Figure 2, there are several different transfer functions, $(1 + gc)^{-1}$, $c(1 + gc)^{-1}$, etc. Associated with these transfer functions one can set up many different $H^2$, $H^\infty$ or mixed optimization problems. Here we have just demonstrated how some of these problems lead to a Sarason or two block problem. It turns out that most of these problems can be transformed to the two-sided Sarason problem discussed in Section I.10. For an in depth discussion of $H^\infty$ control theory see Francis [1], Foias-Osbay-Tannenbaum [1], Green-Limebeer [1], and Zhou-Doyle-Glover [1].

Recall that in the tracking problem if g is rational and c is taken to be the optimal $H^\infty$ stabilizing controller, then the weighted tracking error transfer function satisfies $|w(1 + gc)^{-1}| = |wd(q - nh)| = d_\infty$ a.e. This phenomenon also occurs in the two block disturbance reduction problem. To see this we begin with the following result.

**PROPOSITION 4.2.** *Let* $f_1$ *and* $f_2$ *be two rational functions in* $L^\infty$, *and assume that* $\|f_2\|_\infty < d_\infty$. *Then there exists a unique* $H^\infty$ *optimal solution* $h_o$ *to the two block optimization problem in (4.10). Moreover, in this case* $|f_1 - h_o|^2 + |f_2|^2 = d_\infty^2$, *a.e.*

**PROOF.** As before, let A be the operator from $H^2$ into $K^2 \oplus L^2$ defined by $Ah = P_- f_1 h \oplus f_2 h$. Now let $\psi$ be the outer function in $H^\infty$ defined by $|\psi|^2 = d_\infty^2 - |f_2|^2$ a.e. Clearly $\psi$ is an invertible rational outer function. Let $\Gamma$ be the Hankel operator from $H^2$ into $K^2$ defined by $\Gamma h = P_- f_1 \psi^{-1} h$. Because $f_1 \psi^{-1}$ is rational, $\Gamma$ has finite rank. Recall that $D_A$ is the positive square root of $\gamma^2 I - A^* A$. For h in $H^2$ and $\gamma = d_\infty$ we have

$$0 \leq \|D_A h\|^2 = d_\infty^2 \|h\|^2 - \|P_- f_1 h\|^2 - \|f_2 h\|^2 = \|\psi h\|^2 - \|P_- f_1 h\|^2 = \|\psi h\|^2 - \|\Gamma \psi h\|^2 . \quad (4.14)$$

Since $\psi$ is an invertible outer function, $\Gamma$ is a contraction. In particular, the previous formula shows that $\|D_A h\| = \|D_{\Gamma,1} \psi h\|$, where $D_{\Gamma,1}$ is the positive square root of $I - \Gamma^* \Gamma$. If $h_n$ is a sequence of unit vectors such that $\|Ah_n\|$ approaches $\|A\|$ as n approaches infinity, then $\|D_{\Gamma,1} \psi h_n\|$ approaches zero as n approaches infinity. Since $\psi$ is an invertible outer function, this readily implies that 0 is in the spectrum of $D_{\Gamma,1}$. Because $f_1$ is rational, $\Gamma$ has finite rank, and thus ker $D_{\Gamma,1}$ is not zero. Using $\|D_A h\| = \|D_{\Gamma,1} \psi h\|$ we see that ker $D_A$ is also nonzero. Hence, A attains its norm. According to Corollary IV.2.8, the operator A admits a unique intertwining lifting B of A satisfying $\|B\| = \|A\|$. Therefore there exists a unique solution $h_o$ in $H^\infty$ to the $H^\infty$ optimization problem in (4.10). By consulting Corollary IV.2.8 once again we also see that $|f_1 - h_o|^2 + |f_2|^2 = d_\infty$ a.e. This completes the proof.

Notice that if there exists a unique solution to the $H^\infty$ two block optimization problem in (4.10), then it does not necessarily follow that $\|f_2\| < d_\infty$. For example, if $f_1 = 0$ and $f_2 = 1$, then $d_\infty = 1$ and $h_o = 0$ is the only $H^\infty$ optimal solutions to (4.10). On the other hand, if $\|f_2\|_\infty = d_\infty$, then there can be many $H^\infty$ optimal solution to (4.10). For instance, let $f_1 = 0$ and $f_2$ be any rational function in $L^\infty$ such that $\|f_2\|_\infty = 1$ and $|f_2| \neq 1$ a.e. Then $d_\infty = 1$. Moreover, if $h_o = \phi\psi$ where $\phi$ is any function in $H^\infty$ satisfying $\|\phi\|_\infty \leq 1$ and as before $\psi$ is the outer factor for $d_\infty^2 - |f_2|^2$, then $h_o$ is a $H^\infty$ optimal solution for the two block optimization problem in (4.10).

Now assume that g is a rational function and $g = n/d$ is a coprime factorization of g where n, d, p and q are rational functions satisfying the Bezout identity $np + dq = 1$. Since n, p, d and q are rational, $f_1$ and $f_2$ are rational functions. If $\|f_2\| < d_\infty$, which occurs quite often, then the previous lemma shows that there exists a unique $H^\infty$ optimal solution $h_o$ in $H^\infty$ to (4.10). Since the conversion from the Sarason optimization problem (4.4) to the two block optimization problem (4.10) preserves the norm, it follows that $h = (h_o + P_+ a)/\theta_o$ is the unique $H^\infty$ solution to the disturbance reduction problem in (4.4). Moreover, because $|f_1 - h_o|^2 + |f_2|^2 = d_\infty^2$ a.e. the disturbance reduction error satisfies

$$|w(1+gc)^{-1}|^2 + |w_o gc(1+gc)^{-1}|^2 = |wd(q-nh)|^2 + |w_o n(p+dh)|^2 = d_\infty^2, \text{ a.e.} \quad (4.15)$$

Finally, it is noted that unlike the tracking problem, the $H^\infty$ optimal solution to the disturbance reduction problem in (4.4) do not generically produce controllers $c \notin H^\infty$.

## VII.5 THE MULTIVARIABLE CASE

In this section some of the previous control problems are generalized to the multidimensional setting. First we present the Youla parameterization for all stabilizing controllers for transfer functions G with values in $\mathcal{L}(\mathcal{U}, \mathcal{Y})$. Then we use this result to show how the model matching problem discussed in Section I.10 naturally occurs in control theory.

In the multidimensional setting G is a *transfer function*, if G is analytic in some neighborhood of the origin and G has values in $\mathcal{L}(\mathcal{U}, \mathcal{Y})$. Consider the feedback control system given in Figure 6 where the transfer function G has values in $\mathcal{L}(\mathcal{U}, \mathcal{Y})$. Here we are looking for a compensator C is which is analytic in some neighborhood of the origin, that is, a transfer function, with values in $\mathcal{L}(\mathcal{Y}, \mathcal{U})$ to internally stabilize (see below for the definition) this feedback control system and at the same time satisfy some performance criteria. In this case the input $(r_{1,n})_0^\infty$ and $(r_{2,n})_0^\infty$ are sequences with values in $\mathcal{Y}$ and $\mathcal{U}$, respectively, while the output sequence $(y_n)_0^\infty$ has values in $\mathcal{Y}$. The error sequence $(e_n)_0^\infty$ has values in $\mathcal{Y}$, and both sequences $(v_n)_0^\infty$ and $(u_n)_0^\infty$ have values in $\mathcal{U}$. Finally, it is noted that the output sequence $(y_n)_0^\infty = T_G(u_n)_0^\infty$ and $(v_n)_0^\infty = T_C(e_n)_0^\infty$, where $T_G$ and $T_C$ are the block Toeplitz matrices generated by the Taylor coefficients of $\lambda^n$ computed from G and C, respectively.

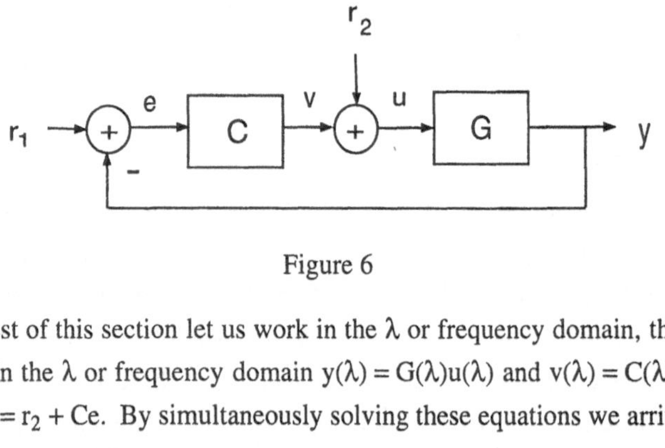


Figure 6

Throughout the rest of this section let us work in the $\lambda$ or frequency domain, that is, $u = \Sigma u_n \lambda^n$, $y = \Sigma y_n \lambda^n$, etc. In the $\lambda$ or frequency domain $y(\lambda) = G(\lambda)u(\lambda)$ and $v(\lambda) = C(\lambda)e(\lambda)$. Moreover, $e = r_1 - Gu$ and $u = r_2 + Ce$. By simultaneously solving these equations we arrive at

$$\begin{bmatrix} e \\ u \end{bmatrix} = \begin{bmatrix} (I+GC)^{-1} & -G(I+CG)^{-1} \\ C(I+GC)^{-1} & (I+CG)^{-1} \end{bmatrix} \begin{bmatrix} r_1 \\ r_2 \end{bmatrix}. \tag{5.1}$$

The $2 \times 2$ matrix function in (5.1) is the transfer function $\Omega$ from $r_1 \oplus r_2$ to $e \oplus u$. We say that the closed loop system in Figure 6 is *internally stable* if this transfer function $\Omega$ is in $H^\infty(\mathcal{Y} \oplus \mathcal{U}, \mathcal{Y} \oplus \mathcal{U})$. So this system is internally stable if and only if all the entries in the $2 \times 2$ matrix in (5.1) are in the appropriate $H^\infty$ spaces. Furthermore, C is a *stabilizing controller* for G if C is the analytic in some neighborhood of the origin and the closed loop system generated by C is internally stable, that is, $\Omega$ is in $H^\infty(\mathcal{Y} \oplus \mathcal{U}, \mathcal{Y} \oplus \mathcal{U})$. This readily leads to the multivariable version of the tracking problem discussed in Sections 1 and 3

$$d_j = \inf \{\|W(I+GC)^{-1}\|_j : C \text{ is a stabilizing controller for } G\} \qquad (\text{for } j = 2 \text{ or } \infty). \quad (5.2)$$

Here $d_j$ is the tracking error and W in $H^\infty(\mathcal{Y}, \mathcal{E})$ is a weighting function chosen to minimize the tracking error over critical frequency ranges and directions or components in $\mathcal{Y}$.

To eliminate some technical problems throughout the rest of this section we assume that $\mathcal{U} \oplus \mathcal{Y}$ is finite dimensional. We say that a transfer function G with values in $\mathcal{L}(\mathcal{U}, \mathcal{Y})$ admits a *coprime factorization* if $G = ND^{-1} = D_0^{-1}N_0$, where N, D, $D_0$ and $N_0$ are functions in the appropriate $H^\infty$ spaces, and there exists $H^\infty$ functions P, Q, $Q_0$ and $P_0$ satisfying

$$\begin{bmatrix} P & Q \\ D_0 & -N_0 \end{bmatrix} \begin{bmatrix} N & Q_0 \\ D & -P_0 \end{bmatrix} = \begin{bmatrix} I_{\mathcal{U}} & 0 \\ 0 & I_{\mathcal{Y}} \end{bmatrix}. \qquad (5.3)$$

The first $2 \times 2$ matrix in (5.3) is a function in $H^\infty(\mathcal{Y} \oplus \mathcal{U}, \mathcal{U} \oplus \mathcal{Y})$ while the second $2 \times 2$ matrix in (5.3) is a function in $H^\infty(\mathcal{U} \oplus \mathcal{Y}, \mathcal{Y} \oplus \mathcal{U})$. (Recall that $I_{\mathcal{H}}$ is the identity on $\mathcal{H}$.) Here (5.3) holds for all $|\lambda| < 1$ and $G = ND^{-1} = D_0^{-1}N_0$ holds for all $\lambda$ where both $D(\lambda)$ and $D_0(\lambda)$ are invertible. Because $G(\lambda)$ is analytic in some neighborhood of the origin both $D(0)$ and $D_0(0)$ are invertible. To see this assume that $D(0)x = 0$ for some nonzero x in $\mathcal{U}$. Then the Bezout identity $PN + QD = I$, implies that $P(0)N(0)x = x$. However, this contradicts the fact that at $\lambda = 0$ we have $0 = G(0)D(0)x = N(0)x$ is nonzero. Hence $D(0)$ is invertible. A similar argument shows that $D_0(0)$ is also invertible. In particular, both $D(\lambda)$ and $D_0(\lambda)$ are invertible in some neighborhood of the origin. Clearly det $[D(0)]$ and det $[D_0(0)]$ are both nonzero. (The determinant of an operator A on $\mathcal{H}$ with respect to any basis is denoted by det [A].) Recall that a nonzero function f in $H^\infty$ is a.e. nonzero on the unit circle; see Hoffman [1]. Hence det $[D(\lambda)]$ and det $[D_0(\lambda)]$ are both nonzero a.e. on the unit circle. Thus both D and $D_0$ are a.e. invertible on the unit circle. Therefore $G = ND^{-1} = D_0^{-1}N_0$ is Lebesgue measurable on the unit circle. If $G(\lambda)$ is rational, then $G(\lambda)$ admits a coprime factorization. In this case one can use state space techniques or polynomial matrix methods to compute a coprime factorization $ND^{-1} = D_0^{-1}N_0$ of G and the $H^\infty$ functions P, Q, $D_0$ and $P_0$ satisfying (5.3); see Francis [1], Green-Limebeer [1], and Zhou-Doyle-Glover [1] for further details. This sets the stage for the following Youla parameterization of all multidimensional stabilizing controllers.

**THEOREM 5.1.** *Assume that the transfer function* G *with values in* $\mathcal{L}(\mathcal{U}, \mathcal{Y})$ *admits a coprime factorization of the form* $G = ND^{-1} = D_0^{-1}N_0$ *where* N, D, $D_0$, $N_0$, P, Q, $Q_0$ *and* $P_0$ *are all* $H^\infty$ *functions on the appropriate spaces satisfying (5.3) and* $\mathcal{U} \oplus \mathcal{Y}$ *is finite dimensional. Then the set of all coprime stabilizing compensators* C *for* G *is given by*

$$C = (P_o + DH)(Q_o - NH)^{-1} = (Q - HN_o)^{-1}(P + HD_o), \tag{5.4}$$

*where* H *is in* $H^\infty(\mathcal{Y}, \mathcal{U})$ *and both* $Q_o - NH$ *and* $Q - HN_o$ *are invertible at* $\lambda = 0$.

**PROOF**. Assume that C admits a factorization of the form (5.4). Then using $G = D_0^{-1}N_o$ along with (5.3) we have

$$(I + GC)^{-1} = [I + D_0^{-1}N_o(P_o + DH)(Q_o - NH)^{-1}]^{-1} =$$
$$[D_0^{-1}[D_0(Q_o - NH) + N_oP_o + N_oDH](Q_o - NH)^{-1}]^{-1} = Q_oD_o - NHD_o. \tag{5.5}$$

Notice that $N_oD - D_0N = 0$ because $G = ND^{-1} = D_0^{-1}N_o$. Equation (5.5) and (5.4) also show that $C(I + GC)^{-1} = (P_o + DH)D_0$. A similar calculation yields

$$\Omega(\lambda) = \begin{bmatrix} Q_oD_o - NHD_o & NHN_o - NQ \\ P_oD_o + DHD_o & DQ - DHN_o \end{bmatrix}, \tag{5.6}$$

where $\Omega(\lambda)$ is the $2 \times 2$ matrix in (5.1). Therefore $\Omega(\lambda)$ is in $H^\infty(\mathcal{Y} \oplus \mathcal{U}, \mathcal{Y} \oplus \mathcal{U})$ and the corresponding feedback system is internally stable. Now let

$$X = P_o + DH, \; Y = Q_o - NH, \; Y_o = Q - HN_o \text{ and } X_o = P + HD_o. \tag{5.7}$$

Since (5.4) holds, $C = XY^{-1} = Y_o^{-1}X_o$. Using $Y_oX = X_oY$ along with (5.3) we arrive at

$$\begin{bmatrix} -X_o & Y_o \\ D_0 & N_o \end{bmatrix}\begin{bmatrix} -N & Y \\ D & X \end{bmatrix} = \begin{bmatrix} I & 0 \\ 0 & I \end{bmatrix}. \tag{5.8}$$

By rearranging terms we see that (5.8) corresponds to (5.3). Hence C admits a coprime factorization. The hypothesis guarantees that both $Y(0)$ and $Y_o(0)$ are invertible and thus $C(\lambda)$ is analytic in some neighborhood of the origin. Therefore C is a stabilizing controller.

Now assume that C is a stabilizing controller and C admits a coprime factorization of the form $C = XY^{-1} = Y_o^{-1}X_o$ where X, Y, $Y_o$ and $X_o$ are functions in the appropriate $H^\infty$ spaces. Using $G = ND^{-1} = D_0^{-1}N_o$ along with the following identities $C(I + GC)^{-1} = (I + CG)^{-1}C$ and $G(I + CG)^{-1} = (I + GC)^{-1}G$ we have

$$
\begin{aligned}
&(I+GC)^{-1} = [I + D_0^{-1}N_0XY^{-1}]^{-1} = Y(D_0Y+N_0X)^{-1}D_0 \\
&(I+CG)^{-1} = [I + Y_0^{-1}X_0ND^{-1}]^{-1} = D(Y_0D+X_0N)^{-1}Y_0 \\
&C(I+GC)^{-1} = X(D_0Y+N_0X)^{-1}D_0 = D(Y_0D+X_0N)^{-1}X_0 \\
&G(I+CG)^{-1} = N(Y_0D+X_0N)^{-1}Y_0 = Y(D_0Y+N_0X)^{-1}N_0 \\
&I-(I+GC)^{-1} = G(I+CG)^{-1}C = N(Y_0D+X_0N)^{-1}X_0 \\
&I-(I+CG)^{-1} = C(I+GC)^{-1}G = X(D_0Y+N_0X)^{-1}N_0 \,.
\end{aligned}
\tag{5.9}
$$

Because C is a stabilizing controller, all the functions in (5.9) are in the appropriate $H^\infty$-spaces. Now let A, B, $B_0$ and $A_0$ be functions in the appropriate $H^\infty$ spaces satisfying the coprime condition

$$
\begin{bmatrix} -X_0 & Y_0 \\ B & A \end{bmatrix}\begin{bmatrix} -A_0 & Y \\ B_0 & X \end{bmatrix} = \begin{bmatrix} I_{\mathcal{U}} & 0 \\ 0 & I_{\mathcal{Y}} \end{bmatrix} . \tag{5.10}
$$

By virtue of (5.3) and (5.10) we have

$$
[P,\, Q]\begin{bmatrix} N(Y_0D+X_0N)^{-1}Y_0 & N(Y_0D+X_0N)^{-1}X_0 \\ D(Y_0D+X_0N)^{-1}Y_0 & D(Y_0D+X_0N)^{-1}X_0 \end{bmatrix}\begin{bmatrix} B_0 \\ A_0 \end{bmatrix} = (Y_0D+X_0N)^{-1} . \tag{5.11}
$$

Notice that all the entries in the previous $2\times 2$ matrix appear in (5.9). Since C is a stabilizing controller, all the entries in (5.11) are in the appropriate $H^\infty$ space. Therefore $(Y_0D+X_0N)^{-1}$ is also in $H^\infty(\mathcal{U}, \mathcal{U})$. In particular, $\Phi_0 = Y_0D+X_0N$ is an invertible outer function, that is, both $\Phi_0$ and $\Phi_0^{-1}$ are in $H^\infty(\mathcal{U}, \mathcal{U})$.

By using the four terms involving $(D_0Y+N_0X)^{-1}$ in (5.9) and performing a calculation similar to (5.11) with the A, B terms on the left and $P_0$, $Q_0$ on the right of the corresponding $2\times 2$ matrix, we see that $(D_0Y+N_0X)^{-1}$ is in $H^\infty(\mathcal{Y}, \mathcal{Y})$. Hence $\Phi = D_0Y+N_0X$ is an invertible outer function in $H^\infty(\mathcal{Y}, \mathcal{Y})$. Thus

$$
D_0Y\Phi^{-1} + N_0X\Phi^{-1} = I \text{ and } \Phi_0^{-1}X_0N + \Phi_0^{-1}Y_0D = I , \tag{5.12}
$$

where $Y\Phi^{-1}$, $X\Phi^{-1}$, $\Phi_0^{-1}X_0$ and $\Phi_0^{-1}Y_0$ are all functions in the appropriate $H^\infty$ spaces. Now let $\mathcal{N}_0$ and $\mathcal{N}$ be subspaces of the appropriate $H^\infty$ spaces defined by

$$\mathcal{N}_o = \{[H_1, H_2]^{tr} \in H^\infty(\mathcal{Y}, \mathcal{Y} \oplus \mathcal{U}) : D_o H_1 + N_o H_2 = I\}$$
$$\mathcal{N} = \{[H_3, H_4] \in H^\infty(\mathcal{Y} \oplus \mathcal{U}, \mathcal{U}) : H_3 N + H_4 D = I\}\,. \tag{5.13}$$

Since $\mathcal{U} \oplus \mathcal{Y}$ is finite dimensional, equation (5.3) shows that the first two matrices in (5.3) are both invertible outer functions. In particular,

$$\begin{bmatrix} P(\lambda) & Q(\lambda) \\ D_o(\lambda) & -N_o(\lambda) \end{bmatrix}^{-1} = \begin{bmatrix} N(\lambda) & Q_o(\lambda) \\ D(\lambda) & -P_o(\lambda) \end{bmatrix}. \tag{5.14}$$

From this we can show that

$$\mathcal{N}_o = \mathcal{N}_1 := \begin{bmatrix} Q_o \\ P_o \end{bmatrix} + \begin{bmatrix} -N \\ D \end{bmatrix} H^\infty(\mathcal{Y}, \mathcal{U})\,. \tag{5.15}$$

Clearly if $[H_1, H_2]^{tr}$ is in $\mathcal{N}_1$, then (5.3) shows that $D_o H_1 + N_o H_2 = I$ and thus $\mathcal{N}_1 \subseteq \mathcal{N}_o$. On the other hand, if $D_o H_1 + N_o H_2 = I$, then

$$\begin{bmatrix} P & Q \\ D_o & -N_o \end{bmatrix} \begin{bmatrix} H_1 \\ -H_2 \end{bmatrix} = \begin{bmatrix} -H \\ I \end{bmatrix}, \tag{5.16}$$

where H is in $H^\infty(\mathcal{Y}, \mathcal{U})$. By employing the inverse in (5.14) we see that $[H_1, H_2]^{tr}$ is in $\mathcal{N}_1$. Hence $\mathcal{N}_o \subseteq \mathcal{N}_1$ and thus $\mathcal{N}_o = \mathcal{N}_1$. A similar argument shows that

$$\mathcal{N} = [P, Q] + H^\infty(\mathcal{Y}, \mathcal{U})[D_o, -N_o]\,. \tag{5.17}$$

Equation (5.12) shows that $[Y\Phi^{-1}, X\Phi^{-1}]^{tr}$ is in $\mathcal{N}_o$ and $[\Phi_o^{-1}X_o, \Phi_o^{-1}Y_o]$ is in $\mathcal{N}$. Because both D and $D_o$ are a.e. invertible on the unit circle, there exists unique functions $H_o$ and H in $H^\infty(\mathcal{Y}, \mathcal{U})$ satisfying

$$X\Phi^{-1} = P_o + DH_o,\; Y\Phi^{-1} = Q_o - NH_o,\; \Phi_o^{-1}X_o = P + HD_o \text{ and } \Phi_o^{-1}Y_o = Q - HN_o\,. \tag{5.18}$$

Since $C = XY^{-1} = Y_o^{-1}X_o$, we obtain

$$C = (P_o + DH_o)(Q_o - NH_o)^{-1} = (Q - HN_o)^{-1}(P + HD_o)\,. \tag{5.19}$$

Using $(Q - HN_o)(P_o + DH_o) = (P + HD_o)(Q_o - NH_o)$ along with (5.3) and $D_o N = N_o D$ we see that $H = H_o$. Therefore C is given by the formula in (5.4). Finally, because C is a stabilizing controller, C is analytic in some neighborhood of the origin. Since $C = XY^{-1} = Y_o^{-1}X_o$ is a coprime factorization of C, this implies that both Y(0) and $Y_o(0)$ are invertible. Equation (5.18) with $H = H_o$ shows that both $Q_o - NH$ and $Q - HN_o$ are invertible at $\lambda = 0$. This completes the proof.

Now assume that the transfer function G admits a coprime factorization $G = ND^{-1} = D_0^{-1}N_o$, where N, D, $D_0$ and $N_o$ are functions in the appropriate $H^\infty$ spaces. As before assume that P, Q, $Q_o$ and $P_o$ are functions in the appropriate $H^\infty$ spaces satisfying (5.3). Then the set of all stabilizing coprime controllers is given by C in (5.4), where H is in $H^\infty(\mathcal{Y}, \mathcal{U})$ and $Q_o - NH$ and $Q - HN_o$ are invertible at $\lambda = 0$. According to (5.1) and (5.5) the tracking error transfer function from $r_1$ to e is given by $(I + GC)^{-1} = Q_oD_o - NHD_o$. Therefore the multivariable tracking problem for coprime compensators in (5.2) now becomes

$$d_j = \inf \{\|W(I + GC)^{-1}\|_j : C \text{ is a coprime stabilizing controller for } G\} =$$
$$= \inf \|WQ_oD_o - WNHD_o\|_j : H \in H^\infty(\mathcal{Y}, \mathcal{U})\}, \tag{5.20}$$

where $j = 2$ or $\infty$. Here W is a weighting function in $H^\infty(\mathcal{Y}, \mathcal{E})$ chosen by the engineer to minimize the tracking error over certain critical frequency ranges and components or directions in the space $\mathcal{Y}$. As before W can be chosen to be the identity. Clearly the $H^2$ and $H^\infty$ optimization problem in (5.20) is a generalization of the scalar valued $H^2$ and $H^\infty$ optimization problems discussed in Section 3. The $H^\infty$ and $H^2$ optimization problem in (5.20) is a special case of the model matching problem discussed in Section I.10. In particular, if $\Theta_2$ is the inner part of WN and $\Theta_1$ is the co-inner part of $D_0$, then $d_\infty = \|A\|$ where A is the operator from $\mathcal{H}_1$ into $\mathcal{H}_2$ defined by $A = P_{\mathcal{H}_2}M_F \mid \mathcal{H}_1$ in Corollary I.10.2 with $F = WQ_oD_o$. (Here $WN = \Theta_2\Theta_{2o}$ where $\Theta_2$ is inner, $\Theta_{2o}$ is outer and $D_0 = \Theta_{1*}\Theta_1$ where $\Theta_{1*}$ is co-outer and $\Theta_1$ is co-inner.) So if H is a solution to the optimization problem in (5.20), i.e., $d_j = \|WQ_oD_o - WNHD_o\|_j$, then the corresponding optimal compensator is computed according to (5.4). In particular, one can use the results in Section IV.9 to compute the optimal $H^2$ controller. The central solution presented in Section IV.9 will also give a stabilizing controller which satisfies the corresponding mixed $H^\infty$ and $H^2$ bounds with the tolerance $\gamma = \delta d_\infty$ for $\delta > 1$. Finally, it is noted that especially in the multidimensional case, a stabilizing controller obtained from the Youla parameterization can be sensitive to numerical errors. One can construct simple examples where a slight perturbation in a stabilizing controller will produce an unstable closed loop system.

Now consider the multivariable disturbance reduction problem corresponding to Figure 5 where c is replaced by C in $L^\infty(\mathcal{Y}, \mathcal{U})$ and g is replaced by G in $L^\infty(\mathcal{U}, \mathcal{Y})$. As before, we assume that G admits a coprime factorization $G = ND^{-1} = D_0^{-1}N_o$ where N, D, $D_0$, $N_o$, P, Q, $Q_o$ and $P_o$ are all function in the appropriate $H^\infty$ spaces satisfying (5.3). The transfer function from the disturbance $r_3$ to $e \oplus y$ is given by $-[(I + GC)^{-1}, GC(I + GC)^{-1}]$. So if C is a coprime stabilizing controller for G, then C is given by Youla parameterization in (5.4) for some H in $H^\infty(\mathcal{Y}, \mathcal{U})$. So according to (5.4) we have

$$GC(I+GC)^{-1} = G(I+CG)^{-1}C = GD(Q-HN_o)C = N(P+HD_o)\,.$$

The second equality follows from the fact that $(I+CG)^{-1}$ is given by the $\Omega_{2,2}$ entry of $\Omega$ in (5.6). (Recall that $\Omega$ is precisely the $2\times 2$ matrix in (5.1).) This along with (5.4) readily yield the following multivariable disturbance reduction control problem

$$d_j = \inf\left\{ \left\| \begin{bmatrix} W(I+GC)^{-1} \\ W_o GC(I+GC)^{-1} \end{bmatrix} \right\|_j : \text{C is a coprime stabilizing controller for G} \right\} = \tag{5.21}$$

$$\inf\left\{ \left\| \begin{bmatrix} WQ_oD_o \\ W_oNP \end{bmatrix} - \begin{bmatrix} WN \\ -W_oN \end{bmatrix} HD_o \right\|_j : H \in H^\infty(\mathcal{Y}, \mathcal{U}) \right\},$$

where $j = 2$ or $\infty$. Here W in $H^\infty(\mathcal{Y}, \mathcal{E})$ and $W_o$ in $H^\infty(\mathcal{Y}, \mathcal{E}_1)$ are weighting function chosen by the designer to minimize the effect of the disturbance $r_3$ on the error e and the output y over certain critical frequency ranges and components in $\mathcal{Y}$. As before $W_o$ or W can be zero or the identity. If $W_o = 0$, then the disturbance reduction problem is precisely the tracking problem. Clearly the disturbance reduction problems in (5.21) is a special case of the model matching problem discussed in Section I.10. In particular, if $\Theta_2$ is the inner part of $[WN, -W_oN]^{tr}$ and $\Theta_1$ is the co-inner part of $D_0$, then $d_\infty = \|A\|$, where A is the operator from $\mathcal{H}_1$ into $\mathcal{H}_2$ defined by $A = P_{\mathcal{H}_2} M_F \mid \mathcal{H}_1$ in Corollary I.10.2 with $F = [WQ_oD_o, W_oNP]^{tr}$. The optimal $H^2$ error $d_2$ is given by equation (IV.9.6). Moreover one can use the results in Section IV.9 to obtain an H in $H^\infty(\mathcal{Y}, \mathcal{U})$ corresponding to the central solution for the commutant lifting theorem with tolerance $\gamma = \delta d_\infty$ and $\delta > 1$, and this H will satisfy the mixed $H^2$ and $H^\infty$ bounds in Theorem IV.9.1. So if H is a solution to any one of these optimization problems in (5.21) or H is the central solution with tolerance $\delta d_\infty$, and $Q_o - NH$ and $Q - HN_o$ are both invertible at $\lambda = 0$, then the corresponding stabilizing controller C is given by (5.4). Finally, it is noted that one can state many other $H^\infty$ control problems associated with the block diagram in Figures 2, 5, or 6 including filtering and robust control problems. However, almost all of these problems can be converted to the model matching problem discussed in Section I.10. For a more in depth account of $H^\infty$ control theory see Francis [1], Foias-Osbay-Tannenbaum [1], Green-Limebeer [1] and Zhou-Doyle-Glover [1].

## *Notes to Chapter VII:*

This chapter belongs to a well-established branch of control theory which is referred to as $H^\infty$ control and was initiated by Zames [1], [2], Zames-Francis [1] and Francis-Helton-Zames [1] at the beginning of the eighties. An early introduction to the subject is Francis [1]. The present chapter contains by now standard material which in different forms can be found in, for example, the books Basar-Bernhard [1], Chui-Chen [1], Doyle-Francis-Tannenbaum [1], Foias-Osbay-Tannenbaum [1], Green-Limebeer [1], Zhou-Doyle-Glover [1], and Part VI of Ball-Gohberg-Rodman [1].

# PART B

## Nonstationary Interpolation and Time-Varying Systems

This part treats norm constrained interpolation problems for doubly infinite lower triangular operator matrices. The problems considered are the nonstationary analogues of the $H^\infty$ interpolation problems studied in Part A. The solutions are obtained in two ways. First, by converting the nonstationary interpolation problems to an infinite dimensional time-invariant setting. The second way is based on a new completion result, called the three chains completion theorem, which may be viewed as a time-variant version of the commutant lifting theorem. The three chains completion theorem is analyzed from different points of view. As in Part A, connections with systems provide motivation and insight in the explicit state space formulas for the solutions.

## CHAPTER VIII

# NONSTATIONARY INTERPOLATION THEOREMS

The interpolation problems introduced in this chapter are nonstationary versions of the ones appearing in Chapter I. In the present chapter the interpolants are not $H^\infty$-functions but lower triangular doubly infinite matrices (with operator entries) and the interpolation data depend on an additional quantity which may be viewed as a discrete time parameter. In the next chapter time-varying systems will be used to give a motivation and a further interpretation of the role of this additional time parameter.

This chapter concentrates on the necessary and sufficient conditions for the existence of solutions for various nonstationary interpolation problems. Included are the nonstationary versions of the interpolation problems of Nevanlinna-Pick, Hermit-Fejér, Nehari, Sarason and Nudelman. Proofs of the existence theorems will be given in Chapter XI using the technique of reduction from nonstationary to stationary presented in Chapter X.

## VIII.I. NONSTATIONARY NEVANLINNA-PICK INTERPOLATION

For the nonstationary left Nevanlinna-Pick interpolation problem the given data are bounded linear operators

$$Z_{j,k} : \mathcal{X}_{k+1} \to \mathcal{X}_k \quad \text{and} \quad \tilde{B}_{j,k} : \mathcal{Y}_k \to \mathcal{X}_k \quad (k \in \mathbb{Z}, j = 1, \dots, N) \tag{1.1}$$

acting between Hilbert spaces. For convenience let us introduce the *(state) transition operator* $\Phi_j$ for the operators $\{Z_{j,k}\}$ by

$$\Phi_j(k, m) = Z_{j,k} Z_{j,k+1} \cdots Z_{j,m-1} : \mathcal{X}_m \to \mathcal{X}_k \qquad (\text{when } k < m). \tag{1.2}$$

Notice that $\Phi_j(k, m)\Phi_j(m, n) = \Phi_j(k, n)$, and by definition $\Phi_j(k, k) = I$, where I is the identity operator on $\mathcal{X}_k$. Throughout we always assume that the data (1.1) satisfy the following conditions

$$\begin{aligned} &\sup_{k \in \mathbb{Z}} \{\|Z_{j,k}\|, \|\tilde{B}_{j,k}\|\} < \infty \qquad (j = 1, \dots, N), \\ &\limsup_{\nu \to \infty} \left[\sup_{k \in \mathbb{Z}} \|\Phi_j(k, k+\nu)\|\right]^{1/\nu} < 1 \qquad (j = 1, \dots, N). \end{aligned} \tag{1.3}$$

The last condition in (1.3) is a nonstationary version of the spectral radius condition placed on the data for the classical Nevanlinna-Pick problem in Chapter I. To see this simply choose $Z_{j,k} = Z_j$ for all k, then the last condition is (1.3) is equivalent to $r_{spec}(Z_j) < 1$ for all j. For later purposes, recall that $\vec{\mathcal{X}}$ and $\vec{\mathcal{Y}}$ are the Hilbert spaces defined by

$$\vec{\mathcal{X}} = \bigoplus_{k=-\infty}^{\infty} \mathcal{X}_k \quad \text{and} \quad \vec{\mathcal{Y}} = \bigoplus_{k=-\infty}^{\infty} \mathcal{Y}_k . \tag{1.4}$$

If $\mathcal{X}_k = \mathcal{X}$ for all k, then $\vec{\mathcal{X}} = \ell^2(\mathcal{X})$. Also, recall that $\ell^2_N(\mathcal{X}_k)$ stands for the Hilbert space direct sum of N copies of the space $\mathcal{X}_k$.

Given data as above the *nonstationary (left) Nevanlinna-Pick interpolation* problem with tolerance $\gamma$ is to find operators $f_{j,k}$ in $\mathcal{L}(\mathcal{Y}_k, \mathcal{X}_j)$ for $-\infty < k \le j < \infty$ satisfying the interpolation conditions

$$\sum_{\nu=0}^{\infty} \Phi_j(k, k+\nu) f_{k+\nu,k} = \tilde{B}_{j,k} \qquad (k \in \mathbb{Z} \text{ and } j = 1, \ldots, N), \tag{1.5}$$

and such that the block lower triangular operator

$$F = \begin{bmatrix} \ddots & & & & \\ \ddots & f_{-1,-1} & & & \\ \ddots & f_{0,-1} & \underline{f_{0,0}} & & \\ \ddots & f_{1,-1} & f_{1,0} & f_{1,1} & \\ \ddots & \ddots & \ddots & \ddots & \ddots \end{bmatrix} : \vec{\mathcal{Y}} \to \vec{\mathcal{X}} \tag{1.6}$$

is bounded in operator norm by $\gamma$, that is, $\|F\| \le \gamma$. (As before the blank entries in (1.6) are zero and the underline on $f_{0,0}$ indicates the (0, 0) entry.) Notice that the nonstationary left Nevanlinna-Pick problem includes the classical Nevanlinna-Pick problem in Chapter I by relaxing the constraint on $Z_{j,k}$ to $Z_{j,k} = Z_j$ for all k and by imposing the additional constraint that F is a lower triangular Laurent operator.

To state our main theorem we need the following operators in $\mathcal{L}(\mathcal{X}_k, \mathcal{X}_k)$:

$$\Delta_{i,j}(k) = \sum_{\nu=0}^{\infty} \Phi_i(k, k+\nu) \left[\gamma^2 I - \tilde{B}_{i,k+\nu} \tilde{B}^*_{j,k+\nu}\right] \Phi_j(k, k+\nu)^* : \mathcal{X}_k \to \mathcal{X}_k . \tag{1.7}$$

Here i, j = 1, ..., N and $k \in \mathbb{Z}$. For $\nu = 0$ the $\nu$-th term in the right hand side of (1.7) reduces to $\gamma^2 I - \tilde{B}_{i,k} \tilde{B}^*_{j,k}$.

**THEOREM 1.1.** *The nonstationary left Nevanlinna-Pick interpolation problem for the data (1.1) and with tolerance* $\gamma$ *has a solution if and only if for each* $k \in \mathbb{Z}$ *the* $N \times N$ *operator matrix* $(\Delta_{i,j}(k))_{i,j=1}^N$ *induces a positive operator* $(\ge 0)$ *on* $\ell^2_N(\mathcal{X}_k)$.

## VIII.2 NONSTATIONARY TANGENTIAL NEVANLINNA-PICK INTERPOLATION

For the nonstationary tangential Nevanlinna-Pick interpolation problem the given data are bounded linear operators

$$Z_{j,k} : \mathcal{X}_{j,k+1} \to \mathcal{X}_{j,k} , \ B_{j,k} : \mathcal{U}_k \to \mathcal{X}_{j,k} , \ \tilde{B}_{j,k} : \mathcal{Y}_k \to \mathcal{X}_{j,k} \quad (j = 1, ..., N , k \in \mathbb{Z}) , \tag{2.1}$$

which act between Hilbert spaces. These operators are assumed to satisfy the following conditions

$$\begin{aligned} &\sup_{k \in \mathbb{Z}} \{\|Z_{j,k}\| , \|B_{j,k}\| , \|\tilde{B}_{j,k}\|\} < \infty \qquad (j = 1, ..., N) , \\ &\limsup_{\nu \to \infty} \left[\sup_{k \in \mathbb{Z}} \|\Phi_j(k, k+\nu)\|\right]^{1/\nu} < 1 \qquad (j = 1, ..., N) , \end{aligned} \tag{2.2}$$

where $\Phi_j$ is the state transition matrix for $\{Z_{j,k}\}$ defined in (1.2) with $\mathcal{X}_m$ and $\mathcal{X}_k$ replaced by $\mathcal{X}_{j,m}$ and $\mathcal{X}_{j,k}$, respectively.

Given data as above the *nonstationary (left) tangential Nevanlinna-Pick interpolation* problem with tolerance $\gamma$ is to find operators $f_{j,k}$ in $\mathcal{L}(\mathcal{Y}_k, \mathcal{U}_j)$ for $-\infty < k \le j < \infty$ satisfying the interpolation conditions

$$\sum_{\nu=0}^{\infty} \Phi_j(k, k+\nu) B_{j,k+\nu} f_{k+\nu,k} = \tilde{B}_{j,k} \qquad (k \in \mathbb{Z} \ \text{ and } \ j = 1, ..., N) , \tag{2.3}$$

and such that the block lower triangular operator

$$F = \begin{bmatrix} \ddots & & & & \\ \ddots & f_{-1,-1} & & & \\ \ddots & f_{0,-1} & \underline{f_{0,0}} & & \\ \ddots & f_{1,-1} & f_{1,0} & f_{1,1} & \\ \ddots & \ddots & \ddots & \ddots & \ddots \end{bmatrix} : \vec{\mathcal{Y}} \to \vec{\mathcal{U}} \tag{2.4}$$

has operator norm $\|F\| \le \gamma$.

To formulate our main theorem we need the following operators associated with the data (2.1)

$$\Omega_{i,j}(k) = \sum_{\nu=0}^{\infty} \Phi_i(k, k+\nu)[\gamma^2 B_{i,k+\nu} B_{j,k+\nu}^* - \tilde{B}_{i,k+\nu} \tilde{B}_{j,k+\nu}^*] \Phi_j(k, k+\nu)^* : \mathcal{X}_{j,k} \to \mathcal{X}_{i,k} . \tag{2.5}$$

Here i, j = 1, ..., N and $k \in \mathbb{Z}$. For $\nu = 0$ the $\nu$-th term in the right hand side of (2.5) reduces to

$\gamma^2 B_{i,k} B_{j,k}^* - \tilde{B}_{i,k} \tilde{B}_{j,k}^*$.

**THEOREM 2.1.** *The nonstationary left tangential Nevanlinna-Pick interpolation problem for the data (2.1) and with tolerance* $\gamma$ *has a solution if and only if for each* $k \in \mathbb{Z}$ *the* $N \times N$ *operator matrix* $(\Omega_{i,j}(k))_{i,j=1}^N$ *induces a positive operator on* $\mathcal{X}_{1,k} \oplus \cdots \oplus \mathcal{X}_{N,k}$.

Theorem 2.1 yields Theorem 1.1 as a corollary. Indeed, if in (2.1) we take $\mathcal{X}_{j,k} = \mathcal{X}_k$ and $B_{j,k} = I$ for $j = 1, \ldots, N$ and $k \in Z$, then Theorem 2.1 is just Theorem 1.1.

The *standard nonstationary (left) Nevanlinna-Pick problem* is the nonstationary left Nevanlinna-Pick problem for the case when $N = 1$. To establish some convenient notation, the data for the standard nonstationary (left) Nevanlinna-Pick problem are operators of the form

$$Z_k : \mathcal{X}_{k+1} \to \mathcal{X}_k , \ B_k : \mathcal{U}_k \to \mathcal{X}_k \ \text{ and } \ \tilde{B}_k : \mathcal{Y}_k \to \mathcal{X}_k \qquad (\text{for } k \in \mathbb{Z}) \tag{2.6}$$

As before, we assume that

$$\begin{aligned} &\sup \{\|Z_k\|, \|B_k\|, \|\tilde{B}_k\| : k \in \mathbb{Z}\} < \infty , \\ &\limsup_{\nu \to \infty} \left[ \sup_{k \in \mathbb{Z}} \|\Phi(k, k+\nu)\| \right]^{1/\nu} < 1 , \end{aligned} \tag{2.7}$$

where $\Phi(m, n)$ is the state transition operator associated with the operators $\{Z_k\}$, i.e., $\Phi(n, n)$ is the identity operator on $\mathcal{X}_n$ and $\Phi(m, n) = Z_m Z_{m+1} \cdots Z_{n-1}$ for $m < n$. Notice that $\Phi(m, n)$ maps $\mathcal{X}_n$ into $\mathcal{X}_m$. The standard nonstationary left Nevanlinna-Pick problem is to find (if possible) a set of operators $f_{j,k}$ for $-\infty < k \leq j < \infty$ satisfying the interpolation conditions

$$\sum_{\nu=0}^{\infty} \Phi(k, k+\nu) B_{k+\nu} f_{k+\nu,k} = \tilde{B}_k \qquad (\text{for } k \in \mathbb{Z}) , \tag{2.8}$$

and such that the block lower triangular operator F from $\vec{\mathcal{Y}}$ to $\vec{\mathcal{U}}$ generated by $\{f_{j,k}\}$ is bounded in norm by $\gamma$.

As expected from the stationary case (see Section I.3), the nonstationary left Nevanlinna-Pick problem can be reduced to the standard nonstationary problem. To see this, let $\mathcal{X}_k = \mathcal{X}_{1,k} \oplus \mathcal{X}_{2,k} \oplus \cdots \oplus \mathcal{X}_{N,k}$. Now let $Z_k$, $B_k$ and $\tilde{B}_k$ be the operators of the form (2.6) given by

$$\begin{aligned} &Z_k = \operatorname{diag} [Z_{1,k}, Z_{2,k}, \ldots, Z_{N,k}] , \\ &B_k = [B_{1,k}, B_{2,k}, \ldots, B_{N,k}]^{tr} \ \text{ and } \ \tilde{B}_k = [\tilde{B}_{1,k}, \tilde{B}_{2,k}, \ldots, \tilde{B}_{N,k}]^{tr} . \end{aligned} \tag{2.9}$$

Then the interpolating conditions in (2.3) and (2.8) are equivalent. Therefore the nonstationary Nevanlinna-Pick problem reduces to the standard nonstationary Nevanlinna-Pick problem.

To provide some further insight into the nonstationary standard Nevanlinna-Pick problem, recall that $\vec{\mathcal{U}}_{(k,\infty)}$ and $\vec{\mathcal{Y}}_{(k,\infty)}$ are the Hilbert space direct sums defined by

$$\vec{\mathcal{U}}_{(k,\infty)} = \bigoplus_{n=k}^{\infty} \mathcal{U}_n \quad \text{and} \quad \vec{\mathcal{Y}}_{(k,\infty)} = \bigoplus_{n=k}^{\infty} \mathcal{Y}_n . \tag{2.10}$$

Let $W_k$ from $\vec{\mathcal{U}}_{(k,\infty)}$ to $\mathcal{X}_k$ and $\tilde{W}_k$ from $\vec{\mathcal{Y}}_{(k,\infty)}$ to $\mathcal{X}_k$ be the *controllability operators* defined by

$$\begin{aligned} W_k &= [B_k,\ \Phi(k, k+1)B_{k+1},\ \Phi(k, k+2)B_{k+2},\ \ldots] \quad \text{and} \\ \tilde{W}_k &= [\tilde{B}_k,\ \Phi(k, k+1)\tilde{B}_{k+1},\ \Phi(k, k+2)\tilde{B}_{k+2},\ \ldots] \end{aligned} \tag{2.11}$$

for $k \in \mathbb{Z}$. The conditions in (2.7) guarantee that $W_k$ and $\tilde{W}_k$ are bounded operators for all k. Now let $\Pi_k$ be the orthogonal projection from $\vec{\mathcal{U}}$ onto $\vec{\mathcal{U}}_{(k,\infty)}$ and F the lower triangular operator from $\vec{\mathcal{Y}}$ to $\vec{\mathcal{U}}$ given by (2.4). Then the interpolating condition in (2.8) is equivalent to

$$W_k \Pi_k F \,|\, \vec{\mathcal{Y}}_{(k,\infty)} = \tilde{W}_k \qquad (\text{for all } k \in \mathbb{Z}) . \tag{2.12}$$

Here we view $\vec{\mathcal{Y}}_{(k,\infty)}$ as a subspace of $\vec{\mathcal{Y}}$ in the usual way. So the standard nonstationary left Nevanlinna-Pick problem in equivalent to finding a lower triangular block matrix F from $\vec{\mathcal{Y}}$ to $\vec{\mathcal{U}}$ satisfying (2.12) and $\|F\| \le \gamma$. In this case we say that F (or equivalently its lower triangular entries $\{f_{j,k}\}$) is a *solution* to the standard nonstationary left Nevanlinna-Pick problem with tolerance $\gamma$.

The *controllability grammians* $P_k$ and $\tilde{P}_k$ at time k associated with the pairs $\{Z_j, B_j\}_{j\in\mathbb{Z}}$ and $\{Z_j, \tilde{B}_j\}_{j\in\mathbb{Z}}$ are given by $P_k = W_k W_k^*$ and $\tilde{P}_k = \tilde{W}_k \tilde{W}_k^*$, respectively. Notice that if F is a solution to the standard nonstationary left Nevanlinna-Pick problem, then $W_k \Pi_k F \,|\, \vec{\mathcal{Y}}_{(k,\infty)} = \tilde{W}_k$ and $\|F\| \le \gamma$ readily yields $\gamma^2 P_k \ge \tilde{P}_k$ for all k. This proves the necessary part of the following result. The sufficiency part will be obtained in Section XI.1 by reduction to the stationary case.

**THEOREM 2.2.** *Let* $Z_k$, $B_k$ *and* $\tilde{B}_k$, $k \in \mathbb{Z}$, *be the data for the standard nonstationary left Nevanlinna-Pick problem. Let* $P_k$ *and* $\tilde{P}_k$ *for* $k \in \mathbb{Z}$, *be the controllability grammians at time* k *for* $\{Z_j, B_j\}_{j\in\mathbb{Z}}$ *and* $\{Z_j, \tilde{B}_j\}_{j\in\mathbb{Z}}$, *respectively. Then there exists a solution to the nonstationary left Nevanlinna-Pick problem with tolerance* $\gamma$ *if and only if* $\gamma^2 P_k \ge \tilde{P}_k$ *for all* k *in* $\mathbb{Z}$.

## VIII.3 NONSTATIONARY TANGENTIAL HERMITE-FEJER INTERPOLATION

For the nonstationary (left) tangential Hermite-Fejér interpolation problem the given data are bounded linear operators

$$Z_{j,k} : \mathcal{X}_{j,k+1} \to \mathcal{X}_{j,k}, \quad B_{i,j,k} : \mathcal{U}_k \to \mathcal{X}_{j,k+i}, \quad \text{and} \quad \tilde{B}_{i,j,k} : \mathcal{Y}_k \to \mathcal{X}_{j,k+i}, \tag{3.1}$$

which act between Hilbert spaces. In (3.1) the index k is an arbitrary integer, $i = 0, \ldots, r_j - 1$, and $j = 1, \ldots, N$. The operators in (3.1) are assumed to be uniformly bounded, that is,

$$\sup_{k \in \mathbb{Z}} \{\|Z_{j,k}\|, \|B_{i,j,k}\|, \|\tilde{B}_{i,j,k}\|\} < \infty \qquad (i = 0, \ldots, r_{j-1} \text{ and } j = 1, \ldots, N). \tag{3.2}$$

Furthermore, we require that

$$\limsup_{\nu \to \infty} \left\{ \sup_{k \in \mathbb{Z}} \|\Phi_j(k, k+\nu)\| \right\}^{1/\nu} < 1 \qquad (j = 1, \ldots, N). \tag{3.3}$$

Here $\Phi_j$ is the state transition operator for the operators $\{Z_{j,k}\}$ (see (1.2)).

Given data as above the *nonstationary (left) tangential Hermite-Fejér interpolation problem* with tolerance $\gamma$ is to find operators $f_{j,k}$ in $\mathcal{L}(\mathcal{Y}_k, \mathcal{U}_j)$ for $-\infty < k \le j < \infty$, satisfying the interpolation conditions

$$\sum_{k=0}^{i} \sum_{\nu=0}^{\infty} \begin{bmatrix} \nu+k \\ \nu \end{bmatrix} \Phi_j(q+i, q+i+\nu) B_{i-k,j,q+\nu+k} f_{q+\nu+k,q}$$
$$= \tilde{B}_{i,j,q} \qquad (q \in \mathbb{Z}, i = 0, \ldots, r_{j-1}, j = 1, \ldots, N), \tag{3.4}$$

and such that the block lower triangular operator

$$F = \begin{bmatrix} \ddots & & & & \\ \ddots & f_{-1,-1} & & & \\ \ddots & f_{0,-1} & \underline{f_{0,0}} & & \\ \ddots & f_{1,-1} & f_{1,0} & f_{1,1} & \\ \ddots & \vdots & \vdots & \vdots & \ddots \end{bmatrix} : \vec{\mathcal{Y}} \to \vec{\mathcal{U}} \tag{3.5}$$

is bounded in norm by $\gamma$, that is, $\|F\| \le \gamma$.

To state the solution of the nonstationary Hermite-Fejér problem we need the following auxiliary operators. For $k \in \mathbb{Z}$ and $j = 1, \ldots, N$ put

$$\mathcal{E}_{j,k} = \mathcal{X}_{j,k} \oplus \mathcal{X}_{j,k+1} \oplus \cdots \oplus \mathcal{X}_{j,k+r_j-1}\ , \tag{3.6}$$

and define operators

$$M_{j,k} : \mathcal{E}_{j,k+1} \to \mathcal{E}_{j,k},\quad R_{j,k} : \mathcal{U}_k \to \mathcal{E}_{j,k},\quad \tilde{R}_{j,k} : \mathcal{Y}_k \to \mathcal{E}_{j,k} \tag{3.7}$$

by setting

$$M_{j,k} \begin{bmatrix} x_{j,k+1} \\ x_{j,k+2} \\ \vdots \\ x_{j,k+r_j} \end{bmatrix} = \begin{bmatrix} Z_{j,k} x_{j,k+1} \\ x_{j,k+1} + Z_{j,k+1} x_{j,k+2} \\ \vdots \\ x_{j,k+r_j-1} + Z_{j,k+r_j-1} x_{j,k+r_j} \end{bmatrix}, \tag{3.8}$$

and

$$R_{j,k} u_k = \begin{bmatrix} B_{0,j,k} \\ B_{1,j,k} \\ \vdots \\ B_{r_j-1,j,k} \end{bmatrix} u_k\ ,\quad \tilde{R}_{j,k} y_k = \begin{bmatrix} \tilde{B}_{0,j,k} \\ \tilde{B}_{1,j,k} \\ \vdots \\ \tilde{B}_{r_j-1,j,k} \end{bmatrix} y_k\ . \tag{3.9}$$

With the operators $\{M_{j,k}\}$ we associate the state transition operator $\Phi_{M,j}(n, m)$ from $\mathcal{E}_{j,m}$ into $\mathcal{E}_{j,n}$. In other words, $\Phi_{M,j}(n, n)$ is the identity operator on $\mathcal{E}_{j,n}$ and

$$\Phi_{M,j}(n, m) = M_{j,n} M_{j,n+1} \cdots M_{j,m-1} \quad (m > n)\ . \tag{3.10}$$

Finally, for $i, j = 1, \ldots, N$ and $k \in \mathbb{Z}$ consider the operator

$$\Xi_{i,j}(k) = \sum_{\nu=0}^{\infty} \Phi_{M,i}(k, k+\nu)[\gamma^2 R_{i,k+\nu} R_{j,k+\nu}^* - \tilde{R}_{i,k+\nu} \tilde{R}_{j,k+\nu}^*] \Phi_{M,j}(k, k+\nu)^*\ .$$

Notice that $\Xi_{i,j}(k)$ maps $\mathcal{E}_{j,k}$ into $\mathcal{E}_{i,k}$. We are now ready to state the main result of this section.

**THEOREM 3.1.** *The nonstationary left tangential Hermite-Fejeŕ interpolation problem for the data (3.1) and with tolerance $\gamma$ has a solution if and only if for each $k \in \mathbb{Z}$ the operator matrix $(\Xi_{i,j}(k))_{i,j=1}^{N}$ induces a positive operators on $\mathcal{E}_{1,k} \oplus \cdots \oplus \mathcal{E}_{N,k}$.*

**PROOF** (by reduction to Theorem 2.1). Fix $q \in \mathbb{Z}$ and $1 \le j \le N$. For each $n \ge 0$ we have

$$\Phi_{M,j}(q, q+n) : \begin{bmatrix} \mathcal{X}_{j,q+n} \\ \vdots \\ \mathcal{X}_{j,q+n+r_j-1} \end{bmatrix} \to \begin{bmatrix} \mathcal{X}_{j,q} \\ \vdots \\ \mathcal{X}_{j,q+r_j-1} \end{bmatrix},$$

and hence we can represent $\Phi_{M,j}(q, q+n)$ by an $r_j \times r_j$ operator matrix, as follows

$$\Phi_{M,j}(q, q+n) = (A_{s,t}(n))_{s,t=0}^{r_j-1}.$$

From the definition of $M_{j,k}$ in (3.8) it follows that $A_{s,t}(n)$ is a zero operator for $t > s$. We claim that

$$A_{s,t}(n) = \begin{bmatrix} n \\ n-(s-t) \end{bmatrix} \Phi_j(q+s, q+n+t) \qquad (\text{for } 0 \le t \le s \le r_j - 1). \tag{3.11}$$

Here $\begin{bmatrix} n \\ p \end{bmatrix}$ is usual binomial coefficient for $0 \le p \le n$ and $\begin{bmatrix} n \\ p \end{bmatrix} = 0$ for $p < 0$. Furthermore, $\Phi_j(m, n)$ is the state transition operator associated with the operators $\{Z_{j,k}\}$. For $n = 0$ the identities in (3.11) are trivially true, because $\Phi_{M,j}(q, q)$ is the identity operator on $\mathcal{E}_{j,q}$. To prove (3.11) for $n > 0$ we proceed by induction. Assume (3.11) holds for some $n \ge 0$. Then, by (3.10), we have for $t \le s$

$$\begin{aligned} A_{s,t}(n+1) &= A_{s,t}(n)Z_{j,q+n+t} + A_{s,t+1}(n) \\ &= \left\{ \begin{bmatrix} n \\ n-(s-t) \end{bmatrix} + \begin{bmatrix} n \\ n-(s-t)+1 \end{bmatrix} \right\} \Phi_j(q+s, q+n+t+1) \\ &= \begin{bmatrix} n+1 \\ n+1-(s-t) \end{bmatrix} \Phi_j(q+s, q+n+t+1), \end{aligned}$$

which proves (3.11) for $n+1$ in place of $n$.

Next, let F be as in (3.5). By using (3.11) one computes that

$$\sum_{n=0}^{\infty} \Phi_{M,j}(q, q+n) R_{j,q+n} f_{q+n,q} = \tilde{R}_{j,q} \qquad (q \in \mathbb{Z},\ j = 1, \dots, N) \tag{3.12}$$

is equivalent to (3.4). Indeed, for $i = 0, \dots, r_j - 1$ the i-th block entry of the left hand side of (3.12) is equal to

$$\sum_{n=0}^{\infty} \sum_{t=0}^{i} A_{i,t}(n) B_{t,j,q+n} f_{q+n,q} = \sum_{t=0}^{i} \sum_{n=i-t}^{\infty} \begin{bmatrix} n \\ n-(i-t) \end{bmatrix} \Phi_j(q+i, q+n+t) B_{t,j,q+n} f_{q+n,q}.$$

Here we used that $\begin{bmatrix} n \\ p \end{bmatrix} = 0$ for $p < 0$. Now, replace $i - t$ by $k$ and $n - k$ by $\nu$. We see that for

$i = 0, \ldots, r_j - 1$ the i-th entry of the left hand side of (3.12) is equal to

$$\sum_{k=0}^{i} \sum_{\nu=0}^{\infty} \binom{\nu+k}{\nu} \Phi_j(q+i,\, q+\nu+i) B_{i-k,j,q+\nu+k} f_{q+\nu+k,q} \,.$$

Thus (3.12) and (3.4) are equivalent.

From the equivalence of (3.12) and (3.4) we conclude that F in (3.5) is a solution of the nonstationary left tangential Hermite-Fejér interpolation problem for the data (3.1) and tolerance $\gamma$ if and only if F in (3.5) is a solution of the nonstationary left tangential Nevanlinna-Pick interpolation problem for the data (3.7) and tolerance $\gamma$. By Theorem 2.1 the necessary and sufficient condition for the solvability of the latter Nevanlinna-Pick problem is the requirement that the operator matrix $(\Xi_{i,j}(k))_{i,j=1}^{N}$ induces a positive operator on $\mathcal{E}_{1,k} \oplus \cdots \oplus \mathcal{E}_{N,k}$ for each $k \in \mathbb{Z}$. This completes the proof of Theorem 3.1.

As an application of Theorem 3.1 let us consider the following nonstationary version of the Schur interpolation problem. Assume we have given operators

$$A_{j,k} \in \mathcal{L}(\mathcal{Y}_k, \mathcal{U}_j), \quad -\infty < k \le j \le k+r-1 < \infty . \tag{3.13}$$

Here r is a fixed positive integer. The *nonstationary Schur interpolation problem* associated with the data (3.13) and tolerance $\gamma$ is to find operators $A_{j,k}$ in $\mathcal{L}(\mathcal{Y}_k, \mathcal{U}_j)$ for $-\infty \le k+r \le j < \infty$ such that the resulting doubly infinite block lower triangular operator matrix

$$A = \begin{bmatrix} \ddots & & & & \\ \cdots & A_{-1,-1} & & & \\ \cdots & A_{0,-1} & \underline{A_{0,0}} & & \\ \cdots & A_{1,-1} & A_{1,0} & A_{1,1} & \\ \vdots & \vdots & \vdots & \vdots & \ddots \end{bmatrix} \tag{3.14}$$

induces a bounded linear operator from $\vec{\mathcal{Y}}$ into $\vec{\mathcal{U}}$ of norm at most $\gamma$. In order for this nonstationary Schur problem to be solvable, it is clear for each $k \in \mathbb{Z}$ the operator $T_k$ from $\mathcal{Y}_k \oplus \mathcal{Y}_{k+1} \oplus \cdots \oplus \mathcal{Y}_{k+r-1}$ into $\mathcal{U}_k \oplus \mathcal{U}_{k+1} \oplus \cdots \oplus \mathcal{U}_{k+r-1}$ defined by

$$T_k = \begin{bmatrix} A_{k,k} & & & \\ A_{k+1,k} & A_{k+1,k+1} & & \\ \vdots & \vdots & \ddots & \\ A_{k+r-1,k} & A_{k+r-1,k+1} & \cdots & A_{k+r-1,k+r-1} \end{bmatrix} \tag{3.15}$$

must be bounded in norm by $\gamma$. As usual, the blank spots in (3.15) denote zero entries. We shall use Theorem 3.1 to show that the condition $\|T_k\| \le \gamma$ for each $k \in \mathbb{Z}$ is not only necessary but also sufficient.

**COROLLARY 3.2.** *The nonstationary Schur interpolation problem associated with the data (3.13) and tolerance* $\gamma$ *is solvable if and only if for each* $k \in \mathbb{Z}$ *the operator* $T_k$ *in (3.15) is bounded in norm by* $\gamma$.

**PROOF.** Let us associate with the data (3.13) operators

$$Z_k : \mathcal{U}_{k+1} \to \mathcal{U}_k,\ B_{i,k} : \mathcal{U}_k \to \mathcal{U}_{k+i},\ \tilde{B}_{i,k} : \mathcal{Y}_k \to \mathcal{U}_{k+i} \qquad (k \in \mathbb{Z},\ i = 0, \ldots, r-1) \quad (3.16)$$

in the following way. The operator $Z_k$ is the zero operator, $B_{0,k}$ is the identity operator on $\mathcal{U}_k$ and $B_{i,k}$ is zero for $i = 1, \ldots, r-1$, while $\tilde{B}_{i,k} = A_{k+i,k}$. (Here $N = 1$ and the index j is dropped from the notation in (3.1).) Notice that for $v > 0$ the state transition operator $\Phi(k, k+v)$ associated with the operators $\{Z_j\}$ is equal to zero. It follows that A in (3.14) is a solution of the nonstationary Schur problem associated with the data (3.13) and tolerance $\gamma$ if and only if A is a solution of the nonstationary Hermite-Fejér problem associated with the operators (3.16) and $\gamma$. By Theorem 3.1 the latter problem is solvable if and only if for each $k \in \mathbb{Z}$ a certain operator $\Xi(k)$ is positive. This operator $\Xi(k)$ is given by the formula

$$\Xi(k) = \sum_{v=0}^{\infty} \Phi_M(k, k+v)[\gamma^2 R_{k+v} R_{k+v}^* - \tilde{R}_{k+v} \tilde{R}_{k+v}^*] \Phi_M(k, k+v)^* ,$$

where

$$M_k = \begin{bmatrix} 0 & \cdots & 0 & 0 \\ I_{\mathcal{U}_{k+1}} & & & 0 \\ & \ddots & & \vdots \\ 0 & & I_{\mathcal{U}_{k+r-1}} & 0 \end{bmatrix}, \quad R_k = \begin{bmatrix} I_{\mathcal{U}_k} \\ 0 \\ \vdots \\ 0 \end{bmatrix}, \quad \tilde{R}_k = \begin{bmatrix} A_{k,k} \\ A_{k+1,k} \\ \vdots \\ A_{k+r-1,k} \end{bmatrix},$$

and

$$\Phi_M(k, k) = I_{\mathcal{E}_k}, \quad \Phi_M(k, k+v) = M_k M_{k+1} \cdots M_{k+v-1} \quad (v > 0) .$$

Notice that $T_k$ in (3.15) may also be written as a one block operator row, namely

$$T_k = [T_{k,0}, T_{k,1}, \ldots, T_{k,r-1}] ,$$

where

$$T_{k,j} = \Phi_M(k, k+j)\tilde{R}_{k+j} \qquad (j = 0, 1, \ldots, r-1) .$$

Since $\Phi_M(k, k+j) = 0$ for $j \geq r$, it follows that operator $\Xi(k)$ is equal to $\gamma^2 I_k - T_k T_k^*$, where $I_k$ is the identity operator on $\mathcal{U}_k \oplus \cdots \oplus \mathcal{U}_{k+r-1}$. Thus $\Xi(k)$ is positive if and only $\|T_k\| \le \gamma$, which completes the proof.

## VIII.4 NONSTATIONARY NEHARI INTERPOLATION

For the nonstationary Nehari extension problem the given *data* are bounded linear operators

$$f_{j,k} : \mathcal{U}_k \to \mathcal{Y}_j \qquad (-\infty < j < k < \infty). \tag{4.1}$$

A bounded linear operator G from $\vec{\mathcal{U}}$ to $\vec{\mathcal{Y}}$ with entries $g_{j,k}$ in $\mathcal{L}(\mathcal{U}_k, \mathcal{Y}_j)$ is said to be an *interpolant* of the data $\{f_{j,k}\}$ in (4.1) if

$$g_{j,k} = f_{j,k} \qquad (\text{for } -\infty < j < k < \infty). \tag{4.2}$$

The smallest possible operator norm of such an interpolant is of particular interest. If G is an interpolant and $\|G\| \le \gamma < \infty$, then we refer to G as a *solution* of the nonstationary Nehari problem for the data $\{f_{j,k}\}$ and with tolerance $\gamma$.

Notice that G is an interpolant of the data in (4.1) if and only if G admits an operator matrix representation of the form

$$G = \begin{bmatrix} & \vdots & \vdots & \vdots & \\ \cdots & f_{-2,-1} & f_{-2,0} & f_{-2,1} & \cdots \\ \cdots & g_{-1,-1} & f_{-1,0} & f_{-1,1} & \cdots \\ \cdots & g_{0,-1} & \underline{g_{0,0}} & f_{0,1} & \cdots \\ \cdots & g_{1,-1} & g_{1,0} & g_{1,1} & \cdots \\ & & \vdots & \vdots & \end{bmatrix} : \vec{\mathcal{U}} \to \vec{\mathcal{Y}}. \tag{4.3}$$

The operator

$$H_k := \begin{bmatrix} \cdot & \cdot & \cdot & \cdots \\ \cdot & \cdot & \cdot & \cdots \\ \cdot & \cdot & \cdot & \cdots \\ f_{k-3,k} & f_{k-3,k+1} & f_{k-3,k+2} & \cdots \\ f_{k-2,k} & f_{k-2,k+1} & f_{k-2,k+2} & \cdots \\ f_{k-1,k} & f_{k-1,k+1} & f_{k-1,k+2} & \cdots \end{bmatrix} : \bigoplus_{n=k}^{\infty} \mathcal{U}_n \to \bigoplus_{n=-\infty}^{k-1} \mathcal{Y}_k \tag{4.4}$$

is the compression of G to the appropriate spaces. In particular, $\|H_k\| \le \|G\|$ for all k. By introducing the unitary flip operator which rearranges the rows of $H_k$ we obtain part of the following result.

**THEOREM 4.1.** *The operators* $\{f_{j,k}\}$ *in (4.1) have an interpolant if and only if for each* k *the operator*

$$\Gamma_k = \begin{bmatrix} f_{k-1,k} & f_{k-1,k+1} & f_{k-1,k+2} & \cdots \\ f_{k-2,k} & f_{k-2,k+1} & f_{k-2,k+2} & \cdots \\ f_{k-3,k} & f_{k-3,k+1} & f_{k-3,k+2} & \cdots \\ \vdots & \vdots & \vdots & \end{bmatrix} : \bigoplus_{n=k}^{\infty} \mathcal{U}_n \to \bigoplus_{n=k-1}^{-\infty} \mathcal{Y}_n \tag{4.5}$$

*is well-defined and its operator norm has a finite upper bound independent of* k. *Moreover, in this case*

$$d_\infty = \min \{\|G\| : G \text{ is an interpolant of } (4.1)\} \tag{4.6}$$

*exists and* $d_\infty = \sup \{\|\Gamma_k\| : k \in \mathbb{Z}\}$.

To complete this section let us show how we can use the nonstationary Nehari extension theorem to derive the Parrott Lemma for $2 \times 2$ block matrices. (In Section XII.2 we follow a reverse direction and will use the Parrott Lemma to prove the three chains completion theorem, which in turn is used to prove Theorem 4.1.) To this end, let T be an operator of the form

$$T = \begin{bmatrix} A & B \\ C & D \end{bmatrix} : \begin{bmatrix} \mathcal{U}_1 \\ \mathcal{U}_2 \end{bmatrix} \to \begin{bmatrix} \mathcal{Y}_0 \\ \mathcal{Y}_1 \end{bmatrix} . \tag{4.7}$$

Now let $\mu$ be the scalar defined by

$$\mu = \max \{\|[A, B]\|, \|[B^*, D^*]\|\} . \tag{4.8}$$

Then obviously $\|T\| \geq \mu$ for all C in $\mathcal{L}(\mathcal{U}_1, \mathcal{Y}_1)$. The Parrott Lemma states that there exists an optimal C satisfying $\|T\| = \mu$. In other words, if A, B and D are all specified operators, then one can always find an operator C in $\mathcal{L}(\mathcal{U}_1, \mathcal{Y}_1)$ such that $\|T\| = \mu$. To see this, simple choose $f_{0,1} = A$, $f_{0,2} = B$ and $f_{1,2} = D$ and set all spaces $\mathcal{U}_{j+1}$ and $\mathcal{Y}_j$ with $j \notin \{0, 1\}$ equal to $\{0\}$. Then according to the nonstationary Nehari theorem (Theorem 4.1), there exists an optimal interpolant G in $\mathcal{L}(\vec{\mathcal{U}}, \vec{\mathcal{Y}})$ of $\{f_{j,k}\}$ satisfying $\|G\| = d_\infty$. Moreover, in this case $\mu = d_\infty$. So if we set $C = g_{1,1}$, then $\mu \leq \|T\| \leq \|G\| = \mu$. Therefore $\|T\| = \mu$ and this proves the Parrott Lemma.

## VIII.5. NONSTATIONARY SARASON INTERPOLATION

For the nonstationary Sarason interpolation problem the given data are a block lower triangular operator F from $\vec{\mathcal{U}}$ to $\vec{\mathcal{Y}}$ and a block lower triangular isometry $\Theta$ from $\vec{\mathcal{E}}$ to $\vec{\mathcal{Y}}$ where

$\vec{\mathcal{E}} = \oplus\{\mathcal{E}_n : n \in \mathbb{Z}\}$. By block lower triangular we mean that the operator entries $f_{j,k}$ for F and $\theta_{j,k}$ for $\Theta$ are zero for all $-\infty < j < k < \infty$. For these operators F and $\Theta$ the *nonstationary Sarason interpolation* problem with tolerance $\gamma$ is to determine a block lower triangular operator H from $\vec{\mathcal{U}}$ to $\vec{\mathcal{E}}$ such that $\|F - \Theta H\| \le \gamma$. In particular, we are interested in the smallest $\gamma$ for which such an interpolant H exists.

**THEOREM 5.1.** *Let* F *from* $\vec{\mathcal{U}}$ *to* $\vec{\mathcal{Y}}$ *and* $\Theta$ *from* $\vec{\mathcal{E}}$ *to* $\vec{\mathcal{Y}}$ *be block lower triangular operators, and assume* $\Theta$ *to be an isometry. For each* $k \in \mathbb{Z}$ *put*

$$\mathcal{M}_k = \{\Theta\vec{x} : \vec{x} = (x_j)_{j=-\infty}^{\infty} \in \vec{\mathcal{E}} \ \text{ and } \ x_j = 0 \ \text{ for } \ j < k\}\,. \tag{5.1}$$

*Then there exists a block lower triangular operator* H *from* $\vec{\mathcal{U}}$ *to* $\vec{\mathcal{E}}$ *such that* $\|F - \Theta H\| \le \gamma$ *if and only if*

$$\gamma \ge d_\infty := \sup\{\|(I - P_{\mathcal{M}_k})FP_k\| : k \in \mathbb{Z}\}\,. \tag{5.2}$$

*Here* $P_{\mathcal{M}_k}$ *is the orthogonal projection of* $\vec{\mathcal{Y}}$ *onto* $\mathcal{M}_k$, *and* $P_k$ *is the orthogonal projection on* $\vec{\mathcal{U}}$ *defined by*

$$(P_k\vec{x})_i = \begin{cases} 0 & \text{for } \ i < k\,, \\ x_i & \text{for } \ i \ge k\,. \end{cases} \tag{5.3}$$

*In particular, the number* $d_\infty$ *in the right hand side of (5.2) is the smallest tolerance for which the nonstationary Sarason interpolation problem is solvable.*

## VIII.6. NONSTATIONARY NUDELMAN INTERPOLATION

For the nonstationary Nudelman interpolation problem the given data are operators

$$\begin{gathered} Z_k : \mathcal{X}_{k+1} \to \mathcal{X}_k\,,\ B_k : \mathcal{U}_k \to \mathcal{X}_k\,,\ \tilde{B}_k : \mathcal{Y}_k \to \mathcal{X}_k \\ \Lambda_k : \mathcal{F}_k \to \mathcal{F}_{k-1}\,,\ C_k : \mathcal{F}_k \to \mathcal{Y}_k\,,\ \tilde{C}_k : \mathcal{F}_k \to \mathcal{U}_k,\ \Gamma_k : \mathcal{F}_{k-1} \to \mathcal{X}_k \end{gathered} \tag{6.1}$$

which act between Hilbert spaces. These operators are assumed to be uniformly bounded in the operator norm with respect to k and

$$\limsup_{\nu\to\infty} \left[\sup_{k\in\mathbb{Z}} \|\Phi(k, k+\nu)\|\right]^{1/\nu} < 1 \ \text{ and } \ \limsup_{\nu\to\infty} \left[\sup_{k\in\mathbb{Z}} \|\Psi(k-\nu, k)\|\right]^{1/\nu} < 1\,. \tag{6.2}$$

Here $\Phi$ and $\Psi$ are the (state) transition operators corresponding to the families $\{Z_k\}$ and $\{\Lambda_{k+1}\}$, that is, for m < n we have

$$\Phi(m, n) = Z_m Z_{m+1} \cdots Z_{n-1} \quad \text{and} \quad \Psi(m, n) = \Lambda_{m+1}\Lambda_{m+2} \cdots \Lambda_n . \tag{6.3}$$

Notice that $\Phi(m, n)$ starts with $Z_m$ on the left and ends with $Z_{n-1}$ on the right, while $\Psi(m, n)$ starts with $\Lambda_{m+1}$ on the left and ends with $\Lambda_n$. Clearly $\Phi(m, n)$ maps $\mathcal{X}_n$ into $\mathcal{X}_m$, while $\Psi(m, n)$ maps $\mathcal{F}_n$ into $\mathcal{F}_m$. As before, $\Phi(m, m) = I$ and $\Psi(m, m) = I$.

Given operators as above the *nonstationary Nudelman interpolation problem* with tolerance $\gamma$ is to find operators $f_{j,k}$ in $\mathcal{L}(\mathcal{Y}_k, \mathcal{U}_j)$ for $-\infty < k \leq j < \infty$ satisfying the interpolation conditions

$$\sum_{\nu=0}^{\infty} \Phi(k, k+\nu)B_{k+\nu}f_{k+\nu,k} = \tilde{B}_k \qquad (k \in \mathbb{Z}), \tag{6.4a}$$

$$\sum_{\nu=0}^{\infty} f_{k,k-\nu}C_{k-\nu}\Psi(k-\nu, k) = \tilde{C}_k \qquad (k \in \mathbb{Z}), \tag{6.4b}$$

$$\sum_{j,k=0}^{\infty} \Phi(n+1, n+j+1)B_{n+j+1}f_{n+j+1,n-k}C_{n-k}\Psi(n-k, n) = \Gamma_{n+1} \qquad (n \in \mathbb{Z}), \tag{6.4c}$$

and such that the block lower triangular operator

$$F = \begin{bmatrix} \ddots & & & & \\ \cdots & f_{-1,-1} & & & \\ \cdots & f_{0,-1} & \underline{f_{0,0}} & & \\ \cdots & f_{1,-1} & f_{1,0} & f_{1,1} & \\ \vdots & \vdots & \vdots & \vdots & \ddots \end{bmatrix} : \vec{\mathcal{Y}} \to \vec{\mathcal{U}} \tag{6.5}$$

has operator norm $\|F\| \leq \gamma$.

Assume that there exists a block lower triangular operator F as in (6.5) which satisfies the interpolation conditions (6.4a)-(6.4c). Then using (6.4a)-(6.4c) we have

$$\Gamma_n\Lambda_n = \sum_{j,k=1}^{\infty} \Phi(n, n+j)B_{n+j}f_{n+j,n-k}C_{n-k}\Psi(n-k, n) + B_n\tilde{C}_n - B_nf_{n,n}C_n ,$$

and

$$Z_n\Gamma_{n+1} = \sum_{j,k=1}^{\infty} \Phi(n, n+j)B_{n+j}f_{n+j,n-k}C_{n-k}\Psi(n-k, n) + \tilde{B}_nC_n - B_nf_{n,n}C_n .$$

Hence for the nonstationary Nudelman interpolation problem to be solvable the following compatibility condition must be satisfied:

$$Z_n\Gamma_{n+1} - \Gamma_n\Lambda_n = \tilde{B}_nC_n - B_n\tilde{C}_n \qquad (n \in \mathbb{Z}). \tag{6.6}$$

Therefore in what follows we assume that the time varying Sylvester equation in (6.6) is fulfilled. The following theorem provides a necessary and sufficient condition for the existence of a solution to the Nudelman interpolation problem.

**THEOREM 6.1.** *Consider the Nudelman interpolation data (6.1), and assume that (6.6) holds. For each* $k \in \mathbb{Z}$ *put*

$$M_k = \sum_{\nu=0}^{\infty} (\Psi(k-\nu, k)^*(\gamma^2 C_{k-\nu}^* C_{k-\nu} - \tilde{C}_{k-\nu}^* \tilde{C}_{k-\nu})\Psi(k-\nu, k), \tag{6.7}$$

$$N_k = \sum_{\nu=0}^{\infty} \Phi(k, k+\nu)(\gamma^2 B_{k+\nu} B_{k+\nu}^* - \tilde{B}_{k+\nu}\tilde{B}_{k+\nu}^*)\Phi(k, k+\nu)^*. \tag{6.8}$$

*Then the nonstationary Nudelman interpolation problem for the data (6.1) and tolerance* $\gamma$ *has a solution if and only if for each* $k \in \mathbb{Z}$ *the operator*

$$\Xi_k = \begin{bmatrix} M_k & \gamma\Gamma_{k+1}^* \\ \gamma\Gamma_{k+1} & N_{k+1} \end{bmatrix} : \mathcal{F}_k \oplus \mathcal{X}_{k+1} \to \mathcal{F}_k \oplus \mathcal{X}_{k+1} \tag{6.9}$$

*is positive* ($\geq 0$).

## VIII.7. NONSTATIONARY TWO-SIDED SARASON INTERPOLATION

For the nonstationary two-sided Sarason interpolation problem the given data are operators

$$F : \vec{\mathcal{U}} \to \vec{\mathcal{Y}}, \quad L_1 : \vec{\mathcal{U}} \to \vec{\mathcal{E}}, \text{ and } L_2 : \vec{\mathcal{F}} \to \vec{\mathcal{Y}} \tag{7.1}$$

acting between doubly infinite Hilbert space direct sums. (Recall that the space $\vec{\mathcal{H}}$ stands for the Hilbert space direct sum $\vec{\mathcal{H}} = \oplus\{\mathcal{H}_n : n \in \mathbb{Z}\}$.) We assume additionally that

(i) $L_1$ is a block lower triangular co-isometry;

(ii) $L_2$ is a block lower triangular isometry.

By block lower triangular we mean that the operator entries $(L_2)_{i,j}$ and $(L_1)_{i,j}$ are all zero for $i < j$. For F, $L_1$ and $L_2$ as above and a given tolerance $\gamma \geq 0$ the *nonstationary two-sided Sarason problem* is to determine block lower triangular operators R from $\vec{\mathcal{E}}$ to $\vec{\mathcal{F}}$ such that $\|F - L_2RL_1\| \leq \gamma$. In particular, we are interested in the smallest $\gamma$ for which such an interpolant

R exists.

Notice that this two-sided Sarason problem is a nonstationary version of the two-sided Sarason problem considered in Section I.10. The following theorem provides a necessary and sufficient condition for the existence of a solution.

**THEOREM 7.1.** *Let* F *from* $\vec{\mathcal{U}}$ *to* $\vec{\mathcal{Y}}$ *and* $L_1$ *from* $\vec{\mathcal{U}}$ *to* $\vec{\mathcal{E}}$ *and* $L_2$ *from* $\vec{\mathcal{F}}$ *to* $\vec{\mathcal{Y}}$ *be operators, where* $L_2$ *is a block lower triangular isometry and* $L_1$ *is a block lower triangular co-isometry. For each* $k \in \mathbb{Z}$ *put*

$$\mathcal{H}_k = \left\{ \vec{u} \in \vec{\mathcal{U}} : (L_1\vec{u})_j = 0 \text{ for } j < k \right\} \quad \textit{and} \quad \mathcal{M}_k = \left\{ L_2\vec{f} : \vec{f} \in \vec{\mathcal{F}} \text{ and } f_j = 0 \text{ for } j < k \right\}. \tag{7.2}$$

*Then there exists a block lower triangular operator* R *from* $\vec{\mathcal{E}}$ *to* $\vec{\mathcal{F}}$ *such that* $\|F - L_2RL_1\| \le \gamma$ *if and only if*

$$\gamma \ge d_\infty = \sup\{\|(I - P_{\mathcal{M}_k})F \mid \mathcal{H}_k\| : k \in \mathbb{Z}\}\,. \tag{7.3}$$

*Here* $P_{\mathcal{M}_k}$ *is the orthogonal projection of* $\vec{\mathcal{Y}}$ *onto* $\mathcal{M}_k$*. In particular, the number* $d_\infty$ *defined by the right hand side of (7.2) is the smallest tolerance for which the nonstationary two-sided Sarason problem is solvable.*

With $L_1$ equal to the identity operator on $\vec{\mathcal{U}}$ and F block lower triangular, the nonstationary two-sided Sarason problem reduces to the one-sided problem. Notice that in this case the space $\mathcal{H}_k$ is just the range of the orthogonal projection $P_k$ defined by (5.3).

## *Notes to Chapter VIII:*

The left hand side of formula (1.5) may be seen as a generalization of point evaluation in the time-variant setting. It appears in Alpay-Dewilde [1] in the role of W-transforms and has been developed further in Alpay-Dewilde-Dym [1] for triangular operators with matrix or operator coefficients, in connection with time-invariant lossless inverse scattering problems. The nonstationary Nevanlinna-Pick problem in Section 1 with matrix entries has been treated in Dewilde [1]. The description of all solutions of the left tangential Nevanlinna-Pick problem with operator entries (as in Section 2) appears in Dewilde-Dym [1]. In these papers the main tools came from an appropriate generalization of the reproducing kernel space method. In Ball-Gohberg-Kaashoek [1], [4], [7] solutions to the left tangential Nevanlinna-Pick problem and two-sided Nudelman problem, both with matrix entries, are given. The results of these papers also have continuous analogues (see Ball-Gohberg-Kaashoek [2], [5]). Recursive methods for obtaining solutions of the nonstationary Nevanlinna-Pick problem have been obtained in Sayed-Constantinescu-Kailath [2]. Further developments and applications to completion problems may be found in Kos [2]. The nonstationary tangential Hermite-Fejér problem has also been treated in Dewilde-Dym [1], Sayed-Constantinescu-Kailath [1], and Kos [1]. The nonstationary Nehari problem was first treated in Gohberg-Kaashoek-Woerdeman [1], [2] by using the band method, see also Constantinescu [2] and Dewilde-Dym [1]. The nonstationary Sarason problem of Section 5 seems too new. For collections of papers on nonstationary interpolation problems we refer to the books Gohberg [1] and Dewilde-Kaashoek-Verhaegen [1]. Most of the material in this chapter is taken from Foias-Frazho-Gohberg-Kaashoek [1].

CHAPTER IX

# NONSTATIONARY SYSTEMS AND POINT EVALUATION

This chapter introduces discrete time-varying input output systems, and various notions connected with them. The chapter also serves as a further motivation for the interpolation problems considered in the previous chapter. Special attention is paid to the interpretation of the nonstationary version of point evaluation. It will also be shown how one may convert a time-varying system into an infinite dimensional time-invariant system, which gives a first insight into the reduction techniques that will be used later. Finally, a nonstationary version of the filtering problem is connected to a nonstationary Sarason problem.

## IX.1. TIME VARYING SYSTEMS

In this section we will introduce time varying systems. To begin, a state space (discrete) time varying system (moving in forward time) is given by a set of equations of the form

$$x(n+1) = A_n x(n) + B_n u(n) \quad \text{and} \quad y(n) = C_n x(n) + D_n u(n) . \tag{1.1}$$

The system coefficients are Hilbert spaces operators, $A_n$ is an operator from $\mathcal{X}_n$ to $\mathcal{X}_{n+1}$ and $B_n$ maps $\mathcal{U}_n$ into $\mathcal{X}_{n+1}$ while $C_n$ maps $\mathcal{X}_n$ into $\mathcal{Y}_n$ and the external coefficient $D_n$ maps $\mathcal{U}_n$ into $\mathcal{Y}_n$. The index n can run from $-\infty$ to $\infty$. The space $\mathcal{X}_n$ is called the *state space* at time n, and the vector x(n) in $\mathcal{X}_n$ is the *state vector* at time n. The sequence of vectors (u(n)) is the *input* and (y(n)) is the *output*. In the sequel we will denote the system (1.1) by $\{A_n, B_n, C_n, D_n\}$.

For the moment let us assume that we start operating the system at time k. So assume that we have given a sequence of input vectors $(u(n))_k^\infty$ and that the state x(k) at time k is specified. By recursively solving for the state x(n) for $n \geq k+1$ we have

$$x(n) = \Phi(n, k)x(k) + \sum_{i=k}^{n-1} \Phi(n, i+1)B_i u(i) . \tag{1.2}$$

Here $\Phi(n, m)$ is the *state transition operator* for the state space operators $\{A_j\}$ defined by

$$\Phi(n, m) = A_{n-1}A_{n-2} \cdots A_m \quad (\text{if } n > m) \quad \text{and} \quad \Phi(n, n) = I . \tag{1.3}$$

Notice that the state transition operator $\Phi(n, m)$ starts with $A_m$ on the right and ends with $A_{n-1}$ on the left. In particular, $\Phi(n, m)$ maps $\mathcal{X}_m$ into $\mathcal{X}_n$. Substituting (1.2) into the expression for y(n) in (1.1) we obtain

$$y(n) = C_n\Phi(n, k)x(k) + D_n u(n) + \sum_{i=k}^{n-1} C_n\Phi(n, i+1)B_i u(i) \quad (n \ge k) . \tag{1.4}$$

For $n = k$ formula (1.4) reduces to $y(k) = C_k x(k) + D_k u(k)$. Writing (1.4) out in matrix form yields

$$\begin{bmatrix} y(k) \\ y(k+1) \\ y(k+2) \\ \vdots \end{bmatrix} = \begin{bmatrix} C_k \\ C_{k+1}\Phi(k+1, k) \\ C_{k+2}\Phi(k+2, k) \\ \vdots \end{bmatrix} x(k) + \begin{bmatrix} F_{k,k} & 0 & 0 & \cdots \\ F_{k+1,k} & F_{k+1,k+1} & 0 & \cdots \\ F_{k+2,k} & F_{k+2,k+1} & F_{k+2,k+2} & \cdots \\ \vdots & \vdots & \vdots & \end{bmatrix} \begin{bmatrix} u(k) \\ u(k+1) \\ u(k+2) \\ \vdots \end{bmatrix} \tag{1.5}$$

where $F_{n,m}$ are the operators from $\mathcal{U}_m$ into $\mathcal{Y}_n$ defined by

$$F_{n,n} = D_n \quad \text{and} \quad F_{n,m} = C_n\Phi(n, m+1)B_m \quad \text{if } n > m . \tag{1.6}$$

The block column matrix

$$W_{ok} = \begin{bmatrix} C_k \\ C_{k+1}\Phi(k+1, k) \\ C_{k+2}\Phi(k+2, k) \\ \vdots \end{bmatrix} \tag{1.7}$$

appearing in (1.5) is called the *forward* (or *causal*) *observability operator* of the system (1.1) at time k. We view $W_{ok}$ as a linear map of the state space $\mathcal{X}_k$ to the space consisting of all sequences $[y_k, y_{k+1}, \ldots]^{tr}$ where $y_j \in \mathcal{Y}_j$. Since $W_{ok}$ does not depend on the coefficients $B_n$ and $D_n$, we also refer to $W_{ok}$ as the observability operator of the pairs $\{C_n, A_n\}_{n\in\mathbb{Z}}$ at time k. The finite block lower triangular matrix in (1.5), which will be denoted by $T_k$, is a section of the doubly infinite block lower triangular operator matrix

$$T = \begin{bmatrix} \ddots & & & & \\ \cdots & F_{-1,-1} & & & \\ \cdots & F_{0,-1} & F_{0,0} & & \\ \cdots & F_{1,-1} & F_{1,0} & F_{1,1} & \\ \vdots & \vdots & \vdots & \vdots & \ddots \end{bmatrix}, \tag{1.8}$$

of which the operator entries $F_{n,m} (n \ge m)$ are given by (1.6). We shall refer to T as the *input output map* of system (1.1), and to $T_k$ as the k-th *section* of the input output map. From the previous discussion we see that for a system that starts operating at time k the output $(y(n))_k^\infty$ decomposes into two sequences. One of the sequences is given by the action of the observability operator at time k on the state vector x(k) and the other is obtained by applying the k-th section of the input output map to the input $(u(n))_k^\infty$.

The system (1.1) or its sequence of state operators $\{A_n\}$ is said to be *forward stable* if

$$\limsup_{\nu\to\infty} \left[\sup_{k\in\mathbb{Z}} \|\Phi(k+\nu, k)\|\right]^{1/\nu} < 1 . \tag{1.9}$$

If $A_n = A$ for all n, then the stability condition in (1.8) is equivalent to $r_{spec}(A) < 1$. Finally, we will say that a set of operators $\{G_k : k \in \mathbb{Z}\}$ are *uniformly bounded* if there exists a finite scalar M such that $\|G_k\| \le M$ for all $k \in \mathbb{Z}$. This sets the state for the following stability result.

**PROPOSITION 1.1**. *Let* $\{A_n, B_n, C_n, D_n\}$ *be a forward stable system of the form (1.1), and assume that* $\{B_n\}$, $\{C_n\}$ *and* $\{D_n\}$ *are all uniformly bounded. Let* $T_k$ *be the* k–th *section of the input output map and* $W_{ok}$ *the forward observability operator at time* k *of* $\{A_n, B_n, C_n, D_n\}$. *Then the operator* $T_k$ *is a bounded operator from* $\vec{\mathcal{U}}_{(k,\infty)} = \oplus\{\mathcal{U}_n : n \ge k\}$ *into* $\vec{\mathcal{Y}}_{(k,\infty)} = \oplus\{\mathcal{Y}_n : n \ge k\}$, *and* $W_{ok}$ *is a bounded operator from* $\mathcal{X}_k$ *into* $\vec{\mathcal{Y}}_{(k,\infty)}$. *In particular, the forward stable system* $\{A_n, B_n, C_n, D_n\}$ *in (1.1) defines a bounded operator from* $\mathcal{X}_k \oplus \vec{\mathcal{U}}_{(k,\infty)}$ *into* $\vec{\mathcal{Y}}_{(k,\infty)}$ *by*

$$y = W_{ok}x + T_k u \qquad (x \in \mathcal{X}_k,\ u = (u(n))_k^\infty \ \text{ and } \ y = (y(n))_k^\infty) . \tag{1.10}$$

*Furthermore, the input output map* T *of the system* $\{A_n, B_n, C_n, D_n\}$ *defines a bounded operator from* $\vec{\mathcal{U}}$ *into* $\vec{\mathcal{Y}}$.

**PROOF**. From (1.9) it follows that there exist constants $0 \le \alpha < 1$ and $\beta \ge 0$ such that

$$\|\Phi(k+\nu, k)\| \le \beta\alpha^\nu \qquad (k \in \mathbb{Z},\ \nu \ge 0) .$$

Since the coefficients of (1.1) are uniformly bounded in norm with respect to n, we can find a constant $\tilde{\beta}$ such that

$$\|C_{k+i}\Phi(k+i, k)\| \le \tilde{\beta}\alpha^i \quad (k \in \mathbb{Z},\ i \ge 0) ,$$

$$\|F_{m+i,m}\| \le \tilde{\beta}\alpha^i \quad (n \in \mathbb{Z},\ i \ge 0) .$$

Here $F_{n,m}$ $(n \ge m)$ is given by (1.6). From these inequalities, the boundedness of the operators $W_{ok}$, $T_k$ and T is clear, which completes the proof.

In many applications to nonstationary interpolation problems the controllability and observability operators correspond to systems that move backwards in time. (This phenomenon is hidden in stationary interpolation problems.) The state space equations for a system $\{Z_n, B_n, C_n, D_n\}$ moving backwards in time (or anticausal) are

$$x(n-1) = Z_n x(n) + B_n u(n) \quad \text{and} \quad y(n) = C_n x(n) + D_n u(n)\,. \tag{1.11}$$

As before, the vector x(n) is the state at time n, the input is (u(n)) and the sequence (y(n)) is the output. Here $Z_n$ maps $\mathcal{X}_n$ into $\mathcal{X}_{n-1}$ and $B_n$ maps $\mathcal{U}_n$ into $\mathcal{X}_{n-1}$ while $C_n$ maps $\mathcal{X}_n$ into $\mathcal{Y}_n$ and $D_n$ maps $\mathcal{U}_n$ into $\mathcal{Y}_n$, and as before the underlying spaces are Hilbert spaces. Now assume that the input $u(j) = 0$ for $j > k$ and the initial condition x(k) is specified. By recursively solving for the state x(n), and then substituting this into the output equation for y(n) we obtain for $n \le k$

$$y(n) = C_n \Phi(n, k) x(k) + D_n u(n) + \sum_{i=n+1}^{k} C_n \Phi(n, i-1) B_i u(i)\,. \tag{1.12}$$

Here $\Phi(n, m)$ is the state transition operator for $\{Z_n\}$ moving backward in time defined by

$$\Phi(n, m) = Z_{n+1} Z_{n+2} \cdots Z_m \ \ (\text{if } n < m) \quad \text{and} \quad \Phi(n, n) = I\,. \tag{1.13}$$

This state transition operator starts with $Z_m$ on the right and ends with $Z_{n+1}$ on the left. In particular, $\Phi(n, m)$ maps $\mathcal{X}_m$ into $\mathcal{X}_n$. (Notice that this state transition operator is defined slightly different from the one in Section VIII.2, because in the latter section $Z_n$ maps $\mathcal{X}_{n+1}$ into $\mathcal{X}_n$. In defining state transition operators we always follow the convention that $\Phi(n, m)$ maps the state space $\mathcal{X}_m$ at time m into the state space $\mathcal{X}_n$ at time n.) Notice that the sum appearing in the right hand side of (1.12) is zero if $n = k$. Rewriting the input-output map in (1.12) in matrix form gives

$$\begin{bmatrix} \cdot \\ \cdot \\ \cdot \\ y(k-2) \\ y(k-1) \\ y(k) \end{bmatrix} = \begin{bmatrix} \cdot \\ \cdot \\ \cdot \\ C_{k-2}\Phi(k-2, k) \\ C_{k-1}\Phi(k-1, k) \\ C_k \end{bmatrix} x(k) + \begin{bmatrix} \cdots & \cdot & \cdot & \cdot \\ \cdots & \cdot & \cdot & \cdot \\ \cdots & \cdot & \cdot & \cdot \\ \cdots & G_{k-2,k-2} & G_{k-2,k-1} & G_{k-2,k} \\ \cdots & 0 & G_{k-1,k-1} & G_{k-1,k} \\ \cdots & 0 & 0 & G_{k,k} \end{bmatrix} \begin{bmatrix} \cdot \\ \cdot \\ \cdot \\ u(k-2) \\ u(k-1) \\ u(k) \end{bmatrix}. \tag{1.14}$$

The operators $G_{n,m}$ from $\mathcal{U}_m$ into $\mathcal{Y}_n$ are defined by

$$G_{n,n} = D_n \quad \text{and} \quad G_{n,m} = C_n \Phi(n, m-1) B(m) \ \ \text{if } n < m\,. \tag{1.15}$$

The block column matrix appearing immediately to the right of the equal sign in (1.14), which will be denoted by $W_{oka}$, is called the *backward* (or *anticausal*) *observability operator at time k* for the pairs $\{C_n, Z_n\}_{n \in \mathbb{Z}}$ or of the system $\{Z_n, B_n, C_n, D_n\}$. The infinite block upper triangular matrix in (1.14) will be denoted by $T_{ka}$. We may view $T_{ka}$ as a section of the *input output map* $T_a$ of the system (1.11), which, by definition, is the doubly infinite block upper triangular operator matrix

$$T_a = \begin{bmatrix} \ddots & \vdots & \vdots & \vdots & \vdots \\ & G_{-1,-1} & G_{-1,0} & G_{-1,1} & \cdots \\ & & G_{0,0} & G_{0,1} & \cdots \\ & & & G_{1,1} & \cdots \\ & & & & \ddots \end{bmatrix}, \tag{1.16}$$

of which the operator entrices $G_{n,m}$ $(n \le m)$ are defined by (1.15). We shall refer to $T_{ka}$ as the k-th *backward section* of $T_a$.

The system (1.11) or its sequence of state operators $\{Z_n\}$ is said to be *backward stable* if

$$\lim_{v \to \infty} \sup \left[ \sup_{k \in \mathbb{Z}} \|\Phi(k, k+v)\|^{1/v} \right] < 1 . \tag{1.17}$$

Using this definition we now state the anticausal version of Proposition 1.1.

**PROPOSITION 1.2.** *Let* $\{Z_n, B_n, C_n, D_n\}$ *be a backward stable system of the form (1.11), and assume that* $\{B_n\}$, $\{C_n\}$ *and* $\{D_n\}$ *are all uniformly bounded. Let* $T_{ka}$ *be the* k–th *backward section of the input output map, and let* $W_{oka}$ *be the backward observability operator at time* k *of the system* $\{Z_n, B_n, C_n, D_n\}$. *Then* $T_{ka}$ *is a bounded operator from* $\vec{\mathcal{U}}_{(-\infty,k)} = \oplus \{\mathcal{U}_n : n \le k\}$ *into* $\vec{\mathcal{Y}}_{(-\infty,k)} = \oplus \{\mathcal{Y}_n : n \le k\}$, *and* $W_{oka}$ *is a bounded operator from* $\mathcal{X}_k$ *into* $\vec{\mathcal{Y}}_{(-\infty,k)}$. *In particular, the backward stable system* $\{Z_n, B_n, C_n, D_n\}$ *in (1.11) defines a bounded operator from* $\mathcal{X}_k \oplus \vec{\mathcal{U}}_{(-\infty,k)}$ *into* $\vec{\mathcal{Y}}_{(-\infty,k)}$ *by*

$$y = W_{oka}x + T_{ka}u \qquad (x \in \mathcal{X}_k,\ u = (u(n))_{-\infty}^{k} \ \text{and} \ y = (y(n))_{-\infty}^{k}) . \tag{1.18}$$

*Furthermore, the input output map* $T_a$ *of the system* $\{Z_n, B_n, C_n, D_n\}$ *defines a bounded operator from* $\vec{\mathcal{U}}$ *into* $\vec{\mathcal{Y}}$.

## IX.2. NONSTATIONARY CONTROLLABILITY AND OBSERVABILITY

In this section we will discuss controllability and observability for nonstationary systems. As in the stationary case, let us begin with observability, and for the moment assume that all the state spaces $\mathcal{X}_n$ are finite dimensional. The forward system $\{A_n, B_n, C_n, D_n\}$ in (1.1) is *forward observable at time* k if given the input $u = (u(n))_k^\infty$ and output $y = (y(n))_k^\infty$ one can uniquely determine the state $x = (x(n))_k^\infty$. The forward system $\{A_n, B_n, C_n, D_n\}$ is called *forward observable* if it is forward observable for all k. However, because the state x is uniquely determined by the initial condition x(k) and the input u, the forward system $\{A_n, B_n, C_n, D_n\}$ is observable at time k if given the input $u = (u(n))_k^\infty$ and output $y = (y(n))_k^\infty$, then one can uniquely determine the initial state x(k). According to (1.10) the output $y = W_{ok}x(k) + T_k u$, where $W_{ok}$ is the forward observability operator for the pairs $\{C_n, A_n\}_{n\in\mathbb{Z}}$ at time k and $T_k$ is the k-th section of the input output map of $\{A_n, B_n, C_n, D_n\}$ defined in (1.6). Then obviously the forward system is observable at time k if and only if $W_{ok}$ is one to one. Moreover, the forward system $\{A_n, B_n, C_n, D_n\}$ is forward observable if and only if $W_{ok}$ is one to one for all k. As in the stationary case observability has nothing to do with $\{B_n\}$ and $\{D_n\}$. So we say that the family of pairs $\{C_n, A_n\}_{n\in\mathbb{Z}}$ is *forward observable* at time k if and only if $W_{ok}$ is one to one. Summing up this analysis leads to the following result.

**PROPOSITION 2.1.** *Let* $\{A_n, B_n, C_n, D_n\}$ *be a forward finite dimensional linear system of the form (1.1). Then the following statements are equivalent*

*(i)* *The system* $\{A_n, B_n, C_n, D_n\}$ *is forward observable at time k.*

*(ii)* *The observability operator* $W_{ok}$ *generated by* $\{C_n, A_n\}_{n\in\mathbb{Z}}$ *is one to one.*

*(iii)* $\mathcal{X}_k = \bigvee \{\Phi(k+v, k)^* C_{k+v}^* \mathcal{Y}_{k+v} : v \geq 0\}$.

*Finally, if* $\{A_n\}$ *is forward stable and* $\{C_n\}$ *is uniformly bounded, then the previous three conditions are equivalent to* ran $W_{ok}^* = \mathcal{X}_k$.

Now assume that the operators $\{A_n\}$ are forward stable and $\{C_n\}$ are uniformly bounded. Then all $W_{ok}$ are bounded operators from $\mathcal{X}_k$ into $\vec{\mathcal{Y}}_{(k,\infty)} = \oplus\{\mathcal{Y}_n : n \geq k\}$. In this case, the *forward observability grammian* for the family of pairs $\{C_n, A_n\}_{n\in\mathbb{Z}}$ at time k is defined by $Q_k = W_{ok}^* W_{ok}$. Obviously $Q_k \geq 0$. Moreover, the family of pairs $\{C_n, A_n\}$ is forward observable at time k if and only if $Q_k > 0$. Notice that by (1.3) and (1.7) we have

$$W_{ok} = \begin{bmatrix} C_k \\ W_{o(k+1)} A_k \end{bmatrix}. \tag{2.1}$$

Therefore the observability grammian satisfies the following nonstationary Lyapunov difference equation

$$Q_k = A_k^* Q_{k+1} A_k + C_k^* C_k . \tag{2.2}$$

Now assume that the input $u = (u(n))_k^\infty$ and the output $y = (y(n))_k^\infty$ are known. Then using $y = W_{ok}x(k) + T_k u$ it follows that $W_{ok}^* y = W_{ok}^* W_{ok} x(k) + W_{ok}^* T_k u$. So if the family of pairs $\{C_n, A_n\}_{n\in\mathbb{Z}}$ is observable at time k, then the initial condition at time k can be computed by

$$x(k) = Q_k^{-1}(W_{ok}^* y - W_{ok}^* T_k u) . \tag{2.3}$$

If $\mathcal{X}_k$ is infinite dimensional, and $\{A_n\}$ is forward stable and $\{C_n\}$ is uniformly bounded, then $W_{ok}$ is a bounded operator. In this case we say that the family of pairs $\{C_n, A_n\}_{n\in\mathbb{Z}}$ is *forward observable* at time k if $\|W_{ok}x\| \geq \delta\|x\|$ for some $\delta > 0$ and all x in $\mathcal{X}_k$. Clearly, $\{C_n, A_n\}_{n\in\mathbb{Z}}$ is forward observable at time k if and only if $Q_k := W_{ok}^* W_{ok}$ is strictly positive. Moreover, because $\{A_n\}$ is forward stable and $\{C_n\}$ is uniformly bounded, the observability grammians $Q_k$ are well defined and satisfy the Lyapunov difference equation in (2.2). Finally, we say that this forward stable infinite dimensional system is forward observable if $\{C_n, A_n\}_{n\in\mathbb{Z}}$ is forward observable for all k.

Once again let us assume that $\mathcal{X}_k$ is finite dimensional. Then the system $\{Z_n, B_n, C_n, D_n\}$ in (1.11) moving backwards in time is *backward observable at time* k if given the input $u = (u(n))_{-\infty}^k$ and output $y = (y(n))_{-\infty}^k$ one can uniquely determine the state $x = (x(n)_{-\infty}^k$, or, equivalently, the initial condition x(k). As expected, we call the system in (1.11) *backward observable* if the system is backward observable for all k. By mimicking our previous analysis for forward observable systems we obtain the following result.

**PROPOSITION 2.2.** *Let* $\{Z_n, B_n, C_n, D_n\}$ *be the finite dimensional backward system given in (1.11). Then the following statements are equivalent*

*(i)* *The system* $\{Z_n, B_n, C_n, D_n\}$ *is backward observable at time k.*

*(ii)* *The backward observability operator* $W_{oka}$ *is one to one.*

*(iii)* $\mathcal{X}_k = \bigvee \{\Phi(k-\nu, k)^* C_{k-\nu}^* \mathcal{Y}_{k-\nu} : \nu \geq 0\}$.

*Finally, if* $\{Z_n\}$ *is backward stable and* $\{C_n\}$ *is uniformly bounded, then the previous three conditions are equivalent to* ran $W_{oka}^* = \mathcal{X}_k$.

Now assume that the operators $\{Z_n\}$ are backward stable and $\{C_n\}$ are uniformly bounded. Then the backward observability operator $W_{oka}$ is a bounded operator from $\mathcal{X}_k$ into $\vec{\mathcal{Y}}_{(-\infty,k)} = \oplus\{\mathcal{Y}_n : n \leq k\}$. In this case, the *backward observability grammian* for $\{C_n, Z_n\}_{n\in\mathbb{Z}}$ at time k is defined by $Q_{ka} = W_{oka}^* W_{oka}$. Clearly $Q_{ka} \geq 0$. Moreover, $Q_{ka} > 0$ if and only if the family of pairs $\{C_n, Z_n\}_{n\in\mathbb{Z}}$ is backward observable at time k. Using

$$W_{oka} = \begin{bmatrix} W_{o(k-1)a} Z_k \\ C_k \end{bmatrix} \tag{2.4}$$

it follows that the backward observability grammian satisfies the following nonstationary

Lyapunov difference equation

$$Q_{ka} = Z_k^* Q_{(k-1)a} Z_k + C_k^* C_k \ . \tag{2.5}$$

Finally, assume that the family of pairs $\{C_n, Z_n\}_{n \in \mathbb{Z}}$ is backward observable at time k (i.e., the system (1.11) is backward observable at time k) and the input $u = (u(n))_{-\infty}^{k}$ and output $y = (y(n))_{-\infty}^{k}$ are known. Then following our previous analysis the initial state x(k) can be computed by

$$x(k) = Q_{ka}^{-1}(W_{oka}^* y - W_{oka}^* T_{ka} u) \ . \tag{2.6}$$

Now assume that $\mathcal{X}_k$ is infinite dimensional, $\{Z_n\}$ is backward stable and $\{C_n\}$ is uniformly bounded. Then obviously the observability grammians $Q_{ka} := W_{oka}^* W_{oka}$ are well defined and satisfy the nonstationary Lyapunov equation in (2.5). In this stable infinite dimensional setting we say that the family of pairs $\{C_n, Z_n\}_{n \in \mathbb{Z}}$ is *backward observable* at time k if $\|W_{oka} x\| \geq \delta \|x\|$ for some $\delta > 0$ and all x in $\mathcal{X}_k$. Clearly, $\{C_n, Z_n\}_{n \in \mathbb{Z}}$ is backward observable at time k if and only if $Q_{ka}$ is strictly positive. As in the finite dimensional setting, the backward stable system $\{A_n, B_n, C_n, D_n\}$ is backward observable if $\{C_n, Z_n\}_{n \in \mathbb{Z}}$ is backward observable at time k for all k.

To discuss controllability for simplicity of presentation let us assume that the state space $\mathcal{X}_n$ is finite dimensional for all n. The forward system $\{A_n, B_n, C_n, D_n\}$ in (1.1) is called *forward controllable at time* k if for each $x_k \in \mathcal{X}_k$ there exists a sequence of input vectors u(r), u(r + 1), ..., u(k − 1) (depending on $x_k$) such that starting from x(r) = 0 this input sequence drives the system to the state vector $x(k) = x_k$ at time k. By recursively solving for $x_k = x(k)$ in (1.1) or using (1.2) we see that the state $x(k) = x_k$ at time k if and only if

$$x_k = \sum_{i=r}^{k-1} \Phi(k, i+1) B_i u(i) \ . \tag{2.7}$$

It follows that the system $\{A_n, B_n, C_n, D_n\}$ in (1.1) is forward controllable at time k if and only if

$$\mathcal{X}_k = \bigvee_{\nu \geq 0} \operatorname{ran} \Phi(k, k-\nu) B_{k-\nu-1} \ . \tag{2.8}$$

The system $\{A_n, B_n, C_n, D_n\}$ is said to be *forward controllable* if it is forward controllable for all k. Since controllability involves only the state and input coefficient, we shall also speak about forward controllability at time k of the family of pairs $\{A_n, B_n\}_{n \in \mathbb{Z}}$. By definition the *forward controllability operator* $W_{ck}$ is given by the infinite operator row

$$W_{ck} = [B_{k-1}, \Phi(k, k-1)B_{k-2}, \Phi(k, k-2)B_{k-3}, \dots] . \tag{2.9}$$

We view $W_{ck}$ as a linear map defined on the space consisting of all sequences $(u_{k-1}, u_{k-2}, \dots)$, where $u_j \in \mathcal{U}_j$, such that only a finite number of $u_j$'s is nonzero. Such input sequences are said to have *finite support.* The range space of $W_{ck}$ is contained in $\mathcal{X}_k$. So we see from (2.8) that the system in (1.1) is forward controllable at time k if and only if the map $W_{ck}$ is onto $\mathcal{X}_k$. Clearly, forward controllability only depends on the pairs $\{A_n, B_n\}$ with $n \le k-1$. Therefore we shall also speak about forward controllability of the pairs $\{A_n, B_n\}_{n\in\mathbb{Z}}$ at time k.

**PROPOSITION 2.3.** *Let* $\{A_n, B_n, C_n, D_n\}$ *be a forward finite dimensional system of the form (1.1). Then the following statements are equivalent*

*(i)* *The forward system* $\{A_n, B_n, C_n, D_n\}$ *is controllable at time* k.

*(ii)* *The forward controllability operator* $W_{ck}$ *is onto* $\mathcal{X}_k$.

*(iii)* *The operator* $W_{ck}^* := [B_{k-1}^*, B_{k-2}^*\Phi(k, k-1)^*, \dots]^{tr}$ *is one to one.*

*(iv)* *The family of pairs* $\{A_n, B_n\}$ *is forward controllable at time k.*

*(v)* *The family of pairs* $\{B_n^*, A_n^*\}$ *is backward observable at time* $k-1$.

Now assume that $\{A_n\}$ is forward stable and $\{B_n\}$ is uniformly bounded. Then $W_{ck}$ defines a bounded operator from $\mathcal{U}_{(-\infty,k-1)} = \oplus\{\mathcal{U}_n : n < k\}$ into $\mathcal{X}_k$. In this case the *forward controllability grammian at time* k for the family of pairs $\{A_n, B_n\}_{n\in\mathbb{Z}}$ is defined by $P_k = W_{ck}W_{ck}^*$. Obviously $P_k \ge 0$. Moreover, $P_k > 0$ if and only if the family of pairs $\{A_n, B_n\}_{n\in\mathbb{Z}}$ is controllable at time k. Using $W_{c(k+1)} = [B_k, A_kW_{ck}]$ it follows that $P_k$ satisfies the following nonstationary Lyapunov difference equation

$$P_{k+1} = A_kP_kA_k^* + B_kB_k^* . \tag{2.10}$$

Notice that this is the dual of the backward observability Lyapunov difference equation (2.5) for the anticausal observability grammian. Finally, if $u = (u(n))_{-\infty}^{k-1}$ from $\vec{\mathcal{U}}_{(-\infty,k-1)}$ is the input to the forward system (1.1) with no initial condition, then $x(k) = W_{ck}u$. So if the family of pairs $\{A_n, B_n\}_{n\in\mathbb{Z}}$ is forward controllable at time k, we can use standard pseudo inversion techniques to show that the input u which drives the forward system (1.1) to the state $x_k$ at time k is given by

$$u = W_{ck}^*(W_{ck}W_{ck}^*)^{-1}x_k = W_{ck}^*P_k^{-1}x_k . \tag{2.11}$$

In fact substituting (2.11) into $x(k) = W_{ck}u$ yields $x(k) = x_k$.

Now assume that $\mathcal{X}_k$ is infinite dimensional, $\{A_n\}$ is forward stable and $\{B_n\}$ is uniformly bounded. Then the controllability grammians $P_k = W_{ck} W_{ck}^*$ are well defined and satisfy the Lyapunov difference equation in (2.10). In this stable infinite dimensional setting we say that the family of pairs $\{A_n, B_n\}_{n \in \mathbb{Z}}$ is *forward controllable at time* k if $W_{ck}$ is onto all of $\mathcal{X}_k$, or equivalently, $P_k$ is strictly positive. As in the finite dimensional setting $\{A_n, B_n\}_{n \in \mathbb{Z}}$ is forward controllable if $\{A_n, B_n\}_{n \in \mathbb{Z}}$ is forward controllable at time k for all k.

Once again let us assume that $\mathcal{X}_k$ is finite dimensional. In this case the anticausal system $\{Z_n, B_n, C_n, D_n\}$ in (1.11) is *backward controllable at time* k if given any $x_k$ in $\mathcal{X}_k$, then there exists an input sequence $u = (u(n))_{k+1}^{\infty}$ with $u_j \in \mathcal{U}_j$ non-zero for a finite number of j's only and such that $x_k = x(k)$. This system is *backward controllable* if it is backward controllable for all k in $\mathbb{Z}$. By recursively solving for x(k) in (1.10) we see that $x(k) = W_{cka} u$, where $W_{cka}$ is the *backward controllability operator at time* k defined by

$$W_{cka} = [B_{k+1}, \Phi(k, k+1)B_{k+2}, \Phi(k, k+2)B_{k+3}, \ldots] . \tag{2.12}$$

Here we view $W_{cka}$ as a linear map defined on the space consisting of all sequences $(u_{k+1}, u_{k+2}, \ldots)$ with $u_j \in \mathcal{U}_j$ non-zero for a finite number of j's. Obviously this system is backward controllable at time k if and only if $W_{cka}$ is onto. By adapting Proposition 2.3 to the backward setting we have.

**PROPOSITION 2.4**. *Let* $\{Z_n, B_n, C_n, D_n\}$ *be a finite dimensional backward system of the form (1.11). Then the following statements are equivalent*

*(i)* *The anticausal system* $\{Z_n, B_n, C_n, D_n\}$ *in (1.11) is backward controllable at time* k.

*(ii)* *The backward controllability operator* $W_{cka}$ *is onto.*

*(iii)* *The operator* $W_{cka}^* := [B_{k+1}^*, B_{k+2}^*\Phi(k, k+1)^*, \ldots]^{tr}$ *is one to one.*

*(iv)* *The family of pairs* $\{Z_n, B_n\}$ *is backward controllable at time* k.

*(v)* *The family of pairs* $\{B_n^*, Z_n^*\}$ *is forward observable at time* k + 1.

Now assume that $\{Z_n\}$ is backward stable and $\{B_n\}$ is uniformly bounded. Then $W_{cka}$ is a bounded operator from $\vec{\mathcal{U}}_{(k+1,\infty)}$ to $\mathcal{X}_k$. In this case *backward controllability grammian at time* k for the family of pairs $\{Z_n, B_n\}_{n \in \mathbb{Z}}$ is defined by $P_{ka} = W_{cka} W_{cka}^*$. Obviously $P_{cka} \geq 0$. Moreover, $P_{cka} > 0$ if and only if the family of pairs $\{Z_n, B_n\}_{n \in \mathbb{Z}}$ is backward controllable at time k. Using $W_{c(k-1)a} = [B_k, Z_k W_{cka}]$ it follows that $P_{ka}$ satisfies the discrete time Lyapunov equation

$$P_{(k-1)a} = Z_k P_{ka} Z_k^* + B_k B_k^* . \tag{2.13}$$

Clearly this difference equation is the dual of the causal observability Lyapunov difference equation in (2.2). Now assume that $\mathcal{X}_k$ is infinite dimensional, $\{Z_n\}$ is backward stable and $\{B_n\}$ is uniformly bounded. Then $P_{ka}$ is well defined and satisfies the Lyapunov difference equation in (2.13). In this setting we say that $\{Z_n, B_n\}_{n\in\mathbb{Z}}$ is *backward controllable at time* k if $W_{cka}$ is onto all of $\mathcal{X}_k$, or equivalently, $P_{ka}$ is strictly positive. As in the finite dimensional setting $\{Z_n, B_n\}_{n\in\mathbb{Z}}$ is backward controllable, if $\{Z_n, B_n\}_{n\in\mathbb{Z}}$ is backward controllable at time k for all k.

## IX.3. POINT EVALUATION

Motivated by our previous discussion on nonstationary systems, we will now introduce the concept of point evaluation for nonstationary systems. Recall that in the stationary case, point or operator evaluation for a function $F(\lambda)$ in $H^\infty(\mathcal{Y}, \mathcal{U})$ with respect to the operator values $\{Z, B\}$ is defined by

$$(BF)(Z)_{\text{left}} = \sum_0^\infty Z^n B F_n = W_c T_F \Pi_0^* \quad \text{where } \Pi_0 = [I, 0, 0, \ldots ] . \tag{3.1}$$

Here $\{F_n\}_0^\infty$ are the Fourier coefficients for $F(\lambda)$, and Z is a stable operator on $\mathcal{X}$ while B maps $\mathcal{U}$ into $\mathcal{X}$. Moreover, $W_c$ is the (stationary) controllability operator $[B, ZB, Z^2B, \ldots ]$ from $\ell_+^2(\mathcal{U})$ into $\mathcal{X}$ and $T_F$ is the lower triangular Toeplitz operator generated by F. By choosing F in $H^\infty$ and setting $B = 1$ with $Z = \alpha$, a complex number in the open unit disc, then (3.1) reduces to $F(\alpha)$; the evaluation of F at the point $\alpha$. So point evaluation in the stationary case can be viewed as $W_c T_F \Pi_0^*$ where $W_c$ is a controllability operator.

From our previous analysis we can deduce that point evaluation in the nonstationary case should be defined as $W_c T \Pi_0^*$ where $W_c$ is a nonstationary controllability operator and T is a lower triangular block matrix. However, what controllability operator should we use, the forward or backward. It turns out that the backward controllability operator is the most natural one to use in nonstationary point evaluation. To see this let $\{A_n, B_n, C_n, D_n\}$ be a forward stable system of the form (1.1). According to (1.10) we have

$$W_{ok}^* T_k u = W_{ok}^* W_{ok} x(k) - W_{ok}^* y \tag{3.2}$$

where the output $y = (y(n))_k^\infty$ and the input $u = (u(n))_k^\infty$. Here $W_{ok}$ is the forward observability operator generated by the family of pairs $\{C_n, A_n\}$ at time k and $T_k$ is the block lower triangular matrix generated by $\{F_{n,m} : n \geq k \text{ and } m \geq k\}$ defined in (1.6). Since $W_{ok}^*$ can be viewed as a backward controllability operator, the term $W_{ok}^* T_k$ in (3.2) suggests that point evaluation in the nonstationary case should be of the form $W_c T \Pi_o^*$ where $W_c$ is a backward controllability operator.

Following this analysis we are now ready to define point evaluation for nonstationary systems. Let $\{Z_n : \mathcal{X}_n \to \mathcal{X}_{n-1}, B_n\}$ be a backward stable system where $\{B_n : \mathcal{U}_n \to \mathcal{X}_{n-1}\}$ is uniformly bounded. Let $\mathcal{F} = \{F_{n,m} \in \mathcal{L}(\mathcal{Y}_m, \mathcal{U}_n) : n, m \in \mathbb{Z}\}$ be a set of operators where $F_{n,m} = 0$ if $m > n$. Finally, let $T_k$ from $\vec{\mathcal{Y}}_k = \oplus\{\mathcal{Y}_n : n \geq k\}$ into $\vec{\mathcal{U}}_k = \oplus\{\mathcal{U}_n : n \geq k\}$ be the block lower triangular operators generated by $\{F_{n,m} : n \geq k \text{ and } m \geq k\}$ and assume that $\{T_k\}$ are uniformly bounded. (Obviously $\{T_k\}$ are uniformly bounded if and only if the lower triangular operator L with entries $\{F_{n,m}\}$ determines a bounded linear operator from $\oplus\{\mathcal{Y}_n : n \in \mathbb{Z}\}$ into $\oplus\{\mathcal{U}_n : n \in \mathbb{Z}\}$.) Then the left nonstationary *point evaluation of $\mathcal{F}$ at time k with respect to the family of pairs* $\{Z_n, B_n\}$ is defined by

$$(\{B_n\}\mathcal{F}\{Z_n\})_{k,\,\text{left}} = \sum_{\nu=0}^{\infty} \Phi(k-1, k+\nu-1) B_{k+\nu} F_{k+\nu,k} =$$

$$[B_k, \Phi(k-1,k)B_{k+1}, \Phi(k-1,k+1)B_{k+2}, \ldots]T_k \Pi_o^* = W_k T_k \Pi_o^* \tag{3.3}$$

where $\Phi(n, m) = Z_{n+1}Z_{n+2}\ldots Z_m$ for $n < m$ and $\Phi(n, n) = I$ is the state transition operator generated by $\{Z_n\}$. Here $\Pi_o$ is the operator from $\vec{\mathcal{Y}}_k$ onto $\mathcal{Y}_k$ defined by $\Pi_o = [I, 0, 0, \ldots]$. Notice that the definition of $\Pi_o$ depends upon k. However, for convenience the emphasis on k is not expressed in the notation. (Right nonstationary point evaluation is defined in an analogous way.) Obviously the operator $W_k$ appearing before $T_k$ is the backward controllability grammian for the family of pairs $\{Z_n, B_n\}$ at time $k-1$. For convenience of notation only we shifted the index by one in the backward controllability operator $W_k$ for our definition of nonstationary point evaluation. To see how this reduces to point evaluation in the stationary case for the moment assume that $B = B_n$ and $Z = Z_n$ for all n where $r_{spec}(Z) < 1$. Furthermore, assume that the $\{F_{n,m}\}$ are stationary, that is, $F_{n,m} = F_{n-m}$ for all n and m, and of course $F_j = 0$ if $j < 0$. In this case $T_k = T_F$ where F is now the symbol for the Toeplitz operator $T_F$ generated by $(F_n)_0^\infty$. Then $(\{B_n\}\mathcal{F}\{Z_n\})_{k,\,\text{left}} = (BF)(Z)_{\text{left}}$ for all k.

As in the time invariant case nonstationary point evaluation naturally occur when one chooses the appropriate input for the nonstationary system. To be more precise, let $\{A_n, B_n, C_n, D_n\}$ be a forward stable system of the form (1.1). Recall that the output $y = (y(n))_{-\infty}^{\infty}$ for this causal system with no initial condition and input $u = (u(n))_{-\infty}^{\infty}$ is given by $y = Lu$ where L is now the block lower triangular (causal) matrix from $\vec{\mathcal{U}} = \oplus\{\mathcal{U}_n : n \in \mathbb{Z}\}$ into $\vec{\mathcal{Y}} = \{\oplus \mathcal{Y}_n : n \in \mathbb{Z}\}$ generated by $\mathcal{F} = \{F_{n,m} : n, m \in \mathbb{Z}\}$ in (1.6). Now let us choose an input u of the form $u(n) = 0$ if $n > k$ while $u(k) = E_k e$ and $u(n) = E_n e z_{n+1} z_{n+2} \ldots z_k$ for $n < k$

where $\{E_n\}$ are uniformly bounded operators from $\mathcal{E}$ into $\mathcal{U}_n$ for all n and e is in $\mathcal{E}$. Furthermore, assume that $\{z_n\}$ are stable scalars (that is, $x_{n+1} = z_n x_n$ is a stable system). Then the output y(k) at time k is

$$y(k) = F_{k,k}E_k e + \sum_{v=-\infty}^{-1} F_{k,k+v}E_{k+v}z_{k+v+1} \cdots z_k e = (\{E\}\mathcal{F}\{Z_n\})_{k,\text{right}} . \tag{3.4}$$

Therefore the output y(k) at time k is precisely the right point evaluation at k acting on the vector e.

To complete this we will present some sample examples to provide some insight into point evaluation for nonstationary systems.

**EXAMPLE 3.1.** Probably the simplist example of point evaluation is when $Z_j = 0$ and $B_j = 1$ for all j. Then the controllability operator becomes $W_k = [1, 0, 0, 0, \ldots]$. So if $T_k$ is the lower triangular matrix on $\ell^2_+$ generated by $\{f_{m,n} : m \geq k \text{ and } n \geq k\}$, then the point evaluation of $\{f_{m,n}\}$ at $\{Z_j, B_j\}$ at time k is given by $W_k T_k \Pi_o^* = f_{k,k}$ for all k. In this case, point evaluation simply picks the diagonal entries $f_{k,k}$ from the set $\{f_{m,n}\}$.

**EXAMPLE 3.2.** Let $a(\lambda)$ and $b(\lambda)$ be two functions in $H^\infty$ whose Fourier coefficients are $\{a_n\}_0^\infty$ and $\{b_n\}_0^\infty$, respectively. Let L be the lower triangular operator on $\ell^2$ generated by $\{f_{m,n}\}$ defined by

$$\begin{aligned} f_{m,n} &= a_{m-n} \quad \text{if} \quad m \geq n \quad \text{and} \quad m \leq 0 \\ &= b_{m-n} \quad \text{if} \quad m \geq n \quad \text{and} \quad m > 0 \\ &= 0 \qquad \text{otherwise} \end{aligned} \tag{3.5}$$

Notice that the operator L is precisely the lower triangle operator obtained by placing $a_j$ on the $j-$th diagonal below the main diagonal above the main row, otherwise we put $b_j$ on the $j-$th subdiagonal, that is,

$$L = \begin{bmatrix} & \vdots & \vdots & \vdots & \\ \cdots & a_0 & 0 & 0 & \cdots \\ \cdots & a_1 & \underline{a_0} & 0 & \cdots \\ \cdots & b_2 & b_1 & b_0 & \cdots \\ & \vdots & \vdots & \vdots & \end{bmatrix} . \tag{3.6}$$

In this case the operator $T_k$ on $\ell^2_+$ generated by $\{f_{m,n} : m \geq k \;\text{ and }\; n \geq k\}$ is given by for $k \leq 0$ and $q = -k$

$$T_k = \begin{bmatrix} a_0 & 0 & 0 & \cdots \\ a_1 & a_0 & 0 & \cdots \\ \vdots & \vdots & \vdots & \\ a_q & a_{q-1} & a_{q-2} & \cdots \\ b_{q+1} & b_q & b_{q-1} & \cdots \\ b_{q+2} & b_{q+1} & b_q & \cdots \\ \vdots & \vdots & \vdots & \end{bmatrix} . \tag{3.7}$$

Moreover, $T_k = T_b$ the Toeplitz operator on $\ell_+^2$ generated by b if $k > 0$. Now assume that $B_j = 1$ and $Z_j = \alpha$ for all j where $\alpha$ is a scalar in the open unit disc. In this case the (backward) controllability operator becomes

$$W_k = [1, \alpha, \alpha^2, \alpha^3, \ldots] . \tag{3.8}$$

As before, $\Pi_o = [1, 0, 0, \ldots]$ So by using (3.7) and (3.8) we see that the point evaluation of $\{f_{m,n}\}$ with respect to $\{Z_j, B_j\}$ is given by

$$W_k T_k \Pi_o^* = a_o + a_1\alpha + \cdots + a_q\alpha^q + \sum_{j=q+1}^{\infty} b_j\alpha^j \tag{3.9}$$

when $-q = k \le 0$. In particular, we see that the point evaluation $W_k T_k \Pi_o^*$ approaches $a(\alpha)$ as k approaches $-\infty$. On the other hand, the point evaluation of $\{f_{m,n}\}$ at time k for $k > 0$ is defined by $W_k T_k \Pi_o^* = W_k T_b \Pi_o^* = b(\alpha)$. As expected this is precisely the stationary point evaluation of the function $b(\lambda)$ at $\alpha$.

**EXAMPLE 3.3**. For our next example consider the case when $B_j = 1$ for all j and $Z_j = \alpha$ for $j < 0$ and $Z_j = \beta$ for $j \ge 0$ where $\alpha$ and $\beta$ are two complex numbers in the open unit disc. Now let L be the Laurent operator on $\ell^2$ generated by the function a in $H^\infty$ with Fourier coefficients $\{a_n\}_0^\infty$. Then $T_k = T_a$ is the Toeplitz matrix on $\ell_+^2$ with symbol a. In this case if $-q = k < 0$, then the backward controllability operator becomes

$$W_k = [1, \alpha, \alpha^2, \ldots, \alpha^q, \alpha^q\beta, \alpha^q\beta^2, \alpha^q\beta^3, \ldots] . \tag{3.10}$$

So if $k < 0$, then the evaluation of $\{f_{m,n} = a_{m-n}\}$ at the points $\{Z_j, 1\}$ at time k is given by

$$W_k T_k \Pi_o^* = W_k T_a \Pi_o^* = a_0 + a_1\alpha + \cdots + a_q\alpha^q + \alpha^q \sum_{j=q+1}^{\infty} a_j\beta^{(j-q)} . \tag{3.11}$$

On the other hand, if $k \ge 0$, then $W_k = [1, \beta, \beta^2, \ldots]$, and thus the point evaluation at time $k \ge 0$ is given by $W_k T_k \Pi_o^* = W_k T_a \Pi_o^* = a(\beta)$.

**EXAMPLE 3.4.** In this example we will show how nonstationary periodic point evaluation can be viewed as stationary point evaluation. To this end, let $\{a_j\}$ and $\{b_j\}$ be a sequence of complex numbers satisfying $a_j = 0$ and $b_j = 0$ for $j < 0$. Now let $\{f_{i,j}\}$ be the sequence defined by

$$\begin{aligned} f_{i,j} &= a_{i-j} \quad \text{if } j \text{ is even} \\ &= b_{i-j} \quad \text{if } j \text{ is odd}\,. \end{aligned} \tag{3.12}$$

Notice that $\{f_{i,j}\}$ defined a periodic sequence of order two, that is, $f_{i,j} = f_{i+2,j+2}$. In particular, for even k the lower triangular block matrix $T_k$ becomes

$$T_k = \begin{bmatrix} a_0 & 0 & 0 & 0 & \cdots \\ a_1 & b_0 & 0 & 0 & \cdots \\ a_2 & b_1 & a_0 & 0 & \cdots \\ a_3 & b_2 & a_1 & b_0 & \cdots \\ \cdot & \cdot & \cdot & \cdot & \cdots \\ \cdot & \cdot & \cdot & \cdot & \cdots \\ \cdot & \cdot & \cdot & \cdot & \cdots \end{bmatrix} . \tag{3.13}$$

For odd k the operator $T_k$ is given by (3.13) where $b_j$ replaces $a_j$ and $a_j$ replaces $b_j$. Throughout we assume that the operator $T_k$ is bounded for some k and thus all k. Now let $B_j = 1$ for all j and set $Z_j = \alpha$ for even j and $Z_j = \beta$ for odd j where $\alpha$ and $\beta$ are complex numbers in the open unit disc. For this choice of $\{Z_j, B_j\}$ the controllability operator becomes

$$\begin{aligned} W_k &= [1,\ \alpha,\ \alpha\beta,\ \alpha\beta\alpha,\ \alpha\beta\alpha\beta,\ \ldots\,] \qquad \text{(for even k)} \\ &= [1,\ \beta,\ \beta\alpha,\ \beta\alpha\beta,\ \beta\alpha\beta\alpha,\ \ldots\,] \qquad \text{(for odd k)}\,. \end{aligned} \tag{3.14}$$

Therefore the point evaluation of $\{f_{i,j}\}$ with respect to $\{Z_j, B_j\}$ is given by

$$\begin{aligned} a(\alpha, \beta) &:= W_k T_k \Pi_o^* = a_0 + a_1\alpha + a_2\alpha\beta + a_3\alpha\beta\alpha + \cdots \qquad \text{(for even k)} \\ b(\alpha, \beta) &:= W_k T_k \Pi_o^* = b_0 + b_1\beta + b_2\beta\alpha + b_3\beta\alpha\beta + \cdots \qquad \text{(for odd k)}\,. \end{aligned} \tag{3.15}$$

To show how this can be viewed as stationary point evaluation, let F(z) be the function whose values are linear operators on $C^2$ defined by

$$F(z) = \begin{bmatrix} a_0 & b_0 \\ a_1 & b_1 \end{bmatrix} + \begin{bmatrix} a_2 & b_2 \\ a_3 & b_3 \end{bmatrix} z + \begin{bmatrix} a_4 & b_4 \\ a_5 & b_5 \end{bmatrix} z^2 + \cdots \tag{3.16}$$

Throughout we assume that F(z) is in $H^\infty(\mathbb{C}^2, \mathbb{C}^2)$. Now let Z and B be the matrices defined by

$$Z = \begin{bmatrix} \alpha\beta & 0 \\ 0 & \beta\alpha \end{bmatrix} \quad \text{and} \quad B = \begin{bmatrix} 1 & \alpha \\ 1 & \beta \end{bmatrix}. \tag{3.17}$$

Then the controllability operator $W = [B, ZB, Z^2B, \dots]$ generated by the pair $\{Z, B\}$ is given by

$$W = \begin{bmatrix} 1 & \alpha & \alpha\beta & \alpha\beta\alpha & \dots \\ 1 & \beta & \beta\alpha & \beta\alpha\beta & \dots \end{bmatrix} = \begin{bmatrix} W_0 \\ W_1 \end{bmatrix}. \tag{3.18}$$

Here $W_0$ and $W_1$ are the nonstationary controllability operator defined in (3.14) for $k = 0, 1$. Notice that the left point evaluation of F(z) with respect to $\{Z, B\}$ is given by

$$(BF)(Z)_{\text{left}} = WT_F\Pi_o^* = \begin{bmatrix} a(\alpha, \beta) & * \\ * & b(\alpha, \beta) \end{bmatrix} \tag{3.19}$$

where $a(\alpha, \beta)$ and $b(\alpha, \beta)$ are defined in (3.15). Therefore in this periodic case we can obtain the nonstationary point evaluation by computing a special stationary point evaluation. Here we have concentrated on the case when the period is two. Obviously this can be generalized to the case when the period $\tau$, that is, when $f_{i,j} = f_{i+\tau,j+\tau}$ and $Z_j = Z_{j+\tau}$. The details are left to the reader as a simple exercise.

## IX.4. FROM NONSTATIONARY SYSTEMS TO STATIONARY SYSTEMS

In this section we will show how one can convert a nonstationary system into an infinite dimensional stationary system. This technique will play a fundamental role in solving many nonstationary interpolation problems.

To begin, let $\{A_n, B_n, C_n, D_n\}$ be a forward stable system of the form (1.1). Moreover, assume that $\{A_n\}$, $\{B_n\}$, $\{C_n\}$ and $\{D_n\}$ are all uniformly bounded. To convert this system to an infinite dimensional, stable, stationary system, let $\vec{\mathcal{U}}$, $\vec{\mathcal{X}}$ and $\vec{\mathcal{Y}}$ be the spaces defined by

$$\vec{\mathcal{U}} = \bigoplus_{-\infty}^{\infty} \mathcal{U}_n \quad \text{and} \quad \vec{\mathcal{X}} = \bigoplus_{-\infty}^{\infty} \mathcal{X}_n \quad \text{and} \quad \vec{\mathcal{Y}} = \bigoplus_{-\infty}^{\infty} \mathcal{Y}_n. \tag{4.1}$$

Now let us introduce the operators

$$\mathcal{A} : \vec{\mathcal{X}} \to \vec{\mathcal{X}},\ (\mathcal{A}\vec{x})_j = A_{j-1}x_{j-1} \quad (j \in \mathbb{Z})\ ;$$

$$\mathcal{B} : \vec{\mathcal{U}} \to \vec{\mathcal{X}},\ (\mathcal{B}\vec{u})_j = B_{j-1}u_{j-1} \quad (j \in \mathbb{Z})\ ;$$
$$\tag{4.2}$$
$$\mathcal{C} : \vec{\mathcal{X}} \to \vec{\mathcal{Y}},\ (\mathcal{C}\vec{x})_j = C_jx_j \quad (j \in \mathbb{Z})\ ;$$

$$\mathcal{D} : \vec{\mathcal{U}} \to \vec{\mathcal{Y}},\ (\mathcal{D}\vec{u})_j = D_ju_j \quad (j \in \mathbb{Z})\ .$$

Since the families $\{A_n\}$, $\{B_n\}$, $\{C_n\}$ and $\{D_n\}$ are uniformly bounded, the operators $\mathcal{A}$, $\mathcal{B}$, $\mathcal{C}$ and $\mathcal{D}$ are well-defined bounded linear operators. Notice that $\mathcal{A}$ and $\mathcal{B}$ are block weighted forward shifts, while $\mathcal{C}$ and $\mathcal{D}$ are block diagonal operators. By using these operators it is easy to show that the forward system $\{A_n, B_n, C_n, D_n\}$ in (1.1) is equivalent to

$$\vec{x} = \mathcal{A}\vec{x} + \mathcal{B}\vec{u} \quad \text{and} \quad \vec{y} = \mathcal{C}\vec{x} + \mathcal{D}\vec{u} \tag{4.3}$$

where $\vec{u} = (u(n))_{-\infty}^{\infty}$ is the input and $\vec{y} = (y(n))_{-\infty}^{\infty}$ the output.

Notice that for $n \geq 1$

$$\|\mathcal{A}^n\| = \sup\ \{\|A_{k-1}A_{k-2} \cdots A_{k-n}\| : k \in \mathbb{Z}\}\ .$$

Since the system (1.1) is assumed to be forward stable it follows that $r_{spec}(\mathcal{A}) < 1$. So $\mathcal{A}$ is a stable operator. In particular, $I - \mathcal{A}$ has a bounded inverse. So by solving for $\vec{y}$ in terms of $\vec{u}$ in (4.3), we readily obtain that

$$\vec{y} = \mathcal{C}(I - \mathcal{A})^{-1}\mathcal{B}\vec{u} + \mathcal{D}\vec{u}\ . \tag{4.4}$$

Therefore the bounded input-output map L of the stable forward system $\{A_n, B_n, C_n, D_n\}$ is given by

$$L = \mathcal{C}(I - \mathcal{A})^{-1}\mathcal{B} + \mathcal{D}\ . \tag{4.5}$$

Using $(I - \mathcal{A})^{-1} = \Sigma\mathcal{A}^n$ along with the fact that $\mathcal{A}$ is a block weighted forward shift, we see that $(I - \mathcal{A})^{-1}$ is block lower triangular. Since $\mathcal{B}$ is a block weighted forward shift while $\mathcal{C}$ and $\mathcal{D}$ are block diagonal, we conclude that the operator L in (4.5) is a block lower triangular operator from $\vec{\mathcal{U}}$ into $\vec{\mathcal{Y}}$. This is another way of showing that the stable forward system $\{A_n, B_n, C_n, D_n\}$ generates a causal bounded input output map. Finally, notice that the block diagonal entrices of L coincide with those of $\mathcal{D}$. Furthermore, the n-th block diagonal of L below its main diagonal is precisely the n-th block diagonal in the lower triangular part of $\mathcal{C}\mathcal{A}^{n-1}\mathcal{B}$. It follows that for $n \geq m$ the (n, m)-th entry of our operator L in (4.5) is precisely the operator $F_{n,m}$ defined in (1.6).

Now consider the time invariant system generated by the enlarged stable linear system $\{\mathcal{A}, \mathcal{B}, \mathcal{C}, \mathcal{D}\}$, that is,

$$\vec{x}(n+1) = \mathcal{A}\vec{x}(n) + \mathcal{B}\vec{u}(n) \quad \text{and} \quad \vec{y}(n) = \mathcal{C}\vec{x}(n) + \mathcal{D}\vec{u}(n) . \tag{4.6}$$

Here the input $\vec{u}(n)$ at time n is a vector in $\vec{\mathcal{U}}$ and the output $\vec{y}(n)$ at time n is a vector in $\vec{\mathcal{Y}}$. So for the enlarged system (4.6) the input-output map is of the form $y = L_F u$ where the input $u = (\vec{u}(n))_{-\infty}^{\infty}$ and output $y = (\vec{y}(n))_{-\infty}^{\infty}$. Moreover, $L_F$ is a bounded block lower triangular Laurent operator from $\ell^2(\vec{\mathcal{U}})$ into $\ell^2(\vec{\mathcal{Y}})$ whose symbol F is the transfer function in $H^\infty(\vec{\mathcal{U}}, \vec{\mathcal{Y}})$ given by

$$F(\lambda) = \mathcal{D} + \lambda\mathcal{C}(I - \lambda\mathcal{A})^{-1}\mathcal{B} . \tag{4.7}$$

By consulting equation (III.1.6), the Taylor coefficients $\{F_n\}$ of F are given by

$$F_0 = \mathcal{D} \quad \text{and} \quad F_n = \mathcal{C}(\mathcal{A})^{n-1}\mathcal{B} \qquad (\text{if } n \geq 1) . \tag{4.8}$$

Recall that $F_n$ is an operator from $\vec{\mathcal{U}}$ into $\vec{\mathcal{Y}}$. Using the special form of $\mathcal{A}$, $\mathcal{B}$, $\mathcal{C}$ and $\mathcal{D}$, it follows that all the entries of the block operator matrix representation of $F_n$ off the $n-$th subdiagonal are zero. More precisely

$$F_n = (\delta_{n,j-k}F_{j,k})_{j,k=-\infty}^{\infty} , \tag{4.9}$$

where $\delta_{m,n}$ is the Kronecker delta, $F_{j,k} = 0$ for $j < k$ and $F_{j,k}$ for $j \geq k$ is defined in (1.6). Therefore the block Laurent operator $L_F$ generated by the enlarged system $\{\mathcal{A}, \mathcal{B}, \mathcal{C}, \mathcal{D}\}$ has a special structure. This structure will be analyzed further in Chapter X, and will be used to solve many nonstationary interpolation problems.

## IX.5. A NONSTATIONARY FILTERING PROBLEM

This section is the nonstationary version of Section I.11. Here we will show how a right-sided nonstationary Sarason problem naturally arises in estimation theory. To this end, let $x = (x(n))_{-\infty}^{\infty}$ and $y = (y(n))_{-\infty}^{\infty}$ be two scalar valued random processes, and assume that the covariance operators $R_x$ and $R_y$ are bounded on $\ell^2$. This readily implies that the covariance operator $R_z$ for $z = [x, y]^{tr}$ defined by

$$R_z = \begin{bmatrix} R_x & R_{xy} \\ R_{yx} & R_y \end{bmatrix} \tag{5.1}$$

is a bounded, positive operator on $\ell^2 \oplus \ell^2$; see Section I.11. Throughout this section we always assume that the covariance operator $R_z$ is invertible. In particular, this assumption requires that both $R_x$ and $R_y$ are invertible positive operators on $\ell^2$. In the nonstationary case, we say that the process $\hat{x} = (\hat{x}(n))_{-\infty}^{\infty}$ is an *estimate* of the process x from y if $\hat{x} = L_c y$ where $L_c$ is a lower triangular bounded operator on $\ell^2$. In this case, the operator $L_c$ is called the *filter* for the estimate $\hat{x}$ of x. So our filtering problem is to construct a causal filter such that the process $\hat{x} = L_c y$ extracted from y comes as close to the process x as possible, that is, we want to construct a causal filter $L_c$ to make the error processes $(e(n))_{-\infty}^{\infty} = e = x - L_c y$ as small as possible. (By causal we mean that $L_c$ is lower triangular.) This leads to the following nonstationary optimization problem

$$d_\infty^2 = \inf \{\|R_e\| : e = x - L_c y \text{ and } L_c \text{ is a bounded causal operator on } \ell^2\} . \tag{5.2}$$

To obtain a solution to the optimization problem in (5.2), let $L_c$ be any bounded lower triangular operator on $\ell^2$. Then using $e = x - L_c y$ we obtain

$$R_e = E(ee^*) = R_x - L_c R_{yx} - R_{xy} L_c^* + L_c R_y L_c^* . \tag{5.3}$$

Since $R_y$ is a bounded positive invertible operator on $\ell^2$, it admits a factorization of the form $R_y = L_\theta L_\theta^*$ where $L_\theta$ is invertible, and both $L_\theta$ and $L_\theta^{-1}$ are (bounded) lower triangular operators on $\ell^2$. We call this operator $L_\theta$ a *nonstationary co-outer factor of* $R_y$. Now let $L_h$ be the lower triangular operator defined by $L_h = L_c L_\theta$. Using $R_{xy} = R_{yx}^*$ and $R_y = L_\theta L_\theta^*$ we have

$$\begin{aligned} R_e &= R_x - L_h L_\theta^{-1} R_{yx} - R_{xy} L_\theta^{-*} L_h^* + L_h L_h^* = \\ &R_x - R_{xy} R_y^{-1} R_{yx} + (R_{xy} L_\theta^{-*} - L_h)(R_{xy} L_\theta^{-*} - L_h)^* . \end{aligned} \tag{5.4}$$

Since $R_z$ is a positive invertible operator, equation (I.11.4) shows that its Schur complement $R = R_x - R_{xy} R_y^{-1} R_{yx}$ is a bounded positive invertible operator. Therefore R admits a nonstationary co-outer factorization $R = L_o L_o^*$ where $L_o$ is a bounded, invertible causal operator on $\ell^2$. Substituting $R = L_o L_o^*$ into (5.4) readily yields

$$R_e = L_o L_o^* + (R_{xy} L_\theta^{-*} - L_h)(R_{xy} L_\theta^{-*} - L_h)^* . \tag{5.5}$$

Using the fact that $\|A\| = \|A^*\|$ we have

$$\|R_e\| = \|[R_{xy} L_\theta^{-*} - L_h, L_o]\|^2 . \tag{5.6}$$

Therefore the filtering minimization problem in (5.2) is equivalent to minimizing the norm of $\|[R_{xy} L_\theta^{-*} - L_h, L_o]\|$ over the set of all bounded lower triangular operators $L_h$ on $\ell^2$. This is precisely a nonstationary Sarason problem which will be solved in Chapter XI. Summarizing the previous analysis we readily obtain the following result.

**PROPOSITION 5.1.** *Let* $x = (x(n))_{-\infty}^{\infty}$ *and* $y = (y(n))_{-\infty}^{\infty}$ *be two random processes such that the covariance* $R_z$ *for* $z = [x, y]^{tr}$ *defined in (5.1) is a bounded invertible (positive) operator on* $\ell^2 \oplus \ell^2$. *Let* $L_\theta$ *and* $L_o$ *be two bounded invertible lower triangular operators with a lower triangular inverse satisfying*

$$R_y = L_\theta L_\theta^* \quad \text{and} \quad R_x - R_{xy} R_y^{-1} R_{yx} = L_o L_o^* . \tag{5.7}$$

*Then the filtering optimization problem in (5.2) is equivalent to the following nonstationary right-hand side Sarason problem:*

$$d_\infty = \inf \{ \| [R_{xy} L_\theta^{-*} - L_h ,\ L_o] \|^2 : L_h \text{ is a bounded causal operator on } \ell^2 \} . \tag{5.8}$$

*Moreover, if* $L_h$ *is an optimal solution to (5.8), then an optimal filter solving the nonstationary filtering optimization problem in (5.2) is given by* $L_c = L_h L_\theta^{-1}$.

## *Notes to Chapter IX:*

Sections 1 and 2 contain standard material about time-varying systems which can be found in many textbooks (see, e.g., Kailath [1] and Rugh [1]). The concept of point evaluation for doubly infinite block lower triangular matrices appears explicitly in Ball-Gohberg-Kaashoek [1], and has its roots in Alpay-Dewilde [1]. The system interpretation of this time-varying point evaluation has been explored in Ball-Gohberg-Kaashoek [8], and the latter paper served as a basis for the material in Section 3. The idea of transforming time-varying systems into time-invariant ones (described in Section 4) seems to be known in system theory, and has been used in nonstationary interpolation problems (see, e.g., Ball-Gohberg-Kaashoek [1] and Dewilde-Dym [1]). The nonstationary filtering problem in Section 5 is a natural analogue of the stationary one considered in Section I.11. Nonstationary filtering problems are treated in Caines [1].

## CHAPTER X

# REDUCTION TECHNIQUES: FROM NONSTATIONARY TO STATIONARY AND VICE VERSA

This chapter presents the reduction technique that will allow us to convert nonstationary interpolation problems into stationary ones. This technique is based on a transformation (and its inverse) which maps a doubly infinite operator matrix $F = (f_{j,k})_{j,k=-\infty}^{\infty}$ into a doubly infinite block Laurent matrix $\hat{F} = ([F]_{j-k})_{j,k=-\infty}^{\infty}$, where $[F]_n$ is the matrix which one obtains from F if all (operator) entries in F are set to zero except those on the n-th diagonal which are left unchanged. To be more precise,

$$[F]_n = (\delta_{j-k,n} f_{j,k})_{j,k=-\infty}^{\infty} ,$$

where $\delta_{m,n}$ is the Kronecker delta. This transformation $F \to \hat{F}$, which is called the diagonally sparse transform, is left invertible. One particular left inverse is defined as follows. Let $G = (G_{j-k})_{j,k=-\infty}^{\infty}$ be an arbitrary block Laurent operator, where $G_n$ is a double infinite operator matrix for each n, then the inverse diagonally sparse transform $\check{G}$ of G is the doubly infinite operator matrix whose n-th diagonal is equal to the n-th diagonal of $G_n$ for each n. In other words,

$$\check{G} = \sum_{n=-\infty}^{\infty} [G_n]_n .$$

These two operations are studied in the second section of this chapter. The first, which has a preliminary character, introduces the spatial features underlying these operations.

## X.1 SPATIAL FEATURES

Let $\mathcal{X}_j$, for $j \in \mathbb{Z}$, be a Hilbert space, and consider the Hilbert space direct sum

$$\vec{\mathcal{X}} = \bigoplus_{j=-\infty}^{\infty} \mathcal{X}_j = \left\{ \vec{x} = (x_j)_{j=-\infty}^{\infty} : x_j \in \mathcal{X}_j \text{ and } \sum_{j=-\infty}^{\infty} \|x_j\|^2 < \infty \right\}. \tag{1.1}$$

In this section we study different representations of the space $\ell^2(\vec{\mathcal{X}})$, with $\vec{\mathcal{X}}$ as in (1.1). Let $\tau_k$ be the canonical embedding of $\mathcal{X}_k$ into $\vec{\mathcal{X}}$, that is,

$$\tau_k : \mathcal{X}_k \to \vec{\mathcal{X}} \quad \text{and} \quad \tau_k x = (\delta_{j,k} x)_{j=-\infty}^{\infty} . \tag{1.2}$$

Here $\delta_{j,k}$ is the Kronecker delta. Thus all elements of the sequence $\tau_k x$ are zero except the k-th which is equal to x. Obviously, $\tau_k^*$ maps $\vec{\mathcal{X}}$ into $\mathcal{X}_k$ and $\tau_k^* \vec{x} = x_k$, where $x_k$ is the k-th component of $\vec{x}$. Now consider the operator $\tau$ from $\vec{\mathcal{X}}$ to $\ell^2(\vec{\mathcal{X}})$ defined by

$$\tau[\ldots, x_{-1}, \underline{x_0}, x_1, \ldots]^{tr} = [\ldots, \tau_{-1} x_{-1}, \underline{\tau_0 x_0}, \tau_1 x_1, \ldots]^{tr} . \tag{1.3}$$

We shall refer to $\tau\vec{\mathcal{X}}$ as the *sparse embedding* of $\vec{\mathcal{X}}$ into $\ell^2(\vec{\mathcal{X}})$, and we shall call $\tau$ the *canonical sparsing operator* corresponding to the spaces $\mathcal{X}_k$ for $k \in \mathbb{Z}$. The operator $\tau$ is an isometry. Indeed,

$$\tau^*[\ldots, \vec{x}_{-1}, \vec{x}_0, \vec{x}_1, \ldots]^{tr} = [\ldots, \tau_{-1}^* \vec{x}_{-1}, \tau_0^* \vec{x}_0, \tau_1^* \vec{x}_1, \ldots]^{tr}$$

and hence $\tau^* \tau$ is the identity operator.

Let V be the block forward shift on $\ell^2(\vec{\mathcal{X}})$. Since the range of $\tau_p$ is orthogonal to the range of $\tau_q$ for $p \neq q$, we see that the spaces $V^k \tau \vec{\mathcal{X}}$, for $k \in \mathbb{Z}$, are mutually orthogonal. (In particular $\tau\vec{\mathcal{X}}$ is a wandering subspace for V.) Their closed linear hull is precisely $\ell^2(\vec{\mathcal{X}})$, which leads to following proposition.

**PROPOSITION 1.1**. *Let $\vec{\mathcal{X}}$ be the Hilbert space direct sum defined in (1.1), and let $\tau$ from $\vec{\mathcal{X}}$ to $\ell^2(\vec{\mathcal{X}})$ be the corresponding canonical sparsing operator in (1.3). Then the map* U *on $\ell^2(\vec{\mathcal{X}})$ given by the one row operator matrix*

$$U = [\ldots, V^{*2}\tau, V^*\tau, \underline{\tau}, V\tau, V^2\tau, \ldots] \tag{1.4}$$

*is well-defined and unitary.*

**PROOF**. Since V and $\tau$ are isometries, we have $\|V^j \tau \vec{x}_j\| = \|\vec{x}_j\|$. It follows that

$$\sum_{j=-\infty}^{\infty} \|V^j \tau \vec{x}_j\|^2 = \sum_{j=-\infty}^{\infty} \|\vec{x}_j\|^2 = \|(\vec{x}_j)_{j=-\infty}^{\infty}\|^2 .$$

The vectors $V^j \tau \vec{x}_j$, for $j \in \mathbb{Z}$, are mutually orthogonal. Thus the above identities imply that the map U is well-defined and an isometry. Obviously, the range of U contains the linear hull of the spaces ran $V^j \tau$ for $j \in \mathbb{Z}$. Thus ran $U = \ell^2(\vec{\mathcal{X}})$, and U is unitary. This completes the proof.

We shall refer to the operator U in (1.4) as the *unitary operator associated with the sparse embedding* of $\vec{\mathcal{X}}$ into $\ell^2(\vec{\mathcal{X}})$. Using $VV^* = I$ in (1.4), it readily follows that $VU = UV$, and hence U is a Laurent operator. This operator U may also be written in the form

$$U = \sum_{j=-\infty}^{\infty} V^j \tau \pi_j . \tag{1.5}$$

Here for each j the map $\pi_j$ is the operator defined by picking out the $j-$th component from $\ell^2(\vec{\mathcal{X}})$, that is,

$$\pi_j : \ell^2(\vec{\mathcal{X}}) \to \vec{\mathcal{X}} \quad \text{and} \quad \pi_j((\vec{x}_k)_{k=-\infty}^{\infty}) = \vec{x}_j . \tag{1.6}$$

The convergence in (1.5) is in the strong operator topology. Since $\pi_j V = \pi_{j-1}$ for each j, the representation in (1.5) for U also shows that U and the block forward shift V commute. Indeed,

$$VU = \sum_{j=-\infty}^{\infty} V^{j+1} \tau \pi_j = \sum_{j=-\infty}^{\infty} V^j \tau \pi_{j-1} = UV .$$

Therefore once again one sees that U is a Laurent operator. According to (1.5) the column with index zero is given by $\tau$. Hence U admits a matrix representation of the form

$$U = \begin{bmatrix} & \vdots & \vdots & \vdots & \\ \cdots & \pi_0\tau & \pi_{-1}\tau & \pi_{-2}\tau & \cdots \\ \cdots & \pi_1\tau & \underline{\pi_0\tau} & \pi_{-1}\tau & \cdots \\ \cdots & \pi_2\tau & \pi_1\tau & \pi_0\tau & \cdots \\ \cdots & \pi_3\tau & \pi_2\tau & \pi_1\tau & \cdots \\ & \vdots & \vdots & \vdots & \end{bmatrix} \text{ on } \ell^2(\vec{\mathcal{X}}) \tag{1.7}$$

with block entries $U_{i,j} = \pi_{i-j}\tau$ on $\vec{\mathcal{X}}$.

As usual we identify $\ell^2_+(\vec{\mathcal{X}})$ with the subspace of $\ell^2(\vec{\mathcal{X}})$ consisting of all sequences $(\vec{x}_j)_{j=-\infty}^{\infty}$ in $\ell^2(\vec{\mathcal{X}})$ with $\vec{x}_j = 0$ for $j < 0$.

**PROPOSITION 1.2.** *Let* $P_+$ *be the orthogonal projection of* $\ell^2(\vec{\mathcal{X}})$ *onto* $\ell^2_+(\vec{\mathcal{X}})$, *and let* U *be the unitary operator defined by (1.4) or (1.5). Then*

$$U^{-1}P_+U = \begin{bmatrix} \ddots & & & \\ & P_1 & & \\ & & \underline{P_0} & \\ & & & P_{-1} & \\ & & & & \ddots \end{bmatrix} : \ell^2(\vec{\mathcal{X}}) \to \ell^2(\vec{\mathcal{X}}), \tag{1.8}$$

*where* $P_j$ *is the orthogonal projection on* $\vec{\mathcal{X}}$ *defined by*

$$(P_j\vec{x})_i = \begin{cases} 0 & \text{for } i < j, \\ x_i & \text{for } i \ge j, \end{cases} \tag{1.9}$$

*and blank spots in the operator matrix in (1.8) denote zero entries.*

**PROOF.** For $\vec{x} = [\ldots, x_{-1}, \underline{x_0}, x_1, \ldots]^{tr}$ in $\vec{\mathcal{X}}$, we have

$$\begin{aligned} P_+V^j\tau\vec{x} &= P_+V^j[\ldots, \tau_{-1}x_{-1}, \underline{\tau_0 x_0}, \tau_1 x_1, \ldots]^{tr} \\ &= P_+[\ldots, \tau_{-j-1}x_{-j-1}, \underline{\tau_{-j}x_{-j}}, \tau_{-j+1}x_{-j+1}, \ldots]^{tr} \\ &= [\ldots, 0, 0, \underline{\tau_{-j}x_{-j}}, \tau_{-j+1}x_{-j+1}, \ldots]^{tr} = V^j\tau P_{-j}\vec{x}. \end{aligned} \tag{1.10}$$

Let $\hat{P}$ be the diagonal operator defined in the right-hand side of (1.8). Then using $P_+V^j\tau = V^j\tau P_{-j}$ along with the form of U in (1.4) gives

$$P_+U = [\ldots, P_+V^*\tau, \underline{P_+\tau}, P_+V\tau, \ldots] = [\ldots, V^*\tau P_1, \underline{\tau P_0}, V\tau P_{-1}, \ldots] = U\hat{P}.$$

Therefore $U^*P_+U = \hat{P}$, and this completes the proof.

By using the representation for U in (1.4) or (1.5), we can rewrite (1.8) is a more concise form as follows:

$$P_+ = \sum_{j=-\infty}^{\infty} V^j\tau P_{-j}\tau^* V^{-j}, \tag{1.11}$$

with convergence in the strong operator topology. To see this, using (1.5) first note that $U^{-1} = \sum_{j=-\infty}^{\infty} \pi_j^*\tau^*V^{-j}$, where

$$\pi_j^* : \vec{\mathcal{X}} \to \ell^2(\vec{\mathcal{X}}) \quad \text{and} \quad \pi_j^*\vec{x} = (\delta_{k,j}\vec{x})_{k=-\infty}^{\infty}. \tag{1.12}$$

Now, let $\hat{P}$ be the diagonal operator in the right hand side of (1.8). Then $\pi_j\hat{P}\pi_k^* = 0$ for $j \neq k$ and

$\pi_j \hat{P} \pi_j^* = P_{-j}$. It follows that $P_+ = U\hat{P}U^{-1}$ is given by the right hand side of (1.11).

## X.2 OPERATOR FEATURES

Throughout this section $\vec{\mathcal{X}}$ is the Hilbert space direct sum in (1.1) and $\vec{\mathcal{Y}}$ is the Hilbert space direct sum $\oplus\{\mathcal{Y}_k : k \in \mathbb{Z}\}$. Let F be a bounded linear operator from $\vec{\mathcal{X}}$ into $\vec{\mathcal{Y}}$. Then F has the operator matrix representation

$$F = (f_{j,k})_{j,k=-\infty}^{\infty} : \vec{\mathcal{X}} \to \vec{\mathcal{Y}}, \tag{2.1}$$

where $f_{j,k}$ in $\mathcal{L}(\mathcal{X}_k, \mathcal{Y}_j)$ is the operator given by $(\tau_j')^* F \tau_k$. Here $\tau_k$ from $\mathcal{X}_k$ to $\vec{\mathcal{X}}$ is the embedding operator defined in (1.2), and $\tau_k'$ from $\mathcal{Y}_k$ to $\vec{\mathcal{Y}}$ is the corresponding embedding operator with $\vec{\mathcal{X}}$ replaced by $\vec{\mathcal{Y}}$. For each $n \in \mathbb{Z}$ we denote by $[F]_n$ the operator

$$[F]_n = (\delta_{j-k,n} f_{j,k})_{j,k=-\infty}^{\infty}, \tag{2.2}$$

which acts as a bounded linear operator from $\vec{\mathcal{X}}$ into $\vec{\mathcal{Y}}$. Notice that $[F]_n$ for $n \geq 0$ (for $n < 0$) is the block matrix obtained by extracting the n-th diagonal below (above) the main diagonal from F and then setting all the other entries equal to zero, respectively. Next, we build with the operators $[F]_n$, for $n \in \mathbb{Z}$, the doubly infinite Laurent operator matrix $\hat{F}$ by

$$\hat{F} = ([F]_{j-k})_{j,k=-\infty}^{\infty}. \tag{2.3}$$

In other words, $\hat{F}$ is derived from F by replacing each entry $f_{j,k}$ by the doubly infinite block matrix which one obtains by setting all entries in F to zero except those on the diagonal containing $f_{j,k}$ which remain unchanged. We shall refer to $\hat{F}$ as the *diagonally sparse transform* of F. The next theorem shows that $\hat{F}$ acts as a bounded linear operator from $\ell^2(\vec{\mathcal{X}})$ into $\ell^2(\vec{\mathcal{Y}})$.

**PROPOSITION 2.1.** *Let* F *be a bounded linear operator from* $\vec{\mathcal{X}}$ *into* $\vec{\mathcal{Y}}$, *and let* $\hat{F}$ *be its diagonally sparse transform. Then* $\hat{F}$ *acts as a bounded linear operator from* $\ell^2(\vec{\mathcal{X}})$ *into* $\ell^2(\vec{\mathcal{Y}})$, *and*

$$(U')^{-1}\hat{F}U = \begin{bmatrix} \ddots & & & \\ & F & & \\ & & F & \\ & & & F \\ & & & & \ddots \end{bmatrix} : \ell^2(\vec{\mathcal{X}}) \to \ell^2(\vec{\mathcal{Y}}), \tag{2.4}$$

*where* U *on* $\ell^2(\vec{\mathcal{X}})$ *is the unitary operator defined by (1.5) and* U′ *on* $\ell^2(\vec{\mathcal{Y}})$ *is the corresponding unitary operator for* $\vec{\mathcal{Y}}$ *in place of* $\vec{\mathcal{X}}$. *The blank spots in the operator matrix in (2.4) denote zero entries. In particular,* F *and* $\hat{F}$ *have the same norm.*

**PROOF.** Let $\Delta_F$ be the block diagonal operator defined by the operator matrix in the right-hand side of (2.4), and set $G = U'\Delta_F U^{-1}$, where U and U′ are as in (2.4). Obviously, $G : \ell^2(\vec{\mathcal{X}}) \to \ell^2(\vec{\mathcal{Y}})$ is a bounded linear operator. It suffices to show that G is a block Laurent operator, $G = (G_{j-k})_{j,k=-\infty}^{\infty}$, with $G_n = [F]_n$ for each $n \in \mathbb{Z}$.

Let V and V′ be the block forward shifts on $\ell^2(\vec{\mathcal{X}})$ and $\ell^2(\vec{\mathcal{Y}})$, respectively. Note that $V'\Delta_F = \Delta_F V$. Since V commutes with U, and V′ commutes with U′, we obtain $V'G = GV$, and hence G is a block Laurent operator.

Let $\pi_j$ be as in (1.6), and let $\pi_j'$ be the corresponding operator for $\vec{\mathcal{Y}}$ in place of $\vec{\mathcal{X}}$. It remains to show that $\pi_n' G \pi_0^* = [F]_n$ for each $n \in \mathbb{Z}$. By using the representation (1.5) and its analogue for U′ with $\vec{\mathcal{Y}}$ in place of $\vec{\mathcal{X}}$ we see that

$$G = \sum_{j,k=-\infty}^{\infty} (V')^j \tau' \pi_j' \Delta_F \pi_k^* \tau^* V^{-k} = \sum_{j=-\infty}^{\infty} (V')^j \tau' F \tau^* V^{-j} .$$

Here $\tau$ and $\tau'$ are the canonical sparsing operators corresponding to the spaces $\mathcal{X}_k$ and $\mathcal{Y}_k$ for $k \in \mathbb{Z}$, respectively. The above sums converge in the strong operator topology. Since $\pi_j' V' = \pi_{j-1}'$ and $\pi_j V = \pi_{j-1}$ for each j, we have

$$\pi_n' G \pi_0^* = \sum_{j=-\infty}^{\infty} \pi_{n-j}' \tau' F \tau^* \pi_{-j}^* ,$$

again with convergence in the strong operator topology. For each $j \in \mathbb{Z}$ the operator $\pi_j \tau$ on $\vec{\mathcal{X}}$ is the orthogonal projection $Q_j$ of $\vec{\mathcal{X}}$ which assigns to the sequence $\vec{x} = (x_n)_{n=-\infty}^{\infty}$ the sequence $(\delta_{j,n} x_n)_{n=-\infty}^{\infty}$, where $\delta_{j,n}$ is the Kronecker delta. Similarly, $\pi_j' \tau' = Q_j'$, where $Q_j' : \vec{\mathcal{Y}} \to \vec{\mathcal{Y}}$ is defined as $Q_j$ with $\mathcal{Y}$ in place of $\mathcal{X}$. We conclude that

$$G_n = \pi_n' G \pi_0^* = \sum_{j=-\infty}^{\infty} Q_{n-j}' F Q_{-j} = [F]_n ,$$

which completes the proof.

From the proof of Proposition 2.1 we see that (2.4) may be rewritten in a more concise form as follows:

$$\hat{F} = \sum_{j=-\infty}^{\infty} (V')^j \tau' F \tau^* V^{-j} , \tag{2.5}$$

with convergence in the strong operator topology. Since $\tau$ is an isometry and $\tau^* V^k \tau = 0$ for each $k \neq 0$, this implies that $\hat{F}\tau = \tau' F$, and thus $\tau'^* \hat{F} \tau = F$. From Proposition 2.1 it also follows that F and its diagonally sparse transforms $\hat{F}$ have the same operator norm.

Consider the block Toeplitz operator $\hat{F}_+$ associated with the diagonally sparse transform of F, that is,

$$\hat{F}_+ = ([F]_{j-k})_{j,k=0}^{\infty} : \ell_+^2(\vec{\mathcal{X}}) \to \ell_+^2(\vec{\mathcal{Y}}) . \tag{2.6}$$

Let $P_+$ be the orthogonal projection of $\ell^2(\vec{\mathcal{X}})$ onto $\ell_+^2(\vec{\mathcal{X}})$, and let $P'_+$ be the corresponding projection for $\ell^2(\vec{\mathcal{Y}})$. We have

$$\hat{F}_+ = P'_+ \hat{F} \,|\, \ell_+^2(\vec{\mathcal{X}}) = \sum_{j=-\infty}^{\infty} (V')^j \tau' P'_{-j} F P_{-j} \tau^* V^{-j} \,|\, \ell_+^2(\vec{\mathcal{X}}) , \tag{2.7}$$

where $P_j$ is the orthogonal projection defined by (1.9) and $P'_j$ is the analogous projection for $\mathcal{Y}$ in place $\mathcal{X}$. Formula (2.7) follows from (2.5) and $P_+ V^j \tau = V^j \tau P_{-j}$; see (1.10). Note that Proposition 2.1 implies that F, $\hat{F}$ and $\hat{F}_+$ all have the same norm.

In the remaining part of this section we consider an operation which (to a certain extent) is the reverse of the diagonally sparse transform. Consider the block Laurent operator

$$G = (G_{j-k})_{j,k=-\infty}^{\infty} : \ell^2(\vec{\mathcal{X}}) \to \ell^2(\vec{\mathcal{Y}}) . \tag{2.8}$$

Each block entry $G_n$ is a bounded linear operator from $\vec{\mathcal{X}}$ into $\vec{\mathcal{Y}}$, and hence admits an operator matrix representation

$$G_n = (g_{j,k}^{(n)})_{j,k=-\infty}^{\infty} : \vec{\mathcal{X}} \to \vec{\mathcal{Y}} , \tag{2.9}$$

where $g_{j,k}^{(n)} : \mathcal{X}_k \to \mathcal{Y}_j$ is a bounded linear operator for each j and k. Now, let

$$\check{G} = (g_{j,k}^{(j-k)})_{j,k=-\infty}^{\infty} : \vec{\mathcal{X}} \to \vec{\mathcal{Y}} . \tag{2.10}$$

Thus $\check{G}$ is the doubly infinite matrix whose n-th diagonal is precisely equal to the n-th diagonal of $G_n$ for each $n \in \mathbb{Z}$. In other words, $\check{G}$ is obtained from G by replacing each block entry $G_{j-k}$ in G by the (j, k)-th entry of $G_{j-k}$ (or equivalently, $\check{G}_{n+k,k} = g_{n+k,k}^{(n)}$ for all n, k in $\mathbb{Z}$). We shall refer to $\check{G}$ in (2.10) as the *diagonal compression* of the block Laurent operator G in (2.8). If G is the diagonally sparse transform $\hat{F}$ of the operator F in (2.1), then F is precisely the diagonal compression of G, that is $(\hat{F})^{\vee} = F$ (and in this sense diagonal compression is the inverse of the diagonally sparse transform). The next proposition shows that the operator $\check{G}$ in (2.10) is a well-defined bounded operator.

**PROPOSITION 2.2.** *Let* G *be the block Laurent operator in (2.8). Then the diagonal compression* $\check{G}$ *of* G *is the operator* $\tau'^* G\tau$, *where* $\tau$ *is the canonical sparsing operator corresponding to the spaces* $\mathcal{X}_k$ *for* $k \in \mathbb{Z}$, *and* $\tau'$ *is the analogous operator with* $\mathcal{Y}_k$ *in place of* $\mathcal{X}_k$. *In particular,* $\|\check{G}\| \leq \|G\|$.

**PROOF.** Let $\tau_k$ be the operator defined in (1.2), and let $\tau_k'$ be the corresponding operator for $\mathcal{Y}$ in place of $\mathcal{X}$. It suffices to show that

$$(\tau'\tau_j')^* G\tau\tau_k = g_{j,k}^{(j-k)} \qquad (\text{for } j, k \in \mathbb{Z}). \tag{2.11}$$

Now $\tau\tau_k = \pi_k^*\tau_k$ and $\tau'\tau_k' = (\pi_k')^*\tau_k'$, where $\pi_k$ is defined by (1.6) and $\pi_k'$ is the analogous operator for $\mathcal{Y}$ in place of $\mathcal{X}$. It follows that the left hand side of (2.11) is equal to $(\tau_j')^*\pi_j' G\pi_k^*\tau_k$. On the other hand,

$$\pi_j' G\pi_k^* = G_{j-k} \quad \text{and} \quad \tau_j'^* G_{j-k}\tau_k = g_{j,k}^{(j-k)},$$

Therefore (2.11) holds and this completes the proof.

Proposition 2.2 may also be obtained as a corollary of the following more general result.

**PROPOSITION 2.3.** *Let* G *be the block Laurent operator in (2.8) of which the block entries are given by (2.9). Put* $\tilde{G} = (U')^{-1}GU$, *where* U *is the unitary operator on* $\ell^2(\vec{\mathcal{X}})$ *in (1.5) associated with the sparse embedding of* $\vec{\mathcal{X}}$ *into* $\ell^2(\vec{\mathcal{X}})$, *and* U′ *is the corresponding unitary operator with* $\mathcal{X}$ *replaced by* $\mathcal{Y}$. *Then* $\tilde{G}$ *is a block Laurent operator,* $\tilde{G} = (\tilde{G}_{j-k})_{j,k=-\infty}^{\infty}$, *with*

$$\tilde{G}_n = (g_{j,k}^{(n+j-k)})_{j,k=-\infty}^{\infty}. \tag{2.12}$$

*In particular, the diagonal entries* $\tilde{G}_0$ *of* $\tilde{G}$ *are equal to the diagonal compression* $\check{G}$ *of* G.

**PROOF.** We use the notation introduced in the proof of Proposition 2.1. Since G is block Laurent, we have $V'G = GV$. Recall that U and V commute, and that the same holds true for U′ and V′. Thus $V'\tilde{G} = \tilde{G}V$, and hence $\tilde{G}$ is a block Laurent operator.

Let us compute $\tilde{G}_n$. We have $\tilde{G}_n = \pi_n'\tilde{G}\pi_0^*$, where according to (1.5)

$$\tilde{G} = \sum_{j,k=-\infty}^{\infty} \pi_j'^{*}\tau'^{*}V'^{*j}GV^{k}\tau\pi_k ,$$

with convergence in the strong operator topology. Now, $\pi_n'(\pi_j')^* = \delta_{n,j}I$ and $\pi_k\pi_0^* = \delta_{k,0}I$. Thus

$$\tilde{G}_n = \tau'^{*}(V'^{*})^nGV^0\tau = \tau'^{*}(V'^{*})^nG\tau .$$

Thus the (j, k)-th block entry of $\tilde{G}_n$ is given by

$$\begin{aligned}(\tilde{G}_n)_{j,k} &= (\tau_j')^*\tilde{G}_n\tau_k = (\tau_j')^*(\tau')^*(V')^{-n}G\tau\tau_k = (\tau_j')^*\pi_j'(V')^{-n}G\pi_k^*\tau_k\\ &= (\tau_j')^*\pi_{j+n}'G\pi_k^*\tau_k = (\tau_j')^*G_{j+n-k}\tau_k = g_{j,k}^{(n+j-k)} ,\end{aligned}$$

which proves the proposition.

Let us conclude this section with the following useful result.

**COROLLARY 2.4.** *Let* $\hat{F}$ *from* $\ell^2(\vec{\mathcal{X}})$ *into* $\ell^2(\vec{\mathcal{Y}})$ *be the diagonally sparse transform of* F *in (2.1), and let* G *from* $\ell^2(\vec{\mathcal{E}})$ *to* $\ell^2(\vec{\mathcal{X}})$ *be a block Laurent operator, where* $\vec{\mathcal{E}}$ *is the Hilbert space direct* sum $\oplus\{\mathcal{E}_k : k \in \mathbb{Z}\}$. *Then the diagonal compression of* $\hat{F}G$ *is* $F\check{G}$, *where* $\check{G}$ *is the diagonal compression of* G.

**PROOF.** Let U, U′, U″ be the unitary operators associated with the sparse embeddings of $\vec{\mathcal{X}}$ into $\ell^2(\vec{\mathcal{X}})$, of $\vec{\mathcal{Y}}$ in $\ell^2(\vec{\mathcal{Y}})$, and of $\vec{\mathcal{E}}$ into $\ell^2(\vec{\mathcal{E}})$, respectively. Then

$$(U')^{-1}\hat{F}GU'' = (U')^{-1}\hat{F}UU^{-1}GU'' .$$

By Proposition 2.1, the operator $(U')^{-1}\hat{F}U$ is a diagonal operator with diagonal entries equal to F. Hence the diagonal entries of $(U')^{-1}\hat{F}GU''$ are equal to $F\check{G}$, where $\check{G}$ is the diagonal compression of G. Thus Proposition 2.3 shows that the diagonal compression of $\hat{F}G$ is the operator $F\check{G}$. This completes the proof.

**REMARK 2.5.** Let $F = (f_{j,k})_{j,k=-\infty}^{\infty}$ be a bounded linear operator from $\vec{\mathcal{X}}$ into $\vec{\mathcal{Y}}$. We say that F belongs to the nonstationary Wiener class $NSW(\vec{\mathcal{X}}, \vec{\mathcal{Y}})$ if

$$\sum_{n=-\infty}^{\infty} \sup_{j-k=n} \|f_{j,k}\| < \infty . \tag{2.13}$$

Notice that condition (2.13) is equivalent to the requirement that

$$\sum_{n=-\infty}^{\infty} \|[F]_n\| < \infty . \tag{2.14}$$

Hence in this case $F = \Sigma_{n=-\infty}^{\infty} [F]_n$, and the latter series is (absolutely) convergent in operator norm. Thus, if F belongs to the nonstationary Wiener class $NSW(\vec{\mathcal{X}}, \vec{\mathcal{Y}})$, then the symbol $G(\cdot)$ of the diagonally sparse transform $\hat{F}$ of F, that is,

$$G(\lambda) = \sum_{n=-\infty}^{\infty} \lambda^n [F]_n ,$$

is absolutely convergent for $|\lambda| = 1$, and $G(1) = F$. It follows that in this case the diagonal compression of $\hat{F}$ is obtained by evaluating the symbol of $\hat{F}$ at the point $\lambda = 1$.

## *Notes to Chapter X:*

The material in this chapter is taken from the paper Foias-Frazho-Gohberg-Kaashoek [1].

CHAPTER XI

# PROOFS OF THE NONSTATIONARY INTERPOLATION THEOREMS BY REDUCTION TO THE STATIONARY CASE

In this chapter the nonstationary theorems stated in Chapter VIII are proved by using the reduction technique developed in the previous chapter. In each case the main point is to show that the nonstationary interpolation problem is equivalent to a stationary one, and hence can be solved by using the corresponding result of the stationary case.

## XI.1. THE STANDARD NONSTATIONARY NEVANLINNA-PICK INTERPOLATION THEOREM

In this section we prove the standard nonstationary version of the left tangential Nevanlinna-Pick interpolation theorem (Theorem VIII.2.1 for $N = 1$), by reduction to its time-invariant version (Theorem I.3.1). As before, throughout this chapter $\vec{\mathcal{H}}$ denotes the infinite direct sum of Hilbert spaces $\oplus \{\mathcal{H}_n : n \in \mathbb{Z}\}$.

Let $Z_k$ from $\mathcal{X}_{k+1}$ to $\mathcal{X}_k$ and $B_k$ from $\mathcal{U}_k$ to $\mathcal{X}_k$ and $\tilde{B}_k$ from $\mathcal{Y}_k$ into $\mathcal{X}_k$ for $k \in \mathbb{Z}$ be the data associated with the standard nonstationary left Nevanlinna-Pick problem in Section VIII.2. Let $Z$ be the operator on $\vec{\mathcal{X}}$ defined by

$$Z = \begin{bmatrix} \ddots & \ddots & & & \\ & 0 & Z_{-1} & & \\ & & \underline{0} & Z_0 & \\ & & & 0 & Z_1 \\ & & & & \ddots & \ddots \end{bmatrix} \text{ on } \vec{\mathcal{X}}. \tag{1.1}$$

Let $B$ from $\vec{\mathcal{U}}$ into $\vec{\mathcal{X}}$ and $\tilde{B}$ from $\vec{\mathcal{Y}}$ into $\vec{\mathcal{X}}$ be the operators defined by

$$B = \operatorname{diag}(B_k)_{-\infty}^{\infty} \quad \text{and} \quad \tilde{B} = \operatorname{diag}(\tilde{B}_k)_{-\infty}^{\infty}. \tag{1.2}$$

The first condition in (VIII.2.7) implies that $Z$, $B$, and $\tilde{B}$ are all well-defined bounded linear operators. For $\nu \geq 1$ the $(k, m)$-th block entry of $Z^\nu$ is given by

$$(Z^{\nu})_{k,m} = \begin{cases} \Phi(k, k+\nu) \ , & \text{if } m = k+\nu \\ 0 \ , & \text{otherwise} \end{cases} \tag{1.3}$$

where $\Phi(k, n) = Z_k Z_{k+1} \cdots Z_{n-1}$, $k < n$, is the *(state) transition operator* for the operators $\{Z_n\}_{n\in\mathbb{Z}}$. (Recall that by definition $\Phi(k, k) = I$.) It follows that

$$\|Z^{\nu}\| = \sup\{\|\Phi(k, k+\nu)\| : k \in \mathbb{Z}\} \tag{1.4}$$

and hence the second condition in (VIII.2.7) implies that $r_{spec}(Z) < 1$. Thus the operators

$$Z : \vec{\mathcal{X}} \to \vec{\mathcal{X}}, \qquad B : \vec{\mathcal{U}} \to \vec{\mathcal{X}}, \qquad \tilde{B} : \vec{\mathcal{Y}} \to \vec{\mathcal{X}} \tag{1.5}$$

may be considered as the data for a standard stationary Nevanlinna-Pick problem, namely the problem of finding $G \in H^{\infty}(\vec{\mathcal{Y}}, \vec{\mathcal{U}})$ such that $\|G\|_{\infty} \leq \gamma$ and $(BG)(Z)_{left} = \tilde{B}$. We shall refer to this latter problem as the *standard stationary Nevanlinna-Pick problem associated with the nonstationary data* $\{Z_k, B_k, \tilde{B}_k\}$ and the bound $\gamma$. The next theorem shows that these two interpolation problems are equivalent.

**THEOREM 1.1.** *The standard nonstationary Nevanlinna-Pick problem for the data* $\{Z_k, B_k, \tilde{B}_k\}$ *and tolerance* $\gamma$ *is solvable if and only if the standard stationary Nevanlinna-Pick problem associated with the nonstationary data* $\{Z_k, B_k, \tilde{B}_k\}$ *and tolerance* $\gamma$ *is solvable. More precisely, if* F *is a solution of the standard nonstationary Nevanlinna-Pick problem associated with the data* $\{Z_k, B_k, \tilde{B}_k\}$ *and the tolerance* $\gamma$, *then the symbol* G *in* $H^{\infty}(\vec{\mathcal{Y}}, \vec{\mathcal{U}})$ *of the diagonally sparse transform* $L_G = \hat{F}$ *of* F *is a solution of the associated standard stationary Nevanlinna-Pick problem and* $\|G\|_{\infty} = \|F\| \leq \gamma$. *Conversely, if* $G \in H^{\infty}(\vec{\mathcal{Y}}, \vec{\mathcal{U}})$ *is a solution of the standard stationary Nevanlinna-Pick problem associated with the nonstationary data* $\{Z_k, B_k, \tilde{B}_k\}$ *and the tolerance* $\gamma$, *then the diagonal compression* F *of the block Laurent operator* $L_G$ *is a solution of the original standard nonstationary Nevanlinna-Pick problem for* $\{Z_k, B_k, \tilde{B}_k\}$ *with tolerance* $\gamma$ *and* $\|F\| \leq \|G\|_{\infty} \leq \gamma$.

**PROOF.** Let F be a solution of the nonstationary Nevanlinna-Pick problem with tolerance $\gamma$ associated with the data $\{Z_k, B_k, \tilde{B}_k\}$, and let $\hat{F}$ be the diagonally sparse transform of F. We know (see the previous chapter) that $\hat{F}$ is a block Laurent operator from $\ell^2(\vec{\mathcal{Y}})$ to $\ell^2(\vec{\mathcal{U}})$, and $\|\hat{F}\| = \|F\|$ (see Proposition X.2.1). Let G be the symbol for $\hat{F}$, that is, $\hat{F} = L_G$ where $G \in H^{\infty}(\vec{\mathcal{Y}}, \vec{\mathcal{U}})$. Then $\|G\|_{\infty} = \|\hat{F}\| = \|F\| \leq \gamma$. By definition the n-th Taylor coefficient of G is equal to $[F]_n$, where $[F]_n$ is the doubly infinite operator matrix which one obtains if all entries in F are set to zero except those on the n-th diagonal below the main diagonal, which remain unchanged. A straightforward calculation (using (1.3)) shows that the interpolation conditions

$$B_k f_{k,k} + \sum_{\nu=1}^{\infty} \Phi(k, k+\nu) B_{k+\nu} f_{k+\nu,k} = \tilde{B}_k, \qquad (k \in \mathbb{Z}) \tag{1.6}$$

imply that $(BG)(Z)_{\text{left}} = \tilde{B}$, where Z, B, and $\tilde{B}$ are defined by (1.1) and (1.2). Thus G solves the stationary Nevanlinna-Pick interpolation problem with tolerance $\gamma$ associated with the data (1.5).

To prove the converse statement assume that there exists $G \in H^\infty(\vec{\mathcal{Y}}, \vec{\mathcal{U}})$ such that $\|G\|_\infty \le \gamma$ and $(BG)(Z)_{\text{left}} = \tilde{B}$. Let $L_G$ be the block Laurent operator induced by G, and define F to be the diagonal compression of $L_G$. Then F is a block lower triangular operator from $\vec{\mathcal{Y}}$ to $\vec{\mathcal{U}}$ satisfying $\|F\| \le \|L_G\|$ (see Proposition X.2.2). Hence $\|F\| \le \|L_G\| = \|G\|_\infty \le \gamma$. Let $G_n = (g_{j,k}^{(n)})_{j,k=-\infty}^{\infty}$ be the n-th Taylor coefficient of the operator function G. Then, by definition,

$$F = (f_{j,k})_{j,k=-\infty}^{\infty} = (g_{j,k}^{(j-k)})_{j,k=-\infty}^{\infty} .$$

In particular, $g_{k+\nu,k}^{(\nu)} = f_{k+\nu,k}$. It follows that the (k, k)-th block entry of the operator $(BG)(Z)_{\text{left}}$ is precisely equal to

$$((BG)(Z)_{\text{left}})_{k,k} = B_k f_{k,k} + \sum_{\nu=1}^{\infty} \Phi(k, k+\nu) B_{k+\nu} f_{k+\nu,k} .$$

Since $\tilde{B}$ is block diagonal, we conclude that $(BG)(Z)_{\text{left}} = \tilde{B}$ implies (1.6), and hence F is a solution of the nonstationary Nevanlinna-Pick problem with tolerance $\gamma$ associated with the data $\{Z_k, B_k, \tilde{B}_k\}$. This completes the proof.

**PROOF OF THEOREM VIII.2.1.** It suffices to consider the case $N = 1$. Consider the nonstationary interpolation data $Z_k$, $B_k$, and $\tilde{B}_k$ for $k \in \mathbb{Z}$, and let Z, B, and $\tilde{B}$ be the operators defined by (1.1) and (1.2). Introduce the generalized Pick operators

$$\Omega_k := \sum_{\nu=0}^{\infty} \Phi(k, k+\nu)[\gamma^2 B_{k+\nu} B_{k+\nu}^* - \tilde{B}_{k+\nu} \tilde{B}_{k+\nu}^*] \Phi(k, k+\nu)^* \qquad (k \in \mathbb{Z}),$$

$$\Omega := \sum_{\nu=0}^{\infty} Z^\nu [\gamma^2 BB^* - \tilde{B}\tilde{B}^*](Z^\nu)^* .$$

From the special form of Z, B, and $\tilde{B}$ in (1.1) and (1.2) it follows that $\Omega$ is a block diagonal operator on $\vec{\mathcal{X}}$, and for each $k \in \mathbb{Z}$ the k-th block entry of $\Omega$ is precisely equal to $\Omega_k$. Therefore $\Omega$ is positive if and only if $\Omega_k$ is positive for each $k \in \mathbb{Z}$. But then we can apply Theorems 1.1 and I.3.1 to finish the proof.

As before, let $\{Z_k, B_k, \tilde{B}_k\}$ be the data for the nonstationary standard left Nevanlinna-Pick problem. The above analysis lead to the following optimization problem

$$d_\infty = \inf\{\|F\| : F \text{ in } \mathcal{L}(\vec{\mathcal{Y}}, \vec{\mathcal{U}}) \text{ is lower triangular and (1.6) holds}\}. \tag{1.7}$$

We shall refer to $d_\infty$ as the *optimal error* associated with the nonstationary data $\{Z_k, B_k, \tilde{B}_k\}$. Now let Z, B and $\tilde{B}$ be the operators defined in (1.1) and (1.2). Let W from $\ell_+^2(\vec{\mathcal{U}})$ to $\vec{\mathcal{X}}$ and $\tilde{W}$ from $\ell_+^2(\vec{\mathcal{Y}})$ to $\vec{\mathcal{X}}$ be the controllability operators generated by the pairs $\{Z, B\}$ and $\{Z, \tilde{B}\}$ respectively; see (I.4.1). Then it follows that $G \in H^\infty(\vec{\mathcal{Y}}, \vec{\mathcal{U}})$ satisfies $(BG)(Z)_{\text{left}} = \tilde{B}$ if and only if $WT_G = \tilde{W}$ where $T_G$ is the block lower triangular Toeplitz operator from $\ell_+^2(\vec{\mathcal{Y}})$ to $\ell_+^2(\vec{\mathcal{U}})$ generated by G. Theorem 1.1 shows that the optimization problem in (1.7) is equivalent to the following optimization problem

$$d_\infty = \inf\{\|G\|_\infty : G \in H^\infty(\vec{\mathcal{Y}}, \vec{\mathcal{U}}) \text{ and } WT_G = \tilde{W}\}. \tag{1.8}$$

In particular, $d_\infty$ in (1.7) and (1.8) is the same optimal error.

According to the results in Section II.2, the error $d_\infty$ is finite if and only if the relation $WA = \tilde{W}$ defines a bounded operator A from $\ell_+^2(\vec{\mathcal{Y}})$ into $\mathcal{H}' = \overline{\operatorname{ran} W^*}$. Moreover, in this case the operator A is uniquely determined and $d_\infty = \|A\|$. Furthermore, there exists an optimal G in $H^\infty(\vec{\mathcal{Y}}, \vec{\mathcal{U}})$ satisfying the interpolation condition $WT_G = \tilde{W}$ and $d_\infty = \|G\|_\infty$. So if F is the diagonal compression of $L_G$, then F is a lower triangular operator satisfying the nonstationary interpolation conditions in (1.6) and $d_\infty = \|F\|$. If the controllability grammian $P = WW^*$ for $\{Z, B\}$ is invertible, then $d_\infty$ is finite. In this case the bounded operator A is given by $A = W^*P^{-1}\tilde{W}$. Furthermore,

$$d_\infty^2 = r_{\text{spec}}(\tilde{P}P^{-1}) \tag{1.9}$$

where $\tilde{P} = \tilde{W}\tilde{W}^*$ is the controllability grammian for $\{Z, \tilde{B}\}$; see Section II.2.

As in Chapter VIII, let $\vec{\mathcal{U}}_{(k,\infty)}$ and $\vec{\mathcal{Y}}_{(k,\infty)}$ be the Hilbert space direct sums given by

$$\vec{\mathcal{U}}_{(k,\infty)} = \bigoplus_{n=k}^{\infty} \mathcal{U}_n \quad \text{and} \quad \vec{\mathcal{Y}}_{(k,\infty)} = \bigoplus_{n=k}^{\infty} \mathcal{Y}_k, \tag{1.10}$$

and define $W_k$ from $\vec{\mathcal{U}}_{(k,\infty)}$ into $\mathcal{X}_k$ and $\tilde{W}_k$ from $\vec{\mathcal{Y}}_{(k,\infty)}$ into $\mathcal{X}_k$ to be the nonstationary controllability operators at time k for the families of pairs $\{Z_n, B_n\}_{n\in\mathbb{Z}}$ and $\{Z_n, \tilde{B}_n\}_{n\in\mathbb{Z}}$, that is,

$$W_k = [B_k, \Phi(k, k+1)B_{k+1}, \Phi(k, k+2)B_{k+2}, \ldots] \text{ and}$$
$$\tilde{W}_k = [\tilde{B}_k, \Phi(k, k+1)\tilde{B}_{k+1}, \Phi(k, k+2)\tilde{B}_{k+2}, \ldots], \tag{1.11}$$

where $\Phi(m, n) = Z_m Z_{m+1} \cdots Z_{n-1}$ for $m < n$ and $\Phi(n, n)$ is the identity operator on $\mathcal{X}_n$. The operators $W_k$ and $\tilde{W}_k$, $k \in \mathbb{Z}$, and the controllability operators $W$ and $\tilde{W}$ associated with the pairs $\{Z, B\}$ and $\{Z, \tilde{B}\}$ are closely related. To describe this relation (see formula (1.12) below) we need same additional notation. Let $P_+$ and $P'_+$ be (respectively) the orthogonal projections of $\ell^2(\vec{\mathcal{U}})$ onto $\ell^2_+(\vec{\mathcal{U}})$ and of $\ell^2(\vec{\mathcal{Y}})$ onto $\ell^2_+(\vec{\mathcal{Y}})$, and let $P_{\mathcal{U},k}$ and $P_{\mathcal{Y},k}$ be the orthogonal projections of $\vec{\mathcal{U}}$ onto $\vec{\mathcal{U}}_{(k,\infty)}$ and of $\vec{\mathcal{Y}}$ onto $\vec{\mathcal{Y}}_{(k,\infty)}$, respectively. By $U$ we denote the unitary operator on $\ell^2(\vec{\mathcal{U}})$ associated with the sparse embedding of $\vec{\mathcal{U}}$ into $\ell^2(\vec{\mathcal{U}})$, and $U'$ is the analogous operator $\mathcal{Y}$ in place of $\mathcal{U}$ (see Section X.1). Finally, let $\pi_j$ and $\pi'_j$ be the respective operators from $\ell^2(\vec{\mathcal{U}})$ to $\vec{\mathcal{U}}$ and from $\ell^2(\vec{\mathcal{Y}})$ to $\vec{\mathcal{Y}}$ obtained by picking out the j-th component of the appropriate $\ell^2$-sequences, and, as in Section X.1, let $\tau_j$ be the canonical embedding of $\mathcal{X}_j$ into $\vec{\mathcal{X}}$ (see (X.1.2)). Then using (1.3) one obtains

$$WP_+U = \sum_{j=-\infty}^{\infty} \tau_{-j} W_{-j} P_{\mathcal{U},-j} \pi_j \text{ and } \tilde{W}P'_+U' = \sum_{j=-\infty}^{\infty} \tau_{-j} \tilde{W}_{-j} P_{\mathcal{Y},-j} \pi'_j . \tag{1.12}$$

Thus $WP_+U$ and $\tilde{W}P'_+U'$ may be viewed a block diagonal operators. Since $U$ and $U'$ are unitary operators, the identities in (1.12) imply that the controllability grammians $P$ and $\tilde{P}$ on $\vec{\mathcal{X}}$ associated with the pairs $\{Z, B\}$ and $\{Z, \tilde{B}\}$ are given by

$$P = WW^* = \operatorname{diag}(P_j)_{-\infty}^{\infty} \text{ and } \tilde{P} = \tilde{W}\tilde{W}^* = \operatorname{diag}(\tilde{P}_j)_{-\infty}^{\infty}, \tag{1.13}$$

where $P_j = W_j W_j^*$ and $\tilde{P}_j = \tilde{W}_j \tilde{W}_j^*$ are the nonstationary controllability grammians at time j corresponding to the family of pairs $\{Z_n, B_n\}_{n\in\mathbb{Z}}$ and $\{Z_n, \tilde{B}_n\}_{n\in\mathbb{Z}}$, respectively. Recall that $P_j$ and $\tilde{P}_j$ can be recursively computed from the nonstationary Lyapunov difference equations

$$P_k = Z_k P_{k+1} Z_k^* + B_k B_k^* \text{ and } \tilde{P}_k = Z_k \tilde{P}_{k+1} Z_k^* + \tilde{B}_k \tilde{B}_k^* \quad (k \in \mathbb{Z}). \tag{1.14}$$

If $T = \operatorname{diag}(T_k)_{-\infty}^{\infty}$ on $\mathcal{X}$ where $T_k$ on $\mathcal{X}_k$ are uniformly bounded operators, then $r_{spec}(T) \geq d := \sup\{r_{spec}(T_k) : k \in \mathbb{Z}\}$. Following Problem 98 in Halmos [1], to demonstrate that we do not have equality in general, let $T_k$ be the lower shift on $\mathbb{C}^{|k|}$ for $k \in \mathbb{Z}$, that is, one's appear immediately below the main diagonal and zeroes elsewhere. Then $\|T^n\| = 1$ for all $n \geq 0$ and thus $r_{spec}(T) = 1 > d = 0$. Clearly if all the operators $T_k$ are self-adjoint, then $r_{spec}(T) = d$. Moreover, if $T_k = R_k Q_k$ where $R_k$ is self-adjoint and $Q_k$ is positive, then we also have $r_{spec}(T) = d$. To see this let $R = \operatorname{diag}(R_k)_{-\infty}^{\infty}$ and $Q = \operatorname{diag}(Q_k)_{-\infty}^{\infty}$. Then $T = RQ$ and the self-adjoint operator $Q^{1/2}RQ^{1/2}$ have the same spectral radius. Since $R_kQ_k$ and $Q_k^{1/2}R_kQ_k^{1/2}$ also have the same spectral radius, it follows that $r_{spec}(T) = d$. In particular, $r_{spec}(T) = r_{spec}(\tilde{P}P^{-1}) = d$ when $T_k = \tilde{P}_k P_k^{-1}$. From the remarks made in the three preceding paragraphs and Theorem 1.1 it is straightforward to derive the following corollary.

**COROLLARY 1.2.** *Let* $W_k$ *from* $\vec{\mathcal{U}}_{(k,\infty)}$ *to* $\mathcal{X}_k$ *and* $\tilde{W}_k$ *from* $\vec{\mathcal{Y}}_{(k,\infty)}$ *to* $\mathcal{X}_k$ *be the controllability operators at time* k *for the family of pairs* $\{Z_n, B_n\}_{n\in\mathbb{Z}}$ *and* $\{Z_n, \tilde{B}_n\}_{n\in\mathbb{Z}}$*, respectively. Then the optimal error* $d_\infty$ *associated with the nonstationary interpolation data* $\{Z_n, B_n, \tilde{B}_n\}$ *is finite if and only if for each* $k \in \mathbb{Z}$ *there exists a bounded linear operator* $A_k$ *from* $\vec{\mathcal{Y}}_{(k,\infty)}$ *to the closure of* ran $W_k$ *such that*

$$W_k A_k = \tilde{W}_k \quad \text{and} \quad \|A_k\| \le M < \infty \qquad (k \in \mathbb{Z}), \tag{1.15}$$

*where* M *is a constant not depending on* k. *In this case each* $A_k$ *is uniquely determined and*

$$d_\infty = \sup\{\|A_k\| : k \in \mathbb{Z}\}. \tag{1.16}$$

*Furthermore, if* $P_k = W_k W_k^* \ge \delta > 0$, *then* $A_k = W_k^* P_k^{-1} \tilde{W}_k$ *and*

$$d_\infty^2 = \sup\{r_{spec}(P_k^{-1}\tilde{P}_k) : k \in \mathbb{Z}\}, \tag{1.17}$$

*where* $\tilde{P}_k = \tilde{W}_k \tilde{W}_k^*$ *for each* $k \in \mathbb{Z}$.

Let $G_\gamma$ in $H^\infty(\vec{\mathcal{Y}}, \vec{\mathcal{U}})$ be the central interpolant with tolerance $\gamma$ for the standard stationary Nevanlinna-Pick problem associated with the nonstationary data $\{Z_k, B_k, \tilde{B}_k\}$ and tolerance $\gamma$, and let F be the solution to the original nonstationary Nevanlinna-Pick problem with tolerance $\gamma$ obtained by taking the diagonal compression of the block Laurent operator $L_{G_\gamma}$; see Theorem 1.1. Then we call this F the *central interpolant* for the standard nonstationary Nevanlinna-Pick problem with data $\{Z_k, B_k, \tilde{B}_k\}$ and tolerance $\gamma$. Finally, we say that the family of pairs $\{Z_k, B_k\}_{k\in\mathbb{Z}}$ is *uniformly controllable* if there exists an $\varepsilon > 0$ such that $P_j = W_j W_j^* \ge \varepsilon I$ for all j in $\mathbb{Z}$, or equivalently, diag $(P_k)_{-\infty}^{\infty}$ is a strictly positive operator on $\vec{\mathcal{X}}$. The following result gives an explicit state space formula to compute the central interpolant for the nonstationary Nevanlinna-Pick problem when the family of pairs $\{Z_k, B_k\}_{k\in\mathbb{Z}}$ is uniformly controllable.

**THEOREM 1.3.** *Let* $\{Z_k, B_k, \tilde{B}_k\}$ *be the data for the nonstationary standard Nevanlinna-Pick problem with tolerance* $\gamma > d_\infty$ *and assume that the family of pairs* $\{Z_k, B_k\}_{k\in\mathbb{Z}}$ *is uniformly controllable. Let* $C_k$ *from* $\mathcal{X}_k$ *into* $\mathcal{U}_k$ *and* $M_k$ *from* $\mathcal{X}_k$ *into* $\mathcal{X}_{k+1}$ *be the operators defined by*

$$C_k = \gamma^2 B_k^*(\gamma^2 P_k - Z_k \tilde{P}_{k+1} Z_k^*)^{-1} \text{ and} \tag{1.18}$$

$$M_k = (\gamma^2 P_{k+1} - \tilde{P}_{k+1}) Z_k^* (\gamma^2 P_k - Z_k \tilde{P}_{k+1} Z_k^*)^{-1} \quad (k \in \mathbb{Z}) .$$

*Finally, let* $\Psi(m, n)$ *from* $\mathcal{X}_n$ *into* $\mathcal{X}_m$ *be the state transition operator for* $\{M_k\}$ *defined by*

$$\Psi(m, n) = M_{m-1} M_m \cdots M_{n+1} M_n \text{ if } m > n \text{ and } \Psi(n, n) = I . \tag{1.19}$$

*Then the central interpolant* F *with tolerance* $\gamma$ *to this nonstationary Nevanlinna-Pick interpolation problem is given by*

$$\begin{aligned} F_{j,k} &= C_j \Psi(j, k) \tilde{B}_k && (\text{if } j \geq k) \\ F_{j,k} &= 0 && (\text{if } j < k) . \end{aligned} \tag{1.20}$$

*Moreover, the sequence* $\{M_n\}$ *is forward stable, that is*

$$\limsup_{\nu \to \infty} (\sup \{\|\Psi(k+\nu, k)\| : k \in \mathbb{Z}\})^{1/\nu} < 1 . \tag{1.21}$$

PROOF. To prove this result we combine Theorem 1.1 with the formula for the central solution for the stationary Nevanlinna-Pick problem in Theorem V.1.2. As before, let Z, B and $\tilde{B}$ be the operators defined in (1.1) and (1.2). Recall that P is the controllability grammian for $\{Z, B\}$, while $\tilde{P}$ is the controllability grammian for $\{Z, \tilde{B}\}$. Since Z is a block weighted backward shift and B, $\tilde{B}$ are block diagonal operators it follows that $P = \text{diag}\,(P_k)_{-\infty}^{\infty}$ and $\tilde{P} = \text{diag}\,(\tilde{P}_k)_{-\infty}^{\infty}$. Moreover, because the family of pairs $\{Z_k, B_k\}$ is uniformly controllable, P is strictly positive. By employing Theorem V.1.2 we see that the central solution $G_\gamma$ in $H^\infty(\vec{\mathcal{Y}}, \vec{\mathcal{U}})$ with tolerance $\gamma$ for the stationary Nevanlinna-Pick problem associated with the nonstationary data $\{Z_k, B_k, \tilde{B}_k\}$ is given by

$$G_\gamma(\lambda) = C(I - \lambda M)^{-1} \tilde{B} \tag{1.22}$$

where C from $\vec{\mathcal{X}}$ into $\vec{\mathcal{U}}$ and M on $\vec{\mathcal{X}}$ are the operators defined by

$$C = \gamma^2 B^* (\gamma^2 P - Z\tilde{P}Z^*)^{-1} \text{ and } M = (\gamma^2 P - \tilde{P}) Z^* (\gamma^2 P - Z\tilde{P}Z^*)^{-1} . \tag{1.23}$$

Since B, P and $Z\tilde{P}Z^*$ are all block diagonal operators, it follows that $C = \text{diag}\,(C_k)_{-\infty}^{\infty}$ where $C_k$ from $\mathcal{X}_k$ into $\mathcal{U}_k$ is the operator defined in (1.18). Using the fact that $Z^*$ is a block forward weighted shift and all the other matrices in the definition of M in (1.23) are block diagonal matrices, it follows that M is the block weighted forward shift defined by

$$M = \begin{bmatrix} \ddots & & & & \\ & 0 & & & \\ & M_{-1} & \underline{0} & & \\ & & M_0 & 0 & \\ & & & M_1 & \ddots \\ & & & & \ddots \end{bmatrix} \text{ on } \vec{\mathcal{X}} \tag{1.24}$$

where the operators $M_k$ from $\mathcal{X}_k$ into $\mathcal{X}_{k+1}$ are given by (1.18) and all the unspecified entries in (1.24) are zero. Moreover, for $\nu \geq 0$ we have

$$(M^\nu)_{j,k} = \begin{cases} \Psi(k+\nu, k) & \text{if } j = k+\nu\,, \\ 0 & \text{if } j \neq k+\nu\,. \end{cases} \tag{1.25}$$

Theorem V.1.2 shows that M is stable, that is, $r_{spec}(M) < 1$. According to (1.25) we have $\|M^\nu\| = \sup\{\|\Psi(k+\nu, k)\| : k \in \mathbb{Z}\}$. Therefore the state transition operators $\Psi$ satisfy the stability condition in (1.21).

Recall that the central interpolant F with tolerance $\gamma$ for the nonstationary Nevanlinna-Pick problem is precisely the diagonal compression of $L_{G_\gamma}$. The diagonal compression of $L_{G_\gamma}$ places the $\nu$-th subdiagonal of $G_{\gamma,\nu}$ onto the $\nu$-th subdiagonal of F where $G_{\gamma,\nu}$ is the Taylor coefficient of $\lambda^\nu$ for $G_\gamma$. To be precise, $F_{k+\nu,k} = (G_{\gamma,\nu})_{k+\nu,k}$ for all $\nu \geq 0$ and k in $\mathbb{Z}$; see (X.2.10). Since $G_\gamma(\lambda) = C(I-\lambda M)^{-1}\tilde{B}$ the $\nu$-th Taylor coefficient for $G_\gamma$ is given by $G_{\gamma,\nu} = CM^\nu\tilde{B}$. So using (1.25) along with the fact that C and $\tilde{B}$ are diagonal matrices we have

$$F_{k+\nu,k} = (G_{\gamma,\nu})_{k+\nu,k} = (CM^\nu\tilde{B})_{k+\nu,k} = C_{k+\nu}\Psi(k+\nu, k)\tilde{B}_k\,. \tag{1.26}$$

Therefore the (j, k)-entry of F is given by (1.20). This completes the proof.

**REMARK 1.4**. In this remark we show that the diagonally sparse transform of the solution F constructed in Theorem 1.3 is equal to $L_{G_\gamma}$, where $G_\gamma$ is the central solution of the associated stationary problem. As before, let $\{Z_k, B_k, \tilde{B}_k\}$ be the data for the nonstationary standard Nevanlinna-Pick problem with tolerance $\gamma > d_\infty$, and assume that the family of pairs $\{Z_k, B_k\}_{k\in\mathbb{Z}}$ is uniformly controllable. Let M on $\vec{\mathcal{X}}$ and C from $\vec{\mathcal{X}}$ into $\vec{\mathcal{U}}$ be the operators defined by (1.23), where $P = \operatorname{diag}(P_k)_{-\infty}^{\infty}$ and $\tilde{P} = \operatorname{diag}(\tilde{P}_k)_{-\infty}^{\infty}$. Then M is stable and the central interpolant F with tolerance $\gamma$ for the nonstationary data $\{Z_k, B_k, \tilde{B}_k\}$ is given by $F = G_\gamma(1)$, that is,

$$F = C(I-M)^{-1}\tilde{B}\,. \tag{1.27}$$

To see this notice that because M is stable, $(I-M)^{-1}$ exists as a bounded operator. So using

(1.25) along with $(I-M)^{-1} = I + M + M^2 + \cdots$ we see that the (j, k)-th entry of $(I-M)^{-1}$ is given by

$$((I-M)^{-1})_{j,k} = \begin{cases} \Psi(j,k) & \text{if } j \geq k , \\ 0 & \text{if } j < k . \end{cases} \tag{1.28}$$

By combining this along with the fact that C and $\tilde{B}$ are block diagonal operators it readily follows that the (j, k)-th entry of $C(I-M)^{-1}\tilde{B}$ is given by the right hand side of (1.20), and thus $C(I-M)^{-1}\tilde{B}$ is the central interpolant with tolerance $\gamma$.

The above calculation shows that the operator $[F]_n = (\delta_{j-k,n} f_{j,k})_{j,k=-\infty}^{\infty}$, where $\delta_{p,n}$ is the Kronecker delta, is equal to $CM^n\tilde{B}$ for $n \geq 0$, and $[F]_n = 0$ otherwise. Since M is stable, it follows that F belongs to the nonstationary Wiener class $NSW(\vec{\mathcal{Y}}, \vec{\mathcal{U}})$; see Remark X.2.5. Furthermore, the symbol $G(\cdot)$ of the diagonally sparse transform $\hat{F}$ of F is given by

$$G(\lambda) = \sum_{n=0}^{\infty} \lambda^n CM^n \tilde{B} = C(I - \lambda M)^{-1}\tilde{B} ,$$

and hence the symbol of the diagonally sparse transform of F is precisely the central interpolant $G_\gamma$ of the associated stationary Nevanlinna-Pick problem.

## XI.2 THE NONSTATIONARY VERSION OF NEHARI'S THEOREM

In this section we prove the nonstationary version of the Nehari extension theorem (Theorem VIII.4.1). Recall that for the nonstationary Nehari problem the given data are bounded linear operators $f_{j,k} \in \mathcal{L}(\mathcal{U}_k, \mathcal{Y}_j)$ with $-\infty < j < k < \infty$. With these data we associate for each $k \in \mathbb{Z}$ the infinite operator matrix

$$\Gamma_k = \begin{bmatrix} f_{k-1,k} & f_{k-1,k+1} & f_{k-1,k+2} & \cdots \\ f_{k-2,k} & f_{k-2,k+1} & f_{k-2,k+2} & \cdots \\ f_{k-3,k} & f_{k-3,k+1} & f_{k-3,k+2} & \cdots \\ \vdots & \vdots & \vdots & \end{bmatrix} . \tag{2.1}$$

Now assume that G in $\mathcal{L}(\vec{\mathcal{U}}, \vec{\mathcal{Y}})$ is an interpolant of the data $\{f_{j,k}\}$, that is, G can be represented by a doubly infinite operator matrix of the form (VIII.4.3). Since after a reordering of the rows $\Gamma_k$ can be obtained as a compression of G, we have

$$d_\infty := \sup\{\|\Gamma_k\| : k \in \mathbb{Z}\} \leq \|G\| . \tag{2.2}$$

In the sequel we assume that the error in the left hand side of (2.2) is finite. It remains to show that we can find an interpolant $G \in \mathcal{L}(\vec{\mathcal{U}}, \vec{\mathcal{Y}})$ of the data $\{f_{j,k}\}$ such that $\|G\| = d_\infty$. For this purpose we shall employ the stationary version of the Nehari extension theorem for possibly nonseparable Hilbert spaces (Theorem I.6.2). Our assumption on $\Gamma_k$ for $k \in \mathbb{Z}$ implies that

$$\sup\{\|f_{j,k}\| : -\infty < j < k < \infty\} < \infty . \tag{2.3}$$

Hence for each $n = -1, -2, \ldots$ the operator

$$[F]_n = (\delta_{j-k,n} f_{j,k})_{j,k=-\infty}^{\infty} : \vec{\mathcal{U}} \to \vec{\mathcal{Y}} \tag{2.4}$$

is well-defined and bounded. (Here we assume that $f_{j,k} = 0$ for $j \geq k$.) Consider the block Hankel matrix

$$\Gamma = \begin{bmatrix} [F]_{-1} & [F]_{-2} & [F]_{-3} & \cdots \\ [F]_{-2} & [F]_{-3} & [F]_{-4} & \cdots \\ [F]_{-3} & [F]_{-4} & [F]_{-5} & \cdots \\ \vdots & \vdots & \vdots & \end{bmatrix} . \tag{2.5}$$

We shall prove (see Lemma 2.1 below) that our condition $d_\infty < \infty$ implies that $\Gamma$ defines a bounded linear operator from $\ell_+^2(\vec{\mathcal{U}})$ to $\ell_+^2(\vec{\mathcal{Y}})$ and $\|\Gamma\| = d_\infty$. Assume that these facts have been established. Then we can apply Theorem I.6.2 to show that there exists a block Laurent operator

$$L = (L_{i-j})_{i,j=-\infty}^{\infty} : \ell^2(\vec{\mathcal{U}}) \to \ell^2(\vec{\mathcal{Y}}) \tag{2.6}$$

such $\|L\| = \|\Gamma\| = d_\infty$ and

$$L_{-n} = [F]_{-n} \qquad (n = 1, 2, 3, \ldots) . \tag{2.7}$$

Now, let $G = \check{G}$ be the diagonal compression of L. Then $\|G\| \leq \|L\| = d_\infty$ by Proposition X.2.2. By definition the (j, k)-th entry of G is equal to the (j, k)-th entry of $L_{j-k}$. Thus (2.7) shows that for $-\infty < j < k < \infty$ the (j, k)-th entry of G is equal to $f_{j,k}$. In other words, G is a nonstationary interpolant of the data $\{f_{j,k}\}$ with $\|G\| \leq d_\infty$. By combining this with (2.2), we see that $\|G\| = d_\infty$ and Theorem VIII.4.1 is proved.

It remains to show that the block Hankel matrix $\Gamma$ in (2.5) induces a bounded linear operator with norm $d_\infty$ (where $d_\infty$ is as in (2.2)). For this purpose we need the following lemma.

**LEMMA 2.1**. *Let* $f_{j,k} \in \mathcal{L}(\mathcal{U}_k, \mathcal{Y}_j)$ *for* $-\infty < j < k < \infty$ *be given, and for each* $k \in \mathbb{Z}$ *let* $\tilde{H}_k$ *be the doubly infinite operator matrix* $(h_{i,j}^{(k)})_{i,j=-\infty}^{\infty}$ *with*

$$h_{i,j}^{(k)} = \begin{cases} f_{i,j} & \textit{for } i \le k-1 \textit{ and } j \ge k, \\ 0 & \textit{otherwise}. \end{cases}$$

*Assume that* $\tilde{H}_k$ *induces a bounded linear operator from* $\vec{\mathcal{U}}$ *to* $\vec{\mathcal{Y}}$ *and*

$$\tilde{d}_\infty = \sup\{\|\tilde{H}_k\| : k \in \mathbb{Z}\} < \infty. \tag{2.8}$$

*Consider the block diagonal operator*

$$D = \operatorname{diag}(\tilde{H}_{-k})_{k=-\infty}^{\infty} : \ell^2(\vec{\mathcal{U}}) \to \ell^2(\vec{\mathcal{Y}}),$$

*and put* $\tilde{H} = U'DU^{-1}$, *where* U *on* $\ell^2(\vec{\mathcal{U}})$ *and* U′ *on* $\ell^2(\vec{\mathcal{Y}})$ *are the unitary operators associated with the sparse embeddings of* $\vec{\mathcal{U}}$ *into* $\ell^2(\vec{\mathcal{U}})$ *and of* $\vec{\mathcal{Y}}$ *into* $\ell^2(\vec{\mathcal{Y}})$, *respectively. Then* $\tilde{H}$ *is a bounded linear operator from* $\ell^2(\vec{\mathcal{U}})$ *into* $\ell^2(\vec{\mathcal{Y}})$ *of norm* $\tilde{d}_\infty$ *and its* (r, s)*-th operator entry is given by*

$$\tilde{H}_{r,s} = \begin{cases} [F]_{r-s} & \textit{for } r \le -1 \textit{ and } s \ge 0, \\ 0 & \textit{otherwise}, \end{cases}$$

*where* $[F]_n$ *is as in (2.4).*

Notice that the operator matrix $\Gamma_k$ in (2.1) may be obtained from the operator matrix $\tilde{H}_k$ in Lemma 2.1 by deleting in $\tilde{H}_k$ the rows with index $i \ge k$ and the columns with index $j \le k-1$ and by reordering the remaining rows. In particular, the numbers $d_\infty$ in (2.2) and $\tilde{d}_\infty$ in (2.8) are equal. Furthermore, the operator matrix Γ in (2.5) is obtained from the operator matrix $\tilde{H}$ in Lemma 2.1 by deleting in $\tilde{H}$ the rows with index $i \ge 0$ and the columns with index $j \le -1$ and by reordering the remaining columns. Thus, by Lemma 2.1, the condition $d_\infty < \infty$ implies that Γ induces a bounded linear operator and $\|\Gamma\| = d_\infty$.

**PROOF OF LEMMA 2.1.** Using the notation introduced in Section X.2 (adjusted for the present setting, i.e., with $\mathcal{U}_j$ in place of $\mathcal{X}_j$), we have

$$\tilde{H}_{r,s} = \pi_r' U' D U^* \pi_s^* = \pi_r' \left[ \sum_{j,k=-\infty}^{\infty} (V')^j \tau' \pi_j' D \pi_k^* \tau^* V^{-k} \right] \pi_s^*$$

$$= \pi_r' \left[ \sum_{j=-\infty}^{\infty} (V')^j \tau' \tilde{H}_{-j} \tau^* V^{-j} \right] \pi_s^* = \sum_{j=-\infty}^{\infty} \pi_{r-j}' \tau' \tilde{H}_{-j} \tau^* \pi_{s-j}^*$$

with convergence in the strong operator topology. Now recall that $\pi_k \tau$ on $\vec{\mathcal{U}}$ is the orthogonal projection $Q_k$ of $\vec{\mathcal{U}}$ which assigns to the sequence $\vec{u} = (u_n)_{n=-\infty}^{\infty}$ the sequence $(\delta_{k,n}u_n)_{n=-\infty}^{\infty}$, where $\delta_{k,n}$ is the Kronecker delta. The projection $Q'_k = \pi'_k\tau'$ is defined in an analogous way with $\mathcal{Y}$ in place of $\mathcal{U}$. Thus $\pi'_{r-j}\tau'\tilde{H}_{-j}\tau^*\pi^*_{s-j}$ is the doubly infinite operator matrix of which all entries are zero except the $(r-j, s-j)$-th entry which is equal to the $(r-j, s-j)$-th entry of $\tilde{H}_{-j}$. But then we can use the definition of $\tilde{H}_{-j}$ to show that $\pi'_{r-j}\tau'\tilde{H}_{-j}\tau^*\pi^*_{s-j}$ is zero for $r \geq 0$ or $s \leq -1$, and

$$\pi'_{r-j}\tau'\tilde{H}_{-j}\tau^*\pi^*_{s-j} = (\delta_{r-j,p}\delta_{s-j,q}f_{r-j,s-j})_{p,q=-\infty}^{\infty}$$

otherwise. We conclude that $\tilde{H}_{r,s} = 0$ for $r \geq 0$ or $s \leq -1$, and

$$\tilde{H}_{r,s} = (\delta_{r-s,j-k}f_{j,k})_{j,k=-\infty}^{\infty} = [F]_{r-s} \quad (r \leq -1,\ s \geq 0),$$

which completes the proof.

From Lemma 2.1 and our analysis in the first two paragraphs of this section it follows that solutions of the nonstationary Nehari problem may be obtained as diagonal compressions of solutions of the associated stationary Nehari problem. The next theorem shows that all solutions are obtained in this way.

**THEOREM 2.2.** *Let* $f_{j,k} \in \mathcal{L}(\mathcal{U}_k, \mathcal{Y}_j)$ *for* $-\infty < j < k < \infty$ *be given. Assume that (2.3) holds, and for* $n = -1, -2, \ldots$ *let* $[F]_n \in \mathcal{L}(\vec{\mathcal{U}}, \vec{\mathcal{Y}})$ *be defined by (2.4). Then the nonstationary Nehari problem for the data* $\{f_{j,k}\}$ *and tolerance* $\gamma$ *is solvable if and only if the stationary Nehari problem for the associated data* $\{[F]_n\}$ *and tolerance* $\gamma$ *is solvable. More precisely, if* $G \in \mathcal{L}(\vec{\mathcal{U}}, \vec{\mathcal{Y}})$ *is a solution of the nonstationary Nehari problem for the data* $\{f_{j,k}\}$, *then the diagonally sparse transform* $\hat{G}$ *of* $G$ *is a solution of the stationary Nehari problem for the data* $\{[F]_n\}$ *and* $\|\hat{G}\| = \|G\|$. *Conversely, if the block Laurent operator* $L$ *from* $\ell^2(\vec{\mathcal{U}})$ *to* $\ell^2(\vec{\mathcal{Y}})$ *is a solution of the stationary Nehari problem for the data* $\{[F]_n\}$ *and tolerance* $\gamma$, *then the diagonal compression* $\check{L}$ *of* $L$ *is a solution of the nonstationary Nehari problem for the data* $\{f_{j,k}\}$ *and* $\|\check{L}\| \leq \|L\|$.

**PROOF.** To prove the theorem it only remains to show that the diagonally sparse transform $\hat{G}$ is a solution of the stationary Nehari problem whenever $G$ is a solution of the nonstationary problem. Put $G = (g_{j,k})_{j,k=-\infty}^{\infty}$, and recall that $\hat{G} = ([G]_{j-k})_{j,k=-\infty}^{\infty}$, where

$$[G]_n = (\delta_{j-k,n}g_{j,k})_{j,k=-\infty}^{\infty}.$$

Since $G$ is a solution of the nonstationary problem, $g_{j,k} = f_{j,k}$ for $j < k$, and thus $[G]_n = [F]_n$ for $n < 0$, which completes the proof.

To complete this section we will use Theorem 2.2 to solve a nonstationary Nehari problem whose data is given by a nonstationary state space system. To this end, let $Z_k$ from $\mathcal{X}_{k+1}$ into $\mathcal{X}_k$, $B_k$ from $\mathcal{U}_k$ into $\mathcal{X}_k$ and $C_k$ from $\mathcal{X}_{k+1}$ into $\mathcal{Y}_k$ be a set of Hilbert space operators for k in $\mathbb{Z}$. The state transition operator $\Phi(m, n)$ mapping $\mathcal{X}_n$ into $\mathcal{X}_m$ is defined by $\Phi(m, n) = Z_m Z_{m+1} \cdots Z_{n-1}$ if $m < n$ and $\Phi(n, n) = I$. Throughout we always assume that the families of operators $\{B_k\}$ and $\{C_k\}$ are uniformly bounded and that the family $\{Z_k\}$ is backward stable, that is,

$$\limsup_{\nu \to \infty} (\sup \{\|\Phi(k-\nu, k)\| : k \in \mathbb{Z}\})^{1/\nu} < 1 . \tag{2.9}$$

We say that $\{Z_k, B_k, C_k\}$ is a *nonstationary anticausal stable realization* of the data $\{f_{j,k}\}$ if

$$f_{j,k} = C_j \Phi(j+1, k) B_k \qquad (\text{for } -\infty < j < k < \infty) . \tag{2.10}$$

Without loss of generality we always assume that $f_{j,k} = 0$ if $j \geq k$.

In this setting the *controllability grammian* $P_k$ at time k for the family of pairs $\{Z_n, B_n\}_{n \in \mathbb{Z}}$ is the positive operator defined by

$$P_k = \sum_{\nu=0}^{\infty} \Phi(k, k+\nu) B_{k+\nu} B_{k+\nu}^* \Phi(k, k+\nu)^* . \tag{2.11}$$

The stability condition in (2.9) along with the fact that $\{B_n\}$ is uniformly bounded guarantees that $P_k$ is a bounded positive operator on $\mathcal{X}_k$. We say that the family of pairs $\{Z_n, B_n\}_{n \in \mathbb{Z}}$ is *uniformly controllable* if there exists an $\varepsilon > 0$ such that $P_k \geq \varepsilon I$ for all k in $\mathbb{Z}$. The *observability grammian* $Q_k$ at time k for the family of pairs $\{C_n, Z_n\}_{n \in \mathbb{Z}}$ is the bounded positive operator on $\mathcal{X}_k$ defined by

$$Q_k = \sum_{\nu=1}^{\infty} \Phi(k+1-\nu, k)^* C_{k-\nu}^* C_{k-\nu} \Phi(k+1-\nu, k) . \tag{2.12}$$

Once again the stability condition in (2.9) along with the facts that the family $\{C_n\}$ is uniformly bounded guarantees that $Q_k$ is a positive bounded operator for all k. The family of pairs $\{C_n, Z_n\}_{n \in \mathbb{Z}}$ is *uniformly observable* if there exists an $\varepsilon > 0$ such that $Q_k \geq \varepsilon I$ for all k in $\mathbb{Z}$. Finally, using the fact that $\Phi(m, n) = Z_m \Phi(m+1, n)$ and $\Phi(m, n) = \Phi(m, n-1) Z_{n-1}$ when $m < n$, it follows that $P_k$ and $Q_k$ satisfy the following nonstationary Lyapunov difference equations

$$P_k = Z_k P_{k+1} Z_k^* + B_k B_k^* \text{ and } Q_k = Z_{k-1}^* Q_{k-1} Z_{k-1} + C_{k-1}^* C_{k-1} . \tag{2.13}$$

Let $f_{j,k}$ for $j < k$ be given by (2.10). Then condition (2.3) is fulfilled, and for $n \leq -1$ we may consider the operators $[F]_n$ defined by (2.4). Let L be the central interpolant for the stationary Nehari problem associated with the data $\{[F]_n\}$ and tolerance $\gamma$, and let $G = \check{L}$ in $\mathcal{L}(\vec{\mathcal{U}}, \vec{\mathcal{Y}})$ be the solution to the original nonstationary Nehari problem for the data $\{f_{j,k}\}$ with tolerance $\gamma$ obtained by taking the diagonal compression of L; see Theorem 2.2. Then we call this G the *central interpolant for the nonstationary Nehari problem with tolerance* $\gamma$. The following result gives a state space description for the central Nehari interpolant when the data is given in state space form.

**THEOREM 2.3.** *Let* $\{Z_k, B_k, C_k\}$ *be a nonstationary anticausal stable, uniformly controllable and uniformly observable realization of the data* $\{f_{j,k}\}$, *and assume that both* $\{B_k\}$ *and* $\{C_k\}$ *are uniformly bounded. Then the error* $d_\infty$ *in the nonstationary Nehari problem is given by*

$$d_\infty^2 = \sup \{r_{spec}(Q_k P_k) : k \in \mathbb{Z}\}, \tag{2.14}$$

*where* $P_k$ *is the controllability grammian at time* k *for the pairs* $\{Z_n, B_n\}_{n\in\mathbb{Z}}$, *and* $Q_k$ *is the observability grammian at time* k *for the pairs* $\{C_n, Z_n\}_{n\in\mathbb{Z}}$. *Now assume that* $\gamma > d_\infty$ *and let* $M_k$ *from* $\mathcal{X}_k$ *into* $\mathcal{X}_{k+1}$ *be the operators defined by*

$$M_k = (\gamma^2 I - Z_k^* Q_k Z_k P_{k+1})^{-1} Z_k^* (\gamma^2 I - Q_k P_k) \quad (k \in \mathbb{Z}). \tag{2.15}$$

*Let* $\Psi(m, n) = M_{m-1} M_m \cdots M_n$ *if* $m > n$ *and* $\Psi(n, n) = I$ *be the state transition operator from* $\mathcal{X}_n$ *into* $\mathcal{X}_m$ *generated by* $\{M_k\}$. *Then the central interpolant* G *in* $\mathcal{L}(\vec{\mathcal{U}}, \vec{\mathcal{Y}})$ *to the nonstationary Nehari interpolation problem for the data* $\{f_{j,k}\}$ *with tolerance* $\gamma$ *is given by*

$$\begin{aligned} G_{j,k} &= -C_j P_{j+1} \Psi(j+1, k+1)(\gamma^2 I - Z_k^* Q_k Z_k P_{k+1})^{-1} Z_k^* Q_k B_k \qquad && \text{if } j \ge k \\ G_{j,k} &= f_{j,k} = C_j \Phi(j+1, k) B_k && \text{if } k > j. \end{aligned} \tag{2.16}$$

*Moreover, the sequence* $\{M_n\}$ *is forward stable, that is,*

$$\limsup_{\nu\to\infty} (\sup \{\|\Psi(k+\nu, k)\| : k \in \mathbb{Z}\})^{1/\nu} < 1. \tag{2.17}$$

**PROOF.** Let Z be the backward block weighted shift on $\vec{\mathcal{X}}$ associated with the operators $\{Z_k\}$, defined as in (1.1), and $B = \operatorname{diag}(B_k)_{-\infty}^{\infty}$ mapping $\vec{\mathcal{U}}$ into $\vec{\mathcal{X}}$. Let C be the backward block weighted shift from $\vec{\mathcal{X}}$ into $\vec{\mathcal{Y}}$ defined by

$$C = \begin{bmatrix} \ddots & \ddots & & & & \\ & 0 & C_{-1} & & & \\ & & \underline{0} & C_0 & & \\ & & & 0 & C_1 & \\ & & & & \ddots & \ddots \end{bmatrix} : \vec{\mathcal{X}} \to \vec{\mathcal{Y}}, \tag{2.18}$$

where all the unspecified entries are zero. Since $\{B_k\}$ and $\{C_k\}$ are uniformly bounded, B and C are bounded linear operators. Notice that the entries of $Z^\nu$ are all zero except those on the ν-th diagonal above the main diagonal, that is,

$$(Z^\nu)_{j,k} = \begin{cases} \Phi(j, j+\nu) & \text{if } k = j+\nu, \\ 0 & \text{if } k \neq j+\nu. \end{cases} \tag{2.19}$$

Hence $\|Z^\nu\| = \sup\{\|\Phi(j, j+\nu)\| : j \in \mathbb{Z}\}$ and the stability condition (2.9) implies that Z is stable, that is, $r_{spec}(Z) < 1$. By combining (2.4), (2.10) and (2.19) with $\nu > 0$ we have

$$\begin{gathered}([F]_{-\nu})_{j,j+\nu} = f_{j,j+\nu} = C_j\Phi(j+1, j+\nu)B_{j+\nu} = \\ C_j\Phi(j+1, j+1+\nu-1)B_{j+\nu} = C_j(Z^{\nu-1})_{j+1,j+\nu}B_{j+\nu}\,.\end{gathered} \tag{2.20}$$

Since C and Z are both block weighted forward shifts and B is a diagonal matrix, the only possible nonzero entries of $CZ^{\nu-1}B$ appear in the ν-th diagonal above the main diagonal. Therefore (2.20) shows that

$$[F]_{-\nu} = CZ^{\nu-1}B \qquad (\text{for } \nu \geq 1). \tag{2.21}$$

This means that {Z, B, C} is an anticausal stable realization of $\{[F]_{-\nu}\}_1^\infty$; see III.5. In particular,

$$C(\lambda I - Z)^{-1}B = \sum_{\nu=1}^{\infty} \lambda^{-\nu}[F]_{-\nu} \qquad (|\lambda| > 1). \tag{2.22}$$

Notice that $F(\lambda) = C(\lambda I - Z)^{-1}B$ is a bounded analytic operator-valued function on $|\lambda| > 1$.

The above analysis shows that the stationary Nehari interpolation problem associated with the nonstationary Nehari interpolation problem with data $\{f_{j,k}\}$ given by (2.10) and tolerance γ, is: Given the function $F(\lambda) = C(\lambda I - Z)^{-1}B$, find a function H in $H^\infty(\vec{\mathcal{U}}, \vec{\mathcal{Y}})$ such that $\|F - H\|_\infty \leq \gamma$. Moreover, by Theorem 2.2, if $G = F - H$, where H is in $H^\infty(\vec{\mathcal{U}}, \vec{\mathcal{Y}})$ and $\|F - H\|_\infty \leq \gamma$, then the diagonal compression $G = (L_G)^\vee$ of the Laurent operator $L_G$ is a solution to the original nonstationary Nehari problem with tolerance γ. Theorem 2.2 also shows that the error $d_\infty = \|\Gamma\|$ in the nonstationary Nehari problem (see (2.2)) equals the error in the corresponding stationary Nehari problem, that is,

$$\|\Gamma\| = d_\infty = \inf\{\|F - H\|_\infty : H \in H(\vec{\mathcal{U}}, \vec{\mathcal{Y}})\}\,. \tag{2.23}$$

So to complete the proof we will use the results in Section V.5 to compute the central interpolant $G_\gamma$ in $L^\infty(\vec{\mathcal{U}}, \vec{\mathcal{Y}})$ for the associated stationary Nehari interpolation problem with data F in (2.22) and tolerance $\gamma$. Then we will obtain the central interpolant G for the nonstationary Nehari problem by computing the diagonal compression of $L_{G_\gamma}$.

Let P be the controllability grammian for $\{Z, B\}$ and Q be the observability grammian for $\{C, Z\}$. Because Z is stable, P and Q are given by (see Section III.2)

$$P = \sum_{\nu=0}^{\infty} Z^\nu B^* B^* Z^{*\nu} \quad \text{and} \quad Q = \sum_{\nu=0}^{\infty} Z^{*\nu} C^* C Z^\nu\,. \tag{2.24}$$

Since Z and C are backward block weighted shifts and B is a block diagonal operator, it follows that both P and Q are block diagonal operators on $\vec{\mathcal{X}}$. In fact, using (2.19) it follows that $P = \operatorname{diag}(P_k)_{-\infty}^{\infty}$ and $Q = \operatorname{diag}(Q_k)_{-\infty}^{\infty}$, where $P_k$ and $Q_k$ are given by (2.11) and (2.12), respectively. Because the nonstationary system $\{Z_k, B_k, C_k\}$ is uniformly controllable and observable, P and Q are both strictly positive. Hence $\{Z, B, C\}$ is a controllable and observable, stable anticausal realization of F. According to Theorem V.5.1, the error $d_\infty$ (see 2.23) in the associated stationary Nehari interpolation problem is given by $d_\infty^2 = r_{spec}(QP)$. Since P and Q are diagonal operators and $\|\Gamma\| = d_\infty$, we obtain the formula for $d_\infty$ in (2.14). Finally, it is noted that one can also obtain the time varying Lyapunov equations for $P_k$ and $Q_k$ in (2.13) by computing the diagonal components in the corresponding stationary Lyapunov equation

$$P = ZPZ^* + BB^* \quad \text{and} \quad Q = Z^* QZ + C^* C\,. \tag{2.25}$$

According to Theorem V.5.2 the central interpolant with tolerance $\gamma$ for the stationary Nevanlinna-Pick problem with the data F is given by $G_\gamma = F - H$, where H is now the function in $H^\infty(\vec{\mathcal{U}}, \vec{\mathcal{Y}})$ defined by

$$\begin{aligned} H(\lambda) &= CP(I - \lambda M)^{-1}(\gamma^2 I - Z^* QZP)^{-1} Z^* QB, \quad \text{where} \\ M &= (\gamma^2 I - Z^* QZP)^{-1} Z^* (\gamma^2 I - QP)\,. \end{aligned} \tag{2.26}$$

Using the fact that $Z^*$ is a block forward weighted shift and P and Q are diagonal matrices it follows that M is a block forward weighted shift of the form

$$M = \begin{bmatrix} \ddots & & & & \\ \ddots & 0 & & & \\ & M_{-1} & \underline{0} & & \\ & & M_0 & 0 & \\ & & & M_1 & \ddots \\ & & & & \ddots \end{bmatrix} \tag{2.27}$$

where the operators $M_k$ from $\mathcal{X}_k$ into $\mathcal{X}_{k+1}$ are given by (2.15) and all the unspecified entries in (2.27) are zero. Moreover, for $\nu \geq 0$, the j, k entry of $M^\nu$ is given by

$$(M^\nu)_{j,k} = \Psi(k+\nu,\ k)\delta_{j,k+\nu} \tag{2.28}$$

where $\delta_{m,n}$ is the Kronecker delta. Theorem V.1.2 shows that M is stable. So according to (2.28) we have $\|M^\nu\| = \sup\{\|\Psi(k+\nu,\ k)\| : k \in \mathbb{Z}\}$. Therefore the stability condition in (2.17) holds.

Recall that the central interpolant $G_\gamma$ with tolerance $\gamma$ for the nonstationary Nehari problem is precisely the diagonal compression of $L_{G_\gamma}$. The diagonal compression of $L_{G_\gamma}$ places the $\nu$-th diagonal of $G_{\gamma,\nu}$ onto the $\nu$-th diagonal of G where $G_{\gamma,\nu}$ is the Fourier coefficient of $\lambda^\nu$ for $G_\gamma$. To be precise, $G_{k+\nu,k} = (G_{\gamma,\nu})_{k+\nu,k}$ for all $\nu$ and k in $\mathbb{Z}$; see (X.2.10). Since $G_\gamma = F - H$, where F is in $K^\infty(\vec{\mathcal{U}}, \vec{\mathcal{Y}})$ and H in $H^\infty(\vec{\mathcal{U}}, \vec{\mathcal{Y}})$ has the state space representation in (2.26), the $\nu$-th Fourier coefficient of $G_\gamma$ is given by

$$G_{\gamma,\nu} = -CPM^\nu(\gamma^2 I - Z^*QZP)^{-1}Z^*QB \qquad (\text{for } \nu \geq 0). \tag{2.29}$$

So using (2.28) along with the fact that $Z^*$ and M are block forward weighted shifts and all the other operators are block diagonal matrices we have for $\nu \geq 0$

$$-G_{k+\nu,k} = (CPM^\nu(\gamma^2 I - Z^*QZP)^{-1}Z^*QB)_{k+\nu,k} =$$

$$C_{k+\nu}P_{k+\nu+1}\Psi(k+\nu+1,\ k+1)(\gamma^2 I - Z_k^*Q_kZ_kP_{k+1})^{-1}Z_k^*Q_kB_k\ .$$

Therefore the j, k entry of G is given by (2.16). This completes the proof.

**REMARK 2.4**. In this remark we show that the diagonally sparse transform of the central interpolant G constructed in Theorem 2.2 is equal to $L_{G_\gamma}$, where $G_\gamma$ is central solution to the associated stationary problem. As before, let $\{Z_k, B_k, C_k\}$ be a stable, uniformly controllable and uniformly observable, anticausal realization for the data $\{f_{j,k}\}$ in the nonstationary Nehari interpolation problem with tolerance $\gamma > d_\infty$. Let M on $\vec{\mathcal{X}}$ be the stable operator defined in (2.27), with $M_k$ as in (2.15), and set $P = \text{diag}\,(P_k)_{-\infty}^{\infty}$ and $Q = \text{diag}\,(Q_k)_{-\infty}^{\infty}$. Then the central interpolant G with tolerance $\gamma$ for this nonstationary Nehari interpolation problem is given by $G = G_\gamma(1)$, that is,

$$G = C(I-Z)^{-1}B - CP(I-M)^{-1}(\gamma^2 I - Z^* QZP)^{-1}Z^* QB . \tag{2.30}$$

To see this recall that the central interpolant $G_\gamma$ for the stationary Nehari interpolation problem associated with the nonstationary Nehari problem with tolerance $\gamma$ is given by $G_\gamma = F - H$, where $F = C(\lambda I - Z)^{-1}B$ is in $K^\infty(\vec{\mathcal{U}}, \vec{\mathcal{Y}})$ and H in $H^\infty(\vec{\mathcal{U}}, \vec{\mathcal{Y}})$ is given by (2.26). By combining (2.21) with (2.29) we have

$$G_{\gamma,\nu} = \begin{cases} [F]_\nu = CZ^{|1+\nu|}B & \text{if } \nu \le -1 , \\ -CPM^\nu(\gamma^2 I - Z^* QZP)^{-1}Z^* QB & \text{if } \nu \ge 0 . \end{cases} \tag{2.31}$$

Notice that because, Z, M and C are block weighted shifts while P, Q and B are block diagonal operators, it follows that all the entries of $G_{\gamma,\nu}$ are zero except possibly those on $\nu$-th block diagonal below the main diagonal for $\nu \ge 0$ and above the main diagonal for $\nu < 0$. Hence $G_{\gamma,\nu} = [G_{\gamma,\nu}]_\nu$. Since the diagonal compression extracts the $\nu$-th diagonal from $G_{\gamma,\nu}$ and places it in the $\nu$-diagonal of G and $G = (L_{G_\gamma})^\vee$, it follows that the central interpolant

$$G = \sum_{\nu=-\infty}^{\infty} G_{\gamma,\nu} = \sum_{\nu=1}^{\infty} G_{\gamma,-\nu} + \sum_{\nu=0}^{\infty} G_{\gamma,\nu} . \tag{2.32}$$

Substituting (2.31) into (2.32) along with the fact that $(I-T)^{-1} = I + T + T^2 + \cdots$ when $r_{spec}(T) < 1$, readily yields (2.30).

Since Z is a backward block weighted shift with $r_{spec}(Z) < 1$ and M is a forward block weighted shift with $r_{spec}(M) < 1$, formula (2.30) implies that G belongs to the nonstationary Wiener class $NSW(\vec{\mathcal{U}}, \vec{\mathcal{Y}})$; see Remark X.2.5. The calculations in the previous paragraph show that the symbol of the diagonally sparse transform of G is given by

$$G_\gamma(\lambda) = C(\lambda I - Z)^{-1}B - CP(I - \lambda M)^{-1}(\gamma^2 I - Z^* QZP)^{-1}Z^* QB .$$

In other words, the symbol of the diagonally sparse transform of the central interpolant for the nonstationary Nehari problem with tolerance $\gamma$ is precisely the central interpolant of the associated stationary Nehari problem.

## XI.3 THE NONSTATIONARY SARASON INTERPOLATION THEOREM

In this section we prove the nonstationary Sarason interpolation theorem (Theorem VIII.5.1). Recall that in the nonstationary Sarason problem the given data are bounded linear operators

$$F : \vec{\mathcal{U}} \to \vec{\mathcal{Y}} \text{ and } \Theta : \vec{\mathcal{E}} \to \vec{\mathcal{Y}}, \tag{3.1}$$

where F is a block lower triangular operator, and $\Theta$ is a block lower triangular isometry. (Recall also that $\vec{\mathcal{H}} = \oplus \{\mathcal{H}_k : k \in \mathbb{Z}\}$.) Given such operators and a tolerance $\gamma$, the nonstationary Sarason problem is to determine a block lower triangular operator H from $\vec{\mathcal{U}}$ to $\vec{\mathcal{E}}$ such that

$$\|F - \Theta H\| \leq \gamma. \tag{3.2}$$

To solve this problem we consider an associated stationary version of the Sarason problem. For this purpose we need the diagonally sparse transforms $\hat{F}$ and $\hat{\Theta}$ of F and $\Theta$, respectively. Recall that

$$\hat{F} : \ell^2(\vec{\mathcal{U}}) \to \ell^2(\vec{\mathcal{Y}}) \text{ and } \hat{\Theta} : \ell^2(\vec{\mathcal{E}}) \to \ell^2(\vec{\mathcal{Y}}), \tag{3.3}$$

and these operators are block Laurent operators. Furthermore, the operators $\hat{F}$ and $\hat{\Theta}$ are both block lower triangular. By using Proposition X.2.1, one sees that $\hat{\Theta}$ is an isometry. Thus we may consider (cf., Section I.7) the problem of finding a block lower triangular Laurent operator K from $\ell^2(\vec{\mathcal{U}})$ to $\ell^2(\vec{\mathcal{E}})$ such that

$$\|\hat{F} - \hat{\Theta} K\| \leq \gamma. \tag{3.4}$$

We shall refer to the latter problem as the *stationary Sarason problem associated with the data* (3.1) *and the bound* $\gamma$. The following result provides the key to the proof of Theorem VIII.5.1.

**THEOREM 3.1**. *The nonstationary Sarason problem for the data (3.1) and tolerance* $\gamma$ *is solvable if and only if the associated stationary Sarason problem is solvable with tolerance* $\gamma$. *More precisely, if* H *is a solution of the nonstationary Sarason problem (3.2), then the diagonally sparse transform* $\hat{H}$ *of* H *is a solution of the associated stationary problem (3.4) and* $\|\hat{H}\| = \|H\|$. *Conversely, if* K *is a solution of the associated stationary Sarason problem (3.2), then the diagonal compression* H *of* K *is a solution of the original nonstationary Sarason problem (3.2) and* $\|H\| \leq \|K\|$.

**PROOF**. First we introduce some notation. Given the Hilbert space direct sum $\vec{\mathcal{H}} = \oplus_{j=-\infty}^{\infty} \mathcal{H}_j$, we write $U_{\vec{\mathcal{H}}}$ for the unitary operator on $\ell^2(\vec{\mathcal{H}})$ associated with the sparse embedding of $\vec{\mathcal{H}}$ into $\ell^2(\vec{\mathcal{H}})$; see Section X.1. Furthermore, given a second Hilbert space direct sum $\vec{\mathcal{K}} = \oplus_{j=-\infty}^{\infty} \mathcal{K}_j$ and a bounded linear operator $T : \vec{\mathcal{H}} \to \vec{\mathcal{K}}$, we use the symbol $\Delta(T)$ to denote the diagonal operator

$$\Delta(T) := \begin{bmatrix} \ddots & & & \\ & T & & \\ & & T & \\ & & & T \\ & & & & \ddots \end{bmatrix} : \ell^2(\vec{\mathcal{H}}) \to \ell^2(\vec{\mathcal{K}}) . \tag{3.5}$$

Using this notation we see that Proposition X.2.1 tells us that

$$\hat{F} = U_{\vec{\mathcal{Y}}} \Delta(F) U^*_{\vec{\mathcal{U}}} \text{ and } \hat{\Theta} = U_{\vec{\mathcal{Y}}} \Delta(\Theta) U^*_{\vec{\mathcal{E}}} . \tag{3.6}$$

Now, assume that H from $\vec{\mathcal{U}}$ to $\vec{\mathcal{E}}$ is a solution to the nonstationary Sarason problem associated with the data (3.1) and the tolerance $\gamma$. Let $\hat{H}$ be the diagonally sparse transform of H. Thus $\hat{H} = U_{\vec{\mathcal{E}}} \Delta(H) U^*_{\vec{\mathcal{U}}}$. Since the norm of the operator $\Delta(T)$ in (3.5) is equal to $\|T\|$, we can use (3.2) and (3.6) to show that

$$\gamma \geq \|F - \Theta H\| = \|\Delta(F - \Theta H)\| = \|\Delta(F) - \Delta(\Theta)\Delta(H)\|$$

$$= \|U^*_{\vec{\mathcal{Y}}}(\hat{F} - \hat{\Theta}\hat{H})U_{\vec{\mathcal{U}}}\| .$$

But $U_{\vec{\mathcal{U}}}$ and $U_{\vec{\mathcal{Y}}}$ are unitary operators. Thus (3.4) is fulfilled with $K = \hat{H}$. Since H from $\vec{\mathcal{U}}$ to $\vec{\mathcal{E}}$ is block lower triangular, the same is true for $\hat{H}$ from $\ell^2(\vec{\mathcal{U}})$ to $\ell^2(\vec{\mathcal{E}})$. Thus $\hat{H}$ is a solution of the associated stationary Sarason problem and $\|\hat{H}\| = \|H\|$, by Proposition X.2.1.

Conversely, assume K from $\ell^2(\vec{\mathcal{U}})$ to $\ell^2(\vec{\mathcal{E}})$ is a solution of the associated stationary Sarason problem. Thus K is block lower triangular, and (3.4) holds. Let H be the diagonal compression of K. Then H is a bounded linear operator from $\vec{\mathcal{U}}$ into $\vec{\mathcal{E}}$, and H is block lower triangular. Recall (see the paragraph preceding Proposition X.2.2) that F is the diagonal compression of $\hat{F}$, and $\Theta$ is the diagonal compression of $\hat{\Theta}$. Furthermore, by applying Corollary X.2.4, we see that the diagonal compression of $\hat{\Theta}K$ is equal to $\Theta H$. Thus $F - \Theta H$, is precisely the diagonal compression of $\hat{F} - \hat{\Theta}K$. Since the diagonal compression does not increase the operator norm (by Proposition X.2.2) we have $\|H\| \leq \|K\|$,

$$\|F - \Theta H\| \leq \|\hat{F} - \hat{\Theta}K\| \leq \gamma ,$$

and hence H is a solution of our original nonstationary model matching problem. This completes the proof.

**PROOF OF THEOREM VIII.5.1.** First recall that $P_j$ is the orthogonal projection on $\vec{\mathcal{U}}$ defined by

$$(P_j\vec{u})_i = \begin{cases} 0 & \text{for} \quad i<j, \\ u_i & \text{for} \quad i\geq j. \end{cases} \tag{3.7}$$

Furthermore, we write $P_+$ for the orthogonal projection of $\ell^2(\vec{\mathcal{U}})$ onto $\ell^2_+(\vec{\mathcal{U}})$. Moreover, $P_{\mathcal{E},j}$ denotes the orthogonal projection of $\vec{\mathcal{E}}$ defined according to (3.7) where $\vec{\mathcal{E}}$ replaces $\vec{\mathcal{U}}$, and $P_{\mathcal{E}+}$ is the orthogonal projection from $\ell^2(\vec{\mathcal{E}})$ onto $\ell^2_+(\vec{\mathcal{E}})$. Clearly, the spaces $\mathcal{M}_k$ appearing in Theorem VIII.5.1 are given by

$$\mathcal{M}_k = \operatorname{ran} \Theta P_{\mathcal{E},k} \quad (k\in \mathbb{Z}). \tag{3.8}$$

Now let $\hat{\mathcal{M}}$ be the subspace in $\ell^2(\vec{\mathcal{Y}})$ defined by $\hat{\mathcal{M}} = \operatorname{ran} \hat{\Theta} P_{\mathcal{E}+}$. As before, $\hat{F}$ and $\hat{\Theta}$ are the diagonally sparse transforms of F and $\Theta$, which also appear in (3.6). According to Theorem I.7.1 (and Theorem I.10.1 for the nonseparable case) the stationary Sarason problem associated with the data (3.1) and the bound $\gamma$ is solvable if and only if

$$\gamma \geq \hat{d}_\infty := \|(I - P_{\hat{\mathcal{M}}})\hat{F} \mid \ell^2_+(\vec{\mathcal{U}})\|. \tag{3.9}$$

But then we see from Proposition 3.1 that in order to prove Theorem VIII.5.1 it suffices to show that

$$\hat{d}_\infty = d_\infty := \sup \{\|(I - P_{\mathcal{M}_k})FP_k\| k \in \mathbb{Z}\}. \tag{3.10}$$

To prove $\hat{d}_\infty = d_\infty$, we first show that

$$U^*_{\vec{\mathcal{Y}}}\hat{\Theta}P_{\mathcal{E}+}U_{\vec{\mathcal{E}}} = \operatorname{diag}\,(\Theta P_{\mathcal{E},-k})_{k=-\infty}^{\infty}. \tag{3.11}$$

Here, as in the proof of Theorem 3.1, the symbol $U_{\vec{\mathcal{K}}}$ denotes the unitary operator on $\ell^2(\vec{\mathcal{K}})$ associated with the sparse embedding of $\vec{\mathcal{K}}$ into $\ell^2(\vec{\mathcal{K}})$. To prove (3.11) one uses Proposition X.1.2, Proposition X.2.1 and the identities in (3.6). Indeed,

$$U^*_{\vec{\mathcal{Y}}}\hat{\Theta}P_{\mathcal{E}+}U_{\vec{\mathcal{E}}} = \Delta(\Theta)U^*_{\vec{\mathcal{E}}}P_{\mathcal{E}+}U_{\vec{\mathcal{E}}} = \operatorname{diag}\,(\Theta P_{\mathcal{E},-k})_{k=-\infty}^{\infty}.$$

Hence (3.11) holds.

Proposition X.1.2 shows that

$$\operatorname{diag}\,(P_{-k})_{k=-\infty}^{\infty} = U^*_{\vec{\mathcal{U}}}P_+U_{\vec{\mathcal{U}}}. \tag{3.12}$$

Now using $\hat{\mathcal{M}} = \operatorname{ran} \hat{\Theta}P_{\mathcal{E}+}$ we see that $P_{\hat{\mathcal{M}}} = \hat{\Theta}P_{\mathcal{E}+}\hat{\Theta}^*$ and similarly $P_{\mathcal{M}_k} = \Theta P_{\mathcal{E},k}\Theta^*$. This and (3.11) gives

$$\operatorname{diag}(P_{\mathcal{M}_{-k}})_{k=-\infty}^{\infty} = U_{\vec{\mathcal{Y}}}^* P_{\hat{\mathcal{M}}} U_{\vec{\mathcal{Y}}} . \tag{3.13}$$

Using (3.12) and (3.13) we may rewrite the right-hand side of (3.10) in the following way

$$\begin{aligned} d_\infty &= \| \operatorname{diag} \left[ (I - P_{\mathcal{M}_k}) F P_k \right]_{k=-\infty}^{\infty} \| \\ &= \| U_{\vec{\mathcal{Y}}}^* (I - P_{\hat{\mathcal{M}}}) U_{\vec{\mathcal{Y}}} \Delta(F) U_{\vec{\mathcal{U}}}^* P_+ U_{\vec{\mathcal{U}}} \| \\ &= \| (I - P_{\hat{\mathcal{M}}}) \hat{F} P_+ \| , \end{aligned}$$

where for the last equality we used the first identity in (3.6) and the fact that $U_{\vec{\mathcal{U}}}$ and $U_{\vec{\mathcal{Y}}}$ are unitary operators. The above calculation yields $d_\infty = \hat{d}_\infty$, and the proof is complete.

## XI.4 THE NONSTATIONARY VERSION OF NUDELMAN'S THEOREM

In this section we prove the nonstationary Nudelman theorem (Theorem VIII.6.1). As before, $\vec{\mathcal{H}}$ is the Hilbert space formed by the orthogonal direct sum $\vec{\mathcal{H}} = \oplus \{\mathcal{H}_k : k \in \mathbb{Z}\}$. With the interpolation data in (VIII.6.1) we associate operators

$$\begin{aligned} &Z : \vec{\mathcal{X}} \to \vec{\mathcal{X}}, \quad B : \vec{\mathcal{U}} \to \vec{\mathcal{X}}, \quad \tilde{B} : \vec{\mathcal{Y}} \to \vec{\mathcal{X}}, \\ &\Lambda : \vec{\mathcal{F}} \to \vec{\mathcal{F}}, \quad C : \vec{\mathcal{F}} \to \vec{\mathcal{Y}}, \quad \tilde{C} : \vec{\mathcal{F}} \to \vec{\mathcal{U}}, \ \Gamma : \vec{\mathcal{F}} \to \vec{\mathcal{X}}, \end{aligned} \tag{4.1}$$

by setting

$$\begin{aligned} &Z = (\delta_{j,k-1} Z_j)_{j,k=-\infty}^{\infty}, \quad B = \operatorname{diag}(B_k)_{-\infty}^{\infty}, \quad \tilde{B} = \operatorname{diag}(\tilde{B}_k)_{-\infty}^{\infty}, \\ &\Lambda = (\delta_{j,k-1} \Lambda_{j+1})_{j,k=-\infty}^{\infty}, \quad C = \operatorname{diag}(C_k)_{-\infty}^{\infty}, \quad \tilde{C} = \operatorname{diag}(\tilde{C}_k)_{-\infty}^{\infty}, \\ &\Gamma = (\delta_{j,k+1} \Gamma_j)_{j,k=-\infty}^{\infty} . \end{aligned} \tag{4.2}$$

Here $\delta_{p,q}$ denotes the Kronecker delta, that is, $\delta_{p,q} = 1$ if $p = q$ and zero otherwise. It follows that $Z$ and $\Lambda$ are block weighted backward shifts, while $\Gamma$ can be viewed as a block weighted forward shift from $\vec{\mathcal{F}}$ into $\vec{\mathcal{X}}$. For $\nu \geq 1$ the $(j, k)$ – th entries of $Z^\nu$ and $\Lambda^\nu$ are given by

$$(Z^\nu)_{j,k} = \delta_{-\nu,j-k} \Phi(j, j+\nu) = \delta_{-\nu,j-k} \Phi(k-\nu, k) , \tag{4.3}$$

$$(\Lambda^\nu)_{j,k} = \delta_{-\nu,j-k} \Psi(j, j+\nu) = \delta_{-\nu,j-k} \Psi(k-\nu, k) , \tag{4.4}$$

where $\Phi$ and $\Psi$ are the state transition operators associated with $\{Z_n\}$ and $\{\Lambda_n\}$, respectively, that is,

$$\Phi(m, n) = Z_m Z_{m+1} \cdots Z_{n-1} \qquad (m < n, \text{ and } \Phi(m, m) = I_{\mathcal{X}_m}), \tag{4.5}$$

$$\Psi(m, n) = \Lambda_{m+1} \Lambda_m \cdots \Lambda_n \qquad (m < n, \text{ and } \Psi(m, m) = I_{\mathcal{F}_m}). \tag{4.6}$$

We conclude that the growth conditions in (VIII.6.2) are equivalent to the requirement that the spectral radii of Z and $\Lambda$ are strictly less than one.

Thus, given the bound $\gamma$, we may consider the problem of finding $G \in H^\infty(\vec{\mathcal{Y}}, \vec{\mathcal{U}})$ such that $\|G\|_\infty \leq \gamma$ and

(a) $(BG)(Z)_{\text{left}} = \tilde{B}$,

(b) $(GC)(\Lambda)_{\text{right}} = \tilde{C}$,

(c) $\Gamma = \Sigma_{j,k=0}^{\infty} Z^j B G_{j+k+1} C \Lambda^k$,

where $G_0$, $G_1$, $G_2$, ... are the Taylor coefficients of G. We shall refer to this problem as the *stationary Nudelman interpolation problem associated with the nonstationary data* (VIII.6.1) *and bound* $\gamma$.

**THEOREM 4.1.** *The nonstationary Nudelman interpolation problem for the nonstationary data (VIII.6.1) and tolerance $\gamma$ is solvable if and only of the stationary Nudelman interpolation problem associated with the data (VIII.6.1) and the bound $\gamma$ is solvable. More precisely, if* F *is a solution of the nonstationary Nudelman interpolation problem for data (VIII.6.1) and $\gamma$, then the symbol* G *of the diagonally sparse transform* $\hat{F}$ *of* F *is a solution of the associated stationary Nudelman interpolation problem, and* $\|G\|_\infty = \|F\| \leq \gamma$. *Conversely, if* G *in* $H^\infty(\vec{\mathcal{Y}}, \vec{\mathcal{U}})$ *is a solution of the associated stationary Nudelman interpolation problem with bound $\gamma$, then the diagonal compression* F *of the block Laurent operator* $L_G$ *defined by* G *is a solution of the original nonstationary Nudelman interpolation problem and* $\|F\| \leq \|G\|_\infty \leq \gamma$.

**PROOF.** Let F be as in (VIII.6.5), and assume that F is a solution of the nonstationary Nudelman interpolation problem for the data (VIII.6.1) and with tolerance $\gamma$. Let G be the symbol of the diagonally sparse transform $\hat{F}$ of F. Thus $G \in H^\infty(\vec{\mathcal{Y}}, \vec{\mathcal{U}})$, the n – th Taylor coefficient of G is given by

$$G_n = [F]_n = (\delta_{j-k,n} f_{j,k})_{j,k=-\infty}^{\infty} \qquad (n \geq 0), \tag{4.7}$$

and $\|G\|_\infty = \|\hat{F}\| = \|F\|$. Let us show that G satisfies the interpolation conditions (a), (b) and (c) appearing in the paragraph preceding the present theorem. For each $n \geq 0$ the operator $Z^n B G_n$ is a block diagonal operator. To see this one uses the fact that all block entries of $Z^n$ are zero except those in the $n$ – th diagonal above the main diagonal and that all block entries of $G_n$ are zero except those on the $n$ – th diagonal below the main diagonal. Since B is a block diagonal operator, these facts imply that $Z^n B G_n$ is also block diagonal. It follows that

$$(BG)(Z)_{\text{left}} = \sum_{n=0}^{\infty} Z^n B G_n$$

is also a block diagonal operator. We shall compute its $k$ – th diagonal entry. From the special shifted diagonal forms of $Z^n$ and $G_n$ it follows that

$$\left[(BG)(Z)_{\text{left}}\right]_{k,k} = \sum_{n=0}^{\infty} (Z^n B G_n)_{k,k}$$

$$= \sum_{n=0}^{\infty} (Z^n)_{k,n+k} B_{n+k} (G_n)_{n+k,k} = \sum_{n=0}^{\infty} \Phi(k,\, n+k) B_{n+k} f_{n+k,k} = \tilde{B}_k \,.$$

Here we used (4.3) and (4.7), and the interpolation condition (VIII.6.4a). Thus (a) is fulfilled. In a similar way one checks that (b) holds. To prove (c), we first note that all block entries of the operator $Z^j B G_{j+k+1} C \Lambda^k$ are zero except may be those in the first diagonal below the main diagonal. The latter block entries are given by

$$(Z^j B G_{j+k+1} C \Lambda^k)_{i,i-1} = (Z^j)_{i,i+j} B_{i+j} (G_{j+k+1})_{i+j,i-k-1} C_{i-k-1} (\Lambda^k)_{i-k-1,i-1}$$

$$= \Phi(i,\, i+j) B_{i+j} f_{i+j,i-k-1} C_{i-k-1} \Psi(i-k-1,\, i-1)\,.$$

This holds for each $i \in \mathbb{Z}$ and $j,\, k = 0,\, 1,\, 2,\, \ldots$. But then we can use (VIII.6.4c) to show that

$$\left[\sum_{j,k=0}^{\infty} Z^j B G_{j+k+1} C \Lambda^k\right]_{i,i-1} = \Gamma_i \qquad (i \in \mathbb{Z})\,,$$

and (c) holds. We conclude that G is a solution of the associated stationary Nudelman interpolation problem.

To prove the reverse implication, assume that $G \in H^\infty(\vec{\mathcal{Y}},\, \vec{\mathcal{U}})$ is a solution of the stationary Nudelman interpolation problem associated with the data (VIII.6.1) and the tolerance $\gamma$. Let F be the diagonal compression of the block Laurent operator $L_G$ defined by G. Then F is a block lower triangular operator from $\vec{\mathcal{Y}}$ into $\vec{\mathcal{U}}$ of the form (VIII.6.5), and according to Proposition X.2.2 we have $\|F\| \le \|L_G\| = \|G\|_\infty \le \gamma$. We have to check that the entries $f_{j,k}$ of F satisfy the identities (VIII.6.4a)-(VIII.6.4c). Let $G_n$ be the $n$ – th Taylor coefficient of G, and let $g_{j,k}^{(n)} : \mathcal{Y}_k \to \mathcal{U}_j$ be the $(j,\, k)$ – th block entry of $G_n$. Then the block entries in the lower triangular part of the diagonal compression F of $L_G$ are given by

$$f_{j,k} = g_{j,k}^{(j-k)} \qquad (-\infty < k \le j < \infty). \tag{4.8}$$

Since $(BG)(Z)_{\text{left}} = \tilde{B}$, we see that

$$\tilde{B}_k = \sum_{n=0}^{\infty} (Z^n B G_n)_{k,k} = \sum_{n=0}^{\infty} (Z^n)_{k,k+n} B_{k+n} (G_n)_{n+k,k} .$$

But then we can use the first equality in (4.3) and (4.8) to conclude that

$$\tilde{B}_k = \sum_{n=0}^{\infty} \Phi(k, k+n) B_{n+k} f_{n+k,k} \qquad (k \in \mathbb{Z}),$$

and hence (VIII.6.4a) is fulfilled. In a similar way using that $(GC)(\Lambda)_{\text{right}} = \tilde{C}$, one proves that (VIII.6.4b) holds. To prove (VIII.6.c), we use the fact that G satisfies the interpolation condition (c) appearing in the paragraph preceding the present theorem. It follows that

$$\begin{aligned}
\Gamma_{n+1} &= \sum_{j,k=0}^{\infty} (Z^j B G_{j+k+1} C \Lambda^k)_{n+1,n} \\
&= \sum_{j,k=0}^{\infty} (Z^j)_{n+1,n+1+j} B_{n+1+j} (G_{j+k+1})_{n+1+j,n-k} C_{n-k} (\Lambda^k)_{n-k,n} \\
&= \sum_{j,k=0}^{\infty} \Phi(n+1, n+j+1) B_{n+j+1} f_{n+j+1,n-k} C_{n-k} \Psi(n-k, n),
\end{aligned}$$

and thus (VIII.6.4c) is satisfied, which completes the proof.

Notice that the compatibility condition (VIII.6.6) is equivalent to the requirement that

$$Z\Gamma - \Gamma\Lambda = \tilde{B}C - B\tilde{C}, \tag{4.9}$$

which is precisely (cf., formula (I.9.4)) the compatibility condition for the associated stationary Nudelman interpolation problem.

Next, we consider for each $k \in \mathbb{Z}$ the operators $M_k$ on $\mathcal{F}_k$, and $N_k$ on $\mathcal{X}_k$, and $\Xi_k$ on $\mathcal{F}_k \oplus \mathcal{X}_{k+1}$ defined by (VIII.6.7), (VIII.6.8), and (VIII.6.9), respectively. Put

$$M = \operatorname{diag}(M_k)_{-\infty}^{\infty}, \; N = \operatorname{diag}(N_k)_{-\infty}^{\infty}, \; \Xi = \operatorname{diag}(\Xi_k)_{-\infty}^{\infty} .$$

Using (4.3) and (4.4) a straightforward computation shows that

$$M = \sum_{\nu=0}^{\infty} (\Lambda^\nu)^* (\gamma^2 C^* C - \tilde{C}^* \tilde{C}) \Lambda^\nu \quad \text{and} \quad N = \sum_{\nu=0}^{\infty} Z^\nu (\gamma^2 B B^* - \tilde{B}\tilde{B}^*)(Z^\nu)^* . \tag{4.10}$$

To analyze the diagonal operator $\Xi$ we use the map

$$J : \bigoplus_{k=-\infty}^{\infty} (\mathcal{F}_k \oplus \mathcal{X}_{k+1}) \to \vec{\mathcal{F}} \oplus \vec{\mathcal{X}} \text{ defined by } J\left(\begin{pmatrix} f_k \\ e_{k+1} \end{pmatrix}_{k=-\infty}^{\infty}\right) = \begin{pmatrix} (f_k)_{-\infty}^{\infty} \\ (e_k)_{-\infty}^{\infty} \end{pmatrix}. \tag{4.11}$$

Obviously, J is a unitary operator. A direct computation yields

$$J \Xi J^{-1} = \begin{pmatrix} M & \gamma\Gamma^* \\ \gamma\Gamma & N \end{pmatrix} : \vec{\mathcal{F}} \oplus \vec{\mathcal{X}} \to \vec{\mathcal{F}} \oplus \vec{\mathcal{X}}. \tag{4.12}$$

Indeed,

$$J\,\Xi\left(\begin{pmatrix} f_k \\ e_{k+1} \end{pmatrix}_{k=-\infty}^{\infty}\right) = J\left(\left(\Xi_k \begin{pmatrix} f_k \\ e_{k+1} \end{pmatrix}\right)_{k=-\infty}^{\infty}\right)$$

$$= J\left(\begin{pmatrix} M_k f_k + \gamma\Gamma_{k+1}^* e_{k+1} \\ \gamma\Gamma_{k+1} f_k + N_{k+1} e_{k+1} \end{pmatrix}_{k=-\infty}^{\infty}\right) = \begin{pmatrix} (M_k f_k + \gamma\Gamma_{k+1}^* e_{k+1})_{k=-\infty}^{\infty} \\ (\gamma\Gamma_k f_{k-1} + N_k e_k)_{k=-\infty}^{\infty} \end{pmatrix}.$$

Moreover,

$$\begin{pmatrix} M & \gamma\Gamma^* \\ \gamma\Gamma & N \end{pmatrix} J\left(\begin{pmatrix} f_k \\ e_{k+1} \end{pmatrix}_{k=-\infty}^{\infty}\right) = \begin{pmatrix} M & \gamma\Gamma^* \\ \gamma\Gamma & N \end{pmatrix}\begin{pmatrix} (f_k)_{k=-\infty}^{\infty} \\ (e_k)_{k=-\infty}^{\infty} \end{pmatrix} = \begin{pmatrix} (M_k f_k + \gamma\Gamma_{k+1}^* e_{k+1})_{k=-\infty}^{\infty} \\ (\gamma\Gamma_k f_{k-1} + N_k e_k)_{k=-\infty}^{\infty} \end{pmatrix},$$

and hence (4.12) holds. We are now ready to prove Theorem VIII.6.1.

**PROOF OF THEOREM VIII.6.1.** Consider the nonstationary data (VIII.6.1), and assume that (VIII.6.6) holds. By the remark made in the first paragraph after the proof of Theorem 4.1, the latter assumption implies that (4.9) is fulfilled. So we may apply Theorem I.9.1 to the stationary Nudelman interpolation problem associated with the data (VIII.6.1) and the tolerance $\gamma$. By Theorem 4.1 our nonstationary Nudelman interpolation problem is solvable if and only if the associated stationary version is solvable. According to Theorem I.9.1 the latter happens if and only if the operator

$$\begin{pmatrix} M & \gamma\Gamma^* \\ \gamma\Gamma & N \end{pmatrix} \tag{4.13}$$

is positive ($\geq 0$). Here M and N are given by (4.10). Since the map J defined by (4.11) is unitary, formula (4.12) shows that the operator in (4.13) is positive if and only if for each $k \in \mathbb{Z}$ the operator $\Xi_k$ is positive. Together these remarks complete the proof.

## XI.5 THE NONSTATIONARY TWO-SIDED SARASON INTERPOLATION THEOREM

In this section we prove the nonstationary two-sided Sarason interpolation theorem (Theorem VIII.7.1). Recall that in the nonstationary two-sided Sarason problem the given data are bounded linear operators

$$F : \vec{\mathcal{U}} \to \vec{\mathcal{Y}}, L_1 : \vec{\mathcal{U}} \to \vec{\mathcal{E}}, L_2 : \vec{\mathcal{F}} \to \vec{\mathcal{Y}}, \tag{5.1}$$

where $L_1$ is a block lower triangular co-isometry, and $L_2$ is a block lower triangular isometry. Given such operators and a tolerance $\gamma$ the nonstationary two-sided Sarason problem is to determine a block lower triangular R from $\vec{\mathcal{E}}$ to $\vec{\mathcal{F}}$ such that

$$\|F - L_2RL_1\| \le \gamma. \tag{5.2}$$

To solve this problem we consider an associated stationary version. For this purpose we need the diagonally sparse transforms $\hat{F}, \hat{L}_1, \hat{L}_2$ of the operators in (5.1). Recall that

$$\hat{F} : \ell^2(\vec{\mathcal{U}}) \to \ell^2(\vec{\mathcal{Y}}), \hat{L}_1 : \ell^2(\vec{\mathcal{U}}) \to \ell^2(\vec{\mathcal{E}}), \hat{L}_2 : \ell^2(\vec{\mathcal{F}}) \to \ell^2(\vec{\mathcal{Y}}), \tag{5.3}$$

and these operators are block Laurent operators. Furthermore, the operators $\hat{L}_1$ and $\hat{L}_2$ are both block lower triangular, and by using Proposition X.2.1, one sees that $\hat{L}_1$ is a co-isometry and $\hat{L}_2$ is an isometry. Thus we may consider (cf., Section I.10) the problem of finding a block lower triangular Laurent operator K from $\ell^2(\vec{\mathcal{E}})$ to $\ell^2(\vec{\mathcal{F}})$ such that

$$\|\hat{F} - \hat{L}_1K\hat{L}_2\| \le \gamma. \tag{5.4}$$

We shall refer to the latter problem as the *stationary two-sided Sarason problem associated with the data* (5.1) *and the bound* $\gamma$. The following result provides the key to the proof of Theorem VIII.7.1.

**THEOREM 5.1**. *The nonstationary two-sided Sarason problem for the data (5.1) and tolerance $\gamma$ is solvable if and only if the associated stationary two-sided Sarason problem with tolerance $\gamma$ is solvable. More precisely, if* R *is a solution of the nonstationary two-sided Sarason problem (5.2), then the diagonally sparse transform* $K = \hat{R}$ *of* R *is a solution of the associated stationary problem (5.4) and* $\|\hat{R}\| = \|R\|$. *Conversely, if* K *is a solution of the associated stationary two-sided Sarason problem (5.4), then the diagonal compression* R *of* K *is a solution of the original nonstationary two-sided Sarason problem (5.2) and* $\|R\| \le \|K\|$.

**PROOF.** First we introduce some notation. Given the Hilbert space direct sum $\vec{\mathcal{H}} = \oplus_{-\infty}^{\infty} \mathcal{H}_j$, we write $U_{\vec{\mathcal{H}}}$ for the unitary operator on $\ell^2(\vec{\mathcal{H}})$ associated with the sparse embedding of $\vec{\mathcal{H}}$ into $\ell^2(\vec{\mathcal{H}})$. Furthermore, given a second Hilbert space direct sum $\vec{\mathcal{K}} = \oplus_{j=-\infty}^{\infty} \mathcal{K}_j$ and a bounded linear operator $T : \vec{\mathcal{H}} \to \vec{\mathcal{K}}$, we use the symbol $\Delta(T)$ to denote the diagonal operator

$$\Delta(T) := \begin{bmatrix} \ddots & & & \\ & T & & \\ & & T & \\ & & & T \\ & & & & \ddots \end{bmatrix} : \ell^2(\vec{\mathcal{H}}) \to \ell^2(\vec{\mathcal{K}}). \tag{5.5}$$

Using this notation we see that Proposition X.2.1 tells us that

$$\hat{F} = U_{\vec{\mathcal{Y}}} \Delta(F) U_{\vec{\mathcal{U}}}^*, \; \hat{L}_1 = U_{\vec{\mathcal{E}}} \Delta(L_1) U_{\vec{\mathcal{U}}}^*, \; \hat{L}_2 = U_{\vec{\mathcal{Y}}} \Delta(L_2) U_{\vec{\mathcal{F}}}^*. \tag{5.6}$$

Now, assume that R from $\vec{\mathcal{E}}$ to $\vec{\mathcal{F}}$ is a solution of the nonstationary two-sided Sarason problem associated with the data (5.1) and the tolerance $\gamma$. Let $\hat{R}$ be the diagonally sparse transform of R. Thus $\hat{R} = U_{\vec{\mathcal{F}}} \Delta(R) U_{\vec{\mathcal{E}}}^*$. Since the norm of the operator $\Delta(T)$ in (5.5) is equal to $\|T\|$, we can use (5.2) and (5.4) to show that

$$\begin{aligned} \gamma &\geq \|F - L_2 R L_1\| = \|\Delta(F - L_2 R L_1)\| \\ &= \|\Delta(F) - \Delta(L_2)\Delta(R)\Delta(L_1)\| \\ &= \|U_{\vec{\mathcal{Y}}}^*(\hat{F} - \hat{L}_2 \hat{R} \hat{L}_1) U_{\vec{\mathcal{U}}}\|. \end{aligned}$$

But $U_{\vec{\mathcal{U}}}$ and $U_{\vec{\mathcal{Y}}}$ are unitary operators. Thus (5.4) is fulfilled with $K = \hat{R}$. Since R from $\vec{\mathcal{E}}$ to $\vec{\mathcal{F}}$ is block lower triangular, the same is true for $\hat{R}$ from $\ell^2(\vec{\mathcal{E}})$ to $\ell^2(\vec{\mathcal{F}})$. Thus $\hat{R}$ is a solution of the associated stationary two-sided Sarason problem (5.4) and $\|\hat{R}\| = \|R\|$, by Proposition X.2.1.

Conversely, assume K from $\ell^2(\vec{\mathcal{E}})$ to $\ell^2(\vec{\mathcal{F}})$ is a solution of the associated stationary two-sided Sarason problem. Thus K is block lower triangular, and (5.4) holds. Let R be the diagonal compression of K. Then R is a bounded linear operator from $\vec{\mathcal{E}}$ into $\vec{\mathcal{F}}$, and R is block lower triangular. Recall (see the paragraph preceding Proposition X.2.2) that F is the diagonal compression of $\hat{F}$, the operator $L_1$ is the diagonal compression of $\hat{L}_1$, and $L_2$ is the diagonal compression of $\hat{L}_2$. Furthermore, by applying Corollary X.2.4 twice, we see that the diagonal compression of $\hat{L}_2 K \hat{L}_1$ is equal to $L_2 R L_1$. Thus $F - L_2 R L_1$, is precisely the diagonal compression of $\hat{F} - \hat{L}_2 K \hat{L}_1$. Since the diagonal compression does not increase the operator norm (by Proposition X.2.2) we have $\|R\| \le \|K\|$ and

$$\|F - L_2RL_1\| \le \|\hat{F} - \hat{L}_2K\hat{L}_1\| \le \gamma .$$

Hence R is a solution of our original nonstationary two-sided Sarason problem (5.2). This completes the proof.

**PROOF OF THEOREM VIII.7.1.** First we introduce some additional notation. Given a Hilbert space direct sum $\vec{\mathcal{K}} = \oplus_{-\infty}^{\infty} \mathcal{K}_j$ we let $P_{\mathcal{K},j}$ be the orthogonal projection on $\vec{\mathcal{K}}$ defined by

$$(P_{\mathcal{K},j}\vec{e})_i = \begin{cases} 0 & \text{for} \quad i < j , \\ e_i & \text{for} \quad i \ge j . \end{cases} \tag{5.7}$$

Furthermore, we write $P_{\vec{\mathcal{K}}+}$ for the orthogonal projection of $\ell^2(\vec{\mathcal{K}})$ onto $\ell^2_+(\vec{\mathcal{K}})$. Using the projections (5.7) for $\vec{\mathcal{E}}$ and $\vec{\mathcal{F}}$ in place of $\vec{\mathcal{K}}$, we see that the spaces $\mathcal{H}_k$ and $\mathcal{M}_k$ appearing in Theorem VIII.7.1 are given by

$$\mathcal{H}_k = \ker (I - P_{\mathcal{E},k})L_1 \ \text{ and } \ \mathcal{M}_k = \operatorname{ran} L_2P_{\mathcal{F},k} \quad (k \in \mathbb{Z}) . \tag{5.8}$$

Next, consider the spaces

$$\hat{\mathcal{H}} = \ker (I - P_{\vec{\mathcal{E}}+})\hat{L}_1 \ \text{ and } \ \hat{\mathcal{M}} = \operatorname{ran} \hat{L}_2P_{\vec{\mathcal{F}}+} . \tag{5.9}$$

Here $\hat{L}_1$ and $\hat{L}_2$ are the diagonally sparse transforms of $L_1$ and $L_2$, which also appear in (5.4). According to Theorem I.10.1 the stationary two-sided Sarason problem associated with the data (5.3) and the bound $\gamma$ is solvable if and only if

$$\gamma \ge \hat{d}_\infty := \|(I - P_{\hat{\mathcal{M}}})\hat{F} \,|\, \hat{\mathcal{H}}\| . \tag{5.10}$$

But then we see from Theorem 5.1 that in order to prove Theorem VIII.7.1 it suffices to show that

$$\hat{d}_\infty = d_\infty := \sup \{\|(I - P_{\mathcal{M}_k})F \,|\, \mathcal{H}_k\| : k \in \mathbb{Z}\} . \tag{5.11}$$

To prove $\hat{d}_\infty = d_\infty$, we first show that

$$U_{\vec{\mathcal{E}}}^*(I - P_{\vec{\mathcal{E}}+})\hat{L}_1U_{\vec{\mathcal{U}}} = \operatorname{diag} \Big[(I - P_{\mathcal{E},-k})L_1\Big]_{k=-\infty}^{\infty} , \tag{5.12}$$

$$U_{\vec{\mathcal{Y}}}^*\hat{L}_2P_{\vec{\mathcal{F}}+}U_{\vec{\mathcal{F}}} = \operatorname{diag} (L_2P_{\mathcal{F},-k})_{k=-\infty}^{\infty} . \tag{5.13}$$

Here, as in the proof of Theorem 5.1, the symbol $U_{\vec{\mathcal{K}}}$ denotes the unitary operator on $\ell^2(\vec{\mathcal{K}})$

associated with the sparse embedding of $\vec{\mathcal{K}}$ into $\ell^2(\vec{\mathcal{K}})$. To prove (5.12) and (5.13) one uses Proposition X.1.2 and the identities in (5.6). Indeed,

$$\begin{aligned} U_{\vec{\mathcal{E}}}^*(I - P_{\vec{\mathcal{E}}+})\hat{L}_1 U_{\vec{\mathcal{U}}} &= U_{\vec{\mathcal{E}}}^*(I - P_{\vec{\mathcal{E}}+})U_{\vec{\mathcal{E}}} U_{\vec{\mathcal{E}}}^* \hat{L}_1 U_{\vec{\mathcal{U}}} \\ &= \operatorname{diag}\,(I - P_{\mathcal{E},-k})_{k=-\infty}^{\infty}\, \Delta(L_1) \\ &= \operatorname{diag}\left[(I - P_{\mathcal{E},-k})L_1\right]_{k=-\infty}^{\infty}, \end{aligned}$$

and similarly for (5.13).

From (5.12), the first identity in (5.8), and the first identity in (5.9) we see that

$$\operatorname{diag}\,(P_{\mathcal{H}_{-k}})_{k=-\infty}^{\infty} = U_{\vec{\mathcal{U}}}^* P_{\hat{\mathcal{H}}} U_{\vec{\mathcal{U}}}\ . \tag{5.14}$$

Similarly, (5.13), the second identity in (5.8), and the second identity in (5.9) yield

$$\operatorname{diag}\,(P_{\mathcal{M}_{-k}})_{k=-\infty}^{\infty} = U_{\vec{\mathcal{Y}}}^* P_{\hat{\mathcal{M}}} U_{\vec{\mathcal{Y}}}\ . \tag{5.15}$$

Using (5.14) and (5.15) we may rewrite the right-hand side of (5.11) in the following way

$$\begin{aligned} d_\infty &= \|\operatorname{diag}\left[(I - P_{\mathcal{M}_k})FP_{\mathcal{H}_k}\right]_{k=-\infty}^{\infty}\| \\ &= \|U_{\vec{\mathcal{Y}}}^*(I - P_{\hat{\mathcal{M}}})U_{\vec{\mathcal{Y}}}\Delta(F)U_{\vec{\mathcal{U}}}^* P_{\hat{\mathcal{H}}} U_{\vec{\mathcal{U}}}\| \\ &= \|(I - P_{\hat{\mathcal{M}}})\hat{F}P_{\hat{\mathcal{H}}}\|, \end{aligned}$$

where for the last equality we used the first identity in (5.6) and the fact that $U_{\vec{\mathcal{U}}}$ and $U_{\vec{\mathcal{Y}}}$ are unitary operators. The above calculation yields $d_\infty = \hat{d}_\infty$, and the proof is complete.

## *Notes to Chapter XI:*

The proofs given in Sections 1 and 4 are taken from Foias-Frazho-Gohberg-Kaashoek [1]. The proofs in the other sections follow the same line of arguments. The connections between the nonstationary interpolation problems and the associated stationary ones described in this chapter can be used to derive state space formulas for solutions of these nonstationary interpolation problems. For the nonstsationary problems of Nevanlinna-Pick and Nehari such formulas are given in the present chapter; for the other nonstationary interpolation problems similar results can be obtained. In particular, from the formulas for the central interpolants of the associated stationary problems one can (by diagonally compressing) obtain formulas for the central interpolants of the original nonstationary problems.

CHAPTER XII

# A GENERAL COMPLETION THEOREM

In this chapter a general completion theorem, which may be viewed as a time-varying version of the commutant lifting theorem, is presented. Three different proofs are given. One proof uses the reduction techniques of Chapter X to convert the three chains completion theorem to a standard commutant lifting setup. The second proof goes by one step extensions, using Parrott's lemma. The third proof gives an explicit formula for a solution which is the analogue of the central intertwining lifting. A nonstationary maximum entropy principle is given. Finally, as a first application, the three chains completion theorem is used to give a new proof of the Carswell-Schubert theorem.

## XII.1. THE THREE CHAINS COMPLETION THEOREM

The following theorem, which is the main result of this section, is a time-varying version of the commutant lifting theorem. We shall refer to this theorem as the *three chains completion theorem.*

**THEOREM 1.1.** *Let $\mathcal{U}$ and $\mathcal{Y}$ be Hilbert spaces, and let $\mathcal{H}_k \subset \mathcal{U}$ and $\mathcal{M}_k \subset \mathcal{K}_k \subset \mathcal{Y}$, where $k \in \mathbb{Z}$, be subspaces satisfying*

$$\mathcal{H}_{k-1} \subset \mathcal{H}_k\ ,\ \mathcal{M}_{k-1} \subset \mathcal{M}_k \ \text{ and }\ \mathcal{K}_{k-1} \subset \mathcal{K}_k \qquad (k \in \mathbb{Z})\ . \tag{1.1}$$

*Furthermore, for each $k \in \mathbb{Z}$, let $A_k : \mathcal{H}_k \to \mathcal{K}_k \ominus \mathcal{M}_k$ be a bounded linear operator. In order that there exists $B \in \mathcal{L}(\mathcal{U}, \mathcal{Y})$ such that*

$$B\mathcal{H}_k \subset \mathcal{K}_k \ \text{ and }\ (I - P_{\mathcal{M}_k})B \,|\, \mathcal{H}_k = A_k \qquad (k \in \mathbb{Z}) \tag{1.2}$$

*it is necessary and sufficient that*

(i) $(I - P_{\mathcal{M}_k})A_{k-1} = A_k \,|\, \mathcal{H}_{k-1}$ *for all* $k \in \mathbb{Z}$,

(ii) $d_\infty := \sup\{\|A_k\| : k \in \mathbb{Z}\} < \infty$.

*In this case there exists a $B \in \mathcal{L}(\mathcal{U}, \mathcal{Y})$ satisfying (1.2) with the additional property $\|B\| = d_\infty$.*

In the sequel any $B \in \mathcal{L}(\mathcal{U}, \mathcal{Y})$ satisfying (1.2) will be called an *interpolant* of the operators $A_k : \mathcal{H}_k \to \mathcal{K}_k \ominus \mathcal{M}_k$ for $k \in \mathbb{Z}$. In this case, the spaces $\mathcal{H}_k \subset \mathcal{U}$ and $\mathcal{M}_k \subset \mathcal{K}_k \subset \mathcal{Y}$, for $k \in \mathbb{Z}$, are assumed to satisfy (1.1). For any interpolant B we have

$$\sup \{\|A_k\| : k \in \mathbb{Z}\} \le \|B\| . \tag{1.3}$$

If we have equality in (1.3), then the interpolant B is said to be *optimal*. Notice that the two conditions in (1.2) are not independent. In fact, since $\mathcal{M}_k \subset \mathcal{K}_k$ and $A_k$ maps $\mathcal{H}_k$ into $\mathcal{K}_k$, the first part of (1.2) follows from the second part of (1.2).

**PROOF**. Assume an interpolant $B \in \mathcal{L}(\mathcal{U}, \mathcal{Y})$ exists. Take $u_{k-1} \in \mathcal{H}_{k-1}$. The second inclusion in (1.1) implies that $P_{\mathcal{M}_k} P_{\mathcal{M}_{k-1}} = P_{\mathcal{M}_{k-1}}$. Therefore, since (1.2) holds, we have

$$\begin{aligned}(I - P_{\mathcal{M}_k})A_{k-1}u_{k-1} &= (I - P_{\mathcal{M}_k})(I - P_{\mathcal{M}_{k-1}})Bu_{k-1} \\ &= (I - P_{\mathcal{M}_k})Bu_{k-1} = A_k u_{k-1} ,\end{aligned}$$

because $u_{k-1} \in \mathcal{H}_{k-1} \subset \mathcal{H}_k$. So condition (i) is fulfilled. From (1.3) we know that (ii) is also satisfied.

Conversely, let us assume that (i) and (ii) are fulfilled. We have to prove the existence of an optimal interpolant. This we will do by applying the commutant lifting theorem. Introduce the following spaces:

$$\begin{aligned}\mathcal{H} &= \{\vec{u} = (u_k)_{-\infty}^{\infty} \in \ell^2(\mathcal{U}) : u_k \in \mathcal{H}_k \text{ for } k \in \mathbb{Z}\} ,\\ \mathcal{M}' &= \{\vec{y} = (y_k)_{-\infty}^{\infty} \in \ell^2(\mathcal{Y}) : y_k \in \mathcal{M}_k \text{ for } k \in \mathbb{Z}\} ,\\ \mathcal{K}' &= \{\vec{y} = (y_k)_{-\infty}^{\infty} \in \ell^2(\mathcal{Y}) : y_k \in \mathcal{K}_k \text{ for } k \in \mathbb{Z}\} .\end{aligned} \tag{1.4}$$

Then $\mathcal{M}' \subset \mathcal{K}'$, and we set $\mathcal{H}' = \mathcal{K}' \ominus \mathcal{M}'$. Let A be the diagonal operator from $\mathcal{H}$ to $\mathcal{H}'$ defined by

$$A((h_k)_{k=-\infty}^{\infty}) = (A_k h_k)_{k=-\infty}^{\infty} \qquad (\vec{h} = (h_k)_{-\infty}^{\infty} \in \mathcal{H}) . \tag{1.5}$$

Condition (ii) implies that A is a bounded linear operator and $\|A\| = d_\infty$.

Let $V_1$ and $V_2$ be the block forward shifts on $\ell^2(\mathcal{U})$ and $\ell^2(\mathcal{Y})$, respectively. From the inclusions in (1.1) it follows that $\mathcal{H}$ is invariant under $V_1$ and that the spaces $\mathcal{M}'$ and $\mathcal{K}'$ are invariant under $V_2$. Let T be the isometry on $\mathcal{H}$ and T′ the contraction on $\mathcal{H}'$ defined by $T = V_1 | \mathcal{H}$ and $T' = P_{\mathcal{H}'} V_2 | \mathcal{H}'$, respectively. Moreover, let U′ be the isometry on $\mathcal{K}'$ defined by $U' = V_2 | \mathcal{K}'$. Then U′ is an isometric lifting of T′. The compatibility condition (i) implies that A intertwines T and T′. Indeed, for $\vec{h} = (h_k)_{k=-\infty}^{\infty}$ we have

$$AT\vec{h} = AV_1\vec{h} = A((h_{k-1})_{k=-\infty}^{\infty}) = (A_k h_{k-1})_{k=-\infty}^{\infty}$$

$$= ((I - P_{\mathcal{M}_k})A_{k-1}h_{k-1})_{k=-\infty}^{\infty} = P_{\mathcal{H}'}((A_{k-1}h_{k-1})_{k=-\infty}^{\infty})$$

$$= P_{\mathcal{H}'}V_2((A_k h_k)_{k=-\infty}^{\infty}) = P_{\mathcal{H}'}V_2 A\vec{h} = T'A\vec{h}\,.$$

So we may apply the commutant lifting theorem (Theorem II.1.1) to show that there exists $B' : \mathcal{H} \to \mathcal{K}'$ such that $B'$ intertwines $T = V_1 \mid \mathcal{H}$ with $U' = V_2 \mid \mathcal{K}'$, the norm $\|B'\| = \|A\| = d_\infty$ and $P_{\mathcal{H}'}B' = A$. According to Lemma II.3.1 there exists an operator $\hat{B}$ from $\ell^2(\mathcal{U})$ to $\ell^2(\mathcal{Y})$ extending $B'$ (that is, $\hat{B} \mid \mathcal{H} = B'$) preserving the norm of $B'$ and intertwining $V_1$ with $V_2$. Since $V_2\hat{B} = \hat{B}V_1$, it follows that $\hat{B}$ is a block Laurent operator, $\hat{B} = (B_{j-k})_{j,k=-\infty}^{\infty}$.

Take $B : \mathcal{U} \to \mathcal{Y}$ to be the operator $B_0$ appearing in the main diagonal of $\hat{B}$. We claim that B has the desired properties. Fix $k \in \mathbb{Z}$, and take $h_k \in \mathcal{H}_k$. Put $\vec{h} = (\delta_{j,k}h_k)_{j=-\infty}^{\infty}$. Then $\hat{B}\vec{h} = (B_{j-k}h_k)_{j=-\infty}^{\infty}$. The definition of $\mathcal{M}'$ in (1.4) along with $P_{\mathcal{H}'}\hat{B} \mid \mathcal{H} = A$, implies that $\hat{B}\vec{h} = A\vec{h} + \vec{m}$, where $\vec{m} = (m_j)_{j=-\infty}^{\infty}$ with $m_j \in \mathcal{M}_j$ for $j \in \mathbb{Z}$. It follows that

$$Bh_k = B_0 h_k = (\hat{B}\vec{h})_k = (A\vec{h})_k + m_k = A_k h_k + m_k\,.$$

Thus $B\mathcal{H}_k \subset A_k\mathcal{H}_k + \mathcal{M}_k \subset \mathcal{K}_k$. Furthermore, $(I - P_{\mathcal{M}_k})Bh_k = A_k h_k$. So the two statements in (1.2) hold true. Finally, note that $\|B\| = \|B_0\| \le \|\hat{B}\| = d_\infty$. On the other hand, by the second part of (1.2),

$$\|B\| \ge \|(I - P_{\mathcal{M}_k})B \mid \mathcal{H}_k\| = \|A_k\|\,,$$

for all $k \in \mathbb{Z}$ and thus $\|B\| = d_\infty$. This completes the proof.

By replacing k by $-k$ in Theorem 1.1 we obtain the following reformulation of Theorem 1.1 (which we also shall refer to as the *three chains completion theorem*).

**THEOREM 1.1a.** *Let $\mathcal{U}$ and $\mathcal{Y}$ be Hilbert spaces, and let $\mathcal{H}_k \subset \mathcal{U}$ and $\mathcal{M}_k \subset \mathcal{K}_k \subset \mathcal{Y}$, where $k \in \mathbb{Z}$, be subspaces satisfying*

$$\mathcal{H}_{k+1} \subset \mathcal{H}_k,\ \mathcal{K}_{k+1} \subset \mathcal{K}_k,\ \mathcal{M}_{k+1} \subset \mathcal{M}_k \quad (k \in \mathbb{Z})\,. \tag{1.6}$$

*Furthermore, for each $k \in \mathbb{Z}$, let $A_k : \mathcal{H}_k \to \mathcal{K}_k \ominus \mathcal{M}_k$ be a bounded linear operator. In order that there exists $B \in \mathcal{L}(\mathcal{U}, \mathcal{Y})$ such that*

$$B\mathcal{H}_k \subset \mathcal{K}_k \text{ and } (I - P_{\mathcal{M}_k})B \mid \mathcal{H}_k = A_k \quad (k \in \mathbb{Z}) \tag{1.7}$$

*it is necessary and sufficient that*

(j) $(I - P_{\mathcal{M}_k})A_{k+1} = A_k \,|\, \mathcal{H}_{k+1}$ for *all* $k \in \mathbb{Z}$,

(jj) $d_\infty := \sup \{\|A_k\| : k \in \mathbb{Z}\} < \infty$.

*In this case there exists* $B \in \mathcal{L}(\mathcal{U}, \mathcal{Y})$ *satisfying (1.7) with the additional property* $\|B\| = d_\infty$.

If the operators $A_k$ in Theorem 1.1 satisfy some additional intertwining relation, then we can improve the properties of the interpolants as is shown by the next two results.

**PROPOSITION 1.2.** *Let* $B \in \mathcal{L}(\mathcal{U}, \mathcal{Y})$ *be an interpolant of the operators* $A_k$ *from* $\mathcal{H}_k$ *to* $\mathcal{K}_k \ominus \mathcal{M}_k$ *for* $k \in \mathbb{Z}$, *where* $\mathcal{H}_k$, $\mathcal{M}_k$ *and* $\mathcal{K}_k$ *satisfy (1.1). Let* T *on* $\mathcal{U}$ *and* R *on* $\mathcal{Y}$ *be bounded linear operators satisfying*

$$T\mathcal{H}_k \subset \mathcal{H}_{k+1}\,,\ R^*\mathcal{K}_{k+1} \subset \mathcal{K}_k \quad \textit{and} \quad R(\mathcal{K}_k \ominus \mathcal{M}_k) \subset \mathcal{K}_{k+1} \ominus \mathcal{M}_{k+1} \qquad (k \in \mathbb{Z})\,. \tag{1.8}$$

*Then* $R^*BT$ *is also an interpolant if and only if*

$$(I - P_{\mathcal{M}_k})R^*A_{k+1}T \,|\, \mathcal{H}_k = A_k \qquad (k \in \mathbb{Z})\,. \tag{1.9}$$

**PROOF.** From the first part of (1.2) it follows that for $k \in \mathbb{Z}$

$$R^*BT\mathcal{H}_k \subset R^*B\mathcal{H}_{k+1} \subset R^*\mathcal{K}_{k+1} \subset \mathcal{K}_k\,.$$

The third inclusion of (1.8) yields $P_{\mathcal{M}_{k+1}}R(I - P_{\mathcal{M}_k})\mathcal{K}_k = \{0\}$ and by taking the adjoint

$$P_{\mathcal{K}_k}(I - P_{\mathcal{M}_k})R^*P_{\mathcal{M}_{k+1}} = 0 \qquad (k \in \mathbb{Z})\,. \tag{1.10}$$

Assume now that (1.9) holds. Take $h_k \in \mathcal{H}_k$, and set $h'_{k+1} = Th_k$. Then $h'_{k+1} \in \mathcal{H}_{k+1}$, and $R^*Bh'_{k+1} \in \mathcal{K}_k$. It follows that $(I - P_{\mathcal{M}_k})R^*BTh_k \in \mathcal{K}_k$. Since $R^*BT\mathcal{H}_k \subset \mathcal{K}_k$, we can use (1.10) to show that

$$\begin{aligned}(I - P_{\mathcal{M}_k})R^*BTh_k &= P_{\mathcal{K}_k}(I - P_{\mathcal{M}_k})R^*Bh'_{k+1} \\ &= P_{\mathcal{K}_k}(I - P_{\mathcal{M}_k})R^*(I - P_{\mathcal{M}_{k+1}})Bh'_{k+1} \\ &= P_{\mathcal{K}_k}(I - P_{\mathcal{M}_k})R^*A_{k+1}Th_k = P_{\mathcal{K}_k}A_kh_k = A_kh_k\end{aligned}$$

for all k in $\mathbb{Z}$. Thus $R^*BT$ is an interpolant.

Conversely, assume that $R^*BT$ is an interpolant. Take $h_k \in \mathcal{H}_k$. Then

$$A_kh_k = P_{\mathcal{K}_k}(I - P_{\mathcal{M}_k})A_kh_k = P_{\mathcal{K}_k}(I - P_{\mathcal{M}_k})R^*BTh_k\,.$$

But then we can use (1.10) and the fact that B is an interpolant to show that

$$A_k h_k = P_{\mathcal{K}_k}(I - P_{\mathcal{M}_k})R^*(I - P_{\mathcal{M}_{k+1}})BTh_k$$

$$= P_{\mathcal{K}_k}(I - P_{\mathcal{M}_k})R^* A_{k+1} Th_k = (I - P_{\mathcal{M}_k})R^* A_{k+1} Th_k \ ,$$

hence (1.9) holds. This completes the proof.

If $R \mid (\mathcal{K}_n \ominus \mathcal{M}_n)$ in Proposition 1.2 is an isometry for each $n \in \mathbb{Z}$, and if $A_{n+1} T \mid \mathcal{H}_n = RA_n$ for each $n \in \mathbb{Z}$, then the compatibility condition (1.9) is fulfilled.

**COROLLARY 1.3**. *Assume the operators* $A_k : \mathcal{H}_k \to \mathcal{K}_k \ominus \mathcal{M}_k$ *for* $k \in \mathbb{Z}$, *with* $\mathcal{H}_k$, $\mathcal{M}_k$, $\mathcal{K}_k$ *satisfying (1.1) have an interpolant. Let* T *on* $\mathcal{U}$ *and* R *on* $\mathcal{Y}$ *be contractions satisfying (1.8) and (1.9). Then the operators* $A_k$ *for* $k \in Z$ *have an optimal interpolant* B *with the additional property that*

$$B = R^* BT \tag{1.11}$$

**PROOF**. Let $B_0$ be an optimal interpolant of the operators $A_k$ for $k \in \mathbb{Z}$. By Proposition 1.2, for each $j \geq 0$ the operator $B_j = (R^*)^j B_0 T^j$ is again an interpolant. Since T and R are contractions, we have $\|B_j\| \leq \|B_0\|$, and thus $B_j$ is an optimal interpolant. Now, let $\mathrm{LIM}_{j\to\infty}$ be any Banach generalized limit on $\ell^\infty$ (see Conway [1], Section III.7), and define $B : \mathcal{U} \to \mathcal{Y}$ by

$$<Bu, y> = \mathrm{LIM}_{j\to\infty} <B_j u, y> . \tag{1.12}$$

Since $\|B_j\| = d_\infty$, where $d_\infty$ is defined by (ii) in Theorem 1.1, the operator B is well-defined and bounded. In fact, $\|B\| = d_\infty$. Since $R^* B_j T = B_{j+1}$ for each $j \geq 0$, formula (1.12) shows that B satisfies (1.11). Furthermore, since (1.2) holds for each $B_j$, formula (1.12) implies that (1.2) holds for B. Thus B is an optimal interpolant with the additional property (1.11). This completes the proof.

We conclude this section by showing that the commutant lifting theorem (Theorem II.1.1) may be derived as a corollary of Theorem 1.1. To this end, recall that an operator $\hat{U}$ on $\hat{\mathcal{K}}$ is a *unitary extension* of an isometry U on $\mathcal{K}$, if $\hat{U}$ is a unitary operator and $\mathcal{K}$ is an invariant subspace for $\hat{U}$ satisfying $\hat{U} \mid \mathcal{K} = U$. Furthermore, $\hat{U}$ is a *minimal unitary extension* of U if $\hat{U}$ is a unitary extension of U and

$$\hat{\mathcal{K}} = \bigvee_{-\infty}^{\infty} \hat{U}^n \mathcal{K} . \tag{1.13}$$

Any isometry U admits a minimal unitary extension. To see this, simply use the Sz.-Nagy-

Schäffer construction, and let $\hat{U}$ be the operator on $\hat{\mathcal{K}} = \ell^2_-(\mathcal{L}) \oplus \mathcal{K}$ defined by

$$\hat{U} = \begin{bmatrix} \cdots & \cdot & \cdot & \cdot & \cdot \\ \cdots & \cdot & \cdot & \cdot & \cdot \\ \cdots & \cdot & \cdot & \cdot & \cdot \\ \cdots & 0 & 0 & 0 & 0 \\ \cdots & I & 0 & 0 & 0 \\ \cdots & 0 & I & 0 & 0 \\ \cdots & 0 & 0 & \tau_{\mathcal{L}} & U \end{bmatrix} \tag{1.14}$$

where $\mathcal{L} = \mathcal{K} \ominus U\mathcal{K}$ and $\tau_{\mathcal{L}}$ is the canonical embedding of $\mathcal{L}$ into $\mathcal{K}$. Clearly $\mathcal{K}$ is an invariant subspace for $\hat{U}$. It is easy to show that both $\hat{U}$ and $\hat{U}^*$ are isometries. So $\hat{U}$ is a unitary extension of U. Since $\hat{\mathcal{K}}$ is spanned by $\hat{U}^{*n}\mathcal{K}$ for $n \geq 0$, it follows that $\hat{U}$ is a minimal unitary extension of U. It turns out that the minimal unitary extension $\hat{U}$ of U is unique up to an isomorphism; see Sz.-Nagy-Foias [3], Chapter I (or Gohberg-Goldberg-Kaashoek [2], Chapter XXVII) for further details.

Recall that if U on $\mathcal{K}$ is an isometric lifting of a contraction T on $\mathcal{H}$ and $\hat{U}$ on $\hat{\mathcal{K}}$ is a unitary extension of U, then $\hat{U}$ is called a *unitary dilation* of T. Using the fact that $\mathcal{H}$ is an invariant subspace for $U^*$ and $\mathcal{K}$ is invariant for $\hat{U}$, it follows that $\hat{U}$ admits a matrix representation of the form

$$\hat{U} = \begin{bmatrix} * & 0 & 0 \\ * & T & 0 \\ * & * & * \end{bmatrix} \text{ on } (\hat{\mathcal{K}} \ominus \mathcal{K}) \oplus \mathcal{H} \oplus (\mathcal{K} \ominus \mathcal{H}). \tag{1.15}$$

Here the stars denote unspecified entries. Now we are ready to prove the commutant lifting theorem.

**PROOF OF THEOREM II.1.1** (Using Theorem 1.1). In what follows we use the notation introduced in Theorem II.1.1. To prove the commutant lifting theorem, let A be an operator from $\mathcal{H}$ to $\mathcal{H}'$ intertwining the isometry T on $\mathcal{H}$ with the contraction T′ on $\mathcal{H}'$. As before, let U′ on $\mathcal{K}'$ be an isometric lifting of T′. Let $\hat{T}$ on $\mathcal{U}$ be a minimal unitary extension of U = T, and let $\hat{U}'$ on $\mathcal{Y}$ be a minimal unitary extension of U′. In particular,

$$\mathcal{U} = \bigvee_{n=0}^{\infty} \hat{T}^{-n}\mathcal{H} \quad \text{and} \quad \mathcal{Y} = \bigvee_{n=0}^{\infty} (\hat{U}')^{-n}\mathcal{K}'.$$

For each $k \in \mathbb{Z}$ we consider the subspaces

$$\mathcal{H}_k = \hat{T}^{-k}\mathcal{H}\,,\ \mathcal{K}_k = (\hat{U}')^{-k}\mathcal{K}' \ \text{ and } \ \mathcal{M}_k = (\hat{U}')^{-k}(\mathcal{K}' \ominus \mathcal{H}')\,.$$

Furthermore, define $A_k$ from $\mathcal{H}_k$ to $\mathcal{K}_k \ominus \mathcal{M}_k$ by setting

$$A_k = (I - P_{\mathcal{M}_k})(\hat{U}')^{-k} A \hat{T}^k \,|\, \mathcal{H}_k\,. \tag{1.16}$$

From the construction of a minimal unitary extension it follows that $\hat{T}$ and $\hat{U}'$ admit the following partitionings:

$$\hat{T} = \begin{bmatrix} T & * \\ 0 & * \end{bmatrix} : \mathcal{H} \oplus (\mathcal{U} \ominus \mathcal{H}) \to \mathcal{H} \oplus (\mathcal{U} \ominus \mathcal{H})\,, \tag{1.17}$$

$$\hat{U}' = \begin{bmatrix} * & 0 & 0 \\ * & T' & 0 \\ * & * & * \end{bmatrix} : (\mathcal{Y} \ominus \mathcal{K}') \oplus \mathcal{H}' \oplus (\mathcal{K}' \ominus \mathcal{H}') \to (\mathcal{Y} \ominus \mathcal{K}') \oplus \mathcal{H}' \oplus (\mathcal{K}' \ominus \mathcal{H}')\,. \tag{1.18}$$

Here the stars denote unspecified entries. The $2 \times 2$ block matrix in the right lower corner of the block partitioning of $\hat{U}'$ is precisely $U'$. From the block matrix representations (1.17) and (1.18) we see that $\mathcal{H}$ is invariant under $\hat{T}$ and the spaces $\mathcal{K}'$ and $\mathcal{K}' \ominus \mathcal{H}'$ are invariant under $\hat{U}'$. These invariance properties imply that the spaces $\mathcal{H}_k$, $\mathcal{K}_k$ and $\mathcal{M}_k$ satisfy the inclusion relations (1.1).

Next use the intertwining relation $AT = T'A$ to show that condition (i) in Theorem 1.1 holds. Take $h_{k-1} = \hat{T}^{-k+1} h$ in $\mathcal{H}_{k-1}$, where h is in $\mathcal{H}$. By (1.17), we have $\hat{T}h = Th$. Thus

$$A_k h_{k-1} = (I - P_{\mathcal{M}_k})(\hat{U}')^{-k} A \hat{T} h = (I - P_{\mathcal{M}_k})(\hat{U}')^{-k} T' A h\,.$$

Note that the orthogonal complement of the subspace $\mathcal{M}_k$ in $\mathcal{K}_k$ is $(\hat{U}')^{-k}\mathcal{H}'$ and thus $I - P_{\mathcal{M}_k} = (\hat{U}')^{-k} P_{\mathcal{H}'} (\hat{U}')^k$. Furthermore, the block partitioning of $\hat{U}'$ in (1.18) implies that $P_{\mathcal{H}'}\hat{U}'Ah = T'Ah$. We conclude that

$$\begin{aligned} A_k h_{k-1} &= (I - P_{\mathcal{M}_k})(\hat{U}')^{-k} P_{\mathcal{H}'} \hat{U}' A h \\ &= (I - P_{\mathcal{M}_k})(I - P_{\mathcal{M}_k})(\hat{U}')^{-k+1} A h = (I - P_{\mathcal{M}_k})(\hat{U}')^{-k+1} A h\,. \end{aligned}$$

Since $\mathcal{M}_{k-1} \subset \mathcal{M}_k$, we have $(I - P_{\mathcal{M}_k})(I - P_{\mathcal{M}_{k-1}}) = (I - P_{\mathcal{M}_k})$. Therefore,

$$(I - P_{\mathcal{M}_k}) A_{k-1} h_{k-1} = (I - P_{\mathcal{M}_k})(I - P_{\mathcal{M}_{k-1}})(\hat{U}')^{-k+1} A h = (I - P_{\mathcal{M}_k})(\hat{U}')^{-k+1} A h = A_k h_{k-1}\,,$$

which shows that condition (i) is fulfilled. Since $\|A_k\| \le \|A\|$ for each k, condition (ii) in Theorem 1.1 is also fulfilled. Hence there exists an interpolant $\hat{B} \in \mathcal{L}(\mathcal{U}, \mathcal{Y})$ of the operators $A_k$ for $k \in \mathbb{Z}$.

Next, we apply Corollary 1.3 with $\hat{T}^{-1}$ in place of T and with $(\hat{U}')^*$ in place of R. For the spaces and operators considered here the inclusions (1.8) hold. Note that, by (1.16),

$$\begin{aligned} A_k &= (I - P_{\mathcal{M}_k})\hat{U}'\{(\hat{U}')^{-k-1}A\hat{T}^{k+1}\}\hat{T}^{-1} \,|\, \mathcal{H}_k \\ &= (I - P_{\mathcal{M}_k})\hat{U}'\{(I - P_{\mathcal{M}_{k+1}})(\hat{U}')^{-k-1}A\hat{T}^{k+1}\}\hat{T}^{-1} \,|\, \mathcal{H}_k \\ &= (I - P_{\mathcal{M}_k})\hat{U}'A_{k+1}\hat{T}^{-1} \,|\, \mathcal{H}_k\,, \end{aligned}$$

and thus (1.9) is also satisfied in the present context. Therefore we may assume that our interpolant satisfies $\hat{U}'\hat{B}\hat{T}^{-1} = \hat{B}$.

Finally, set $B = \hat{B} \,|\, \mathcal{H}$. Since $\mathcal{H} = \mathcal{H}_0$ and $\mathcal{K}' = \mathcal{K}_0$, we have $B \in \mathcal{L}(\mathcal{H}, \mathcal{K}')$. Furthermore,

$$BT = \hat{B}\hat{T} \,|\, \mathcal{H} = \hat{U}'\hat{B} \,|\, \mathcal{H} = U'B\,,$$

$$P_{\mathcal{H}'}B \,|\, \mathcal{H} = (I - P_{\mathcal{M}_0})\hat{B} \,|\, \mathcal{H}_0 = A_0 = A\,.$$

From $\|B\| \le \|\hat{B}\| \le \|A\|$ and $P_{\mathcal{H}'}B \,|\, \mathcal{H} = A$, we conclude that $\|B\| = \|A\|$. Thus $B : \mathcal{H} \to \mathcal{K}$ has the desired lifting properties. This completes the proof.

## XII.2. PROOF BY ONE STEP EXTENSIONS

In this section we prove the three chains completion theorem by one step extensions, using Parrott's lemma, which we have already met in Section VIII.4. We first recall the Parrott lemma in a form which will be convenient for the present section. For a proof of Parrott's lemma see Section IV.3 in Foias-Frazho [4].

**LEMMA 2.1.** *Given Hilbert spaces $\mathcal{E}_0 \subset \mathcal{E}$ and $\mathcal{F}_0 \subset \mathcal{F}$, and contractions $K : \mathcal{E}_0 \to \mathcal{F}$ and $L : \mathcal{E} \to \mathcal{F}_0$, there exists a contraction $C : \mathcal{E} \to \mathcal{F}$ such that $C \,|\, \mathcal{E}_0 = K$ and $P_{\mathcal{F}_0}C = L$ if and only if the following compatibility condition is fulfilled:*

$$L \,|\, \mathcal{E}_0 = P_{\mathcal{F}_0}K\,. \tag{2.1}$$

*Here $P_{\mathcal{F}_0}$ is the orthogonal projection of $\mathcal{F}$ onto $\mathcal{F}_0$.*

Let us now prove the three chains completion theorem (Theorem 1.1) using Lemma 2.1. Since the necessity of the conditions (i) and (ii) is simple and has already been proved in the previous section, we concentrate on the construction of an interpolant. Therefore in what follows $\mathcal{H}_k \subset \mathcal{U}$ and $\mathcal{M}_k \subset \mathcal{K}_k \subset \mathcal{Y}$, where $k \in \mathbb{Z}$, are Hilbert spaces satisfying the inclusion relations (1.1), and

$$A_k : \mathcal{H}_k \to \mathcal{K}_k \ominus \mathcal{M}_k \qquad (k \in \mathbb{Z}), \tag{2.2}$$

are operators satisfying the conditions (i) and (ii) in Theorem 1.1. Without loss of generality we may take the number $d_\infty$ in condition (ii) to be one, that is, the operators $A_k$ are assumed to be contractions.

Using Lemma 2.1 we shall construct by induction a sequence of contractions $(C_j)_{j=0}^{\infty}$ such that

$$C_j : \mathcal{H}_j \to \mathcal{K}_j, \tag{2.3a}$$

$$C_{j+1} | \mathcal{H}_j = C_j, \tag{2.3b}$$

$$(I - P_{\mathcal{M}_j})C_j = A_j, \tag{2.3c}$$

for $j = 0, 1, 2, \ldots$. To this end, set $C_0 = A_0$. Then (2.3a) and (2.3c) are trivially satisfied for $j = 0$. Next, assume that contractions $C_0, \ldots, C_k$ have been constructed so that (2.3a) and (2.3c) hold for $0 \le j \le k$, and so that (2.3b) is fulfilled for $0 \le j \le k-1$. To construct the next element $C_{k+1}$ we apply Lemma 2.1 with

$$\mathcal{E} = \mathcal{H}_{k+1},\ \mathcal{E}_0 = \mathcal{H}_k,\ \mathcal{F} = \mathcal{K}_{k+1},\ \mathcal{F}_0 = \mathcal{K}_{k+1} \ominus \mathcal{M}_{k+1},\ K = C_k,\ L = A_{k+1}.$$

Notice that K and L are contractions. Furthermore, by condition (i) in Theorem 1.1 and formula (2.3c) with $j = k$ we have

$$\begin{aligned} L | \mathcal{E}_0 = A_{k+1} | \mathcal{H}_k &= (I - P_{\mathcal{M}_{k+1}})A_k \\ &= (I - P_{\mathcal{M}_{k+1}})(I - P_{\mathcal{M}_k})C_k \\ &= (I - P_{\mathcal{M}_{k+1}})C_k = P_{\mathcal{F}_0}K. \end{aligned}$$

For the penultimate equality, we used that $\mathcal{M}_k \subset \mathcal{M}_{k+1}$ and for the last equality we used (2.3a) with $j = k$. It follows that we may apply Lemma 2.1 to obtain a contraction $C_{k+1}$ from $\mathcal{H}_{k+1}$ into $\mathcal{K}_{k+1}$ such that

$$C_{k+1} | \mathcal{H}_k = C_k \quad \text{and} \quad (I - P_{\mathcal{M}_{k+1}})C_{k+1} = A_{k+1}.$$

Hence (2.3a) and (2.3c) are now satisfied for $0 \le j \le k+1$, and (2.3b) holds for $0 \le j \le k$. So, by induction, there exists a sequence of contractions $(C_j)_{j=0}^{\infty}$ such that (2.3a)-(2.3c) hold for each $j \ge 0$.

Let $\mathcal{H} = \bigvee \{\mathcal{H}_j : j \geq 0\}$, and let $\mathcal{K} = \bigvee \{\mathcal{K}_j : j \geq 0\}$. Since each $C_j$ is a contraction, (2.3b) implies that there exists a contraction $C_\infty$ from $\mathcal{H}$ into $\mathcal{K}$ such that $C_\infty | \mathcal{H}_j = C_j$ for each $j \geq 0$. In fact,

$$C_\infty x = \lim_{j\to\infty} C_j P_{\mathcal{H}_j} x \qquad (x \in \mathcal{H}) . \tag{2.4}$$

Notice that

$$C_\infty \mathcal{H}_j \subset \mathcal{K}_j \quad \text{and} \quad (I - P_{\mathcal{M}_j}) C_\infty | \mathcal{H}_j = A_j \qquad (j \geq 0) . \tag{2.5}$$

Indeed, since $C_\infty | \mathcal{H}_j = C_j$, the formulas (2.5) follow directly from (2.3a) and (2.3c), respectively.

Set $B_0 = C_\infty$. The operator $B_0$ is a contraction from $\mathcal{H}$ to $\mathcal{K}$ and has the following properties:

$$B_0 \mathcal{H}_n \subset \mathcal{K}_n \quad \text{and} \quad (I - P_{\mathcal{M}_n}) B_0 | \mathcal{H}_n = A_n \qquad (\text{for } n \geq 0) . \tag{2.6}$$

By induction, using Lemma 2.1, we construct a sequence of contractions $B_0, B_{-1}, B_{-2}, \ldots$, acting from $\mathcal{H}$ into $\mathcal{K}$ such that

$$B_{-j} \mathcal{H}_n \subset \mathcal{K}_n \qquad (\text{for } n \geq -j) , \tag{2.7a}$$

$$(I - P_{\mathcal{M}_n}) B_{-j} | \mathcal{H}_n = A_n \qquad (\text{for } n \geq -j) , \tag{2.7b}$$

$$(I - P_{\mathcal{M}_{-j}})(B_{-j-1} - B_{-j}) = 0 , \tag{2.7c}$$

for $j = 0, 1, 2, \ldots$. For $j = 0$ the formulas (2.7a) and (2.7b) coincide with (2.6). Therefore, we fix $k \geq 0$, and assume that contractions $B_0, \ldots, B_{-k}$ have been constructed such that (2.7a) and (2.7b) are fulfilled for $0 \leq j \leq k$, and (2.7c) for $0 \leq j \leq k-1$. To construct the next element $B_{-k-1}$ we apply Lemma 2.1 with

$$\mathcal{E} = \mathcal{H}, \ \mathcal{E}_0 = \mathcal{H}_{-k-1}, \ \mathcal{F} = \mathcal{K}, \ \mathcal{F}_0 = \mathcal{K} \ominus \mathcal{M}_{-k}, \ K = A_{-k-1}, \ L = (I - P_{\mathcal{M}_{-k}}) B_{-k} .$$

Since $\mathcal{H}_{-k-1} \subset \mathcal{H}_{-k}$, we can use (2.7b) with $n = -k$ and condition (i) in Theorem 1.1 to show that

$$L | \mathcal{E}_0 = (I - P_{\mathcal{M}_{-k}}) B_{-k} | \mathcal{H}_{-k-1} = A_{-k} | \mathcal{H}_{-k-1} = (I - P_{\mathcal{M}_{-k}}) A_{-k-1} = P_{\mathcal{F}_0} K .$$

So for the case considered here the compatibility condition (2.1) is satisfied, and by Lemma 2.1, there exists a contraction $B_{-k-1}$ from $\mathcal{H}$ into $\mathcal{K}$ such that

$$B_{-k-1} \mid \mathcal{H}_{-k-1} = A_{-k-1} \quad \text{and} \quad (I - P_{\mathcal{M}_{-k}})B_{-k-1} = (I - P_{\mathcal{M}_{-k}})B_{-k} . \tag{2.8}$$

From the first identity in (2.8) we conclude that (2.7a) and (2.7b) hold for $j = -k-1$ and $n = -k-1$. For $n > -k-1$ we use (2.7b) with $j = k$ and the second identity in (2.8) to show that

$$\begin{aligned}(I - P_{\mathcal{M}_n})B_{-k-1} \mid \mathcal{H}_n &= (I - P_{\mathcal{M}_n})(I - P_{\mathcal{M}_{-k}})B_{-k-1} \mid \mathcal{H}_n \\ &= (I - P_{\mathcal{M}_n})(I - P_{\mathcal{M}_{-k}})B_{-k} \mid \mathcal{H}_n \\ &= (I - P_{\mathcal{M}_n})B_{-k} \mid \mathcal{H}_n = A_n .\end{aligned}$$

Thus (2.7b) holds for $j = k+1$. From (2.7b) and the fact that $\mathcal{M}_n \subset \mathcal{K}_n$ we see that (2.7a) holds. The second identity in (2.8) is the same as (2.7c) for $j = k$. So, by induction, there exist contractions $B_0, B_{-1}, B_{-2}, \ldots$, acting from $\mathcal{H}$ into $\mathcal{K}$, such that (2.7a)-(2.7c) are satisfied for $j = 0, 1, 2, \ldots$.

Put $\mathcal{M}_\infty = \cap\{\mathcal{M}_n : n \in \mathbb{Z}\}$, and let Q be the orthogonal projection of $\mathcal{Y}$ onto $\mathcal{K} \ominus \mathcal{M}_\infty$. Define $\tilde{B}_{-j} = QB_{-j}P_{\mathcal{H}}$ for $j \geq 0$. For each $j \geq 0$ the operator $\tilde{B}_{-j}$ is a contraction and (2.7a)-(2.7c) hold for $\tilde{B}_{-j}$ in place of $B_{-j}$. In particular,

$$\tilde{B}^*_{-j-1} \mid \mathcal{K} \ominus \mathcal{M}_{-j} = \tilde{B}^*_{-j} \mid \mathcal{K} \ominus \mathcal{M}_{-j} \qquad (j \geq 0) . \tag{2.9}$$

The latter identities imply that there exists a contraction $C : \mathcal{Y} \to \mathcal{U}$ such that $C(I - Q) = 0$ and

$$C \mid \mathcal{K} \ominus \mathcal{M}_{-j} = \tilde{B}^*_{-j} \mid \mathcal{K} \ominus \mathcal{M}_{-j} \qquad (j \geq 0) .$$

In fact $Cy = \lim_{j\to\infty} \tilde{B}^*_{-j}y$ for each $y \in \mathcal{Y}$. Put $B = C^*$. Then B is a contraction, and

$$B = \text{weak} \lim_{j\to\infty} \tilde{B}_{-j} . \tag{2.10}$$

But then we can use (2.7a) and (2.7b), with $\tilde{B}_{-j}$ in place of $B_{-j}$, to show that B satisfies (1.2). Thus B is an interpolant of the operators in (2.2), and Theorem 1.1 is proved.

## XII.3. AN EXPLICIT SOLUTION OF THE THREE CHAINS COMPLETION PROBLEM

In this section we present an explicit solution of the three chains completion problem. This provides a new, direct and straightforward proof of Theorem 1.1 (the three chains completion theorem). The construction that will be given here is motivated by the central solution in the commutant lifting theorem which has been presented in Chapter IV. Namely, we will identify in the first proof of Theorem 1.1 (given in Section 1), the interpolant which corresponds to the central solution in the commutant lifting theorem.

First let us recall the three chains completion problem. In what follows we assume that $\gamma \geq d_\infty$ where $d_\infty$ is the number in condition (ii) of Theorem 1.1. Recall that the given data are operators

$$A_k : \mathcal{H}_k \to \mathcal{K}_k \ominus \mathcal{M}_k \quad (k \in \mathbb{Z}), \tag{3.1}$$

where $\mathcal{H}_k \subset \mathcal{U}$ and $\mathcal{M}_k \subset \mathcal{K}_k \subset \mathcal{Y}$ for $k \in \mathbb{Z}$ are Hilbert spaces satisfying the (increasing) inclusion relations (1.1), and $\|A_k\| \leq \gamma$ for all k. The three chains completion problem consists in finding an operator B from $\mathcal{U}$ into $\mathcal{Y}$ such that $\|B\| \leq \gamma$,

$$B\mathcal{H}_k \subset \mathcal{K}_k \quad \text{and} \quad (I - P_{\mathcal{M}_k})B \mid \mathcal{H}_k = A_k \quad (k \in \mathbb{Z}). \tag{3.2}$$

According to Theorem 1.1 we know that for this problem to be solvable it is necessary and sufficient that the operators $A_k$ for $k \in \mathbb{Z}$ satisfy the following compatibility conditions

$$(I - P_{\mathcal{M}_k})A_{k-1} = A_k \mid \mathcal{H}_{k-1} \quad (k \in \mathbb{Z}). \tag{3.3}$$

Therefore, in what follows we assume that (3.3) is fulfilled.

Let us recall the commutant lifting setting used in the proof of Theorem 1.1 given in Section 1. Let the Hilbert spaces $\mathcal{H}$, $\mathcal{M}'$ and $\mathcal{K}'$ be as defined by (1.4), let $\mathcal{H}' = \mathcal{K}' \ominus \mathcal{M}'$ and let A be the operator defined by (1.5). Let $V_1$ and $V_2$ be the block forward shifts on $\ell^2(\mathcal{U})$ and $\ell^2(\mathcal{Y})$ respectively, and set $T = V_1 \mid \mathcal{H}$ and $T' = P_{\mathcal{H}'}V_2 \mid \mathcal{H}'$. Then $T'A = AT$ and $V_2 \mid \mathcal{K}'$ on $\mathcal{K}'$ is an isometric lifting of $T'$. Let $\hat{B}_\gamma$ from $\mathcal{H}$ to $\mathcal{K}'$ be the central lifting of A with tolerance $\gamma \geq d_\infty$, as defined in Section IV.2 (see Theorem IV.2.3). Note that by putting together formula (IV.2.15), Theorem IV.2.3 and Proposition IV.1.3 we obtain

$$\hat{B}_\gamma h = Ah + \sum_{n=0}^{\infty} WU'^n \Pi' \omega P_{\mathcal{F}} (\Pi_A \omega P_{\mathcal{F}})^n D_A h \qquad (h \in \mathcal{H}). \tag{3.4}$$

Here $U'$ is the Sz.-Nagy-Schäffer minimal isometric lifting of $T'$ and W is the unique isometry from $\mathcal{H}' \oplus H^2(\mathcal{D}_{T'})$ into $\mathcal{K}'$ defined by $WU' = V_2W$ and $W \mid \mathcal{H}' = I$ (use Proposition IV.1.3 with $U_1 = V_2 \mid \mathcal{K}'$). In this section by a slight abuse of terminology $D_{T'} = D'$ is the positive square root of $I - T'^*T'$ while $\mathcal{D}' = \mathcal{D}_{T'}$ is the closure of the range of $D_{T'}$. As before, $\mathcal{F} = (D_A T \mathcal{H})^-$ and the operators $\omega$, $\Pi'$ and $\Pi_A$ are defined in (IV.2.8) and (IV.2.10), respectively. Now let $\Psi$ be the isometry from $\mathcal{D}_{T'}$ into $\mathcal{K}'$ defined by $\Psi = W \mid \mathcal{D}_{T'}$ where $\mathcal{D}_{T'}$ is viewed as the subspace of constant functions in $\mathcal{H}^2(\mathcal{D}_{T'})$. Then using $V_2W = WU'$ we have

$$\hat{B}_\gamma h = Ah + \sum_{n=0}^{\infty} V_2^n \Psi \Pi' \omega P_{\mathcal{F}} (\Pi_A \omega P_{\mathcal{F}})^n D_A h \qquad (h \in \mathcal{H}). \tag{3.5}$$

The formula (IV.1.9) shows that

$$\Psi D_{T'} h = (V_2 - T')h = P_{\mathcal{M}'} V_2 h \quad (h \in \mathcal{H}'). \tag{3.6}$$

To give an explicit form to (3.5) we need to compute the various operators appearing in the right hand side of (3.5). For this purpose we introduce some auxiliary spaces and operators. Throughout $D_k$ is the positive square root of $\gamma^2 I - A_k^* A_k$, and we write $\mathcal{D}_k$ for the closure of ran $D_k$. For each $k \in \mathbb{Z}$, set $\mathcal{F}_k = \overline{D_k \mathcal{H}_{k-1}}$ and

$$\mathcal{F}_k' = \text{closure}\left\{ \begin{bmatrix} D_{k-1} h_{k-1} \\ P_{\mathcal{M}_k} A_{k-1} h_{k-1} \end{bmatrix} \in \mathcal{U} \oplus \mathcal{Y} : h_{k-1} \in \mathcal{H}_{k-1} \right\}. \tag{3.7}$$

Fix $h_{k-1} \in \mathcal{H}_{k-1}$. From (3.3) and (3.1) it follows that $A_{k-1} h_{k-1}$ is the orthogonal direct sum of the vectors $A_k h_{k-1}$ and $P_{\mathcal{M}_k} A_{k-1} h_{k-1}$, and hence

$$\|D_k h_{k-1}\|^2 = \|D_{k-1} h_{k-1}\|^2 + \|P_{\mathcal{M}_k} A_{k-1} h_{k-1}\|^2 .$$

It follows that the map

$$\omega_k (D_k h_{k-1}) = \begin{bmatrix} D_{k-1} h_{k-1} \\ P_{\mathcal{M}_k} A_{k-1} h_{k-1} \end{bmatrix} \qquad (h_{k-1} \in \mathcal{H}_{k-1}) \tag{3.8}$$

extends to an isometry (also denoted by $\omega_k$) from $\mathcal{F}_k$ into $\mathcal{F}_k'$. For each $k \in \mathbb{Z}$ we also need the operators

$$\pi_k : \mathcal{F}_k' \to \mathcal{D}_{k-1} \text{ and } \pi_k' : \mathcal{F}_k' \to \mathcal{M}_k , \tag{3.9}$$

which are defined by

$$\pi_k \begin{bmatrix} D_{k-1} h_{k-1} \\ P_{\mathcal{M}_k} A_{k-1} h_{k-1} \end{bmatrix} = D_{k-1} h_{k-1} \text{ and } \pi_k' \begin{bmatrix} D_{k-1} h_{k-1} \\ P_{\mathcal{M}_k} A_{k-1} h_{k-1} \end{bmatrix} = P_{\mathcal{M}_k} A_{k-1} h_{k-1} . \tag{3.10}$$

Obviously, $\pi_k$ and $\pi_k'$ are contractions for each $k \in \mathbb{Z}$. Next, for each $k \in \mathbb{Z}$ define the contraction $\rho_k$ from $\mathcal{D}_k$ into $\mathcal{D}_{k-1}$ by

$$\rho_k = \pi_k \omega_k P_{\mathcal{F}_k} . \tag{3.11}$$

Finally, let $\Phi_\rho(m, n)$ from $\mathcal{D}_n$ into $\mathcal{D}_m$ be the (state ) transition operator generated by the operators $\{\rho_k\}$, that is, for $n > m$

$$\Phi_\rho(m, n) = \rho_{m+1}\rho_{m+2} \cdots \rho_n \text{ and } \Phi_\rho(m, m) = I_{\mathcal{D}_m} . \tag{3.12}$$

Notice that for $h = (h_k)_{-\infty}^{\infty} \in \mathcal{H}$ and $h' = (h'_k)_{-\infty}^{\infty} \in \mathcal{H}'$ we have

$$(D_A h)_k = D_k h_k, \quad (\Psi D_{T'} h')_k = P_{\mathcal{M}_k} h_{k-1} \qquad (k \in \mathbb{Z}) .$$

Thus

$$D_A h = (D_k h_k)_{k=-\infty}^{\infty} \qquad (\text{for } h \in (h_k)_{k=-\infty}^{\infty} \in \mathcal{H}) \tag{3.13}$$

and

$$\Psi D_{T'} h' = (P_{\mathcal{M}_k} h'_{k-1})_{k=-\infty}^{\infty} \qquad (\text{for } h' \in (h'_k)_{k=-\infty}^{\infty} \in \mathcal{H}') . \tag{3.14}$$

Recall that $\mathcal{F} = (D_A T \mathcal{H})^-$ and $\mathcal{F}_k = \overline{D_k \mathcal{H}_{k-1}}$ for each $k \in \mathbb{Z}$. Since $D_A Th = (D_k h_{k-1})_{k=-\infty}^{\infty}$ for each $h = (h_k)_{k=-\infty}^{\infty}$ in $\mathcal{H}$, it follows that

$$\mathcal{F} = \{f = (f_k)_{k=-\infty}^{\infty} : f_k \in \mathcal{F}_k \quad (k \in \mathbb{Z})\} . \tag{3.15}$$

For $h = (h_k)_{k=-\infty}^{\infty} \in \mathcal{H}$ we have

$$\Psi \Pi' \omega D_A Th = \Psi D_{T'} Ah = (P_{\mathcal{M}_k} A_{k-1} h_{k-1})_{k=-\infty}^{\infty} = (\text{diag}\,(\pi'_k \omega_k P_{\mathcal{F}_k})) D_A Th ,$$

$$V_2 \Pi_A \omega D_A Th = V_2 D_A h = V_2 (D_k h_k)_{k=-\infty}^{\infty} = (D_{k-1} h_{k-1})_{k=-\infty}^{\infty} =$$

$$= (\text{diag}\,(\pi_k \omega_k P_{\mathcal{F}_k})) D_A Th .$$

Here $\text{diag}\,(\pi'_k \omega_k P_{\mathcal{F}_k})$ is the block diagonal operator $\oplus_{-\infty}^{\infty} \pi'_k \omega_k P_{\mathcal{F}_k}$ from $\oplus_{-\infty}^{\infty} \mathcal{D}_k$ into $\oplus_{-\infty}^{\infty} \mathcal{M}_k$ and $\text{diag}\,(\pi_k \omega_k P_{\mathcal{F}_k})$ equals $\oplus_{-\infty}^{\infty} \pi_k \omega_k P_{\mathcal{F}_k}$ from $\oplus_{-\infty}^{\infty} \mathcal{D}_k$ into $\oplus_{-\infty}^{\infty} \mathcal{D}_{k-1}$. So the above calculation yields

$$\Psi \Pi' \omega P_{\mathcal{F}} = (\text{diag}\,(\pi'_k \omega_k P_{\mathcal{F}_k})) , \tag{3.16}$$

$$V_2 \Pi_A \omega P_{\mathcal{F}} = (\text{diag}\,(\pi_k \omega_k P_{\mathcal{F}_k})) . \tag{3.17}$$

From (3.17) we see that

$$(\Pi_A \omega P_{\mathcal{F}})^n = (V_2^{-1} \text{ diag}\,(\pi_k \omega_k P_{\mathcal{F}_k}))^n . \tag{3.18}$$

Introducing (3.16) and (3.18) into (3.5), we obtain

$$\hat{B}_\gamma h = Ah + \sum_{n=0}^{\infty} V_2^n (\text{diag}\,(\pi'_k \omega_k P_{\mathcal{F}_k}))(V_2^{-1} (\text{diag}\,(\pi_k \omega_k P_{\mathcal{F}_k}))^n W_A D_A h \qquad (h \in \mathcal{H}) . \tag{3.19}$$

Notice that $\hat{B}_\gamma$ in (3.19) is a block diagonal operator from $\oplus_{-\infty}^{\infty} \mathcal{H}_k$ into $\oplus_{-\infty}^{\infty} \mathcal{K}_k$. Moreover, using (3.11) and (3.12) we can write (3.19) component wise as follows

$$(\hat{B}_\gamma h)_k = A_k h_k + \sum_{v=0}^{\infty} \pi'_{k-v}\omega_{k-v}P_{\mathcal{F}_{k-v}}\Phi_\rho(k-v,\, k)D_k h_k \quad (h = (h_k)_{-\infty}^{\infty} \in \mathcal{H}). \tag{3.20}$$

We are now ready to state the main result of this section.

**THEOREM 3.1.** *Let* $A_k$ *from* $\mathcal{H}_k$ *to* $\mathcal{K}_k \ominus \mathcal{M}_k$, *for* $k \in \mathbb{Z}$, *satisfy the compatibility condition (3.3), and assume that* $\|A_k\| \le \gamma$ *for all* k. *Let* $B_\gamma$ *from* $\mathcal{U}$ *to* $\mathcal{Y}$ *be the operator defined by*

$$B_\gamma h = A_k h + \sum_{v=0}^{\infty} \pi'_{k-v}\omega_{k-v}P_{\mathcal{F}_{k-v}}\Phi_\rho(k-v,\, k)D_k h \qquad (h \in \mathcal{H}_k), \tag{3.21}$$

*and* $B_\gamma x = 0$ *if* x *is orthogonal to all* $\mathcal{H}_n$ *for* $n \in \mathbb{Z}$. *Then* $B_\gamma$ *is a well-defined operator,* $\|B_\gamma\| \le \gamma$, *and* $B_\gamma$ *is a solution to the three chains completion problem associated with the operators* $A_k$ *for* $k \in \mathbb{Z}$.

The operator $B_\gamma$ defined in Theorem 3.1 will be referred to as the *central interpolant* (or *central solution*) of the three chains completion problem associated with the operators $A_k$ for $k \in \mathbb{Z}$ and tolerance $\gamma$.

**PROOF.** From (3.20) and the arguments used in the last paragraph of the proof of Theorem 1.1 in Section 1 we may conclude that $B_\gamma$ has the desired properties. However here below we will follow another route and shall give a self-contained proof without any reference to the commutant lifting theorem. The main problem is to show that $B_\gamma$ is a well-defined operator satisfying $\|B_\gamma\| \le \gamma$. Indeed, if $B_\gamma : \mathcal{U} \to \mathcal{Y}$ is a bounded operator of the form (3.21), then $B_\gamma$ has the properties in (3.2). To see this, notice that ran $\pi'_{k-v} \subset \mathcal{M}_{k-v} \subset \mathcal{M}_k$ for $v \ge 0$. Thus given $h \in \mathcal{H}_k$ formula (3.21) shows that $B_\gamma h = A_k h + m$, where $m \in \mathcal{M}_k$. This implies that $B_\gamma h \in \mathcal{K}_k$ and $(I - P_{\mathcal{M}_k})B_\gamma h = A_k h$ and hence (3.2) is fulfilled.

To prove that $B_\gamma$ is a well-defined operator, we consider the operators $B_{k,N}$ from $\mathcal{H}_k$ into $\mathcal{Y}$ defined by

$$B_{k,N}h_k = A_k h_k + \sum_{v=0}^{N} \pi'_{k-v}\omega_{k-v}P_{\mathcal{F}_{k-v}}\Phi_\rho(k-v,\, k)D_k h_k\,.$$

Here N is an arbitrary non-negative integer. First we show that for each $k \in \mathbb{Z}$ and each $N \ge 0$

$$\|B_{k,N}h_k\| \le \gamma\|h_k\| \qquad (h_k \in \mathcal{H}_k). \tag{3.22}$$

From (3.10) and (3.1) we see that ran $\pi'_k$ is a subspace of $\mathcal{M}_k \ominus \mathcal{M}_{k-1}$ for each k. But then the second inclusion relation (1.1) shows that ran $\pi'_j$ is orthogonal to ran $\pi'_k$ for all $j \neq k$. Also, ran $A_k$ is orthogonal to ran $\pi'_{k-v}$ for $v \geq 0$. In particular, for $h_k$ in $\mathcal{H}_k$

$$\|B_{k,0}h_k\|^2 = \|A_k h_k\|^2 + \|\pi'_k \omega_k P_{\mathcal{F}_k} D_k h_k\|^2 \leq \|A_k h_k\|^2 + \|D_k h_k\|^2 = \gamma^2 \|h_k\|^2 .$$

Thus (3.22) holds for $N = 0$ and each k. We proceed by induction. Assume (3.22) holds for some $N \geq 0$ and for all $k \in \mathbb{Z}$. Put

$$C_{k,N+1} = \sum_{v=0}^{N+1} \pi'_{k-v} \omega_{k-v} P_{\mathcal{F}_{k-v}} \Phi_\rho(k-v, k) . \tag{3.23}$$

To prove (3.22) for $N+1$ in place of N is suffices to show that

$$\|C_{k,N+1} f_k\| \leq \|f_k\| \qquad (f_k \in \mathcal{D}_k), \tag{3.24}$$

for each $k \in \mathbb{Z}$. Indeed, since ran $A_k$ is orthogonal to ran $\pi'_{k-v}$ for $v \geq 0$, we have

$$\|B_{k,N+1} h_k\|^2 = \|A_k h_k\|^2 + \|C_{k,N+1} D_k h_k\|^2 ,$$

for $h_k \in \mathcal{H}_k$ and thus, by (3.24),

$$\|B_{k,N+1} h_k\|^2 \leq \|A_k h_k\|^2 + \|D_k h_k\|^2 = \gamma^2 \|h_k\|^2 \qquad (h_k \in \mathcal{K}_k) .$$

Notice that $C_{k,N+1}$ is zero on $\mathcal{D}_k \ominus \mathcal{F}_k$; thus to prove (3.24) we may without loss of generality assume that $f_k = D_k h_{k-1}$ for some $h_{k-1} \in \mathcal{H}_{k-1}$. But then we can use

$$\rho_k D_k h_{k-1} = D_{k-1} h_{k-1} \qquad (h_{k-1} \in \mathcal{H}_{k-1}) \tag{3.25}$$

to show that

$$C_{k,N+1} f_k = C_{k,N+1} D_k h_{k-1} = \pi'_k \omega_k D_k h_{k-1} + \sum_{v=1}^{N+1} \pi'_{k-v} \omega_{k-v} P_{\mathcal{F}_{k-v}} \Phi_\rho(k-v, k-1) D_{k-1} h_{k-1}$$

$$= P_{\mathcal{M}_k} A_{k-1} h_{k-1} + \sum_{v=0}^{N} \pi'_{k-1-v} \omega_{k-1-v} P_{\mathcal{F}_{k-v-1}} \Phi_\rho(k-v-1, k-1) D_{k-1} h_{k-1} .$$

Let us now use (3.23) and the fact that ran $A_k$ is orthogonal to ran $\pi'_{k-v}$ for $v \geq 0$, to obtain

$$\|C_{k,N+1}D_k h_{k-1}\|^2 = \|A_{k-1}h_{k-1}\|^2 - \|(I - P_{\mathcal{M}_k})A_{k-1}h_{k-1}\|^2 +$$

$$+ \sum_{v=0}^{N} \|\pi'_{k-1-v}\omega_{k-1-v}P_{\mathcal{F}_{k-v-1}}\Phi_\rho(k-v-1, k-1)D_{k-1}h_{k-1}\|^2$$

$$= \|C_{k-1,N}D_{k-1}h_{k-1}\|^2 + \|A_{k-1}h_{k-1}\|^2 - \|(I - P_{\mathcal{M}_k})A_{k-1}h_{k-1}\|^2$$

$$= \|B_{k-1,v}h_{k-1}\|^2 - \|(I - P_{\mathcal{M}_k})A_{k-1}h_{k-1}\|^2 .$$

Next, we use our induction hypothesis (which says that (3.23) holds for the given N and each k) and the compatibility condition (3.3) to conclude that

$$\|C_{k,N+1}D_k h_{k-1}\|^2 \le \gamma^2\|h_{k-1}\|^2 - \|A_k h_{k-1}\|^2 = \|D_k h_{k-1}\|^2 ,$$

which establishes (3.24). So (3.22) holds with $N+1$ in place of N, and therefore, by induction for all N.

From (3.22) and the fact that the range of $\pi'_j$ is orthogonal to the range of $\pi'_k$ for all $j \neq k$, it follows that the series in the right hand side of (3.21) converges. Thus for each $h \in \mathcal{H}_k$ the vector $B_k h$ as given by the right hand side of (3.21) is well-defined, and

$$B_k h = \lim_{N\to\infty} B_{k,N}h \qquad (h \in \mathcal{H}_k) .$$

Since the operators $B_{k,N}$ are all bounded by $\gamma$, it follows that the same holds true for $B_k$.

To complete the proof it remains to show that for $h_{k-1} \in \mathcal{H}_{k-1}$

$$B_k h_{k-1} = B_{k-1}h_{k-1} \qquad (k \in \mathbb{Z}) . \tag{3.26}$$

Indeed, if (3.26) has been established, then

$$B_\gamma u = \lim_{k\to\infty} B_k P_{\mathcal{H}_k} u \qquad (u \in \mathcal{U}) ,$$

defines an operator $B_\gamma : \mathcal{U} \to \mathcal{Y}$ for which (3.21) holds, $\|B_\gamma\| \le \gamma$, and such that $B_\gamma x = 0$ if x is orthogonal to all $\mathcal{H}_n$ (for $n \in \mathbb{Z}$).

To prove (3.26), we first use the compatibility condition in (3.3) and the definition of $\omega_k$ to show that

$$A_k h_{k-1} + \pi'_k\omega_k P_{\mathcal{F}_k}D_k h_{k-1} = A_k h_{k-1} + P_{\mathcal{M}_k}A_{k-1}h_{k-1} = A_{k-1}h_{k-1} .$$

Next, by (3.25)

$$\sum_{v=1}^{\infty} \pi'_{k-v}\omega_{k-v}P_{\mathcal{F}_{k-v}}\Phi_\rho(k-v,\,k)D_k h_{k-1} = \sum_{v=1}^{\infty} \pi'_{k-v}\omega_{k-v}P_{\mathcal{F}_{k-v}}\Phi_\rho(k-v,\,k-1)D_{k-1}h_{k-1}$$

$$= \sum_{v=0}^{\infty} \pi'_{k-1-v}\omega_{k-1-v}P_{\mathcal{F}_{k-v-1}}\Phi_\rho(k-v-1,\,k-1)D_{k-1}h_{k-1}\,.$$

Taking all terms together yields (3.26). This completes the proof.

**REMARK 3.2.** Define the block diagonal operator $\hat{B}$ from $\ell^2(\mathcal{U})$ into $\ell^2(\mathcal{Y})$ by

$$\hat{B}(u_n)_{n=-\infty}^{\infty} = (B_\gamma u_n)_{n=-\infty}^{\infty} \qquad (\text{for } (u_n)_{n=-\infty}^{\infty} \in \ell^2(\mathcal{U}))\,, \tag{3.27}$$

where $B_\gamma$ is the operator defined in Theorem 3.1. Then, by comparing (3.20) with (3.21) we infer that $\hat{B}_\gamma = \hat{B}\,|\,\mathcal{H}$.

## XII.4. MAXIMUM ENTROPY

In this section we shall present the maximum principle for the three chains completion theorem which is the analogue of the maximum principle for the commutant lifting theorem studied in Chapter IV (see especially Section IV.7). As an instructive example we will explicate the maximum principle for the Parrott's completion theorem, and then present an illuminating application to the three chains completion theorem connecting the results of Sections 2 and 3.

To start, we will use the notation and concepts introduced in the preceding section. In particular, the setting for the commutant lifting theorem will be precisely that described in the third paragraph of Section 3. Then the maximum principle given in Theorem IV.7.1 states that if C is any intertwining lifting of A with tolerance $\gamma$, that is, C is an operator from $\mathcal{H}$ into $\mathcal{K}'$ such that

$$CT = V_2C,\ \|C\| \le \gamma \text{ and } P_{\mathcal{H}'}C = A$$

we have (see (IV.7.11))

$$D_C P_{\mathcal{G}(C\cdot T)} D_C \le D_{\hat{B}_\gamma} P_{\mathcal{G}(\hat{B}_\gamma\cdot T)} D_{\hat{B}_\gamma} = D_A P_{\mathcal{G}(A\cdot T)} D_A\,. \tag{4.1}$$

(Here $\hat{B}_\gamma$ is the central intertwining lifting of A given in (3.5) or (3.19).) Recall that for an operator X from $\mathcal{H}$ into $\mathcal{K}'$ whose norm is bounded by $\gamma$, the space $\mathcal{G}(X\cdot T)$ is defined by

$$\mathcal{G}(X\cdot T) = \mathcal{D}_X \ominus (D_X T\mathcal{H})^-\,, \tag{4.2}$$

where $D_X = (\gamma^2 I - X^*X)^{1/2}$ and $\mathcal{D}_X$ is the closure of the range of $D_X$. Moreover, if $V_2\,|\,\mathcal{K}'$ is the minimal isometric lifting of T′, then we have equality in (4.1) if and only if $C = \hat{B}_\gamma$.

*In order to avoid minor technical difficulties we shall assume throughout this section that the linear span of the spaces* $\mathcal{H}_n (n \in \mathbb{Z})$ *is dense in* $\mathcal{U}$, *that is,* $\mathcal{U} = \bigvee \{\mathcal{H}_n : n \in \mathbb{Z}\}$.

Now for an interpolant B in the three chains completion theorem, let $\hat{B}$ denote the operator from $\mathcal{H} = \oplus \{\mathcal{H}_n : n \in \mathbb{Z}\}$ to $\ell^2(\mathcal{Y})$ defined by

$$\hat{B} = \operatorname{diag}(\dots, B, B, B, \dots) \,|\, \mathcal{H} . \tag{4.3}$$

Clearly $\hat{B}$ is an intertwining lifting of A with tolerance $\gamma$ in the above setting of the commutant lifting theorem. So, from (4.1) we infer that

$$D_{\hat{B}} P_{\mathcal{G}(\hat{B}\cdot T)} D_{\hat{B}} \le D_{\hat{B}_\gamma} P_{\mathcal{G}(\hat{B}_\gamma \cdot T)} D_{\hat{B}_\gamma} = D_A P_{\mathcal{G}(A\cdot T)} D_A . \tag{4.4}$$

Note that if we denote by $B_\gamma$ the interpolant in the three chains completion theorem provided by Theorem 3.1, then using Remark 3.2 we have $(B_\gamma)^\wedge = \hat{B}_\gamma$. Therefore, in order to express the maximum principle only in terms of the three chains completion theorem, we must first give explicit forms to $D_{\hat{B}} P_{\mathcal{G}(\hat{B}\cdot T)} D_{\hat{B}}$ and $D_A P_{\mathcal{G}(A\cdot T)} D_A$. For this purpose, we recall that

$$\mathcal{D}_A = \bigoplus_{n=-\infty}^{\infty} \mathcal{D}_n \subset \ell^2(\mathcal{U}) \text{ and } (D_A T \mathcal{H})^- = \mathcal{F} = \bigoplus_{n=-\infty}^{\infty} \mathcal{F}_n \subset \ell^2(\mathcal{U}) ; \tag{4.5}$$

see (3.13), (3.15). Therefore

$$\mathcal{G}(A \cdot T) = (\mathcal{D}_A \ominus \mathcal{F}) = \left[ \bigoplus_{n=-\infty}^{\infty} \mathcal{D}_n \right] \ominus \left[ \bigoplus_{n=-\infty}^{\infty} \mathcal{F}_n \right] = \bigoplus_{n=-\infty}^{\infty} (\mathcal{D}_n \ominus \mathcal{F}_n) ,$$

which readily implies that

$$\mathcal{G}(A \cdot T) = \bigoplus_{n=-\infty}^{\infty} \mathcal{G}_n \text{ where } \mathcal{G}_n = \mathcal{D}_n \ominus \mathcal{F}_n \quad (\text{for all } n \in \mathbb{Z}) . \tag{4.6}$$

It follows that, for all $h = (h_n)_{n=-\infty}^{\infty}$ and $\tilde{h} = (\tilde{h}_n)_{n=-\infty}^{\infty}$ in $\mathcal{H}$

$$(D_A P_{\mathcal{G}(A\cdot T)} D_A h, \tilde{h}) = (P_{\mathcal{G}(A\cdot T)} D_A h, D_A \tilde{h}) = \sum_{n=-\infty}^{\infty} (P_{\mathcal{G}_n} D_n h_n, D_n \tilde{h}_n)$$

and consequently

$$D_A P_{\mathcal{G}(A\cdot T)} D_A h = (D_n P_{\mathcal{G}_n} D_n h_n)_{n=-\infty}^{\infty} \qquad (\text{for all } h = (h_n)_{n=-\infty}^{\infty} \text{ in } \mathcal{H}) . \tag{4.7}$$

Similarly if we define

$$B_n = B \mid \mathcal{H}_n\,,\ D_{B_n} = (\gamma^2 I - B_n^* B_n)^{1/2}\,,\ \mathcal{G}_n(B) = \mathcal{D}_{B_n} \ominus (D_{B_n}\mathcal{H}_{n-1})^- \tag{4.8}$$

for $n \in \mathbb{Z}$, then

$$D_{\hat{B}} P_{\mathcal{G}(\hat{B}\cdot T)} D_{\hat{B}} h = (D_{B_n} P_{\mathcal{G}_n(B)} D_{B_n} h_n)_{n=-\infty}^{\infty} \qquad (\text{for all } h \in (h_n)_{n=-\infty}^{\infty} \text{ in } \mathcal{H})\,. \tag{4.9}$$

With this preliminary completed we can now state the maximum principle for the three chains completion theorem.

**THEOREM 4.1.** *Let* $B_\gamma$ *be the central interpolant for the three chains completion problem associated with the operators* $\{A_k\}$ *and tolerance* $\gamma$, *and assume that* $\bigvee \{\mathcal{H}_n : n \in \mathbb{Z}\} = \mathcal{U}$. *If* B *is any interpolant in this three chains completion problem, with tolerance* $\gamma$, *then*

$$D_{B_n} P_{\mathcal{G}_n(B)} D_{B_n} \le D_{(B_\gamma)n} P_{\mathcal{G}_n(B_\gamma)} D_{(B_\gamma)n} = D_n P_{\mathcal{G}_n} D_n \quad (n \in \mathbb{Z})\,, \tag{4.10}$$

*where* $(B_\gamma)_n = B_\gamma \mid \mathcal{H}_n$ *and the spaces* $\mathcal{G}_n(B)$ *and* $\mathcal{G}_n$ *are defined in (4.8) and (4.6), respectively. Moreover if*

$$\bigcap_{v \in \mathbb{Z}} \mathcal{M}_v = \{0\} \ \text{ and } \ \mathcal{K}_n^{\perp} \cap \mathcal{M}_{n+1} = \{0\} \ \ (n \in \mathbb{Z}) \tag{4.11}$$

*then equality holds in (4.10) if and only if* $B = B_\gamma$.

The proof of Theorem 4.1 trivially follows from Theorem IV.7.1 once we established the following result.

**LEMMA 4.2.** *The operator* $U' = V_2 \mid \mathcal{K}'$ *is a minimal isometric lifting of* $T'$ *if and only if the conditions (4.11) hold.*

**PROOF.** Notice that $U'$ has the required minimality condition if and only if

$$\mathcal{K}' \ominus \mathcal{H}' = \bigvee_{n=0}^{\infty} U'^n (U' - T')\mathcal{H}'\,. \tag{4.12}$$

Indeed, since the right hand side of (4.12) is contained in $\bigvee \{U'^n \mathcal{H}' : n \ge 0\}$, the equality (4.12) implies minimality. To obtain the converse first note that (4.12) holds when $U'$ is the Sz.-Nagy-Schäffer minimal isometric dilation of $T'$. Using the unitary equivalence of all minimal isometric dilations of $T'$, we arrive at (4.12). Observe that

$$(U' - T')\mathcal{H}' = \bigoplus_{-\infty}^{\infty} P_{\mathcal{M}_k}(\mathcal{K}_{k-1} \oplus \mathcal{M}_{k-1}) .$$

Since $P_{\mathcal{M}_n}(\mathcal{K}_{n-1} \ominus \mathcal{M}_{n-1}) = P_{\mathcal{M}_n \ominus \mathcal{M}_{n-1}} \mathcal{K}_{n-1}$ by (1.1) it is easy to check that (4.12) is equivalent to

$$\mathcal{M}_k = \bigvee_{v=0}^{\infty} P_{\mathcal{M}_{k-v} \ominus \mathcal{M}_{k-v-1}} \mathcal{K}_{k-v-1} \qquad (\text{for all } k \in \mathbb{Z}) . \tag{4.13}$$

Clearly, the spaces $P_{\mathcal{M}_{k-v} \ominus \mathcal{M}_{k-v-1}} \mathcal{K}_{k-v-1}$ for $v = 0, 1, 2, \ldots$ are mutually orthogonal. Because the right hand side of (4.13) is (for fixed k) orthogonal to

$$\left[\bigcap_{n \in \mathbb{Z}} \mathcal{M}_n\right] \text{ and } \mathcal{K}_{k-1}^{\perp} \cap (\mathcal{M}_k \ominus \mathcal{M}_{k-1}) = \mathcal{K}_{k-1}^{\perp} \cap \mathcal{M}_k$$

it follows that (4.11) is necessary for the minimality of $U'$. Conversely, if the second condition in (4.11) holds, then

$$(P_{\mathcal{M}_{k-v} \ominus \mathcal{M}_{k-v-1}} \mathcal{K}_{k-v-1})^{-} = \mathcal{M}_{k-v} \ominus \mathcal{M}_{k-v-1} .$$

So, assuming the first condition in (4.11) also holds, we see that the right hand side of (4.13) (with k fixed) equals

$$\bigvee_{v=0}^{\infty} (\mathcal{M}_{k-v} \ominus \mathcal{M}_{k-v-1}) = \mathcal{M}_k \ominus \left[\bigcap_{n=-\infty}^{\infty} \mathcal{M}_n\right] = \mathcal{M}_k$$

and therefore (4.13) holds. Consequently $U'$ is minimal. This completes the proof of the lemma and Theorem 4.1.

We want now to apply Theorem 4.1 to the particular case of Parrott's completion theorem stated in Lemma 2.1 above (see also Sections IV.3 and IV.8 in Foias-Frazho [4]). According to this theorem for an operator matrix B (acting from $\mathcal{E} = \mathcal{E}_0 \oplus \mathcal{E}_1$ into $\mathcal{N} = \mathcal{N}_0 \oplus \mathcal{N}_1$) of the form

$$B = \begin{bmatrix} E & F \\ G & H \end{bmatrix} \text{ with } \left\| \begin{bmatrix} E \\ G \end{bmatrix} \right\| \le \gamma, \ \|[E \ \ F]\| \le \gamma \tag{4.14}$$

one can find an entry H such that norm of the whole operator B is bounded by $\gamma$. In fact, the operator H is of the form (see Theorem IV.3.1 in Foias-Frazho [4])

$$H = -YE^*X + \gamma(I - YY^*)^{1/2}\Gamma(I - X^*X)^{1/2} , \tag{4.15}$$

where Y from $\mathcal{E}_0$ to $\mathcal{N}_1$ and X from $\mathcal{E}_1$ to $\mathcal{N}_0$ are the contractions that are uniquely determined by the following properties:

$$G = Y(\gamma^2 I - E^*E)^{1/2}, \quad Y \mid \ker(\gamma^2 I - E^*E)^{1/2} = 0, \tag{4.16a}$$

$$F = (\gamma^2 I - EE^*)^{1/2} X, \quad X^* \mid \ker(\gamma^2 I - EE^*)^{1/2} = 0. \tag{4.16b}$$

Furthermore, $\Gamma$ in (4.15) is an arbitrary contraction from $\mathcal{D}_X$, the closure of the range of $D_X = (I - X^*X)^{1/2}$, into $\mathcal{D}_{Y^*}$, the closure of the range of $(I - YY^*)^{1/2}$. In what follows $D_E = (\gamma^2 I - E^*E)^{1/2}$ and $D_{E^*} = (\gamma^2 I - EE^*)^{1/2}$, while $\mathcal{D}_E$ and $\mathcal{D}_{E^*}$ are equal to the closures of the ranges of $D_E$ and $D_{E^*}$, respectively.

To apply Theorem 4.1 to this case we define the operators

$$A_0 = \begin{bmatrix} E \\ G \end{bmatrix} \text{ and } A_1 = [E \ \ F] \tag{4.17}$$

from $\mathcal{E}_0$ into $\mathcal{N}$ and from $\mathcal{E}$ into $\mathcal{N}_0$, respectively. Next, to reformulate the Parrott's completion as a three chains completion problem, we introduce the spaces

$$\mathcal{H}_n = \begin{cases} \{0\} & n < 0 \\ \mathcal{E}_0 & n = 0 \\ \mathcal{E} & n \geq 1 \end{cases}, \quad \mathcal{K}_n = \begin{cases} \{0\} & n < 0 \\ \mathcal{N} & n \geq 0 \end{cases}, \quad \mathcal{M}_n = \begin{cases} \{0\} & n \leq 0 \\ \mathcal{N}_1 & n = 1 \\ \mathcal{N} & n > 1 \end{cases} \tag{4.18}$$

and the operators

$$A_n = \begin{cases} 0 & \text{if } n \neq 0, 1 \\ A_n & \text{if } n = 0, 1. \end{cases} \tag{4.19}$$

Then using (4.17), (4.18) and (4.19) we have

$$(I - P_{\mathcal{M}_1})A_0 = A_1 \mid \mathcal{H}_0 \text{ and } (I - P_{\mathcal{M}_k})A_{k-1} = 0 = A_k \mid \mathcal{H}_{k-1} \text{ (for all } k \neq 1), \tag{4.20}$$

and thus the compatibility conditions in Theorem 1.1 are satisfied. It is also clear that $d_\infty = \sup\{\|A_n\| : n \in \mathbb{Z}\} \leq \gamma$ where $\gamma$ is as in (4.14). Consequently applying Theorem 1.1 we obtain an interpolant B with tolerance $\gamma$, and it is obvious that this B has the $2 \times 2$ operator matrix representation in (4.14).

However, our real interest here is to apply Theorem 4.1. First note that the conditions (4.11) hold since $\mathcal{M}_0 = \{0\}$ and either $\mathcal{K}_n^\perp = \{0\}$ or $\mathcal{M}_{n+1} = \{0\}$. So in this case (4.10) holds with equality only for $B = B_\gamma$. Notice that, for $n \neq 0, 1$ we clearly have

$$D_{B_n} P_{\mathcal{G}_n(B)} D_{B_n} = 0 = D_{(B_\gamma)_n} P_{\mathcal{G}_n(B_\gamma)} D_{(B_\gamma)_n} = D_n P_{\mathcal{G}_n} D_n \tag{4.21}$$

because the spaces $\mathcal{G}_n(B)$ and $\mathcal{G}_n$ reduce to $\{0\}$. For $n = 0$ we have $\mathcal{G}_0(B) = \mathcal{D}_{B_0}$ and

$\mathcal{G}_0 = \mathcal{D}_0 = \mathcal{D}_{A_0}$. Furthermore,

$$A_o = (I - P_{\mathcal{M}_o})B \mid \mathcal{H}_o = B \mid \mathcal{H}_o = B_o .$$

Thus we have

$$D_{B_0} P_{\mathcal{G}_0(B)} D_{B_0} = D_0 P_{\mathcal{G}_0} D_0 \tag{4.22}$$

for all interpolants B of tolerance $\gamma$. In particular, this holds for $B_\gamma = B$. Therefore for the case considered here (3.8) reduces to

$$D_{B_1} P_{\mathcal{G}_1(B)} D_{B_1} \le D_{(B_\gamma)_1} P_{\mathcal{G}_1(B_\gamma)} D_{(B_\gamma)_1} = D_1 P_{\mathcal{G}_1} D_1 \tag{4.23}$$

and equality holds in (4.23) if and only if it holds in (4.10). Since $B_1 = B$ and $B_\gamma = (B_\gamma)_1$, we come to the following result.

**THEOREM 4.3**. *Let* B *be a block matrix of the form*

$$B = \begin{bmatrix} E & F \\ G & H \end{bmatrix} : \begin{bmatrix} \mathcal{E}_0 \\ \mathcal{E}_1 \end{bmatrix} \to \begin{bmatrix} \mathcal{N}_0 \\ \mathcal{N}_1 \end{bmatrix}, \tag{4.24}$$

*and assume that* $\|B\| \le \gamma$. *Then*

$$D_B P_{\mathcal{G}(B)} D_B \le D_{A_1} P_{\mathcal{G}_1} D_{A_1} , \tag{4.25}$$

*where* $A_1 = [E \ \ F]$ *and*

$$\mathcal{G}(B) = \mathcal{D}_B \ominus (D_B \mathcal{E}_0)^- \ \text{ and } \ \mathcal{G}_1 = \mathcal{D}_{A_1} \ominus (D_{A_1} \mathcal{E}_0)^- . \tag{4.26}$$

*Equality in (4.25) holds if and only if* $B = B_\gamma$, *where*

$$B_\gamma = \begin{bmatrix} E & F \\ G & -YE^*X \end{bmatrix} \tag{4.27}$$

*and* Y *from* $\mathcal{E}_0$ *to* $\mathcal{N}_1$ *and* X *from* $\mathcal{E}_1$ *to* $\mathcal{N}_0$ *are the unique contractions determined by (4.16a) and (4.16b), respectively.*

**PROOF**. It remains to prove the statement concerning the equality in (4.25). From the discussion preceding the theorem we know that we have equality in (4.25) if and only if $B = B_\gamma$, where $B_\gamma$ is the central interpolant in (3.21) of Theorem 3.1 corresponding to the three chains completion problem determined by (4.17) to (4.19). So it suffices to show that this $B_\gamma$ is also given by the right hand side of (4.27).

Since $\mathcal{H}_1 = \mathcal{E}_0 \oplus \mathcal{E}_1$, Theorem 3.1 shows that $B_\gamma$ is given by (3.21) with $k = 1$. Because $\mathcal{H}_{n-1} = \{0\}$ for $n \leq 0$, it follows that $\mathcal{F}_n = D_n \mathcal{H}_{n-1}$ is zero for all $n \leq 0$. So for $k = 1$, equation (3.21) simplifies to

$$B_\gamma = A_1 + \pi_1' \omega_1 P_{\mathcal{F}_1} D_1 \,. \tag{4.28}$$

To show that $B_\gamma$ is given by (4.27) we need to compute $\pi_1' \omega_1 P_{\mathcal{F}_1} D_1$. Recall that $D_E$, and $D_{E^*}$ are the positive square roots of $\gamma^2 I - E^* E$, and respectively $\gamma^2 I - EE^*$. Let $D_X$ be the positive square root of $I - X^* X$ where X is the unique contraction in (4.16b) satisfying $F = D_{E^*} X$. Now let $\Omega$ be the operator defined by

$$\Omega = \begin{bmatrix} E & D_{E^*} X \\ D_E & -E^* X \\ 0 & D_X \end{bmatrix} : \begin{bmatrix} \mathcal{E}_0 \\ \mathcal{E}_1 \end{bmatrix} \rightarrow \begin{bmatrix} \mathcal{N}_0 \\ \mathcal{D}_E \\ \mathcal{D}_X \end{bmatrix} .$$

Recall that $E^* D_{E^*} = D_E E^*$. So $E^*$ maps $\mathcal{D}_{E^*}$ into $\mathcal{D}_E$. Since the range of X is contained in $\mathcal{D}_{E^*}$, it follows that the range of $-E^* X$ is contained in $\mathcal{D}_E$. Therefore $\Omega$ is well defined. Using $D_{E^*} E = E D_E$ once again a simple calculation shows that $\Omega^* \Omega = I$, and thus $\Omega$ is an isometry.

Now let $h = e_o \oplus e_1$ be a vector in $\mathcal{E}_0 \oplus \mathcal{E}_1$. Because $A_1$ is the first row of $\Omega$ and $\Omega$ is an isometry, we have

$$\|D_1 h\|^2 = \|h\|^2 - \|A_1 h\|^2 = \|\Omega h\|^2 - \|A_1 h\|^2 = \|D_E e_o - E^* X e_1\|^2 + \|D_X e_1\|^2 .$$

This implies that there exists a unitary operator $W_1$ from $\mathcal{D}_1$ onto $\mathcal{D}_E \oplus \mathcal{D}_X$ satisfying

$$W_1 D_1 = \begin{bmatrix} D_E & -E^* X \\ 0 & D_X \end{bmatrix} .$$

Finally, it is noted that $W_1 D_1 (e_o \oplus 0) = D_E e_o \oplus 0$. Since $\mathcal{F}_1 = \overline{D_1 \mathcal{H}_0}$ and $\mathcal{H}_0 = \mathcal{E}_0$, it follows that $W_1 \mathcal{F}_1 = \mathcal{D}_E \oplus \{0\}$. Hence $W_1 P_{\mathcal{F}_1} = P_{\mathcal{D}_E} W_1$.

In our setting, $\pi_1'$ is the orthogonal projection from $\mathcal{F}_1' \subseteq (\mathcal{D}_0 \oplus \mathcal{N}_1)$ into $\mathcal{N}_1$. Moreover, the isometry $\omega_1$ in (3.8) is now given by

$$\omega_1 D_1 \begin{bmatrix} e_o \\ 0 \end{bmatrix} = \begin{bmatrix} D_0 e_o \\ P_{\mathcal{M}_1} A_0 e_o \end{bmatrix} = \begin{bmatrix} D_0 e_o \\ G e_o \end{bmatrix} = \begin{bmatrix} D_0 e_o \\ Y D_E e_o \end{bmatrix} .$$

So using $P_{\mathcal{F}_1} W_1^* = W_1^* P_{\mathcal{D}_E}$ and taking g in $\mathcal{D}_X$ we have

$$\pi_1'\omega_1 P_{\mathcal{F}_1} W_1^* \begin{bmatrix} D_E e_o \\ g \end{bmatrix} = \pi'\omega_1 W_1^* \begin{bmatrix} D_E e_o \\ 0 \end{bmatrix} = \pi_1'\omega_1 D_1 \begin{bmatrix} e_o \\ 0 \end{bmatrix} = \begin{bmatrix} 0 \\ YD_E e_o \end{bmatrix}.$$

Since $D_E \mathcal{E}_0$ is dense in $\mathcal{D}_E$, this shows that

$$\pi_1'\omega_1 P_{\mathcal{F}_1} W_1^*(d \oplus g) = (0 \oplus Yd) \quad ((d \oplus g) \in \mathcal{D}_E \oplus \mathcal{D}_X).$$

Using $e_o \oplus e_1$ in $\mathcal{E}_0 \oplus \mathcal{E}_1$ we obtain

$$\pi_1'\omega_1 P_{\mathcal{F}_1} D_1 \begin{bmatrix} e_o \\ e_1 \end{bmatrix} = \pi_1'\omega_1 P_{\mathcal{F}_1} W_1^* W_1 D_1 \begin{bmatrix} e_o \\ e_1 \end{bmatrix} =$$

$$\pi_1'\omega_1 P_{\mathcal{F}_1} W_1^* \begin{bmatrix} D_E e_o - E^* X e_1 \\ D_X e_1 \end{bmatrix} = \begin{bmatrix} 0 \\ YD_E e_o - YE^* X e_1 \end{bmatrix} = \begin{bmatrix} 0 \\ Ge_o - YE^* X e_1 \end{bmatrix}.$$

Combining this with the formula for $B_\gamma$ in (4.28) we obtain that $B_\gamma$ is given by (4.27). This completes the proof.

**REMARK 4.4.** We shall now present an alternative proof of Theorem 4.3 based on the description of all solutions of the Parrott problem in (4.14) and (4.15). Without loss of generality we can assume that $\gamma = 1$. (To see this simply divide the operators E, F, G and B by $\gamma$.) It is not hard to prove that (see for instance Foias-Frazho [4], Corollary IV.1.5 and Corollary IV.3.2) that there exist unitary operators $W_1$ from $\mathcal{D}_{A_1}$ onto $\mathcal{D}_E \oplus \mathcal{D}_X$ and W from $\mathcal{D}_B$ (where B is given in (4.14) to (4.16)) onto $\mathcal{D}_Y \oplus \mathcal{D}_\Gamma$ defined by

$$W_1 D_{A_1} = \begin{bmatrix} D_E & -E^* X \\ 0 & D_X \end{bmatrix} \text{ and } WD_B = \begin{bmatrix} D_Y D_E & -(D_Y E^* X + Y^* \Gamma D_X) \\ 0 & D_\Gamma D_X \end{bmatrix}.$$

Here $A_1$, B, E, X, Y and $\Gamma$ are contractions and $D_C = (I - C^* C)^{1/2}$ for C equal to any of the previous contractions. Clearly, $W_1 D_{A_1} e_0 = D_E e_0 \oplus 0$ for $e_0 \in \mathcal{E}_0$, and thus $W_1 \mathcal{G}_1 = W_1(\mathcal{D}_{A_1} \ominus (D_{A_1}\mathcal{H}_0)^-) = \{0\} \oplus \mathcal{D}_X$.

It follows that $W_1 P_{\mathcal{G}_1} W_1^{-1} = 0 \oplus I_{\mathcal{D}_X}$ and thus

$$D_{A_1} P_{\mathcal{G}_1} D_{A_1} = D_{A_1} W_1^* W_1 P_{\mathcal{G}_1} W_1^* W_1 D_{A_1} = \begin{bmatrix} 0 & 0 \\ 0 & D_X^2 \end{bmatrix}. \tag{4.29}$$

Similarly

$$D_B P_{\mathcal{G}(B)} D_B = \begin{bmatrix} 0 & 0 \\ 0 & D_X D_\Gamma^2 D_X \end{bmatrix}. \tag{4.30}$$

Therefore (4.25) reduces to

$$D_X D_\Gamma^2 D_X \le D_X^2 . \tag{4.31}$$

Equality in (4.31) is equivalent to $D_X \Gamma^* \Gamma D_X = 0$ which in turn (since $\Gamma$ is defined on $\mathcal{D}_X = (\operatorname{ran} D_X)^-$) is equivalent to $\Gamma = 0$. Due to the parameterization of all H's in (4.15), we conclude that equality holds in (4.25) if and only if $B = B_\gamma$, with $B_\gamma$ being given by (4.27).

As an application of the maximum principles in Theorems 4.1 and 4.3 we will establish the connection between Sections 2 and 3. Consider the three chains completion problem associated with the operators $A_n : \mathcal{H}_n \to \mathcal{K}_n \ominus \mathcal{M}_n$ for $n \in \mathbb{Z}$. First, without loss of generality, we shall assume that $\mathcal{K}_n = \mathcal{Y}$ (for all $n \in \mathbb{Z}$). Then the relations (4.11) reduce to

$$\bigcap_{n \in \mathbb{Z}} \mathcal{M}_n = \{0\}$$

which we will also assume throughout the remainder of this section. Let now $B_\gamma$ be the central interpolant in the three chains completion problem for the given operators, and fix some $k \in \mathbb{Z}$. Set

$$A'_k = (I - P_{\mathcal{M}_{k-1}}) B_\gamma \mid \mathcal{H}_k .$$

Then

$$A'_k \mid \mathcal{H}_{k-1} = A_{k-1}, \; (I - P_{\mathcal{M}_k}) A'_k = A_k \; \text{ and } \; \|A'_k\| \le \gamma .$$

So $A'_k$ is a completion in the Parrott problem with data

$$\begin{bmatrix} E \\ G \end{bmatrix} = A_{k-1} : \mathcal{E}_0 = \mathcal{H}_{k-1} \to \mathcal{N} = (I - P_{\mathcal{M}_{k-1}}) \mathcal{K} \tag{4.32}$$

$$[E \;\; F] = A_k : \mathcal{E} = \mathcal{H}_k \to \mathcal{N}_0 = (I - P_{\mathcal{M}_k}) \mathcal{K} . \tag{4.33}$$

Moreover, for all $h_k \in \mathcal{H}_k$ we have

$$\|D_{A'_k} h_k\|^2 = \|h_k\|^2 - \|A'_k h_k\|^2 = \|h_k\|^2 - \|B_\gamma h_k\|^2 + \|P_{\mathcal{M}_{k-1}} B_\gamma h_k\|^2 \ge \|D_{B_\gamma} h_k\|^2 .$$

Then, using the maximum principle for the Parrott completion problem to obtain the first inequality below, we see that

$$\|P_{\mathcal{G}_k} D_{A_k} h_k\|^2 \geq \|P_{\mathcal{G}(A'_k)} D_{A'_k} h_k\|^2 = \inf_{h_{k-1} \in \mathcal{H}_{k-1}} \|D_{A'_k}(h_k - h_{k-1})\|^2 \geq$$

$$\geq \inf_{h_{k-1} \in \mathcal{H}_{k-1}} \|D_{B_\gamma}(h_k - h_{k-1})\|^2 = \|P_{\mathcal{G}_k(B_\gamma)} D_{(B_\gamma)_k} h_k\|^2 = \|P_{\mathcal{G}_k} D_{A_k} h_k\|^2 \tag{4.34}$$

where the last equality follows from Theorem 4.1 using the fact that $B_\gamma$ is the central interpolant for the operators $\{A_n\}_{n \in \mathbb{Z}}$. It follows that

$$\|P_{\mathcal{G}(A'_k)} D_{A'_k} h_k\|^2 = \|P_{\mathcal{G}_k} D_{A_k} h_k\|^2 \qquad (\text{for all } h_k \in \mathcal{H}_k)\,. \tag{4.35}$$

Since $\mathcal{G}_k = \mathcal{G}(A_k)$, Theorem 4.3 implies that $A'_k$ is the central completion in the Parrott problem with data (4.32), (4.33). We can now pass to the following useful result.

**LEMMA 4.5**. *Let* $B_\gamma$ *in (3.21) be the central interpolant with tolerance* $\gamma$ *for the three chains problem associated with the operators* $A_n : \mathcal{H}_n \to \mathcal{K}_n \ominus \mathcal{M}_n$ *for* $n \in \mathbb{Z}$, *where* $\mathcal{K}_n = \mathcal{Y}$ *for all* $n \in \mathbb{Z}$ *and* $\cap\, \mathcal{M}_n = \{0\}$. *Fix* $k \in \mathbb{Z}$, *and let* $A'_k$ *be the central completion with tolerance* $\gamma$ *to the Parrott completion problem corresponding to the data (4.32), (4.33). Consider the new* ($\mathrm{v}$) *three chains completion problem corresponding to the following new data:*

$$\mathcal{H}_n^{\mathrm{v}} = \mathcal{H}_n\,, \quad \mathcal{K}_n^{\mathrm{v}} = \mathcal{K}\,, \quad \mathcal{M}_n^{\mathrm{v}} = \begin{cases} \mathcal{M}_n & \text{if } n \neq k\,, \\ \mathcal{M}_{n-1} & \text{if } n = k\,, \end{cases} \tag{4.36}$$

*and*

$$A_{n,\mathrm{v}} = \begin{cases} A_n & \text{if } n \neq k\,, \\ A'_n & \text{if } n = k\,. \end{cases} \tag{4.37}$$

*Then* $B_\gamma$ *is the central interpolant with tolerance* $\gamma$ *in this new three chain completion problem.*

**PROOF**. By virtue of the discussion preceding Lemma 4.5, it follows that $B_\gamma$ is also a interpolant with tolerance $\gamma$ of the new three chains problem. Also, by setting $\mathcal{G}_n^{\mathrm{v}} = \mathcal{D}_{A_{n,\mathrm{v}}} \ominus (D_{A_{n,\mathrm{v}}} \mathcal{H}_{n-1}^{\mathrm{v}})^-$ for all $n \in \mathbb{Z}$ we have, for $n \neq k$,

$$D_{(B_\gamma)_n} P_{\mathcal{G}_n(B_\gamma)} D_{(B_\gamma)_n} = D_{A_n} P_{\mathcal{G}_n} D_{A_n} = D_{A_{n,\mathrm{v}}} P_{\mathcal{G}_n^{\mathrm{v}}} D_{A_{n,\mathrm{v}}} \tag{4.38}$$

because $B_\gamma$ is the central solution of the old problem. On the other hand for $n = k$ we have

$$D_{(B_\gamma)_k} P_{\mathcal{G}_k(B_\gamma)} D_{(B_\gamma)_k} = D_{A_k} P_{\mathcal{G}_k} D_{A_k} = D_{A'_k} P_{\mathcal{G}(A'_k)} D_{A'_k} = D_{A_{k,\mathrm{v}}} P_{\mathcal{G}_k^{\mathrm{v}}} D_{A_{k,\mathrm{v}}} \tag{4.39}$$

where the equalities follow from the the fact that $B_\gamma$ is the central interpolant of the old problem and (4.35). The equalities (4.38) and (4.39) show that $B_\gamma$ yields equality in all the relations (4.10) corresponding to the new three chain completion problem. Therefore by virtue of Theorem 4.1, the operator $B_\gamma$ is also the central interpolant with tolerance $\gamma$ of that problem. The proof is now complete.

To put Lemma 4.5 in a more general context, we introduce the notion of a partial completion. Consider a second three chains completion problem which is associated to operators $A'_n : \mathcal{H}_n \to \mathcal{Y} \ominus \mathcal{M}'_n$. We call the operators $\{A'_n\}_{n\in\mathbb{Z}}$ a *partial completion* of our original problem with data $A_n : \mathcal{H}_n \to \mathcal{Y} \ominus \mathcal{M}_n$ for $n \in \mathbb{Z}$ if $\|A'_n\| \le \gamma$ for all n, the new chain $\{\mathcal{M}'_n\}_{n\in\mathbb{Z}}$ satisfies $\mathcal{M}'_n \subseteq \mathcal{M}_n$ for $n \in \mathbb{Z}$, and

$$(I - P_{\mathcal{M}_n})A'_n = A_n \quad (n \in \mathbb{Z}) . \tag{4.40}$$

Notice that $\mathcal{M}'_n \subseteq \mathcal{M}_n$ implies that $(I - P_{\mathcal{M}_n})(I - P_{\mathcal{M}'_n}) = I - P_{\mathcal{M}_n}$. Therefore, if B from $\mathcal{U}$ to $\mathcal{Y}$ is an interpolant in the three chains completion problem for the partial completion $\{A'_n\}_{n\in\mathbb{Z}}$, then B is an interpolant for the original problem. Clearly Lemma 4.5 yields such a partial completion, which we will call a *one step central Parrott completion* of the original problem. It is also clear that any iteration of one step central Parrott completions produces a partial completion $\{A'_n, \mathcal{M}'_n\}_{n\in\mathbb{Z}}$. By virtue of Lemma 4.5, we have

$$A'_n = (I - P_{\mathcal{M}'_n})B_\gamma | \mathcal{H}_n \quad (n \in \mathbb{Z}) , \tag{4.41}$$

where $B_\gamma$ is the central interpolant with tolerance $\gamma$ of the original problem.

**LEMMA 4.6.** *Let $B_\gamma$ in (3.21) be the central interpolant with tolerance $\gamma$ for the three chains completion problem associated with the operators $A_n : \mathcal{H}_n \to \mathcal{Y} \ominus \mathcal{M}_n$ for $n \in \mathbb{Z}$, where $\cap \mathcal{M}_n = \{0\}$. Then $B_\gamma$ is the central interpolant with tolerance $\gamma$ of any partial completion $A'_n : \mathcal{H}_n \to \mathcal{Y} \ominus \mathcal{M}'_n$ with $n \in \mathbb{Z}$ satisfying (4.41).*

**PROOF.** Since $B_\gamma$ is the central interpolant of the operators $\{A_n\}_{n\in\mathbb{Z}}$, we can use (4.10) to show that

$$\|P_{\mathcal{G}_n(B_\gamma)} D_{(B_\gamma)_n} h_n\|^2 = \|P_{\mathcal{G}_n} D_{A_n} h_n\|^2 =$$

$$= \inf_{h_{n-1}\in\mathcal{H}_{n-1}} \|D_{A_n}(h_n - h_{n-1})\|^2 \ge \inf_{h_{n-1}\in\mathcal{H}_{n-1}} \|D_{A'_n}(h_n - h_{n-1})\|^2 =$$

$$= \|P_{\mathcal{G}(A'_n)} D_{A'_n} h_n\|^2 \ge \inf_{h_{n-1}\in\mathcal{H}_{n-1}} \|D_{(B_\gamma)_n}(h_n - h_{n-1})\|^2 = \|P_{\mathcal{G}_n(B_\gamma)} D_{(B_\gamma)_n} h_n\|^2 .$$

The first inequality in the above calculation is a consequences of (4.40), and the second follows from (4.41) and the maximum principle (4.10). We have now proved that

$$D_{(B_\gamma)_n} P_{\mathcal{G}_n(B_\gamma)} D_{(B_\gamma)_n} = D_{A'_n} P_{\mathcal{G}(A'_n)} D_{A'_n} \quad (\text{for } n \in \mathbb{Z}) .$$

By Theorem 4.1, the operator $B_\gamma$ is also the central interpolant with tolerance $\gamma$ of the partial completion. The proof is now complete.

Let $\{\{A_n^{(\beta)}, \mathcal{M}_n^{(\beta)}\}_{n\in\mathbb{Z}}\}_{\beta\in\mathcal{B}}$ be any net of partial completions satisfying (4.41) such that for all $\beta_1, \beta_2$ in $\mathcal{B}$ there exists $\beta_3 \in \mathcal{B}$ such that the $\beta_3$-completion is also a completion of both the $\beta_1$ and $\beta_2$-completions. Then clearly (4.41) with

$$\mathcal{M}'_n = \bigcap_{\beta\in\mathcal{B}} \mathcal{M}_n^{(\beta)} \tag{4.42}$$

yields a partial completion.

Lemmas 4.5 and 4.6 and the remark made in the previous paragraph show that by taking one step central Parrott completions as well as limits of sequences of such completions, we always end up with partial completions which satisfy (4.41). As long as some spaces $\mathcal{M}'_n$ are not equal to zero, we can continue to apply the one step central Parrott completion at the juncture k where $\mathcal{M}'_{k-1} \neq \mathcal{M}'_k$; note that since $\bigcap_{n\in\mathbb{Z}}\mathcal{M}'_n = \{0\}$, such a k must exist. Therefore if the whole successive application of one step central Parrott extensions (including the limits of such completions) is exhaustive we must end up with a three chains problem in which $\mathcal{M}'_n = \{0\}$ for all n. Therefore all the corresponding operator data $A'_n$ are given by $A'_n = B_\gamma | \mathcal{H}_n$. Thus $B_\gamma$ equals the strong limit of $A'_n P_{\mathcal{H}_n}$ as n approaches infinity. This discussion can be summed up as follows.

**THEOREM 4.7.** *An exhaustive successive application of one step central Parrott completions (including limits if necessary) yields* $B_\gamma$, *the central interpolant with tolerance* $\gamma$ *of the initial three chain completion problem.*

**REMARK 4.8.** Assume that $\mathcal{H} = \bigvee \{\mathcal{H}_n : n \in \mathbb{Z}\}$, $\mathcal{K}_n = \mathcal{Y}$ for all n in $\mathbb{Z}$ and $\bigcap_{n\in\mathbb{Z}}\mathcal{M}_n = \{0\}$. Let us apply the above discussion to the construction in Section 2. Then we conclude that if in that construction in each application of the Parrott Lemma, we select the central completion, then the interpolant constructed in Section 2 coincides with the central interpolant introduced in Section 3.

## XII.5. A QUOTIENT FORMULA FOR THE CENTRAL INTERPOLANT

In this section we will present a formula for the central interpolant for the three chains completion problem which is the analogue of the quotient formula for the central solution in the commutant lifting theorem discussed in Section IV.4. As before, let $A_k$ from $\mathcal{H}_k$ into $\mathcal{K}_k \ominus \mathcal{M}_k$ be the data for the three chains completion problem where the spaces $\mathcal{H}_k$, $\mathcal{M}_k$ and $\mathcal{K}_k$ satisfy the increasing property in (1.1) and the compatibility condition in (i) of Theorem 1.1 holds. Moreover, assume that $d_\infty = \sup\{\|A_k\| : k \in \mathbb{Z}\} < \gamma$,

$$\mathcal{U} = \bigvee\{\mathcal{H}_k : k \in \mathbb{Z}\} \text{ and } \bigcap\{\mathcal{H}_k : k \in \mathbb{Z}\} = \{0\}. \tag{5.1}$$

Now let $\mathcal{L}_k$ be the Hilbert space defined by $\mathcal{L}_k = \mathcal{H}_k \ominus \mathcal{H}_{k-1}$. From the two conditions in (5.1) it follows that $\mathcal{U} = \oplus\{\mathcal{L}_k : k \in \mathbb{Z}\}$ and that $\mathcal{H}_k = \oplus\{\mathcal{L}_m : m \le k\}$. Let $E_k$ be the isometric embedding of $\mathcal{L}_k$ into $\mathcal{H}_k$. Finally, let L from $\mathcal{U}$ into $\mathcal{Y}$ and R on $\mathcal{U}$ be the linear maps defined by

$$\begin{aligned} L &= [\ldots, A_{-1}D_{-1}^{-2}E_{-1}, A_0D_0^{-2}E_0, A_1D_1^{-2}E_1, \ldots], \\ R &= [\ldots, D_{-1}^{-2}E_{-1}, D_0^{-2}E_0, D_1^{-2}E_1, \ldots]. \end{aligned} \tag{5.2}$$

Clearly, these maps are well defined on a dense set in $\mathcal{U}$. This sets the stage for the following result.

**THEOREM 5.1.** *Let $A_k$ from $\mathcal{H}_k$ into $\mathcal{K}_k \ominus \mathcal{M}_k$ for $k \in \mathbb{Z}$ be the data for the three chains completion problem where the spaces satisfy the increasing property in (1.1) and the conditions in (5.1). Assume the compatibility condition (i) holds, and let $\gamma > d_\infty$. Then $R^{-1}$ exists as a bounded operator on $\mathcal{U}$ and $B_\gamma = LR^{-1}$ is the central interpolant for the three chains completion problem with tolerance $\gamma$. Furthermore, if $\Theta$ is the operator on $\mathcal{U}$ defined by*

$$\Theta = \operatorname{diag}\,[((E_k^* D_k^{-2} E_k)^{1/2})_{-\infty}^{\infty}]R^{-1}, \tag{5.3}$$

*then the range of $\Theta$ is dense in $\mathcal{U}$ and $\gamma^2 I - B_\gamma^* B_\gamma = \Theta^*\Theta$.*

To give a precise meaning to the inverse of R in Theorem 5.1 and Theorem 5.1a below, let $\mathcal{U}_0$ be the subspace consisting of the set of all sequences in $\mathcal{U}$ with finite support. Then $\mathcal{N} = R\,\mathcal{U}_0$ is dense in $\mathcal{U}$ and the inverse $R^{-1}$ of $R \mid \mathcal{U}_0$ is a bounded operator from $\mathcal{N}$ into $\mathcal{U}$. Therefore $R^{-1}$ can be uniquely extended by continuity to a bounded operator on all of $\mathcal{U}$ which we also denoted by $R^{-1}$. So without loss of generality we say that $R^{-1}$ exists as a bounded operator on $\mathcal{U}$. Finally, as in Remark IV.4.5 the equality $B_\gamma = LR^{-1}$ has to be understood in a special way, namely $B_\gamma \mid \mathcal{N} = LR^{-1} \mid \mathcal{N}$ is well defined and bounded on $\mathcal{N}$, and its continuous extension to $\mathcal{U}$ is $B_\gamma$.

**PROOF**. To prove that the range of R is dense in $\mathcal{U}$ let A be the operator from $\mathcal{H}$ into $\mathcal{K}' \ominus \mathcal{M}'$ defined by (1.4) and (1.5). Recall also that $T'A = AT$ where $T'$ is a contraction, and T on $\mathcal{H}$ is the isometry defined by $T = V_1 \mid \mathcal{H}$ where $V_1$ is the bilateral shift on $\ell^2(\mathcal{U})$. Using the second condition in (5.1) it follows that $\cap \{T^n \mathcal{H} : n \geq 0\} = \{0\}$, and thus T is a unilateral shift. Clearly, $\mathcal{L} = \ker T^* = \oplus_{-\infty}^{\infty} \mathcal{L}_k$ where $\mathcal{L}_k \subseteq \mathcal{H}_k$ lives in the k-th component of $\ell^2(\mathcal{U})$. Since $\|A\| = d_\infty < \gamma$, Lemma IV.4.3 shows that span $\{T^n D_A^{-2} \mathcal{L} : n \geq 0\}$ is dense in $\mathcal{H}$. Because A is a diagonal operator, this implies that

$$\mathcal{H}_k = \bigvee_{\nu \geq 0} D_{k-\nu}^{-2} \mathcal{L}_{k-\nu} \,. \tag{5.4}$$

Notice that $D_k^{-2} E_k$ is precisely the k-th column in R. So (5.4) and the first condition in (5.1) shows that the range of R is dense in $\mathcal{H}$.

To show that the inverse of R is bounded observe that the compatibility condition in part (i) of Theorem 1.1 gives

$$P_{\mathcal{H}_\nu} A_k^* = A_\nu^* (I - P_{\mathcal{M}_k}) \qquad (\text{for all } \nu \leq k)\,. \tag{5.5}$$

Now let $f = (f_n)_{-\infty}^{\infty}$ be any sequence in $\oplus \mathcal{L}_k$ with compact support. Then using (5.5) and $D_k^{-2} E_k f_k = D_k^{-2} f_k$ we have

$$\gamma^2 \|Rf\|^2 - \|Lf\|^2 = \sum_{n=-\infty}^{\infty} \sum_{m=-\infty}^{\infty} \gamma^2 (D_n^{-2} f_n, D_m^{-2} f_m) - (A_n D_n^{-2} f_n, A_m D_m^{-2} f_m) =$$

$$\sum_{n \geq m} ((\gamma^2 I - A_m^* (I - P_{\mathcal{M}_n}) A_n) D_n^{-2} f_n, D_m^{-2} f_m) + \sum_{m > n} (D_n^{-2} f_n, (\gamma^2 I - A_n^* (I - P_{\mathcal{M}_m}) A_m) D_m^{-2} f_m) =$$

$$\sum_{n \geq m} (P_{\mathcal{H}_m} D_n^2 D_n^{-2} f_n, D_m^{-2} f_m) + \sum_{m > n} (D_n^{-2} f_n, P_{\mathcal{H}_n} D_m^2 D_m^{-2} f_m) = \sum_{n=-\infty}^{\infty} (D_n^{-2} f_n, f_n)\,.$$

The last equality follows because $f_k$ is orthogonal to $\mathcal{H}_\nu$ when $k > \nu$. Since $d_\infty < \gamma$ the positive operator $D_k^{-2} \geq \varepsilon^2 I$ for some $\varepsilon > 0$ and all k. Therefore the previous equation shows that

$$\gamma^2 \|Rf\|^2 - \|Lf\|^2 = \sum_{n=-\infty}^{\infty} (D_n^{-2} f_n, f_n) = \sum_{n=-\infty}^{\infty} \|(E_n^* D_n^{-2} E_n)^{1/2} f_n\|^2 \geq \varepsilon^2 \|f\|^2\,. \tag{5.6}$$

for some $\varepsilon > 0$. This readily implies that $(\gamma/\varepsilon)\|Rf\| \geq \|f\|$ for all f on a dense set in $\oplus \mathcal{L}_k$. Recall that the range of R is dense in $\mathcal{U}$. Therefore R admits a bounded inverse $R^{-1}$ on a dense set and by continuity can be extended to a bounded operator (also denoted by $R^{-1}$) on all of $\mathcal{U}$.

Moreover, $\|R^{-1}\| \le \gamma/\varepsilon$.

Since $\|A_k\| < \gamma$, it follows that $\mathcal{F}_k = D_k\mathcal{H}_{k-1}$. We claim that $\mathcal{G}_k := \mathcal{D}_k \ominus \mathcal{F}_k$ equals $D_k^{-1}\mathcal{L}_k$. To see this notice that g is in $\mathcal{G}_k$ if and only if g is in $\mathcal{D}_k$ and g is orthogonal to $D_k\mathcal{H}_{k-1}$, or equivalently, $D_k g$ is orthogonal to $\mathcal{H}_{k-1} = \mathcal{H}_k \ominus \mathcal{L}_k$. So this g is in $\mathcal{G}_k$ if and only if $D_k g$ is in $\mathcal{L}_k$, or equivalently, g is in $D_k^{-1}\mathcal{L}_k$ which proves our claim that $\mathcal{G}_k = D_k^{-1}\mathcal{L}_k$. Now let $f_k$ be in $\mathcal{L}_k$, then clearly $D_k^{-1}f_k$ is in $\mathcal{G}_k$ and thus $P_{\mathcal{F}_k}D_k^{-1}f_k = 0$. By consulting (3.11) and (3.12) this shows that $\Phi_p(k-\nu, k)D_kD_k^{-2}f_k = 0$ for all $\nu > 0$. Applying $D_k^{-2}f_k$ to the central interpolant $B_\gamma$ for the three chains completion problem displayed in (3.21) shows that $B_\gamma D_k^{-2}f_k = A_kD_k^{-2}f_k$. Because $D_k^{-2}E_k$ respectively $A_kD_k^{-2}E_k$ is the k-th column of R respectively L, it follows that $B_\gamma R = L$. Hence $B_\gamma = LR^{-1}$.

To show that $\gamma^2 I - B_\gamma^* B_\gamma = \Theta^*\Theta$, let $h = Rf$ where $f = (f_n)_{-\infty}^{\infty}$ is a sequence in $\oplus\,\mathcal{L}_k$ with compact support. By using $B_\gamma h = B_\gamma Rf = Lf$ along with (5.6) we have

$$((\gamma^2 I - B_\gamma^* B_\gamma)h, h) = \gamma^2\|h\|^2 - \|B_\gamma h\|^2 = \gamma^2\|Rf\|^2 - \|Lf\|^2 =$$

$$\|\operatorname{diag}[((E_k^* D_k^{-2}E_k)^{1/2})_{-\infty}^{\infty}]f\|^2 = \|\Theta h\|^2 = (\Theta^*\Theta h, h)\,.$$

Since this equality holds on a dense set, by continuity, it holds for all h in $\mathcal{U}$. Therefore $\gamma^2 I - B_\gamma^* B_\gamma = \Theta^*\Theta$. This completes the proof.

Now let $\{A_n\}$ be the data for the three chains completion problem given by $2\times 2$ Parrott completion problem in (4.17) to (4.19) where $\|A_k\| < \gamma$ for $k = 0, 1$. Then using Theorem 5.1 it is easy to check that the central interpolant for this three chains problem is given by

$$B_\gamma = \begin{bmatrix} E & F \\ G & -GE^*(\gamma^2 I - EE^*)^{-1}F \end{bmatrix}.$$

As expected, this is precisely the central interpolant given by (4.27), (4.28) in Theorem 4.3.

In many applications it may be convenient to compute the central interpolant for the three chains completion problem corresponding to the data given in Theorem 1.1a. As noted in Section 1 we an obtain a solution to this three chains completion problem by replacing k by $-k$ and then applying our previous results. So by proceeding in this fashion we say that $B_\gamma$ is the central interpolant with tolerance $\gamma$ corresponding to the three chains completion problem in Theorem 1.1a, if $B_\gamma$ is the corresponding central interpolant for the three chains completion problem where k is replaced $-k$. By adapting Theorem 5.1 to the setting of Theorem 1.1a, we readily obtain the following result.

**THEOREM 5.1a.** *Let* $A_k$ *from* $\mathcal{H}_k$ *into* $\mathcal{K}_k \ominus \mathcal{M}_k$ *for* $k \in \mathbb{Z}$ *be the data for the three chains completion problem where the spaces satisfy the decreasing property in (1.6) and the* *conditions in (5.1). Assume the compatibility condition* (j) *holds, and let* $\gamma > d_\infty$. *Put* $\mathcal{L}_k = \mathcal{H}_k \ominus \mathcal{H}_{k+1}$ *and let* $E_k$ *be the isometry which embeds* $\mathcal{L}_k$ *into* $\mathcal{U} = \oplus\{\mathcal{L}_k : k \in \mathbb{Z}\}$. *Finally, let* L *from* $\mathcal{U}$ *into* $\mathcal{Y}$ *and* R *on* $\mathcal{U}$ *be the linear maps defined by (5.2). Then* $R^{-1}$ *exists as a bounded operator on* $\mathcal{U}$ *and* $B_\gamma = LR^{-1}$ *is the central interpolant for this three chains completion problem with tolerance* $\gamma$. *Furthermore, if* $\Theta$ *is the operator on* $\mathcal{U}$ *defined as in (5.3), then the range of* $\Theta$ *is dense in* $\mathcal{U}$ *and* $\gamma^2 I - B_\gamma^* B_\gamma = \Theta^* \Theta$.

To complete this section let us demonstrate how one can use the previous theorem to compute the central solution to the nonstationary standard Nevanlinna-Pick problem discussed in Section VIII.2.1. To this end, recall that the data for this Nevanlinna-Pick problem is the set of Hilbert space operators

$$Z_k : \mathcal{X}_{k+1} \to \mathcal{X}_k, \; B_k : \mathcal{U}_k \to \mathcal{X}_k \text{ and } \tilde{B}_k : \mathcal{Y}_k \to \mathcal{X}_k . \tag{5.7}$$

Throughout we always assume that $\{Z_k\}$, $\{B_k\}$ and $\{\tilde{B}_k\}$ are uniformly bounded. The state transition operator $\Phi(m, n)$ mapping $\mathcal{X}_n$ into $\mathcal{X}_m$ is defined by

$$\Phi(m, n) = Z_m Z_{m+1} \cdots Z_{n-1} \text{ if } m < n \text{ and } \Phi(n, n) = I .$$

Moreover, we assume that

$$\limsup_{\nu \to \infty} (\sup\{\|\Phi(k, k+\nu)\| : k \in \mathbb{Z}\})^{1/\nu} < 1 . \tag{5.8}$$

Associated with the Nevanlinna-Pick problem are the controllability operators $W_k$ from $\vec{\mathcal{U}}_{(k,\infty)} = \oplus\{\mathcal{U}_n : n \geq k\}$ into $\mathcal{X}_k$ and $\tilde{W}_k$ from $\vec{\mathcal{Y}}_{(k,\infty)} = \oplus\{\mathcal{Y}_n : n \geq k\}$ into $\mathcal{X}_k$ defined by

$$\begin{aligned} W_k &= [B_k, \Phi(k, k+1)B_{k+1}, \Phi(k, k+2)B_{k+2} \ldots] \\ \tilde{W}_k &= [\tilde{B}_k, \Phi(k, k+1)\tilde{B}_{k+1}, \Phi(k, k+2)\tilde{B}_{k+2}, \ldots] . \end{aligned} \tag{5.9}$$

We say that F is *an interpolant for* the nonstationary Nevanlinna-Pick problem if F is a lower block triangular matrix from $\vec{\mathcal{Y}} = \oplus\{\mathcal{Y}_n : n \in \mathbb{Z}\}$ into $\vec{\mathcal{U}} = \oplus\{\mathcal{U}_n : n \in \mathbb{Z}\}$ satisfying $W_k \Pi_k F | \vec{\mathcal{Y}}_{(k,\infty)} = \tilde{W}_k$ for all $k \in \mathbb{Z}$. Here $\Pi_k$ is the orthogonal projection from $\vec{\mathcal{U}}$ onto $\vec{\mathcal{U}}_{(k,\infty)}$. Furthermore, F is a solution to the nonstationary Nevanlinna-Pick problem with tolerance $\gamma$, if F is an interpolant for the nonstationary Nevanlinna-Pick problem and $\|F\| \leq \gamma$.

Recall that $P_k = W_k W_k^*$ is the controllability grammian for the family of pairs $\{Z_k, B_k\}$ while $\tilde{P}_k = \tilde{W}_k \tilde{W}_k^*$ is the controllability grammian for the family of pairs $\{Z_k, \tilde{B}_k\}$. Moreover, these operators satisfy the following time varying Lyapunov equations

$$P_k = Z_k P_{k+1} Z_k^* + B_k B_k^* \text{ and } \tilde{P}_k = Z_k \tilde{P}_{k+1} Z_k^* + \tilde{B}_k \tilde{B}_k^* . \tag{5.10}$$

According to the discussion below or the results in Section VIII.2, the nonstationary Nevanlinna-Pick problem with tolerance $\gamma$ is solvable if and only if $\gamma^2 P_k \geq \tilde{P}_k$ for all $k \in \mathbb{Z}$.

Assume that $P_k$ is invertible for $k \in \mathbb{Z}$. The time varying Nevanlinna-Pick problem can be converted to the framework of the three chains completion problem by setting

$$A_k = W_k^* P_k^{-1} \tilde{W}_k : \mathcal{H}_k \to \mathcal{K}_k \ominus \mathcal{M}_k \text{ where}$$
$$\mathcal{H}_k = \vec{\mathcal{Y}}_{(k,\infty)},\ \mathcal{K}_k = \vec{\mathcal{U}}_{(k,\infty)} \text{ and } \mathcal{M}_k = \mathcal{K}_k \cap \ker W_k \quad (k \in \mathbb{Z}) . \tag{5.11}$$

Using $P_k = W_k W_k^*$ it follows that $W_k A_k = \tilde{W}_k$ for all k in $\mathbb{Z}$. Moreover, since $W_k \mid \mathcal{K}_k \ominus \mathcal{M}_k$ is one to one, the operator $A_k = W_k^* P_k^{-1} \tilde{W}_k$ is the only operator from $\mathcal{H}_k$ into $\mathcal{K}_k \ominus \mathcal{M}_k$ satisfying $W_k A_k = \tilde{W}_k$. Clearly, the spaces $\mathcal{H}_k$, $\mathcal{M}_k$ and $\mathcal{K}_k$ satisfy the decreasing property in (1.6). To obtain the compatibility condition (j) in Theorem 1.1a, notice that $\Phi(k, n) = Z_k \Phi(k+1, n)$ and $\mathcal{K}_k \ominus \mathcal{M}_k \subseteq \mathcal{K}_k \ominus \mathcal{M}_{k+1}$ gives

$$W_k(I - P_{\mathcal{M}_k}) A_{k+1} = W_k A_{k+1} = [B_k, Z_k W_{k+1}] A_{k+1} =$$
$$Z_k W_{k+1} A_{k+1} = Z_k \tilde{W}_{k+1} = \tilde{W}_k \mid \mathcal{H}_{k+1} = W_k A_k \mid \mathcal{H}_{k+1} .$$

Because $W_k \mid \mathcal{K}_k \ominus \mathcal{M}_k$ is one to one, the compatibility condition $(I - P_{\mathcal{M}_k}) A_{k+1} = A_k \mid \mathcal{H}_{k+1}$ holds. Therefore all the conditions in Theorem 1.1a hold.

We claim that F is an interpolant for the three chains completion problem with data $\{A_k\}$ in (5.11) if and only if F is an interpolant for the corresponding nonstationary Nevanlinna-Pick problem. If F is an interpolant for $\{A_k\}$, then applying $A_k = (I - P_{\mathcal{M}_k}) F \mid \mathcal{H}_k$ we have $W_k \Pi_k F \mid \mathcal{H}_k = W_k A_k = \tilde{W}_k$ for all k in $\mathbb{Z}$. Since F maps $\mathcal{H}_k$ into $\mathcal{K}_k$ for k in $\mathbb{Z}$, this F is lower triangular. Hence F is an interpolant for the nonstationary Nevanlinna-Pick problem. On the other hand, if F is a nonstationary Nevanlinna-Pick interpolant, then $W_k A_k = \tilde{W}_k = W_k \Pi_k F \mid \mathcal{H}_k$ for all k in $\mathbb{Z}$. Because $W_k \mid \mathcal{K}_k \ominus \mathcal{M}_k$ is one to one, $A_k = (I - P_{\mathcal{M}_k}) F \mid \mathcal{H}_k$ for all k in $\mathbb{Z}$. Since F is a lower triangular matrix, $F\mathcal{H}_k \subseteq \mathcal{K}_k$ and thus F is an interpolant for the three chains completion problem with data $\{A_k\}$. This proves our claim.

By implementing the three chains completion Theorem 1.1a on the data in (5.11), the above analysis shows that there exists a solution F to the nonstationary Nevanlinna-Pick problem with tolerance $\gamma$ if and only if $\gamma \geq d_\infty$ where $d_\infty = \sup \{\|A_k\| : k \in \mathbb{Z}\}$. Moreover, because F is an interpolant for the three chains completion problem with data $\{A_k\}$ if and only if F is a Nevanlinna-Pick interpolant it follows that

$$d_\infty = \sup\{\|A_k\| : k \in \mathbb{Z}\} = \inf\{\|F\| : F \text{ is a nonstationary Nevanlinna–Pick interpolant}\} .$$

Finally, it is noted that by replacing A, W and $\tilde{W}$ respectively by $A_k$, $W_k$ and $\tilde{W}_k$ in Remark II.1.4 we arrive at $\|A_k\|^2 = r_{spec}(\tilde{P}_k P_k^{-1})$. Therefore

$$d_\infty^2 = \sup\{r_{spec}(\tilde{P}_k P_k^{-1}) : k \in \mathbb{Z}\} .$$

Notice $\tilde{P}_k P_k^{-1}$ is similar to $P_k^{-1/2}\tilde{P}_k P_k^{-1/2}$. To see this observe that $P_k^{-1/2}(\tilde{P}_k P^{-1})P_k^{1/2} = P_k^{-1/2}\tilde{P}_k P^{-1/2}$. Hence $d_\infty^2 = \sup\{r_{spec}(P_k^{-1/2}\tilde{P}_k P^{-1/2}) : k \in \mathbb{Z}\}$. So there exists a solution to the nonstationary Nevanlinna-Pick problem with tolerance $\gamma$ if and only if $\gamma^2 P_k \geq \tilde{P}_k$ for all k in $\mathbb{Z}$.

If $F = B_\gamma$ where $\gamma \geq d_\infty$ is the central interpolant F with tolerance $\gamma$ for the three chains completion problem with data $\{A_k\}$ in (5.11), then this F solves the Nevanlinna-Pick problem with tolerance $\gamma$ and *is called the central Nevanlinna-Pick interpolant with tolerance* $\gamma$. Recall that the family of pairs $\{Z_k, B_k\}$ is *uniformly controllable*, if there exists an $\varepsilon > 0$ such that $P_k \geq \varepsilon I$ for all k in $\mathbb{Z}$. (Because $\{B_k\}$ is uniformly bounded and $\{Z_k\}$ satisfy the stability condition in (5.8), there exists a finite $\delta$ such that $\delta I \geq P_k$ for all k in $\mathbb{Z}$.) Now we are ready to use Theorem 5.1a to compute the central nonstationary Nevanlinna-Pick interpolant F with tolerance $\gamma > d_\infty$ for uniformly controllable systems. As expected, this F is precisely the central interpolant with tolerance $\gamma$ computed in Theorem XI.1.3 by the reduction techniques in Chapter XI, and can be viewed as the time varying analogue of the central interpolant for the stationary Nevanlinna-Pick problem computed in Theorem V.1.2.

**THEOREM 5.2.** *Let* $\{Z_k, B_k, \tilde{B}_k\}$ *be the data for the nonstationary standard Nevanlinna-Pick problem with tolerance* $\gamma > d_\infty$, *and assume that the family of pairs* $\{Z_k, B_k\}$ *is uniformly controllable. Let* $C_k$ *from* $\mathcal{X}_k$ *into* $\mathcal{U}_k$ *and* $M_k$ *from* $\mathcal{X}_k$ *into* $\mathcal{X}_{k+1}$ *be the opertors defined by*

$$C_k = \gamma^2 B_k^*(\gamma^2 P_k - \tilde{P}_k + \tilde{B}_k\tilde{B}_k^*)^{-1}$$

$$\text{and } M_k = (\gamma^2 P_{k+1} - \tilde{P}_{k+1})Z_k^*(\gamma^2 P_k - \tilde{P}_k + \tilde{B}_k\tilde{B}_k^*)^{-1} \quad (k \in \mathbb{Z}) . \tag{5.12}$$

*Finally, let* $\Psi(m, n)$ *from* $\mathcal{X}_n$ *into* $\mathcal{X}_m$ *be the state transition operator for* $\{M_k\}$ *defined by*

$$\Psi(m, n) = M_{m-1}M_m \cdots M_{n+1}M_n \quad \text{if } m > n \text{ and } \Psi(n, n) = I . \tag{5.13}$$

*Then the central interpolant* $F = (F_{j,k})_{j,k=-\infty}^{\infty}$ *with tolerance* $\gamma$ *to this nonstationary Nevanlinna-Pick problem is given by*

$$F_{j,k} = 0 \quad (j < k), \quad \textit{and} \quad F_{j,k} = C_j \Psi(j, k)\tilde{B}_k \qquad (j \geq k) \tag{5.14}$$

*Moreover, the state transition operator* $\Psi(m, n)$ *is stable, that is,*

$$\lim_{v \to \infty} \sup \left(\sup \{\|\Psi(k + v, k)\| : k \in \mathbb{Z}\}\right)^{1/v} < 1 . \tag{5.15}$$

**PROOF**. Clearly Theorem 5.2 is equivalent to Theorem XI.1.3. To obtain a proof of this theorem is based on Theorem 5.1a, notice that $A_k = W_k^* P_k^{-1} \tilde{W}_k$ and $P_k = W_k W_k^*$ gives $D_k^2 = \gamma^2 I - \tilde{W}_k^* P_k^{-1} \tilde{W}_k$. By consulting (V.1.12a) to (V.1.14) with A, P, W, $\tilde{P}$ and $\tilde{W}$ replaced by $A_k$, $P_k$, $W_k$, $\tilde{P}_k$ and $\tilde{W}_k$ respectively we obtain

$$D_k^{-2} = \gamma^{-2} I + \gamma^{-2} \tilde{W}_k^* (\gamma^2 P_k - \tilde{P}_k)^{-1} \tilde{W}_k \;\text{ and }\; A_k D_k^{-2} = W_k^* (\gamma^2 P_k - \tilde{P}_k)^{-1} \tilde{W}_k . \tag{5.16}$$

Clearly $\mathcal{L}_k = \mathcal{H}_k \ominus \mathcal{H}_{k+1} = \mathcal{Y}_k$ and $E_k$ is simply the operator embedding $\mathcal{Y}_k$ into $\vec{\mathcal{Y}}_{(k,\infty)} = \mathcal{H}_k$ defined by $E_k = [I, 0, 0, \ldots]^*$. Since $\tilde{W}_k E_k = \tilde{B}_k$ equation (5.16) shows that the linear maps R and L in (5.2) become

$$R = \gamma^{-2} I + \gamma^{-2} [\ldots, \tilde{W}_{-1}^* (\gamma^2 P_{-1} - \tilde{P}_{-1})^{-1} \tilde{B}_{-1}, \tilde{W}_0^* (\gamma^2 P_0 - \tilde{P}_0)^{-1} \tilde{B}_0, \tilde{W}_1 (\gamma^2 P_1 - \tilde{P}_1)^{-1} \tilde{B}_1, \ldots]$$

$$L = [\ldots, W_{-1}^* (\gamma^2 P_{-1} - \tilde{P}_{-1})^{-1} \tilde{B}_{-1}, W_0^* (\gamma^2 P_0 - \tilde{P}_0)^{-1} \tilde{B}_0, W_1 (\gamma^2 P_1 - \tilde{P}_1)^{-1} \tilde{B}_1, \ldots] . \tag{5.17}$$

As in Section XI.1, see (XI.1.1), let Z be the backward weighted shift on $\vec{\mathcal{X}} = \oplus \{\mathcal{X}_k : k \in \mathbb{Z}\}$ defined by placing $\{Z_k\}$ immediately above the main diagonal and zeros elsewhere, that is the (j, k)-entry of Z is given by $Z_{j,k} = Z_{k-1} \delta_{j,k-1}$ where $\delta_{m,n}$ is the Kronecker delta. The stability condition in (5.8) implies that $r_{spec}(Z) = d_\infty < 1$. So $(I - Z^*)^{-1}$ exists. Using $(I - Z^*)^{-1} = I + Z^* + Z^{*2} + \ldots$ we obtain the lower triangular matrix

$$(I - Z^*)^{-1} = \begin{bmatrix} \cdots & \cdot & \cdot & \cdot & \cdot \\ \cdots & \cdot & \cdot & \cdot & \cdot \\ \cdots & \cdot & \cdot & \cdot & \cdot \\ \cdots & \Phi(-1, -1)^* & 0 & 0 & \cdots \\ \cdots & \Phi(-1, 0)^* & \Phi(0, 0)^* & 0 & \cdots \\ \cdots & \Phi(-1, 1)^* & \Phi(0, 1)^* & \Phi(1, 1)^* & \cdots \\ \cdots & \Phi(-1, 2)^* & \Phi(0, 2)^* & \Phi(1, 2)^* & \cdots \\ \vdots & \vdots & \vdots & \vdots & \vdots \end{bmatrix} \tag{5.18}$$

where $\Phi(0, 0)^*$ appears on the (0, 0)-entry. Let B from $\vec{\mathcal{U}}$ into $\vec{\mathcal{X}}$, $\tilde{B}$ from $\vec{\mathcal{Y}}$ into $\vec{\mathcal{X}}$, P on $\vec{\mathcal{X}}$ and $\tilde{P}$ on $\vec{\mathcal{X}}$ be the diagonal operators defined by

$$B = \operatorname{diag}\,(B_k)_{-\infty}^{\infty}\,,\quad \tilde{B} = \operatorname{diag}\,(\tilde{B}_k)_{-\infty}^{\infty}\,,\quad P = \operatorname{diag}\,(P_k)_{-\infty}^{\infty} \text{ and } \tilde{P} = \operatorname{diag}\,(\tilde{P}_k)_{-\infty}^{\infty}\,. \tag{5.19}$$

Because the family of pairs $\{Z_k, B_k\}$ is uniformly controllable P and its inverse exists as a bounded operator. Moreover, $d_\infty^2 = r_{spec}\,(\tilde{P}P^{-1})$. By combining (5.9), (5.17) and (5.19) we see that the operators R and L can be expressed as

$$R = \gamma^{-2}I + \gamma^{-2}\tilde{B}^*(I - Z^*)^{-1}(\gamma^2 P - \tilde{P})^{-1}\tilde{B} \text{ and } L = B^*(I - Z^*)^{-1}(\gamma^2 P - \tilde{P})^{-1}\tilde{B}\,. \tag{5.20}$$

Since $\gamma > d_\infty$ the inverse of $\gamma^2 P - \tilde{P}$ exists and is a strictly positive bounded operator. According to Theorem 5.1a, the central intertwining lifting is given by $B_\gamma = LR^{-1}$. Therefore

$$\begin{aligned}
B_\gamma &= \gamma^2 B^*(I - Z^*)^{-1}(\gamma^2 P - \tilde{P})^{-1}\tilde{B}[I + \tilde{B}^*(I - Z^*)^{-1}(\gamma^2 P - \tilde{P})^{-1}\tilde{B}]^{-1} = \\
&\gamma^2 B^*[I + (I - Z^*)^{-1}(\gamma^2 P - \tilde{P})^{-1}\tilde{B}\tilde{B}^*]^{-1}(I - Z^*)^{-1}(\gamma^2 P - \tilde{P})^{-1}\tilde{B} = \\
&\gamma^2 B^*[(\gamma^2 P - \tilde{P})(I - Z^*) + \tilde{B}\tilde{B}^*]^{-1}\tilde{B} = \\
&\gamma^2 B^*[\gamma^2 P - \tilde{P} + \tilde{B}\tilde{B}^* - (\gamma^2 P - \tilde{P})Z^*]^{-1}\tilde{B} = \\
&\gamma^2 B^*[(I - (\gamma^2 P - \tilde{P})Z^*(\gamma^2 P - \tilde{P} + \tilde{B}\tilde{B}^*)^{-1})(\gamma^2 P - \tilde{P} + \tilde{B}\tilde{B}^*)]^{-1}\tilde{B} = \\
&\gamma^2 B^*(\gamma^2 P - \tilde{P} + \tilde{B}\tilde{B}^*)^{-1}[I - (\gamma^2 P - \tilde{P})Z^*(\gamma^2 P - \tilde{P} + \tilde{B}\tilde{B}^*)^{-1}]^{-1}\tilde{B}\,.
\end{aligned} \tag{5.21}$$

Therefore the central interpolant $F = B_\gamma$ with tolerance $\gamma$ is given by

$$F = B_\gamma = C(I - M)^{-1}\tilde{B} \tag{5.22}$$

where C is the diagonal operator from $\vec{\mathcal{X}}$ into $\vec{\mathcal{U}}$ and M on $\vec{\mathcal{X}}$ is the block forward shift defined by

$$C = \gamma^2 B^*(\gamma^2 P - \tilde{P} + \tilde{B}\tilde{B}^*)^{-1} \text{ and } M = (\gamma^2 P - \tilde{P})Z^*(\gamma^2 P - \tilde{P} + \tilde{B}\tilde{B}^*)^{-1}\,. \tag{5.23}$$

Because P is the controllability grammian for $\{Z, B\}$ and Z is stable Proposition V.1.7 shows that M is stable, that is, $r_{spec}\,(M) < 1$. Hence $(I - M)^{-1}$ exists as a bounded operator. Notice that C is a block diagonal matrix because B, P, $\tilde{P}$ and $\tilde{B}$ are all diagonal matrices. Thus $C = \operatorname{diag}\,(C_k)_{-\infty}^{\infty}$ where $C_k$'s are given in (5.12). The matrix M is (5.23) is a block forward shift because $Z^*$ is a block forward shift and all the other matrices are diagonal matrices. The j, k entry of M is given by

$$M_{j,k} = \begin{cases} (\gamma^2 P_{k+1} - \tilde{P}_{k+1})Z_k^*(\gamma^2 P_k - \tilde{P}_k + \tilde{B}_k\tilde{B}_k^*)^{-1} = M_k & \text{if } j = k+1\,, \\ 0 & \text{if } j \neq k+1\,, \end{cases} \tag{5.24}$$

where $M_k$ is the operator from $\mathcal{X}_k$ into $\mathcal{X}_{k+1}$ defined in (5.12). By computing $I + M + M^2 + \cdots$ it is easy to show that the inverse of $I - M$ is given by

$$(I-M)^{-1} = \begin{bmatrix} \cdots & \cdot & \cdot & \cdot & \cdots \\ \cdots & \cdot & \cdot & \cdot & \cdots \\ \cdots & \cdot & \cdot & \cdot & \cdots \\ \cdots & \Psi(-1,-1) & 0 & 0 & \cdots \\ \cdots & \Psi(0,-1) & \Psi(0,0) & 0 & \cdots \\ \cdots & \Psi(1,-1) & \Psi(1,0) & \Psi(1,1) & \cdots \\ \cdots & \cdot & \cdot & \cdot & \cdots \\ \cdots & \cdot & \cdot & \cdot & \cdots \\ \cdots & \cdot & \cdot & \cdot & \cdots \end{bmatrix}. \tag{5.25}$$

Using this in (5.22) along with $C = \operatorname{diag}(C_k)_{-\infty}^{\infty}$ we see that the j, k entry of $B_\gamma$ is given in (5.14).

Finally, to complete the proof recall that M is stable. Since M is a block forward shift with $\{M_k\}$ on the subdiagonal it follows that $\{\Psi(k+\nu, k)\}$ appear on the $\nu$-th subdiagonal of $M^\nu$ with all the other entries of $M^\nu$ being zero. Therefore $r_{spec}(M)$ is given by the left hand side of (5.15). Because M is stable, (5.15) holds. This completes the proof.

**REMARK 5.3.** As before, let F be the central interpolant with tolerance $\gamma > d_\infty$ for the nonstationary Nevanlinna-Pick problem with data $\{Z_k, B_k, \tilde{B}_k\}$. Furthermore, assume that the family of pairs $\{Z_k, B_k\}$ is uniformly controllability. Now let $N_k$ be the positive operators on $\mathcal{Y}_k$ defined by

$$N_k = (I + \tilde{B}_k^*(\gamma^2 P_k - \tilde{P}_k)^{-1}\tilde{B}_k)^{1/2} \qquad (k \in \mathbb{Z}), \tag{5.26}$$

and let $\Theta$ be the block lower triangular matrix on $\vec{\mathcal{Y}}$ given by

$$\Theta_{j,k} = \begin{cases} \gamma N_j^{-1} & \text{if } j = k, \\ -\gamma N_j \tilde{B}_j^* (\gamma^2 P_j - \tilde{P}_j + \tilde{B}_j \tilde{B}_j^*)^{-1}\Psi(j,k)\tilde{B}_k & \text{if } j > k, \\ 0 & \text{if } j < k. \end{cases} \tag{5.27}$$

Then $\Theta$ is a lower triangular invertible matrix with dense range satisfying $\gamma^2 I - F^*F = \Theta^*\Theta$. In particular, $\|F\| < \gamma$.

To prove this fact we show that $\Theta = \operatorname{diag}[((E_k^* D_k^{-2} E_k)_k^{1/2})_{-\infty}^{\infty}]R^{-1}$ is precisely the lower triangular matrix in (5.27); see Theorem 5.1a. To see this notice that $\tilde{W}_k E_k = \tilde{B}_k$ and (5.16) shows that

$$E_k^* D_k^{-2} E_k = \gamma^{-2} I + \gamma^{-2} \tilde{B}_k^* (\gamma^2 P_k - \tilde{P}_k)^{-1} \tilde{B}_k = \gamma^{-2} N_k^2 \qquad (k \in \mathbb{Z}). \tag{5.28}$$

Using the inversion formula $(I + bc^{-1}d)^{-1} = I - b(db + c)^{-1}d$ on the operator R in (5.20) we obtain $N_k^{-2} = I - \tilde{B}_k^*(\gamma^2 P_k - \tilde{P}_k + \tilde{B}_k \tilde{B}_k^*)^{-1} \tilde{B}_k$ and

$$\begin{aligned} \gamma^{-2} R^{-1} &= (I + \tilde{B}^* (I - Z^*)^{-1} (\gamma^2 P - \tilde{P})^{-1} \tilde{B})^{-1} = \\ &I - \tilde{B}^* (\tilde{B}\tilde{B}^* + (\gamma^2 P - \tilde{P})(I - Z^*))^{-1} \tilde{B} = \\ &I - \tilde{B}^* (\gamma^2 P - \tilde{P} + \tilde{B}\tilde{B}^* - (\gamma^2 P - \tilde{P}) Z^*)^{-1} \tilde{B} = \\ &I - \tilde{B}^* (\gamma^2 P - \tilde{P} + \tilde{B}\tilde{B}^*)^{-1} (I - M)^{-1} \tilde{B} \end{aligned} \tag{5.29}$$

where M is defined in (5.23). This readily implies that $\Theta$ in (5.3) is given by

$$\Theta = \gamma \operatorname{diag} [(N_k)_{-\infty}^{\infty}][I - \tilde{B}^* (\gamma^2 P - \tilde{P} + \tilde{B}\tilde{B}^*)^{-1} (I - M)^{-1} \tilde{B}]. \tag{5.30}$$

By employing the formula for $(I - M)^{-1}$ in (5.25) along with the fact that P, $\tilde{P}$ and $\tilde{B}$ are all diagonal matrices we readily obtain the formula for $\Theta$ in (5.27). According to Theorem 5.1a this $\Theta$ has dense range and $\gamma^2 I - F^* F = \Theta^* \Theta$.

Because the family of pairs $\{Z_k, B_k\}$ is uniformly controllable, P admits a bounded inverse. Since $\gamma > d_\infty$, it follows that $\gamma^2 P - \tilde{P}$ has a bounded inverse. The stability condition on $\{Z_k\}$ in (5.8) implies that $r_{spec}(Z) < 1$. In particular, $(I - Z)$ admits a bounded inverse. By assumption $\{\tilde{B}_k\}$ are uniformly bounded and thus $\tilde{B}$ is a bounded operator. Therefore all the operators in the formula for R in (5.20) are bounded and hence R is a bounded. By consulting the definition of $\Theta$ in (5.3), it follows that $\Theta$ admits a bounded inverse. Because $\gamma^2 I - F^* F = \Theta^* \Theta$ with $\Theta$ boundly invertible $\|F\| < \gamma$. This completes the proof.

## XII.6. THE CARSWELL-SCHUBERT THEOREM

In this section we derive the Carswell-Schubert theorem (see Carswell-Schubert [1] and Page [1]) as a corollary of the three chains completion theorem. A proof of the Carswell-Schubert theorem based on the commutant lifting theorem along with some of its applications is also given in Sections VIII.3 and VIII.6 of Foias-Frazho [4]. First we state the Carswell-Schubert theorem.

**THEOREM 6.1.** *Let* U *and* U′ *be isometries acting on Hilbert spaces* $X$ *and* $X'$, *respectively. Let* $\mathcal{H}$ *be a* U*-invariant subspace of* $X$, *and let* $C : \mathcal{H} \to X'$ *be a linear operator satisfying* $\|C\| \le \gamma$. *Let* $P_n$ *on* $X$ *and* $P_n'$ *on* $X'$ *be the orthogonal projections onto the kernels of*

$U^{*n}$ *and* $U'^{*n}$, *respectively, for all* $n \geq 1$. *Furthermore, assume that* $U'$ *is a unilateral shift, that is,* $U'^{*n} \to 0$ *strongly for* $n \to \infty$. *Then there exists* $F \in \mathcal{L}(\mathcal{X}, \mathcal{X}')$ *such that*

$$F|\mathcal{H} = C, \quad FU = U'F \quad and \quad \|F\| \leq \gamma \tag{6.1}$$

*if and only if the following two conditions are fulfilled:*

(i) $CU|\mathcal{H} = U'C$,

(ii) $\|P'_n Ch\| \leq \gamma \|P_n h\| \qquad$ (*for all* $h \in \mathcal{H}$ *and* $n \geq 1$). $\qquad$ (6.2)

It will be convenient to derive Theorem 6.1 from its dual version which may be stated as follows.

**THEOREM 6.2.** *Let* $V$ *and* $V'$ *be co-isometries acting on Hilbert spaces* $\mathcal{Y}$ *and* $\mathcal{Y}'$, *respectively. Let* $\mathcal{H}$ *be a* $(V')^*$*-invariant subspace of* $\mathcal{Y}'$, *and let* $A : \mathcal{Y} \to \mathcal{Y}'$ *be a linear operator such that* $\operatorname{ran} A \subset \mathcal{H}$ *and* $\|A\| \leq \gamma$. *Let* $P_n$ *on* $\mathcal{Y}$ *and* $P'_n$ *on* $\mathcal{Y}'$ *be the orthogonal projections onto* $\ker V^n$ *and* $\ker V'^n$ *for all* $n \geq 1$, *respectively. Furthermore, assume that* $V^n \to 0$ *strongly for* $n \to \infty$. *Then there exists* $B \in \mathcal{L}(\mathcal{Y}, \mathcal{Y}')$ *such that*

$$P_{\mathcal{H}} B = A, \quad BV = V'B \quad and \quad \|B\| \leq \gamma \tag{6.3}$$

*if and only if the following two conditions are fulfilled:*

(a) $P_{\mathcal{H}} V'A = AV$,

(b) $\|P_n A^* h\| \leq \gamma \|P'_n h\| \qquad$ (*for all* $h \in \mathcal{H}$ *and all* $n \geq 1$). $\qquad$ (6.4)

To derive Theorem 6.1 from Theorem 6.2 we apply the latter theorem with

$$\mathcal{Y} = \mathcal{X}', \quad \mathcal{Y}' = \mathcal{X}, \quad V = U'^*, \quad V' = U^* \quad \text{and} \quad A = C^*,$$

where $C^*$ is viewed as a map from $\mathcal{Y}'$ into $\mathcal{Y}$. With these choices conditions (i) and (ii) in Theorem 6.1 are just equivalent to conditions (a) and (b) in Theorem 6.2. Furthermore, $F \in \mathcal{L}(\mathcal{X}, \mathcal{X}')$ satisfies (6.1) if and only if $B = F^*$ satisfies (6.3). So Theorems 6.1 and 6.2 are equivalent.

**PROOF OF THEOREM 6.2.** We split the proof into three parts. In the first part we prepare the notation and rewrite condition (b) in an equivalent form.

Part ($\alpha$). It will be convenient to define $P_n$ and $P'_n$ to be the zero operators on $\mathcal{Y}$ and $\mathcal{Y}'$, respectively, whenever $n \leq 0$. Since $V$ and $V'$ are co-isometries, we have

$$P_n = I_{\mathcal{Y}} - (V^n)^* V^n \text{ and } P'_n = I_{\mathcal{Y}'} - (V'^n)^* V'^n \quad (n \geq 0). \tag{6.5}$$

Next, we introduce the following auxiliary subspaces. For each $n \in \mathbb{Z}$ we set

$$\mathcal{H}_n = \operatorname{ran} P_n, \quad \mathcal{K}_n = \operatorname{ran} P'_n, \quad \mathcal{M}_n = \mathcal{K}_n \cap (\mathcal{Y}' \ominus \mathcal{H}).$$

Obviously, $\mathcal{H}_n$ and $\mathcal{K}_n$ consist of the zero vector only if $n \leq 0$, and for $n > 0$ we have $\mathcal{H}_n = \ker V^n$ and $\mathcal{K}_n = \ker V'^n$. It follows that the spaces $\mathcal{H}_n$, $\mathcal{K}_n$ and $\mathcal{M}_n$ satisfy the following inclusion relations:

$$\mathcal{H}_{n-1} \subset \mathcal{H}_n, \quad \mathcal{K}_{n-1} \subset \mathcal{K}_n, \quad \mathcal{M}_{n-1} \subset \mathcal{M}_n \quad (n \in \mathbb{Z}). \tag{6.6}$$

Note that $y \in \mathcal{M}_n$ if and only if $y \in \mathcal{K}_n$ and $y$ is orthogonal to $P'_n\mathcal{H}$. Hence

$$\mathcal{K}_n \ominus \mathcal{M}_n = \overline{P'_n\mathcal{H}} \quad (n \in \mathbb{Z}). \tag{6.7}$$

For each $n \in \mathbb{Z}$ define $A_n$ from $\mathcal{H}_n$ to $\mathcal{K}_n \ominus \mathcal{M}_n$ by setting

$$A_n^* P'_n h = P_n A^* h \quad (h \in \mathcal{H}). \tag{6.8}$$

Equality (6.7) implies that $A_n$ is a well-defined bounded linear operator. (In fact, condition (b) implies that $\|A_n\| \leq \gamma$.) We claim that

$$A_{n+1} \mid \mathcal{H}_n = (I - P_{\mathcal{M}_{n+1}})A_n \quad (n \in \mathbb{Z}). \tag{6.9}$$

For $n \leq 0$ the operator $A_n = 0$ and the identity (6.9) is trivial. Let us prove (6.9) for $n \geq 1$. Take a vector $h \in \mathcal{H}$. Then

$$P_n A_{n+1}^* P'_{n+1} h = P_n P_{n+1} A^* h = P_n A^* h = A_n^* P'_n h = A_n^* P'_n P'_{n+1} h,$$

and therefore (using (6.7))

$$P_n A_{n+1}^* \mid (\mathcal{K}_{n+1} \ominus \mathcal{M}_{n+1}) = A_n^* P'_n \mid (\mathcal{K}_{n+1} \ominus \mathcal{M}_{n+1}).$$

By taking adjoints we see that

$$A_{n+1} \mid \mathcal{H}_n = P_{\mathcal{K}_{n+1} \ominus \mathcal{M}_{n+1}} P'_n A_n = (I - P_{\mathcal{M}_{n+1}})A_n,$$

which proves (6.9).

We conclude this part of the proof by showing that condition (b) is equivalent the requirement that

$$\sup \{\|A_n\| : n \in \mathbb{Z}\} \leq \gamma. \tag{6.10}$$

Note that $\|A_n\| = 0$ for all $n \leq 0$. Take $n \geq 1$. Then we can use (6.7) to show that $\|A_n\| \leq \gamma$ is equivalent to

$$\|A_n^* P_n' h\| \le \gamma \|P_n' h\| \quad (h \in \mathcal{H}).$$

By using (6.8) the latter inequality can be rewritten as condition (ii). Thus (6.10) is equivalent to condition (b).

Part (β). In this part we prove the necessity of the conditions (a) and (b). So, let us assume that $B \in \mathcal{L}(\mathcal{Y}, \mathcal{Y}')$ satisfies (6.3). Since $V'^*$ leaves $\mathcal{H}$ invariant, $P_{\mathcal{H}} V' = P_{\mathcal{H}} V' P_{\mathcal{H}}$. But then we can use the first and second equality in (6.3) to show that

$$AV = P_{\mathcal{H}} BV = P_{\mathcal{H}} V'B = P_{\mathcal{H}} V' P_{\mathcal{H}} B = P_{\mathcal{H}} V'A,$$

and hence condition (a) must hold.

Next, we prove that condition (b) is satisfied. To do this we show that B is an interpolant of the operators $A_n : \mathcal{H}_n \to \mathcal{K}_n \ominus \mathcal{M}_n$ for $n \in \mathbb{Z}$, that is, we prove

$$(I - P_{\mathcal{M}_n}) B | \mathcal{H}_n = A_n \qquad (n \in \mathbb{Z}), \tag{6.11}$$

$$B\mathcal{H}_n \subset \mathcal{K}_n \qquad (n \in \mathbb{Z}). \tag{6.12}$$

Since $\|B\| \le \gamma$, by (6.3), formula (6.11) implies that (6.10) holds, or equivalently, by the final result of Part (α), condition (b) is fulfilled.

We need (6.12) to prove (6.11). So let us first prove (6.12). For $n \le 0$ the inclusion (6.12) is trivial. Take $n \ge 1$. From the second identity in (6.3) we see that $V'^n B = BV^n$. Since $\operatorname{ran} P_n = \ker V^n$ we have $V'^n B P_n = 0$. So $B\mathcal{H}_n \subset \ker V'^n = \mathcal{K}_n$, which proves (6.12). We also see that

$$P_n' B P_n = B P_n \qquad (n \in \mathbb{Z}). \tag{6.13}$$

Next, we prove (6.11). It suffices to consider the case when $n \ge 1$. So fix $n \ge 1$. Take a vector $h \in \mathcal{H}$. Then, using (6.13), $P_{\mathcal{H}} B = A$ and (6.8), we have

$$P_n B^* P_n' h = P_n B^* h = P_n B^* P_{\mathcal{H}} h = P_n A^* h = A_n^* P_n' h.$$

According to (6.7), the latter implies that $P_n B^* | (\mathcal{K}_n \ominus \mathcal{M}_n) = A_n^*$, which yields (6.11) by taking adjoints.

Part (γ). In this part we assume that conditions (a) and (b) are fulfilled, and we prove that there exists $B \in \mathcal{L}(\mathcal{Y}, \mathcal{Y}')$ satisfying (6.3). To do this we apply the three chains completion theorem and Corollary 1.3 with $T = V^*$ and $R = V'^*$. First, note that (6.5) readily yields

$$V^* P_n = P_{n+1} V^* \quad \text{and} \quad V'^* P_n' = P_{n+1}' V'^* \qquad (n \ge 0). \tag{6.14}$$

It follows that $V^* \mathcal{H}_n \subset \mathcal{H}_{n+1}$ and by taking adjoints on the second equation $V'\mathcal{K}_{n+1} \subset \mathcal{K}_n$ for each $n \in \mathbb{Z}$. So the first two conditions in (1.8) hold. Next, take $h \in \mathcal{H}$. Since $\mathcal{H}$ is invariant under $V'^*$ we have

$$V'^* P'_n h = P'_{n+1} V'^* h \in P'_{n+1} \mathcal{H} .$$

So $V'^*(P'_n \mathcal{H}) \subset P'_{n+1} \mathcal{H}$. But then we can use (6.7) to conclude that for all $n \in \mathbb{Z}$

$$V'^*(\mathcal{K}_n \ominus \mathcal{M}_n) \subset \mathcal{K}_{n+1} \ominus \mathcal{M}_{n+1} .$$

Hence the last condition in (1.8) holds for $R = V'^*$. We proceed by showing that

$$(I - P_{\mathcal{M}_n}) V' A_{n+1} V^* \mid \mathcal{H}_n = A_n \quad (n \in \mathbb{Z}) . \tag{6.15}$$

Put $R_n = (I - P_{\mathcal{M}_n}) V' A_{n+1} V^* P_n$. Take $x = P'_n h$ for some $h \in \mathcal{H}$. Then $(I - P_{\mathcal{M}_n}) x = x$ by (6.7). By using (6.8) and (6.14), we see that

$$R_n^* P'_n h = P_n V A_{n+1}^* V'^* (I - P_{\mathcal{M}_n}) P'_n h = P_n V A_{n+1}^* V'^* P'_n h$$

$$= P_n V A_{n+1}^* P'_{n+1} V'^* h = P_n V P_{n+1} A^* V'^* h ,$$

because $V'^* h$ is in $\mathcal{H}$. Since condition (a) is fulfilled, we have $A^* V'^* h = V^* A^* h$. So, again by (6.8) and (6.14), we obtain

$$R_n^* P'_n h = P_n V P_{n+1} V^* A^* h = P_n V V^* P_n A^* h = P_n A^* h = A_n^* P'_n h .$$

Since $P'_n \mathcal{H}$ is dense in $\mathcal{K}_n \ominus \mathcal{M}_n$ by (6.7), we conclude that $R_n = A_n P_n$, and hence (6.15) holds.

From (6.9) and (6.10) we see that we may apply the three chains interpolation theorem to show the existence of an optimal interpolant of the operators $A_n : \mathcal{H}_n \to \mathcal{K}_n \ominus \mathcal{M}_n$ for $n \in \mathbb{Z}$. In other words there exists $B \in \mathcal{L}(\mathcal{Y}, \mathcal{Y}')$ such that (6.11) and (6.12) hold and $\|B\| \le \gamma$. This together with the result (6.15) of the previous paragraph shows that the conditions of Corollary 1.3 are satisfied (where $T = V^*$ and $R^* = V'$). So we may assume that our interpolant B has the additional property

$$B = V' B V^* . \tag{6.16}$$

Note that $B\mathcal{H}_1 \subset \mathcal{K}_1$, because B satisfies (6.12). Since $\mathcal{K}_1 = \operatorname{Ker} V'$, we see that (6.16) yields $BV = V'BV^*V = V'B - V'BP_1 = V'B$. So we have shown that B satisfies the second and third part of (6.3). It remains to check the first equality in (6.3).

Fix $n \ge 1$, and take $h \in \mathcal{H}$. From (6.12) it follows that $BP_n = P'_n B P_n$, and hence $P_n B^* = P_n B^* P'_n$. Since $P'_n h \in \mathcal{K}_n \ominus \mathcal{M}_n$, we can use (6.11) and (6.8) to show that

$$P_nB^*P'_nh = A_n^*P'_nh = P_nA^*h .$$

We conclude that $P_nB^* \mid \mathcal{H} = P_nA^* \mid \mathcal{H}$, and hence, by duality, $P_{\mathcal{H}}BP_n = AP_n$. The latter identity holds for each $n \geq 1$. Now, use the fact that $V^n \to 0$ strongly for $n \to \infty$. It follows that $P_n \to I$ strongly for $n \to \infty$, and therefore

$$Ay = \lim_{n \to \infty} AP_ny = \lim_{n \to \infty} P_{\mathcal{H}}BP_ny = P_{\mathcal{H}}By ,$$

for each $y \in \mathcal{Y}$. Thus $P_{\mathcal{H}}B = A$. This completes the proof.

Theorem 6.1 yields the following corollary for lower triangular block Toeplitz operators.

**COROLLARY 6.3.** *Assume the Hilbert space operators $\Phi$ from $\mathcal{Y}$ to $\ell_+^2(\mathcal{E})$ and $\Phi'$ from $\mathcal{Y}$ to $\ell_+^2(\mathcal{E}')$ satisfy the intertwining relations*

$$\Phi\Lambda = S\Phi \quad \textit{and} \quad \Phi'\Lambda = S'\Phi' , \tag{6.17}$$

*where $S$ and $S'$ are the block forward shifts on $\ell_+^2(\mathcal{E})$ and $\ell_+^2(\mathcal{E}')$, respectively, and $\Lambda \in \mathcal{L}(\mathcal{Y}, \mathcal{Y})$. Then there exists $F \in H^\infty(\mathcal{E}, \mathcal{E}')$ such that $T_F\Phi = \Phi'$ and $\|F\|_\infty \leq \gamma$ if and only if*

$$\|P'_n\Phi'y\| \leq \gamma\|P_n\Phi y\| \qquad (y \in \mathcal{Y} \;\; \textit{and} \;\; n \geq 1) . \tag{6.18}$$

*Here $P_n$ and $P'_n$ are the projections on $\ell_+^2(\mathcal{E})$ and $\ell_+^2(\mathcal{E}')$, respectively, which assign to the sequence $[x_1, x_2, \ldots]^{tr}$ the sequence $[x_1, \ldots, x_n, 0, 0, \ldots]^{tr}$.*

**PROOF.** First we prove the necessity of the conditions (6.18). So, assume there exists an $F \in H^\infty(\mathcal{E}, \mathcal{E}')$ such that $T_F\Phi = \Phi'$ and $\|F\|_\infty \leq \gamma$. Then the block Toeplitz operator $T_F$ is lower triangular and $\|T_F\| \leq \gamma$. The first property implies that $P'_nT_F = P'_nT_FP_n$, and from the second it follows that $P'_nT_F$ is bounded in norm by $\gamma$. Thus

$$\|P'_n\Phi'y\| = \|P'_nT_FP_n\Phi y\| \leq \gamma\|P_n\Phi y\| \qquad (y \in \mathcal{Y}) .$$

Here n is an arbitrary positive integer. So (6.18) is proved.

To prove the converse, assume (6.18) holds true. Since $P_n \to I$ and $P'_n \to I$ strongly if $n \to \infty$, we see from (6.18) that $\|\Phi'y\| \leq \gamma\|\Phi y\|$ for $y \in \mathcal{Y}$. Hence there exists an operator $C$ from $\mathcal{H} := \overline{\Phi\mathcal{Y}}$ into $\ell_+^2(\mathcal{E}')$ such that $C\Phi = \Phi'$ and $\|C\| \leq \gamma$. Now, apply Theorem 6.1 with $U = S$ and $U' = S'$. In this case, condition (6.2) is equivalent to (6.18). The first identity in (6.17) implies that $\mathcal{H}$ is invariant under $S$. Together the two identities in (6.17) yield

$$CS(\Phi y) = C\Phi\Lambda y = \Phi'\Lambda y = S'\Phi' y = S'C(\Phi y)\,,$$

where y is an arbitrary vector in $\mathcal{Y}$. It follows that $CS\,|\,\mathcal{H} = S'C$, and we can apply Theorem 6.1 to show that there exists an operator $T : \ell^2_+(\mathcal{E}) \to \ell^2_+(\mathcal{E}')$ such that $\|T\| \le \gamma$ and $T\,|\,\mathcal{H} = C$ and $TS = S'T$. The latter identity implies that $T = T_F$ for some $F \in H^\infty(\mathcal{E}, \mathcal{E}')$. The fact that $\|T\| \le \gamma$ implies that $\|F\|_\infty \le \gamma$. Finally, $T\,|\,\mathcal{H} = C$ yields $T_F\Phi = \Phi'$. This completes the proof.

## *Notes to Chapter XII:*

The material of the first three sections is taken from Foias-Frazho-Gohberg-Kaashoek [2]. The three chains completion theorem in Section 1 easily extends to separable nests. It contains as a particular case the commutant lifting theorems by Constantinescu [1] and Anousis-Katsoulis-Moore-Trent [1]. It is also related to the time-variant version of the commutant lifting theorem for finite triangular matrices in Ball-Gohberg [1], and to the nest algebra commutant lifting theorem due to Power-Poulsen-Ward [1]. In fact, for separable nests the two theorems directly imply one another, see Ball [1] for a further discussion of this topic. Section 4 is taken from Foias-Frazho-Gohberg-Kaashoek [4]. The quotient formulas in Theorems 5.1 and 5.1a are time-variant versions of analogous formulas in Foias-Frazho-Li [1] and they are related to the formulas for the central solution in Gohberg-Kaashoek-Woerdeman [2]. Theorem 6.1 is due Carswell-Schubert [1], the proof given here is new.

CHAPTER XIII

# APPLICATIONS OF THE THREE CHAINS COMPLETION THEOREM TO INTERPOLATION

In this chapter the three chains completion theorem is used to give a direct proof of the nonstationary interpolation theorems stated in Chapter VIII. It is also shown that the three chains completion problem is equivalent to a nonstationary four block problem, and the connections between these two problems is discussed.

## XIII.1. ABSTRACT NONSTATIONARY INTERPOLATION

In this section we treat an abstract norm constrained interpolation problem which contains as a special case the nonstationary left tangential Nevanlinna-Pick problem. First, let us describe the given data. Consider the Hilbert space direct sums

$$\vec{\mathcal{Y}} = \bigoplus_{-\infty}^{\infty} \mathcal{Y}_j, \quad \vec{\mathcal{U}} = \bigoplus_{-\infty}^{\infty} \mathcal{U}_j, \quad \vec{\mathcal{X}} = \bigoplus_{-\infty}^{\infty} \mathcal{X}_j, \tag{1.1}$$

and the subspaces

$$\mathcal{H}_k = \{\vec{y} \in \vec{\mathcal{Y}} : y_j = 0 \text{ for } j < k\} \text{ and } \mathcal{K}_k = \{\vec{u} \in \vec{\mathcal{U}} : u_j = 0 \text{ for } j < k\}. \tag{1.2}$$

The given data are bounded linear operators

$$R_k : \vec{\mathcal{U}} \to \mathcal{X}_k, \tilde{R}_k : \vec{\mathcal{Y}} \to \mathcal{X}_k \text{ and } Z_k : \mathcal{X}_{k+1} \to \mathcal{X}_k \quad (k \in \mathbb{Z}) \tag{1.3}$$

which satisfy for each $k \in \mathbb{Z}$ the following identities

$$R_k = R_k P_{\mathcal{K}_k}, \quad Z_k R_{k+1} = R_k P_{\mathcal{K}_{k+1}}; \tag{1.4}$$

$$\tilde{R}_k = \tilde{R}_k P_{\mathcal{H}_k}, \quad Z_k \tilde{R}_{k+1} = \tilde{R}_k P_{\mathcal{H}_{k+1}}. \tag{1.5}$$

The problem is to find a block lower triangular operator

$$F = \begin{bmatrix} \ddots & & & & \\ \cdots & f_{-1,-1} & & & \\ \cdots & f_{0,-1} & \underline{f_{0,0}} & & \\ \cdots & f_{1,-1} & f_{1,0} & f_{1,1} & \\ \vdots & \vdots & \vdots & \vdots & \ddots \end{bmatrix} : \vec{\mathcal{Y}} \to \vec{\mathcal{U}} \tag{1.6}$$

such that

(a) $R_k F P_{\mathcal{H}_k} = \tilde{R}_k$ for $k \in \mathbb{Z}$;

(b) $\|F\| \le \gamma$.

The next theorem is the nonstationary analogue of Theorem II.1.2.

**THEOREM 1.1.** *Let* $R_k$ *and* $\tilde{R}_k$ *for* $k \in \mathbb{Z}$ *be a set of operators of the form (1.3) satisfying (1.4) and (1.5). Then there exists a block lower triangular operator* F *of the form (1.6) such that* (a) *and* (b) *holds if and only if*

$$\gamma^2 R_k R_k^* - \tilde{R}_k \tilde{R}_k^* \ge 0 \qquad (k \in \mathbb{Z}). \tag{1.7}$$

**PROOF.** To prove the necessity of (1.7), assume that F is a block lower triangular operator satisfying the interpolation conditions (a) and (b). Then (a) yields

$$\tilde{R}_k \tilde{R}_k^* = R_k F P_{\mathcal{H}_k} P_{\mathcal{H}_k}^* F^* R_k \qquad (k \in \mathbb{Z}). \tag{1.8}$$

Since $P_{\mathcal{H}_k}$ is an orthogonal projection, $P_{\mathcal{H}_k} P_{\mathcal{H}_k}^* \le I_{\vec{\mathcal{Y}}}$. Furthermore, because of condition (b), we have $FF^* \le \gamma^2 I$. Inserting these inequalities into the right-hand side of (1.8) we obtain (1.7).

To prove the sufficiency, assume (1.7) is satisfied. We shall apply the three chains completion theorem to show that there exists a block lower triangular F satisfying (a), (b). Put

$$\mathcal{M}_k = \{\vec{u} \in \mathcal{K}_k : R_k \vec{u} = 0\}.$$

We have

$$\mathcal{H}_{k+1} \subset \mathcal{H}_k, \ \mathcal{K}_{k+1} \subset \mathcal{K}_k, \ \mathcal{M}_{k+1} \subset \mathcal{M}_k \qquad (k \in \mathbb{Z}). \tag{1.9}$$

The first two inclusions in (1.9) are obvious. To prove the third, take $\vec{u} \in \mathcal{M}_{k+1}$. Then $\vec{u} \in \mathcal{K}_{k+1} \subset \mathcal{K}_k$ and $R_{k+1}\vec{u} = 0$. Now, use the second identity in (1.4). It follows that

$$R_k \vec{u} = R_k P_{\mathcal{K}_{k+1}} \vec{u} = Z_k R_{k+1} \vec{u} = 0.$$

Thus $\vec{u} \in \mathcal{K}_k$ and $R_k\vec{u} = 0$. In other words $\vec{u} \in \mathcal{M}_k$.

Since (1.7) is fulfilled, for each $k \in \mathbb{Z}$ there exists a unique operator $A_k$ from $\mathcal{H}_k$ to $\mathcal{K}_k \ominus \mathcal{M}_k$ such that

$$\tilde{R}_k \mid \mathcal{H}_k = R_k A_k \quad \text{and} \quad \|A_k\| \le \gamma \qquad (k \in \mathbb{Z}). \tag{1.10}$$

Indeed, for each $k \in \mathbb{Z}$ there exists an operator $\Lambda_k$ from the closure of ran $R_k^*$ to the closure of ran $\tilde{R}_k^*$ such that $\Lambda_k R_k^* x = \tilde{R}_k^* x$ for each $x \in \mathcal{X}_k$ and $\|\Lambda_k\| \le \gamma$. According to the first equation in (1.4) we have $\ker R_k \supset \mathcal{K}_k^{\perp}$. It follows that $\ker R_k = \mathcal{K}_k^{\perp} \oplus \mathcal{M}_k$, and hence $\overline{\operatorname{ran} R_k^*} = \mathcal{K}_k \ominus \mathcal{M}_k$. The first identity in (1.5) shows that $\ker \tilde{R}_k \supset \mathcal{H}_k^{\perp}$, and hence the closure of ran $\tilde{R}_k^*$ is contained in $\mathcal{H}_k$. Thus without loss of generality we may view $\Lambda_k$ as an operator from $\mathcal{K}_k \ominus \mathcal{M}_k$ into $\mathcal{H}_k$. From the definition of $\Lambda_k$ we see that

$$\Lambda_k P_{\mathcal{K}_k \ominus \mathcal{M}_k} R_k^* = P_{\mathcal{H}_k} \tilde{R}_k^* \qquad (k \in \mathbb{Z}). \tag{1.11}$$

Here $P_{\mathcal{K}_k \ominus \mathcal{M}_k}$ from $\vec{\mathcal{U}}$ to $\mathcal{K}_k \ominus \mathcal{M}_k$ is the orthogonal projection of $\vec{\mathcal{U}}$ onto $\mathcal{K}_k \ominus \mathcal{M}_k$, and $P_{\mathcal{H}_k}$ from $\vec{\mathcal{Y}}$ to $\mathcal{H}_k$ is the orthogonal projection of $\vec{\mathcal{Y}}$ onto $\mathcal{H}_k$. Now, define $A_k$ from $\mathcal{H}_k$ into $\mathcal{K}_k \ominus \mathcal{M}_k$ to be the adjoint of $\Lambda_k$. Then $\|A_k\| = \|\Lambda_k^*\| = \|\Lambda_k\| \le \gamma$, and by taking adjoints in (1.11) we obtain the identities in the first part of (1.10). Since $\ker R_k = (\mathcal{K}_k \ominus \mathcal{M}_k)^{\perp}$, it is clear that (1.10) determines $A_k$ uniquely.

By using the identities in (1.4) and (1.5) we see that for each $x \in \mathcal{X}_k$ we have

$$P_{\mathcal{H}_{k+1}} \Lambda_k R_k^* x = P_{\mathcal{H}_{k+1}} \tilde{R}_k^* x = \tilde{R}_{k+1}^* Z_k^* x = \Lambda_{k+1} R_{k+1}^* Z_k^* x = \Lambda_{k+1} P_{\mathcal{K}_{k+1}} R_k^* x ,$$

and therefore

$$(P_{\mathcal{H}_{k+1}} \mid \mathcal{H}_k)\Lambda_k = \Lambda_{k+1}(P_{\mathcal{K}_{k+1} \ominus \mathcal{M}_{k+1}} \mid \mathcal{K}_k \ominus \mathcal{M}_k). \tag{1.12}$$

By taking adjoints in (1.12), we obtain

$$A_k \mid \mathcal{H}_{k+1} = P_{\mathcal{K}_k \ominus \mathcal{M}_k} A_{k+1} \qquad (k \in \mathbb{Z}). \tag{1.13}$$

Since $\|A_k\| \le \gamma$ for each k, formula (1.13) allows us to apply the second version of the three chains completion theorem (Theorem XII.1.1a), with $\vec{\mathcal{Y}}$ in place of $\mathcal{U}$ and $\vec{\mathcal{U}}$ in place of $\mathcal{Y}$. So there exists $F \in \mathcal{L}(\vec{\mathcal{Y}}, \vec{\mathcal{U}})$ such that

$$F\mathcal{H}_k \subset \mathcal{K}_k, \ (I - P_{\mathcal{M}_k})F \mid \mathcal{H}_k = A_k \ \text{ and } \ \|F\| \le \gamma. \tag{1.14}$$

From the first inclusion in (1.14) it follows that F has the block lower triangular form (1.6). According to the second part of (1.14), the operator F satisfies condition (a). Indeed, since

$\ker R_k = (\mathcal{K}_k \ominus \mathcal{M}_k)^\perp$, we can use the first part of (1.10) to show that

$$R_k F | \mathcal{H}_k = R_k P_{\mathcal{K}_k \ominus \mathcal{M}_k} F | \mathcal{H}_k = R_k A_k = \tilde{R}_k | \mathcal{H}_k .$$

Thus F satisfies the conditions (a) and (b). This completes the proof.

The proof of Theorem 1.1 also yields the following corollary.

**COROLLARY 1.2.** *Let* $R_k$ *and* $\tilde{R}_k$ *for* $k \in \mathbb{Z}$ *be a set of operators of the form (1.3) satisfying (1.4) and (1.5). Put*

$$\mathcal{M}_k = \{\vec{u} \in \mathcal{K}_k : R_k \vec{u} = 0\} \qquad (k \in \mathbb{Z}) . \tag{1.15}$$

*Then there exists a block lower triangular* $F \in \mathcal{L}(\vec{\mathcal{Y}}, \vec{\mathcal{U}})$ *such that*

$$R_k F P_{\mathcal{H}_k} = \tilde{R}_k \qquad (k \in \mathbb{Z}) . \tag{1.16}$$

*if and only if there exist linear operators*

$$A_k : \mathcal{H}_k \to \mathcal{K}_k \ominus \mathcal{M}_k \qquad (k \in \mathbb{Z}) \tag{1.17}$$

*such that*

$$R_k A_k = \tilde{R}_k | \mathcal{H}_k \qquad (k \in \mathbb{Z}) \textit{ and } \sup \{\|A_k\| : k \in \mathbb{Z}\} < \infty . \tag{1.18}$$

*In this case a block lower triangular operator* $F \in \mathcal{L}(\vec{\mathcal{Y}}, \vec{\mathcal{U}})$ *satisfies (1.16) if and only if* F *is in interpolant of the operators (1.17), that is,* F *is a solution of the three chains completion problem associated with the spaces* $\mathcal{H}_k$, $\mathcal{K}_k$, $\mathcal{M}_k$ *and the operators* $A_k$ *for* $k \in \mathbb{Z}$.

To prove Corollary 1.2 it only remains to show that a block lower triangular operator $F \in \mathcal{L}(\vec{\mathcal{Y}}, \vec{\mathcal{U}})$ satisfies (1.16) implies that F is an interpolant of the operators $A_k$ for $k \in \mathbb{Z}$. In other words, we have to show that for such an operator F we have,

$$F\mathcal{H}_k \subset \mathcal{K}_k, \quad (I - P_{\mathcal{M}_k}) F | \mathcal{H}_k = A_k \quad (k \in \mathbb{Z}) . \tag{1.19}$$

Since F is block lower triangular, the first part of (1.19) is true. To prove the second part, notice that because of (1.16) and $\ker R_k = (\mathcal{K}_k \ominus \mathcal{M}_k)^\perp$ we have

$$\tilde{R}_k | \mathcal{H}_k = R_k \{(I - P_{\mathcal{M}_k}) F | \mathcal{H}_k\} \qquad (k \in \mathbb{Z}) .$$

The latter can only happen (because of the uniqueness of the operators $A_k$) when the second part of (1.19) holds.

## XIII.2. APPLICATION TO NEVANLINNA-PICK INTERPOLATION

In this section we use Theorem 1.1 to give a second proof of the standard nonstationary left tangential Nevanlinna-Pick interpolation theorem (i.e., Theorem VIII.2.2 or Theorem VIII.2.1 with $N = 1$). To this end let us recall that the data for the standard nonstationary Nevanlinna-Pick problem are operators $Z_k$ from $\mathcal{X}_{k+1}$ to $\mathcal{X}_k$ and $B_k$ from $\mathcal{U}_k$ to $\mathcal{X}_k$ and $\tilde{B}_k$ from $\mathcal{Y}_k$ to $\mathcal{X}_k$ for all $k \in \mathbb{Z}$. These operators are assumed to satisfy the following two conditions

$$\sup\{\|Z_k\|, \|B_k\|, \|\tilde{B}_k\| : k \in \mathbb{Z}\} < \infty,$$
$$\limsup_{\nu\to\infty} \left\{\sup_{k\in\mathbb{Z}} \|\Phi(k, k+\nu)\|\right\}^{1/\nu} < 1. \tag{2.1}$$

Here $\Phi(m, n)$ is the state transition operator associated with the operators $\{Z_k\}$, that is, $\Phi(m, n) = Z_m Z_{m+1} \cdots Z_{n-1}$ for $m < n$ and $\Phi(n, n)$ is the identity operator on $\mathcal{X}_n$. The standard nonstationary Nevanlinna-Pick problem associated with the data $\{Z_k, B_k, \tilde{B}_k\}$ is to find operators $f_{j,k}$ in $\mathcal{L}(\mathcal{Y}_k, \mathcal{U}_j)$ for $-\infty < k \le j < \infty$ satisfying the interpolation conditions

$$\sum_{\nu=0}^{\infty} \Phi(k, k+\nu) B_{k+\nu} f_{k+\nu,k} = \tilde{B}_k \quad (\text{for } k \in \mathbb{Z}), \tag{2.2}$$

and such that the block lower triangular operator F from $\vec{\mathcal{Y}}$ to $\vec{\mathcal{U}}$ generated by $\{f_{j,k}\}$ is bounded in norm by $\gamma$.

To treat the standard nonstationary Nevanlinna-Pick problem by using Theorem 1.1, let us introduce operators $R_k$ from $\vec{\mathcal{U}}$ to $\mathcal{X}_k$ and $\tilde{R}_k$ from $\vec{\mathcal{Y}}$ to $\mathcal{X}_k$ by setting

$$R_k \vec{u} = \sum_{\nu=k}^{\infty} \Phi(k, k+\nu) B_{k+\nu} u_{k+\nu} \quad (k \in \mathbb{Z}), \tag{2.3}$$

$$\tilde{R}_k \vec{y} = \sum_{\nu=k}^{\infty} \Phi(k, k+\nu) \tilde{B}_{k+\nu} y_{k+\nu} \quad (k \in \mathbb{Z}). \tag{2.4}$$

Here $\vec{u} = (u_n)_{-\infty}^{\infty}$ is an arbitrary vector in $\vec{\mathcal{U}}$ and $\vec{y} = (y_n)_{-\infty}^{\infty}$ is an arbitrary vector in $\vec{\mathcal{Y}}$. As in the previous section, put

$$\mathcal{H}_k = \{\vec{y} \in \vec{\mathcal{Y}} : y_j = 0 \text{ for } j < k\} \quad \text{and} \quad \mathcal{K}_k = \{\vec{u} \in \vec{\mathcal{U}} : u_j = 0 \text{ for } j < k\}. \tag{2.5}$$

Obviously,

$$R_k P_{\mathcal{K}_k} = R_k \text{ and } \tilde{R}_k P_{\mathcal{H}_k} = \tilde{R}_k \quad (k \in \mathbb{Z}). \tag{2.6}$$

Since $\Phi(k, k+\nu+1) = Z_k \Phi(k+1, k+\nu+1)$ for each $\nu \geq 0$, it is straightforward to check that

$$R_k P_{\mathcal{K}_{k+1}} = Z_k R_{k+1} \text{ and } \tilde{R}_k P_{\mathcal{H}_{k+1}} = Z_k \tilde{R}_{k+1} \quad (k \in \mathbb{Z}). \tag{2.7}$$

Thus the identities (1.4) an (1.5) hold in the present setting.

Next, notice that the operator $R_k$ may be identified with the controllability operator $W_k$ at time k associated with the family of pairs $\{Z_n, B_n\}_{n \in \mathbb{Z}}$. More precisely, $R_k = W_k P_{\mathcal{K}_k}$. Analogously, $\tilde{R}_k = \tilde{W}_k P_{\mathcal{H}_k}$, where $\tilde{W}_k$ is the controllability operator at time k associated with the family of pairs $\{Z_n, \tilde{B}_n\}_{n \in \mathbb{Z}}$. It follows, using (VIII.2.12), that the interpolation conditions in (2.2) are equivalent to

$$R_k F P_{\mathcal{H}_k} = \tilde{R}_k \quad (k \in \mathbb{Z}). \tag{2.7}$$

Furthermore, we see that $\gamma^2 R_k R_k^* - \tilde{R}_k \tilde{R}_k^* \geq 0$ if and only if $\gamma^2 P_k - \tilde{P}_k \geq 0$, where $P_k = W_k W_k^*$ and $\tilde{P}_k = \tilde{W}_k \tilde{W}_k^*$. We conclude that Theorem VIII.2.2 is a direct corollary of Theorem 1.1.

Associated with the standard nonstationary Nevanlinna-Pick interpolation problem is the following optimization problem

$$d_\infty = \inf \{\|F\| : F = (f_{j,k})_{j,k=-\infty}^{\infty} \in \mathcal{L}(\vec{\mathcal{Y}}, \vec{\mathcal{U}}) \text{ is block lower triangular and (2.2) holds}\}. \tag{2.8}$$

The previous analysis and Corollary 1.2 readily yield the following result.

**THEOREM 2.1.** *Let* $Z_k$, $B_k$ *and* $\tilde{B}_k$, $k \in \mathbb{Z}$, *be the data for the standard nonstationary Nevanlinna-Pick problem, and let* $W_k$ *and* $\tilde{W}_k$ *be the controllability operators at time* k *associated with the families of pairs* $\{Z_n, B_n\}_{n \in \mathbb{Z}}$ *and* $\{Z_n, \tilde{B}_n\}_{n \in \mathbb{Z}}$, *respectively. Then* $d_\infty$ *in (2.8) is finite if and only if there exist linear operators*

$$A_k : \mathcal{H}_k \to \mathcal{K}_k \ominus \mathcal{M}_k \quad (k \in \mathbb{Z}) \tag{2.9}$$

*such that* $\sup \{\|A_k\| : k \in \mathbb{Z}\} < \infty$. *Here* $\mathcal{H}_k$ *and* $\mathcal{K}_k$ *are defined by (2.5) and*

$$\mathcal{M}_k = \{\vec{u} \in \mathcal{K}_k : W_k P_{\mathcal{K}_k} \vec{u} = 0\} \quad (k \in \mathbb{Z}).$$

*Moreover, in this case*

$$d_\infty = \sup \{\|A_k\| : k \in \mathbb{Z}\}, \tag{2.10}$$

*and a block lower triangular operator* $F = (f_{j,k})_{j,k=-\infty}^{\infty} \in \mathcal{L}(\vec{\mathcal{Y}}, \vec{\mathcal{U}})$ *satisfies the interpolation conditions (2.2) if and only if* F *is an interpolant of the operators* $A_k$ *for* $k \in \mathbb{Z}$.

Finally, let us notice that if $P_k = W_k W_k^*$ is strictly positive for all k, then an explicit formula for $A_k$ is given by $A_k = W_k^* P_k^{-1} \tilde{W}_k$. In particular, from Corollary XI.1.2, it readily follows that $\|A_k\|^2 = r_{spec}(\tilde{P}_k P_k^{-1})$, where $\tilde{P}_k = \tilde{W}_k \tilde{W}_k^*$. So in this case

$$d_\infty^2 = \sup\{r_{spec}(\tilde{P}_k P_k^{-1}) : k \in \mathbb{Z}\}. \tag{2.11}$$

## XIII.3. APPLICATION TO THE NEHARI PROBLEM

In this section we use the three chains completion theorem (Theorem XII.1.1a) to give a second proof of the main part of Theorem VIII.4.1, which guarantees the existence of an optimal solution of the nonstationary Nehari extension problem. Therefore, in what follows

$$f_{j,k} : \mathcal{U}_k \to \mathcal{Y}_j \qquad (-\infty < j < k < \infty) \tag{3.1}$$

are bounded linear operators acting between Hilbert spaces, and we assume that

$$d_\infty := \sup\{\|\Gamma_k\| : k \in \mathbb{Z}\} < \infty, \tag{3.2}$$

where $\Gamma_k$ is the block operator defined by (VIII.4.5). The problem is to construct an interpolant G of the operators in (3.1) such that $\|G\| = d_\infty$.

Let $\vec{\mathcal{U}}$ and $\vec{\mathcal{Y}}$ be the Hilbert space direct sums $\oplus\{\mathcal{U}_k : k \in \mathbb{Z}\}$ and $\oplus\{\mathcal{Y}_k : k \in \mathbb{Z}\}$, respectively. For each $k \in \mathbb{Z}$ introduce the subspaces $\mathcal{K}_k = \mathcal{Y}$,

$$\mathcal{H}_k = \{\vec{u} \in \mathcal{U} : u_j = 0 \text{ for } j < k\} \quad \text{and} \quad \mathcal{M}_k = \{\vec{y} \in \mathcal{Y} : y_j = 0 \text{ for } j < k\}. \tag{3.3}$$

Obviously these subspaces satisfy the inclusion conditions in (XII.1.6). Next, let $A_k$ be the operator from $\mathcal{H}_k$ to $\mathcal{K}_k \ominus \mathcal{M}_k$ obtained by setting $A_k = H_k$, where $H_k$ is defined in (VIII.4.4). Here we identify $\mathcal{H}_k$ with the space $\vec{\mathcal{U}}_{(k,\infty)} = \oplus\{\mathcal{U}_n : n \geq k\}$ and $\mathcal{K}_k \ominus \mathcal{M}_k$ with the space $\vec{\mathcal{Y}}_{(-\infty,k-1)} = \oplus\{\mathcal{Y}_n : n \leq k-1\}$. The operator $A_k$ is unitarily equivalent to $\Gamma_k$, and hence $\sup\{\|A_k\| : k \in \mathbb{Z}\} = d_\infty$. Furthermore, by consulting the form of $H_k$ in (VIII.4.4) we have

$$A_k \mid \mathcal{H}_{k+1} = (I - P_{\mathcal{M}_k})A_{k+1} \tag{3.4}$$

for each $k \in \mathbb{Z}$. Thus the compatibility condition (j) in Theorem XII.1.1a is fulfilled. Now applying Theorem XII.1.1a, we obtain a bounded linear operator $G : \vec{\mathcal{U}} \to \vec{\mathcal{Y}}$ such that $\|G\| = d_\infty$ and

$$(I - P_{\mathcal{M}_k})G \mid \mathcal{H}_k = A_k \qquad (k \in Z). \tag{3.5}$$

Formula (3.5) is equivalent to the statement that G is an interpolant of the operators in (3.1).

Since $\|G\| = d_\infty$, the proof is complete.

The above analysis also shows that G is an interpolant of the operators in (3.1) if and only if G is a solution of the three chains completion problems associated with the spaces $\mathcal{H}_k = \vec{\mathcal{U}}_{(k,\infty)}$, $\mathcal{K}_k = \mathcal{Y}$ and $\mathcal{M}_k = \vec{\mathcal{Y}}_{(-\infty,k-1)}$ and the operators $A_k = H_k$ for $k \in \mathbb{Z}$.

## XIII.4. APPLICATION TO THE TWO-SIDED SARASON PROBLEM

In this section we use the three chains completion theorem to treat the nonstationary two-sided Sarason interpolation problem. In particular, we shall give a second proof of Theorem VIII.7.1.

Recall that for the nonstationary two-sided Sarason problem the given data is an operator F from $\vec{\mathcal{U}}$ to $\vec{\mathcal{Y}}$, a block lower triangular isometry $L_2$ from $\vec{\mathcal{F}}$ to $\vec{\mathcal{Y}}$, and a block lower triangular co-isometry $L_1$ from $\vec{\mathcal{U}}$ to $\vec{\mathcal{E}}$. As before, we write $\vec{\mathcal{H}}$ for the doubly infinite Hilbert space direct sum $\cdots \oplus \mathcal{H}_{-1} \oplus \mathcal{H}_0 \oplus \mathcal{H}_1 \oplus \cdots$. Given F, $L_1$ and $L_2$ as above and a tolerance $\gamma$, the problem is to determine block lower triangular operators R from $\vec{\mathcal{E}}$ to $\vec{\mathcal{F}}$ such that

$$\|F - L_2 R L_1\| \leq \gamma. \tag{4.1}$$

To put the above problem in the context of the three chains completion theorem we introduce for $k \in \mathbb{Z}$ the spaces

$$\mathcal{H}_k = \{\vec{u} \in \vec{\mathcal{U}} : (L_1\vec{u})_j = 0 \text{ for } j < k\}, \tag{4.2}$$

$$\mathcal{M}_k = \{L_2\vec{f} : \vec{f} \in \vec{\mathcal{F}} \text{ and } f_j = 0 \text{ for } j < k\}, \tag{4.3}$$

$$\mathcal{K}_k = \vec{\mathcal{Y}}, \tag{4.4}$$

and the operators

$$A_k = (I - P_{\mathcal{M}_k})F \,|\, \mathcal{H}_k : \mathcal{H}_k \to \mathcal{K}_k \ominus \mathcal{M}_k. \tag{4.5}$$

Clearly the operators $A_k$ are well-defined. Notice that the spaces $\mathcal{H}_k$, $\mathcal{M}_k$ and $\mathcal{K}_k$ satisfy the inclusion relations in (1.6) of Section XII.1.

**PROPOSITION 4.1.** *The nonstationary two-sided Sarason problem (4.1) has a solution if and only if the operators (4.5) have an interpolant of norm bounded by* $\gamma$, *and in this case all solutions* R *of the problem (4.1) are given by* $R = L_2^*(F - B)L_1^*$, *where* B *from* $\vec{\mathcal{U}}$ *to* $\vec{\mathcal{Y}}$ *has norm bounded by* $\gamma$ *and is an interpolant of the operators (4.5).*

To prove the above proposition we need the following lemma.

**LEMMA 4.2.** *Let $\mathcal{H}_k$ and $\mathcal{M}_k$ be given by (4.2) and (4.3). Assume the operator* T *from* $\vec{\mathcal{U}}$ *to* $\vec{\mathcal{Y}}$ *satisfies* $(I - P_{\mathcal{M}_k})T\mathcal{H}_k = \{0\}$ *for each* $k \in \mathbb{Z}$. *Then there exists a unique* R *from* $\vec{\mathcal{E}}$ *to* $\vec{\mathcal{F}}$ *such that* $T = L_2RL_1$. *This operator* R *is block lower triangular and is given by* $R = L_2^*TL_1^*$.

**PROOF.** Since $L_2$ is injective and $L_1$ is surjective, the uniqueness statement is evident. Put $R = L_2^*TL_1^*$. We have to show that R is block lower triangular and $T = L_2RL_1$.

First, we show that R is block lower triangular. Fix an integer k, and let

$$\vec{e} = (\ldots, 0, 0, e_k, e_{k+1}, \ldots) \in \vec{\mathcal{E}}\,. \tag{4.6}$$

Put $\vec{u} = L_1^*\vec{e}$. Then $L_1\vec{u} = L_1L_1^*\vec{e} = \vec{e}$, and thus $\vec{u} \in \mathcal{H}_k$. But then $TL_1^*\vec{e} \in \mathcal{M}_k$, that is, $TL_1^*\vec{e} = L_2\vec{f}$ for some $\vec{f} \in \vec{\mathcal{F}}$ satisfying $f_j = 0$ for $j < k$. As $L_2^*L_2 = I$, we conclude that $R\vec{e} = \vec{f}$. Thus $(R\vec{e})_j = 0$ for $j < k$, whenever $\vec{e}$ has the form (4.6). Thus R is block lower triangular.

Notice that

$$L_2RL_1 = T - (I - L_2L_2^*)T - L_2L_2^*T(I - L_1^*L_1)\,.$$

Thus, in order to prove that $T = L_2RL_1$ is suffices to show that $T(I - L_1^*L_1)$ and $(I - L_2L_2^*)T$ are zero operators. However, $I - L_1^*L_1$ is the orthogonal projection onto $\ker L_1$ and $I - L_2L_2^*$ is the orthogonal projection onto $\ker L_2^*$. Thus we have to show that

$$\ker L_1 \subset \ker T, \quad \operatorname{ran} T \subset \operatorname{ran} L_2\,. \tag{4.7}$$

Assume $L_1\vec{u} = 0$. Fix $k \in \mathbb{Z}$. Notice that $\vec{u} \in \mathcal{H}_k$. Thus $(I - P_{\mathcal{M}_k})T\vec{u} = 0$ by our hypothesis on T. Therefore, $T\vec{u} = P_{\mathcal{M}_k}T\vec{u} = L_2\vec{f}$ for some $\vec{f} \in \vec{\mathcal{F}}$ satisfying $f_j = 0$ for $j < k$. Since $L_2$ is block lower triangular, we conclude that $(T\vec{u})_j = 0$ for $j < k$. But k is arbitrary. So $T\vec{u} = 0$, and the first inclusion in (4.7) is proved.

To prove the second inclusion in (4.7), let $\vec{u}$ be an arbitrary element of $\vec{\mathcal{U}}$. For each $k \in \mathbb{Z}$ let z(k) be the element of $\vec{\mathcal{U}}$ given by $z(k)_j = u_j$ for $j \geq k$ and $z(k)_j = 0$ otherwise. Then z(k) belongs to $\mathcal{H}_k$ (because $L_1$ is block lower triangular), and hence $(I - P_{\mathcal{M}_k})Tz(k) = 0$. It follows that $Tz(k) \in \mathcal{M}_k \subset \operatorname{ran} L_2$. Notice that z(k) tends to $\vec{u}$ if $k \to -\infty$. Since $\operatorname{ran} L_2$ is closed, we conclude that $T\vec{u} \in \operatorname{ran} L_2$, as desired. This completes the proof.

**PROOF OF PROPOSITION 4.1.** Let R from $\vec{\mathcal{E}}$ to $\vec{\mathcal{F}}$ be a solution of the two-sided Sarason problem (4.1). Put $B = F - L_2RL_1$. Then the norm of B is bounded by $\gamma$. Since R is block lower triangular, $L_2RL_1\mathcal{H}_k$ is contained in $\mathcal{M}_k$ for each $k \in \mathbb{Z}$, and hence

$$(I - P_{\mathcal{M}_k})B \mid \mathcal{H}_k = (I - P_{\mathcal{M}_k})F \mid \mathcal{H}_k = A_k \qquad (k \in \mathbb{Z}).$$

Since $\mathcal{K}_k = \vec{\mathcal{Y}}$, the condition B maps $\mathcal{H}_k$ into $\mathcal{K}_k$ is trivially fulfilled for each k. Thus B is an interpolant of the operators $A_k$ in (4.5). Recall that $L_2^*L_2$ and $L_1L_1^*$ are identity operators. Thus $L_2^*(F - B)L_1^* = L_2^*L_2RL_1L_1^* = R$, as desired.

Conversely, assume that B is an interpolant of the operators (4.5) and $\|B\| \le \gamma$. Put $R = L_2^*(F - B)L_1^*$. Notice that $T = F - B$ satisfies the conditions of Lemma 4.2. Hence R is block lower triangular and $F - B = L_2RL_1$. Since B has norm bounded by $\gamma$, the same is true for $F - L_2RL_1$, and hence R is a solution of the two-sided Sarason problem (4.1). This completes the proof.

**PROOF OF THEOREM VIII.7.1.** Notice that the operators defined by (4.5) satisfy the compatibility condition (j) in Theorem XII.1.1a. Therefore we see from Proposition 4.1 that Theorem VIII.7.1 follows directly by applying this three chains completion theorem to the operators $A_k$ in (4.5). Notice that in this case

$$d_\infty = \sup\{\|A_k\| : k \in \mathbb{Z}\} = \sup\{\|(I - P_{\mathcal{M}_k})F \mid \mathcal{H}_k\| : k \in \mathbb{Z}\},$$

and the latter quantity is precisely the optimal error in the two-sided Sarason problem. This completes the proof.

## XIII.5. APPLICATION TO THE NUDELMAN PROBLEM

In this section we use the three chains completion theorem to treat the nonstationary Nudelman interpolation problem. In particular, we shall give a second proof of Theorem VIII.6.1.

Recall that for the nonstationary Nudelman interpolation problem (Section VIII.6) the given data are bounded linear operators

$$Z_k : \mathcal{X}_{k+1} \to \mathcal{X}_k, \; B_k : \mathcal{U}_k \to \mathcal{X}_k, \; \tilde{B}_k : \mathcal{Y}_k \to \mathcal{X}_k,$$

$$A_k : \mathcal{F}_k \to \mathcal{F}_{k-1}, \; C_k : \mathcal{F}_k \to \mathcal{Y}_k, \tilde{C}_k : \mathcal{F}_k \to \mathcal{U}_k,$$

$$\Gamma_k : \mathcal{F}_{k-1} \to \mathcal{X}_k,$$

which act between Hilbert spaces. These operators are assumed to be uniformly bounded in the operator norm with respect k and

$$\limsup_{\nu\to\infty}\left[\sup_{k\in\mathbb{Z}}\|\Phi(k,k+\nu)\|\right]^{1/\nu}<1 \text{ and } \limsup_{\nu\to\infty}\left[\sup_{k\in\mathbb{Z}}\|\Psi(k-\nu,k)\|\right]^{1/\nu}<1$$

Here $\Phi$ and $\Psi$ are the state transition operators associated with the families $\{Z_k\}$ and $\{\Lambda_{k+1}\}$, i.e.,

$$\Phi(m,n)=Z_m Z_{m+1}\cdots Z_{n-1}\quad(m<n),\qquad \Phi(n,n)=I_{\mathcal{X}_n};$$

$$\Psi(m,n)=\Lambda_{m+1}\Lambda_{m+1}\cdots\Lambda_n\quad(m<n),\qquad \Psi(n,n)=I_{\mathcal{F}_n}.$$

We also assume that the data satisfy the following compatibility condition:

$$Z_k\Gamma_{k+1}-\Gamma_k\Lambda_k=\tilde{B}_kC_k-B_k\tilde{C}_k\quad(k\in\mathbb{Z}). \tag{5.1}$$

Given operators as above and a bound $\gamma$ the nonstationary Nudelman interpolation problem (see Section VIII.6) is to find operators $f_{j,k}$ in $\mathcal{L}(\mathcal{Y}_k,\mathcal{U}_j)$ for $-\infty<k\le j<\infty$ satisfying the interpolation conditions

$$\sum_{\nu=0}^{\infty}\Phi(k,k+\nu)B_{k+\nu}f_{k+\nu,k}=\tilde{B}_k\quad(k\in\mathbb{Z}), \tag{5.2}$$

$$\sum_{\nu=0}^{\infty}f_{k,k-\nu}C_{k-\nu}\Psi(k-\nu,k)=\tilde{C}_k\quad(k\in\mathbb{Z}), \tag{5.3}$$

$$\sum_{j,k=0}^{\infty}\Phi(n,n+j)B_{n+j}f_{n+j,n-k-1}C_{n-k-1}\Psi(n-k-1,n-1)=\Gamma_n\quad(n\in\mathbb{Z}), \tag{5.4}$$

and such that the block lower triangular operator

$$F=\begin{bmatrix}\ddots &&&&\\ \cdots & f_{-1,-1}&&&\\ \cdots & f_{0,-1} & f_{0,0}&&\\ \cdots & f_{1,-1} & f_{1,0} & f_{1,1}&\\ \vdots&\vdots&\vdots&\vdots&\ddots\end{bmatrix}:\vec{\mathcal{Y}}\to\vec{\mathcal{U}} \tag{5.5}$$

has operator norm $\|F\|\le\gamma$. Here, as before, $\vec{\mathcal{Y}}$ is the Hilbert space direct sum $\oplus\{\mathcal{Y}_n : n\in\mathbb{Z}\}$ and $\vec{\mathcal{U}}$ is the Hilbert space direct sum $\oplus\{\mathcal{U}_n : n\in\mathbb{Z}\}$.

We begin by rewriting the interpolation conditions (5.2)-(5.4) using the block lower triangular operator F in (5.5) instead of its operator entries. For this purpose we define for each $k\in\mathbb{Z}$ bounded linear operators

$$R_k : \vec{\mathcal{U}} \to \mathcal{X}_k, \quad \tilde{R}_k : \vec{\mathcal{Y}} \to \mathcal{X}_k,$$

$$K_k : \mathcal{F}_{k-1} \to \vec{\mathcal{Y}}, \quad \tilde{K}_k : \mathcal{F}_{k-1} \to \vec{\mathcal{U}},$$

by setting

$$R_k\vec{u} = \sum_{\nu=0}^{\infty} \Phi(k, k+\nu)B_{k+\nu}u_{k+\nu}, \tag{5.6}$$

$$\tilde{R}_k\vec{y} = \sum_{\nu=0}^{\infty} \Phi(k, k+\nu)\tilde{B}_{k+\nu}y_{k+\nu}, \tag{5.7}$$

$$K_k f = \vec{y}, \quad y_j = \begin{cases} C_j\Psi(j, k-1)f, & j \le k-1, \\ 0, & j > k-1, \end{cases} \tag{5.8}$$

$$\tilde{K}_k f = \vec{u}, \quad u_j = \begin{cases} \tilde{C}_j\Phi(j, k-1)f, & j \le k-1 \\ 0, & j > k-1. \end{cases} \tag{5.9}$$

Here $\vec{y} = (y_j)_{-\infty}^{\infty}$ and $\vec{u} = (u_j)_{-\infty}^{\infty}$ are vectors in $\vec{\mathcal{Y}}$ and $\vec{\mathcal{U}}$, respectively. We shall also need the operators $P_{\vec{\mathcal{Y}},k}$ and $P_{\vec{\mathcal{U}},k}$ which, by definition, are the orthogonal projections of $\vec{\mathcal{Y}}$ onto the subspace $\vec{\mathcal{Y}}_{(k,\infty)} = \oplus\{\mathcal{Y}_n : n \ge k\}$ and of $\vec{\mathcal{U}}$ onto the space $\vec{\mathcal{U}}_{(k,\infty)} = \oplus\{\mathcal{U}_n : n \ge k\}$, respectively. We have the following identities:

$$R_k P_{\vec{\mathcal{U}},k} = R_k, \quad \tilde{R}_k P_{\vec{\mathcal{Y}},k} = \tilde{R}_k, \tag{5.10}$$

$$R_k P_{\vec{\mathcal{U}},k+1} = Z_k R_{k+1}, \quad \tilde{R}_k P_{\vec{\mathcal{Y}},k+1} = Z_k \tilde{R}_{k+1}, \tag{5.11}$$

$$(I - P_{\vec{\mathcal{Y}},k})K_k = K_k, \quad (I - P_{\vec{\mathcal{U}},k})\tilde{K}_k = \tilde{K}_k, \tag{5.12}$$

$$(I - P_{\vec{\mathcal{Y}},k-1})K_k = K_{k-1}\Lambda_{k-1}, \quad (I - P_{\vec{\mathcal{U}},k-1})\tilde{K}_k = \tilde{K}_{k-1}\Lambda_{k-1}. \tag{5.13}$$

Using the auxiliary operators introduced in the previous paragraph it is straightforward to check that our nonstationary Nudelman interpolation problem is to find a block lower triangular bounded linear operator $F$ from $\vec{\mathcal{Y}}$ to $\vec{\mathcal{U}}$ of norm bounded by $\gamma$ such that $F$ satisfies the following identities:

$$R_k F P_{\vec{\mathcal{Y}},k} = \tilde{R}_k \quad (k \in \mathbb{Z}), \tag{5.14}$$

$$(I - P_{\vec{\mathcal{U}},k})FK_k = \tilde{K}_k \quad (k \in \mathbb{Z}), \tag{5.15}$$

$$R_k F K_k = \Gamma_k \quad (k \in \mathbb{Z}). \tag{5.16}$$

In fact, (5.14) is equivalent to (5.2), while (5.15) is equivalent to (5.3), and (5.16) to (5.4).

To put the nonstationary Nudelman problem in the context of the three chains completion theorem we introduce the following spaces:

$$\mathcal{H}_k = \overline{\text{ran } K_k} \oplus \vec{\mathcal{Y}}_{(k,\infty)} \quad (k \in \mathbb{Z}), \tag{5.17}$$

$$\mathcal{M}_k = \{\vec{u} \in \vec{\mathcal{U}}_{(k,\infty)} : R_k\vec{u} = 0\} \quad (k \in \mathbb{Z}), \tag{5.18}$$

$$\mathcal{K}_k = \vec{\mathcal{U}} \quad (k \in \mathbb{Z}). \tag{5.19}$$

Notice that

$$\overline{\text{ran } K_k} \subset \vec{\mathcal{Y}}_{(-\infty,k-1)} = \{\vec{y} \in \vec{\mathcal{Y}} : y_j = 0 \text{ for } j \geq k\}, \tag{5.20}$$

and hence the orthogonal direct sum symbol in (5.17) is justified.

**LEMMA 5.1**. *The spaces* $\mathcal{H}_k \subset \vec{\mathcal{Y}}$ *and* $\mathcal{M}_k \subset \mathcal{K}_k \subset \vec{\mathcal{U}}$, *where* $k \in \mathbb{Z}$, *satisfy the inclusion relations (XII.1.6) and*

$$\mathcal{K}_k \ominus \mathcal{M}_k = \vec{\mathcal{U}}_{(-\infty,k-1)} \oplus \overline{\text{ran } R_k^*} \quad (k \in \mathbb{Z}). \tag{5.21}$$

*Here* $\vec{\mathcal{U}}_{(-\infty,k-1)} = \{\vec{u} \in \vec{\mathcal{U}} : u_j = 0$ *for* $j \geq k\}$.

**PROOF**. Clearly, $\vec{\mathcal{Y}}_{(k+1,\infty)} \subset \vec{\mathcal{Y}}_{(k,\infty)} \subset \mathcal{H}_k$. Hence, in order to prove $\mathcal{H}_{k+1} \subset \mathcal{H}_k$, it suffices to show that ran $K_{k+1} \subset \mathcal{H}_k$. Take $\vec{y} = K_{k+1}f_k \in$ ran $K_{k+1}$. Then, using the first identity in (5.13), we have

$$\begin{aligned}\vec{y} &= (I - P_{\vec{\mathcal{Y}},k})K_{k+1}f_k + P_{\vec{\mathcal{Y}},k}\vec{y} \\ &= K_k\Lambda_k f_k + P_{\vec{\mathcal{Y}},k}\vec{y} \in \text{ran } K_k + \vec{\mathcal{Y}}_{(k,\infty)} \subset \mathcal{H}_k .\end{aligned}$$

Thus $\mathcal{H}_{k+1} \subset \mathcal{H}_k$.

The inclusion $\mathcal{K}_{k+1} \subset \mathcal{K}_k$ is trivially satisfied. To prove $\mathcal{M}_{k+1} \subset \mathcal{M}_k$, take $\vec{u} \in \mathcal{M}_{k+1}$. Then $\vec{u} = P_{\vec{\mathcal{U}},k+1}\vec{u}$ and $R_{k+1}\vec{u} = 0$. But then we can use the first identity in (5.11) to show that

$$R_k\vec{u} = R_kP_{\vec{\mathcal{U}},k+1}\vec{u} = Z_kR_{k+1}\vec{u} = 0 .$$

Thus $R_k\vec{u} = 0$. Since $\vec{u} \in \vec{\mathcal{U}}_{(k+1,\infty)} \subset \vec{\mathcal{U}}_{(k,\infty)}$, we conclude that $\vec{u} \in \mathcal{M}_k$. Thus $\mathcal{M}_{k+1} \subset \mathcal{M}_k$.

Notice that ran $R_k^* \subset \vec{\mathcal{U}}_{(k,\infty)}$, and hence the space in the right hand side of (5.21) is well-defined and closed. It follows that

$$\mathcal{K}_k \ominus \mathcal{M}_k = \mathcal{M}_k^\perp = (\vec{\mathcal{U}}_{(k,\infty)} \cap \ker R_k)^\perp = \vec{\mathcal{U}}_{(-\infty,k-1)} \oplus \overline{\operatorname{ran} R_k^*},$$

which completes the proof.

Now, let us assume that our nonstationary Nudelman interpolation problem has a solution. So there exists a block lower triangular operator F from $\vec{\mathcal{Y}}$ to $\vec{\mathcal{U}}$ such that $\|F\| \le \gamma$ and F satisfies (5.14)-(5.16). For each $k \in \mathbb{Z}$ put

$$A_{c,k} = P_{\overline{\operatorname{ran} R_k^*}} F \,|\, \vec{\mathcal{Y}}_{(k,\infty)}, \quad A_{o,k} = (I - P_{\vec{\mathcal{U}},k}) F \,|\, \overline{\operatorname{ran} K_k}, \quad A_{1,2,k} = P_{\overline{\operatorname{ran} R_k^*}} F \,|\, \overline{\operatorname{ran} K_k}.$$

Then

$$A_{c,k} : \vec{\mathcal{Y}}_{(k,\infty)} \to \overline{\operatorname{ran} R_k^*}, \qquad A_{o,k} : \overline{\operatorname{ran} K_k} \to \vec{\mathcal{U}}_{(-\infty,k-1)}, \tag{5.22}$$

$$A_{1,2,k} : \overline{\operatorname{ran} K_k} \to \overline{\operatorname{ran} R_k^*}, \tag{5.23}$$

and these operators satisfy the identities

$$R_k A_{c,k} = \tilde{R}_k \,|\, \vec{\mathcal{Y}}_{(k,\infty)}, \quad A_{o,k} K_k = (I - P_{\vec{\mathcal{U}},k}) \tilde{K}_k = \tilde{K}_k, \tag{5.24}$$

$$\Gamma_k = R_k A_{1,2,k} K_k. \tag{5.25}$$

Furthermore, the operator $A_k$ from $\mathcal{H}_k$ to $\mathcal{K}_k \ominus \mathcal{M}_k$ defined by

$$A_k = \begin{bmatrix} A_{o,k} & 0 \\ A_{1,2,k} & A_{c,k} \end{bmatrix} : \begin{bmatrix} \overline{\operatorname{ran} K_k} \\ \vec{\mathcal{Y}}_{(k,\infty)} \end{bmatrix} \to \begin{bmatrix} \vec{\mathcal{U}}_{(-\infty,k-1)} \\ \overline{\operatorname{ran} R_k^*} \end{bmatrix} \tag{5.26}$$

satisfies the norm condition

$$\|A_k\| \le \gamma. \tag{5.27}$$

Indeed, since F is block lower triangular,

$$A_k = (I - P_{\mathcal{M}_k}) F \,|\, \mathcal{H}_k, \tag{5.28}$$

and hence (5.27) follows from the fact that $\|F\| \le \gamma$. Notice that for each $k \in \mathbb{Z}$ the operators $A_{c,k}$, $A_{o,k}$ and $A_{1,2,k}$ in (5.22) and (5.23) are uniquely determined by the identities in (5.24) and (5.25).

Next, let us consider the condition of solvability. We have to show (Theorem VIII.6.1) that there exists a block lower triangular operator F from $\vec{\mathcal{Y}}$ to $\vec{\mathcal{U}}$ with $\|F\| \le \gamma$ and satisfying (5.14)-(5.16) if and only if for each $k \in \mathbb{Z}$ the operator

$$\Xi_k = \begin{bmatrix} M_k & \gamma\Gamma_{k+1}^* \\ \gamma\Gamma_{k+1} & N_{k+1} \end{bmatrix} : \mathcal{F}_k \oplus \mathcal{X}_{k+1} \to \mathcal{F}_k \oplus \mathcal{X}_{k+1} \tag{5.29}$$

is positive. Here $M_k$ on $\mathcal{F}_k$ and $N_{k+1}$ on $\mathcal{X}_{k+1}$ are defined by (VIII.6.7) and (VIII.6.8). These operators are also given by

$$M_k = \gamma^2 K_{k+1}^* K_{k+1} - \tilde{K}_{k+1}^* \tilde{K}_{k+1} \quad (k \in \mathbb{Z}), \tag{5.30}$$

$$N_{k+1} = \gamma^2 R_{k+1} R_{k+1}^* - \tilde{R}_{k+1} \tilde{R}_{k+1}^* \quad (k \in \mathbb{Z}). \tag{5.31}$$

We need the following lemma.

**LEMMA 5.2.** *The operator $\Xi_{k-1}$ in (5.29) is positive if and only if there exists operators $A_{c,k}$, $A_{o,k}$ and $A_{1,2,k}$ as in (5.22) and (5.23) satisfying (5.24) and (5.25) and such that the operator $A_k$ from $\mathcal{H}_k$ to $\mathcal{K}_k \ominus \mathcal{M}_k$ defined by (5.26) has norm bounded by $\gamma$.*

**PROOF.** Assume that $\Xi_{k-1}$ is positive. This implies that the operators $M_{k-1}$ and $N_k$ are positive. From the representation (5.30) and the fact that $M_{k-1} \geq 0$ it follows that there exists a unique operator $A_{o,k}$ from $\overline{\operatorname{ran} K_k}$ to $\vec{\mathcal{U}}_{(-\infty,k-1)}$ such that

$$A_{o,k} K_k = (I - P_{\vec{\mathcal{U}},k}) \tilde{K}_k \quad \text{and} \quad \|A_{o,k}\| \leq \gamma. \tag{5.32}$$

Similarly, using (5.31) and $N_k \geq 0$, there exists a unique operator $A_{c,k}$ from $\vec{\mathcal{Y}}_{(k,\infty)}$ to $\overline{\operatorname{ran} R_k^*}$ such that

$$R_k A_{c,k} = \tilde{R}_k \,|\, \vec{\mathcal{Y}}_{(k,\infty)} \quad \text{and} \quad \|A_{c,k}\| \leq \gamma. \tag{5.33}$$

Using (5.32) and (5.33) we can rewrite (5.30) and (5.31), with $k-1$ in place of k, as

$$M_{k-1} = K_k^* D_{o,k}^2 K_k \quad \text{and} \quad N_k = R_k D_{c,k}^2 R_k^*, \tag{5.34}$$

where

$$D_{o,k} = (\gamma^2 I - A_{o,k}^* A_{o,k})^{1/2} \quad \text{and} \quad D_{c,k} = (\gamma^2 I - A_{c,k} A_{c,k}^*)^{1/2}. \tag{5.35}$$

The representations for $M_{k-1}$ and $N_k$ in (5.34) and the fact that $\Xi_{k-1}$ in (5.29) is positive allows us (use Lemma II.5.2) to find a contraction $\Omega_k$ from the closure of the range of $D_{o,k}$ to the closure of the range of $D_{c,k}$ such that

$$\gamma\Gamma_k = R_k D_{c,k} \Omega_k D_{o,k} K_k. \tag{5.36}$$

Now, put

$$A_{1,2,k} = \frac{1}{\gamma} D_{c,k} \Omega_k D_{o,k} : \overline{\operatorname{ran} K_k} \to \overline{\operatorname{ran} R_k^*}. \tag{5.37}$$

Then the operators $A_{c,k}$, $A_{o,k}$ and $A_{1,2,k}$ introduced above satisfy (5.22)-(5.25), and, by Lemma II.5.1, the operator $A_k$ from $\mathcal{H}_k$ to $\mathcal{K}_k \ominus \mathcal{M}_k$ defined by (5.26) has norm bounded by $\gamma$.

Conversely, assume that there exist operators $A_{c,k}$, $A_{o,k}$ and $A_{1,2,k}$ satisfying (5.22)-(5.25) and such that the operator $A_k$ from $\mathcal{H}_k$ to $\mathcal{K}_k \ominus \mathcal{M}_k$ defined by (5.26) has norm bounded by $\gamma$. Since $\|A_k\| \le \gamma$, the operators $A_{o,k}$ and $A_{c,k}$ are bounded in norm by $\gamma$, and we can apply Lemma II.5.1 to show that $\gamma A_{1,2,k} = D_{c,k}\Omega_k D_{o,k}$ for some contraction $\Omega_k$. Here $D_{o,k}$ and $D_{c,k}$ are the operators defined by (5.35). From (5.24), (5.30), (5.31) and the fact that $A_{o,k}$ and $A_{c,k}$ are bounded in norm by $\gamma$, we see that $M_{k-1}$ and $N_k$ are given by (5.34), and these operators are both positive. From (5.25) and the fact that $\gamma A_{1,2,k} = D_{c,k}\Omega_k D_{o,k}$, we see that (5.36) holds, and hence (using (5.34)) we have

$$\Xi_{k-1} = \begin{bmatrix} K_k^* D_{o,k} & 0 \\ R_k D_{c,k} \Omega_k & R_k D_{c,k} \end{bmatrix} \begin{bmatrix} I & 0 \\ 0 & I - \Omega_k \Omega_k^* \end{bmatrix} \begin{bmatrix} D_{o,k} K_k & \Omega_k^* D_{c,k} R_k^* \\ 0 & D_{c,k} R_k^* \end{bmatrix}.$$

Since $\Omega_k$ is a contraction, this shows that $\Xi_{k-1}$ is positive, which completes the proof.

We are now ready to prove the necessity part of Theorem VIII.6.1. Indeed, assume that our nonstationary Nudelman interpolation problem has a solution. So (see the first paragraph after the proof of Lemma 5.1) for each $k \in \mathbb{Z}$ there exist operators $A_{c,k}$, $A_{o,k}$ and $A_{1,2,k}$ satisfying (5.22)-(5.25) such that the operator $A_k$ from $\mathcal{H}_k$ to $\mathcal{K}_k \ominus \mathcal{M}_k$ defined by (5.26) is bounded in norm by $\gamma$. By Lemma 5.2 we may conclude that for each $k \in \mathbb{Z}$ the operator $\Xi_k$ is positive, as desired.

It remains to prove the reverse implication. Therefore, in what follows we assume that $\Xi_k$ is positive for each $k \in \mathbb{Z}$ or, equivalently, we assume (cf., Lemma 5.2) that for each $k \in \mathbb{Z}$ there exist operators $A_{c,k}$, $A_{o,k}$ and $A_{1,2,k}$ satisfying (5.22)-(5.25) such that the operator $A_k$ defined by (5.26) has norm bounded by $\gamma$, that is,

$$d_\infty := \sup \{\|A_k\| : k \in \mathbb{Z}\} \le \gamma < \infty. \tag{5.38}$$

Recall that $A_k$ is an operator from $\mathcal{H}_k$ into $\mathcal{K}_k \ominus \mathcal{M}_k$ for each $k \in \mathbb{Z}$. We shall apply the second version of the three chains completion theorem (Theorem XII.1.1a) to the operators $\{A_k\}$. By (5.38) condition (jj) in Theorem XII.1.1a is fulfilled. The next proposition shows that the operators $\{A_k\}$ also satisfy the compatibility condition (j) in Theorem XII.1.1a.

**PROPOSITION 5.3.** *Let the spaces $\mathcal{H}_k$, $\mathcal{M}_k$ and $\mathcal{K}_k$ for $k \in \mathbb{Z}$ be defined by (5.17)-(5.19), and for each $k \in \mathbb{Z}$ let $A_k$ from $\mathcal{H}_k$ to $\mathcal{K}_k \ominus \mathcal{M}_k$ be the operator defined by (5.26),* *where* $A_{c,k}$, $A_{o,k}$ *and* $A_{1,2,k}$ *are operators satisfying (5.22)-(5.25). Then*

$$(I - P_{\mathcal{M}_k}) | A_{k+1} = A_k | \mathcal{H}_{k+1} \quad (k \in \mathbb{Z}) . \tag{5.39}$$

**PROOF**. Fix $k \in \mathbb{Z}$. To prove (5.39) it suffices to show that for each $\vec{y} \in \mathcal{H}_{k+1}$ we have

$$A_{k+1}\vec{y} - A_k\vec{y} \in \vec{\mathcal{U}}_{(k,\infty)} \text{ and } R_k(A_{k+1}\vec{y} - A_k\vec{y}) = 0 . \tag{5.40}$$

Indeed, (5.40) implies that $A_{k+1}\vec{y} - A_k\vec{y} \in \mathcal{M}_k$, and hence

$$(I - P_{\mathcal{M}_k})A_{k+1}\vec{y} = (I - P_{\mathcal{M}_k})A_k\vec{y} .$$

Since $A_k\vec{y} \in \mathcal{K}_k \ominus \mathcal{M}_k$, the latter equality yields (5.39).

Next, recall that $\mathcal{H}_{k+1} = \overline{\text{ran } K_{k+1}} \oplus \vec{\mathcal{Y}}_{(k+1,\infty)}$. So it suffices to check (5.40) for $\vec{y} = K_{k+1}f$, where f is an arbitrary element of $\mathcal{F}_k$, and for $\vec{y} \in \vec{\mathcal{Y}}_{(k+1,\infty)}$.

Take $\vec{y} = K_{k+1}f$ with $f \in \mathcal{F}_k$. From the first paragraph of the proof of Lemma 5.1 we know that $\vec{y} = K_k\Lambda_k f + P_{\vec{\mathcal{Y}},k}K_{k+1}f$. Thus, by (5.26),

$$A_k\vec{y} = A_{o,k}K_k\Lambda_k f \oplus (A_{1,2,k}K_k\Lambda_k f + A_{c,k}P_{\vec{\mathcal{Y}},k}K_{k+1}f) . \tag{5.41}$$

Furthermore,

$$A_{k+1}\vec{y} = A_{o,k+1}K_{k+1}f \oplus A_{1,2,k+1}K_{k+1}f . \tag{5.42}$$

By using the second parts of (5.12), (5.13) and (5.24) repeatedly we see that

$$\begin{aligned} A_{o,k+1}K_{k+1}f &= (I - P_{\vec{\mathcal{U}},k+1})\tilde{K}_{k+1}f = \tilde{K}_{k+1}f \\ &= (I - P_{\vec{\mathcal{U}},k})\tilde{K}_{k+1}f + P_{\vec{\mathcal{U}},k}\tilde{K}_{k+1}f \\ &= \tilde{K}_k\Lambda_k f + P_{\vec{\mathcal{U}},k}\tilde{K}_{k+1}f \\ &= A_{o,k}K_k\Lambda_k f + P_{\vec{\mathcal{U}},k}\tilde{K}_{k+1}f . \end{aligned}$$

In particular, we see that

$$A_{o,k+1}K_{k+1}f - A_{o,k}K_k\Lambda_k f \in \vec{\mathcal{U}}_{(k,\infty)} . \tag{5.43}$$

Notice that the term between brackets in the right hand side of (5.41) belongs to $\vec{\mathcal{U}}_{(k,\infty)}$. The second term in the right hand side of (5.42) belongs to $\vec{\mathcal{U}}_{(k+1,\infty)}$, and hence this vector is also in the space $\vec{\mathcal{U}}_{(k,\infty)}$. But then we can use (5.43) to show that $A_{k+1}\vec{y} - A_k\vec{y}$ belongs to $\vec{\mathcal{U}}_{(k,\infty)}$.

Again, take $\vec{y} = K_{k+1} f$ with $f \in \mathcal{F}_k$, and let us compute $R_k(A_{k+1}\vec{y} - A_k\vec{y})$. From (5.8) and (5.9), we see that

$$P_{\vec{\mathcal{Y}},k} K_{k+1} f = (\delta_{j,k} C_k f)_{-\infty}^{\infty} \quad \text{and} \quad P_{\vec{\mathcal{U}},k} \tilde{K}_{k+1} f = (\delta_{j,k} \tilde{C}_k f)_{-\infty}^{\infty} , \tag{5.44}$$

where $\delta_{j,k}$ is the Kronecker delta. Thus, using (5.24), and ran $A_{o,k} \subseteq \vec{\mathcal{U}}_{(-\infty,k-1)} \subseteq \ker R_k$, we see that

$$R_k A_{c,k} P_{\vec{\mathcal{Y}},k} K_{k+1} f = \tilde{B}_k C_k f , \quad R_k A_{o,k} K_k \Lambda_k f = 0 ,$$

$$R_k A_{o,k+1} K_{k+1} f = B_k \tilde{C}_k f .$$

Next, use (5.25), the first identity in (5.13) and (5.11) to show that

$$R_k A_{1,2,k} K_k \Lambda_k f = \Gamma_k \Lambda_k f ,$$

$$R_k A_{1,2,k+1} K_{k+1} f = R_k P_{\vec{\mathcal{U}},k+1} A_{1,2,k+1} K_{k+1} f = Z_k R_{k+1} A_{1,2,k+1} K_{k+1} f = Z_k \Gamma_{k+1} f .$$

From this analysis and the one in the previous paragraph we conclude (use (5.4) and (5.42)) that

$$R_k(A_{k+1}\vec{y} - A_k\vec{y}) = B_k \tilde{C}_k f + Z_k \Gamma_{k+1} f - \Gamma_k \Lambda_k f - \tilde{B}_k C_k f = 0 ,$$

because of (5.1). Thus (5.40) has been established for $\vec{y} = K_{k+1} f$.

Next, take $\vec{y} \in \vec{\mathcal{Y}}_{(k+1,\infty)}$. Then

$$A_{k+1}\vec{y} - A_k\vec{y} = A_{c,k+1}\vec{y} - A_{c,k}\vec{y} \in \vec{\mathcal{U}}_{(k,\infty)} .$$

Furthermore, using the first identities in (5.24) and (5.11), we see that

$$\begin{aligned} R_k(A_{k+1}\vec{y} - A_k\vec{y}) &= R_k A_{c,k+1}\vec{y} - R_k A_{c,k}\vec{y} \\ &= R_k P_{\vec{\mathcal{U}},k+1} A_{c,k+1}\vec{y} - \tilde{R}_k\vec{y} \\ &= Z_k \tilde{R}_{k+1}\vec{y} - \tilde{R}_k\vec{y} . \end{aligned}$$

Since $\vec{y} \in \vec{\mathcal{Y}}_{(k+1,\infty)}$, the second identity in (511) implies that $Z_k \tilde{R}_{k+1}\vec{y} = \tilde{R}_k\vec{y}$. Thus (5.40) is also fulfilled for $\vec{y} \in \vec{\mathcal{Y}}_{(k+1,\infty)}$, which completes the proof.

From Proposition 5.3 and formula (5.38) it follows that we can apply Theorem XII.1.1a to show that the operators $A_k$ ($k \in \mathbb{Z}$) defined by (5.26) have an interpolant with norm bounded by $\gamma$, that is, there exists F in $\mathcal{L}(\vec{\mathcal{Y}}, \vec{\mathcal{U}})$ such that

$$\|F\| \le \gamma , \quad (I - P_{\mathcal{M}_k}) F | \mathcal{H}_k = A_k \qquad (k \in \mathbb{Z}) . \tag{5.45}$$

The next theorem shows that the set of all interpolants F in (5.45) coincides with the set of all solutions to our nonstationary Nudelman interpolation problem.

**THEOREM 5.4**. *Let the spaces* $\mathcal{H}_k$, $\mathcal{M}_k$ *and* $\mathcal{K}_k$ *for* $k \in \mathbb{Z}$ *be defined by (5.17)-(5.19), and for each* $k \in \mathbb{Z}$ *let the operator* $A_k$ *from* $\mathcal{H}_k$ *to* $\mathcal{K}_k \ominus \mathcal{M}_k$ *be given by (5.26), where* $A_{c,k}$, $A_{l,k}$ *and* $A_{1,2,k}$ *are operators satisfying (5.22) to (5.25) and assume (5.27) holds. Then there exists* F *in* $\mathcal{L}(\vec{\mathcal{Y}}, \vec{\mathcal{U}})$ *with norm bounded by* $\gamma$ *and such that*

$$(I - P_{\mathcal{M}_k})F \mid \mathcal{H}_k = A_k \qquad (k \in \mathbb{Z}). \tag{5.46}$$

*Furthermore, an operator* F *from* $\vec{\mathcal{Y}}$ *to* $\vec{\mathcal{U}}$ *satisfies (5.46) if and only if* F *is block lower triangular and* F *satisfies the interpolation conditions (5.14)-(5.16).*

**PROOF**. Proposition 5.3, formula (5.27) and Theorem XII.1.1a guarantee the existence of an operator F from $\vec{\mathcal{Y}}$ to $\vec{\mathcal{U}}$ such that $\|F\| \le \gamma$ and F satisfies (5.46).

Next, let $F \in \mathcal{L}(\vec{\mathcal{Y}}, \vec{\mathcal{U}})$ satisfy (5.46). Notice that $\vec{\mathcal{Y}}_{(k,\infty)} \subset \mathcal{H}_k$ and $A_k$ maps $\vec{\mathcal{Y}}_{(k,\infty)}$ into $\vec{\mathcal{U}}_{(k,\infty)}$. Thus, by (5.46),

$$F\vec{\mathcal{Y}}_{(k,\infty)} \subset (I - P_{\mathcal{M}_k})F\vec{\mathcal{Y}}_{(k,\infty)} + \mathcal{M}_k = A_k\vec{\mathcal{Y}}_{(k,\infty)} + \mathcal{M}_k \subset \vec{\mathcal{U}}_{(k,\infty)},$$

because $\mathcal{M}_k$ is also contained in $\vec{\mathcal{U}}_{(k,\infty)}$. Thus F maps $\vec{\mathcal{Y}}_{(k,\infty)}$ into $\vec{\mathcal{U}}_{(k,\infty)}$ for each $k \in \mathbb{Z}$, which implies that F is block lower triangular.

Let F be as in the previous paragraph, and let us check that F satisfies the interpolation conditions (5.14)-(5.16). By (5.46), (5.26), the first identity in (5.24), and the second identity in (5.10), we have

$$\begin{aligned} R_kFP_{\vec{\mathcal{Y}},k} &= R_k((I - P_{\mathcal{M}_k})F \mid \mathcal{H}_k)P_{\vec{\mathcal{Y}},k} \\ &= R_kA_kP_{\vec{\mathcal{Y}},k} = R_kA_{c,k}P_{\vec{\mathcal{Y}},k} = \tilde{R}_kP_{\vec{\mathcal{Y}},k} = \tilde{R}_k \qquad (k \in \mathbb{Z}), \end{aligned}$$

which yields (5.14). Analogously, using (5.46), (5.26) and the properties of $K_k$ and $\tilde{K}_k$, we have

$$\begin{aligned} (I - P_{\vec{\mathcal{U}},k})FK_k &= (I - P_{\vec{\mathcal{U}},k})((I - P_{\mathcal{M}_k})F \mid \mathcal{H}_k)K_k = (I - P_{\vec{\mathcal{U}},k})A_kK_k \\ &= A_{o,k}K_k = (I - P_{\vec{\mathcal{U}},k})\tilde{K}_k = \tilde{K}_k \quad (k \in \mathbb{Z}), \end{aligned}$$

and (5.15) is proved. To get (5.16), we use (5.25) in the following computation:

$$R_k F K_k = R_k((I - P_{\mathcal{M}_k})F \mid \mathcal{H}_k)K_k$$

$$= R_k A_k K_k = R_k A_{1,2,k} K_k = \Gamma_k \qquad (k \in \mathbb{Z}) .$$

In the preceding calculations we used that $R_k$ is zero on $\mathcal{M}_k$, and hence $R_k = R_k(I - P_{\mathcal{M}_k})$ for each $k \in \mathbb{Z}$.

Finally, let F be a block lower triangular operator from $\vec{\mathcal{Y}}$ to $\vec{\mathcal{U}}$ satisfying (5.14)-(5.16). Then we know (see (5.28)) from the analysis in the first paragraph after the proof of Lemma 5.1 and from the uniqueness of the operators $A_{c,k}$, $A_{o,k}$ and $A_{1,2,k}$ ($k \in \mathbb{Z}$) in (5.22)-(5.25) that F satisfies (5.46). This remark completes the proof.

To conclude let us note that the sufficiency part of Theorem VIII.6.1 directly follows from Lemma 5.2 and Theorem 5.4. Indeed, if for each $k \in \mathbb{Z}$ the operator $\Xi_k$ in (5.29) is positive, then the operators $A_k$ from $\mathcal{H}_k$ to $\mathcal{K}_k \ominus \mathcal{M}_k$ ($k \in \mathbb{Z}$) given by (5.26) are well defined and have norm bounded by $\gamma$. By Theorem 5.4 this guarantees the existence of a solution to our nonstationary Nudelman interpolation problem.

## XIII.6. THE THREE CHAINS COMPLETION PROBLEM AND THE FOUR BLOCK PROBLEM

In this section we show that the three chains completion problem is equivalent to a nonstationary four block problem, and we discuss the connections between the two problems.

In the nonstationary case the four block problem is just a bordered nonstationary Nehari problem. To be more precise assume we are given Hilbert space operators

$$f_{j,k} : \mathcal{U}_k \to \mathcal{Y}_j \qquad (-\infty < k < j < \infty) , \tag{6.1a}$$

$$g_j : \mathcal{U}_- \to \mathcal{Y}_j \qquad (-\infty < j < \infty) , \tag{6.1b}$$

$$e_k : \mathcal{U}_k \to \mathcal{Y}_+ \qquad (-\infty < k < \infty) , \tag{6.1c}$$

$$d : \mathcal{U}_- \to \mathcal{Y}_+ . \tag{6.1d}$$

Define

$$\mathcal{U} = \mathcal{U}_- \oplus (\cdots \oplus \mathcal{U}_{-1} \oplus \mathcal{U}_0 \oplus \mathcal{U}_1 \oplus \cdots) , \tag{6.2a}$$

$$\mathcal{Y} = (\cdots \oplus \mathcal{Y}_{-1} \oplus \mathcal{Y}_0 \oplus \mathcal{Y}_1 \oplus \cdots) \oplus \mathcal{Y}_+ . \tag{6.2b}$$

Given a tolerance $\gamma$, we say that $B \in \mathcal{L}(\mathcal{U}, \mathcal{Y})$ is a *solution of the four block problem* associated with data (6.1a)-(6.1d) and the tolerance $\gamma$ if $\|B\| \le \gamma$ and

$$P_{\mathcal{Y}_j} B \mid \mathcal{U}_k = f_{j,k} \qquad (-\infty < k < j < \infty)\,, \tag{6.3a}$$

$$P_{\mathcal{Y}_j} B \mid \mathcal{U}_- = g_j \qquad (-\infty < j < \infty)\,, \tag{6.3b}$$

$$P_{\mathcal{Y}_+} B \mid \mathcal{U}_k = e_k \qquad (-\infty < k < \infty)\,, \tag{6.3c}$$

$$P_{\mathcal{Y}_+} B \mid \mathcal{U}_- = d\,. \tag{6.3d}$$

As before, let $\vec{\mathcal{U}} = \oplus\{\mathcal{U}_k : k \in \mathbb{Z}\}$ and $\vec{\mathcal{Y}} = \oplus\{\mathcal{Y}_k : k \in \mathbb{Z}\}$. In this case we can partition B into a $2 \times 2$ operator matrix,

$$B = \begin{bmatrix} G & F \\ D & E \end{bmatrix} : \begin{bmatrix} \mathcal{U}_- \\ \vec{\mathcal{U}} \end{bmatrix} \to \begin{bmatrix} \vec{\mathcal{Y}} \\ \mathcal{Y}_+ \end{bmatrix},$$

where G, E and D are entirely determined by the data,

$$G = [\ldots, g_{-1}, g_0, g_1, \ldots]^{tr}\,, \quad E = [\ldots, e_{-1}, e_0, e_1, \ldots]\,, \quad D = d\,,$$

and $F = (F_{j,k})_{j,k=-\infty}^{\infty}$ is a doubly finite operator matrix of which the strictly lower triangular part is determined by the data, namely $F_{j,k} = f_{j,k}$ for $j > k$. Thus, if the spaces $\mathcal{U}_-$ and $\mathcal{Y}_+$ are trivial, then the nonstationary four block problem is just a nonstationary Nehari problem.

To describe the solution of the nonstationary four block problem we consider for each $n \in \mathbb{Z}$ the operator matrix

$$A_n = \begin{bmatrix} g_{n+1} & \cdots & f_{n+1,n-1} & f_{n+1,n} \\ g_{n+2} & \cdots & f_{n+2,n-1} & f_{n+2,n} \\ \vdots & & \vdots & \vdots \\ d & \cdots & e_{n-1} & e_n \end{bmatrix}, \tag{6.4}$$

and the auxiliary spaces

$$\mathcal{H}_n = \mathcal{U}_- \oplus (\cdots \oplus \mathcal{U}_{n-1} \oplus \mathcal{U}_n)\,, \tag{6.5a}$$

$$\mathcal{H}'_n = (\mathcal{Y}_{n+1} \oplus \mathcal{Y}_{n+2} \oplus \cdots) \oplus \mathcal{Y}_+\,. \tag{6.5b}$$

Notice that the operator matrix $A_n$ in (6.4) is entirely determined by the data. We shall refer to the operator $A_n$ as the n-th *generalized Hankel operator* corresponding to the nonstationary four block problem associated with the data (6.1a)-(6.1d).

**THEOREM 6.1**. *The nonstationary four block problem associated with the data (6.1a)-(6.1d) and the tolerance $\gamma$ has a solution if and only if for each $n \in \mathbb{Z}$ the operator matrix $A_n$ in*

*(6.4) defines a bounded linear operator from $\mathcal{H}_n$ to $\mathcal{H}'_n$ of norm at most $\gamma$.*

**PROOF**. If a solution B of the four block problem exists, then $A_n$ is a compression of B for each n, and hence $A_n$ is bounded and $\|A_n\| \leq \|B\| \leq \gamma$ for each n.

To prove the reverse implication assume that $A_n : \mathcal{H}_n \to \mathcal{H}'_n$ is a bounded linear operator of norm at most $\gamma$ for each n. For each $n \in \mathbb{Z}$ put $\mathcal{M}_n = \mathcal{Y} \ominus \mathcal{H}'_n$, where $\mathcal{Y}$ is given by (6.2b), and set $\mathcal{K}_n = \mathcal{Y}$. The spaces $\mathcal{H}_n$, $\mathcal{M}_n$ and $\mathcal{K}_n$ for $n \in \mathbb{Z}$, satisfy the conditions (XII.1.1) of the three chains completion theorem (Theorem XII.1.1), and

$$A_n : \mathcal{H}_n \to \mathcal{K}_n \ominus \mathcal{M}_n \qquad (n \in \mathbb{Z}) \tag{6.6}$$

are bounded linear operators such that $d_\infty = \sup\{\|A_n\| : n \in \mathbb{Z}\} \leq \gamma$. Since the operators $A_n$ for $n \in \mathbb{Z}$ are defined by (6.4), the compatibility condition (i) in Theorem XII.1.1 is also fulfilled. Thus, according to Theorem XII.1.1, the operators $A_n$ in (6.6) have an interpolant $B \in \mathcal{L}(\mathcal{U}, \mathcal{Y})$ with $\|B\| \leq \gamma$. The fact that $(I - P_{\mathcal{M}_n})B \,|\, \mathcal{H}_n = A_n$ for each n implies that B satisfies (6.3a)-(6.3d), and thus B is a solution of the nonstationary four block problem. This completes the proof.

Note that the proof of Theorem 6.1 shows that the set of all solutions of the nonstationary four block problem associated with the data (6.1a)-(6.1d) and the tolerance $\gamma$ is precisely equal to the set of all solutions for the three chain completion problem for the operators $A_n$ in (6.4), (6.6) with tolerance $\gamma$.

We shall now show conversely that the three chains completion problem can be restated as a (somewhat more general version of a) nonstationary four block problem. Recall (see Section XII.1) that the data for the three chains completion problem consist of operators

$$A_k : \mathcal{H}_k \to \mathcal{K}_k \ominus \mathcal{M}_k \qquad (k \in \mathbb{Z}) \tag{6.7}$$

where $\mathcal{H}_k \subset \mathcal{U}$ and $\mathcal{M}_k \subset \mathcal{K}_k \subset \mathcal{Y}$ for $k \in \mathbb{Z}$ are Hilbert spaces satisfying the inclusion relations

$$\mathcal{H}_{k-1} \subset \mathcal{H}_k, \ \mathcal{M}_{k-1} \subset \mathcal{M}_k, \text{ and } \mathcal{K}_{k-1} \subset \mathcal{K}_k \qquad (k \in \mathbb{Z}). \tag{6.8}$$

We shall also assume that the operators $A_k$ in (6.7) satisfy the compatibility condition, that is,

$$(I - P_{\mathcal{M}_k})A_{k-1} = A_k \,|\, \mathcal{H}_{k-1} \qquad (k \in \mathbb{Z}). \tag{6.9}$$

Given this data the three chains completion problem is to find all bounded linear operators B from $\mathcal{U}$ into $\mathcal{Y}$ such that

$$B\mathcal{H}_k \subset \mathcal{K}_k \text{ and } (I - P_{\mathcal{M}_k})B \mid \mathcal{H}_k = A_k \qquad (k \in \mathbb{Z}) . \tag{6.10}$$

As before (see Section XII.1) any $B \in \mathcal{L}(\mathcal{U}, \mathcal{Y})$ satisfying (6.10) will be called an *interpolant* of the operators in (6.7).

To restate the above problem as a nonstationary four block problem we decompose the spaces $\mathcal{U}$ and $\mathcal{Y}$ in the following way:

$$\mathcal{U} = \mathcal{L}_- \oplus (\cdots \oplus \mathcal{L}_{-1} \oplus \mathcal{L}_0 \oplus \mathcal{L}_1 \oplus \cdots) \oplus \mathcal{L}_+ , \tag{6.11}$$

$$\mathcal{Y} = \mathcal{D}'_- \oplus (\cdots \oplus \mathcal{D}'_{-1} \oplus \mathcal{D}'_0 \oplus \mathcal{D}'_1 \oplus \cdots) \oplus \mathcal{D}'_+ , \tag{6.12}$$

where

$$\mathcal{L}_- = \bigcap_{n \in \mathbb{Z}} \mathcal{H}_n , \qquad \mathcal{D}'_- = \bigcap_{n \in \mathbb{Z}} \mathcal{M}_n ,$$

$$\mathcal{L}_+ = \mathcal{U} \ominus (\bigvee_{n \in \mathbb{Z}} \mathcal{H}_n) , \qquad \mathcal{D}'_+ = \mathcal{Y} \ominus (\bigvee_{n \in \mathbb{Z}} \mathcal{M}_n) ,$$

$$\mathcal{L}_j = \mathcal{H}_j \ominus \mathcal{H}_{j-1} , \qquad \mathcal{D}'_j = \mathcal{M}_j \ominus \mathcal{M}_{j-1} .$$

Furthermore, we set

$$\mathcal{L} = \bigoplus_{j=-\infty}^{\infty} \mathcal{L}_j \text{ and } \mathcal{D}' = \bigoplus_{j=-\infty}^{\infty} \mathcal{D}'_j . \tag{6.13}$$

Let us remark that the notation is chosen to agree with the notation that will be used later in Chapter XIV. In particular, the primes on the $\mathcal{D}$-spaces will appear naturally in Chapter XIV.

Next, we use the decompositions in (6.11), (6.12) and (6.13) to partition the operators $A_k$ in (6.7). For $-\infty < i < j < +\infty$ we define the following operators

$$A_{j,i} : \mathcal{L}_i \to \mathcal{D}'_j , \qquad A_{j,-\infty} : \mathcal{L}_- \to \mathcal{D}'_j ,$$

$$A_{+\infty,i} : \mathcal{L}_i \to \mathcal{D}'_+ , \qquad A_{+\infty,-\infty} : \mathcal{L}_- \to \mathcal{D}'_+ ,$$

by setting

$$A_{j,i}x = P_{\mathcal{D}'_j} A_i x \qquad (x \in \mathcal{L}_i) \tag{6.14a}$$

$$A_{+\infty,i}x = P_{\mathcal{D}'_+} A_i x \qquad (x \in \mathcal{L}_i) \tag{6.14b}$$

$$A_{j,-\infty}x = P_{\mathcal{D}'_j} A_{j-1} x \qquad (x \in \mathcal{L}_-) \tag{6.14c}$$

$$A_{+\infty,-\infty}x = P_{\mathcal{D}'_+} A_i x \qquad (x \in \mathcal{L}_-) \tag{6.14d}$$

Later (see Corollary 6.4 (ii) below) we shall show that in (6.14d) the choice of the integer i is irrelevant. Now consider the set of operators

$$\mathbb{A}_{\mathcal{L}} = \{A_{j,i} : -\infty \leq i < j \leq +\infty\} . \tag{6.15}$$

We shall refer to $\mathbb{A}_{\mathcal{L}}$ as the *lower triangular array associated to* the operators $A_k$ in (6.7). We view $\mathbb{A}_{\mathcal{L}}$ as a partially given operator matrix (the question marks denote unspecified entries),

$$\left[\begin{array}{c|ccccccc|c}
? & \cdots & ? & ? & ? & ? & ? & \cdots & ? \\
\hline
\vdots & \ddots & \vdots & \vdots & \vdots & \vdots & \vdots & & \vdots \\
A_{-2,-\infty} & & ? & ? & ? & ? & ? & \cdots & ? \\
A_{-1,-\infty} & \cdots & A_{-1,-2} & ? & ? & ? & ? & \cdots & ? \\
A_{0,-\infty} & \cdots & A_{0,-2} & A_{0,-1} & \underline{?} & ? & ? & \cdots & ? \\
A_{1,-\infty} & \cdots & A_{1,-2} & A_{1,-1} & A_{1,0} & ? & ? & \cdots & ? \\
A_{2,-\infty} & \cdots & A_{2,-2} & A_{2,-1} & A_{2,0} & A_{2,1} & ? & & ? \\
\vdots & & \vdots & \vdots & \vdots & \vdots & & \ddots & \vdots \\
\hline
A_{+\infty,-\infty} & \cdots & A_{+\infty,-2} & A_{+\infty,-1} & A_{+\infty,0} & A_{+\infty,1} & & \cdots & ?
\end{array}\right], \tag{6.16}$$

which partitions into nine blocks as indicated by the dotted lines. An operator B from $\mathcal{U}$ into $\mathcal{Y}$ will be called a *completion* of the array $\mathbb{A}_{\mathcal{L}}$ if

$$P_{\mathcal{D}'_j} B \mid \mathcal{L}_i = A_{j,i} \text{ for } -\infty < i < j < \infty , \tag{6.17a}$$

$$P_{\mathcal{D}'_+} B \mid \mathcal{L}_i = A_{+\infty,i} \text{ for } -\infty < i < \infty , \tag{6.17b}$$

$$P_{\mathcal{D}'_j} B \mid \mathcal{L}_- = A_{j,-\infty} \text{ for } -\infty < j < \infty , \tag{6.17c}$$

$$P_{\mathcal{D}'_+} B \mid \mathcal{L}_- = A_{+\infty,-\infty} . \tag{6.17d}$$

Notice that in this case the strictly lower triangular part of the operator matrix of B relative to the decompositions (6.11) and (6.12) is precisely the array $\mathbb{A}_{\mathcal{L}}$.

**THEOREM 6.2.** *Let $\mathbb{A}_{\mathcal{L}}$ be the lower triangular array associated to the operators $A_k$ in (6.7). Then $B \in \mathcal{L}(\mathcal{U}, \mathcal{Y})$ is a completion of $\mathbb{A}_{\mathcal{L}}$ if and only if B is an interpolant of the operators in (6.7).*

Notice that the top block row and the right block column in the operator matrix (6.16) consist entirely of question marks. Thus in order to find a completion B of the array $\mathbb{A}_{\mathcal{L}}$ of norm of most $\gamma$, one solves first the nonstationary four block problem associated with the operators (6.14a)-(6.14d) and tolerance $\gamma$, and next one applies simply two times the Parrott lemma (Lemma XII.2.1) to take care of the top block row and the right block column. From this remark and Theorem 6.2 it follows that the three chains completion theorem can be derived as a corollary of Theorem 6.1. For the proof of Theorem 6.2 we need the following lemma.

**LEMMA 6.3.** *Let* $A_n$ *for* $n \in \mathbb{Z}$ *be the operators defined in (6.7) satisfying the compatibility condition (6.9). Let* $h_k \in \mathcal{H}_k$. *Then*

(i) $P_{\mathcal{D}'_{j+1}} A_j h_k = P_{\mathcal{D}'_{j+1}} A_k h_k$ *for* $j \geq k$;

(ii) $P_{\mathcal{D}'_+} A_j h_k = P_{\mathcal{D}'_+} A_k h_k$ *for* $j \geq k$.

**PROOF.** First we show that

$$A_j h_k = (I - P_{\mathcal{M}_j}) A_k h_k \qquad (j \geq k)\,. \tag{6.18}$$

Write $j = k + n$. Since $A_k$ maps $\mathcal{H}_k$ into the orthogonal complement of $\mathcal{M}_k$, there is nothing to prove for $n = 0$. We proceed by induction on n. Assume (6.18) holds for $j = k + n$, where $n \geq 0$. Consider $j = k + n + 1$. Thus $k \leq j - 1$, and hence $h_k \in \mathcal{H}_{j-1}$. But then we can use the compatibility condition (6.9) to show that

$$A_j h_k = (I - P_{\mathcal{M}_j}) A_{j-1} h_k\,. \tag{6.19}$$

Since $j - 1 = k + n$, our induction hypothesis implies that $A_{j-1} h_k = (I - P_{\mathcal{M}_{j-1}}) A_k h_k$. Recall that $\mathcal{M}_{j-1} \subset \mathcal{M}_j$. Thus $(I - P_{\mathcal{M}_j}) P_{\mathcal{M}_{j-1}}$ is the zero operator, and therefore (6.19) gives

$$A_j h_k = (I - P_{\mathcal{M}_j})(I - P_{\mathcal{M}_{j-1}}) A_k h_k = (I - P_{\mathcal{M}_j}) A_k h_k\,,$$

which proves (6.18).

Notice that $P_{\mathcal{D}'_{j+1}} P_{\mathcal{M}_j}$ and $P_{\mathcal{D}'_+} P_{\mathcal{M}_j}$ are both zero operators. Hence

$$P_{\mathcal{D}'_{j+1}} (I - P_{\mathcal{M}_j}) = P_{\mathcal{D}'_{j+1}} \text{ and } P_{\mathcal{D}'_+} (I - P_{\mathcal{M}_j}) = P_{\mathcal{D}'_+}\,.$$

By using these identities in (6.18), we obtain (i) and (ii) in the lemma, which completes the proof.

Since $\mathcal{L}_- \subset \mathcal{H}_k$ for each $k \in \mathbb{Z}$, the above lemma yields the following corollary.

**CORALLARY 6.4.** *Let* $A_n$ *for* $n \in \mathbb{Z}$ *be the operators in (6.7) satisfying (6.9). Let* $x \in \mathcal{L}_-$. *Then*

(i) $P_{\mathcal{D}'_{j+1}} A_j x = P_{\mathcal{D}'_{j+1}} A_k x$ *whenever* $j \geq k$,

(ii) $P_{\mathcal{D}'_+} A_j x = P_{\mathcal{D}'_+} A_k x$ *for all* j *and* k.

From Corollary 6.4 (ii) it follow that in (6.14d) the particular choice of the integer i is not important, and hence the operator $A_{+\infty,-\infty}$ is well-defined. Also, notice that Corollary 6.4 (i) implies that for each $k < j$

$$A_{j,-\infty} x = P_{\mathcal{D}'_j} A_k x \qquad (x \in \mathcal{L}_-) . \tag{6.20}$$

Next, we represent the operators $A_k$ in (6.7) by operator matrices using the spaces appearing in (6.11) and (6.12). First, notice that

$$\mathcal{H}_k = \mathcal{L}_- \oplus (\cdots \oplus \mathcal{L}_{k-1} \oplus \mathcal{L}_k) , \tag{6.21}$$

$$\mathcal{Y} \ominus \mathcal{M}_k = (\mathcal{D}'_{k+1} \oplus \mathcal{D}'_{k+2} \oplus \cdots) \oplus \mathcal{D}'_+ . \tag{6.22}$$

These spaces are used in the following result.

**PROPOSITION 6.5**. *Relative to the decompositions (6.21) and (6.22) the operator* $A_k$, *which maps* $\mathcal{H}_k$ *into* $\mathcal{Y} \ominus \mathcal{M}_k$, *has the following operator matrix representation*

$$A_k = \begin{bmatrix} A_{k+1,-\infty} & \dots & A_{k+1,k-1} & A_{k+1,k} \\ A_{k+2,-\infty} & \dots & A_{k+2,k-1} & A_{k+2,k} \\ \vdots & & \vdots & \vdots \\ A_{-\infty,+\infty} & \dots & A_{+\infty,k-1} & A_{+\infty,k} \end{bmatrix} , \tag{6.23}$$

*with the entries being given by (6.14a)-(6.14d).*

**PROOF**. Fix $-\infty < i \leq k < j < \infty$, and take $x \in \mathcal{L}_i$. Notice that $x \in \mathcal{H}_i \subset \mathcal{H}_k$. Thus, by applying Lemma 6.3 (i) twice, we see that

$$(A_k)_{j,i} = P_{\mathcal{D}'_j} A_k x = P_{\mathcal{D}'_j} A_{j-1} x = P_{\mathcal{D}'_j} A_i x = A_{j,i} x .$$

Similarly, if $k < j < \infty$ and $x \in \mathcal{L}_-$. Then $x \in \mathcal{H}_k$, and thus by (6.20) we have

$$(A_k)_{j,-\infty} x = P_{\mathcal{D}'_j} A_k x = A_{j,-\infty} x .$$

Next, take $x \in \mathcal{L}_i$, where $-\infty < i \leq k$. By Lemma 6.3 (ii) we have

$$(A_k)_{+\infty,i} = P_{\mathcal{D}'_+} A_k x = P_{\mathcal{D}'_+} A_i x = A_{+\infty,i} x .$$

Finally, if $x \in \mathcal{L}_-$, then, by (6.14d) and the fact that the particular choice of the integer i in

(6.14d) is irrelevant, we have

$$(A_k)_{+\infty,-\infty} = P_{\mathcal{D}'_+} A_k x = P_{\mathcal{D}'_+} A_j x = A_{+\infty,-\infty} x ,$$

which completes the proof.

**PROOF OF THEOREM 6.2.** Assume B is an interpolant of the operators $A_k$ in (6.7). Then $(I - P_{\mathcal{M}_k})B \mid \mathcal{H}_k = A_k$ for each $k \in \mathbb{Z}$. Since the operator matrix of $A_k$ relative to the decompositions (6.21) and (6.22) is given by the right hand side of (6.23), we see that (6.14a)-(6.14d) are fulfilled. This holds for each $k \in \mathbb{Z}$, and thus B is a completion of $\mathbb{A}_{\mathcal{L}}$.

Conversely, assume B is a completion of $\mathbb{A}_{\mathcal{L}}$. Then formulas (6.14a)-(6.14d) and (6.23) imply that $A_k = (I - P_{\mathcal{M}_k})B \mid \mathcal{H}_k$ for each $k \in \mathbb{Z}$. It follows that

$$B\mathcal{H}_k \subset \{(I - P_{\mathcal{M}_k})B + P_{\mathcal{M}_k}B\}\mathcal{H}_k \subset A_k\mathcal{H}_k + \mathcal{M}_k . \tag{6.24}$$

By (6.7) the operator $A_k$ maps $\mathcal{H}_k$ into $\mathcal{K}_k$. Since $\mathcal{M}_k \subset \mathcal{K}_k$, we see from (6.24) that $B\mathcal{H}_k \subset \mathcal{K}_k$. Hence (6.10) is proved. Thus B is an interpolant of the operators in (6.7). This completes the proof.

Notice that (6.23) shows that the operators $A_k$ in (6.7) are precisely the generalized Hankel operators corresponding to the nonstationary four block problem associated with the data (6.14a)-(6.14d). Thus we can use Theorems 6.1 and 6.2 (see also the remark in the paragraph directly after Theorem 6.2) to show that the three chains completion problem for the operators $A_k$ in (6.7) has a solution B of norm bounded by $\gamma$ if and only if $\|A_k\| \le \gamma$ for each $k \in \mathbb{Z}$.

The stationary analogue of Theorem 6.2, with the three chains completion problem replaced by the problem of commutant lifting, can be found in Section VIII.7 of Foias-Frazho [4].

## *Notes to Chapter XIII:*

The material in Sections 1, 2, 4 and 6 is taken from the paper Foias-Frazho-Gohberg-Kaashoek [2]. The other two sections in this chapter use similar techniques.

## CHAPTER XIV

# PARAMETERIZATION OF ALL SOLUTIONS OF THE THREE CHAINS COMPLETION PROBLEM

In this chapter it is shown that the set of all solutions of the three chains completion problem with a prescribed tolerance is parameterized by a natural set of contractive upper triangular operators. The result is used to derive the set of all solutions of a nonstationary Nehari problem with the data being given in state space form.

## XIV.1. MAIN THEOREM

Recall that for the three chains completion problem the given data are bounded linear operators

$$A_k : \mathcal{H}_k \to \mathcal{K}_k \ominus \mathcal{M}_k \qquad (k \in \mathbb{Z}). \tag{1.1}$$

Here $\mathcal{H}_k \subset \mathcal{U}$ and $\mathcal{M}_k \subset \mathcal{K}_k \subset \mathcal{Y}$ for $k \in \mathbb{Z}$ are Hilbert spaces satisfying the inclusion relations

$$\mathcal{H}_{k-1} \subset \mathcal{H}_k\,,\ \ \mathcal{M}_{k-1} \subset \mathcal{M}_k\,,\ \ \mathcal{K}_{k-1} \subset \mathcal{K}_k \qquad (k \in \mathbb{Z}). \tag{1.2}$$

Given the operators (1.1) and a tolerance $\gamma$ the problem is to find an operator B from $\mathcal{U}$ into $\mathcal{Y}$ such that $||B|| \leq \gamma$ and

$$B\mathcal{H}_k \subset \mathcal{K}_k\,,\ \ (I - P_{\mathcal{M}_k})B\,|\,\mathcal{H}_k = A_k \qquad (k \in \mathbb{Z}). \tag{1.3}$$

From Theorem XII.1.1 we know that for an operator B to exist it is necessary and sufficient that the operators $A_k$ for $k \in \mathbb{Z}$ satisfy the compatibility conditions

$$(I - P_{\mathcal{M}_k})A_{k-1} = A_k\,|\,\mathcal{H}_{k-1} \qquad (k \in \mathbb{Z}), \tag{1.4}$$

and that

$$\gamma \geq d_\infty := \sup\,\{||A_k|| : k \in \mathbb{Z}\}\,. \tag{1.5}$$

In this section we assume that the conditions (1.4) and (1.5) are fulfilled, and our aim is to describe the set of all solutions B.

First, let us notice that

$$\operatorname{ran} P_{\mathcal{M}_k} A_{k-1} \subset \mathcal{M}_k \ominus \mathcal{M}_{k-1} \qquad (k \in \mathbb{Z}). \tag{1.6}$$

Indeed, take $h \in \mathcal{H}_{k-1}$. Then, according to (1.1), we have $A_{k-1}h$ is orthogonal to $\mathcal{M}_{k-1}$. Since $\mathcal{M}_{k-1} \subset \mathcal{M}_k$, we may write $A_{k-1}h = x + y$, where $x \perp \mathcal{M}_k$ and $y \in \mathcal{M}_k \ominus \mathcal{M}_{k-1}$. But then $P_{\mathcal{M}_k} A_{k-1} h = y \in \mathcal{M}_k \ominus \mathcal{M}_{k-1}$, as desired.

To state our main result we need to consider a number of auxiliary Hilbert spaces. For each $k \in \mathbb{Z}$ put

$$D_k = (\gamma^2 I - A_k^* A_k)^{1/2}, \qquad \mathcal{D}_k = \overline{\operatorname{ran} D_k}\,; \tag{1.7}$$

$$\mathcal{D}'_k = \mathcal{M}_k \ominus \mathcal{M}_{k-1}\,; \tag{1.8}$$

$$\mathcal{F}_k = \overline{D_k \mathcal{H}_{k-1}}, \qquad \mathcal{G}_k = \mathcal{D}_k \ominus \mathcal{F}_k\,; \tag{1.9}$$

$$\mathcal{F}'_k = \operatorname{cl}\left\{ \begin{bmatrix} D_{k-1} h_{k-1} \\ P_{\mathcal{M}_k} A_{k-1} h_{k-1} \end{bmatrix} : h_{k-1} \in \mathcal{H}_{k-1} \right\}, \tag{1.10}$$

$$\mathcal{G}'_k = \begin{pmatrix} \mathcal{D}_{k-1} \\ \mathcal{D}'_k \end{pmatrix} \ominus \mathcal{F}'_k. \tag{1.11}$$

Here the first in the right hand side of (1.11) denotes the Hilbert space direct sum $\mathcal{D}_{k-1} \oplus \mathcal{D}'_k$. By (1.5) the operator $D_k$ is well defined. Notice that $D_k$ acts on $\mathcal{H}_k$. Since $\mathcal{H}_{k-1} \subset \mathcal{H}_k$, we may consider the restriction of $D_k$ to $\mathcal{H}_{k-1}$, and hence $\mathcal{F}_k$ is a well defined subspace of $\mathcal{D}_k$. It follows that $\mathcal{G}_k$ is also well defined. According to (1.6) the vector $D_{k-1}h_{k-1} \oplus P_{\mathcal{M}_k} A_{k-1} h_{k-1}$ belongs to $\mathcal{D}_{k-1} \oplus \mathcal{D}'_k$ whenever $h_{k-1} \in \mathcal{H}_{k-1}$. Hence $\mathcal{F}'_k$ is a subspace of $\mathcal{D}_{k-1} \oplus \mathcal{D}'_k$, and therefore $\mathcal{G}'_k$ is well defined too.

We also need a number of auxiliary operators. For each $k \in \mathbb{Z}$ we define

$$\pi_k : \begin{pmatrix} \mathcal{D}_{k-1} \\ \mathcal{D}'_k \end{pmatrix} \to \mathcal{D}_{k-1}, \quad \pi_k \begin{pmatrix} d_{k-1} \\ d'_k \end{pmatrix} = d_{k-1}\,; \tag{1.12}$$

$$\pi'_k : \begin{pmatrix} \mathcal{D}_{k-1} \\ \mathcal{D}'_k \end{pmatrix} \to \mathcal{D}'_k, \quad \pi'_k \begin{pmatrix} d_{k-1} \\ d'_k \end{pmatrix} = d'_k\,; \tag{1.13}$$

$$\omega_k : \mathcal{F}_k \to \begin{bmatrix} \mathcal{D}_{k-1} \\ \mathcal{D}'_k \end{bmatrix}, \ \omega_k(D_k h_{k-1}) = \begin{bmatrix} D_{k-1} h_{k-1} \\ P_{\mathcal{M}_k} A_{k-1} h_{k-1} \end{bmatrix}; \tag{1.14}$$

$$\rho_k : \mathcal{D}_k \to \mathcal{D}_{k-1}, \quad \rho_k = \pi_k \omega_k P_{\mathcal{F}_k}. \tag{1.15}$$

In the fourth paragraph of Section XII.3 we have seen that $\omega_k$ is a well-defined isometry from $\mathcal{F}_k$ onto $\mathcal{F}'_k$. Since $\pi_k$ and $\pi'_k$ are orthogonal projections, we conclude that $\rho_k$ is a contraction. In what follows we write $\Phi_\rho(m, n)$ for the state transition matrix generated by $\rho_k$, that is,

$$\Phi_\rho(m, n) = \rho_{m+1}\rho_{m+2} \cdots \rho_n : \mathcal{D}_n \to \mathcal{D}_m \qquad (m < n);$$

$$\Phi_\rho(m, m) = I_{\mathcal{D}_m}.$$

Next, we introduce four upper triangular operators matrices, namely

$$\Theta_{11} = \left[(\Theta_{11})_{j,k}\right]_{j,k=-\infty}^{\infty}, \ (\Theta_{11})_{j,k} : \mathcal{G}'_k \to \mathcal{G}_j,$$

$$\Theta_{12} = \left[(\Theta_{12})_{j,k}\right]_{j,k=-\infty}^{\infty}, \ (\Theta_{12})_{j,k} : \mathcal{D}_k \to \mathcal{G}_j,$$

$$\Theta_{21} = \left[(\Theta_{21})_{j,k}\right]_{j,k=-\infty}^{\infty}, \ (\Theta_{21})_{j,k} : \mathcal{G}'_k \to \mathcal{D}'_j,$$

$$\Theta_{22} = \left[(\Theta_{22})_{j,k}\right]_{j,k=-\infty}^{\infty}, \ (\Theta_{22})_{j,k} : \mathcal{D}_k \to \mathcal{D}'_j,$$

where

$$(\Theta_{11})_{j,k} = \begin{cases} P_{\mathcal{G}_j}\Phi_\rho(j, k-1)\pi_k \mid \mathcal{G}'_k, & j < k, \\ 0, & j \geq k, \end{cases}$$

$$(\Theta_{12})_{j,k} = \begin{cases} P_{\mathcal{G}_j}\Phi_\rho(j, k), & j \leq k, \\ 0, & j > k, \end{cases}$$

$$(\Theta_{21})_{j,k} = \begin{cases} \pi'_j\omega_j P_{\mathcal{F}_j}\Phi_\rho(j, k-1)\pi_k \mid \mathcal{G}'_k, & j < k, \\ \pi'_k \mid \mathcal{G}'_k, & j = k, \\ 0, & j > k, \end{cases}$$

and

$$(\Theta_{22})_{j,k} = \begin{cases} \pi'_j\omega_j P_{\mathcal{F}_j}\Phi_\rho(j, k), & j \leq k, \\ 0, & j > k. \end{cases}$$

Notice that the operator matrix $\Theta_{11}$ is strictly upper triangular.

Now, let $R = (R_{j,k})_{j,k=-\infty}^{\infty}$, where $R_{j,k}$ maps $\mathcal{G}_k$ into $\mathcal{G}'_j$, be an upper triangular operator matrix (that is, $R_{j,k} = 0$ for $j > k$). Given such an R we write C(R) for the upper triangular operator matrix defined by

$$C(R) := \Theta_{22} + \Theta_{21}R(I - \Theta_{11}R)^{-1}\Theta_{12}\,. \tag{1.16}$$

The products and inverse in (1.16) have to be understood algebraically and have to be carried out in the algebra of upper triangular operator matrices. Notice that $\Theta_{11}R$ is strictly upper triangular. Thus $(I - \Theta_{11}R)^{-1}$ is well defined algebraically and is upper triangular. In fact, for each j and k in $\mathbb{Z}$ we have

$$\left[(I - \Theta_{11}R)^{-1}\right]_{j,k} = \sum_{\nu=0}^{\infty} \left[(\Theta_{11}R)^{\nu}\right]_{j,k}, \tag{1.17}$$

and the sum in the right hand side of (1.17) is a finite one.

Consider the doubly infinite Hilbert space direct sums

$$\mathcal{G} = \bigoplus_{-\infty}^{\infty} \mathcal{G}_k\,, \qquad \mathcal{G}' = \bigoplus_{-\infty}^{\infty} \mathcal{G}'_k\,; \tag{1.18}$$

$$\mathcal{D} = \bigoplus_{-\infty}^{\infty} \mathcal{D}_k\,, \qquad \mathcal{D}' = \bigoplus_{-\infty}^{\infty} \mathcal{D}'_k\,. \tag{1.19}$$

In what follows we shall always assume that the operator matrix R in (1.16) is a contractive operator from $\mathcal{G}$ into $\mathcal{G}'$. In this case (see Proposition 1.2 below) the k-th block column of the operator matrix C(R) in (1.16) turns out to be a contractive operator from $\mathcal{D}_k$ into $\mathcal{D}'$ for all k in $\mathbb{Z}$.

We are now ready to state the main result of this section.

**THEOREM 1.1.** *Assume* $A_k$ *from* $\mathcal{H}_k$ *to* $\mathcal{K}_k \ominus \mathcal{M}_k$, $k \in \mathbb{Z}$, *satisfy conditions (1.4) and (1.5). Let* R *from* $\mathcal{G}$ *to* $\mathcal{G}'$ *be a block upper triangular and contractive operator. Then the formula*

$$B(R)h = \begin{cases} A_k h + \sum_{j=-\infty}^{k} C_{j,k}(R)D_k h\,, & h \in \mathcal{H}_k\,, \\ 0\,, & h \perp \bigvee_{n\in\mathbb{Z}} \mathcal{H}_n\,, \end{cases} \tag{1.20}$$

*where*

$$C_{j,k}(R) = (\Theta_{22} + \Theta_{21}R(I - \Theta_{11}R)^{-1}\Theta_{12})_{j,k} \qquad (j \leq k),$$

*defines a linear operator* B(R) *from* $\mathcal{U}$ *to* $\mathcal{Y}$ *such that* $\|B(R)\| \leq \gamma$ *and* B(R) *is a solution of the three chains completion problem associated with the operators* $A_k$ $(k \in \mathbb{Z})$ *and tolerance* $\gamma$.

*Conversely, if* B *is a solution of the three chains completion problem associated with* $A_k$ $(k \in \mathbb{Z})$ *and tolerance* $\gamma$, *then there exists a unique block upper triangular contractive operator* R *from* $\mathcal{G}$ *to* $\mathcal{G}'$ *such that* $Bh = B(R)h$ *for each* $h \in \bigvee \{\mathcal{H}_n : n \in \mathbb{Z}\}$.

If in Theorem 1.1 the free parameter R is taken to be the zero operator, then the resulting interpolant B coincides with the interpolant $B_\gamma$ in Theorem XII.3.1. This fact is another reason for calling $B_\gamma$ the central interpolant.

The proof of Theorem 1.1 will be given in the next two sections. In Section 2 it will be shown that indeed the operator B(R) defined by (1.20) yields a solution of the three chains completion problem for the operators (1.1). The converse implication, which requires the construction of a contractive block upper triangular operator R, will be proved in Section 3. We conclude this section with an alternative representation of the operator matrix C(R) in (1.16) which is often more convenient to work with.

Let $R = (R_{j,k})_{j,k=-\infty}^{\infty}$, where $R_{j,k}$ maps $\mathcal{G}_k$ into $\mathcal{G}_j'$, be an upper triangular operator matrix. Consider

$$W_{j,k} = \begin{pmatrix} \delta_{j,k}\omega_k & 0 \\ 0 & R_{j,k} \end{pmatrix} : \begin{pmatrix} \mathcal{F}_k \\ \mathcal{G}_k \end{pmatrix} \to \begin{pmatrix} \mathcal{F}_j' \\ \mathcal{G}_j' \end{pmatrix}. \tag{1.21}$$

Here $\delta_{j,k}$ is the Kronecker index, and j and k are arbitrary integers. Recall that $\mathcal{D}_k = \mathcal{F}_k \oplus \mathcal{G}_k$ and $\mathcal{D}_{j-1} \oplus \mathcal{D}_j' = \mathcal{F}_j' \oplus \mathcal{G}_j'$. Thus we may define

$$W_{j,k}^{(1)} : \mathcal{D}_k \to \mathcal{D}_j, \quad W_{j,k}^{(1)} = \pi_{j+1} W_{j+1,k}; \tag{1.22}$$

$$W_{j,k}^{(2)} : \mathcal{D}_k \to \mathcal{D}_j', \quad W_{j,k}^{(2)} = \pi_j' W_{j,k}. \tag{1.23}$$

Put

$$W_1(R) = (W_{j,k}^{(1)})_{j,k=-\infty}^{\infty}, \quad W_2(R) = (W_{j,k}^{(2)})_{j,k=-\infty}^{\infty}. \tag{1.24}$$

Then $W_1(R)$ and $W_2(R)$ are both upper triangular operator matrices, and $W_1(R)$ is strictly upper triangular. In particular, $(I - W_1(R))^{-1}$ is well-defined in the algebra of upper triangular operator matrices.

**PROPOSITION 1.2.** *Let* $R = (R_{j,k})_{j,k=-\infty}^{\infty}$, *where* $R_{j,k}$ *maps* $\mathcal{G}_k$ *into* $\mathcal{G}'_j$, *be upper triangular. Then the operator matrix* $C(R)$ *in (1.16) is also given by*

$$C(R) = W_2(R)(I - W_1(R))^{-1} . \tag{1.25}$$

*Furthermore, if* $R$ *is a contraction from* $\mathcal{G}$ *to* $\mathcal{G}'$, *then for each* $k \in \mathbb{Z}$

$$\sum_{j=-\infty}^{k} \|C_{j,k}(R)d_k\|^2 \leq \|d_k\|^2 \qquad (d_k \in \mathcal{D}_k) , \tag{1.26}$$

*where* $C_{j,k}(R)$ *is the* $(j, k)$*-th block entry of* $C(R)$.

**PROOF.** We split the proof into two parts.

Part (a). First we prove (1.25). For this purpose we introduce the following block diagonal doubly infinite operator matrices:

$$\Pi = \operatorname{diag}(\pi_k)_{k=-\infty}^{\infty} , \quad \Pi' = \operatorname{diag}(\pi'_k)_{k=-\infty}^{\infty} , \tag{1.27}$$

$$\Omega = \operatorname{diag}(\omega_k)_{k=-\infty}^{\infty} , \tag{1.28}$$

$$P_{\mathcal{F}} = \operatorname{diag}(P_{\mathcal{F}_k})_{k=-\infty}^{\infty} , \quad P_{\mathcal{G}} = \operatorname{diag}(P_{\mathcal{G}_k})_{k=-\infty}^{\infty} , \tag{1.29}$$

$$E_{\mathcal{G}'} = \operatorname{diag}(E_{\mathcal{G}'_k})_{k=-\infty}^{\infty} , \tag{1.30}$$

where $E_{\mathcal{G}'_k}$ is the canonical embedding of $\mathcal{G}'_k$ into $\mathcal{D}_{k-1} \oplus \mathcal{D}'_k$. Furthermore, consider the backwards block (weighted) shifts on $\mathcal{D}$

$$S^* = (\delta_{j,k-1} I_{\mathcal{D}_{k-1}})_{j,k=-\infty}^{\infty} , \tag{1.31}$$

$$S_\rho = (\delta_{j,k-1} \rho_k)_{j,k=-\infty}^{\infty} . \tag{1.32}$$

Notice that $I - S_\rho$ is invertible in the algebra of upper triangular operator matrices, and

$$((I - S_\rho)^{-1})_{j,k} = \Phi_\rho(j, k) \qquad (j \leq k) .$$

One may use the latter identity to show that the operator matrices $\Theta_{i,j}$, $1 \leq i, j \leq 2$, are given by the following formulas:

$$\Theta_{11} = P_{\mathcal{G}}(I - S_\rho)^{-1} S^* \Pi E_{\mathcal{G}'} , \quad \Theta_{12} = P_{\mathcal{G}}(I - S_\rho)^{-1} , \tag{1.33}$$

$$\Theta_{21} = \Pi' \Omega P_{\mathcal{F}}(I - S_\rho)^{-1} S^* \Pi E_{\mathcal{G}'} + \Pi' E_{\mathcal{G}'} , \tag{1.34}$$

$$\Theta_{22} = \Pi' \Omega P_{\mathcal{F}}(I - S_\rho)^{-1} . \tag{1.35}$$

All factors in these products are upper triangular operator matrices.

Now, let $R = (R_{j,k})_{j,k=-\infty}^{\infty}$, where $R_{j,k}$ maps $\mathcal{G}_k$ into $\mathcal{G}'_j$, be an upper triangular operator matrix, and let C(R) be given by (1.16). Using the algebraic identity

$$(I - cd^{-1}b)^{-1} = I + c(d - bc)^{-1}b\,,$$

together with the first part of (1.33), we see that

$$(I - \Theta_{11}R)^{-1} = I_{\mathcal{G}} + P_{\mathcal{G}}(I - S_\rho - S^*\Pi E_{\mathcal{G}'}RP_{\mathcal{G}})^{-1}S^*\Pi E_{\mathcal{G}'}R\,.$$

Multiplying the latter identity on the right by $P_{\mathcal{G}}(I - S_\rho)^{-1}$ and using

$$S^*\Pi E_{\mathcal{G}'}RP_{\mathcal{G}} = (I - S_\rho) - (I - S_\rho - S^*\Pi E_{\mathcal{G}'}RP_{\mathcal{G}})\,, \tag{1.36}$$

we obtain that

$$(I - \Theta_{11}R)^{-1}\Theta_{12} = P_{\mathcal{G}}(I - S_\rho - S^*\Pi E_{\mathcal{G}'}RP_{\mathcal{G}})^{-1}\,.$$

Using the identity (1.36) a second time, together with the formula for $\Theta_{21}$ in (1.34), we get

$$\Theta_{21}R(I - \Theta_{11}R)^{-1}\Theta_{12} =$$

$$-\Pi'\Omega P_{\mathcal{F}}(I - S_\rho)^{-1} + (\Pi'E_{\mathcal{G}'}RP_{\mathcal{G}} + \Pi'\Omega P_{\mathcal{F}})(I - S_\rho - S^*\Pi E_{\mathcal{G}'}RP_{\mathcal{G}})^{-1}\,.$$

It follows (use (1.35)) that

$$C(R) = (\Pi'E_{\mathcal{G}'}RP_{\mathcal{G}} + \Pi'\Omega P_{\mathcal{F}})(I - S_\rho - S^*\Pi E_{\mathcal{G}'}RP_{\mathcal{G}})^{-1}\,.$$

To get (1.25) notice that the upper triangular operator matrices $W_1(R)$ and $W_2(R)$ defined by (1.21)-(1.24) are also given by

$$W_1(R) = S^*\Pi(\Omega P_{\mathcal{F}} + E_{\mathcal{G}'}RP_{\mathcal{G}})\,,\quad W_2(R) = \Pi'(\Omega P_{\mathcal{F}} + E_{\mathcal{G}'}RP_{\mathcal{G}})\,.$$

Finally, since

$$S_\rho = S^*\Pi\Omega P_{\mathcal{F}}\,, \tag{1.37}$$

these identities yield (1.25).

Part (b). In this part we assume additionally that R is a contraction from $\mathcal{G}$ to $\mathcal{G}'$, and we prove (1.26). Put $W = (W_{j,k})_{j,k=-\infty}^{\infty}$, where $W_{j,k}$ is defined by (1.21). Since R is a contraction, we conclude that W is a contraction. From (1.22)-(1.24) we see that

$$W = \begin{pmatrix} SW_1(R) \\ W_2(R) \end{pmatrix} : \overset{\infty}{\underset{-\infty}{\oplus}} \mathcal{D}_k \to \overset{\infty}{\underset{-\infty}{\oplus}} \begin{pmatrix} \mathcal{D}_{k-1} \\ \mathcal{D}'_k \end{pmatrix}, \tag{1.38}$$

where S is the block forward shift from $\overset{\infty}{\underset{-\infty}{\oplus}} \mathcal{D}_k$ to $\overset{\infty}{\underset{-\infty}{\oplus}} \mathcal{D}_{k-1}$.

Fix $d_k \in \mathcal{D}_k$, and set $d = (\delta_{j,k} d_k)_{j=-\infty}^{\infty}$, where $\delta_{j,k}$ is the Kronecker delta. Put

$$y = (I - W_1(R))^{-1} d . \tag{1.39}$$

Since $(I - W_1(R))^{-1}$ is upper triangular and its i-th block diagonal entry is equal to $I_{\mathcal{D}_i}$, we obtain that $y_j = 0$ for $j > k$ and $y_k = d_k$. From the definition of y in (1.39) and from (1.25) we get

$$W_1(R)y = y - d , \quad W_2(R)y = C(R)d .$$

Thus, using (1.38), we have

$$Wy = \begin{pmatrix} SW_1(R)y \\ W_2(R)y \end{pmatrix} = \begin{pmatrix} S(y-d) \\ C(R)d \end{pmatrix} .$$

By taking finite sections it follows that

$$\begin{pmatrix} W_{k-n+1,k-n+1} & \cdots & W_{k-n+1,k} \\ & \ddots & \vdots \\ & & W_{k,k} \end{pmatrix} \begin{pmatrix} y_{k-n+1} \\ \vdots \\ y_k \end{pmatrix} = \begin{pmatrix} u \\ v \end{pmatrix},$$

where the $n \times n$ operator matrix in the left hand side is block upper triangular and

$$u = \begin{pmatrix} y_{k-n} \\ \vdots \\ y_{k-1} \end{pmatrix}, \quad v = \begin{pmatrix} C_{k-n+1,k}(R)d_k \\ \vdots \\ C_k(R)d_k \end{pmatrix} .$$

The finite sections of W are also contractions. Thus

$$\sum_{j=k-n}^{k-1} \|y_j\|^2 + \sum_{j=k-n+1}^{k} \|C_{j,k}(R)d_k\|^2 \leq \sum_{j=k-n+1}^{k} \|y_j\|^2 .$$

By bringing the first sum in the left hand side to the right we get

$$\sum_{j=k-n+1}^{k} \|C_{j,k}(R)d_k\|^2 \leq \|y_k\|^2 - \|y_{k-n}\|^2 \leq \|y_k\|^2 = \|d_k\|^2 .$$

These inequalities hold for each $n \geq 1$, and thus (1.26) is proved.

## XIV.2. PROOF OF THE MAIN THEOREM (first part)

In this section we prove the first part of Theorem 1.1 which concerns the case when the free parameter R is given. Throughout this section we use the notations and terminology introduced in the preceding section.

We begin with some preparations. Assume $B \in \mathcal{L}(\mathcal{U}, \mathcal{Y})$ is a solution of the three chains completion problem associated with the operators $A_k$ ($k \in \mathbb{Z}$) in (1.1) and the tolerance $\gamma$. Then for each k and h in $\mathcal{H}_k$ we have $Bh = A_k h + P_{\mathcal{M}_k} Bh$. Hence $\|P_{\mathcal{M}_k} Bh\|^2 \leq \gamma^2 \|h\|^2 - \|A_k h\|^2 = \|D_k h\|^2$. So there exists a unique contraction $C_k$ from $\mathcal{D}_k$ to $\mathcal{M}_k$ such that

$$P_{\mathcal{M}_k} B \,|\, \mathcal{H}_k = C_k D_k \,. \tag{2.1}$$

Since $\mathcal{H}_{k-1} \subset \mathcal{H}_k$, the operators $C_k$ and $C_{k-1}$ are related. In fact, using the compatibility conditions in (1.4), we have

$$\begin{aligned} C_k D_k \,|\, \mathcal{H}_{k-1} &= P_{\mathcal{M}_k} B \,|\, \mathcal{H}_{k-1} = (P_{\mathcal{M}_k} - P_{\mathcal{M}_{k-1}}) B \,|\, \mathcal{H}_{k-1} + P_{\mathcal{M}_{k-1}} B \,|\, \mathcal{H}_{k-1} \\ &= P_{\mathcal{M}_k}(I - P_{\mathcal{M}_{k-1}}) B \,|\, \mathcal{H}_{k-1} + C_{k-1} D_{k-1} \\ &= P_{\mathcal{M}_k} A_{k-1} + C_{k-1} D_{k-1} \,, \end{aligned}$$

and therefore

$$C_{k-1} D_{k-1} = C_k D_k \,|\, \mathcal{H}_{k-1} - P_{\mathcal{M}_k} A_{k-1} \quad (k \in \mathbb{Z}) \,. \tag{2.2}$$

The next proposition concerns the reverse implications, and can be viewed as a time varying version of Lemma IV.2.1.

**PROPOSITION 2.1.** *Let* $A_k$ *from* $\mathcal{H}_k$ *to* $\mathcal{K}_k \ominus \mathcal{M}_k$ *for* $k \in \mathbb{Z}$ *satisfy conditions (1.4) and (1.5), and let* $C_k$ *from* $\mathcal{D}_k$ *to* $\mathcal{M}_k$ *be contractions satisfying (2.2). Define* B *from* $\mathcal{U}$ *to* $\mathcal{Y}$ *by setting*

$$Bh = \begin{cases} A_k h + C_k D_k h \,, & h \in \mathcal{H}_k \,, \\ 0 \,, & h \perp \bigvee_{n \in \mathbb{Z}} \mathcal{H}_n \,. \end{cases} \tag{2.3}$$

*Then* B *is a solution of the three chains completion problem for the operators* $A_k$ ($k \in \mathbb{Z}$) *and the tolerance* $\gamma$.

**PROOF**. First we show that B is well-defined. Take $h \in \mathcal{H}_{k-1} \subset \mathcal{H}_k$. Then, using (2.2) and (1.4), we have

$$\begin{aligned} A_{k-1}h + C_{k-1}D_{k-1}h &= A_{k-1}h + C_kD_kh - P_{\mathcal{M}_k}A_{k-1}h \\ &= (I - P_{\mathcal{M}_k})A_{k-1}h + C_kD_kh \\ &= A_kh + C_kD_kh . \end{aligned}$$

Thus B is well-defined.

From the definition of B, we see that $(I - P_{\mathcal{M}_k})B \mid \mathcal{H}_k = A_k$ and that (2.1) holds. Since $C_k$ is a contraction, (2.1) implies that for f in $\mathcal{H}_k$ we have $\|Bf\|^2 \le \|A_kf\|^2 + \|D_kf\|^2 \le \gamma^2\|f\|^2$. Hence $\|B \mid \mathcal{H}_k\| \le \gamma$ for each k. By continuity, it follows that

$$\|B \mid \bigvee_{n\in\mathbb{Z}} \mathcal{H}_n\| \le \gamma . \tag{2.4}$$

Now using the fact that B is the zero operator on the orthogonal complement of $\bigvee_{n\in\mathbb{Z}} \mathcal{H}_n$, we conclude that $\|B\| \le \gamma$.

From (2.3) and $(I - P_{\mathcal{M}_k})B \mid \mathcal{H}_k = A_k$ for $k \in \mathbb{Z}$ it also follows that

$$B\mathcal{H}_k \subset (I - P_{\mathcal{M}_k})B\mathcal{H}_k + \mathcal{M}_k \subset \mathcal{K}_k \qquad (k \in \mathbb{Z}) ,$$

and thus B satisfies (1.3). Hence B is a solution which completes the proof.

The identities in (2.2) may be rewritten in the following equivalent form

$$C_{k-1}\pi_k\omega_k = C_k \mid \mathcal{F}_k - \pi_k'\omega_k \qquad (k \in \mathbb{Z}) . \tag{2.5}$$

This follows directly from the definition of $\mathcal{F}_k$ in (1.9) and from (1.12)-(1.14).

We are now ready to prove the first part of Theorem 1.1. Assume that $R = (R_{j,k})_{j,k=-\infty}^{\infty}$ from $\mathcal{G}$ to $\mathcal{G}'$ (where $\mathcal{G}$ and $\mathcal{G}'$ are defined in (1.18)) is block upper triangular and a contraction. Define C(R) by (1.16) (or equivalently, by (1.25)). As before for each j, k in $\mathbb{Z}$ let $C_{j,k}(R)$ be the (j, k)-th block entry of C(R). Define

$$C_k(R) : \mathcal{D}_k \to \mathcal{M}_k , \; C_k(R)d_k = \sum_{j=-\infty}^{k} C_{j,k}(R)d_k \qquad (k \in \mathbb{Z}) . \tag{2.6}$$

Recall (cf., (1.8)) that $\mathcal{M}_k = \cdots \oplus \mathcal{D}_{k-2}' \oplus \mathcal{D}_{k-1}' \oplus \mathcal{D}_k'$, and hence formula (1.26) in the second part of Proposition 1.2 shows that $C_k(R)$ is a well-defined contraction for each $k \in \mathbb{Z}$.

Now, use that C(R) is also given by (1.25). It follows that

$$C(R) = C(R)W_1(R) + W_2(R),$$

where $W_1(R)$ and $W_2(R)$ are defined by (1.21)-(1.24). Recall that C(R), $W_1(R)$ and $W_2(R)$ are upper triangular operator matrices and $W_1(R)$ is strictly upper triangular. We get (use (1.22)-(1.24) or (1.38)) that

$$C_{k,k}(R) = \pi'_k W_{k,k},\ C_{j,k}(R) = \sum_{v=j}^{k-1} C_{j,v}(R)\pi_{v+1} W_{v+1,k} + \pi'_j W_{j,k} \qquad (j < k).$$

Next (see (1.21)) use that

$$W_{k,k} \mid \mathcal{F}_k = \omega_k\,,\quad W_{j,k} \mid \mathcal{F}_k = 0 \qquad (j < k). \tag{2.7}$$

We conclude that

$$C_{k,k}(R) \mid \mathcal{F}_k = \pi'_k \omega_k\,, \tag{2.8a}$$

$$C_{j,k}(R) \mid \mathcal{F}_k = C_{j,k-1}(R)\pi_k \omega_k \qquad (j < k). \tag{2.8b}$$

These identities hold for each k, and together they are just equivalent to the statement that the contractions $C_k(R)$, $k \in \mathbb{Z}$, satisfy the identities in (2.5), and therefore those in (2.2). But then we can apply Proposition 2.1 to show that the operator B(R) defined by (1.20) is a solution of our three chains completion problem. This completes the proof of the first part of Theorem 1.1.

## XIV.3. PROOF OF THE MAIN THEOREM (second part)

In this section we prove the second part of Theorem 1.1 which deals with the case when a solution B of our three chains completion problem is given. Our aim is to recover the upper triangular contraction R corresponding to B.

So, let B from $\mathcal{U}$ to $\mathcal{Y}$ be a solution of the three chains completion problem associated with the operators $A_k$ ($k \in \mathbb{Z}$) in (1.1) and the tolerance $\gamma$. As we have seen in the previous section, this implies that for each $k \in \mathbb{Z}$ there exists a contraction $C_k$ from $\mathcal{D}_k$ to $\mathcal{M}_k$ such that (2.1) holds, and these contractions satisfy the identities in (2.5). We know that $\mathcal{M}_k$ is equal to the Hilbert space direct sum $\cdots \oplus \mathcal{D}'_{k-2} \oplus \mathcal{D}'_{k-1} \oplus \mathcal{D}'_k$, and hence we may partition $C_k$ as

$$C_k = \begin{pmatrix} \vdots \\ C_{k-1,k} \\ C_{k,k} \end{pmatrix} : \mathcal{D}_k \to \bigoplus_{j=-\infty}^{k} \mathcal{D}'_j\,.$$

Since ran $\pi'_k = \mathcal{D}'_k$ is orthogonal to $\mathcal{M}_{k-1}$, the identities in (2.5) are equivalent to

$$C_{k,k} \mid \mathcal{F}_k = \pi_k' \omega_k \,, \quad C_{j,k} \mid \mathcal{F}_k = C_{j,k-1}\pi_k\omega_k \quad (j < k) \,, \qquad k \in \mathbb{Z} \,.$$

We denote by C the doubly infinite block upper triangular operator matrix whose (j, k)-th entry is equal to $C_{j,k}$ for $j \le k$. Notice that the operators $C_k$, $k \in \mathbb{Z}$, are uniquely determined by B, and hence in order to prove the second part of Theorem 1.1, we have to show that

$$C = C(R) = W_2(R)(I - W_1(R))^{-1} \tag{3.1}$$

for a unique block upper triangular contractive operator R from $\mathcal{G}$ to $\mathcal{G}'$. Here $W_1(R)$ and $W_2(R)$ are the upper triangular operator matrices defined by (1.21)-(1.24) while $\mathcal{G}$ and $\mathcal{G}'$ are the spaces defined in (1.18) .

Thus we have to solve the following inverse problem. Assume that for each $k \in \mathbb{Z}$ we are given operators

$$C_{j,k} : \mathcal{D}_k \to \mathcal{D}_j' \,, \quad j \le k \,, \tag{3.2}$$

satisfying

$$\sum_{j=-\infty}^{k} \|C_{j,k} d_k\|^2 \le \|d_k\|^2 \,, \quad d_k \in \mathcal{D}_k \,, \tag{3.3}$$

$$C_{k,k} \mid \mathcal{F}_k = \pi_k' \omega_k \,, \tag{3.4}$$

$$C_{j,k} \mid \mathcal{F}_k = C_{j,k-1}\pi_k\omega_k \qquad (j < k) \,. \tag{3.5}$$

Our problem is to find a unique set of operators.

$$W_{j,k} : \mathcal{D}_k \to \begin{bmatrix} \mathcal{D}_{j-1} \\ \mathcal{D}_j' \end{bmatrix} \qquad (j, k \in \mathbb{Z}) \,, \tag{3.6}$$

such that

$$W_{j,k} = 0 \qquad (j > k) \,, \tag{3.7}$$

$$W_{k,k} \mid \mathcal{F}_k = \omega_k \,, \tag{3.8}$$

$$C_{k,k} = \pi_k' W_{k,k} \,, \tag{3.9}$$

$$C_{j,k} = \pi_j' W_{j,k} + \sum_{\nu=j}^{k-1} C_{j,\nu}\pi_{\nu+1} W_{\nu+1,k} \qquad (j < k) \,, \tag{3.10}$$

and

$$\left\| \begin{bmatrix} \ddots & \vdots & \vdots & \vdots & \vdots \\ & W_{-1,-1} & W_{-1,0} & W_{-1,1} & \cdots \\ & & W_{0,0} & W_{0,1} & \cdots \\ & & & W_{1,1} & \cdots \\ & & & & \ddots \end{bmatrix} \right\| \leq 1 . \tag{3.11}$$

Indeed, if such operators $W_{j,k}$ ($j, k \in \mathbb{Z}$) have been found, then from (3.7), (3.8), (3.11) and the fact that $\omega_k$ acts as a unitary operator from $\mathcal{F}_k$ onto $\mathcal{F}'_k$ ($k \in \mathbb{Z}$) it easily follows (e.g., use Lemma XXVIII.1.2 in Gohberg-Goldberg-Kaashoek [2]) that each $W_{j,k}$ is of the form

$$W_{j,k} = \begin{pmatrix} \delta_{j,k}\omega_k & 0 \\ 0 & R_{j,k} \end{pmatrix} : \begin{pmatrix} \mathcal{F}_k \\ \mathcal{G}_k \end{pmatrix} \to \begin{pmatrix} \mathcal{F}'_j \\ \mathcal{G}'_j \end{pmatrix}, \tag{3.12}$$

where $\delta_{j,k}$ is the Kronecker delta and $R = (R_{j,k})_{j,k=-\infty}^{\infty}$ is block upper triangular and contractive. In this case (3.9) and (3.10) are just equivalent to (3.1), and hence C is of the form C(R). Furthermore, because of the uniqueness of the operators $W_{j,k}$($j, k \in \mathbb{Z}$), the operator R is unique too.

In what follows we solve the above inverse problem. First we show that the problem can have at most one solution.

**PROPOSITION 3.1.** *Given operators* $C_{j,k}$ ($j \leq k$) *satisfying (3.2)-(3.5), there exists at most one set of operators* $W_{j,k}$ ($j, k \in \mathbb{Z}$) *such that (3.6)-(3.11) are fulfilled.*

**PROOF.** We first show that for each $k \in \mathbb{Z}$ the operator $\pi'_k$ is one-one on $\mathcal{G}'_k$. Indeed, assume

$$\begin{pmatrix} d_{k-1} \\ d_k \end{pmatrix} \in \mathcal{G}'_k \text{ and } \pi'_k \begin{pmatrix} d_{k-1} \\ d_k \end{pmatrix} = 0 . \tag{3.13}$$

The first part of (3.13) means that the following inner product formula holds:

$$(d_{k-1} \oplus d_k,\ D_{k-1}h \oplus P_{\mathcal{M}_k} A_{k-1}h) = 0 \qquad (h \in \mathcal{H}_{k-1}) . \tag{3.14}$$

According to the second part of (3.13) we have $d_k = 0$. Thus (3.14) reduces to $d_{k-1}$ being orthogonal ran $D_{k-1}$. Since $d_{k-1} \in \mathcal{D}_{k-1} = \overline{\text{ran}\, D_{k-1}}$, this can only happen when $d_{k-1} = 0$. So $\pi'_k \mid \mathcal{G}'_k$ is injective.

Next, we assume that the operators $W_{j,k}$ ($j, k \in \mathbb{Z}$) satisfy (3.6)-(3.11). We already know (from formula (3.12)) that for each j and k the restriction of $W_{j,k}$ to $\mathcal{F}_k$ is fixed and that

$W_{j,k}\mathcal{G}_k \subset \mathcal{G}'_j$. Since $\pi'_j \mid \mathcal{G}'_j$ is injective, we conclude that it suffices to show that the operators $\pi'_j W_{j,k} \mid \mathcal{G}_k$ $(j \le k)$ are uniquely determined by the conditions (3.2)-(3.11). For $j = k$ this follows directly from (3.9). For $j < k$ it follows by induction, using

$$\pi'_j W_{j,k} \mid \mathcal{G}_k = C_{j,k} \mid \mathcal{G}_k - \sum_{\nu=j}^{k-1} C_{j,\nu}\pi_{\nu+1} W_{\nu+1,k} \mid \mathcal{G}_k ,$$

with the latter identity being derived from (3.10). With these remarks the proof is completed.

We begin now the construction of operators $W_{j,k}$ $(j \le k)$ satisfying (3.8)-(3.11). The construction goes by induction on $n = k - j \ge 0$ and is given in a number of steps.

*Step 1.* For each $k \in \mathbb{Z}$ we construct an operator $W_{k,k}$ from $\mathcal{D}_k$ to $\mathcal{D}_{k-1} \oplus \mathcal{D}'_k$ such that

$$\|W_{k,k}\| \le 1 , \quad W_{k,k} \mid \mathcal{F}_k = \omega_k , \quad \pi'_k W_{k,k} = C_{k,k} . \tag{3.15}$$

Notice that $\omega_k$ and $C_{k,k}$ are contractions. Since (3.4) holds, we can apply the Parrott lemma (see Lemma XII.2.1) to show that the problem (3.15) has a solution. Indeed, the condition $\pi'_k \omega_k = C_{k,k} \mid \mathcal{F}_k$ allows us to rewrite the problem (3.15) as a completion problem for $2 \times 2$ operator matrices of the following type:

$$\begin{bmatrix} \alpha & X \\ \beta & \delta \end{bmatrix} : \begin{pmatrix} \mathcal{F}_k \\ \mathcal{G}_k \end{pmatrix} \to \begin{pmatrix} \mathcal{D}_{k-1} \\ \mathcal{D}'_k \end{pmatrix} , \tag{3.16}$$

where the unknown operator X has to be constructed in such a way that the resulting $2 \times 2$ operator matrix is a contraction. Since

$$\omega_k = \begin{pmatrix} \alpha \\ \beta \end{pmatrix} , \quad C_{k,k} = [\beta \quad \delta]$$

are contractions, this problem is solvable by virtue of the Parrott lemma.

*Step 2.* We proceed by induction. For this purpose we make the following induction hypotheses. Let n be a nonnegative integer. Assume that for each $k \in \mathbb{Z}$ we have constructed operators

$$W_{j,k} : \mathcal{D}_k \to \begin{pmatrix} \mathcal{D}_{j-1} \\ \mathcal{D}'_j \end{pmatrix} \qquad (k - n \le j \le k) \tag{3.17}$$

such that

$$W_{k,k} \mid \mathcal{F}_k = \omega_k \,, \tag{3.18}$$

$$C_{k,k} = \pi'_k W_{k,k} \,, \tag{3.19}$$

$$C_{j,k} = \pi'_j W_{j,k} + \sum_{\nu=j}^{k-1} C_{j,\nu}\pi_{\nu+1} W_{\nu+1,k} \qquad (k-n \le j < k) \,, \tag{3.20}$$

and

$$\left\| \begin{bmatrix} W_{k-n,k-n} & \cdots & W_{k-n,k} \\ & \ddots & \vdots \\ & & W_{k,k} \end{bmatrix} \right\| \le 1 \,. \tag{3.21}$$

Since (3.18) holds for each k, the operator matrix in (3.21) acts as a unitary operator from the Hilbert space direct sum $\mathcal{F}_{k-n} \oplus \cdots \oplus \mathcal{F}_k$ onto the Hilbert space direct sum $\mathcal{F}'_{k-n} \oplus \cdots \oplus \mathcal{F}'_k$. This fact together with the norm condition in (3.21) implies (e.g., by repeatedly using Lemma XXVIII.1.2 in Gohberg-Goldberg-Kaashoek [2]) that

$$W_{j,k} \mid \mathcal{F}_k = 0 \qquad (k-n \le j < k) \,, \tag{3.22}$$

$$W_{j,k}\mathcal{G}_k \subset \mathcal{G}'_j \qquad (k-n \le j \le k) \,. \tag{3.23}$$

With the operators $W_{j,k}$ $(k-n \le j \le k)$ we define a doubly infinite operator matrix $W(n) = (W(n)_{j,k})_{j,k=-\infty}^{\infty}$ by setting

$$W(n)_{j,k} = \begin{cases} W_{j,k} \,, & k-n \le j \le k \,, \\ 0 \,, & \text{otherwise} \,. \end{cases} \tag{3.24}$$

Furthermore, put

$$E = (E_{j,k})_{j,k=-\infty}^{\infty} = (I - S^* \Pi W(n))^{-1} \,,$$

where $\Pi$ is defined in (1.27) and $S^*$ by (1.31). Notice that $E_{j,k}$ maps $\mathcal{D}_k$ into $\mathcal{D}_j$ and that E is upper triangular. According to the definition of E, we have $E = I + S^*\Pi W(n)E$, and hence

$$E_{k,k} = I_{\mathcal{D}_k} \,, \tag{3.25}$$

$$E_{j,k} = \pi_{j+1} \sum_{\nu=j+1}^{k} W_{j+1,\nu}E_{\nu,k} \qquad (k-n-1 \le j < k) \,. \tag{3.26}$$

In particular we see that in E the first $n+2$ diagonals above the main diagonal (the main diagonal included) are uniquely determined by $W_{j,k}$ for $k-n \le j \le k$. Also, notice that conditions (3.19) and (3.20) imply that in the operator matrix $C - CS^*\Pi W(n) - \Pi' W(n)$ the first n + 1 upper diagonal entries (the main diagonal included) are all zero. Here $\Pi'$ is defined in (1.27). It follows that the same holds true for $C - \Pi' W(n) E$, and hence

$$C_{j,k} = \pi_j' \sum_{\nu=j}^{k} W_{j,\nu} E_{\nu,k} \qquad (k-n \leq j \leq k)\,. \tag{3.27}$$

*Step 3.* We claim that

$$E_{j,k} \mid \mathcal{F}_k = E_{j,k-1} \pi_k \omega_k \qquad (k-n-1 \leq j < k)\,. \tag{3.28}$$

Indeed,

$$E_{k-1,k} \mid \mathcal{F}_k = \pi_k W_{k,k} E_{k,k} \mid \mathcal{F}_k = \pi_k W_{k,k} \mid \mathcal{F}_k = \pi_k \omega_k = E_{k-1,k-1} \pi_k \omega_k\,.$$

Here we used (3.25) for $k-1$ in place of k. We proceed by induction. Take $k-n-1 \leq i < k-1$, and assume that (3.28) has been proved for all j such that $i < j < k$. Since $i+1 < k$, we have $W_{i+1,k} \mid \mathcal{F}_k = 0$, and therefore (using our induction hypothesis)

$$\begin{aligned} E_{i,k} \mid \mathcal{F}_k &= \pi_{i+1} \sum_{\nu=i+1}^{k-1} W_{i+1,\nu} E_{\nu,k} \mid \mathcal{F}_k \\ &= \pi_{i+1} \sum_{\nu=i+1}^{k-1} W_{i+1,\nu} E_{\nu,k-1} \pi_k \omega_k = E_{i,k-1} \pi_k \omega_k\,, \end{aligned}$$

where the last equality comes from (3.26) with $k-1$ in place of k. In this way we establish (3.28) by induction.

Notice that (3.28) may be rewritten in the following equivalent form:

$$E_{j,k} D_k \mid \mathcal{H}_{k-1} = E_{j,k-1} D_{k-1} \qquad (k-n-1 \leq j < k)\,. \tag{3.29}$$

By using these equalities repeatedly one sees that

$$E_{j,k} D_k \mid \mathcal{H}_i = \begin{cases} E_{j,i} D_i\,, & k-n-1 \leq j \leq i \leq k\,, \\ D_j \mid \mathcal{H}_i\,, & k-n-1 \leq i \leq j \leq k\,. \end{cases} \tag{3.30}$$

In particular, we have

$$E_{j,k} D_k \mathcal{H}_i \subset \mathcal{F}_j\,, \quad k-n-1 \leq i < j \leq k\,. \tag{3.31}$$

*Step 4.* We prove the following lemma.

**LEMMA 3.2.** *Let* $Z_\nu : \mathcal{D}_\nu \to \mathcal{D}'_{k-n}$, *for* $k-n \le \nu \le k$, *be operators such that*
($\alpha$) $C_{k-n,k} = Z_{k-n}E_{k-n,k} + \cdots + Z_k E_{k,k}$, *and*
($\beta$) $Z_\nu \mid \mathcal{F}_\nu = 0$ *for* $k-n < \nu \le k$ *in case of* $n > 0$.

*Then* $Z_\nu = \pi'_{k-n} W_{k-n,\nu}$ *for* $k-n \le \nu \le k$.

**PROOF**. Multiply both sides of ($\alpha$) by $D_k \mid \mathcal{H}_{k-n}$ from the right. Using ($\beta$) and (3.31) we see that

$$C_{k-n,k} D_k \mid \mathcal{H}_{k-n} = Z_{k-n} E_{k-n,k} D_k \mid \mathcal{H}_{k-n} . \tag{3.32}$$

Next, we use (3.30) with $j = i = k-n$ and derive

$$C_{k-n,k} D_k \mid \mathcal{H}_{k-n} = Z_{k-n} D_{k-n} . \tag{3.33}$$

Notice that (3.5) is equivalent to $C_{j,k} D_k \mid \mathcal{H}_{k-1} = C_{j,k-1} D_{k-1}$. By repeatedly applying the latter identity, we obtain

$$C_{j,k} D_k \mid \mathcal{H}_i = C_{j,i} D_i \qquad (j \le i \le k) . \tag{3.34}$$

Using (3.34) with $j = i = k-n$ in (3.33) yields

$$C_{k-n,k-n} D_{k-n} = Z_{k-n} D_{k-n} . \tag{3.35}$$

Recall that $\mathcal{D}_{k-n} = \overline{\operatorname{ran} D_{k-n}}$. Therefore (3.35) implies that $C_{k-n,k-n} = Z_{k-n}$. According to (3.19) we have $C_{k-n,k-n} = \pi'_{k-n} W_{k-n,k-n}$, and hence we have shown that $Z_{k-n} = \pi'_{k-n} W_{k-n,k-n}$ as desired.

We proceed by induction. Take $k-n \le i-1 < k$, and assume that $Z_\nu = \pi'_{k-n} W_{k-n,\nu}$ for $k-n \le \nu \le i-1$. Multiply both sides of ($\alpha$) and both sides of (3.27) with $j = k-n$ by $D_k \mid \mathcal{H}_i$ from the right. This yields (use ($\beta$), (3.30), (3.31), and (3.22)) the following two identities:

$$C_{k-n,k} D_k \mid \mathcal{H}_i = \sum_{\nu=k-n}^{i-1} Z_\nu E_{\nu,k} D_k \mid \mathcal{H}_i + Z_i D_i ,$$

$$C_{k-n,k} D_k \mid \mathcal{H}_i = \sum_{\nu=k-n}^{i-1} \pi'_{k-n} W_{k-n,\nu} E_{\nu,k} D_k \mid \mathcal{H}_i + \pi'_{k-n} W_{k-n,i} D_i .$$

Since $Z_\nu = \pi'_{k-n} W_{k-n,\nu}$ for $\nu = k-n, \ldots, i-1$ by our induction hypothesis, we conclude that $Z_i D_i = \pi'_{k-n,i} W_{k-n,i} D_i$. Therefore $Z_i = \pi'_{k-n} W_{k-n,i}$, and the proof is completed by induction.

*Step 5*. Introduce the following spaces:

$$\mathcal{X}_k = \begin{bmatrix} \mathcal{D}_{k-n-1} \\ \mathcal{D}_{k-n} \\ \vdots \\ \mathcal{D}_k \end{bmatrix}, \quad \mathcal{X}_k^o = \begin{bmatrix} E_{k-n-1,k} \\ E_{k-n,k} \\ \vdots \\ E_{k,k} \end{bmatrix} \mathcal{D}_k,$$

$$\mathcal{Y}_k = \begin{bmatrix} \mathcal{D}_{k-n-2} \oplus \mathcal{D}'_{k-n-1} \\ \mathcal{D}_{k-n-1} \oplus \mathcal{D}'_{k-n} \\ \vdots \\ \mathcal{D}_{k-1} \oplus \mathcal{D}'_k \end{bmatrix}, \quad \mathcal{Y}_k^o = \begin{bmatrix} \{0\} \oplus \mathcal{D}'_{k-n-1} \\ \mathcal{D}_{k-n-1} \oplus \mathcal{D}'_{k-n} \\ \vdots \\ \mathcal{D}_{k-1} \oplus \mathcal{D}'_k \end{bmatrix}.$$

We see $\mathcal{X}_k^o$ as a subspace of $\mathcal{X}_k$ and $\mathcal{Y}_k^o$ as a subspace of $\mathcal{Y}_k$. Define

$$T_k^o : \mathcal{X}_k^o \to \mathcal{Y}_k^o, \quad T_k^o \begin{bmatrix} E_{k-n-1,k} \\ E_{k-n,k} \\ \vdots \\ E_{k,k} \end{bmatrix} d_k = \begin{bmatrix} 0 \oplus C_{k-n-1,k} d_k \\ E_{k-n-1,k} d_k \oplus C_{k-n,k} d_k \\ \vdots \\ E_{k-1,k} d_k \oplus C_{k,k} d_k \end{bmatrix}.$$

We claim that the operator $T_k^o$ is a contraction. Indeed, set $x_k = \operatorname{col}(E_{j,k} d_k)_{j=k-n-1}^{k}$. Then

$$\|T_k^o x_k\|^2 = \sum_{j=k-n-1}^{k} \|C_{j,k} d_k\|^2 + \sum_{j=k-n-1}^{k-1} \|E_{j,k} d_k\|^2$$

$$\leq \|d_k\|^2 + \sum_{j=k-n-1}^{k-1} \|E_{j,k} d_k\|^2 = \sum_{j=k-n-1}^{k} \|E_{j,k} d_k\|^2 = \|x_k\|^2 .$$

Here we used that $E_{k,k} = I_{\mathcal{D}_k}$.

*Step 6.* A first completion problem. We seek

$$W'_{k-n-1,j} : \mathcal{D}_j \to \mathcal{D}'_{k-n-1} \qquad (k-n-1 \leq j \leq k) \tag{3.36}$$

such that the upper triangular operator matrix

$$T'_k = \begin{bmatrix} W'_{k-n-1,k-n-1} & W'_{k-n-1,k-n} & \cdots & W'_{k-n-1,k} \\ & W_{k-n,k-n} & \cdots & W_{k-n,k} \\ & & \ddots & \vdots \\ & & & W_{k,k} \end{bmatrix}$$

from $\mathcal{X}_k$ into $\mathcal{Y}_k^o$ is a contraction and $T'_k \mid \mathcal{X}_k^o = T_k^o$.

To solve this problem we apply Parrott's lemma as formulated in Lemma XII.2.1 with K equal to $T_k^o$ and with L equal to the known part of the operator matrix $T_k'$. Notice that L is a contraction by virtue of the induction assumption (3.21). Thus it remains to prove that the data K and L satisfy the compatibility condition (XII.2.1) in Lemma XII.2.1. This follows from formulas (3.26) and (3.27); indeed these formulas can be summarized as

$$\begin{bmatrix} W_{k-n,k-n} & \cdots & W_{k-n,k} \\ & \ddots & \vdots \\ & & W_{k,k} \end{bmatrix} \begin{bmatrix} E_{k-n,k} \\ \vdots \\ E_{k,k} \end{bmatrix} d_k = \begin{bmatrix} E_{k-n-1,k}d_k \oplus C_{k-n,k}d_k \\ \vdots \\ E_{k-1,k}d_k \oplus C_{k,k}d_k \end{bmatrix}, \tag{3.37}$$

which implies that

$$P_{(\mathcal{D}_{k-n-1} \oplus \mathcal{D}'_{k-n}) \oplus \cdots \oplus (\mathcal{D}_{k-1} \oplus \mathcal{D}'_k)} T_k^o = \begin{bmatrix} 0 & W_{k-n,k-n} & \cdots & W_{k-n,k} \\ \vdots & & \ddots & \vdots \\ 0 & & & W_{k,k} \end{bmatrix} | \mathcal{X}_k^o .$$

Thus the data are compatible and Parrott's lemma gives us the desired operators in (3.36).

Notice that for each $k-n \le j \le k$ the operator $W_{j,j} \mid \mathcal{F}_j$ is an isometry. Since $T_k'$ is a contraction, we conclude (e.g., by using Lemma XXVIII.1.2 in Gohberg-Goldberg-Kaashoek [2]) that

$$W'_{k-n-1,j} \mid \mathcal{F}_j = 0 \qquad (k-n \le j \le k) . \tag{3.38}$$

The identity $T_k' \mid \mathcal{X}_k^o = T_k^o$ implies that

$$C_{k-n-1,k} = \sum_{j=k-n-1}^{k} W'_{k-n-1,j} E_{j,k} . \tag{3.39}$$

Let us restrict both sides of (3.39) to $\mathcal{F}_k$. Using (3.5), (3.25), (3.28), and (3.38) we obtain

$$\begin{aligned} C_{k-n-1,k-1} \pi_k \omega_k &= C_{k-n-1,k} \mid \mathcal{F}_k \\ &= \sum_{j=k-n-1}^{k-1} W'_{k-n-1,j} E_{j,k-1} \pi_k \omega_k + W'_{k-n-1,k} \mid \mathcal{F}_k \\ &= \left[ \sum_{j=k-n-1}^{k-1} W'_{k-n-1,j} E_{j,k-1} \right] \pi_k \omega_k . \end{aligned}$$

The range of $\pi_k \omega_k$ is dense in $\mathcal{D}_{k-1}$. So we get

$$C_{k-n-1,k-1} = \sum_{j=k-n-1}^{k-1} W'_{k-n-1,j} E_{j,k-1} \,.$$

Now we can apply Lemma 3.2 with k − 1 in place of k to obtain

$$W'_{k-n-1,j} = \pi'_{k-n-1} W_{k-n-1,j} \qquad (k-n-1 \le j \le k-1)\,. \tag{3.40}$$

*Step 7.* A second completion problem. We seek an operator

$$W_{k-n-1,k} : \mathcal{D}_k \to \begin{bmatrix} \mathcal{D}_{k-n-2} \\ \mathcal{D}'_{k-n-1} \end{bmatrix} \tag{3.41}$$

such that

$$T_k = \begin{bmatrix} W_{k-n-1,k-n-1} & \cdots & W_{k-n-1,k} \\ & \ddots & \vdots \\ & & W_{k,k} \end{bmatrix} \tag{3.42}$$

from $\mathcal{X}_k$ into $\mathcal{Y}_k$ is a contraction and $P_{\mathcal{Y}_k^o} T_k = T'_k$. Again we apply Parrott's lemma. In this case the data are the operator $T'_k$ from $\mathcal{X}_k$ to $\mathcal{Y}_k^o$ constructed in the previous step and the contraction (see (3.21) with k replaced by k − 1)

$$\begin{bmatrix} W_{k-n-1,k-n-1} & \cdots & W_{k-n-1,k-1} \\ & \ddots & \vdots \\ & & W_{k-1,k-1} \\ 0 & \cdots & 0 \end{bmatrix} : \mathcal{X}'_k \to \mathcal{Y}_k\,, \text{ where } \mathcal{X}'_k = \begin{bmatrix} \mathcal{D}_{k-n-1} \\ \vdots \\ \mathcal{D}_{k-1} \\ \{0\} \end{bmatrix} \subset \mathcal{X}_k\,.$$

The compatibility condition is

$$P_{\mathcal{Y}_k^o} \begin{bmatrix} W_{k-n-1,k-n-1} & \cdots & W_{k-n-1,k-1} \\ & \ddots & \vdots \\ & & W_{k-1,k-1} \\ 0 & \cdots & 0 \end{bmatrix} = T'_k \,|\, \mathcal{X}'_k\,,$$

which is precisely equivalent to (3.40). Thus $W_{k-n-1,k}$ can be constructed as desired for each $k \in \mathbb{Z}$.

*Step 8.* Note that in Step 7 we established the existence of operators $W_{k-n-1,k}$, $k \in \mathbb{Z}$, such that for each k the operator matrix (3.21) with n replaced by n + 1 is a contraction. Now we will show that with these operators $W_{k-n-1,k}$ condition (3.20) is fulfilled with $n+1$ in place of n. Notice that

$$P_{\mathcal{Y}_k^o} T_k | \mathcal{X}_k^o = T'_k | \mathcal{X}_k^o = T_k^o , \tag{3.43}$$

which implies that

$$C_{k-n-1,k} = \pi'_{k-n-1} \sum_{v=k-n-1}^{k} W_{k-n-1,v} E_{v,k} . \tag{3.44}$$

Consider the doubly infinite operator matrix

$$W(n+1) = \left[ W(n+1)_{j,k} \right]_{j,k=-\infty}^{\infty} , \quad \text{where}$$

$$W(n+1)_{j,k} = \begin{cases} W_{j,k} , & k-n-1 \le j \le k , \\ 0 , & \text{otherwise} , \end{cases}$$

and set

$$\tilde{E} = (\tilde{E}_{j,k})_{j,k=-\infty}^{\infty} = (I - S^* \Pi W(n+1))^{-1} .$$

Notice that the first $n+1$ diagonals in the upper triangular part of $W(n)$ coincide with the first $n+1$ diagonals in the upper triangular part of $W(n+1)$. This implies that

$$\tilde{E}_{j,k} = E_{j,k} \qquad (k-n-1 \le j \le k) .$$

But then we see from (3.27) and (3.44) that the first $n+2$ diagonals in the upper triangular part of $\Pi' W(n+1)\tilde{E}$ coincide with the first $n+2$ diagonals in the upper triangular part of C. Put $\tilde{C} = \Pi' W(n+1)\tilde{E}$. Then

$$\tilde{C}(I - S^* \Pi W(n+1)) = \Pi' W(n+1) ,$$

which implies that

$$C_{k-n-1,k} = \pi'_{k-n-1,k} W_{k-n-1,k} + \sum_{v=k-n-1}^{k} C_{k-n-1,v} \pi_{v+1} W_{v+1,k} .$$

We conclude that (3.20) is fulfilled for $n+1$ in place of n.

*Step 9.* From the result of the previous step it follows that we can complete the proof by induction. More precisely, by induction we can find operators $W_{j,k}$ ($j, k \in \mathbb{Z}$) satisfying (3.6)-(3.10), and such that (3.21) holds for each nonnegative integer n. The latter implies that also (3.11) holds, which completes the proof of Theorem 1.1.

## XIV.4 THE CASE OF DECREASING SPACES

In this section we will present the reformation of Theorem 1.1 for the case when the three chains of subspaces are decreasing. To this end, recall that for this version the given data are again bounded linear operators

$$A_k : \mathcal{H}_k \to \mathcal{K}_k \ominus \mathcal{M}_k \qquad (k \in \mathbb{Z}) . \tag{4.1}$$

Here $\mathcal{H}_k \subset \mathcal{U}$ and $\mathcal{M}_k \subset \mathcal{K}_k \subset \mathcal{Y}$ for $k \in \mathbb{Z}$ are Hilbert spaces which now satisfy the inclusion relations

$$\mathcal{H}_{k+1} \subset \mathcal{H}_k , \ \mathcal{M}_{k+1} \subset \mathcal{M}_k , \ \mathcal{K}_{k+1} \subset \mathcal{K}_k \qquad (k \in \mathbb{Z}) . \tag{4.2}$$

Given the operators (4.1) and a tolerance $\gamma$ the problem is to find an operator B from $\mathcal{U}$ into $\mathcal{Y}$ such that $||B|| \le \gamma$ and

$$B\mathcal{H}_k \subset \mathcal{K}_k , \ (I - P_{\mathcal{M}_k})B \,|\, \mathcal{H}_k = A_k \qquad (k \in \mathbb{Z}) . \tag{4.3}$$

From Theorem XII.1.1a we know that for an operator B to exist it is necessary and sufficient that the operators $A_k$ for $k \in \mathbb{Z}$ satisfy the compatibility conditions

$$(I - P_{\mathcal{M}_k})A_{k+1} = A_k \,|\, \mathcal{H}_{k+1} \qquad (k \in \mathbb{Z}) , \tag{4.4}$$

and that

$$\gamma \ge d_\infty := \sup \{||A_k|| : k \in \mathbb{Z}\} . \tag{4.5}$$

In this section we assume that the conditions (4.4) and (4.5) are fulfilled, and our aim is to reformulate Theorem 1.1 is this setting.

First, let us notice that

$$\operatorname{ran} P_{\mathcal{M}_k} A_{k+1} \subset \mathcal{M}_k \ominus \mathcal{M}_{k+1} \qquad (k \in \mathbb{Z}) . \tag{4.6}$$

Indeed, take $h \in \mathcal{H}_{k+1}$. Then, according to (4.1), we have $A_{k+1}h$ is orthogonal to $\mathcal{M}_{k+1}$. Since $\mathcal{M}_{k+1} \subset \mathcal{M}_k$, we may write $A_{k+1}h = x + y$, where $x \perp \mathcal{M}_k$ and $y \in \mathcal{M}_k \ominus \mathcal{M}_{k+1}$. But then $P_{\mathcal{M}_k} A_{k+1} h = y \in \mathcal{M}_k \ominus \mathcal{M}_{k+1}$, as desired.

To state our main result we need to consider a number of auxiliary Hilbert spaces. For each $k \in \mathbb{Z}$ put

$$D_k = (\gamma^2 I - A_k^* A_k)^{1/2}, \qquad \mathcal{D}_k = \overline{\operatorname{ran} D_k}\,; \tag{4.7}$$

$$\mathcal{D}'_k = \mathcal{M}_k \ominus \mathcal{M}_{k+1}\,; \tag{4.8}$$

$$\mathcal{F}_k = \overline{D_k \mathcal{H}_{k+1}}\,, \quad \mathcal{G}_k = \mathcal{D}_k \ominus \mathcal{F}_k\,; \tag{4.9}$$

$$\mathcal{F}'_k = \operatorname{cl}\left\{ \begin{bmatrix} D_{k+1} h_{k+1} \\ P_{\mathcal{M}_k} A_{k+1} h_{k+1} \end{bmatrix} : h_{k+1} \in \mathcal{H}_{k+1} \right\}, \tag{4.10}$$

$$\mathcal{G}'_k = \begin{pmatrix} \mathcal{D}_{k+1} \\ \mathcal{D}'_k \end{pmatrix} \ominus \mathcal{F}'_k\,. \tag{4.11}$$

Here the first term in the right hand side of (4.11) denotes the Hilbert space direct sum $\mathcal{D}_{k+1} \oplus \mathcal{D}'_k$. By (4.5) the operator $D_k$ is well defined. Notice that $D_k$ acts on $\mathcal{H}_k$. Since $\mathcal{H}_{k+1} \subset \mathcal{H}_k$, we may consider the restriction of $D_k$ to $\mathcal{H}_{k+1}$, and hence $\mathcal{F}_k$ is a well defined subspace of $\mathcal{D}_k$. It follows that $\mathcal{G}_k$ is also well defined. According to (4.6) the vector $D_{k+1} h_{k+1} \oplus P_{\mathcal{M}_k} A_{k+1} h_{k+1}$ belongs to $\mathcal{D}_{k+1} \oplus \mathcal{D}'_k$ whenever $h_{k+1} \in \mathcal{H}_{k+1}$. Hence $\mathcal{F}'_k$ is a subspace of $\mathcal{D}_{k+1} \oplus \mathcal{D}'_k$, and therefore $\mathcal{G}'_k$ is well defined too.

We also need a number of auxiliary operators. For each $k \in \mathbb{Z}$ we define

$$\pi_k : \begin{pmatrix} \mathcal{D}_{k+1} \\ \mathcal{D}'_k \end{pmatrix} \to \mathcal{D}_{k+1}\,, \quad \pi_k \begin{pmatrix} d_{k+1} \\ d'_k \end{pmatrix} = d_{k+1}\,; \tag{4.12}$$

$$\pi'_k : \begin{pmatrix} \mathcal{D}_{k+1} \\ \mathcal{D}'_k \end{pmatrix} \to \mathcal{D}'_k\,, \quad \pi'_k \begin{pmatrix} d_{k+1} \\ d'_k \end{pmatrix} = d'_k\,; \tag{4.13}$$

$$\omega_k : \mathcal{F}_k \to \begin{pmatrix} \mathcal{D}_{k+1} \\ \mathcal{D}'_k \end{pmatrix}, \quad \omega_k(D_k h_{k+1}) = \begin{pmatrix} D_{k+1} h_{k+1} \\ P_{\mathcal{M}_k} A_{k+1} h_{k+1} \end{pmatrix}; \tag{4.14}$$

$$\rho_k : \mathcal{D}_k \to \mathcal{D}_{k+1}\,, \quad \rho_k = \pi_k \omega_k P_{\mathcal{F}_k}\,. \tag{4.15}$$

By consulting Section XII.3 and replacing k by $-k$ we see that $\omega_k$ is a well defined isometry from $\mathcal{F}_k$ onto $\mathcal{F}'_k$. Since $\pi_k$ and $\pi'_k$ are orthogonal projections, we conclude that $\rho_k$ is a contraction from $\mathcal{D}_k$ into $\mathcal{D}_{k+1}$. In what follows we write $\Phi_\rho(m, n)$ for the state transition matrix generated by $\rho_k$, that is,

$$\Phi_\rho(m, n) = \rho_{m-1} \cdots \rho_{n+1}\, \rho_n : \mathcal{D}_n \to \mathcal{D}_m \qquad (m > n);$$

$$\Phi_\rho(m, m) = I_{\mathcal{D}_m} .$$

Next, we introduce four lower triangular operators matrices, namely

$$\Theta_{11} = \left[(\Theta_{11})_{j,k}\right]_{j,k=-\infty}^{\infty}, \quad (\Theta_{11})_{j,k} : \mathcal{G}'_k \to \mathcal{G}_j ,$$

$$\Theta_{12} = \left[(\Theta_{12})_{j,k}\right]_{j,k=-\infty}^{\infty}, \quad (\Theta_{12})_{j,k} : \mathcal{D}_k \to \mathcal{G}_j ,$$

$$\Theta_{21} = \left[(\Theta_{21})_{j,k}\right]_{j,k=-\infty}^{\infty}, \quad (\Theta_{21})_{j,k} : \mathcal{G}'_k \to \mathcal{D}'_j ,$$

$$\Theta_{22} = \left[(\Theta_{22})_{j,k}\right]_{j,k=-\infty}^{\infty}, \quad (\Theta_{22})_{j,k} : \mathcal{D}_k \to \mathcal{D}'_j ,$$

where

$$(\Theta_{11})_{j,k} = \begin{cases} P_{\mathcal{G}_j}\Phi_\rho(j, k+1)\pi_k \mid \mathcal{G}'_k , & j > k, \\ 0 , & j \le k, \end{cases}$$

$$(\Theta_{12})_{j,k} = \begin{cases} P_{\mathcal{G}_j}\Phi_\rho(j, k) , & j \ge k, \\ 0 , & j < k, \end{cases}$$

$$(\Theta_{21})_{j,k} = \begin{cases} \pi'_j\omega_j P_{\mathcal{F}_j}\Phi_\rho(j, k+1)\pi_k \mid \mathcal{G}'_k , & j > k, \\ \pi'_k \mid \mathcal{G}'_k , & j = k, \\ 0 , & j < k, \end{cases}$$

and

$$(\Theta_{22})_{j,k} = \begin{cases} \pi'_j\omega_j P_{\mathcal{F}_j}\Phi_\rho(j, k) , & j \ge k, \\ 0 , & j < k. \end{cases}$$

Notice that the operator matrix $\Theta_{11}$ is strictly lower triangular.

Now, let $R = (R_{j,k})_{j,k=-\infty}^{\infty}$, where $R_{j,k}$ maps $\mathcal{G}_k$ into $\mathcal{G}'_j$, be a lower triangular operator matrix (that is, $R_{j,k} = 0$ for $j < k$). Given such an $R$ we write $\Gamma(R)$ for the lower triangular operator matrix defined by

$$\Gamma(R) := \Theta_{22} + \Theta_{21}R(I - \Theta_{11}R)^{-1}\Theta_{12} . \tag{4.16}$$

The products and inverse in (4.16) have to be understood algebraically and have to be carried out in the algebra of lower triangular operator matrices. Notice that $\Theta_{11}R$ is strictly lower triangular. Thus $(I - \Theta_{11}R)^{-1}$ is well defined algebraically and is lower triangular. In fact, for each j and k in $\mathbb{Z}$ we have

$$\left[(I-\Theta_{11}R)^{-1}\right]_{j,k} = \sum_{\nu=0}^{\infty} \left[(\Theta_{11}R)^{\nu}\right]_{j,k}, \tag{4.17}$$

and the sum in the right hand side of (4.17) is a finite one.

Consider the doubly infinite Hilbert space direct sums

$$\mathcal{G} = \bigoplus_{-\infty}^{\infty} \mathcal{G}_k, \qquad \mathcal{G}' = \bigoplus_{-\infty}^{\infty} \mathcal{G}'_k; \tag{4.18}$$

$$\mathcal{D} = \bigoplus_{-\infty}^{\infty} \mathcal{D}_k, \qquad \mathcal{D}' = \bigoplus_{-\infty}^{\infty} \mathcal{D}'_k. \tag{4.19}$$

In what follows we shall always assume that the operator matrix R in (4.16) is a contractive operator from $\mathcal{G}$ into $\mathcal{G}'$. In this case (see Proposition 1.2 with k replaced by $-$ k) the k-th block column of the operator matrix $\Gamma(R)$ in (4.16) turns out to be a contractive operator from $\mathcal{D}_k$ into $\mathcal{D}'$ for all k in $\mathbb{Z}$.

By replacing k by $-$ k in Theorem 1.1 we readily obtain the following parameterization for the set of all interpolants in the three chains completion problem.

**THEOREM 4.1**. *Assume* $A_k$ *from* $\mathcal{H}_k$ *to* $\mathcal{K}_k \ominus \mathcal{M}_k$, $k \in \mathbb{Z}$, *satisfy conditions (4.4) and (4.5). Let* R *from* $\mathcal{G}$ *to* $\mathcal{G}'$ *be a block lower triangular and contractive operator. Then the formula*

$$B(R)h = \begin{cases} A_k h + \sum_{j=k}^{\infty} \Gamma_{j,k}(R) D_k h, & h \in \mathcal{H}_k, \\ 0, & h \perp \bigvee_{n\in\mathbb{Z}} \mathcal{H}_n, \end{cases} \tag{4.20}$$

*where*

$$\Gamma_{j,k}(R) = (\Theta_{22} + \Theta_{21}R(I-\Theta_{11}R)^{-1}\Theta_{12})_{j,k} \qquad (j \geq k), \tag{4.21}$$

*defines a linear operator* B(R) *from* $\mathcal{U}$ *to* $\mathcal{Y}$ *such that* $\|B(R)\| \leq \gamma$ *and* B(R) *is a solution of the three chains completion problem associated with the operators* $A_k$ ($k \in \mathbb{Z}$) *and tolerance* $\gamma$.

*Conversely, if* B *is a solution of the three chains completion problem associated with* $A_k$ ($k \in \mathbb{Z}$) *and tolerance* $\gamma$, *then there exists a unique block lower triangular contractive operator* R *from* $\mathcal{G}$ *to* $\mathcal{G}'$ *such that* $Bh = B(R)h$ *for each* $h \in \bigvee \{\mathcal{H}_n : n \in \mathbb{Z}\}$.

## XIV.5 THE NONSTATIONARY NEHARI PARAMETERIZATION

In this section we will use Theorem 4.1 to obtain the set of all solutions to a nonstationary Nehari interpolation problem whose data is described by an anticausal nonstationary realization. To this end, let $f_{j,k} : \mathcal{U}_k \to \mathcal{Y}_j$ for $-\infty < j < k < \infty$ be the data for the nonstationary Nehari interpolation problem with tolerance $\gamma$. Given these data the problem is to find all linear operators $G = (g_{j,k})_{j,k=-\infty}^{\infty}$ mapping $\vec{\mathcal{U}} = \oplus\{\mathcal{U}_k : k \in \mathbb{Z}\}$ into $\vec{\mathcal{Y}} = \oplus\{\mathcal{Y}_k : k \in \mathbb{Z}\}$ of norm bounded by $\gamma$ such that $g_{j,k} = f_{j,k}$ for $-\infty < j < k < \infty$.

Throughout this section we assume that $\{Z_k, B_k, C_k\}$ is an anticausal nonstationary realization (see Section XI.2) of $\{f_{j,k}\}$, that is,

$$f_{j,k} = C_j \Phi(j+1, k) B_k \qquad (-\infty < j < k < \infty) . \tag{5.1}$$

Here $Z_k$ maps $\mathcal{X}_{k+1}$ into $\mathcal{X}_k$ while $B_k$ maps $\mathcal{U}_k$ into $\mathcal{X}_k$ and $C_j$ maps $\mathcal{X}_{j+1}$ into $\mathcal{Y}_j$. As before, $\Phi(m, n)$ from $\mathcal{X}_n$ into $\mathcal{X}_m$ is the state transition operator for $\{Z_n\}$ defined by $\Phi(n, n) = I$ and $\Phi(m, n) = Z_m Z_{m+1} \cdots Z_{n-1}$ if $m < n$. Throughout we always assume that $\{B_k\}$ and $\{C_k\}$ are uniformly bounded, and that $\{Z_k\}$ is backward stable, that is,

$$\limsup_{\nu\to\infty} (\sup\{\|\Phi(k-\nu, k)\| : k \in \mathbb{Z}\})^{1/\nu} < 1 . \tag{5.2}$$

The controllability grammian $P_k$ at time k for the family of pairs $\{Z_k, B_k\}$ is given by

$$P_k = \sum_{\nu=0}^{\infty} \Phi(k, k+\nu) B_{k+\nu} B_{k+\nu}^* \Phi(k, k+\nu)^* . \tag{5.3}$$

The observability grammian $Q_k$ at time k for the family of pairs $\{C_k, Z_k\}$ is defined by

$$Q_k = \sum_{\nu=1}^{\infty} \Phi(k-\nu+1, k)^* C_{k-\nu}^* C_{k-\nu} \Phi(k-\nu+1, k) . \tag{5.4}$$

The stability condition in (5.2) guarantees that both $P_k$ and $Q_k$ are well defined bounded (positive) operators on $\mathcal{X}_k$ for all k in $\mathbb{Z}$. The controllability and observability grammians can be obtained by solving the nonstationary Lyapunov equations in (XI.2.13). Throughout we also assume that $\{Z_k, B_k, C_k\}$ is uniformly controllable and uniformly observable, that is, there exists an $\varepsilon > 0$ such that $P_k \geq \varepsilon I$ and $Q_k \geq \varepsilon I$ for all k in $\mathbb{Z}$.

Recall that our nonstationary Nehari problem is solvable if and only if $\gamma \geq d_\infty$, where $d_\infty$ is given by (XI.2.2). According to Theorem XI.2.3 the quantity $d_\infty$ in this setting is also given by

$$d_\infty^2 = \sup\{r_{spec}(Q_k P_k) : k \in \mathbb{Z}\} . \tag{5.5}$$

Throughout this section we also assume that $\gamma > d_\infty$, and hence we can define operators $N_k$ on

$\mathcal{U}_k$ and $N_{*k}$ on $\mathcal{Y}_k$ by

$$\begin{aligned} N_k &= \gamma^{-1}[I + B_k^*(\gamma^2 I - Q_k P_k)^{-1} Q_k B_k]^{1/2} , \\ N_{*k} &= [I + C_k P_{k+1}(\gamma^2 I - Q_{k+1} P_{k+1})^{-1} C_k^*]^{1/2} . \end{aligned} \tag{5.6}$$

Notice that in (5.6) the operators between square brackets are strictly positive, and hence $N_k$ and $N_{*k}$ are also strictly positive operators. Moreover using (5.2), (5.5), and the uniform boundedness of the coefficients of the system $\{Z_k, B_k, C_k\}$, one sees that the families $\{N_k^{-1}\}$ and $\{N_{*k}^{-1}\}$ are bounded in operator norm.

As in Section XI.2 equation (2.15), let $M_k$ be the operator from $\mathcal{X}_k$ into $\mathcal{X}_{k+1}$ defined by

$$M_k = (\gamma^2 I - Z_k^* Q_k Z_k P_{k+1})^{-1} Z_k^* (\gamma^2 I - Q_k P_k) . \tag{5.7}$$

According to the results in Section XI.2, the operators $\{M_k\}$ are forward stable. In the sequel $\Psi(m, n)$ from $\mathcal{X}_n$ into $\mathcal{X}_m$ denotes the state transition operator for $\{M_k\}$, that is, $\Psi(n, n) = I$ and $\Psi(m, n) = M_{m-1} \cdots M_{n+1} M_n$ if $m > n$.

To obtain a state space description for the set of all solutions to our nonstationary Nehari problem with tolerance $\gamma > d_\infty$ we introduce the bock lower triangular operators

$$\begin{aligned} &\Omega_{11} : \vec{\mathcal{Y}} \to \vec{\mathcal{U}} , \quad \Omega_{12} : \vec{\mathcal{U}} \to \vec{\mathcal{U}} , \\ &\Omega_{21} : \vec{\mathcal{Y}} \to \vec{\mathcal{Y}} , \quad \Omega_{22} : \vec{\mathcal{U}} \to \vec{\mathcal{Y}} , \end{aligned} \tag{5.8}$$

by setting

$$\begin{aligned} (\Omega_{11})_{k,k} &= 0 \\ (\Omega_{11})_{j,k} &= -N_j^{-1} B_j^* \Psi(j, k+1)(\gamma^2 I - Q_{k+1} P_{k+1})^{-1} C_k^* N_{*k}^{-1} \qquad (j > k) \\ (\Omega_{12})_{k,k} &= N_k^{-1} \\ (\Omega_{12})_{j,k} &= -N_j^{-1} B_j^* \Psi(j, k+1)(\gamma^2 I - Z_k^* Q_k Z_k P_{k+1})^{-1} Z_k^* Q_k B_k \qquad (j > k) \\ (\Omega_{21})_{k,k} &= -N_{*k}^{-1} \\ (\Omega_{21})_{j,k} &= C_j P_{j+1} M_j \Psi(j, k+1)(\gamma^2 I - Q_{k+1} P_{k+1})^{-1} C_k^* N_{*k}^{-1} \qquad (j > k) \\ (\Omega_{22})_{j,k} &= C_j P_{j+1} \Psi(j+1, k+1)(\gamma^2 I - Z_k^* Q_k Z_k P_{k+1})^{-1} Z_k^* Q_k B_k \qquad (j \geq k) . \end{aligned} \tag{5.9}$$

Since the operators $\{M_k\}$ are forward stable, the uniform boundedness of the system coefficients $\{Z_k, B_k, C_k\}$, formula (5.5), and the boundedness of the families $\{N_k^{-1}\}$ and $\{N_{*k}^{-1}\}$ imply that the operators in (5.8) are bounded. Now we are ready to state the following description for the set of all solutions to the nonstationary Nehari problem.

**THEOREM 5.1.** *Let* $\{Z_k, B_k, C_k\}$ *be a backward stable, uniformly controllable and uniformly observable anticausal realization of the data* $\{f_{j,k}\}$, *and assume that* $\gamma > d_\infty$. *Let* $\Omega_{ij}$ *for* $i, j = 1, 2$ *be the block lower triangular operators defined in (5.8) and (5.9). Then the set of all solutions* $G = (g_{j,k})_{j,k=-\infty}^{\infty} = G(R)$ *to the nonstationary Nehari interpolation with data* $\{f_{j,k}\}$ *and tolerance* $\gamma$ *is given by* $g_{j,k} = f_{j,k}$ *for* $j < k$ *and* $g_{j,k} = -H_{j,k}$ *for* $j \geq k$, *where*

$$H = H(R) = \Omega_{22} + \Omega_{21} R(I - \Omega_{11} R)^{-1} \Omega_{12} \tag{5.10}$$

*where* $R$ *is a contractive block lower triangular matrix mapping* $\vec{\mathcal{U}}$ *into* $\vec{\mathcal{Y}}$. *Moreover, the mapping from* $R$ *to* $H$ *given by (5.10) is a bijection from the set of all contractive lower triangular operators from* $\vec{\mathcal{U}}$ *into* $\vec{\mathcal{Y}}$ *onto the set of all solutions of the nonstationary Nehari problem with tolerance* $\gamma$.

Notice that if we choose $R = 0$ in the previous theorem, then $H = \Omega_{22}$ and $G(0)$ is precisely the central interpolant with tolerance $\gamma$ for the nonstationary Nehari interpolation problem computed in Section XI.2.

**PROOF.** The proof of this theorem follows by connecting the nonstationary Nehari interpolation problem to a three chain completion problem, and then using state space techniques on Theorem 4.1 to give the state space formulation of all interpolants in (4.20). To this end, let us apply the three chain completion theorem with

$$\begin{aligned} &\mathcal{U} = \vec{\mathcal{U}}, \;\; \mathcal{Y} = \vec{\mathcal{Y}}, \;\; \mathcal{H}_k = \vec{\mathcal{U}}_{(k,\infty)} = \oplus_{m=k}^{\infty} \mathcal{U}_m, \;\; \mathcal{M}_k = \vec{\mathcal{Y}}_{(k,\infty)} = \oplus_{m=k}^{\infty} \mathcal{Y}_m, \\ &\mathcal{K}_k = \vec{\mathcal{Y}}, \;\; \mathcal{K}_k \ominus \mathcal{M}_k = \vec{\mathcal{Y}}_{(-\infty,k-1)} = \oplus_{m=-\infty}^{k-1} \mathcal{Y}_m . \end{aligned} \tag{5.11}$$

Clearly these spaces satisfy the decreasing property in (4.2). Now let $A_k$ be the operator from $\mathcal{H}_k$ into $\mathcal{K}_k \ominus \mathcal{M}_k$ defined by

$$A_k = \begin{bmatrix} \vdots & \vdots & \vdots & \cdots \\ \vdots & \vdots & \vdots & \cdots \\ f_{k-3,k} & f_{k-3,k+1} & f_{k-3,k+2} & \cdots \\ f_{k-2,k} & f_{k-2,k+1} & f_{k-2,k+2} & \cdots \\ f_{k-1,k} & f_{k-1,k+1} & f_{k-1,k+2} & \cdots \end{bmatrix} . \tag{5.12}$$

By consulting Section XIII.3 we see that $G$ from $\vec{\mathcal{U}}$ into $\vec{\mathcal{Y}}$ is an interpolant for the three chain completion problem with data $\{A_k\}$ and tolerance $\gamma$ if and only if $G$ is a solution for the nonstationary Nehari interpolation problem with tolerance $\gamma$. Moreover, in this case $G$ admits a decomposition of the form $G = F - H$, where $F$ mapping $\vec{\mathcal{U}}$ into $\vec{\mathcal{Y}}$ is the strictly upper triangular operator determined by $(F)_{j,k} = f_{j,k}$ for $j < k$ and H is a lower triangular operator from $\vec{\mathcal{U}}$ to $\vec{\mathcal{Y}}$ given by $(H)_{j,k} = -g_{j,k}$ for $j \geq k$.

Now let $W_{ck}$ from $\mathcal{H}_k$ into $\mathcal{X}_k$ be the controllability operator and $W_{ok}$ the observability operator from $\mathcal{X}_k$ into $\vec{\mathcal{Y}}_{(-\infty,k-1)} = \mathcal{K}_k \ominus \mathcal{M}_k$ defined by

$$\begin{aligned} W_{ck} &= [B_k, \Phi(k, k+1)B_{k+1}, \Phi(k, k+2)B_{k+2}, \ldots] \\ W_{ok} &= [\ldots, C_{k-3}\Phi(k-2, k), C_{k-2}\Phi(k-1, k), C_{k-1}]^{tr} . \end{aligned} \tag{5.13}$$

Notice that $P_k = W_{ck}W_{ck}^*$ and $Q_k = W_{ok}^*W_{ok}$. Moreover, using (5.1) and (5.12) it follows that $A_k = W_{ok}W_{ck}$. Furthermore, by consulting the proof of Theorem III.6.1 we see that $\|A_k\|^2 = r_{spec}(Q_kP_k)$. Therefore the error $d_\infty$ in (5.5) is also given by

$$d_\infty = \sup \{\|A_k\| : k \in \mathbb{Z}\} . \tag{5.14}$$

Theorem 4.1 provides a complete parameterization of all interpolants G of $\{A_k\}$ with tolerance $\gamma$. Because G is an interpolant for the three chain completion problem with data $\{A_k = W_{ok}W_{ck}\}$ if and only if G is a nonstationary Nehari interpolant with data $\{f_{j,k}\}$, Theorem 4.1 can also be used to provide a complete characterization of the set of all nonstationary Nehari interpolants with tolerance $\gamma$. So, to complete the proof we use state space techniques to connect the operator $\Theta_{jk}$ in Theorem 4.1 to their corresponding state space representation in (5.8) and (5.9). Then Theorem 5.1 follows immediately from Theorem 4.1.

Throughout the proof we use the notation established in Section 4. Since $\gamma > d_\infty$, the operator $D_k$ is invertible and $\mathcal{F}_k = D_k\mathcal{H}_{k+1}$. If X is the operator from $\mathcal{H}_{k+1}$ into $\mathcal{D}_k$ defined by $X = D_k \mid \mathcal{H}_{k+1}$, then $X^*X$ has a bounded inverse and the range of X is $\mathcal{F}_k$. Therefore the orthogonal projection onto $\mathcal{F}_k$ is given by $X(X^*X)^{-1}X^*$. Hence

$$P_{\mathcal{F}_k} = D_k(P_{\mathcal{H}_{k+1}}D_k^2 \mid \mathcal{H}_{k+1})^{-1}P_{\mathcal{H}_{k+1}}D_k , \tag{5.15}$$

where $P_{\mathcal{H}_{k+1}}D_k^2 \mid \mathcal{H}_{k+1}$ is viewed as an operator on $\mathcal{H}_{k+1}$. Now we are ready to establish the following useful formulas

$$P_{\mathcal{F}_k}D_k \mid \mathcal{U}_k = -D_kW_{c(k+1)}^*(\gamma^2I - Z_k^*Q_kZ_kP_{k+1})^{-1}Z_k^*Q_kB_k = -D_kW_{c(k+1)}^*\hat{B}_k \tag{5.16}$$

where $\hat{B}_k = (\gamma^2I - Z_k^*Q_kZ_kP_{k+1})^{-1}Z_k^*Q_kB_k$. To obtain (5.16) observe that $W_{ck} \mid \mathcal{H}_{k+1} = Z_kW_{c(k+1)}$ and thus $A_k \mid \mathcal{H}_{k+1} = W_{ok}Z_kW_{c(k+1)}$. Recall also that $Q_k = W_{ok}^*W_{ok}$. Since $\mathcal{U}_k$ is orthogonal to $\mathcal{H}_{k+1}$ (in fact $\mathcal{U}_k = \mathcal{H}_k \ominus \mathcal{H}_{k+1}$) we have $P_{\mathcal{H}_{k+1}}\mathcal{U}_k = \{0\}$. This along with the formula for $P_{\mathcal{F}_k}$ in (5.15) yields

$$P_{\mathcal{F}_k} D_k \mid \mathcal{U}_k = D_k (P_{\mathcal{H}_{k+1}} D_k^2 \mid \mathcal{H}_{k+1})^{-1} P_{\mathcal{H}_{k+1}} D_k^2 \mid \mathcal{U}_k$$

$$= - D_k (\gamma^2 I_{\mathcal{H}_{k+1}} - P_{\mathcal{H}_{k+1}} A_k^* A_k \mid \mathcal{H}_{k+1})^{-1} P_{\mathcal{H}_{k+1}} A_k^* A_k \mid \mathcal{U}_k$$

$$= - D_k (\gamma^2 I_{\mathcal{H}_{k+1}} - W_{c(k+1)}^* Z_k^* Q_k Z_k W_{c(k+1)})^{-1} W_{c(k+1)}^* Z_k^* Q_k W_{ck} \mid \mathcal{U}_k$$

$$= - D_k W_{c(k+1)}^* (\gamma^2 I - Z_k^* Q_k Z_k W_{c(k+1)} W_{c(k+1)}^*)^{-1} Z_k^* Q_k B_k = - D_k W_{c(k+1)}^* \hat{B}_k \, .$$

Hence (5.16) holds. Using the definitions of $\pi_k$ and $\omega_k$ in (4.12) and (4.14), equation (5.16) gives

$$\pi_k \omega_k P_{\mathcal{F}_k} D_k \mid \mathcal{U}_k = - D_{k+1} W_{c(k+1)}^* \hat{B}_k \, . \tag{5.17}$$

Equation (5.16) along with $A_{k+1} = W_{o(k+1)} W_{c(k+1)}$ and the definition of $\pi_k'$ and $\mathcal{D}_k' = \mathcal{Y}_k$ in (4.8) and (4.13) yields

$$\pi_k' \omega_k P_{\mathcal{F}_k} D_k \mid \mathcal{U}_k = - P_{\mathcal{D}_k'} P_{\mathcal{M}_k} A_{k+1} W_{c(k+1)}^* \hat{B}_k = - C_k P_{k+1} \hat{B}_k \, . \tag{5.18}$$

Next let us establish the following result where $M_n$ is defined in (5.7);

$$\pi_n \omega_n P_{\mathcal{F}_n} D_n W_{cn}^* = D_{n+1} W_{c(n+1)}^* M_n \, . \tag{5.19}$$

Using $A_n = W_{on} W_{cn}$ and $P_{\mathcal{H}_{n+1}} W_{cn}^* = W_{c(n+1)}^* Z_n^*$ once again along with the formula for $P_{\mathcal{F}_n}$ in (5.15) we have

$$\pi_n \omega_n P_{\mathcal{F}_n} D_n W_{cn}^* = \pi_n \omega_n D_n (P_{\mathcal{H}_{n+1}} (\gamma^2 I - A_n^* A_n) \mid \mathcal{H}_{n+1})^{-1} P_{\mathcal{H}_{n+1}} D_n^2 W_{cn}^*$$

$$= \pi_n \omega_n D_n (\gamma^2 I_{\mathcal{H}_{n+1}} - P_{\mathcal{H}_{n+1}} W_{cn}^* Q_n W_{cn} \mid \mathcal{H}_{n+1})^{-1} \cdot$$

$$\cdot P_{\mathcal{H}_{n+1}} (\gamma^2 I - W_{cn}^* Q_n W_{cn}) W_{cn}^*$$

$$= D_{n+1} (\gamma^2 I - W_{c(n+1)}^* Z_n^* Q_n Z_n W_{c(n+1)})^{-1} P_{\mathcal{H}_{n+1}} W_{cn}^* (\gamma^2 I - Q_n P_n)$$

$$= D_{n+1} (\gamma^2 I - W_{c(n+1)}^* Z_n^* Q_n Z_n W_{c(n+1)})^{-1} W_{c(n+1)}^* Z_n^* (\gamma^2 I - Q_n P_n)$$

$$= D_{n+1} W_{c(n+1)}^* M_n \, .$$

Hence (5.19) holds. By recursively using (5.19) we readily arrive at the following useful result

$$\Phi_p(m, n) D_n W_{cn}^* = D_m W_{cm}^* \Psi(m, n) \qquad (m \geq n) \, . \tag{5.20}$$

Let us establish

$$\pi_j' \omega_j P_{\mathcal{F}_j} D_j W_{cj}^* = C_j P_{j+1} M_j \, . \tag{5.21}$$

To this end, notice that

$$\pi_j'\omega_j P_{\mathcal{F}_j} D_j W_{cj}^* = \pi_j'\omega_j D_j (P_{\mathcal{H}_{j+1}} D_j^2 \mid \mathcal{H}_{j+1})^{-1} P_{\mathcal{H}_{j+1}} D_j^2 W_{cj}^*$$

$$= \pi_j'\omega_j D_j (\gamma^2 I_{\mathcal{H}_{j+1}} - P_{\mathcal{H}_{j+1}} W_{cj}^* Q_j W_{cj} \mid \mathcal{H}_{j+1})^{-1} P_{\mathcal{H}_{j+1}} W_{cj}^* (\gamma^2 I - Q_j P_j)$$

$$= \pi_j'\omega_j D_j (\gamma^2 I - W_{c(j+1)}^* Z_j^* Q_j Z_j W_{c(j+1)})^{-1} W_{c(j+1)}^* Z_j^* (\gamma^2 I - Q_j P_j)$$

$$= \pi_j'\omega_j D_j W_{c(j+1)}^* M_j = P_{\mathcal{D}_j'} A_{j+1} W_{c(j+1)}^* M_j$$

$$= P_{\mathcal{D}_j'} W_{o(j+1)} W_{c(j+1)} W_{c(j+1)}^* M_j = C_j P_{j+1} M_j \ .$$

Let $\Omega_{22}$ be the block lower triangular operator mapping $\vec{\mathcal{U}}$ into $\vec{\mathcal{Y}}$ defined by $\Omega_{22} = -\Theta_{22}\operatorname{diag}(D_k \mid \mathcal{U}_k)_{-\infty}^{\infty}$, where $\Theta_{22}$ is defined in Theorem 4.1. Then according to (5.18) and the definition of $\Theta_{22}$ in Section 4, the (k, k)-th entry of $\Omega_{22}$ is given by

$$(\Omega_{22})_{k,k} = -\pi_k'\omega_k P_{\mathcal{F}_k} D_k \mid \mathcal{U}_k = C_k P_{k+1} \hat{B}_k \ .$$

Moreover, for $j > k$ equations (5.17) to (5.21) yield

$$(\Omega_{22})_{j,k} = -\pi_j'\omega_j P_{\mathcal{F}_j} \Phi_p(j, k) D_k \mid \mathcal{U}_k = -\pi_j'\omega_j P_{\mathcal{F}_j} \Phi_p(j, k+1)\pi_k\omega_k P_{\mathcal{F}_k} D_k \mid \mathcal{U}_k$$

$$= \pi_j'\omega_j P_{\mathcal{F}_j} \Phi_p(j, k+1) D_{k+1} W_{c(k+1)}^* \hat{B}_k = \pi_j'\omega_j P_{\mathcal{F}_j} D_j W_{cj}^* \Psi(j, k+1)\hat{B}_k$$

$$= C_j P_{j+1} M_j \Psi(j, k+1)\hat{B}_k = C_j P_{j+1} \Psi(j+1, k+1)\hat{B}_k \ .$$

This is precisely the state space formula for $\Omega_{22}$ given in (5.9).

To obtain a state space formula for the other operators $\Theta_{ij}$ in Theorem 4.1 we first need to identify $\mathcal{G}_k$ with $\mathcal{U}_k$ and $\mathcal{G}_k'$ with $\mathcal{Y}_k$. Since $\|A_k\| < \gamma$, it follows that $\mathcal{D}_k = \mathcal{H}_k$ and $\mathcal{G}_k := \mathcal{D}_k \ominus \mathcal{F}_k = \mathcal{H}_k \ominus D_k \mathcal{H}_{k+1}$. We claim that $\mathcal{G}_k = D_k^{-1}\mathcal{U}_k$. To verify this notice that a vector g is in $\mathcal{G}_k$ if and only if g is in $\mathcal{H}_k$ and g is orthogonal to $D_k\mathcal{H}_{k+1}$, or equivalently, $D_k g$ is orthogonal to $\mathcal{H}_{k+1}$. Recall that $\mathcal{H}_k = \mathcal{U}_k \oplus \mathcal{H}_{k+1}$. So then g is in $\mathcal{G}_k$ if and only if $D_k g$ is in $\mathcal{U}_k$, or equivalently, g is in $D_k^{-1}\mathcal{U}_k$ which proves our claim.

Let $N_k$ be the operator defined in (5.6). Then

$$\psi_k = D_k^{-1} N_k^{-1} : \mathcal{U}_k \to \mathcal{G}_k \tag{5.22}$$

is unitary. To see this let X be the operator from $\mathcal{U}_k$ into $\mathcal{G}_k$ defined by $X = D_k^{-1} \mid \mathcal{U}_k$. Since $\mathcal{G}_k = D_k^{-1}\mathcal{U}_k$ and $\|A_k\| < \gamma$, it follows that ran $X = \mathcal{G}_k$ and thus $X(X^*X)^{-1/2}$ is a unitary operator from $\mathcal{U}_k$ onto $\mathcal{G}_k$. Let $\tilde{N}_k$ be the positive square root of $X^*X = P_{\mathcal{U}_k} D_k^{-2} \mid \mathcal{U}_k$. By consulting (V.5.16) with $A_k$, $W_{ck}$, $W_{ok}$, $P_k$ and $Q_k$ replacing A, $W_c$, $W_o$, P and Q we have

$$D_k^{-2} = \gamma^{-2} I + \gamma^{-2} W_{ck}^*(\gamma^2 I - Q_k P_k)^{-1} Q_k W_{ck} .$$

Using $B_k = W_{ck} \mid \mathcal{U}_k$ it follows that $\tilde{N}_k = N_k$, where $N_k$ is given by (5.6). Hence $\psi_k$ is unitary.

Let $N_{*k}$ be the positive operator on $\mathcal{Y}_k$ defined in (5.6). Then we claim that

$$\psi_{*k} = \begin{bmatrix} -D_{k+1}^{-1} W_{c(k+1)}^* C_k^* \\ I \end{bmatrix} N_{*k}^{-1} : \mathcal{Y}_k \to \mathcal{G}_k' \tag{5.23}$$

is unitary. To obtain an expression for $\mathcal{G}_k'$ first observe that in our setting $\mathcal{D}_{k+1} = \mathcal{H}_{k+1}$ and $\mathcal{D}_k' = \mathcal{M}_k \ominus \mathcal{M}_{k+1} = \mathcal{Y}_k$. Recall that

$$\mathcal{G}_k' = (\mathcal{D}_{k+1} \oplus \mathcal{D}_k') \ominus \{D_{k+1} h \oplus P_{\mathcal{D}_k'} A_{k+1} h : h \in \mathcal{H}_{k+1}\}$$

$$= (\mathcal{H}_{k+1} \oplus \mathcal{Y}_k) \ominus \{D_{k+1} h \oplus P_{\mathcal{Y}_k} A_{k+1} h : h \in \mathcal{H}_{k+1}\} .$$

Therefore $f \oplus g$ is in $\mathcal{G}_k'$ if and only if $f \oplus g$ is in $\mathcal{H}_{k+1} \oplus \mathcal{Y}_k$ and $f \oplus g$ is orthogonal to $\{D_{k+1} h \oplus P_{\mathcal{Y}_k} A_{k+1} h : h \in \mathcal{H}_{k+1}\}$, or equivalently, $D_{k+1} f + A_{k+1}^* g$ is orthogonal to $\mathcal{H}_{k+1}$. Hence $f \oplus g$ is in $\mathcal{G}_k'$ if and only if $D_{k+1} f + A_{k+1}^* g = 0$, or equivalently, $f = -D_{k+1}^{-1} A_{k+1}^* g$. So $\mathcal{G}_k'$ is the range of the operator Y from $\mathcal{Y}_k$ into $\mathcal{H}_{k+1} \oplus \mathcal{Y}_k$ defined by

$$Y = \begin{bmatrix} -D_{k+1}^{-1} A_{k+1}^* \\ I \end{bmatrix} \mid \mathcal{Y}_k = \begin{bmatrix} -D_{k+1}^{-1} W_{c(k+1)}^* C_k^* \\ I \end{bmatrix} .$$

The equality follows from $A_{k+1} = W_{o(k+1)} W_{c(k+1)}$ and $W_{o(k+1)}^* \mid \mathcal{Y}_k = C_k^*$. Thus $\psi_{*k} = Y(Y^*Y)^{-1/2}$ is a unitary operator from $\mathcal{Y}_k$ onto $\mathcal{G}_k'$. It remains to show that $N_{*k}^2 = Y^*Y$ and hence $\psi_{*k}$ is given by (5.23). Using $P_{k+1} = W_{c(k+1)} W_{c(k+1)}^*$, we have

$$Y^*Y = I + C_k W_{c(k+1)} (\gamma^2 I - A_{k+1}^* A_{k+1})^{-1} W_{c(k+1)}^* C_k^*$$

$$= I + C_k W_{c(k+1)} (\gamma^2 I - W_{c(k+1)}^* Q_{k+1} W_{c(k+1)})^{-1} W_{c(k+1)}^* C_k^*$$

$$= I + C_k W_{c(k+1)} W_{c(k+1)}^* (\gamma^2 I - Q_{k+1} P_{k+1})^{-1} C_k^* = N_{*k}^2 .$$

Therefore $\psi_{*k}$ in (5.23) is unitary.

Let $\psi$ be the unitary operator from $\vec{\mathcal{U}}$ onto $\mathcal{G} = \oplus\{\mathcal{G}_k : k \in \mathbb{Z}\}$ defined by $\psi = \operatorname{diag}(\psi_k)_{-\infty}^{\infty}$. Let $\psi_*$ be the unitary operator from $\vec{\mathcal{Y}}$ onto $\mathcal{G}' = \oplus\{\mathcal{G}_k' : k \in \mathbb{Z}\}$ defined by $\psi_* = \operatorname{diag}(\psi_{*k})_{-\infty}^{\infty}$. Finally, let $\Omega_{11}$ be the strictly lower triangular block operator from $\vec{\mathcal{Y}}$ into $\vec{\mathcal{U}}$ defined by $\psi\Omega_{11} = \Theta_{11}\psi_*$ where $\Theta_{11}$ is the strictly lower triangular block matrix in Theorem 4.1. Then using (5.20) along with the definition of $\psi$ and $\psi_*$ we have for $j > k$

$$
\begin{aligned}
(\Omega_{11})_{j,k} &= (\psi^* \Theta_{11} \psi_*)_{j,k} = \psi_j^* P_{\mathcal{G}_j} \Phi_p(j, k+1) \pi_k \psi_{*k} \\
&= -\psi_j^* P_{\mathcal{G}_j} \Phi_p(j, k+1) D_{k+1} D_{k+1}^{-2} W_{c(k+1)}^* C_k^* N_{*k}^{-1} \\
&= -\psi_j^* P_{\mathcal{G}_j} \Phi_p(j, k+1) D_{k+1} (\gamma^2 I - W_{c(k+1)}^* Q_{k+1} W_{c(k+1)})^{-1} W_{c(k+1)}^* C_k^* N_{*k}^{-1} \\
&= -\psi_j^* P_{\mathcal{G}_j} \Phi_p(j, k+1) D_{k+1} W_{c(k+1)}^* (\gamma^2 I - Q_{k+1} P_{k+1})^{-1} C_k^* N_{*k}^{-1} \\
&= -\psi_j^* P_{\mathcal{G}_j} D_j W_{cj}^* \Psi(j, k+1) (\gamma^2 I - Q_{k+1} P_{k+1})^{-1} C_k^* N_{*k}^{-1} \\
&= -N_j^{-1} P_{\mathcal{U}_j} D_j^{-1} D_j W_{cj}^* \Psi(j, k+1) (\gamma^2 I - Q_{k+1} P_{k+1})^{-1} C_k^* N_{*k}^{-1} \\
&= -N_j^{-1} B_j^* \Psi(j, k+1) (\gamma^2 I - Q_{k+1} P_{k+1})^{-1} C_k^* N_{*k}^{-1} .
\end{aligned}
$$

Therefore $\Omega_{11}$ is given by the first two equations in (5.9).

Let $\Omega_{12}$ be the block lower triangular operator on $\vec{\mathcal{U}}$ defined by $\Omega_{12} = \psi^* \Theta_{12} \operatorname{diag}(D_k \mid \mathcal{U}_k)_{-\infty}^{\infty}$. Then the (k, k)-th entry of $\Omega_{12}$ is given by

$$
(\Omega_{12})_{k,k} = \psi_k^* P_{\mathcal{G}_k} D_k \mid \mathcal{U}_k = N_k^{-1} P_{\mathcal{U}_k} D_k^{-1} D_k \mid \mathcal{U}_k = N_k^{-1} .
$$

For $j > k$, equations (5.17), (5.20) along with the definition of $\psi_j$ gives

$$
\begin{aligned}
(\Omega_{12})_{j,k} &= \psi_j^* (\theta_{12})_{j,k} D_k \mid \mathcal{U}_k = \psi_j^* P_{\mathcal{G}_j} \Phi_p(j, k) D_k \mid \mathcal{U}_k = \\
&= \psi_j^* P_{\mathcal{G}_j} \Phi_p(j, k+1) \pi_k \omega_k P_{\mathcal{F}_k} D_k \mid \mathcal{U}_k = -\psi_j^* P_{\mathcal{G}_j} \Phi_p(j, k+1) D_{k+1} W_{c(k+1)}^* \hat{B}_k \\
&= -\psi_j^* P_{\mathcal{G}_j} D_j W_{cj}^* \Psi(j, k+1) \hat{B}_k \\
&= -N_j^{-1} P_{\mathcal{U}_j} D_j^{-1} D_j W_{cj}^* \Psi(j, k+1) \hat{B}_k \\
&= -N_j^{-1} B_j^* \Psi(j, k+1) \hat{B}_k .
\end{aligned}
$$

Therefore $\Omega_{12}$ is given by (5.9).

Let $\Omega_{21}$ be the block lower triangular operator on $\vec{\mathcal{Y}}$ defined by $\Omega_{21} = -\Theta_{21} \psi_*$. The (k, k)-th entry of $\Omega_{21}$ is given by

$$
(\Omega_{21})_{k,k} = -\pi_k' \psi_{*k} = -N_{*k}^{-1} .
$$

Using (5.20), (5.21) and the definition of $\psi_{*k}$ we have for $j > k$

$$\begin{aligned}(\Omega_{21})_{j,k} &= -(\Theta_{21}\psi_*)_{j,k} = -\pi_j'\omega_j P_{\mathcal{F}_j}\Phi_p(j,k+1)\pi_k\psi_{*k}\\ &= \pi_j'\omega_j P_{\mathcal{F}_j}\Phi_p(j,k+1)D_{k+1}D_{k+1}^{-2}W_{c(k+1)}^*C_k^*N_{*k}^{-1}\\ &= \pi_j'\omega_j P_{\mathcal{F}_j}\Phi_p(j,k+1)D_{k+1}(\gamma^2 I - W_{c(k+1)}^*Q_{k+1}W_{c(k+1)})^{-1}W_{c(k+1)}^*C_k^*N_{*k}^{-1}\\ &= \pi_j'\omega_j P_{\mathcal{F}_j}\Phi_p(j,k+1)D_{k+1}W_{c(k+1)}^*(\gamma^2 I - Q_{k+1}P_{k+1})^{-1}C_k^*N_{*k}^{-1}\\ &= \pi_j'\omega_j P_{\mathcal{F}_j}D_jW_{cj}^*\Psi(j,k+1)(\gamma^2 I - Q_{k+1}P_{k+1})^{-1}C_k^*N_{*k}^{-1}\\ &= C_jP_{j+1}M_j\Psi(j,k+1)(\gamma^2 I - Q_{k+1}P_{k+1})^{-1}C_k^*N_{*k}^{-1}\,.\end{aligned}$$

Therefore $\Omega_{21}$ is given by (5.9).

Let R be a contractive block lower triangular operator mapping $\vec{\mathcal{U}}$ into $\vec{\mathcal{Y}}$. According to the definition of $\Omega_{ij}$ it follows that $\Gamma(\psi_* R\psi^*)$ in (4.16) is given by

$$-\Gamma(\psi_* R\,|\,\psi^*)\ \mathrm{diag}\,(D_k\,|\,\mathcal{U}_k)_{-\infty}^{\infty} = \Omega_{22} + \Omega_{21}R(I-\Omega_{11}R)^{-1}\Omega_{12} = H(R)\,. \tag{5.24}$$

Notice that all the $\Omega_{ij}$ are block lower triangular bounded linear operators. Since $\Omega_{11}$ is strictly block lower triangular and R is a block lower triangular contraction, it follows that H(R) is a block lower triangular operator. Let F be the strictly block upper triangular operator from $\vec{\mathcal{U}}$ into $\vec{\mathcal{Y}}$ defined by

$$F = \begin{bmatrix} \ddots & & & \vdots & \vdots & \\ & 0 & f_{-1,0} & f_{-1,1} & f_{-1,2} & \cdots \\ & & \underline{0} & f_{0,1} & f_{0,2} & \cdots \\ & & & 0 & f_{1,2} & \\ & & & & 0 & \\ & & & & & \ddots \end{bmatrix}.$$

From (5.1) and the stability condition (5.2) it follows that F is a well defined bounded linear operator. Recall that $A_k = (I - P_{\mathcal{M}_k})F\,|\,\vec{\mathcal{U}}_{(k,\infty)}$, where $I - P_{\mathcal{M}_k}$ is the orthogonal projection onto $\vec{\mathcal{Y}}_{(-\infty,k-1)}$. So according to Theorem 4.1, the set of all solutions to our nonstationary Nehari interpolation problem with tolerance $\gamma$ is given by $B(\psi_* R\psi^*) = F - H(R)$ where R is a block lower triangular contractive matrix from $\vec{\mathcal{U}}$ into $\vec{\mathcal{Y}}$. This completes the proof.

We conclude this section with a number of remarks. First we rewrite the coefficients $\Omega_{ij}$ in (5.9) in a closed form. To this end, given a sequence of Hilbert spaces $\{\mathcal{H}_k\}$ put $\vec{\mathcal{H}} = \oplus\{\mathcal{H}_k : k \in \mathbb{Z}\}$. Using this notation let Z on $\vec{\mathcal{X}}$ and C from $\vec{\mathcal{X}}$ into $\vec{\mathcal{Y}}$ be the block weighted shifts defined by

$$Z_{j,k} = \delta_{j,k-1} Z_j, \qquad C_{j,k} = \delta_{j,k-1} C_j \ , \tag{5.25}$$

where $\delta_{j,k}$ is the Kronecker delta. The backward stability condition placed on $\{Z_k\}$ in (5.2) guarantees that $r_{spec}(Z) < 1$. Now let B from $\vec{\mathcal{U}}$ into $\vec{\mathcal{X}}$ and P on $\vec{\mathcal{X}}$ with Q on $\vec{\mathcal{X}}$ be the block diagonal operators defined by

$$B = \text{diag}\,(B_k)_{-\infty}^{\infty}, \quad P = \text{diag}\,(P_k)_{-\infty}^{\infty}, \quad Q = \text{diag}\,(Q_k)_{-\infty}^{\infty} \ . \tag{5.26}$$

The uniform controllability condition placed on $\{P_k\}$ and the uniform observability condition placed on $\{Q_k\}$ guarantees that both P and Q are invertible positive operators. From (5.5) we conclude that $d_\infty^2 = r_{spec}(QP)$. Notice that for $j \le k$ the (j, k)-th entry of $(I - Z)^{-1}$ is $\Phi(j, k)$. This and (5.1) readily implies that

$$((C(I - Z)^{-1} B)_{j,k} = f_{j,k} \qquad (-\infty < j < k < \infty)\ . \tag{5.27}$$

Next let N on $\vec{\mathcal{U}}$ and $N_*$ on $\vec{\mathcal{Y}}$ be the strictly positive operators defined by the block diagonal matrix

$$N = \text{diag}\,(N_k)_{-\infty}^{\infty}, \quad N_* = \text{diag}\,(N_{*k})_{-\infty}^{\infty} \ , \tag{5.28}$$

and let M be the block forward weighted shift on $\vec{\mathcal{X}}$ defined by

$$M_{j,k} = \delta_{j,k+1} M_k \ , \tag{5.29}$$

where $M_k$ is defined in (5.7). We already know that the family of operators $\{M_k\}$ is forward stable, and thus $r_{spec}(M) < 1$. Notice that the (j, k)-th entry of $(I - M)^{-1}$ is given by $\Psi(j, k)$ for $j \ge k$ and is zero otherwise. Using this fact, and (5.9) it follows that the operators $\Omega_{ij}$ in (5.8) are given by

$$\begin{aligned}
\Omega_{11} &= -N^{-1} B^* (I - M)^{-1} (\gamma^2 I - QP)^{-1} C^* N_*^{-1} \ , \\
\Omega_{12} &= N^{-1} - N^{-1} B^* (I - M)^{-1} (\gamma^2 I - Z^* QZP)^{-1} Z^* QB \ , \\
\Omega_{21} &= CPM(I - M)^{-1} (\gamma^2 I - QP)^{-1} C^* N_*^{-1} - N_*^{-1} \ , \\
\Omega_{22} &= CP(I - M)^{-1} (\gamma^2 I - Z^* QZP)^{-1} Z^* QB \ .
\end{aligned} \tag{5.30}$$

The operator $\Omega_{11}$ is strictly lower triangular. Combining this with (5.27) and Theorem 5.1 we see that the set of all nonstationary Nehari interpolants with tolerance $\gamma > d_\infty$ is given by

$$G(R) = C(I - Z)^{-1} B - \Omega_{22} - \Omega_{21} R (I - \Omega_{11} R)^{-1} \Omega_{12} \tag{5.31}$$

where R is a block lower triangular contraction from $\vec{\mathcal{U}}$ into $\vec{\mathcal{Y}}$.

Next, put $F_{-n} = CZ^{n-1}B$ for $n \geq 1$. From the particular form of Z, B and C in (5.25) and (5.26) it follows that

$$(F_{-n})_{j,k} = (CZ^{n-1}B)_{j,k} = \begin{cases} f_{j,k} & \text{for } j-k=-n, \\ 0 & \text{for } j-k \neq -n. \end{cases}$$

Thus $F_{-n}$ coincides with the operator $[F]_{-n}$ in (XI.2.4), and the operators $\{F_{-n}\}_{n\geq 1}$ are precisely the data of the stationary Nehari problem associated with our nonstationary Nehari problem. Furthermore, {Z, B, C} is an anticausal realization for these stationary data. Since P is the controllability grammian of {Z, B} and Q is the observability grammian of {C, Z}, the system {Z, B, C} is a controllable and observable realization of $\{F_{-n}\}_{n\geq 1}$. So we can apply the results of Section VI.8 to the function $F(\lambda) = C(\lambda I - Z)^{-1}B$, where {Z, B, C} is as above. We see that for $1 \leq i, j \leq 2$ the coefficient $\Omega_{ij}$ in (5.30) is given by $\Omega_{ij} = \Phi_{i,j}(1)$, where $\Phi_{i,j}(\lambda)$ is defined in (VI.8.5). Notice that $\Phi_{i,j}(\cdot)$ is analytic in the closed unit disc because $r_{spec}(M) < 1$, and hence we can evaluate $\Phi_{i,j}(\cdot)$ at $\lambda = 1$. In fact, since M is a stable block forward weighted shift, $\Omega_{ij}$ belongs to a nonstationary Wiener class (cf., Remark X.2.5) and the symbol of the diagonally sparce transform of $\Omega_{ij}$ is precisely the function $\Phi_{i,j}(\lambda)$, and thus $\Phi_{i,j}(1) = \Omega_{ij}$.

The phenomenon described in the previous paragraph applies in an analogous way to the other nonstationary interpolation problems discussed in Chapter VIII. For instance, let us sketch how one can obtain all solutions for the nonstationary Nevanlinna-Pick problem with data $\{Z_k, B_k, \tilde{B}_k\}$ as in (VIII.2.6), where $\{Z_k, B_k\}$ is uniformly controllable and the tolerance $\gamma > d_\infty$ (see (XI.1.7)). To get the parameterization of all solutions one simply goes to the associated stationary Nevanlinna-Pick problem with data $\{Z, B, \tilde{B}\}$ described in Section XI.1; see (XI.1.1), (XI.1.2) and (XI.1.5). Then one takes the coefficients $\Phi_{i,j}(\lambda)$ in the parametrization of all solutions corresponding to this stationary problem (use (VI.7.6)), and one evaluates these coefficients at $\lambda = 1$. So with $\Omega_{ij} = \Phi_{i,j}(1)$ for $1 \leq i, j \leq 2$, the set of all solutions to the nonstationary Nevanlinna-Pick problem is given by

$$F(R) = \Omega_{22} + \Omega_{21}R(I - \Omega_{11}R)^{-1}\Omega_{12}$$

where R is a block lower triangular contractive operator from $\vec{\mathcal{Y}}$ into $\vec{\mathcal{U}}$.

## *Notes to Chapter XIV:*

The material in this chapter is taken from Foias-Frazho-Gohberg-Kaashoek [3].

APPENDIX

# ON FACTORIZATION OF MATRIX-VALUED FUNCTIONS

This appendix reviews some results on positive definite and positive semidefinite matrix-valued functions on the unit circle. The emphasis is a constructive methods for spectral factorization based on finite sections of Toeplitz matrices and associated Riccati difference equations. The first section focuses on the general theory. The second section presents explicit state space formula for the rational case. For a general reference about Riccati equations and its applications see Lancaster-Rodman [1].

## A.1 SQUARE OUTER SPECTRAL FACTORIZATIONS

Throughout this section we assume that $\mathcal{U}$ is a finite dimensional vector space. We say that $\Theta$ is a *square outer function* if $\Theta$ is a function in $H^\infty(\mathcal{U}, \mathcal{U})$ satisfying

$$\overline{\Theta H^2(\mathcal{U})} = H^2(\mathcal{U}). \tag{1.1}$$

In this case $\Theta(\lambda)$ is invertible for all $\lambda$ in the open unit disc. To see this recall that $(1-\bar{\alpha}\lambda)^{-1}y$ where $|\alpha| < 1$ and y is in $\mathcal{U}$ has the following reproducing property

$$(f, (1-\bar{\alpha}\lambda)^{-1}y) = (f(\alpha), y) \qquad (f \in H^2(\mathcal{U})).$$

To verify this let $\{f_n\}$ be the Taylor coefficients of $\lambda^n$ for f in $H^2(\mathcal{U})$. Then

$$(f, (1-\bar{\alpha}\lambda)^{-1}y) = (\sum_0^\infty f_n\lambda^n, \sum_0^\infty (\bar{\alpha}\lambda)^n y) = \sum_0^\infty (f_n\alpha^n, y) = (f(\alpha), y)$$

which proves our claim. To show that a square outer $\Theta(\alpha)$ is invertible for all $|\alpha| < 1$, assume that there exists a vector y in $\mathcal{U}$ which is orthogonal to ran $\Theta(\alpha)$ for some $\alpha$ satisfying $|\alpha| < 1$. Then using the reproducing property of the function $(1-\bar{\alpha}\lambda)^{-1}y$ we have

$$(\Theta h, (1-\bar{\alpha}\lambda)^{-1}y) = (\Theta(\alpha)h(\alpha), y) = 0 \qquad (h \in H^2(\mathcal{U})).$$

However, equation (1.1) shows that y must be zero. Hence $\Theta(\lambda)$ is invertible for all $|\lambda| < 1$. In particular, $\Theta(0)^*\Theta(0)$ is strictly positive.

Let R be a function in $L^\infty(\mathcal{U}, \mathcal{U})$ and $T_R$ the block Toeplitz matrix on $\ell_+^2(\mathcal{U})$ generated by R. We say that F is a *spectral factor for* R if F is a function in $H^\infty(\mathcal{U}, \mathcal{Y})$ satisfying

$T_F^* T_F = T_R$, or equivalently, $F^* F = R$ a.e. Here we are interested in finding necessary and sufficient conditions for the existence of a square outer spectral factor $\Theta$ for $T_R$, that is, a square outer function $\Theta$ satisfying $\Theta^* \Theta = R$ a.e. The following classical result shows that R admits a square outer spectral factorization if $T_R$ is strictly positive.

**THEOREM 1.1.** *Let* R *be a function in* $L^\infty(\mathcal{U}, \mathcal{U})$ *and* $T_R$ *a strictly positive Toeplitz operator on* $\ell_+^2(\mathcal{U})$. *Let* $\Pi$ *be the operator from* $\ell_+^2(\mathcal{U})$ *onto* $\mathcal{U}$ *defined by* $\Pi f = f_0$, *where* $f_0$ *is the first element of the sequence* f, *and let* $\mathcal{F}_+$ *the Fourier transform from* $\ell_+^2(\mathcal{U})$ *into* $H^2(\mathcal{U})$. *Then* $\Omega(\lambda) = (\mathcal{F}_+ T_R^{-1} \Pi^*)(\lambda)$ *is a square outer function in* $H^\infty(\mathcal{U}, \mathcal{U})$ *and* $\Omega(\lambda)^{-1}$ *is also in* $H^\infty(\mathcal{U}, \mathcal{U})$. *Moreover the function* $\Theta$ *in* $H^\infty(\mathcal{U}, \mathcal{U})$ *defined by*

$$\Theta(\lambda) = (\Pi T_R^{-1} \Pi^*)^{1/2} \Omega(\lambda)^{-1} = \Omega(0)^{1/2} \Omega(\lambda)^{-1} \tag{1.2}$$

*is an outer spectral factor for* $T_R$.

**PROOF.** Let $T_0$ be the Toeplitz operator on $H^2(\mathcal{U})$ defined by taking the Fourier transform of $T_R$, that is, $T_0 = \mathcal{F}_+ T_R \mathcal{F}_+^* = P_+ M_R \,|\, H^2(\mathcal{U})$ where $P_+$ is the orthogonal projection onto $H^2(\mathcal{U})$. Then $\Omega(\lambda) = (T_0^{-1} \Pi_0^*)(\lambda)$ where $\Pi_0$ is the operator from $H^2(\mathcal{U})$ onto $\mathcal{U}$ defined by $\Pi_0 = P_{\mathcal{U}}$. Here $\mathcal{U}$ is identified with the subspace of constant functions in $H^2(\mathcal{U})$. Now let us show that $\Omega$ is an outer function. Obviously $\Omega(\lambda)$ is analytic in the open unit disc and $\Omega(\lambda)a$ is in $H^2(\mathcal{U})$ for all a in $\mathcal{U}$. To show that $\Omega\mathcal{U}$ is cyclic for the unilateral shift S on $H^2(\mathcal{U})$ assume that x is orthogonal to $S^n T_0^{-1} \mathcal{U}$ for all $n \geq 0$. Let y be the vector satisfying $x = T_0 y$. Then $S^{*n} T_0 y$ is orthogonal to $T_0^{-1} \mathcal{U}$ for all $n \geq 0$. By choosing $n = 0$ we see that y is also orthogonal to $\mathcal{U}$. Thus $y = S S^* y$. This and $S^* T_0 S = T_0$ shows that $S^{*n} T_0 y = S^{*n-1} T_0 S^* y$ is orthogonal to $T_0^{-1} \mathcal{U}$ for all $n \geq 1$. Choosing $n = 1$ shows that $S^* y$ is also orthogonal to $\mathcal{U}$. Thus $y = S^2 S^{*2} y$. Continuing in this fashion $S^{*n} y$ is orthogonal to $\mathcal{U}$ for all $n \geq 0$. Hence, $y = 0$, the vector $x = 0$ and $\Omega\mathcal{U}$ is cyclic for S. Therefore $\Omega$ is an outer function.

Now let N be the positive square root of $\Pi_0 T_0^{-1} \Pi_0^*$, and a, b vectors in $\mathcal{U}$. Then using $S^* T_0 S = T_0$ we have

$$(T_0 S^n \Omega a, S^n \Omega b) = (T_0 T_0^{-1} a, T_0^{-1} b) = (N^2 a, b). \tag{1.3}$$

On the other hand, if $n > m$, then

$$(T_0 S^m \Omega a, S^n \Omega b) = (T_0 T_0^{-1} a, S^{n-m} T_0^{-1} b) = 0. \tag{1.4}$$

A similar calculation shows that (1.4) holds for all $m > n$ and thus for all $n \neq m$. So by combining (1.3) and (1.4) we see that

$$(T_0^{\frac{1}{2}}M_\Omega^+ f,\ T_0^{\frac{1}{2}}M_\Omega^+ g) = (T_0 M_\Omega^+ f,\ M_\Omega^+ g) = (N^2 f,\ g) = (Nf,\ Ng) \tag{1.5}$$

where f and g are operator valued polynomials. In particular, by choosing f = g $\|T_0^{\frac{1}{2}}M_\Omega^+ f\| \le \|N\|\,\|f\|$. Because $T_0^{\frac{1}{2}}$ is invertible, it follows that $M_\Omega^+$ can be extended to a bounded operator on $H^2(\mathcal{U})$. Since this extension commutes with the unilateral shift S, it follows that $\Omega$ is a function in $H^\infty(\mathcal{U},\ \mathcal{U})$. Equation (1.5) shows that

$$(M_\Omega^+)^* T_0 M_\Omega^+ = N^2 I, \text{ or equivalently, } T_0 = (M_\Omega^+)^{-*} NN (M_\Omega^+)^{-1}\ .$$

Therefore $N\Omega(\lambda)^{-1}$ is an outer spectral factor for $T_0$. This completes the proof.

Now let det [A] be the determinant of an operator A on $\mathcal{U}$ with respect to any basis. Then it is well known that $\Theta$ is a square outer function if and only if det $[\Theta]$ is an outer function in $H^\infty$. To prove this, following Section V.6 in Sz.-Nagy-Foias [3], let $\delta(\lambda) = \det[\Theta(\lambda)]$ and $\Omega(\lambda)$ in $H^\infty(\mathcal{U},\ \mathcal{U})$ be the algebraic adjoint of $\Theta(\lambda)$, that is, $\Theta(\lambda)\Omega(\lambda) = \Omega(\lambda)\Theta(\lambda) = \delta(\lambda)I$. If $\delta$ is an outer function, then

$$H^2(\mathcal{U}) = \overline{\delta H^2(\mathcal{U})} = \overline{\Theta\Omega H^2(\mathcal{U})} \subseteq \overline{\Theta H^2(\mathcal{U})}\ .$$

Thus $\overline{\Theta H^2(\mathcal{U})} = H^2(\mathcal{U})$ and $\Theta$ is a square outer function. Conversely, assume that $\Theta$ is square outer function. Let $\delta_i$ be the inner part of $\delta$ and $\Omega = \Omega_i\Omega_o$ be the inner-outer factorization of $\Omega$ where $\Omega_i$ is an inner function in $H^\infty(\mathcal{F},\ \mathcal{U})$. Then using $\delta = \Omega\Theta$ we have

$$\delta_i H^2(\mathcal{U}) = \overline{\delta H^2(\mathcal{U})} = \overline{\Omega_i\Omega_o\Theta H^2(\mathcal{U})} = \Omega_i H^2(\mathcal{F})\ .$$

According to the Beurling-Lax-Halmos Theorem (see Foias-Frazho [4], Sz.-Nagy-Foias [3]), this implies that $\delta_i I$ equals $\Omega_i$ up to a unitary constant on the right. So without loss of generality we can assume that $\Omega_i = \delta_i I$. Hence $\Omega_o\Theta = \Theta\Omega_o = \delta_o I$ where $\delta_o$ is the outer part of $\delta$. By taking the determinant along with $\delta = \det[\Theta]$ we have

$$\delta_i\delta_o \det[\Omega_o] = \det[\Theta]\det[\Omega_o] = \delta_o^n$$

where n is the dimension of $\mathcal{U}$. This shows that $\delta_i$ divides the outer function $\delta_o^n$. Therefore $\delta_i$ is a constant of modulus one (see Foias-Frazho [4], Sz.-Nagy-Foias [3]) and $\det[\Theta] = \delta_o$ is an outer function. This proves that $\Theta$ is outer if and only if det $[\Theta]$ is outer.

Recall (see pg. 62 in Hoffman [1]) that a function g in $H^\infty$ is outer if and only if

$$\frac{1}{2\pi}\int_0^{2\pi} \ln |g(e^{it})|\, dt = \ln|g(0)|\ .$$

So using $g = \det[\Theta]$ we see that a function $\Theta$ in $H^\infty(\mathcal{U},\ \mathcal{U})$ is outer if and only if

$$\frac{1}{2\pi}\int_0^{2\pi} \ln \det [\Theta^*\Theta]dt = \ln \det [\Theta(0)^*\Theta(0)] . \tag{1.6}$$

If R admits a square outer spectral factor, then (1.6) and $\det[\Theta(0)] \neq 0$ show that

$$\frac{1}{2\pi}\int_0^{2\pi} \ln \det [R(e^{it})]dt > -\infty . \tag{1.7}$$

On the other hand, Section V.6 in Sz.-Nagy-Foias [3] shows that if R is a.e. a positive operator and (1.7) holds, then R admits a square outer spectral factor. Therefore R admits a square outer spectral factor if and only if R is a.e. positive and (1.7) holds.

Now let us pursue another method to determine the existence of a square outer spectral factorization for $T_R$. To this end, let $\{R_n\}_{-\infty}^{\infty}$ be the Fourier coefficients for an a.e. self-adjoint operator valued function R in $L^\infty(\mathcal{U}, \mathcal{U})$. By comparing the coefficients of $e^{int}$ and using $R = R^*$ a.e. it follows that $R_n = R_{-n}^*$ for all n. Let $T_{R,n}$ on $\ell_n^2(\mathcal{U})$ be the block Toeplitz matrix generated by $\{R_j : 0 \leq j < n\}$, that is,

$$T_{R,n} = \begin{bmatrix} R_0 & R_1^* & R_2^* & \cdots & R_{n-1}^* \\ R_1 & R_0 & R_1^* & \cdots & R_{n-2}^* \\ R_2 & R_1 & R_0 & \cdots & R_{n-3}^* \\ \cdot & \cdot & \cdot & \cdots & \cdot \\ \cdot & \cdot & \cdot & \cdots & \cdot \\ \cdot & \cdot & \cdot & \cdots & \cdot \\ R_{n-1} & R_{n-2} & R_{n-3} & \cdots & R_0 \end{bmatrix} . \tag{1.8}$$

If $T_R$ admits a square outer spectral factor $\Theta$, then $T_{R,n}$ is strictly positive for all n. Indeed, if $(T_{R,n}x, x) = 0$ for some x in $\ell_n^2(\mathcal{U})$, then using $T_\Theta^* T_\Theta = T_R$ it follows that $(T_R(x \oplus 0), x \oplus 0) = \|T_\Theta(x \oplus 0)\|^2$ is zero. Because $\Theta(0)$ is invertible, this implies that $x = 0$ and $T_{R,n}$ is strictly positive.

If $T_{R,n}$ is strictly positive for all n, then $T_R$ is positive. However, it does not necessarily follow that $T_R$ admits a square outer spectral factorization. For example, let R be the function in $L^\infty$ defined by $R = 1$ if $0 \leq t \leq \pi$ and $R = 0$ for all other t. Then we claim that $T_{R,n}$ is strictly positive for all n and $T_R$ does not admit an outer spectral factor $\Theta$. If $R = |\Theta|^2$ for some $\Theta$ in $H^\infty$, then $\Theta = 0$ a.e. on a set of positive measure and thus $\Theta = 0$. (Recall that if f is a nonzero function in $H^\infty$ or $H^2$, then $f(e^{it}) \neq 0$ a.e.; see Chapter 5 in Hoffman [1].) So R does not admit an outer spectral factor. To complete the proof it remains to show that $T_R$ is positive and $\ker T_R = \{0\}$. Since $R \geq 0$ a.e. the operator $B := P_{H^2} M_R \,|\, H^2$ on $H^2$ is positive. Because B is the Fourier transform of $T_R$ it follows that $T_R$ is positive. To show that $\ker T_R$ is zero, it is sufficient to show that ker B is zero. If $Bh = 0$ for some h in $H^2$, then Rh is in $K^2 = L^2 \ominus H^2$.

This along with the fact that $Rh = 0$ a.e. on a set of positive Lebesgue measure, implies that $Rh = 0$. So $h = 0$ a.e. on a set of positive Lebesgue measure. Hence $h = 0$ and $\ker T_R = \{0\}$. Therefore $T_{R,n}$ is strictly positive for all n and $T_R$ does not admit an outer spectral factor.

Theorem 1.1 shows that if $T_R$ is strictly positive, then $T_R$ admits a square outer spectral factorization. On the other hand, if $T_R$ admits a square outer spectral factor $\Theta$, then $T_R$ may not be invertible. (By invertible we mean that the inverse of $T_R$ exists and is a bounded operator.) To see this choose $\Theta = \lambda + 1$ and set $R = |\Theta|^2$ a.e. Then $\Theta$ is a square outer factor for $T_R$ and $T_R$ is not invertible. To obtain necessary an sufficient conditions in terms of $T_{R,n}$ for $T_R$ to admit a square outer spectral factor, we need the following classical matrix inversion lemma.

**LEMMA 1.2.** *Let* T *be a block self-adjoint matrix of the form*

$$T = \begin{bmatrix} A & X^* \\ X & Y \end{bmatrix} \text{ on } \begin{bmatrix} \mathcal{U} \\ \mathcal{Y} \end{bmatrix}, \tag{1.9}$$

*and assume that* Y *is strictly positive. Then its Schur complement* $\Delta := A - X^*Y^{-1}X$ *is given by the error to the following optimization problem*

$$(\Delta f, f) = \inf\{(Tx, x) : \Pi x = f\}, \tag{1.10}$$

*where* f *is a specified vector in* $\mathcal{U}$ *and* $\Pi$ *is the orthogonal projection from* $\mathcal{U} \oplus \mathcal{Y}$ *onto* $\mathcal{U}$. *Moreover,* T *is strictly positive if and only if* $\Delta$ *is strictly positive. In this case*

$$T^{-1} = \begin{bmatrix} \Delta^{-1} & -\Delta^{-1}X^*Y^{-1} \\ -Y^{-1}X\Delta^{-1} & Y^{-1} + Y^{-1}X\Delta^{-1}X^*Y^{-1} \end{bmatrix} \tag{1.11}$$

*and the Schur complement* $\Delta$ *is given by*

$$\Delta = A - X^*Y^{-1}X = (\Pi T^{-1}\Pi^*)^{-1}. \tag{1.12}$$

**PROOF.** The last part of this lemma follow by observing that T admits a Schur decomposition of the form

$$T = \begin{bmatrix} I & X^*Y^{-1} \\ 0 & I \end{bmatrix} \begin{bmatrix} A - X^*Y^{-1}X & 0 \\ 0 & Y \end{bmatrix} \begin{bmatrix} I & 0 \\ Y^{-1}X & I \end{bmatrix}. \tag{1.13}$$

If $\Delta > 0$, then inverting the T in (1.13) readily gives the form of the inverse of T in (1.11). Equation (1.12) follows from (1.11) and the definition of the Schur complement. To complete the proof it remains to establish that the Schur complement $\Delta$ is given by the error in the optimization problem (1.10). To this end, let $x = f \oplus g$ be in $\mathcal{U} \oplus \mathcal{Y}$ and notice that the Schur factorization of T in (1.13) gives

$$(Tx, x) = (\Delta f, f) + (Y(Y^{-1}Xf + g), Y^{-1}Xf + g) = (\Delta f, f) + \|Y^{-1/2}Xf + Y^{1/2}g\|^2 .$$

The previous equation shows that $(Tx, x) \geq (\Delta f, f)$ and we have equality by choosing $g = -Y^{-1}Xf$. This establishes (1.10) and completes the proof.

Consider the matrix decomposition of the Toeplitz matrix $T_{R,n}$ in (1.8) given by

$$T_{R,n} = \begin{bmatrix} R_0 & X^* \\ X & T_{R,n-1} \end{bmatrix} \tag{1.14}$$

where $X = [R_1, R_2, \ldots, R_{n-1}]^{tr}$. For the moment assume that $T_{R,n-1}$ is strictly positive. Then the *Schur complement* $\Delta_n$ *associate with* $T_{R,n}$ is the operator on $\mathcal{U}$ defined by

$$\Delta_n = R_0 - X^* T_{R,n-1}^{-1} X \quad \text{and} \quad \Delta_1 := R_0 . \tag{1.15}$$

If $T_{R,n-1}$ is not invertible, then the Schur complement $\Delta_n$ is not well defined and for convenience in this case we set $\Delta_n = -\infty$. Finally, if $T_{R,n}$ is strictly positive, then the matrix inversion Lemma 1.2 shows that

$$\Delta_n = (\Pi_n T_{R,n}^{-1} \Pi_n^*)^{-1} \tag{1.16}$$

where $\Pi_n := [I, 0, 0, \ldots, 0]$ maps $\ell_n^2(\mathcal{U})$ onto $\mathcal{U}$. This sets the state for the following result which allows us to determine whether or not $T_{R,n}$ is strictly positive by checking the positivity of the Schur complements.

**LEMMA 1.3**. *Let* $T_{R,n}$ *be the* $n \times n$ *Toeplitz matrix on* $\ell_n^2(\mathcal{U})$ *given by (1.8), and let* $\Delta_j$ *on* $\mathcal{U}$ *be the Schur complement associated with* $T_{R,j}$ *for* $1 \leq j \leq n$. *Then* $T_{R,n}$ *is strictly positive if and only if* $\Delta_j$ *is strictly positive for all* $1 \leq j \leq n$. *Moreover, in this case* $\{\Delta_j\}_1^n$ *forms a decreasing sequence of positive operators.*

**PROOF**. Since $\Delta_1 = R_0$, the lemma is true for $n = 1$. Now let us proceed by induction and assume that the lemma is true for $n - 1$, that is, $\Delta_1, \Delta_2, \ldots, \Delta_{n-1}$ are all strictly positive if and only if $T_{R,n-1}$ is strictly positive. By applying the matrix inversion Lemma 1.2 to the decomposition of $T_{R,n}$ in (1.14) we see that $T_{R,n}$ is strictly positive if and only if $T_{R,n-1}$ and $\Delta_n$ are strictly positive. This completes the proof of the induction.

Now assume that $\{\Delta_j\}_1^n$ are all strictly positive, or equivalently, $T_{R,n}$ is strictly positive. Notice that $T_{R,j}$ is the compression of $T_{R,n}$ to $\ell_j^2(\mathcal{U})$. Therefore the optimization problem in (1.10) shows that

$$(\Delta_j f, f) = \mu_j(f) = \inf \{(T_{R,n}x, x) : x \in \ell_j^2(\mathcal{U}) \oplus \{0\} \text{ and } \Pi_n x = f\}\ , \tag{1.17}$$

where f is in $\mathcal{U}$. Because the infimum for $\mu_{j+1}$ is taken over a larger set $\ell_{j+1}^2(\mathcal{U})$ than the infimum corresponding to $\mu_j$, it follows that $\{\mu_j(f)\}$ form a decreasing sequence. Thus $\{\Delta_j\}$ forms a decreasing sequence of positive operator. This completes the proof.

Assume that $T_{R,n}$ is strictly positive for all n. The previous lemma shows that the Schur complements $\Delta_n$ form a decreasing sequence of strictly positive operators. So $\Delta_n$ converges to a positive operator $\Delta$ on $\mathcal{U}$ as n approaches infinity. Moreover, by consulting the optimization problems in (1.17) we see that

$$(\Delta f, f) = \inf \{(T_R x, x) : x \in \ell_+^2(\mathcal{U}) \text{ and } \Pi x = f\} \tag{1.18}$$

where f is a specified vector in $\mathcal{U}$ and $\Pi$ is the operator from $\ell_+^2(\mathcal{U})$ onto $\mathcal{U}$ which picks out the first component of $\ell_+^2(\mathcal{U})$. Finally, let us notice that if $T_R$ is invertible, then

$$\Delta = (\Pi T_R^{-1} \Pi^*)^{-1}\ . \tag{1.19}$$

This follows by observing that $T_R$ admits a block matrix decomposition of the form

$$T_R = \begin{bmatrix} R_o & X^* \\ X & T_R \end{bmatrix},$$

and $\Delta$ is the Schur complement for $T_R$.

Now assume that the Toeplitz operator $T_R$ admits a square outer spectral factor $\Theta$. Then $T_{R,n}$ is strictly positive for all n and thus $\Delta_n$ converges to a positive operator $\Delta$. We claim that this $\Delta$ is strictly positive. This follows by taking the Fourier transform of $T_\Theta$ and using (1.1) in the optimization problem (1.18), that is, for a specified f in $\mathcal{U}$ we obtain

$$(\Delta f, f) = \inf \{\|T_\Theta x\|^2 : x \in \ell_+^2(\mathcal{U}) \text{ and } \Pi x = f\} = \inf \{\|\Theta f - \lambda \Theta h\|^2 : h \in H^2(\mathcal{U})\} = \tag{1.20}$$
$$\inf \{\|\Theta f - \lambda h\|^2 : h \in H^2(\mathcal{U})\} = \|\Theta(0) f\|^2\ .$$

Therefore $\Delta = \Theta(0)^* \Theta(0)$ is strictly positive. This proves part of the following result.

**THEOREM 1.4.** *Let* $\{R_n\}_{-\infty}^{\infty}$ *be the Fourier coefficients for a function* R *in* $L^\infty(\mathcal{U}, \mathcal{U})$, *where* $R = R^*$ *a.e. and* $\mathcal{U}$ *is finite dimensional. Let* $\Delta_n$ *be the Schur complements associated with* $T_{R,n}$ *for* $n \geq 1$. *Then the Toeplitz operator* $T_R$ *admits a square outer spectral factorization* $\Theta$ *if and only if* $\Delta_n \geq \delta I$ *for all* n *and some* $\delta > 0$. *In this case the sequence* $\{\Delta_n\}$ *is decreasing and converges to the strictly positive operator* $\Delta = \Theta(0)^* \Theta(0)$. *Furthermore, if* F *in* $H^\infty(\mathcal{U}, \mathcal{U})$ *is any other spectral factorization of* $T_R$, *then*

$$\Delta = \Theta(0)^*\Theta(0) \geq F(0)^*F(0)\,, \tag{1.21}$$

*and there is equality in (1.21) if and only if* F *is a square outer spectral factor of* $T_R$. *Finally,*

$$\frac{1}{2\pi}\int_0^{2\pi} \ln\det[R(e^{it})]dt = \ln\det[\Delta] \geq \ln\det[F(0)^*F(0)]\,, \tag{1.22}$$

*and there is equality in (1.22) if and only if* F *is a square outer spectral factor for* $T_R$.

If $T_R$ admits a square outer spectral factor, then $T_{R,n}$ is strictly positive for all n. So without loss of generality one can assume that $T_{R,n}$ is strictly positive for all n. Now the previous theorem shows that $T_R$ admits a square outer spectral factor $\Theta$ if and only if the Schur complements $\Delta_n = (\Pi_n T_{R,n}^{-1}\Pi_n^*)^{-1}$ form a sequence of positive operators satisfying $\Delta_n \geq \delta I$ for all n and for some $\delta > 0$. In this case $\{\Delta_n\}$ is a decreasing sequence which converges to $\Theta(0)^*\Theta(0) = \Delta$. However, it is not necessary to compute the inverse of $T_{R,n}$. One can use the Levinson algorithm (see Remark 1.7 below) or the Riccati difference equation (discussed in the next section) to recursively compute the Schur complements $\Delta_n$ without inverting $T_{R,n}$. These algorithms provide an efficient method to compute the Schur complements. Furthermore, if at any step n the Schur complement $\Delta_n$ is not strictly positive, then $T_{R,n}$ is not strictly positive, and thus $T_R$ does not admit a square outer spectral factorization.

**REMARK 1.5**. In many applications one is given a spectral factor F in $H^\infty(\mathcal{U}, \mathcal{U})$ for a Toeplitz operator $T_R$, where F(0) is invertible. In this case $T_R$ admits a square outer spectral factorization. To see this (via the previous theorem) first notice that because $T_R = T_F^*T_F$ is positive the Schur complements $\{\Delta_n\}$ form a decreasing sequence of positive operators which converges to $\Delta$. So using $T_R = T_F^*T_F$ once again and (1.18) we have for f in $\mathcal{U}$

$$(\Delta f, f) = \inf\{\|T_F x\| : x \in \ell_+^2(\mathcal{U}) \text{ and } \Pi x = f\} = \inf\{\|Ff - \lambda Fh\| : h \in H^2(\mathcal{U})\} \geq$$
$$\inf\{\|Ff - \lambda h\|^2 : h \in H^2(\mathcal{U})\} = \|F(0)f\|^2\,. \tag{1.23}$$

Therefore $\Delta_n \geq \Delta \geq F(0)^*F(0)$ are all strictly positive, and thus $T_R$ admits a square outer spectral factor.

**PROOF OF THEOREM 1.4**. Our proof relies on the Wold decomposition of an isometry U on $\mathcal{H}$; see the end of Section I.1, Chapter I in Sz.-Nagy-Foias [3] or Chapter VI in Foias-Frahzo [4]. To this end, recall that a subspace $\mathcal{L}$ is *wandering* for an isometry U on $\mathcal{H}$ if $U^n\mathcal{L}$ is orthogonal to $U^m\mathcal{L}$ for all $n \neq m$. A unilateral shift $S_o$ on $\mathcal{H}_+$ is an isometry which contains a cyclic wandering subspace $\mathcal{L}$, that is,

$$\mathcal{H}_+ = \bigoplus_0^\infty S_o^n \mathcal{L} \tag{1.24}$$

where $\mathcal{L} = \mathcal{H}_+ \ominus S_o\mathcal{H}_+$. In this case $S_o$ is unitarily equivalent to the unilateral shift S on $\ell_+^2(\mathcal{L})$. To see this simply let $\Phi$ be the unitary operator from $\mathcal{H}_+$ onto $\ell_+^2(\mathcal{L})$ defined by

$$\Phi \bigoplus_0^\infty S_o^n f_n = [f_0, f_1, f_2, \ldots]^{tr} \qquad (f_n \in \mathcal{L}), \tag{1.25}$$

then $S\Phi = \Phi S_o$. Now let U be any isometry on $\mathcal{H}$. According to the Wold decomposition, there exists two unique reducing subspaces $\mathcal{H}_+$ and $\mathcal{H}_u$ for U such that U admits a reducing decomposition of the form $U = S_o \oplus W$ on $\mathcal{H} = \mathcal{H}_+ \oplus \mathcal{H}_u$ where $S_o = U|\mathcal{H}_+$ is a unilateral shift on $\mathcal{H}_+$ and $W = U|\mathcal{H}_u$ is a unitary operator on $\mathcal{H}_u$. Moreover, the subspace $\mathcal{H}_+$ is given by (1.24) where $\mathcal{L} = \mathcal{H} \ominus U\mathcal{H}$. Finally, it is noted that in this decomposition $\mathcal{H}_+$ or $\mathcal{H}_u$ can be the trivial space $\{0\}$.

Now assume the Schur complements form a decreasing sequence of positive operators which converge to the strictly positive operator $\Delta$. Because $\Delta_n$ is strictly positive for all n, Lemma 1.3 shows that $T_{R,n}$ is strictly positive for all n. Hence $T_R$ is positive. We claim that $\ker T_R$ is zero. To see this we proceed by contradiction and assume that $T_R x = 0$ for some nonzero x in $\ell_+^2(\mathcal{U})$. Thus $(T_R x, x) = 0$. By using the band structure of the Toeplitz matrix $T_R$ we can without loss of generality assume that $\Pi x = f$ is nonzero. The optimization problem in (1.18) shows that $0 = (T_R x, x) \geq (\Delta f, f)$. Since $\Delta$ is strictly positive f must be zero which leads to a contradiction. Therefore $\ker T_R$ is zero.

Let $\Gamma$ be the positive square root of $T_R$ and S be the unilateral shift on $\ell_+^2(\mathcal{U})$. Obviously $\Gamma$ is a quasi-affinity, that is, $\Gamma$ is one to one and the range of $\Gamma$ is dense in $\ell_+^2(\mathcal{U})$. Using $S^*T_RS = T_R$ we have for x in $\ell_+^2(\mathcal{U})$

$$\|\Gamma x\|^2 = (T_R x, x) = (S^*T_R Sx, x) = \|\Gamma Sx\|^2 .$$

So there exists an isometry U on $\mathcal{H} = \overline{\operatorname{ran}\Gamma} = \ell_+^2(\mathcal{U})$ satisfying $U\Gamma = \Gamma S$. According to the Wold decomposition, U admits a reducing decomposition of the form $U = S_o \oplus W$ on $\mathcal{H}_+ \oplus \mathcal{H}_u$ where $S_o = U|\mathcal{H}_+$ is a unilateral shift and W is unitary. Moreover, $\mathcal{H}_+$ is given by (1.24) where $\mathcal{L}$ is the wandering subspace for U defined by $\mathcal{L} = \mathcal{H} \ominus U\mathcal{H}$. Notice that $\mathcal{L} = \mathcal{H} \ominus \overline{\Gamma S\mathcal{H}}$. So $\mathcal{L}$ is given by $\operatorname{ran}(\Gamma S)^\perp$, or equivalently, $\mathcal{L} = \ker(S^*\Gamma)$. In particular, $\dim \mathcal{L} \leq \dim \mathcal{U}$ where dim denotes the dimension. Now by employing the projection theorem along with (1.18) we have for f in $\mathcal{U}$

$$\|P_{\mathcal{L}}\Gamma\Pi^* f\|^2 = \inf\{\|\Gamma\Pi^* f - \Gamma S h\|^2 : h \in \ell_+^2(\mathcal{U})\} = \inf\{\|\Gamma x\|^2 : \Pi x = f\} =$$
$$\inf\{(T_R x, x) : \Pi x = f\} = (\Delta f, f)\,. \tag{1.26}$$

Since $\Delta$ is strictly positive $\mathcal{L}$ and $\mathcal{U}$ have the same dimension.

Because $\dim \mathcal{L} = \dim \mathcal{U} < \infty$ and $\Gamma$ is one to one with dense range in $\mathcal{H}$, Lemma 1.6 below shows that S is unitarily equivalent to U. So there exists a unitary operator $\Phi$ from $\mathcal{H}$ onto $\ell_+^2(\mathcal{U})$ satisfying $S\Phi = \Phi U$. Therefore the operator $\Phi\Gamma$ commutes with the unilateral shift S. So there exists a function $\Theta$ in $H^\infty(\mathcal{U}, \mathcal{U})$ satisfying $T_\Theta = \Phi\Gamma$. Since $\Phi\Gamma$ has dense range, equation (1.1) holds and $\Theta$ is a square outer function. Obviously $T_\Theta^* T_\Theta = \Gamma^2 = T_R$, and thus $\Theta$ is a square outer spectral factor for $T_R$.

Now let us show that $\Theta(0)^*\Theta(0) = \Delta$. To this end, notice that

$$\Phi P_{\mathcal{L}} = \Phi(I - UU^*) = (I - SS^*)\Phi = P_{\mathcal{U}}\Phi \tag{1.27}$$

where $P_{\mathcal{U}}$ is the orthogonal projection onto $\mathcal{U}$ the first component of $\ell_+^2(\mathcal{U})$. Using $\Phi P_{\mathcal{L}} = P_{\mathcal{U}}\Phi$ along with a specified f in $\mathcal{U}$ we have

$$\|\Theta(0)f\|^2 = \|P_{\mathcal{U}} T_\Theta \Pi^* f\|^2 = \|P_{\mathcal{U}}\Phi\Gamma\Pi^* f\|^2 = \|P_{\mathcal{L}}\Gamma\Pi^* f\|^2\,.$$

Equation (1.26) shows that $\Theta(0)^*\Theta(0) = \Delta$.

Now assume that F in $H^\infty(\mathcal{U}, \mathcal{U})$ is another spectral factor for $T_R$. Then F admits an inner-outer factorization of the form $F_o = F_i F_o$, where $F_i$ is inner and $F_o$ is outer. Because $F_o$ is also an outer spectral factor for $T_R$ we can assume without loss of generality that $F_o = \Theta$. Thus $F = F_i\Theta$. This readily implies that for f in $\mathcal{U}$

$$\|F(0)f\|^2 = \|F_i(0)\Theta(0)f\|^2 \le \|\Theta(0)f\|^2 \qquad (f \in \mathcal{U})\,. \tag{1.28}$$

Hence (1.21) holds. If $F(0)^*F(0) = \Theta(0)^*\Theta(0)$, then we have equality in (1.28). Using the fact that $\Theta(0)$ is onto $\mathcal{U}$, this implies that $F_i(0)$ is an isometry on $\mathcal{U}$. Since $\mathcal{U}$ is finite dimensional $F_i(0)$ is unitary. Now using the fact that $F_i$ is an inner function $\|F_i(0)f\| = \|f\| = \|F_i f\|$ for all f in $\mathcal{U}$. So all the Fourier coefficients $(F_i)_n$ of $F_i$ for $n \ge 1$ must be zero, and thus $F_i = F_i(0)$ is a constant unitary function. Therefore $F = F_i\Theta$ is also a square outer spectral factor for $T_R$. Finally, the equality in (1.22) follows from (1.6) and (1.21). The statement concerning the inequality in (1.22) follows from the fact that if $A \ge B > 0$, then $A = B$ if and only if $\det[A] = \det[B]$. (To verify this use the fact that $A \ge B$ if and only if there exists a contraction C such that $B = A^{1/2}C^*CA^{1/2}$.) So if we have equality in (1.22), then $F(0)^*F(0) = \Theta(0)^*\Theta(0)$. From our previous discussion F must be a square outer spectral factor for $T_R$. This completes the proof.

Recall that an operator A from $\mathcal{H}$ into $\mathcal{H}'$ is a *quasi-affinity* if A is one to one and the range of A is dense in $\mathcal{H}'$. This sets the stage for the following result.

**LEMMA 1.6.** *Let* S *be the unilateral shift on* $\ell^2_+(\mathcal{U})$ *where* $\mathcal{U}$ *is finite dimensional. Let* U *be an isometry on* $\mathcal{H}$ *whose wandering subspace* $\mathcal{L} := \mathcal{H} \ominus U\mathcal{H}$ *has the same dimension as* $\mathcal{U}$. *Then* S *is unitarily equivalent to* U *if and only if there is a quasi-affinity* $\Gamma$ *from* $\ell^2_+(\mathcal{U})$ *into* $\mathcal{H}$ *satisfying* $\Gamma S = U\Gamma$.

**PROOF.** Let $U = S_0 \oplus W$ on $\mathcal{H}_+ \oplus \mathcal{H}_u$ be the Wold decomposition of U where $S_0$ is the unilateral shift on $\mathcal{H}_+$. In fact, $\mathcal{H}_+$ is given by (1.24). Since $\mathcal{U}$ and $\mathcal{L}$ have the same dimension and $S_0$ and S are both unilateral shifts, there exists a unitary operator $\Phi$ from $\mathcal{H}_+$ onto $\ell^2_+(\mathcal{U})$ satisfying $S\Phi = \Phi S_0$. In fact, one such unitary operator is given by

$$\Phi \bigoplus_0^\infty S_0^n f_n = [\phi f_0, \phi f_1, \phi f_2, \ldots]^{tr}$$

where $\phi$ is any specified unitary operator from $\mathcal{L}$ onto $\mathcal{U}$. To complete the proof it remains to show that $\mathcal{H}_u = \{0\}$ when $\Gamma$ is a quasi-affinity satisfying $U\Gamma = \Gamma S$. Let A be the quasi-affinity from $\mathcal{H} = \ell^2_+(\mathcal{U})$ into $\mathcal{H}' = \ell^2_+(\mathcal{U}) \oplus \mathcal{H}_u$ defined by $A = \Phi P_{\mathcal{H}_+}\Gamma + (I - P_{\mathcal{H}_+})\Gamma$. Clearly, $U\Gamma = \Gamma S$ implies that $(S \oplus W)A = AS$. Now let V be the bilateral shift on $\ell^2(\mathcal{U})$. Then V is an extension of S and $V' := V \oplus W$ on $\mathcal{K}' := \ell^2(\mathcal{U}) \oplus \mathcal{H}_u$ is an extension of $S \oplus W$. (Recall that C is an extension of an operator D on $\mathcal{H}$ if $C|\mathcal{H} = D$.) By consulting Lemma II.3.1 there exists an operator B from $\ell^2(\mathcal{U})$ into $\mathcal{K}'$ extending A and preserving the norm of A (that is, $\|B\| = \|A\|$) and intertwining V with V′. Using $V'B = BV$ along with the fact that the range of A is dense in $\mathcal{H}'$ it is easy to see that the range of B is dense in $\mathcal{K}'$. In fact,

$$\overline{B\ell^2(\mathcal{U})} = \bigvee_{-\infty}^{\infty} BV^n \ell^2_+(\mathcal{U}) = \bigvee_{-\infty}^{\infty} V'^n B\ell^2_+(\mathcal{U}) = \bigvee_{-\infty}^{\infty} V'^n A\ell^2_+(\mathcal{U}) = \mathcal{K}'.$$

Therefore the range of B is dense in $\mathcal{K}'$.

We claim that V is unitarily equivalent to V′. To see this notice that $V'B = BV$ gives $V^*B^* = B^*V'^*$. Because both V and V′ are unitary $B^*V' = VB^*$. Thus $V'BB^* = BB^*V'$. This implies that $V'(BB^*)^n = (BB^*)^n V'$ for all $n \geq 0$. So for any polynomial $p(\lambda)$ we have $V'p(BB^*) = p(BB^*)V'$. Recall that one can choose a sequence of polynomials $p_n(\lambda)$ such that $p_n(BB^*)$ converges strongly to $(BB^*)^{1/2}$, the positive square root of $BB^*$; see Problem 121 in Halmos [1]. Thus $V'(BB^*)^{1/2} = (BB^*)^{1/2}V'$. Since $\ker B^*$ is zero, B admits a polar decomposition of the form $B = (BB^*)^{1/2}\Omega$ where $\Omega$ is a co-isometry from $\ell^2(\mathcal{U})$ onto $\mathcal{K}'$. So using $V'B = BV$ once again

$$(BB^*)^{1/2}\Omega V = BV = V'B = V'(BB^*)^{1/2}\Omega = (BB^*)^{1/2}V'\Omega .$$

Since the kernel of $(BB^*)^{1/2}$ is zero $\Omega V = V'\Omega$ and $\Omega^* V' = V\Omega^*$. Because V is a bilateral shift with finite multiplicity a classical argument shows that $\mathcal{H}_u = \{0\}$ and thus U is a unilateral shift.

Now let us show that $\mathcal{H}_u = \{0\}$. To this end, let $\phi_1, \phi_2, \ldots, \phi_n$ be an orthonormal basis for $\mathcal{U}$. Then clearly $\{V^k\phi_j : 1 \le j \le n \text{ and } k \in \mathbb{Z}\}$ is an orthonormal basis for $\ell^2(\mathcal{U})$. Using Bessel's inequality and Parseval's equality along with the fact that $\Omega$ is a co-isometry we have

$$n = \sum_i \|\phi_i \oplus 0\|^2 = \sum_{i,j,k} |(\Omega^*(\phi_i \oplus 0), V^k\phi_j)|^2 = \sum_{i,j,k} |(\Omega^* V'^{*k}(\phi_i \oplus 0), \phi_j)|^2 \le \sum_j \|\phi_j\|^2 = n. \quad (1.29)$$

So we have equality. This means that $\mathcal{U}$ which equals the span of $\{\phi_j\}_1^n$ must be contained in the span of the orthonormal set $\{\Omega^* V'^k(\phi_i \oplus 0) : 1 \le i \le n \text{ and } k \in \mathbb{Z}\}$. In other words $\mathcal{U} \subseteq \Omega^*(\ell^2(\mathcal{U}) \oplus \{0\})$. Since $\{V^n \mathcal{U} : n \in \mathbb{Z}\}$ spans $\ell^2(\mathcal{U})$ and $V\Omega^* = \Omega^* V'$, we have $\ell^2(\mathcal{U}) \subseteq \Omega^*(\ell^2(\mathcal{U}) \oplus \{0\})$. Therefore $\Omega$ is unitary and $\Omega \ell^2(\mathcal{U}) = \ell^2(\mathcal{U}) \oplus \{0\}$. Because $\Omega$ is unitary $\mathcal{H}_u = \{0\}$ and U is a unilateral shift. This completes the proof.

**REMARK 1.7.** As before, let $T_{R,n}$ be the $n \times n$ block Toeplitz matrix in (1.8). Recall that the Levinson algorithm (see Foias-Frazho [4], Kailath [2]) gives a recursive procedure to compute the operators $\{E_n\}$ and $\{A_{n,j}\}$ on $\mathcal{U}$ satisfying

$$T_{R,n}[I, A_{n,1}, A_{n,2}, \ldots, A_{n,n-1}]^{tr} = [E_n, 0, 0, \ldots, 0]^{tr} \quad (1.30)$$

where tr denotes the transpose. If $T_{R,n-1}$ is strictly positive, then using the decomposition of $T_{R,n}$ in (1.14) it follows that $E_n = \Delta_n$ the Schur complement for $T_{R,n}$. (Moreover, if $T_{R,n}$ is invertible (1.30) also shows that $E_n = (\Pi_n T_{R,n}^{-1} \Pi_n^*)^{-1}$.) In particular, $T_{R,n}$ is strictly positive if and only if $E_1, E_2, \ldots, E_n$ are all strictly positive. So one can use the Levinson algorithm to recursively compute the Schur complements $E_j = \Delta_j$ for $1 \le j \le n$. According to Lemma 1.3 the $n \times n$ Toeplitz operator $T_{R,n}$ is strictly positive if and only if $E_j$ is strictly positive of all $1 \le j \le n$. In this case $\{E_j\}$ forms a decreasing sequence of positive operators.

If $T_{R,n}$ is strictly positive, then it is well known that

$$\Theta_n(\lambda) := E_n^{1/2}[I + A_{n,1}\lambda + A_{n,2}\lambda^2 + \cdots + A_{n,n-1}\lambda^{n-1}]^{-1} =$$
$$(\Pi_n T_{R,n}^{-1} \Pi_n^*)^{1/2}([I, \lambda I, \lambda^2 I, \ldots, \lambda^{n-1} I] T_{R,n}^{-1} \Pi_n^*)^{-1} \quad (1.31)$$

is a square outer function in $H^\infty(\mathcal{U}, \mathcal{U})$ which is analytic in $|\lambda| < 1 + \varepsilon$ for some $\varepsilon > 0$. Moreover, $\{R_j\}_0^{n-1}$ are the first n Fourier coefficients of $\Theta_n^* \Theta_n$. If $T_R$ is strictly positive, then Theorem 1.1 shows that the square outer spectral factor $\Theta$ for $T_R$ is given by $\Theta = (\Pi T_R^{-1} \Pi^*)^{1/2}(\mathcal{F}_+ T_R^{-1} \Pi^*)(\lambda)$ where $\mathcal{F}_+$ is the Fourier transform from $\ell_+^2(\mathcal{U})$ onto $H^2(\mathcal{U})$.

So if $T_R$ is strictly positive, then one can use (1.31) to show that $\Theta_n(\lambda)$ converges to $\Theta$ the square outer spectral factor for $T_R$.

According to Theorem 1.4, the Toeplitz operator $T_R$ admits a square outer spectral factor $\Theta$ if and only if $E_n \geq \delta I$ for all n and some $\delta > 0$. In this case $\{E_n\}$ forms a decreasing sequence of positive operators which converge to $\Delta = \Theta(0)^*\Theta(0)$. Without loss of generality we can always assume that $\Theta(0)$ is strictly positive. (This can always be accomplished by multiplying $\Theta$ by the appropriate unitary constant on the left.) Equation (1.31) shows that $\Theta_n(0) = E_n^{1/2}$ converges to $\Delta^{1/2} = \Theta(0)$ as n approaches infinity. Recall that according to the Vitali Theorem if $f_n$ is a uniformly bounded sequence of analytic functions in a region $\mathcal{D}$ and $f_n(\alpha)$ converges for some $\alpha$ in $\mathcal{D}$, then the subsequence $f_n$ converges uniformly to an analytic function f on compact sets in $\mathcal{D}$. So using the Vitali Theorem along with the fact that $\{R_j\}_0^{n-1}$ are the first n Fourier coefficients of $\Theta_n^*\Theta_n$ it easily follows that $\Theta_n$ converges uniformly on compact sets in $|\lambda| < 1$ to $\Theta$ the square outer spectral factor for $T_R$. Therefore one can use the Levinson algorithm to compute $\Theta_n(\lambda)$ and approximate the outer spectral factor $\Theta$ for $T_R$. Finally, it is noted that one can also use standard fast Fourier transform techniques to compute the Fourier coefficients $(\Theta_n)_j$ for each $\Theta_n$. Then one can apply the Kalman-Ho algorithm (see Kalman-Falb-Arbib [1]) to $(\Theta_n)_j$ to obtain a low order state space realization $\Sigma_n = \{A_n, B_n, C_n, D_n\}$ for $\Theta_n$. (This realization $\Sigma_n$ is not the state space realization obtained from the Levinson algorithm.) If n is chosen to be large enough, then $\{A_n, B_n, C_n, D_n\}$ can be used to approximate the state space realization for $\Theta$ the square outer spectral factor $T_R$.

## A.2 INNER-OUTER FACTORIZATIONS

In this section we will use Theorem 1.1 to obtain a state space realization for the outer spectral factor $\Theta$ for a strictly positive Toeplitz operator with a rational symbol. Motivated, by this result we will present the positive real lemma for Toeplitz matrices and give state space formulas to compute inner-outer factorizations for rational functions in $H^\infty(\mathcal{U}, \mathcal{Y})$. Finally, we will present the bounded real Lemma in control theory, which is used to compute the $H^\infty$ norm for a rational function in $H^\infty$.

To begin, let $\{R_n\}_{-\infty}^{\infty}$ be a sequence of operators on $\mathcal{U}$ satisfying $R_n = R_{-n}^*$ for all n in $\mathbb{Z}$. Recall that the block Toeplitz matrix $T_R$ on $\ell_+^2(\mathcal{U})$ is the operator on $\ell_+^2(\mathcal{U})$ whose (i, j)-th entry is given by $R_{i-j}$. We say that $\{Z \text{ on } \mathcal{X}, B, C\}$ is a realization of $\{R_n\}_1^\infty$ if Z, B and C are operators on the appropriate spaces satisfying $CZ^{n-1}B = R_n$ for all $n \geq 1$. Now consider the following Riccati difference equation

$$Q_{n+1} = Z^*Q_nZ + (C - B^*Q_nZ)^*(R_0 - B^*Q_nB)^{-1}(C - B^*Q_nZ) \tag{2.1}$$

where $Q_n$ is an operator on $\mathcal{X}$ and the initial condition $Q_0 = 0$. We say that Q is a *steady state solution* to the Riccati difference equation in (2.1) if $Q_n$ strongly converges to an operator Q and the operator $R_0 - B^*QB$ is strictly positive. If $\{Q_n\}$ does not converge or $R_0 - B^*Q_nB$ is not strictly positive for some n, then there is no steady state solution to this Riccati difference equation. Now assume that {Z, B} is a stable, controllable pair. Later we will see that $R_0 - B^*Q_nB$ forms a decreasing sequence of positive operators uniformly bounded below by $\delta I$ for some $\delta > 0$, if and only if the Toeplitz matrix $T_R$ admits a square outer spectral factorization. In this case $\{Q_n\}$ form an increasing sequence of positive operators which converge to a steady state solution Q for (2.1). In particular, if $T_R$ is strictly positive, then the Riccati equation in (2.1) admits a steady state solution. Finally, recall that $\Theta$ is an outer spectral factorization for $T_R$ if $\Theta$ is an outer function in $H^\infty(\mathcal{U}, \mathcal{Y})$ satisfying $T_R = T_\Theta^* T_\Theta$. We begin with the following result.

**THEOREM 2.1.** *Assume that the set of operators $\{R_n\}_{-\infty}^{\infty}$ on $\mathcal{U}$ generates a strictly positive Toeplitz operator $T_R$ on $\ell_+^2(\mathcal{U})$, and let $\{Z \text{ on } \mathcal{X}, B, C\}$ be a stable, controllable realization of $\{R_n\}_1^\infty$. Then the set of operators $\{Q_n\}$ obtained from the Riccati difference equation (2.1) forms an increasing sequence of positive operators which converges to the steady state solution Q to (2.1) and $R_0 - B^*QB$ is strictly positive. Moreover, if*

$$D_0 = (R_0 - B^*QB)^{1/2},\ C_o = D_0^{-1}(C - B^*QZ) \quad \textit{and} \quad \Theta = D_0 + \lambda C_o(I - \lambda Z)^{-1}B\,, \tag{2.2}$$

*then $\Theta$ is the outer spectral factor in $H^\infty(\mathcal{U}, \mathcal{U})$ for $T_R$ and*

$$\Theta(\lambda)^{-1} = D_0^{-1} - \lambda D_0^{-1}C_o(I - \lambda(Z - BD_o^{-1}C_o))^{-1}BD_o^{-1}\,. \tag{2.3}$$

*Finally, {Z, B, C} is controllable and observable if and only if $\{Z, B, C_o, D_0\}$ is controllable and observable. In this case, $Z - BD_o^{-1}C_o$ is stable when the state space is finite dimensional.*

Now assume that Q is a steady state solution to the Riccati difference equation (2.1). Then Q satisfies the following algebraic Riccati equation

$$Q = Z^*QZ + (C - B^*QZ)^*(R_0 - B^*QB)^{-1}(C - B^*QZ)\,. \tag{2.4}$$

We say that Q is a *positive solution* to the algebraic Riccati equation (2.4) if Q is a positive operator on $\mathcal{X}$ satisfying (2.4) and $R_0 - B^*QB$ is strictly positive. Moreover, Q is a *minimal or stabilizing* solution to the algebraic Riccati equation (2.4) if Q is a positive solution to (2.4) and $Q \le Q_1$ where $Q_1$ is any other positive solution to the algebraic Riccati equation (2.4). This Q is called the stabilizing solution because the eigenvalues of the state space operator $Z - BD_o^{-1}C_o$ are in the closed unit disc when $D_0$ and $C_o$ are obtained from the minimal Q. If the Toeplitz operator $T_R$ is strictly positive, then the algebraic Riccati equation (2.4) admits a minimal solution. In fact, the steady state solution Q to the Riccati difference equation (2.1) turns out to be the minimal solution to the algebraic Riccati equation (2.4).

Let R be a.e. a positive operator in $L^\infty(\mathcal{U}, \mathcal{U})$. Recall that the Toeplitz matrix $T_R$ on $\ell_+^2(\mathcal{U})$ admits a square outer spectral factor $\Theta$ if there exists an outer function $\Theta$ in $H^\infty(\mathcal{U}, \mathcal{U})$ satisfying $T_\Theta^* T_\Theta = T_R$. Moreover, $T_R$ admits a square outer spectral factorization if and only if

$$\int_0^{2\pi} \ln \det [R] > -\infty \, ; \tag{2.5}$$

see (Section V.7 in Sz.-Nagy-Foias [3] or Helson-Lowdenslager [1], [2]) for further details. This sets the state for the following result known as the positive real lemma in stochastic realization theory; see Caines [1].

**LEMMA 2.2.** *Let* $\{R_n\}_{-\infty}^{\infty}$ *be a set of operators on a finite dimensional space* $\mathcal{U}$ *satisfying* $R_n = R_{-n}^*$ *for all* $n \geq 0$. *Assume that there exists a finite dimensional stable, controllable realization* $\{Z, B, C\}$ *of* $\{R_n\}_1^\infty$. *Then there exists a positive solution to the algebraic Riccati equation (2.4) if and only if the Toeplitz operator* $T_R$ *admits a square outer spectral factor* $\Theta$. *In this case the algebraic Riccati equation (2.4) admits a minimal solution* Q. *Moreover, the square outer spectral factor* $\Theta$ *is given by the third part of (2.2) using this minimal* Q *in the first two parts of (2.2) and its inverse is computed by (2.3). Finally,*

$$\frac{1}{2\pi} \int_0^{2\pi} \ln \det R = \ln \det [\Theta(0)^* \Theta(0)] = \ln \det [R_0 - B^* Q B] \, . \tag{2.6}$$

**REMARK 2.3.** As before, let $\{R_n\}_{-\infty}^{\infty}$ be a sequence of operators on a finite dimensional space $\mathcal{U}$ satisfying $R_n = R_{-n}^*$ for all n, and assume that $\{Z, B, C\}$ is a finite dimensional stable, controllable realization for $\{R_n\}_1^\infty$. Let $Q_n$ be the n-th solution to the Riccati difference equation (2.1). Then the block Toeplitz matrix $T_R$ on $\ell_+^2(\mathcal{U})$ admits a square outer spectral factorization $\Theta$ if and only if $R_0 - B^* Q_n B \geq \delta I$ for all n and some $\delta > 0$. In this case $\{R_0 - B^* Q_n B\}$ are decreasing and converge to the strictly positive operator $\Theta(0)^* \Theta(0)$. Moreover, $\{Q_n\}$ forms an increasing sequence of positive operators which converge to the minimal solution Q for the algebraic Riccati equation (2.4). Equivalently $T_R$ admits a square outer spectral factorization if and only if the Riccati difference equation (2.1) converges to a positive steady state solution Q such that $R_0 - B^* Q B$ is strictly positive. In either case the square outer spectral factor $\Theta$ for $T_R$ is given by (2.2) and its inverse by (2.3). If $T_R$ is strictly positive, then the steady state solution Q to (2.1) or minimal solution Q to (2.4) is given by $Q = W_o^* T_R^{-1} W_o$ where $W_o$ is the observability operator from $\mathcal{X}$ into $\ell_+^2(\mathcal{U})$ defined by $W_1 x = (C^n x)_0^\infty$. Finally, $\{Z, B, C\}$ is controllable and observable if and only if $\{Z, B, C_o, D_0\}$ is controllable and observable.

The following procedure can be used to compute inner-outer factorizations for certain rational functions F in $H^\infty(\mathcal{U}, \mathcal{Y})$.

**PROCEDURE 2.4** (Inner-Outer factorization). Let $\{Z, B, \tilde{C}, D\}$ be a stable, controllable realization for a rational function F in $H^\infty(\mathcal{U}, \mathcal{Y})$ where $\mathcal{U}$ is finite dimensional and assume that $F(e^{it})^*F(e^{it})$ is a.e. invertible, or equivalently, $\det[F^*F]$ is a.e. nonzero. Then condition (2.5) holds for $R = F^*F$ and F admits an inner-outer factorization of the form $F = F_i\Theta$ where $\Theta$ is a square outer function in $H^\infty(\mathcal{U}, \mathcal{U})$ and $F_i$ is an inner function in $H^\infty(\mathcal{U}, \mathcal{Y})$. To compute a state space realization for $F_i$ and $\Theta$, let X be the maximal positive solution to the following algebraic Riccati equation

$$X = Z^*XZ - (D^*\tilde{C} + B^*XZ)^*(D^*D + B^*XB)^{-1}(D^*\tilde{C} + B^*XZ) + \tilde{C}^*\tilde{C}. \tag{2.7}$$

By the *maximal solution* we mean that X is the largest positive operator satisfying (2.7) where $D^*D + B^*XB$ is strictly positive, that is, $X \geq X_1$, where $X_1$ is any self-adjoint solution of (2.7) and $D^*D + B^*X_1B > 0$. (By following an argument similar to the one used to prove (2.35) below, the maximal solution $X \geq 0$.) It turns out that the maximal $X = \tilde{Q} - Q$ where $\tilde{Q}$ is the observability grammian for $\{\tilde{C}, Z\}$ and Q is the steady state solution to the following Riccati difference equation

$$Q_{n+1} = Z^*Q_nZ + (D^*\tilde{C} + B^*(\tilde{Q} - Q_n)Z)^*(D^*D + B^*(\tilde{Q} - Q_n)B)^{-1}(D^*\tilde{C} + B^*(\tilde{Q} - Q_n)Z). \tag{2.8}$$

In fact, Equation (2.7) is obtained by subtracting the algebraic Riccati equation generated by (2.8) from $\tilde{Q} = Z^*\tilde{Q}Z + \tilde{C}^*\tilde{C}$. Finally, let $C_o$, $D_0$, $A_i$, $B_i$, $C_i$ and $D_i$ be the operators defined by

$$D_0 = (D^*D + B^*XB)^{1/2},\ C_o = D_0^{-1}(D^*\tilde{C} + B^*XZ) \quad\text{and}\quad A_i = Z - BD_o^{-1}C_o$$
$$B_i = BD_o^{-1},\ C_i = \tilde{C} - DD_o^{-1}C_o \quad\text{and}\quad D_i = DD_o^{-1}. \tag{2.9}$$

Then $\Sigma_i = \{A_i, B_i, C_i, D_i\}$ and $\Sigma_0 = \{Z, B, C_o, D_0\}$ are controllable realizations for the inner factor $F_i$ and square outer factor $\Theta$ for F respectively. Notice that the realizations $\Sigma_i$ and $\Sigma_0$ may not be controllable and observable. For example, if $F = F_o$ is outer, then $F_i = D_i$ is an isometric constant and thus $\Sigma_i$ is not controllable and observable. However, one can extract the controllable and observable realizations from $\Sigma_i$ and $\Sigma_0$ by standard state space techniques.

The following result known as the bounded real Lemma can be used to determine if a function F in $H^\infty(\mathcal{U}, \mathcal{Y})$ satisfies $\|F\|_\infty \leq \gamma$.

**LEMMA 2.5.** *Let* {Z, B, $\tilde{C}$, D} *be a stable, controllable realization for a rational function* F *in* $H^\infty(\mathcal{U}, \mathcal{Y})$. *Then* $\|F\|_\infty$ *is the infimum over the set of all* $\gamma > 0$ *such that the algebraic Riccati equation*

$$Y = Z^*YZ + (D^*\tilde{C} + B^*YZ)^*(\gamma^2 I - D^*D - B^*YB)^{-1}(D^*\tilde{C} + B^*YZ) + \tilde{C}^*\tilde{C} \tag{2.10}$$

*admits a positive solution. Moreover, if* $\|F\|_\infty < \gamma$ *and* Y *is the minimal solution to the algebraic Riccati equation (2.10), then the square outer spectral factor* $\Theta$ *for the spectral density* $\gamma^2 I - F^*F$ *is given by*

$$\Theta(\lambda) = D_0 + \lambda C_o(I - \lambda Z)^{-1}B \text{ where}$$
$$D_0 = (\gamma^2 I - D^*D - B^*YB)^{1/2} \text{ and } C_o = -D_0^{-1}(D^*\tilde{C} + B^*YZ)\,. \tag{2.11}$$

*Finally, the inverse for* $\Theta$ *is given by (2.3).*

Recall that Y is a positive solution to the algebraic Riccati equation (2.10) if Y is a positive solution to (2.10) and $\gamma^2 I - D^*D - B^*YB$ is strictly positive. The minimal solution to (2.10) is the smallest possible positive solution to (2.10). Actually the minimal solution for (2.10) is given by $Y = \tilde{Q} + Q$ where $\tilde{Q}$ is the observability grammian for {$\tilde{C}$, Z} and Q is the steady state solution to the following Riccati difference equation

$$Q_{n+1} = Z^*Q_nZ +$$
$$(D^*\tilde{C} + B^*(\tilde{Q} + Q_n)Z)^*(\gamma^2 I - D^*D - B^*(\tilde{Q} + Q_n)B)^{-1}(D^*\tilde{C} + B^*(\tilde{Q} + Q_n)Z)\,. \tag{2.12}$$

So one obtains the algebraic Riccati equation in (2.10) by adding $\tilde{Q} = Z^*\tilde{Q}Z + \tilde{C}^*\tilde{C}$ to the algebraic Riccati equation generated by (2.12).

The previous bounded real lemma shows that $\|F\|_\infty$ is the infimum of the set of all $\gamma$ such that the algebraic Riccati equation (2.10) admits a positive solution. So one can compute $\|F\|_\infty$ by iterating on $\gamma$ to find the smallest $\gamma$ such that (2.10) admits a positive solution. Finally, it is noted that one can also use the fast Fourier transform techniques to efficiently compute the $L^\infty$ norm of a rational function.

**REMARK 2.6.** The above lemma can be used to solve the rational Darlington synthesis problem. To see this, let F be a rational function in $H^\infty(\mathcal{U}, \mathcal{Y})$ satisfying $\|F\|_\infty < 1$ where $\mathcal{U}$ and $\mathcal{Y}$ are finite dimensional. Then the Darlington synthesis problem is to find a two-sided inner function $\Phi(\lambda)$ in $H^\infty(\mathcal{U} \oplus \mathcal{Y}, \mathcal{Y} \oplus \mathcal{U})$ of the form

$$\Phi(\lambda)=\begin{bmatrix} F(\lambda) & F_{1,2}(\lambda) \\ F_{2,1}(\lambda) & F_{2,2}(\lambda) \end{bmatrix}.$$

To solve, this problem let $\{Z, B, \tilde{C}, D\}$ be a stable controllable and observable realization of $F(\lambda)$. Then let Y be the minimal solution to the algebraic Riccati equation in (2.10) where $\gamma = 1$. (Recall that $Y = \tilde{Q} + Q$ where $\tilde{Q}$ is the observability grammian for $\{\tilde{C}, Z\}$ and Q is the steady state solution to the Riccati difference equation in (2.12) where $\gamma = 1$.) According to the previous lemma $F_{2,1}(\lambda) := \Theta(\lambda)$ is the square outer spectral factor of $I - F^*F$, where $\Theta$ is defined in (2.11). Therefore $[F, F_{2,1}]$ is an inner function in $H^\infty(\mathcal{U}, \mathcal{Y} \oplus \mathcal{U})$. Moreover, $\Gamma = \{Z, B, [\tilde{C}, C_o]^{tr}, [D, D_o]^{tr}\}$ is a controllable and observable realization of $[F, F_{21}]^{tr}$.

Let P be the observability grammian for the pair $\{Z, [\tilde{C}, C_o]^{tr}\}$. Then P is the solution to the Lyapunov equation

$$P = Z^*PZ + \tilde{C}^*\tilde{C} + C_o^*C_o .$$

Now let $\Gamma_1 = \{A, B_1, [C_1, C_2]^{tr}, [D_{1,1}, D_{2,1}]^{tr}\}$ be the system defined by

$$A = P^{1/2}ZP^{-1/2},\ C_1 = \tilde{C}P^{-1/2},\ C_2 = C_oP^{-1/2}$$

$$B_1 = P^{1/2}B,\ D_{1,1} = D \text{ and } D_{2,1} = D_o .$$

Then $\Gamma$ is similar to $\Gamma_1$ and thus $\Gamma_1$ is a stable controllable and observable realization of the inner function $[F, F_{2,1}]^{tr}$. Because the observability grammian for $\Gamma_1$ is the identity, it follows that $\Gamma_1$ is unitarily equivalent to the restricted backward shift realization; see Remark III.4.5. Since the restricted backward shift realization of an inner function is an isometric system,

$$\begin{bmatrix} A & B_1 \\ C_1 & D_{11} \\ C_2 & D_{21} \end{bmatrix} : \begin{bmatrix} \mathcal{X} \\ \mathcal{U} \end{bmatrix} \to \begin{bmatrix} \mathcal{X} \\ \mathcal{Y} \\ \mathcal{U} \end{bmatrix}$$

is an isometry; see Theorem III.7.10 and its proof. Now let $B_2$, $D_{1,2}$ and $D_{2,2}$ be operators such that

$$\begin{bmatrix} A & B_1 & B_2 \\ C_1 & D_{1,1} & D_{12} \\ C_2 & D_{2,1} & D_{22} \end{bmatrix} : \begin{bmatrix} \mathcal{X} \\ \mathcal{U} \\ \mathcal{Y} \end{bmatrix} \to \begin{bmatrix} \mathcal{X} \\ \mathcal{Y} \\ \mathcal{U} \end{bmatrix}$$

is a unitary operators. Clearly A is stable, $\{[C_1, C_2]^{tr}, A\}$ is observable and $\{A, [B_1, B_2]\}$ is controllable. So according to Theorem III.10.5

$$\Phi(\lambda) = \begin{bmatrix} F_{1,1}(\lambda) & F_{1,2}(\lambda) \\ F_{2,1}(\lambda) & F_{2,2}(\lambda) \end{bmatrix} = \begin{bmatrix} D_{1,1} & D_{1,2} \\ D_{2,1} & D_{2,2} \end{bmatrix} + \lambda \begin{bmatrix} C_1 \\ C_2 \end{bmatrix} (I - \lambda A)^{-1} [B_1, B_2]$$

is a two-sided inner function in $H^\infty(\mathcal{U} \oplus \mathcal{Y}, \mathcal{Y} \oplus \mathcal{U})$. Since $\{A, B_1, C_1, D_{1,1}\}$ is a realization of $F(\lambda) = F_{1,1}(\lambda)$ it follows that $\Phi(\lambda)$ is a Darlington extension of $F(\lambda)$. Therefore the previous method of constructing $A$, $B_1$, $B_2$, $C_1$, $C_2$, $D_{1,1}$, $D_{1,2}$, $D_{2,1}$ and $D_{2,2}$ yields a state space realization procedure for constructing a Darlington extension for a rational $F(\lambda)$ satisfying $\|F\|_\infty < 1$. Finally, by converting back to the original coordinate system

$$Z = P^{-1/2} A P^{1/2}, \quad \tilde{C} = C_1 P^{1/2}, \quad C_o = C_2 P^{1/2} \quad B = P^{-1/2} B_1 \text{ and } B_e = P^{-1/2} B_2$$

$$D_e = \begin{bmatrix} D_{11} & D_{12} \\ D_{21} & D_{22} \end{bmatrix}$$

it follows that $\{Z, [B, B_e], [\tilde{C}, C_o]^{tr}, D_e\}$ is a stable, controllable and observable realization of $\Phi(\lambda)$ which is a Darlington extension of $F(\lambda)$. In particular, one can use the same state space operator $Z$ in the Darlington extension $\Phi(\lambda)$ of $F(\lambda)$.

**PROOF OF THEOREM 2.1.** This proof follows some of the methods in stochastic realization theory; see Caines [1], Faurre [1]. Let $\Pi$ be the operator from $\ell_+^2(\mathcal{U})$ onto $\mathcal{U}$ defined by $\Pi = P_{\mathcal{U}}$ and set $N$ equal to the positive square root of $\Pi T_R^{-1} \Pi^*$. According to Theorem 1.1 the outer spectral factor $\Theta$ in $H^\infty(\mathcal{U}, \mathcal{U})$ for the Toeplitz operator $T_R$ is given by $\Theta(\lambda) = N\Omega(\lambda)^{-1}$ where $\Omega(\lambda) = (\mathcal{F}_+ T_R^{-1} \Pi^*)(\lambda)$ and $\mathcal{F}_+$ denotes the Fourier transform from $\ell_+^2(\mathcal{U})$ onto $H^2(\mathcal{U})$. Now let $S$ be the unilateral shift on $\ell_+^2(\mathcal{U})$. Notice that

$$\Omega(\lambda) = \Pi(I - \lambda S^*)^{-1} T_R^{-1} \Pi^* = \Pi T_R^{-1} \Pi^* + \lambda \Pi (I - \lambda S^*)^{-1} S^* T_R^{-1} \Pi^* =$$
$$N^2 - \lambda \Pi (I - \lambda S^*)^{-1} T_R^{-1} S^* T_R \Pi^* N^2 . \tag{2.13}$$

The last equality follows from $\Pi^*\Pi = I - SS^*$ and $S^* T_R S = T_R$. The last equality also follows from the fact that the Toeplitz operator $T_R$ admits a matrix partition of the form

$$T_R = \begin{bmatrix} R_0 & \Pi T_R S \\ S^* T_R \Pi^* & T_R \end{bmatrix}. \tag{2.14}$$

So using the matrix inversion Lemma 1.2 with $X = S^* T_R \Pi^*$ we see that $S^* T_R^{-1} \Pi^* = -T_R^{-1} S^* T_R \Pi^* \Delta^{-1}$ where the Schur complement $\Delta = R_o - X^* T_R^{-1} X = N^{-2}$. (Notice that $N^2 = \Pi T_R^{-1} \Pi^* = \Delta^{-1}$.) This readily gives the last equality in (2.13).

According to Lemma III.1.2 if $G(\lambda) = D + \lambda C(I - \lambda Z)^{-1} B$, then its inverse $G(\lambda)^{-1}$ is given by

$$G(\lambda)^{-1} = D^{-1} - \lambda D^{-1}C(I - \lambda(Z - BD^{-1}C))BD^{-1}$$

in some neighborhood of the origin. Using this formula for $G = \Omega$ in (2.13) we obtain

$$\Omega(\lambda)^{-1} = N^{-2}[I + \lambda\Pi(I - \lambda(S^* + T_R^{-1}S^*T_R\Pi^*\Pi))^{-1}T_R^{-1}S^*T_R\Pi^*] . \tag{2.15}$$

Now notice that

$$S^* + T_R^{-1}S^*T_R\Pi^*\Pi = S^* + T_R^{-1}S^*T_R(I - SS^*) = T_R^{-1}S^*T_R .$$

Substituting this into (2.15) and using $\Theta = N\Omega^{-1}$, we see that the outer spectral factor $\Theta$ for $T_R$ is given by

$$\Theta = N^{-1} + \lambda N^{-1}\Pi T_R^{-1}(I - \lambda S^*)^{-1}S^*T_R\Pi^* . \tag{2.16}$$

To complete the proof it remains to give a state space formula for $\Theta$ in (2.16). To this end, let $W_o$ be the observability operator from $\mathcal{X}$ into $\ell_+^2(\mathcal{U})$ defined by $W_o x = (CZ^n x)_0^\infty$ for x in $\mathcal{X}$. Obviously $S^*W_o = W_o Z$. Because $CZ^{n-1}B = R_n$ for all $n \geq 1$ and $\{R_n\}_0^\infty$ is the first column of $T_R$ it follows that $S^*T_R\Pi^* = W_o B$. Using this in (2.14) we see that the Toeplitz matrix admits a matrix partition of the form

$$T_R = \begin{bmatrix} R_0 & B^*W_o^* \\ W_o B & T_R \end{bmatrix} . \tag{2.17}$$

Now let $Q = W_o^* T_R^{-1} W_o$. According to the matrix inversion Lemma 1.2

$$N^2 = \Pi T_R^{-1}\Pi^* = \Delta^{-1} = (R_0 - B^*W_o^*T_R^{-1}W_o B)^{-1} = (R_0 - B^*QB)^{-1} := D_o^{-2} . \tag{2.18}$$

In particular $D_o = N^{-1}$. By applying the matrix inversion Lemma to $\Pi T_R^{-1}$ with $T_R$ in (2.17) and $C_o := D_o^{-1}(C - B^*QZ)$, we obtain

$$N^{-1}\Pi T_R^{-1}W_o = N^{-1}[\Delta^{-1}, -\Delta^{-1}B^*W_o^*T_R^{-1}]\begin{bmatrix} C \\ W_o Z \end{bmatrix} = N^{-1}N^2(C - B^*QZ) = C_o . \tag{2.19}$$

Substituting this into our expression for $\Theta$ in (2.16) and using $S^*T_R\Pi^* = W_o B$ along with $S^*W_o = W_o Z$ we have

$$\Theta = N^{-1} + \lambda N^{-1}\Pi T_R^{-1}(I - \lambda S^*)^{-1}W_o B = D_o + \lambda N^{-1}\Pi T_R^{-1}W_o(I - \lambda Z)^{-1}B =$$

$$D_o + \lambda C_o(I - \lambda Z)^{-1}B .$$

This is precisely the state space form for $\Theta$ in (2.2).

Later we will see that $Q = W_o^*T_R^{-1}W_o$ can be obtained as the steady state solution to the Riccati difference equation (2.1). So to complete the proof let us show that $\{C_o, Z\}$ is observable if and only if $\{C, Z\}$ is observable. Since $T_R$ is strictly positive and $\Theta(\lambda) = N\Omega(\lambda)^{-1}$ is its outer factor, it follows that $\Omega(\lambda)$ is an invertible outer function in $H^\infty(\mathcal{U}, \mathcal{U})$. In particular, the Toeplitz matrix $T_\Omega$ on $\ell_+^2(\mathcal{U})$ generated by $\Omega(\lambda)$ is invertible. Because $T_R^{-1}\Pi^*$ is the first block columns of $T_\Omega$, we see that $T_\Omega = [\Omega_0, \Omega_1, \Omega_2, ...]$ where $\Omega_n$ is the block column operator from $\mathcal{U}$ into $\ell_+^2(\mathcal{U})$ defined by $\Omega_n = S^n T_R^{-1}\Pi^*$ for $n = 0, 1, 2, ....$ By consulting (2.19) we see that $C_o = D_0 \Pi T_R^{-1} W_o$. Hence

$$C_o Z^n x = D_0 \Pi T_R^{-1} W_o Z^n x = D_0 \Pi T_R^{-1} S^{*n} W_o x = D_0 \Omega_n^* W_o x .$$

This readily shows that

$$[C_o, C_o Z, C_o Z^2, ...]^{tr} = \text{diag}\,[D_0, D_0, D_0, ...] T_\Omega^* W_o$$

where diag $[D_0, D_0, D_0, ...]$ is the block diagonal matrix on $\ell_+^2(\mathcal{U})$ generated by $D_0$. Because diag $[D_0, D_0, D_0, ...]T_\Omega^*$ is invertible, $\{C_o, Z\}$ is observable if and only if $\{C, Z\}$ is observable. The inversion formula for $\Theta$ in (2.3) is obtained by applying Lemma III.1.2 to $\Theta$ in (2.2). Finally, notice that for finite dimensional systems, if $\{Z, B, C_o, D_0\}$ is a controllable and observable realization for $\Theta(\lambda)$, then the realization for its inverse $\Theta(\lambda)^{-1}$ in (2.3) is also controllable and observable. (To see this compute the span of $(Z - BD_o^{-1}C_o)^n B$ and $(Z^* - C_o^* D_0^{-1} B^*)^n C_o^*$.) Since $\Theta(\lambda)^{-1}$ is in $H^\infty(\mathcal{U}, \mathcal{U})$, Theorem III.4.4 shows that $Z_o - BD_o^{-1}C_o$ is stable. This completes the proof.

To prove Lemma 2.2 we begin with the following useful result.

**LEMMA 2.7.** *Let* $\{Z, B, \tilde{C}, D\}$ *be a stable, realization for a function* F *in* $H^\infty(\mathcal{U}, \mathcal{Y})$. *Let* $T_R$ *be the Toeplitz operator on* $\ell_+^2(\mathcal{U})$ *generated by a sequence of operator* $\{R_n\}_{-\infty}^{\infty}$ *satisfying* $R_n = R_{-n}^*$ *for all* $n \in \mathbb{Z}$. *Then* F *is a spectral factor for* $T_R$ *if and only if*

$$D^*D + B^*\tilde{Q}B = R_0 \text{ and } (D^*\tilde{C} + B^*\tilde{Q}Z)Z^{n-1}B = R_n \quad (\text{for } n \geq 1) \tag{2.20}$$

*where* $\tilde{Q}$ *is the observability grammian for the pair* $\{\tilde{C}, Z\}$.

**PROOF.** Since $T_R$ is self adjoint Toeplitz operator, it follows that F is a spectral factor for $T_R$ if and only if the first column of $T_F^* T_F$ is $T_R \Pi^* = (R_n)_0^\infty$. Recall that the Fourier coefficients $\{F_n\}$ for F are given by $F_0 = D$ and $F_n = \tilde{C}Z^{n-1}B$ for $n \geq 1$. So the first column of $T_F^* T_F$ equals $T_R \Pi^*$ if and only if

$$\begin{bmatrix} D^* & B^*\tilde{C}^* & B^*Z^*\tilde{C}^* & \cdots \\ 0 & D^* & B^*\tilde{C}^* & \cdots \\ 0 & 0 & D^* & \cdots \\ \cdot & \cdot & \cdot & \cdot \\ \cdot & \cdot & \cdot & \cdot \\ \cdot & \cdot & \cdot & \cdot \end{bmatrix} \begin{bmatrix} D \\ \tilde{C}B \\ \tilde{C}ZB \\ \cdot \\ \cdot \\ \cdot \end{bmatrix} = \begin{bmatrix} R_0 \\ R_1 \\ R_2 \\ \cdot \\ \cdot \\ \cdot \end{bmatrix}. \tag{2.21}$$

Using $\tilde{Q} = \sum Z^{*n}\tilde{C}^*\tilde{C}Z^n$ we readily arrive at (2.20). This completes the proof.

**PROOF OF LEMMA 2.2**. Here we will prove both Lemma 2.2 and of Remark 2.3. To this end, assume that $\tilde{Q}$ is a positive solution to the algebraic Riccati equation (2.4). Set

$$D = (R_0 - B^*\tilde{Q}B)^{1/2} \quad \text{and} \quad \tilde{C} = D^{-1}(C - B^*\tilde{Q}Z). \tag{2.22}$$

By substituting D and $\tilde{C}$ into (2.4) we see that $\tilde{Q}$ is also the observability grammian for the pair $\{\tilde{C}, Z\}$. Now let $\{Z, B, \tilde{C}, D\}$ be a state space realization for a function F in $H^\infty(\mathcal{U}, \mathcal{U})$. Then Lemma 2.7 and the fact that $\{Z, B, C\}$ is a realization of $\{R_n\}_1^\infty$ implies that F is a spectral factor for $T_R$. Because $F(0) = D$ is invertible, $T_R$ admits a square outer spectral factor; see Remark 1.5.

Now let us establish some results concerning the steady state solution to the Riccati difference equation and the algebraic Riccati equation. To this end, $W_n$ be the operator from $\mathcal{X}$ into $\ell_n^2(\mathcal{U})$ defined by

$$W_n = [C, CZ, CZ^2, \cdots, CZ^{n-1}]^{tr}. \tag{2.23}$$

As before, let $T_{R,n}$ be the $n \times n$ Toeplitz matrix given by (1.8). Then using $R_j = CZ^{j-1}B$, it follows that $T_{R,n+1}$ admits a partition of the form

$$T_{R,n+1} = \begin{bmatrix} R_0 & B^*W_n^* \\ W_nB & T_{R,n} \end{bmatrix}. \tag{2.24}$$

For the moment assume that $T_{R,n}$ is strictly positive and set $Q_n = W_n^*T_{R,n}^{-1}W_n$. Then Schur complement $\Delta_{n+1}$ for $T_{R,n+1}$ is given by

$$\Delta_{n+1} = R_0 - B^*Q_nB \text{ where } Q_n = W_n^*T_{R,n}^{-1}W_n. \tag{2.25}$$

So according Theorem 1.4 the Toeplitz operator $T_R$ admits a square outer spectral factorization $\Theta$ if and only if

$$\Delta_{n+1} = R_0 - B^*Q_nB \geq \delta I \quad \text{(for all n and some } \delta > 0). \tag{2.26}$$

In this case, $\{R_0 - B^* Q_n B\}$ forms a decreasing sequence of positive operators converging to $\Delta = \Theta(0)^* \Theta(0)$. Moreover, $T_{R,n}$ is strictly positive for all n. Now let us assume that (2.26) holds. So using (2.23) in the matrix inversion Lemma 1.2 on the partition for $T_{R,n+1}$ in (2.24) along with $Q_{n+1} = W_{n+1}^* T_{R,n+1}^{-1} W_{n+1}$ we obtain

$$Q_{n+1} = [C^*, Z^* W_n^*] \begin{bmatrix} \Delta^{-1} & -\Delta^{-1} B^* W_n^* T_{R,n}^{-1} \\ -T_{R,n}^{-1} W_n B \Delta^{-1}, & T_{R,n}^{-1} + T_{R,n}^{-1} W_n B \Delta^{-1} B^* W_n^* T_{R,n}^{-1} \end{bmatrix} \begin{bmatrix} C \\ W_n Z \end{bmatrix}$$

$$= Z^* Q_n Z + (C - B^* Q_n Z)^* (R_0 - B^* Q_n B)^{-1} (C - B^* Q_n Z) \tag{2.27}$$

where $\Delta = \Delta_{n+1} = R_0 - B^* Q_n B$. This is precisely the Riccati difference equation in (2.1).

Now let us show that $\{Q_n\}$ forms an increasing sequence when (2.26) holds. To this end, notice that $T_{R,n+1}$ admits a decomposition of the form

$$T_{R,n+1} = \begin{bmatrix} T_{R,n} & X \\ X^* & R_0 \end{bmatrix} \tag{2.28}$$

where $X^* = [R_n, R_{n-1}, \cdots, R_1]$. Because $T_{R,n}$ is invertible, $T_{R,n+1}$ admits a Schur factorization of the form

$$T_{R,n+1} = \begin{bmatrix} I & 0 \\ X^* T_{R,n}^{-1} & I \end{bmatrix} \begin{bmatrix} T_{R,n} & 0 \\ 0 & R_0 - X^* T_{R,n}^{-1} X \end{bmatrix} \begin{bmatrix} I & T_{R,n}^{-1} X \\ 0 & I \end{bmatrix}. \tag{2.29}$$

Since $T_{R,n+1}$ is strictly positive, the Schur complement $\Lambda = R_0 - X^* T_{R,n}^{-1} X$ is also strictly positive. In particular, the inverse of $T_{R,n+1}$ is given by

$$T_{R,n+1}^{-1} = \begin{bmatrix} I & -T_{R,n}^{-1} X \\ 0 & I \end{bmatrix} \begin{bmatrix} T_{R,n}^{-1} & 0 \\ 0 & \Lambda^{-1} \end{bmatrix} \begin{bmatrix} I & 0 \\ -X^* T_{R,n}^{-1} & I \end{bmatrix}. \tag{2.30}$$

So using $W_{n+1} = [W_n, CZ^n]^{tr}$ along with x in the state space $\mathcal{X}$ we have

$$\begin{aligned} &(Q_{n+1} x, x) = (T_{R,n+1}^{-1} W_{n+1} x, W_{n+1} x) = \\ &(T_{R,n+1}^{-1} (W_n x \oplus CZ^n x), W_n x \oplus CZ^n x) = \\ &(T_{R,n}^{-1} W_n x, W_n x) + \|\Lambda^{-1/2} (-X^* T_{R,n}^{-1} W_n + CZ^n) x\|^2 \geq (Q_n x, x) . \end{aligned} \tag{2.31}$$

Therefore $Q_n \leq Q_{n+1}$.

Now let us show that $\{Q_n\}$ are uniformly bounded. Recall that $R_0 - B^* Q_n B$ form a decreasing sequence which converge to the strictly positive operator $\Theta(0)^* \Theta(0)$. Hence $\{B^* Q_n B\}$ forms an increasing sequence of positive operators which converge to a bounded operator $\Omega$. By consulting the Riccati difference equation (2.1) we see that $B^*Q_{n+1}B \geq B^*Z^*Q_nZB$. In particular, $B^*Z^*Q_nZB \leq \Omega$. Continuing in this fashion $B^*Z^{*k}Q_nZ^kB \leq \Omega$ for all k and n. This readily implies that the diagonal entries of the positive operator

$$P_n = [B, ZB, \cdots, Z^{m-1}B]^*Q_n[B, ZB, \cdots, Z^{m-1}B] \tag{2.32}$$

are all bounded by $\Omega$. Because $P_n$ is positive and its diagonal entries are bounded by $\Omega$, it follows that $\|P_n\| \leq \gamma < \infty$ for all n. In fact, $\|P_n\| \leq m \operatorname{trace} \Omega$. So if we choose m to be the dimension of the state space $\mathcal{X}$, then the controllability operator $[B, \cdots, Z^{m-1}B]$ in (2.32) is onto. Therefore $Q_n \leq \alpha I$ for some finite $\alpha$. (In all of our applications of the positive real lemma to the two block problem the corresponding Toeplitz operator $T_R$ is strictly positive. In this $Q_n$ converges to the bounded operator $W_o^*T_R^{-1}W_o$ and thus we do not have to assume that the state space is finite dimensional when $T_R$ is invertible.)

Because $\{Q_n\}$ are increasing and uniformly bounded, $Q_n$ converges to a positive operator Q. Moreover, this Q is precisely the steady state solution to the Riccati difference equation (2.1). This Q is also a positive solution to the algebraic Riccati equation (2.4). So if we define $\Theta$ according to (2.2), then Lemma 2.7 or our previous analysis shows that $\Theta$ is a spectral factor for $T_R$. However, $\Theta(0)^*\Theta(0) = R_0 - B^*QB = \Delta$ which is the limit of the $\{\Delta_n\}$; see (2.26). So by Theorem 1.4. this $\Theta$ is precisely the square outer factor for $T_R$.

To complete the proof it remains to show that if $\tilde{Q}$ is any positive solution to the algebraic Riccati equation (2.4), then $Q \leq \tilde{Q}$ where Q is the steady state solution to the Riccati difference equation (2.1). To this end, let D and $\tilde{C}$ be the operators defined according to (2.22). Then using (2.4) follows that $\tilde{Q}$ is the observability grammian for $\{\tilde{C}, Z\}$. Lemma 2.7 shows that the function F generated by the state space realization $\{Z, B, \tilde{C}, D\}$ is a spectral factor for $T_R$. Moreover, the square outer spectral factor $\Theta$ for $T_R$ is given by (2.2), where Q is the observability grammian for the pair $\{C_o, Z\}$. Now consider the Hankel operators $\Gamma_o$ and $\Gamma_1$ from $K^2(\mathcal{U}) = L^2(\mathcal{U}) \ominus H^2(\mathcal{U})$ into $H^2(\mathcal{U})$ defined by

$$\Gamma_o = P_+M_\Theta | K^2(\mathcal{U}) \quad \text{and} \quad \Gamma_1 = P_+M_F | K^2(\mathcal{U}) \tag{2.33}$$

where $P_+$ is the orthogonal projection onto $H^2(\mathcal{U})$. Notice that $S^*\Gamma_o = \Gamma_o S_1$ and $S^*\Gamma_1 = \Gamma_1 S_1$ where S is now the unilateral shift on $H^2(\mathcal{U})$ and $S_1$ is the unilateral shift on $K^2(\mathcal{U})$, that is, $S_1 f = e^{-it}f$ for all f in $K^2(\mathcal{U})$. Now let $\hat{W}_o$ and $W_1$ from $\mathcal{X}$ into $H^2(\mathcal{U})$ and $W_c$ from $\mathcal{X}$ onto $K^2(\mathcal{U})$ be the observability and controllability operators defined by

$$\hat{W}_o x = C_o(I - \lambda Z)^{-1}x, \; W_1 = \tilde{C}(I - \lambda Z)^{-1}x \quad \text{and} \quad W_c^* = B^*(\lambda I - Z^*)^{-1}x \qquad (x \in \mathcal{X}). \tag{2.34}$$

The observability grammians Q and $\tilde{Q}$ are given by $Q = W_o^*W_o$ and $\tilde{Q}_o = W_1^*W_1$.

We claim that $\Gamma_o = \hat{W}_o W_c$ and $\Gamma_1 = W_1 W_c$. To verify this first notice that $S^* \hat{W}_o = \hat{W}_o Z$. This follows from the fact that $S^* h = (h - h(0))/\lambda$ for h in $H^2(\mathcal{U})$, that is,

$$S^* \hat{W}_o = \frac{C(I - \lambda Z)^{-1} - C}{\lambda} = C(I - \lambda Z)^{-1} Z = \hat{W}_o Z .$$

Moreover, using

$$W_c f = \sum_{n=0}^{\infty} Z^n B f_{-(n+1)} \qquad \left( f = \sum_{n=1}^{\infty} f_{-n} e^{-int} \in K^2(\mathcal{U}) \right)$$

it follows that $ZW_c = W_c S_1$. Now notice that for all u in $\mathcal{U}$ we have

$$\Gamma_o e^{-it} u = P_+ e^{-it} \Theta u = C_o (I - \lambda Z)^{-1} Bu = \hat{W}_o W_c e^{-it} u .$$

This along with $S^* \hat{W}_o = \hat{W}_o Z$ and $ZW_c = W_c S_1$ readily gives for all $n \geq 0$

$$\hat{W}_o W_c S_1^n e^{-it} u = \hat{W}_o Z^n W_c e^{-it} u = S^{*n} \hat{W}_o W_c e^{-it} u = S^{*n} \Gamma_o e^{-it} u = \Gamma_o S_1^n e^{-it} u .$$

The last equality follows from $S^* \Gamma_o = \Gamma_o S_1$. Because $\{S_1^n e^{-it} \mathcal{U} : n \geq 0\}$ spans $K^2(\mathcal{U})$, it follows that $\Gamma_o = \hat{W}_o W_c$. A similar argument shows that $\Gamma_1 = W_1 W_c$.

Because F and $\Theta$ are both spectral factors for $T_R$, the function F admits an inner-outer factorization of the form $F = F_i \Theta$ where $F_i$ is an inner function. We claim that $\|P_+ F_i f\| \geq \|P_+ f\|$ for all f in $L^2(\mathcal{U})$. To see this let $P_-$ be the orthogonal projection onto $K^2(\mathcal{U})$. Then using $P_- M_{F_i} = P_- M_{F_i} P_-$ we have

$$\|P_+ F_i f\|^2 = \|F_i f\|^2 - \|P_- F_i f\|^2 = \|f\|^2 - \|P_- M_{F_i} P_- f\|^2 \geq \|f\|^2 - \|P_- f\|^2 = \|P_+ f\|^2 .$$

Hence $\|P_+ f\| \leq \|P_+ F_i f\|$. So for g in $K^2(\mathcal{U})$ we obtain

$$(\Gamma_o^* \Gamma_o g, g) = \|\Gamma_o g\|^2 = \|P_+ \Theta g\|^2 \leq \|P_+ F_i \Theta g\|^2 = \|\Gamma_1 g\|^2 = (\Gamma_1^* \Gamma_1 g, g) . \tag{2.35}$$

Since $\Gamma_o^* \Gamma_o = W_c^* \hat{W}_o^* \hat{W}_o W_c = W_c^* Q W_c$ and $\Gamma_1^* \Gamma_1 = W_c^* \tilde{Q} W_c$, we have $W_c^* Q W_c \leq W_c^* \tilde{Q} W_c$. Because the pair $\{Z, B\}$ is controllable, $W_c$ is onto and $Q \leq \tilde{Q}$. This completes the proof.

**PROOF OF PROCEDURE 2.4.** According to Lemma 2.7 the function F is a spectral factor for the Toeplitz operator $T_R$ where $\{R_n\}_0^\infty$ are generated by (2.20) and $R_n = R_{-n}^*$ for $n < 0$. In this case $\tilde{Q}$ is the observability grammian for $\{\tilde{C}, Z\}$. Moreover, if we set

$$C = D^* \tilde{C} + B^* \tilde{Q} Z \quad \text{and} \quad R_0 = D^* D + B^* \tilde{Q} B , \tag{2.36}$$

then $\{Z, B, C\}$ is a realization of $\{R_n\}_1^\infty$ and $R_0$ is specified in terms of D, B and $\tilde{Q}$. Because F is a rational function and $R = F^*F$ is a.e. invertible, condition (2.5) holds, and thus $T_R$ admits a square outer spectral factor $\Theta$. In particular, $F^*F = \Theta^*\Theta$ and F admits an inner-outer factorization of the form $F = F_i\Theta$. According to the positive real Lemma 2.2 this outer function $\Theta$ is given by (2.2) where $R_0$ and C are specified by (2.36) and Q is the minimal solution for the corresponding algebraic Riccati equation (2.4). In fact, substituting (2.36) into (2.4) we see that Q is the minimal solution to the following algebraic Riccati equation

$$Q = Z^*QZ + (D^*\tilde{C} + B^*(\tilde{Q} - Q)Z)^*(D^*D + B^*(\tilde{Q} - Q)B)^{-1}(D^*\tilde{C} + B^*(\tilde{Q} - Q)Z) . \quad (2.37)$$

So by setting $X = \tilde{Q} - Q$ and subtracting (2.37) from the Lyapunov equation $\tilde{Q} = Z^*\tilde{Q}Z + \tilde{C}^*\tilde{C}$ we arrive at the algebraic Riccati equation for X in (2.7). It is easy to show that Q is the minimal solution for the algebraic Riccati equation (2.37) if and only if $X = \tilde{Q} - Q$ is the maximal solution for (2.7). Finally, using $X = \tilde{Q} - Q$ and the expressions for C and $R_0$ in (2.36) in the formula for $\Theta$ in (2.2) we arrive at the state space realization $\{Z, B, C_o, D_o\}$ for $\Theta$ in (2.9).

The inverse of $\Theta$ is given by (2.3). Therefore

$$\begin{aligned}
&F_i(\lambda) = F(\lambda)\Theta(\lambda)^{-1} = (D + \lambda\tilde{C}(I - \lambda Z)^{-1}B)(D_0^{-1} - \lambda D_0^{-1}C_o(I - \lambda(Z - BD_o^{-1}C_o))^{-1}BD_o^{-1} = \\
&DD_o^{-1} - \lambda DD_o^{-1}C_o(I - \lambda(Z - BD_o^{-1}C_o))^{-1}BD_o^{-1} + \\
&\lambda\tilde{C}(I - \lambda Z)^{-1}[I - \lambda BD_o^{-1}C_o(I - \lambda(Z - BD_o^{-1}C_o))^{-1}]BD_o^{-1} = \qquad (2.38)\\
&DD_o^{-1} - \lambda DD_o^{-1}C_o(I - \lambda(Z - BD_o^{-1}C_o))^{-1}BD_o^{-1} + \lambda\tilde{C}(I - \lambda(Z - BD_o^{-1}C_o))^{-1}BD_o^{-1} = \\
&DD_o^{-1} + \lambda(\tilde{C} - DD_o^{-1}C_o)(I - \lambda(Z - BD_o^{-1}C_o))^{-1}BD_o^{-1} .
\end{aligned}$$

The is precisely the state space formula for the inner function $F_i$ given in (2.9). The proof is now complete.

**PROOF OF LEMMA 2.5.** If $\|F\|_\infty < \gamma$ then the Toeplitz matrix $T_R = \gamma^2 I - T_F^* T_F$ is strictly positive, and hence $T_R$ admits a square outer spectral factorization. On the other hand, if this Toeplitz matrix $T_R$ admits a square outer spectral factorization, then $\gamma^2 I - T_F^* T_F$ is positive and thus $\|F\|_\infty \le \gamma$. Therefore $\|F\|_\infty$ is the infimum over the set of all $\gamma > 0$ such that the Toeplitz matrix $T_R$ admits a square outer spectral factorization. According to Lemma 2.7 the entries $(R_{i-j})$ of $T_R$ are given by

$$R_0 = \gamma^2 I - D^*D - B^*\tilde{Q}B \quad \text{and} \quad R_n = CZ^{n-1}B \text{ for } n \ge 1 \text{ where } C = -(D^*\tilde{C} + B^*\tilde{Q}Z) . \quad (2.39)$$

As before, $\tilde{Q}$ is the observability grammian for $\{\tilde{C}, Z\}$. In particular, $\{Z, B, C\}$ is a stable, controllable realization for $\{R_n\}_1^\infty$.

Now we can use the positive real Lemma to determine whether or not $T_R$ admits a square outer spectral factor. Using the expressions for $R_0$ and C in (2.39) the algebraic Riccati equation for Q in (2.4) becomes

$$Q = Z^*QZ + (D^*\tilde{C} + B^*(\tilde{Q}+Q)Z)^*(\gamma^2 I - D^*D - B^*(\tilde{Q}+Q)B)^{-1}(D^*\tilde{C} + B^*(\tilde{Q}+Q)Z)\,. \quad (2.40)$$

In particular, the Toeplitz operator $T_R = \gamma^2 I - T_F^* T_F$ admits a square outer spectral factorization if and only if the algebraic Riccati equation (2.40) admits a positive solution. Moreover, if we add $\tilde{Q} = Z^*\tilde{Q}Z + \tilde{C}^*\tilde{C}$ to (2.40) and set $Y = \tilde{Q} + Q$ we arrive at the algebraic Riccati equation for Y in (2.10). It is easy to show that Q is a positive solution to (2.40) if and only if $Y = \tilde{Q} + Q$ is a positive solution to (2.10). In particular, Q is the minimal solution to (2.40) if and only if $Y = \tilde{Q} + Q$ is the minimal solution to the algebraic Riccati equation in (2.10). So according to the positive real Lemma 2.2, the Toeplitz operator $T_R = \gamma^2 I - T_F^* T_F$ admits a square outer spectral factor if and only there exists a positive solution Y to the algebraic Riccati equation for Y in (2.10). Therefore $\|F\|_\infty$ is the infimum over the set of all $\gamma > 0$ such that the algebraic Riccati equation for Y in (2.10) admits a positive solution.

To complete the proof assume that $\|F\|_\infty < \gamma$ and $Y = \tilde{Q} + Q$ is the minimal solution to the algebraic Riccati equation (2.10). Then according to Lemma 2.2 the outer spectral factor $\Theta$ for $\gamma^2 I - F^*F$ is given by (2.2) where $R_0$ and C are now specified by (2.39). Using this $R_0$ and C in (2.2) along with $Y = \tilde{Q} + Q$ we arrive at the state space realization for $\Theta$ in (2.11). This completes the proof.

## *Notes to Appendix:*

Section 1 of this appendix contains a number of results about factorization that are used in the main text. These results are by now standard and they appear in different sources; see, for instance, Helson-Lowdenslager [1], [2], Sz.-Nagy-Foias [3], and Caines [1]. Other directions of factorization can be found in the theory of Wiener-Hopf and singular integral equations (see Gohberg-Goldberg-Kaashoek [1]), and the related state space theory developed in Bart-Gohberg-Kaashoek [1]. The material in Section 2 is also standard, and here we mostly follow the state space factorization method developed in stochastic realization theory (see Caines [1]). For the Darlington synthesis problem see Arov [3] and the references therein.

# REFERENCES


Adamjan, V.A., Arov, D.Z. and M.G. Krein,

[1] Infinite Hankel matrices and generalized problems of Carathéodory - Fejér and I. Schur, *Functional Anal. i Prilozen,* **2** (1968) pp. 1-19 (Russian).

[2] On bounded operators commuting with contractions of class $C_{00}$ of unit rank of non-unitarity class, *Funkcional. Anal. i Prilozen,* **3** (1969) pp. 86-87 (Russian).

[3] Analytic properties of Schmidt pairs for a Hankel operator and the generalized Schur-Takagi problem, *Math USSR Sbornick,* **15** (1971) pp. 31-73.

[4] Infinite Hankel block matrices and related extension problems, *Izv. Akad. Nauk. Armjan SSR, Matematika,* **6** (1971) pp. 87-112 (English Translation *Amer. Math. Soc. Transl.,* **III** (1978) pp. 133-156.)

Alpay, D. and P. Dewilde,

[1] Time-varying signal approximation and estimation, in: *Signal processing, scattering and operatory theory, and numerical methods*, Proceedings of the international symposium MTNS-89, Vol. III (eds. M.A. Kaashoek, J.H. van Schuppen and A.C.M. Ran), Birkhäuser Verlag, Boston, 1990, pp. 1-22.

Alpay, D., Dewilde, P. and H. Dym,

[1] Lossless scattering and reproducing kernels for upper triangular operators, in: *Extension and interpolation of linear operators and matrix functions* (Ed. I. Gohberg), OT **47**, Birkhäuser Verlag, Basel, 1990, pp. 61-135.

Anousis, M.,

[1] Interpolation in nest algebras, Proc. Amer. Math. Soc., **114** (1992) pp. 707-710.

Anousis, M., Katsoulis, E.G., Moore, R.L. and T.T. Trent

[1] Interpolation on problems for Hilbert-Schmidt operators in reflexive algebras, *Houston J. Math.*, **19** (1993) pp. 63-73.

Arocena, R.,

[1] Unitary colligations and parametrization formulas, Ukrainskii Mat. Zhurnal, **46** (3) (1994) pp. 147-154.

Arov, D.Z.,

[1] On unitary coupling with losses (scattering theory with losses), Functional Anal. Appl., **8** (4) (1974) pp. 280-294.

[2] Passive linear steady-state dynamical systems, Sibirsk. Mat. Zh., **20** (2) (1979) pp. 211-228 (Russian).

[3] Stable dissipative linear stationary dynamical scattering systems, J. Operator Theory, **2** (1979) pp. 95-126 (Russian).

[4] Regular J-inner matrix-functions and related continuation problems, in: Linear Operators and Function Spaces (eds. H. Helson, B.Sz.-Nagy, F.-H. Vasilescy), OT **43**, Birkhäuser Verlag, Basel, 1990, pp. 63-87.

Arov, D.Z. and M.G. Krein,

[1] Calculation of entropy functionals and of their minima, Acta Sci. Math. Szeged, **45** (1983) pp. 33-50 (Russian).

Arsene, Gr., Ceausescu, Z. and C. Foias,

[1] On intertwining dilations VII, *Proc. Coll. Complex Analysis, Joensuu, Lecture Notes in Math.*, **747** (1979) pp. 24-45.

[2] On intertwining dilations VIII, *J. Operator Theory,* **4** (1980) pp. 55-91.

Arveson, W.B.,

[1] Interpolation problems in nest algebras, *J. Functional Analysis,* **20** (1975) pp. 208-233.

Bakonyi, M. and T. Constantinescu,

[1] *Schur's algorithm and several applications*, Putman Series **261**, Longman, Harlow, 1992.

Ball, J.A.,

[1] Commutant lifting and interpolation: the time-varying case, *Integral Equations and Operator Theory*, **25** (4) (1996) pp. 377-405.

Ball, J.A. and I. Gohberg,

[1] A commutant lifting theorem for triangular matrices with diverse applications, *Intergal Equations and Operator Theory,* **8** (1985) pp. 205-267.

Ball, J.A., Gohberg, I. and M.A. Kaashoek,

[1] Nevanlinna-Pick interpolation for time-varying input-output maps: The discrete case, in: *Time-variant systems and interpolation*, (Ed. I. Gohberg), OT **56**, Birkhäuser Verlag, Basel, 1992, pp. 1-51.

[2] Nevanlinna-Pick interpolation for time-varying input-output maps: The continuous time case, in: *Time-variant systems and interpolation*, (Ed. I. Gohberg), OT **56**, Birkhäuser Verlag, Basel, 1992, pp. 52-89.

[3] Time-varying systems: Nevanlinna-Pick interpolation and sensitivity minimization, in: *Recent Advances in Mathematical Theory of Systems, Control, Networks and Signal Processing I, Proceedings MTNS-91* (eds. H. Kimura, S. Kodama) Mita Press, Tokyo, 1992; pp. 53-58.

[4] Bitangential interpolation for input-output operators of time-varying systems: The discrete time case, in: *New aspects in interpolation and completion theories*, (Ed. I. Gohberg), OT **64**, Birkhäuser Verlag, Basel, 1993, pp. 33-72.

[5] Bitangential interpolation for input-output maps of time-varying systems: the continuous time case, *Integral Equations and Operator Theory* 20 (1994) pp. 1-43.

[6] $H_\infty$-control and interpolation for time-varying systems, in: *Systems and Networks: Mathematical Theory and Applications*, Vol. I (Eds. U. Helmke, R. Mennicken, J. Saurer), Akademie Verlag, Berlin, 1994, pp. 33-38.

[7] Two-sided Nudelman interpolation for input-output operators of discrete time-varying systems, *Integral Equations and Operator Theory* **21** (1995) pp. 174-211.

[8] A frequency response function for linear, time-varying systems, *Math. Control Signals Systems*, **8** (1995) pp. 334-351.

Ball, J.A., Gohberg, I. and L. Rodman,

[1] *Interpolation for Rational Matrix Functions*, OT 45, Birkhäuser, 1990.

Ball, J.A. and J.W. Helton,

[1] A Beurling-Lax Theorem for the Lie group U(m,n) which contains most classical interpolation theory, *J. Operator Theory*, **9** (1983) pp. 107-142.

Ball, J.A. and A.C.M. Ran,

[1] Hankel norm approximation of a rational matrix function in terms of its realization, *Modeling Identification and Robust Control*, Eds. E. I. Byrnes and A. Lindquist, Elsevier Science Publishers B.V. North Holland (1986) pp. 285-296.

[2] Optimal Hankel norm model reductions and Wiener-Hopf Factorizations I: The canonical case, *SIAM J. Contr. and Optimization,* **25** (1987) pp. 362-382.

[3] Optimal Hankel norm model reductions and Wiener-Hopf factorizations II: The non-canonical case, *Integral Equations and Operator Theory,* **10** (1987) pp. 416-436.

Ball, J.A. and V. Vinnikov,

[1] Zero-pole interpolation for meromorphic matrix functions on an algebraic curve and transfer functions of 2D systems, *Acta Applicandae Mathematicae*, **45** (1996) pp. 239-316.

Bart, H., Gohberg, I. and M. A. Kaashoek,

[1] *Minimal Factorization of Matrix and Operator Functions*, OT **1**, Birkhäuser Verlag, Basel, 1979.

Basar, T. and P. Bernhard,

[1] *$\mathcal{H}_\infty$-Optimal Control and Related Minimax Design Problems: A Dynamic Game Approach*. Systems and Control: Foundations and Applications. Birkhäuser, Boston, 1991.

Bercovici, H.,

[1] *Operator Theory and Arithmetic in* $H^\infty$, American Mathematical Society, Providence, Rhode Island, 1988.

Biswas, A.

[1] A harmonic-type maximal principle in commutant lifting, to appear in *Integral Equations and Operator Theory.*

Brodskii, M.S.,

[1] *Triangular and Jordan Representations of Linear Operators*, Transl. Math. Monographs, **32**, Amer. Math. Soc., Providence, R.I., 1970.

[2] Unitary operator colligations and their characteristic functions, *Uspekhi Math. Nauk*, **33** (4) (1978) pp. 141-178 (Russian); English Transl., *Russian Math. Surveys*, **33** (4) (1987) pp. 159-191.

Caines, P. E.,

[1] *Linear Stochastic Systems*, John Wiley & Sons, Inc., Montreal, 1988.

Carathéodory, C.,

[1] Über den Variabilitätsbereich der Koeffizienten von Potenzreihen, die gegebene Werte nicht annehmen, *Math. Ann.*, **64** (1907) pp. 95-115.

[2] Über den Variabilitätsbereich der Fourierschen Konstanten von positiven harmonischen Funktionen, *Rend. Circ. Mat. Palermo,* **32** (1911) pp. 193-217.

Carswell, J.G.W. and C. F. Schubert,

[1] Lifting of operators that commute with shifts, *Michigan Math. J.,* **22** (1975) pp. 65-69.

Chu, C.C., Doyle, J.C. and E.B. Lee,

[1] The general distance problem in $\mathcal{H}_\infty$ optimal control theory, *International Journal of Control*, **44** (1986) pp. 565-596.

Chui, C.K. and G. Chen,

[1] *Signal Processing and Systems Theory*, Springer-Verlag, Berlin, 1992.

Clancey, K. and I. Gohberg,

[1] *Factorization of Matrix Functions and Singular Integral Operators,* OT **3**, Birkhäuser, Basel, 1981.

Constantinescu, T.,

[1] *Some aspects of nonstationarity I*, Acta Sci. Math. (Szedged) **54** (1990) pp. 379-389.

[2] *Some aspects of nonstationarity II*, Mathematica Balkanica **4** (1990) pp. 211-235.

[3] *Schur parameters, factorization and dilation problems*, OT **82**, Birkhäuser Verlag, Basel, 1996.

Conway, J. B.,

[1] *A Course in Functional Analysis*, Springer Verlag, Berlin, 1985.

Dewilde, P.,

[1] A course on the algebraic Schur and Nevanlinna-Pick interpolation problems, in: *Algorithms and parallel VLSI architectures*, Vol. A: Tutorials (eds. E.F. Deprettere and A.-J. van der Veen, Elsevier, Amsterdam, 1991, pp. 13-69.

Dewilde, P. and H. Dym,

[1] Interpolation for upper triangular operators, in: *Time-variant systems and interpolation* (Ed. I. Gohberg), OT **56**, Birkhäuser Verlag 1992, pp. 153-260.

Dewilde, P., Kaashoek, M.A. and M. Verhaegen (Eds.),

[1] *Challenges of a generalized system theory*, Koninklijke Nederlandse Akademie van Wetenschappen, Verhandelingen, Afd. Natuurkunde, Eerste reeks, deel **40**, North-Holland Publ. Co., Amsterdam, 1993.

Doyle, J.C. and B.A. Francis,

[1] Linear control theory with an $H_\infty$ optimality criterion, *SIAM J. Control and Optimization,* **25** (1987) pp. 815-844.

Doyle, J.C., Francis, B.A. and A.R. Tannenbaum,

[1] *Feedback control theory*, Macmillan, New York, 1992.

Dubovoj, V.K., Fritzsche, B. and B. Kirstein,

[1] *Matricial version of the classical Schur problem*, Teubner-Texte zur Mathematik, **129**, Teubner, Leipzig, 1992.

Dym, H.,

[1] *J. Contractive Matrix Functions, Reproducing Kernel Hilbert Spaces and Interpolation,* CBMS Regional Conference series, **71**, American Mathematical Society, Providence, Rhode Island, 1989.

Dym, H. and I. Gohberg,

[1] Extension of kernels of Fredholm operators *J. d Analyse Math.*, **42** (1982) pp. 83-125.

[2] Unitary interpolants, factorization indices and infinite block Hankel matrices, *J. Functional Analysis,* **54** (1983) pp. 229-289.

[3] A maximum entropy principle for contractive interpolants, *J. Functional Analysis,* **65** (1986) pp. 83-125.

[4] A new class of contractive interpolants and maximum entropy principles, *Topics in Operator Theory and Interpolation*, OT **29**, Ed. I. Gohberg, Birkhäuser Verlag, Basel (1988) pp. 117-150.

Faurre, P.L.,

[1] Stochastic realization algorithms, in *System Identification: Advances and Case Studies*, R.K. Mehra and D. F. Lainiotis, Eds., Academic Press, New York (1976) pp. 1-23.

Fedcina, I.P.,

[1] A criterion for the solvability of the Nevanlinna-Pick tangent problem, *Mat. Issled.,* **7** (1972) pp. 213-227.

[2] The tangential Nevanlinna-Pick problem with multiple points, *Akad. Nauk Armjan. SSR Dokl.*, **61** (1975) pp. 214-218.

Feintuch, A. and B.A. Francis,

[1] Distance formulas for operator algebras arising in optimal control problems, in: *Topics in Operator Theory and Interpolation*; OT **29**, Ed. I Gohberg, Birkhäuser Verlag, Basel (1988) pp. 151-170.

Foias, C. and A.E. Frazho,

[1] Redheffer products and the lifting of contractions on Hilbert space, *J. Operator Theory,* **11** (1984) pp. 193-196.

[2] On the Schur representation in the Commutant Lifting Theorem I, in: *Schur Methods in Operator Theory and signal processing*, OT **18**, Ed. I. Gohberg, Birkhäuser Verlag, Basel (1986) pp. 207-217.

[3] On the Schur representation in the commutant lifting theorem II, in: *Topics in Operator Theory and Interpolation*; OT **29**, Ed. I Gohberg, Birkhäuser Verlag, Basel (1988) pp. 171-179.

[4] *The commutant lifting approach to interpolation problems*, OT **44**, Birkhäuser Verlag, Basel, 1990.

[5] Commutant lifting and simultaneous $H^\infty$ and $L^2$ suboptimization, *SIAM J. Math. Analysis*, **23**, 1992, pp. 984-994.

[6] Constructing the Schur contraction in the commutant lifting theorem, *Acta Sci. Math.* (Szeged), **61** (1995) pp. 425-442.

Foias, C., Frazho, A.E. and I. Gohberg,

[1] Central intertwining lifting, maximum entropy and their performanance, *Integral Equations and Operator Theory*, **18** (1994) pp. 166-201.

Foias, C., Frazho, A.E., Gohberg, I. and M.A. Kaashoek,

[1] Discrete time-invariant interpolation as classical interpolation with an operator argument, *Integral Equations and Operator Theory*, **26** (1996) pp. 371-403.

[2] A time-variant version of the commutant lifting theorem and nonstationary interpolation problems, *Integral Equations and Operator Theory*, **28** (1997) pp. 158-190.

[3] Parameterization of all solutions of the three chains completion problem, *Integral Equations and Operator Theory*, to appear.

[4] The maximum principle for the three chains completion problem, *Integral Equations and Operator Theory*, to appear.

Foias, C., Frazho, A.E. and W.S. Li,

[1] The exact $H^2$ estimate for the central $H^\infty$ interpolant, in: *New aspects in interpolation and completion theories* (Ed. I. Gohberg), OT **64**, Birkhäuser Verlag, Basel (1993) pp. 119-156.

[2] On $H^2$ minimization for the Carathéodory-Schur interpolation problem, *Integral Equations and Operator Theory*, **21** (1995) pp. 24-32.

Foias, C., Özbay, H. and A.R. Tannenbaum,

[1] *Robust Control of Infinite Dimensional Systems*, Springer-Verlag, London, 1996.

Francis, B.A.,

[1] *A Course in* $H^\infty$ *Control Theory,* Lecture Notes in Control and Information Science, Springer, New York, 1987.

Francis, B.A., Helton, J.W. and G. Zames,

[1] $H^\infty$-optimal feedback controllers for linear multivariate systems, IEEE Trans. Auto. Control AC-**29**, (1984) pp. 888-900.

Frazho, A.E.,

[1] A four block approach to central commutant lifting, *Indiana University Mathematics Journal*, **42**, No. 3, 1993, pp. 821-838.

Frazho, A.E. and S.M. Kherat,

[1] On mixed $H^2 - H^\infty$ tangential interpolation, in: *New aspects in interpolation and completion theories* (Ed. I. Gohberg), OT **64**, Birkhäuser Verlag, Basel (1993) pp. 157-202.

Frazho, A.E. and M.A. Rotea,

[1] A remark on mixed $L^2/L^\infty$ bounds, *Integral Equations and Operator Theory*, **15** (2) (1992) pp. 343-348.

Fritzsche, B. and B. Kirstein,

[1] *Ausgewählte Arbeiten zu den Ursprüngen der Schur-Analysis*, Gewidmet dem groBen Mathematiker Issai Schur (1875-1941), B. G. Teubner Verlagsgesellschaft, Stuttgart-Leipzig, 1991.

Fuhrmann, P.A.,

[1] *Linear Systems and Operators in Hilbert Space*, McGraw-Hill, Inc., 1981.

Gantmacher, F.R.,

[1] *The Theory of Matrices*, Chelsea Publishing Co., New York, 1977.

Gilbert, E.,

[1] Controllability and observability in multivariable control systems, *SIAM J. Control*, **1** (1963) pp. 128-151.

Glover, K.,

[1] All optimal Hankel-norm approximations of linear multivariable systems and their $L_\infty$-error bounds, *Int.J. Cont.,* **39** (1984) pp. 1115-1193.

Gohberg, I. (Ed.),

[1] *Time-variant systems and interpolation*, OT **56**, Birkhäuser Verlag, Basel, 1992.

Gohberg, I., Goldberg, S. and M.A. Kaashoek,

[1] *Classes of linear operators*, Vol. I, OT **49**, Birkhäuser Verlag, Basel, 1990.

[2] *Classes of linear operators*, Vol. II, OT **63**, Birkhäuser Verlag, Basel, 1993.

Gohberg, I., Kaashoek, M.A., and F. van Schagen,

[1] Rational contractive and unitary interpolants in realized form, *Integral Equations and Operator Theory*, **11** (1988) pp. 105-127.

Gohberg, I., Kaashoek, M.A. and H.J. Woerdeman,

[1] The band method for positive and contractive extension problems, *J. Operator Theory*, **22** (1989) pp. 109-155.

[2] The band method for positive and contractive extension problems: An alternative version and new applications, *Integral Equations and Operator Theory*, **12** (1989) pp. 343-382.

[3] A maximum entropy principle in the general framework of the band method, *J. Funct. Anal.*, **95** (1991) pp. 231-254.

[4] The time variant versions of the Nehari and four block problems, in: *$H_\infty$-control theory* (Mosca, E., Pandolfi, L., eds.), Springer Lecture Notes in Math **1496**, Springer-Verlag, Berlin, (1991) pp. 309-323.

Gohberg, I. and L.A. Sakhnovich (Eds.),

[1] *Matrix and operator valued functions* (The Vladimir Petrovich Potapov Memorial Volume), OT **72**, Birkhäuser Verlag, Basel, 1994.

Green, M. an D.J.N. Limebeer,

[1] *Linear Robust Control*, Prentice-Hall, Inc., New Jersey, 1995.

Halanay, A., and V. Ionescu,

[1] *Time-varying discrete linear systems*, OT **68**, Birkhäuser Verlag, Basel, 1996.

Halmos, Paul R.,

[1] *A Hilbert Space Problem Book*, Springer-Verlag, New York Inc., 1982.

Helmke, U., Mennicken, R. and J. Saurer,

[1] *Systems and Networks: Mathematical Theory and Applications*, Vol. I, Akademie Verlag, Berling, 1994.

Helson, H. and D. Lowdenslager,

[1] Prediction theory and Fourier series in several variables, *Acta Math.,* **99** (1958) pp. 165-202.

[2] Prediction theory and Fourier series in several variables II, *Acta Math.,* **106** (1961) pp. 175-213.

Helton, J.W.,

[1] The distance of a Function to $H^\infty$ in the Poincare´ Metric; Electrical Power Transfer, *J. Functional Analysis*, **38** (1980) pp. 273-314.

[2] Worst case analysis in the frequency-domain; an $H_\infty$ approach to control, *IEEE Trans. Auto. Control,* **30** (1985) pp. 1154-1170.

[3] *Operator Theory, Analytic Functions, Matrices, and Electrical Engineering*, CBMS Regional Conference Series in Math., **68**, Amer. Math. Soc., Providence, Rhode Island, 1987.

Hoffman, K.,

[1] *Banach Spaces of Analytic Functions,* Englewood Cliffs, Prentice Hall, New Jersey, 1962.

Kaashoek, M.A.,

[1] State space theory of rational matrix functions and applications, in: *Lectures on Operator Theory and its Applications*, (Ed. P. Lancaster), Fields Institute Monographs, **3**, Amer. Math. Soc., 1996, pp. 233-333.

Kaftal, V., Larson, D. and G. Weiss,

[1] Quasitriangular subalgebras of semifinite Von Neumann algebras are closed, *J. Functional Anal.*, **107** (2) (1992) pp. 387-401.

Kailath, T.,

[1] *Linear Systems,* Englewood Cliffs: Prentice-Hall, Inc., New Jersey, 1980.

[2] A theorem of I. Schur and its impact on modern signal processing, in: *I. Schur Methods in Operator Theory and Signal Processing*; Ed. I. Gohberg, OT **18**, Birkhäuser Verlag, Basel (1986) pp. 9-30.

Kailath, T. and A.H. Sayed,

[1] Displacement structure: theory and applications, *SIAM Review*, **37** (3) (1995) pp. 297-386.

Kalman, R.E.,

[1] Mathematical description of linear dynamical systems, *SIAM J. Control,* **1** (2) (1963) pp. 152-192.

Kalman, R.E., Falb, P.L. and M.A. Arbib,

[1] *Topics in Mathematical System Theory*, McGraw-Hill, New York, 1969.

Kheifets, A.Y.,

[1] The generalized bitangential Scur-Nevanlinna-Pick problem and the related Parseval equality, Teor. Funktsii, Funktsional'nyi Analizi i ikh Prilozhen, **54** (1990) pp. 89-96 (Russian); English transl. in: J. Sov. Math., **58** (4) (1992) pp. 358-364.

Kimura, H.,

[1] Robust stabilization for a class of transfer functions, *IEEE Trans. Auto. Cont.,* **29** (1984) pp. 788-793.

[2] On interpolation-minimization problems in $H_\infty$, *Control-Theory and Advanced Technology,* **2** (1986) pp. 1-25.

Kos, J.,

[1] Higher order time-varying Nevanlinna-Pick interpolation, in: *Challenges of a generalized system theory*, Koninklijke Nederlandse Akademie van Wetenschappen, Verhandelingen, Afd. Natuurkunde, Eerste reeks, deel **40**, North-Holland Publ. Co., Amsterdam, 1993, pp. 59-71.

[2] *Time-dependent problems in linear operator theory*, Ph.D. Thesis, Vrije Universiteit, Amsterdam, 1995.

Lancaster, P. and L. Rodman,

[1] *Algebraic Riccati Equations*, Clarendon Press, Oxford, 1995.

Livšic, M.S.,

[1] On a class of linear operators in a Hilbert space, *Mat. Sbornik*, **19** (61) (1946) pp. 239-262 (Russian); English Transl., *Amer. Math. Soc. Transl.* (Series 2), **13** (1960) pp. 1-63.

[2] On the spectral resolution of linear non-selfadjoint operators, *Mat. Sbornik*, **34** (76) (1954) pp. 145-199 (Russian); English Transl., *Amer. Math. Soc. Transl* (Series 2), **5** (1957) pp. 67-114.

Livšic, M.S., Kravitsky, N., Markus, A.S. and V. Vinnikov,

[1] *Theory of commuting nonselfadjoint operators*, Mathematics and its Applications, **332**, Kluwer, Dordrecht, 1995.

Luenberger, David G.,

[1] *Optimization by Vector Space Methods*, John Wiley & Sons, Inc., New York, 1969.

Lowdenslager, D.B.,

[1] On factoring matrix valued functions, *Ann. of Math.,* **78** (1963) pp. 450-454.

Mustafa, D. and K. Glover,

[1] *Minimum Entropy $H_\infty$ Control, Lecture notes in Control and Information Sciences*, Springer-Verlag, New York, 1990.

Nehari, Z.,

[1] On bounded bilinear forms, *Ann. of Math.,* **65** (1957) pp. 153-162.

Nevanlinna, R.,

[1] Über beschränkte Funktionen, die in gegebenen Puntken vorgeschriebene Werte annehmen, *Ann. Acad. Sci. Fenn,* **13:1** (1919), 71 pp.

Nikolskii, N.K.,

[1] *Treatise on the Shift Operator,* Springer-Verlag, New York, 1986.

Nudelman, A.A.,

[1] On a new problem of moment type, *Dokl. Akad. Nauk SSSR,* **233** (1977) pp. 792-795; English Transl., *Soviet Math. Dokl.,* **18** (1977) pp. 507-510.

[2] On a generalization of classical interpolation problems, *Dokl. Akad. Nauk SSR,* **256** (1981) pp. 790-793, *Soviet Math. Dokl.,* **23** (1981) pp. 125-128.

Ogata, K.,

[1] *Modern Control Engineering*, Prentice-Hall, Englewood Cliffs, NJ, 1970.

Page, L.B.,

[1] Operator that commute with a unilateral shift on an invariant subspace, *Pacific J. Math.*, **36** (1971) pp. 787-794.

Pick, G.,

[1] Über die Beschränkungen analytischer Functionen, welche durch vorgegebene Funktionswerte bewirkt sind., *Math. Ann.*, **77** (1916) pp. 7-23.

[2] Über die Beschränkungen analytischer Funktionen, welche durch vorgegebene, funktionswerte bewirkt sind, *Math. Ann.*, **78** (1918) pp. 270-275.

Parrott, S.,

[1] On a quotient norm and the Sz.-Nagy-Foias lifting theorem, *J. Functional Analysis*, **30** (1978) pp. 311-328.

Paulsen, V.I., Power, S. and J. Ward,

[1] Semi-discreteness and dilation theory for next algebras, *J. Functional Anal.*, **80** (1988) pp. 76-87.

Power, S. C.,

[1] *Hankel Operators on Hilbert Space,* Pitman, London, 1982.

Redheffer, R.M.,

[1] On a certain linear fractional transformation, *J. Math. Phys.,* **39** (1960) pp. 269-286.

[2] On the relation of transmission - line theory to scattering and transfer, *J. Math. Phys.*, **41** (1962) pp. 1-41.

Rosenblum, M. and J. Rovnyak,

[1] An operator-theoretic approach to theorems of the Pick-Nevanlinna and Loewner types I, *Integral Equations and Operator Theory,* **3** (1980) pp. 408-436.

[2] An operator-theoretic approach to theorems of the Pick-Nevanlinna and Loewner types II, *Integral Equations and Operator Theory,* **5** (1982) pp. 870-887.

[3] *Hardy Classes and Operator Theory,* Oxford Univ. Press, New York, 1985.

Rotea, M.A. and A.E. Frazho,

[1] Bounds on solutions to $H_\infty$ algebraic Riccati equations and $H_2$ properties of $H_\infty$ central solutions, *Sys. Control Letters*, **19** (1992) pp. 341-352.

Rugh, W.J.,

[1] *Linear System Theory*, Prentice-Hall, Inc., New Jersey, 1993.

Sakhnovich, L.A.,

[1] Method of operator identities and problems of analysis, *St. Petersburg Math. J.*, **5** (1) (1994) pp. 1-69.

Sarason, D.,

[1] Generalized interpolation in $H^\infty$, *Trans. American Math. Soc.,* **127** (1967) pp. 179-203.

Sayed, A.H.,

[1] *Displacement structure in signal processing and mathematics*, Ph.D. Thesis, Department of Electrical Engineering, Information Systems Laboratory, Stanford University, Stanford, CA, August 1992, pp. 116-121.

Sayed, A.H., Constantinescu, T. and T. Kailath,

[1] Lattice structures of time-invariant interpolation problems, in: *Proc. 31-st IEEE Conf. on Decision and Control*, (Tuscon, AZ), Dec. 1992, pp. 116-121.

Schäffer, J.J.,

[1] On unitary dilations of contractions, *Proc. Amer. Math. Soc.*, **6** (1955) p. 322.

Schur, I.,

[1] On power series which are bounded in the interior of the unit circle I, *J. für die Reine und Angewandte Mathematik,* **147** (1917) pp. 205-232, English translation in *I. Schur Methods in Operator Theory and Signal Processing*; Ed. I. Gohberg, OT **18**, Birkhäuser Verlag, Basel (1986) pp. 31-59.

[2] On power series which are bounded in the interior of unit circle II., *J.* für *die Reine and Angewandte Mathematik*, **148** (1918) pp. 122-145 (German). English translation in *I. Schur Methods in Operator Theory and Signal Processing*; Ed. I. Gohberg, OT **18**, Birkhäuser Verlag, Basel (1986) pp. 61-88.

Skelton, R.E., Iwasaki, T. and K.M. Grigoriadis,

[1] *A Unified Algebraic Approach to Control Design*, Taylor & Francis, London, 1997.

Smith, M.J. and A.E. Frazho,

[1] $H^2$–$H^\infty$ control system synthesis via lifting techniques, OT, to appear.

Stoorvogel, A.,

[1] *The $H_\infty$ Control Problem*, Prentise-Hall, United Kingdom, 1992.

Sz.-Nagy, B. and C. Foias,

[1] Dilatation des commutants d'opérateurs, *C.R. Acad. Sci. Paris, Serie A,* **266** (1968) pp. 493-495.

[2] Commutants de certains operatéurs, *Acta Sci. Math.,* **29** (1968) pp. 1-17.

[3] *Harmonic analysis of operators on Hilbert space,* North Holland Publishing Co., Amsterdam-Budapest, 1970.

Sz.-Nagy, B. and A. Koranyi,

[1] Relations d'un problème de Nevanlinna et Pick avec la theorie des opérateurs de l'espace Hilbertien, *Acta Sci. Math.,* **7** (1956) pp. 295-302.

[2] Operator theoretische Behandlung and Verallgemeinerung eines Problemkreises in der komplexen funktionentheorie, *Acta Math.,* **100** (1958) pp. 171-202.

Tannenbaum, A.R.,

[1] Feedback stablization of linear dynamical plants with uncertainty in the gain factor, *Int. J. Control,* **32** (1980) pp. 1-16.

[2] Modified Nevanlinna-Pick interpolation of linear plants with uncertainty in the gain factor, *Int. J. Control,* **36** (1982) pp. 331-336.

Teodorescu, R.,

[1] Intertwining liftings and Schur contractions, to appear *Integral Equations and Operator Theory.*

Veen, A.-J. van der,

[1] *Time-varying system theory and computational modeling*, Ph.D. Thesis, Department of Electrical Engineering, Delft University of Technology, The Netherlands, June 1993.

Vidyasagar, M.,

[1] *Control System Synthesis*, The MIT Press, Cambridge, Massachusetts, 1985.

Weiss, L.,

[1] On the structure theory of linear differential systems, *Siam J. Control*, **6** (1968) pp. 659-680.

Woerdeman, H.J.,

[1] *Matrix and Operator Extension*, Ph.D. Thesis, Vrije Universiteit te Amsterdam, CWI Tract **68**, Centre for Mathematics and Computer Science, Amsterdam, 1989.

Youla, D.C.,

[1] The synthesis of linear dynamical systems from prescribed weighting patterns, *SIAM J. Appl. Math.*, **14** (1966) pp. 527-549.

Youla, D.C., Jabr, H.A. and J.J. Bongiorno, Jr.,

[1] Modern Wiener-Hopf design of optimal controllers: part II, *IEEE Trans. Auto. Cont.*, vol. AC-**21**, pp. 319-338.

Zames, G.,

[1] Optimal sensitivity and feedback: weighted seminorms, approximate inverses, and plant invariant schemes, *Proc. Allerton Conf.*, 1979.

[2] Feedback and optimal sensitivity: model reference transformations, multiplicative seminorms, and approximate inverses, *IEEE Trans. Auto. Cont.,* **26** (1981) pp. 301-320.

Zames, G. and B. A. Francis,

[1] Feedback, minimax sensitivity, and optimal robustness, *IEEE Trans. Auto. Cont.,* **28** (1983) pp. 585-601.

Zhou, K., Doyle, J.C. and K. Glover,

[1] *Robust and Optimal Control*, Prentice-Hall, Inc., 1996.

# LIST OF SYMBOLS

| | |
|---|---|
| $A^*$ | the adjoint of the operator A |
| $A^{1/2}$ | the positive square root of the positive operator A |
| $\det[A]$ | the determinant of A |
| $(a_{i,j})_{1,1}^{n,m}$ | a n by m matrix with entries $a_{i,j}$ |
| $A \mid \mathcal{M}$ | the restriction of the operator A to the subspace $\mathcal{M}$ |
| $\mathrm{cl}\mathcal{M}$ | the closure of the set $\mathcal{M}$, also denoted by $\overline{\mathcal{M}}$ or $\mathcal{M}^-$ |
| $\mathbb{C}^n$ | the n-dimensional linear space of complex n-tuples |
| $\dim \mathcal{M}$ | the dimension of the subspace $\mathcal{M}$ |
| $\mathbb{D}$ | the open unit disc |
| $D_A$ | the positive square root of $\gamma^2 I - A^* A$ |
| $\mathcal{D}_A$ | the closure of the range of $D_A$ |
| $D_{A^*}$ | the positive square root of $\gamma^2 I - AA^*$ |
| $\mathcal{D}_{A^*}$ | the closure of the range of $D_{A^*}$ |
| $D'$ | the positive square root of $I - T'^* T'$ |
| $\mathcal{D}'$ | the closure of the range of $D'$ |
| $\mathcal{F}_{\mathcal{U}}$ | the Fourier transform from $\ell^2(\mathcal{U})$ onto $L^2(\mathcal{U})$ |
| $\mathcal{F}_{\mathcal{U}+}$ | the Fourier transform from $\ell^2_+(\mathcal{U})$ onto $H^2(\mathcal{U})$ |
| $\mathcal{G}(A \cdot T)$ | equals $\mathcal{D}_A \ominus \overline{D_A T \mathcal{H}}$ |

| | |
|---|---|
| $H^2$ | the Hardy space of analytic functions in $\mathbb{D}$ with square summable Taylor coefficients |
| $H^2(\mathcal{U})$ | the Hardy space of analytic functions in $\mathbb{D}$ with values in the Hilbert space $\mathcal{U}$ and square summable Taylor coefficients |
| $H^\infty$ | the space of all uniformally bounded analytic functions on $\mathbb{D}$ |
| $H^\infty(\mathcal{U}, \mathcal{Y})$ | the space of all uniformally bounded analytic functions on $\mathbb{D}$ with values in $\mathcal{L}(\mathcal{U}, \mathcal{Y})$ |
| $\mathcal{H}(\Theta)$ | $H^2(\mathcal{Y}) \ominus \Theta H^2(\mathcal{U})$ for an inner $\Theta$ in $H^\infty(\mathcal{U}, \mathcal{Y})$ |
| $H^2(\mathcal{U}, \mathcal{Y})$ | the $H^2$ Hardy space in $\mathbb{D}$ whose values are Hilbert Schmidt operators from $\mathcal{U}$ to $\mathcal{Y}$ |
| $\vec{\mathcal{H}}$ | the Hilbert space direct sum $\bigoplus_{-\infty}^{\infty} \mathcal{H}_k = \oplus \{\mathcal{H}_k : k \in \mathbb{Z}\}$ |
| $\vec{\mathcal{H}}_{(a,b)}$ | the Hilbert space direct sum $\bigoplus_{k=a}^{b} \mathcal{H}_k$ |
| $I$ | identity operator |
| $I_{\mathcal{H}}$ | the identity operator on $\mathcal{H}$ |
| ker A | the kernel (null space) of A |
| $K^2$ | $L^2 \ominus H^2$ |
| $K^2(\mathcal{U})$ | $L^2(\mathcal{U}) \ominus H^2(\mathcal{U})$ |
| $K^\infty$ | the set of all f in $L^\infty$ of the form $f = \sum_1^\infty f_n e^{-int}$ |
| $K^\infty(\mathcal{U}, \mathcal{Y})$ | the set of all F in $L^\infty(\mathcal{U}, \mathcal{Y})$ of the form $F = \sum_1^\infty F_n e^{-int}$ |

| | |
|---|---|
| $\mathcal{L}(\mathcal{U}, \mathcal{Y})$ | the set of all operators from $\mathcal{U}$ to $\mathcal{Y}$ |
| $L^2$ | the set of all square integrable Lebesgue measurable functions in $[0, 2\pi)$ |
| $L^2(\mathcal{U})$ | the set of all square integrable Lebesgue measurable functions in $[0, 2\pi)$ with values in $\mathcal{U}$ |
| $L^2(\mathcal{U}, \mathcal{Y})$ | the $L^2$ space consisting of all square integrable functions whose values are Hilbert Schmidt operators from $\mathcal{U}$ to $\mathcal{Y}$ |
| $L^\infty$ | the set of all Lebesgue measurable function uniformally bounded a.e. in $[0, 2\pi)$ |
| $L^\infty(\mathcal{U}, \mathcal{Y})$ | the set of all strongly Lebesgue measurable functions uniformally bounded a.e. in $[0, 2\pi)$ with values in $\mathcal{L}(\mathcal{U}, \mathcal{Y})$ |
| $\ell^2_+$ | the set of all square summable unilateral sequences $[x_0, x_1, x_2, ...]^{tr}$ |
| $\ell^2_+(\mathcal{U})$ | the set of all square summable unilateral sequences $[x_0, x_1, x_2, ...]^{tr}$ with values in $\mathcal{U}$, that is , $x_i \in \mathcal{U}$ for all i |
| $\ell^2_n(\mathcal{U})$ | $\bigoplus_1^n \mathcal{U}$ which is identified with the set of all vectors of the form $[x_1, x_2, ..., x_n]^{tr}$ where $x_i \in \mathcal{U}$ for all i |
| $\ell^2$ | the set of all square summable bilateral sequences |
| $\ell^2(\mathcal{U})$ | the set of all square summable bilateral sequences with values in $\mathcal{U}$ |
| $\ell^2_-(\mathcal{U})$ | the subspace $\ell^2(\mathcal{U}) \ominus \ell^2_+(\mathcal{U})$ |
| $L_F$ | the Laurent operator from $\ell^2(\cdot)$ into $\ell^2(\cdot)$ for F in $L^\infty(\cdot, \cdot)$ |

| | |
|---|---|
| $\overline{\mathcal{M}}$ or $\mathcal{M}^-$ | the closure of the set $\mathcal{M}$ |
| $M_F$ | the operator from $L^2(\cdot)$ to $L^2(\cdot)$ defined by $M_F x = Fx$ when F is in the appropriate $L^\infty(\cdot, \cdot)$ space and x is in $L^2(\cdot)$ |
| $M_F^+$ | the operator from $H^2(\cdot)$ to $H^2(\cdot)$ defined by $M_F^+ h = Fh$ when F is in the appropriate $H^\infty(\cdot, \cdot)$ space and h is in $H^2(\cdot)$ |
| $P_{\mathcal{H}}$ | the orthogonal projection onto the subspace $\mathcal{H}$ |
| $P_+$ | the orthogonal projection onto the appropriate $H^2(\cdot)$ or $\ell_+^2(\cdot)$ space |
| $P_-$ | the orthogonal projection onto the appropriate $K^2(\cdot)$ or $\ell_-^2(\cdot)$ space |
| ran A | the range (image) of the operator A |
| $r_{spec}(Z)$ | the spectral radius of Z |
| $S_{\mathcal{U}}$ | the unilateral shift on the appropriate $H^2(\mathcal{U})$ or $\ell_+^2(\mathcal{U})$ space |
| $T_F$ | the Toeplitz operator from $\ell_+^2(\cdot)$ into $\ell_+^2(\cdot)$ for F in $L^\infty(\cdot, \cdot)$ |
| tr | transpose |
| $\mathbb{Z}$ | the integers ... $-2, -1, 0, 1, 2,$ ... |
| $V_{\mathcal{U}}$ | the bilateral shift on the appropriate $L^2(\mathcal{U})$ or $\ell^2(\mathcal{U})$ space |
| $(x_i)_m^n$ | a vector in $\bigoplus_m^n \mathcal{X}_i$ identified with $[x_m, x_{m+1}, ..., x_n]^{tr}$ |
| $[A_1, A_2, ..., A_n]^{tr}$ | equals $[A_1^*, A_2^*, ..., A_n^*]^*$ when the $A_i'^s$ are operators |
| $\delta_{j,k}$ | the Kronecker delta |

| | |
|---|---|
| $(x, y)$ | the inner product between x and y |
| $\|x\|$ | the norm of x |
| $\|F\|_\infty$ | the $L^\infty(\cdot, \cdot)$ norm of F |
| $\|f\|_2$ | the $L^2(\cdot)$ norm of f |
| $\|A\|_{HS}$ | the Hilbert Schmidt norm of A |
| $\oplus$ | the orthogonal direct sum |
| $\overset{n}{\underset{m}{\oplus}} x_i \in \overset{n}{\underset{m}{\oplus}} \mathcal{X}_i$ | the vector $x_i$ is in the Hilbert space $\mathcal{X}_i$ for $m \le i \le n$ and the space $\mathcal{X}_i$ is orthogonal to $\mathcal{X}_j$ for all $i \neq j$; the vector $\overset{n}{\underset{m}{\oplus}} x_i$ is identified with $[x_m, x_{m+1}, \ldots, x_n]^{tr}$. |
| $\mathcal{K} \ominus \mathcal{M}$ | the orthogonal complement of $\mathcal{M}$ in $\mathcal{K}$. |
| $\mathcal{M}^\perp$ | the orthogonal complement of $\mathcal{M}$ |
| $x \perp y$ | x is orthogonal to y |
| $\begin{pmatrix} n \\ k \end{pmatrix}$ | the binomial coefficient n!/k! (n-k)! |
| $\bigvee \mathcal{M}$ | the closed linear span of the set $\mathcal{M}$ |

# INDEX